Nordic Landscapes

THE UNIVERSITY OF MINNESOTA PRESS gratefully acknowledges the generosity of the Swedish Council for Planning and Coordination of Research, which contributed to this book.

Nordic Landscapes

Region and Belonging on the Northern Edge of Europe

Michael Jones *and* Kenneth R. Olwig, Editors

Published in cooperation with
the Center for American Places, Chicago, Illinois

University of Minnesota Press
Minneapolis · London

Chapter 8 was originally published in Swedish as "Dalarna—svenskt ideal?" in *Med landskapet i centrum: Kulturgeografisk perspektiv på nutida och historiska landskapet*, ed. Ulf Jansson, Meddelande 119, Kulturgeografiska institutionen (Stockholm: Stockholm University, 2003), 157–86. Portions of chapter 18 were previously published as "Layered Landscapes," in *Postcolonialism, Multitude, and the Politics of Nature: On the Changing Geographies of the European North* (Lanham, Md.: University Press of America, 2006); copyright 2006 by the University Press of America; reprinted with permission.

Excerpts in chapter 18 from Ilmari Kianto, *Punainen viiva* (Helsinki: Otava Publishing House, 1946), are reprinted with kind permission from Otava Publishing House; copyright 1946 Otava Publishing House.

Copyright 2008 by the Regents of the University of Minnesota

All rights reserved. No part of this publication may be reproduced, stored in a retrieval system, or transmitted, in any form or by any means, electronic, mechanical, photocopying, recording, or otherwise, without the prior written permission of the publisher.

Published by the University of Minnesota Press
111 Third Avenue South, Suite 290
Minneapolis, MN 55401-2520
http://www.upress.umn.edu

Library of Congress Cataloging-in-Publication Data

Jones, Michael.
　Nordic landscapes : region and belonging on the northern edge of Europe / Michael Jones and Kenneth R. Olwig, Editors.
　　　p. cm.
　Includes bibliographical references and index.
　ISBN-13: 978-0-8166-3914-4
　ISBN-10: 0-8166-3914-0
　ISBN-13: 978-0-8166-3915-1 (pb : alk. paper)
　ISBN-10: 0-8166-3915-9 (pb : alk. paper)
　　1. Scandinavia—Historical geography. 2. Regionalism—Scandinavia. 3. Group identity—Scandinavia. 4. Landscape—Scandinavia. I. Olwig, Kenneth. II. Title.
　DL6.7.J66 2008
　914.8—dc22

2007033556

Printed in the United States of America on acid-free paper

The University of Minnesota is an equal-opportunity educator and employer.

Contents

Introduction: Thinking Landscape and Regional Belonging
on the Northern Edge of Europe ... ix
KENNETH R. OLWIG *and* MICHAEL JONES

Denmark

1. Danish Landscapes ... 3
   KENNETH R. OLWIG

2. The Jutland Cipher: Unlocking the Meaning and Power of a
   Contested Landscape ... 12
   KENNETH R. OLWIG

The North Atlantic

3. Icelandic Topography and the Sense of Identity ... 53
   KIRSTEN HASTRUP

4. Land Divisions, Land Rights, and Landownership in the
   Faeroe Islands ... 77
   ARNE THORSTEINSSON

5. Perceiving Landscapes in Greenland ... 106
   BO WAGNER SØRENSEN

Sweden

6. The Swedish Landscape: The Regional Identity of
   Historical Sweden   141
   Ulf Sporrong

7. The South of the North: Images of an (Un)Swedish Landscape   157
   Tomas Germundsson

8. The Province of Dalecarlia (Dalarna): Heartland or Anomaly? 192
   Ulf Sporrong

9. Selma Lagerlöf's Värmland: A Swedish *Landskap* in Thought
   and Practice   220
   Gabriel Bladh

10. The Swedish Agropastoral *Hagmark* Landscape: An Approach to
    Integrated Landscape Analysis   251
    Margareta Ihse *and* Helle Skånes

Norway

11. The "Two Landscapes" of North Norway and the "Cultural
    Landscape" of the South   283
    Michael Jones

12. The Landscape in the Sign, the Sign in the Landscape: Periphery
    and Plurality as Aspects of North Norwegian Regional Identity   300
    Venke Åsheim Olsen

13. Changes in the Land and the Regional Identity of Western
    Norway: The Case of Sandhåland, Karmøy   344
    Anders Lundberg

14. The "Fjordscape" of Inner Sogn, Western Norway   372
    Ingvild Austad *and* Leif Hauge

15. The Agropastoral Mountain Landscape in Southern Norway:
    Museum or Living Landscape?   401
    Ann Norderhaug

Finland

16. Reflections on the Historical Landscapes of Finland   421
    W. R. Mead

17. Landscape Territory, Autonomy, and Regional Identity:
    The Åland Islands in a Cultural Perspective   440
    NILS STORÅ

18. Landscapes of Domination: Living in and off the Forests in
    Eastern Finland   458
    ARI AUKUSTI LEHTINEN

19. A Kaleidoscopic Nation: The Finnish National Landscape Imagery   483
    MAUNU HÄYRYNEN

20. Finnish Landscape as Social Practice: Mapping Identity and Scale   511
    ANSSI PAASI

NORDEN

21. The Nordic Countries: A Geographical Overview   543
    MICHAEL JONES *and* JENS CHRISTIAN HANSEN

22. Features of Nordic Physical Landscapes: Regional Characteristics   568
    ULF SPORRONG

    Contributors   585
    Index   589

Introduction: Thinking Landscape and Regional Belonging on the Northern Edge of Europe

Kenneth R. Olwig *and* Michael Jones

Globalization has, in an apparent paradox, been accompanied by a revived interest in the particularities of place as expressed in the landscapes of historical regions. There are concrete reasons for this. The emergence, for example, of globally oriented transnational bodies such as the European Union has opened the way for a reexamination of regional place identities that previously had been suppressed under the political and ideological hegemony of the nation-state. The rise of the European Union and the fall of the Soviet bloc have created what is perceived to be a realistic foundation for the reconstitution of historical regional identities, sometimes crossing national boundaries. Such regional identities have been inhibited since the rise of the nation-state. This situation is by no means limited to continental Europe. From Britain to Canada, national and regional identities are in flux, and even in the United States, which has always had an open federate structure, the particularity of place and regional landscape is a subject of growing interest.

Norden, literally "the North," comprises the states of Denmark, Iceland, Finland, Norway, and Sweden. Denmark, Norway, and Sweden are the three nation-states of Scandinavia (although in the English-speaking world "Scandinavia" sometimes refers to all the Nordic countries). Finland, to the east, was historically once part of Sweden and includes the internally autonomous, Swedish-speaking island territory of Åland. In the North Atlantic to the west are Iceland, once belonging to Norway and later to Denmark but now an independent state, and the Faeroe Islands and Greenland, which are internally autonomous territories under Denmark. Historically, other territories once within the cultural and political sphere of the Scandinavian countries included Slesvig-Holsten (Schleswig-Holstein in German), Orkney and Shetland, and Estonia on the Baltic.

Although lying on the northern edge of Western Europe, bordering the former Soviet

Union to the east and the polar wilderness to the north, Norden is by no means marginal to the construction of Western cultural identity. Borderlands, where cultures meet, often provide a site for cultural definition against an external other. They can also, however, function as liminal zones where, far from the hegemony established at the social and political core, alternative cultural forms can continue to thrive and perhaps later provide inspiration for changes at the core. The regional landscapes of Norden have been affected historically by both sorts of identity-shaping factors. On the one hand, Norden has thus recently been perceived as being a bulwark of Western liberty against the East, while, on the other, reformers, going back at least to Montesquieu, have drawn inspiration from what has been believed to be ancient northern traditions for representative democracy (Olwig 1992). There is a good portion of mythmaking in such constructions. In this mythmaking, the imagination is aided by Norden's distant location on the northern and eastern edges of the West. There is, nonetheless, a historical basis for such ideas.

In the context of the revived interest in the identities of place, landscape, and region, Norden is of note because of the ways Nordic national and regional identities have been constructed in text and image at the interstices of East, West, and North. Norden, however, also fascinates because of the way that it has preserved historical regional landscape identities, which are often but a distant memory elsewhere.

Landscape, Region, and Place Identity

A burgeoning contemporary literature exists concerning place and regional identity as spatial phenomena. There is also a growing literature concerned with landscape as the visual and literary expression of place and region and as a cultural construction. Finally, there is an interest in landscape as the expression of the historical interaction between society and nature. Oddly, it is rare that these discourses are linked in contemporary studies of place, even though they clearly interpenetrate one another (Olwig 2001). One reason for this is that before World War II it was common, particularly in connection with German nation-building, to approach the landscape as a regional unity determined ultimately by nature. This environmental determinism resulted in a "blood-and-soil" nationalistic ideology that led the study of the synergism of society and nature to fall out of favor in the postwar years (Olwig 1996a). The late twentieth-century breakdown of nationalism's hegemony, however, has opened the way for a critical interest in landscape both as a reflection of ideas concerning the relation of society to nature and as an expression of social spatial practice through time. The identity of places emerges, on the one hand, in the interstices between the social processes in which the ideas of region and landscape are created and, on the other, in the practices by which given societies make a place habitable through dwelling while creating the sense of community identity that is necessary to sustain a place through time.

Identity can be held by individuals and groups, and it can be felt in relation to, among other things, places and regions. Individual identity is a matter of self-awareness, an idea of who one is. Identification of oneself occurs in interaction with identification by others, initially the group within which one is born, grows up, and lives. An individual's identity is not given once and for all—one may have varying identities over time and in relation to several different social groups. Group identity refers to a set of common characteristics of a group of people. These may be chosen partly by the members of the group, partly acquired through socialization and practice, and partly through labeling by others. Group identities, like individual identities, are not immutable but change as members of the group change and interactions with other groups change. Group identity may be ascribed on the basis of many types of characteristics, although gender, social class, language, and nationality are among the most enduring. Identification with place for many is deep-seated. Place identity involves a sense of belonging to a place (or even to several places) and involves a sense of familiarity and a perception that the place one is attached to is different from other places. Place identity is frequently articulated as features of the place in question with which individuals and groups particularly identify, perceived as different in relation to other groups and places. Gillian Rose (1990, 89–96) notes that place identity involves identifying with places that one feels one belongs to, identifying against places that one feels different from, and sometimes not identifying at all—as in the case of strangers who feel they do not belong to or even feel excluded from a place.

The scale of place identity varies from the local to the global. The region is frequently regarded as existing at an intermediate scale, for example, between local and national or between national and global. Hence regional place identity is a sense of belonging and attachment to an identified region, distinguished from other regions. A region is socially constructed through the identification of certain characteristics seen as particular to that region, such as customs, language or dialect, way of life, or features of the physical environment. Landscape is an integral part of this. Such characteristics are often said to constitute the personality of a region. Literature, visual arts, music, and other aspects of cultural heritage are generally important means by which regional identity is constituted and maintained. Anssi Paasi (1986) has shown how regions are the product of social processes in which both individual awareness and institutions play a role in establishing regional identity: these processes involve the territorial delineation of the region, its conceptual or symbolic formulation, including giving it a name, its institutional establishment, and its acceptance in social consciousness as an entity with its own regional identity. Once established, regional identity may be maintained by rituals, policies, educational institutions, museums, and other forms of practice. Such processes inevitably involve power relations. Certain groups—the politically powerful, the economically strong, artistic elites—are more influential than others in the constitution of regional identity. However, less powerful groups may contest and influence particular notions of regional identity. To be long-lasting, regional identity depends on mutual identification by those living in the region in question and outsiders.

Norden is a fascinating area within which to pursue the ever-evolving meaning of landscape and region as place. It is the site of recent and powerful national constructions of place, particularly Norway, which first gained full independence in 1905, and Finland, which proclaimed full independence in 1917. At the same time, however, anomalous landscape-regional identities, predating nationalistic constructions, have survived in many places. In such areas, the word "landscape" does not signify a monolithic unity of environment and culture determined by nature, but rather the place of a polity constituted through human law and custom. In a world in which regionalism and the ideas of communitarianism are gaining ground, the concrete experience of these places is instructive. As an expression of society–environment relations, these landscape regions are also of great interest from a natural-science perspective because many still preserve, in their material fabric, the memory of how society and nature have interacted in the sustained generation of environments of both social and ecological value.

New Ideas of Landscape

Nordic Landscapes was initially inspired by Michael Conzen's wonderful anthology, *The Making of the American Landscape* (Conzen 1990). What made this anthology special was that it was not only about the scenic landscape surface; it was also about landscape as place and region. The authors were all deeply knowledgeable about the places they wrote about and no attempt was made to force each landscape into the procrustean bed of a particular theoretical approach to landscape. The result was a book that was both lively and personal, as any book about place ought to be. Landscape and place are important to people with a broad interest in language, art, literature, and history as well as to people with a more narrow, scientific, economic, or political interest. This book is intended to appeal to this broad spectrum of interests and to provide pathways, and the incitement, to more specialized pursuits of a particular topic.

One change that has occurred since the publication of *The Making of the American Landscape* is that the concept of landscape itself has become the subject of intense interest. This debate was provoked in large measure by the publication of an inspired collection of articles from various disciplines titled *The Iconography of Landscape,* edited by the geographers Denis Cosgrove and Stephen Daniels. The term "iconography" comes from art history, and this is an important clue to the slant the editors take on landscape. In the introduction, the two editors made the provocative claim that "a landscape is a cultural image, a pictorial way of representing or symbolizing surroundings" (Daniels and Cosgrove 1988, 1). This approach led to a spate of books, mostly by geographers, on landscape, semiotics, and textuality (Duncan 1990; Barnes and Duncan 1992).

Daniels's approach to landscape is more complex than that presented in tandem with Cosgrove. In an article that won much attention, he counterpoised the outsider perspective of the person standing apart from a place and viewing it in space as a landscape scene with the insider perspective of those who live and act within a place (Daniels 1989). The debate on landscape that Cosgrove and Daniels, singly and jointly, helped inspire led to new thinking in fields such as anthropology and archaeology regarding how these differing dimensions of the concept of landscape might be operationalized in the interpretation of social activity (Hirsch and O'Hanlon 1995). There has also been a counterreaction to the earlier emphasis on vision that has transferred emphasis to a phenomenological approach concerned with the existential experience of the insider (Tilley 1994). This phenomenological reaction is particularly notable in the influential work of the anthropologist Tim Ingold, who has pronounced, with regard to the statement by Cosgrove and Daniels quoted above, that he does "not share this view." He continues, "To the contrary, I reject the distinction between inner and outer worlds—respectively of mind and matter, meaning and substance—upon which such distinction rests." For him, "as the familiar domain of our dwelling," landscape "is *with* us not *against* us, just as we are part of it" (Ingold 1993, 154; see also Ingold 2000, 191). In some recent phenomenologically inspired British work on landscape, the scenic conception of landscape has simply disappeared (Cloke and Jones 2001).

Much of the polarized thinking about landscape that has followed in the wake of *The Iconography of Landscape* has tended either to emphasize a conception of landscape as a spatially distanced scene, to be read and interpreted semiotically as a text, or to focus on the personal existential experience of landscape. Yet, an alternative approach has also emerged that is less concerned with the insider-outsider dichotomy, but rather focuses on the political implications of landscape (Bender 1993), or, more specifically, with landscape as the expression of a polity (Olwig 1996b, 2002). The political landscape focuses on the actions of people as political beings who neither stand alone as individual spectators of a spatially distant scene nor, alternatively, submerge themselves as individual existential insiders in a world of unreflected concrete experience of the authentic phenomena of the lived world—such as old English apple orchards (Cloke and Jones 2001). In the political landscape, people act together in polities to shape the landscape.

The Making of the Nordic Landscape

While this book seeks to emulate the various concerns with particular places that characterize *The Making of the American Landscape*, it also seeks to incorporate impulses from the ongoing debate concerning the meaning of landscape as a concept. This effort is particularly challenging because the Nordic idea of landscape has had a different history from that of the British, which has dominated recent academic discourse on landscape. By contrast, the Nordic idea of landscape may bear some similarity to the notions of landscape that inspire Conzen's anthology of American landscape.[1] In the Scandinavian languages, the word "landscape" retains a historically rooted notion of belonging to region, place, or polity that gets lost in the more purely scenic approach to landscape taken by Cosgrove and Daniels (Olwig 2002).[2] As Kenneth Olwig's chapter on Jutland (Jylland) in Denmark shows, the historically primary meaning of landscape in the Scandinavian languages referred to the legal and territorial domain of a polity. This means that even now, when ancient territorial entities may no longer have political significance, the identification of landscape with region and place remains strong, and also with regard to the construction of new regions and places.

The Structure of the Book

Several of the chapters in the first three parts of the book, encompassing Denmark, the North Atlantic, and Sweden, take their point of departure in ancient Nordic landscape polities and places that to this day have maintained a regional identity, roughly comparable in scale to that of an English county, a Dutch province, or a Swiss canton. They were lands shaped by the customs and laws of a people and demarcated by the domain of that law. The chapter on Jutland is concerned with the constitution of the ancient Nordic concept of landscape as expressed in Denmark and with the subsequent transformations of the landscape concept, thus providing a useful introduction to the different meanings of landscape. The book continues with the island landscape territories of Iceland and the Faeroes. Along with Greenland—the world's largest island, the site of an ancient Viking colony, and the home of the Inuit Greenlanders—these island countries had long-standing ties to Denmark's historical North Sea imperium and earlier to the medieval Norwegian realm. The third part, on Sweden, is devoted to different historical landscape regional territories, as well as with a characteristic landscape environment. The fourth section contains Norwegian examples showing landscape identities at various scales. The mountainous, fjord-indented territory of Norway is the repository of some of Norden's most venerable local landscapes. Modern Norwegian landscape identity was shaped at a time when Norway's spectacular, and difficult, scenic environment played an important role in defining the internal geographical parameters of the new nation-state. The fifth section concentrates on the landscapes of Finland, Norden's newest nation-state, where the exigencies of twentieth-century nation-building at the troubled political and cultural interstices of East and West generated a powerful need for the construction, and reconstruction, of national and regional identities. Finland had long-standing cultural ties to Sweden, and hence also familiarity with the Swedish notion of *landskap* as regional polity, as exemplified by the internally autonomous Åland Islands, the official title of which is *Landskapet Åland*. While Åland is Swedish-speaking, the Finns speak a non-Scandinavian, non-Indo-European, Finno-Ugric language. Through its role in the construction of national identity, the place of landscape in Finland is, to an inordinate degree, a modern construction, however rooted it may be in interpretations of Finnish history. For this reason, the two final chapters on Finland are among the most reflective, but at the same time they provide lively insight into the landscape identity of Finland. Finally, two chapters at the end of the book provide a brief, general introduction to the physical and human geography of Norden as a whole, providing a frame of reference for the rest of the book.

Denmark

In his introductory chapter, "Danish Landscapes," Kenneth Olwig discusses how the Danish realm has been defined differently at different periods in history. Denmark comprises the peninsula of Jutland, the adjacent islands, of which Zealand (Sjælland) is the largest, including Denmark's capital city of Copenhagen, and the internally autonomous North Atlantic island provinces of the Faeroes and Greenland. Two hundred and fifty years ago the Danish realm included the whole of Schleswig-Holstein and Norway, as well as colonial possessions in the West Indies, West Africa, and India. In the late fourteenth century and much of the fifteenth century, during the period of the Kalmar Union, Danish monarchs ruled over the whole of what is now termed Norden, from Greenland in the west to Finland in the east. Copenhagen, once the geographical center and still the political center of the Danish realm, is located at the crossroads of east and west, north and south, lying between the Baltic Sea to the east and the North Sea and Atlantic Ocean to the west, and between the Scandinavian peninsula to the north and Germany and the rest of continental Europe to the south. The personality of Denmark has been shaped by its dual character as a land of farmers and a land of seafarers. As a land, Denmark has its origin in the bonding together of culturally similar landscape polities, that is, regional "political landscapes," such as Jutland, Funen (Fyn), Lolland, Zealand, Scania (Skåne), lost to Sweden in 1658, and others. Later, the Danish monarchy absorbed other landscape polities, such as those of Norway, united with Denmark in 1381. While Norway passed to the Swedish crown in 1814, the formerly Norwegian island landscapes of Iceland (until 1944), the Faeroes, and Greenland remained under Danish rule. Landscape in the sense of "land shaped by polity," that is, lands with ancient laws rooted in the custom of the people, is one of two meanings of the term "landscape" that Olwig identifies. The other is landscape as scenery, a new meaning that began to gain currency during the Renaissance. Landscapes became understood as naturally bordered areas of soil rather than lands shaped by historically constituted polities, and in part were the product of a cartographic imagination concerned with the mapping of national territory. At present, however, old landscape polities such as Jutland and Scania are reasserting themselves as regions increasingly act independently of the nation-state and reconstitute their place identities with the help of historical memories.

JUTLAND

The ancient landscape polity and place of Jutland, Denmark, provides the point of departure for Kenneth Olwig's chapter, "The Jutland Cipher: Unlocking the Meaning and Power of a Contested Landscape." Olwig seeks to present Jutland as understood and perceived by different actors who helped shape the perception and reality of the region as a polity and place defined by the domain of their law and custom, or, alternatively, shaped its perception as a scenic surface defined by the "natural" borders of the sea and by a topography shaped by physical processes. The two senses are deeply interwoven, so that the shape of the present-day physical landscape cannot be disentangled from the contested political and cultural identity of the peninsula. The interpretation of the landscape of Jutland thus requires an understanding of the struggle over the symbolic landscape of Jutland as the arena of a human community shaping a distinctive environment. Over the centuries, this community has gradually been incorporated into the larger state of Denmark. As Jutland lost its independent legal and political identity, the meaning of the name was reduced to simply signifying the name of a peninsula. Despite its integration into the Danish nation-state, however, Jutland remains at the spatial and social margin of the kingdom, which is centered on Copenhagen in the east. It has also retained strong links to continental Europe to the south and Atlantic Europe to the west and north. Jutland thus remains a contested site in which memories of a distinctive culture set in a distinctive environment play a role in shaping an independent Jutland regional identity.

THE NORTH ATLANTIC

ICELAND

The landscape of Iceland appears "wild," yet for Icelanders it is a landscape deeply marked by history and meaning. The significance of landscape for Icelandic identity is the theme of Kirsten Hastrup's chapter, "Icelandic Topography and the Sense of Identity." Like the Faeroes and Greenland, Iceland was settled by Vikings. The period of settlement *(landnámsöld)* in the eighth and early ninth centuries is celebrated as the source of Icelandic identity. Iceland was also the stepping stone for the European discovery of Greenland and North America. According to legend, Iceland was named for the drift ice that the first settlers encountered in the fjords. With the adoption in the tenth century of a lawbook in accordance with Norse practice, Iceland became a landscape in the meaning of both place and polity. The landscape was constituted by the land and the law. The writing down of the sagas in the twelfth century in what was becoming recognized as a distinctive language reinforced the sense of Icelandic identity and history. Thus by the mid-twelfth century, geography, law, history, and language provided a multiple definition of Icelandicness. Hastrup argues that history is not simply the material imprint on the landscape of the Icelanders' ancestors, but that even a landscape that looks empty is full of meaning and memories. Landscape is not simply a surface or stage on which people play their role but is an integral part of social space. Everyday spatial practice is imbued with feelings derived from age-old notions and events. The meanings of the present-day landscape derive from the age of the first settlement. Previous inhabitants of the landscape, trolls as well as historical beings, have left their mark in legends and in place-names. Icelanders share their knowledge of history through the names given to places in the landscape. The landscape is the site of social memory. Naming is a means of appropriation that is more significant than the modest physical traces. Remembering takes place in a landscape of words as much as of concrete things.

THE FAEROE ISLANDS

The origin of the Danish landscape polities is shrouded in the proverbial mists of time. The misty Faeroe Islands, by contrast, were not inhabited until the Viking Age. Hence the establishment of the landscape polities of the islands is of relatively recent historical origin, as Arne Thorsteinsson shows in his chapter, "Land Divisions, Land Rights, and Landownership in the Faeroe Islands." The rugged terrain permits habitation only on small fringes of land along the coast, especially toward the east, and at the bottom of firths and bays, leaving the inland areas as hill pastures and the mountainous western and northern coasts mainly as bird cliffs. The historic settlement consists of small hamlets *(býlingur)* organized into townships *(bygd)*. The *bygd* is the district for the *grannastevna*, the local body in which proprietors or tenants met to decide on matters of common interest, including the community's regulation of individual activities in the landscape. Some historians have thought that the first settlements were made by individuals exercising *landnám*, the taking into possession of whole islands or large districts, and that such landed possessions were the point of origin of the traditional settlement pattern, the original estates supposedly having been split up gradually into ever smaller plots of land. Except for church and crown land, the splitting up of landed property by inheritance was certainly the rule from the Middle Ages until recently. Thorsteinsson argues, however, that the landscape of the Faeroes should not be understood as the outcome of the division of initially large private estates, but rather as large commons that were free land in which the rights of exploitation were shared by the owners of cultivated land on individual, inheritable farmsteads. This interpretation helps explain the extraordinary continuity and stability of Faeroese settlement in an environment that would seem to defy human habitation.

GREENLAND

Like Iceland and the Faeroe Islands, Greenland, the world's largest island, was the site of ancient Viking settlement. But unlike the other North Atlantic islands, there was a native Inuit population living in northern Greenland at the time of the Norse settlement. Norse settlement eventually died out, but Greenland remained, like Iceland and the Faeroes, the colonial possession first of Norway and then of Denmark. Like Iceland and the Faeroes, Greenland is in the midst of a process of establishing a modern postcolonial identity as a land while seeking to retain a sense of its cultural heritage as a hunting and fishing society. Bo Wagner Sørensen's chapter, "Perceiving Landscapes in Greenland," concentrates on the ethnopolitical aspects of landscape and territorial identity in the interstices between a Greenlandic landscape—landscape defined as culturally invested or informed nature—and a collective or national identity. In today's Greenland, nature/landscape also seems to be part and parcel of an "identity complex," encompassing the Greenlandic language, local foods, and certain more elusive Greenlandic "values" and "ways of thinking."

SWEDEN

Introducing Sweden in "The Swedish Landscape: The Regional Identity of Historical Sweden," Ulf Sporrong describes some characteristic features of the physical landscape. Covering the bedrock of granites and gneisses over most of the country are the glacial landforms resulting from the last glaciation, which ended about ten thousand years ago. The constantly changing coastline, due to shore displacement resulting from land uplift, also has its origin in the glaciation, when the ice sheet depressed the earth's crust, which is still rising after the release of the weight of the ice. Sweden was settled by people who followed the retreating ice. The best agricultural areas lie below the highest marine limit at the end of the Ice Age on land that has emerged from the sea. The exploitation of other natural resources, forest and iron ore in particular, have left their stamp on the physical landscape. Political landscapes emerged during the Middle Ages as the various provinces *(landskap)* developed self-governing institutions with their own lawbooks, until these were codified in a common body of law as Sweden became unified. Despite a highly centralized government since the sixteenth century, most Swedes today still associate their origins and identity with these historical provinces rather than with the modern administrative units. Landscape also has a prominent place in Swedish literature, and Sporrong illustrates this with passages from three internationally known Swedish authors: Carl von Linné (Linnaeus), the eighteenth-century botanist and traveler, who described landscapes such as Scania; Selma Lagerlöf, a teacher and novelist who in the late 1800s described the geography and landscapes of Sweden for children and went on to win the Nobel Prize in literature; and Astrid Lindgren, whose children's stories imprinted the rural landscape of her home area of

Småland on the minds of several generations of children in the latter part of the twentieth century. Other records of landscape and people from the sixteenth century onward—parish registers, official statistics, tax records, and cadastral records of landed property—help document the profound changes that have affected the Swedish landscape as a result of land reorganization, emigration, and migration to the towns as Sweden has been transformed from a rural to an urban society.

SCANIA (SKÅNE)

Scania, or Skåne, was a landscape territory under the Danish monarchy until it was forcibly annexed by Sweden in 1658. Despite a thorough program of acculturation by Swedish authorities, Scania's regionalists claim that the area still maintains a special historically grounded cultural identity—with corresponding cultural landscape—that warrants greater regional autonomy. In the late 1990s, a controversial bridge linking Denmark and Scania was constructed, and social scientists, working with the bridgebuilders, sought to construct a new functional regional identity linking Danish Zealand (Sjælland) and Scania. Tomas Germundsson's chapter, "The South of the North: Images of an (Un)Swedish Landscape," explores Danish-Swedish regional identity politics in an era of globalization and the European Union. Scania serves as a case study in which different concepts of landscape can give different insights: landscape as a territory with a perceived internal coherence, landscape as scenery, and landscape as changing juxtapositions of assorted entities flowing in time and space. Both historically and geographically, the chapter on Scania provides an excellent entry to the landscapes further north, which have a more clear-cut historical Swedish identity.

DALECARLIA (DALARNA)

Dalecarlia, or Dalarna, is an ancient Swedish *landskap* with a legendary populace of independent, freedom-minded farmers that is widely seen to form the heartland of Sweden. In the hands of artists and poets, its physical landscape has become a symbol for Sweden as a whole. Yet Dalecarlia is in many ways an anomaly, as Ulf Sporrong shows in his chapter, "The Province of Dalecarlia (Dalarna): Heartland or Anomaly?" Dalecarlia's characteristic landscape identity as a region of small farms linked by strong bonds of kin and community owes much to the importance of legal customs, allowing for joint family inheritance. Such customs are rare in Sweden, though they are found to the west in Norway and the North Atlantic. Dalecarlia exemplifies how the regional landscape identities of Norden can both contradict and reinforce national identities.

VÄRMLAND

Värmland is an ancient but economically and culturally heterogeneous landscape territory. It has an industrialized southern lakeshore and forested borders with the remainder of Sweden to the east and Norway to the west. In the seventeenth century, Finnish colonists settled the forests of northern Värmland. Despite its heterogeneity, Värmland maintains a strong sense of regional identity. This is due in no small measure to the work of the geographically inclined author Selma Lagerlöf (1858–1940) and the regionalist author Carl-Axel Gottlund (1796–1875). Gabriel Bladh's chapter, "Selma Lagerlöf's Värmland: A Swedish *Landskap* in Thought and Practice," explores the changing relationships between physical landscape, regional identity, and the process of industrialization and modernization.

THE *HAGMARK*

Place identity in Sweden is strongly linked to the ancient territorial *landskap* polities, but several of these places share a common environmental feature, the woodland meadow known as the *hagmark,* which has come to be seen as a vitally Swedish characteristic of the landscape. The *hagmark* is wooded grassland that is largely the outcome of the grazing, mowing, and occasional cultivation of clearings in the woods by farmer-foresters. It is an open grassy environment enclosed by trees, and in forested northern Europe it provided the closest equivalent of the Mediterranean image of a pastoral paradise. The *hagmark* is an environment prized both for its cultural-historical value and its ecological diversity. The chapter by Margareta Ihse and Helle Skånes, "The Swedish Agropastoral *Hagmark* Landscape: An Approach to Integrated Landscape Analysis," focuses on the physical geography of this environment as it reflects ecological and social factors. The authors make clear that even though characteristic landscape-regional environments may reflect human use in addition to cultural values, they are, when seen from a natural-science perspective, highly complex and are not readily amenable to superficial forms of management.

NORWAY

Introducing Norway, Michael Jones presents "The 'Two Landscapes' of North Norway and the 'Cultural Landscape' of the South." Specific types of resource use are often made to represent regional identity through the concept of landscape. The "two landscapes" of the north are distinguished by common property regimes, seasonal movements between different resource niches, and landscape knowledge related to coastal fishing and Saami (Lapp) reindeer herding. The historical significance of a third landscape of subsistence agriculture, in which women had the main responsibility for farming, is neglected in the discourse. The "cultural landscape of agriculture," however, is a view of the landscape by planners and environmental managers defined in terms of government payments for keeping agricultural land in use, and, Jones argues, it has a South Norwegian bias. In a national inventory of agricultural landscapes, the environmental agencies have identified more than one hundred "valuable cultural landscapes" that are important for their biodiversity and cultural heritage. Although only

9 percent of Norway's agricultural land is found in North Norway, almost one-quarter of the nationally important "valuable cultural landscapes" were identified there. These areas often were in the process of going out of agricultural production and were identified principally on the basis of their botanical values by experts based largely in the south, rather than by local people. Agricultural land comprises, however, less than 3 percent of Norway's land area, and the significance of other types of rural landscape has been neglected. Norway has extensive forestry landscapes, especially in South Norway. The south, furthermore, has its own landscapes of fishing and reindeer herding. The main urban and industrial landscapes are also found in South Norway. Yet, in the south and north alike, the majority of people work in service occupations and live in urbanized landscapes.

NORTH NORWAY

In her chapter, "The Landscape in the Sign, the Sign in the Landscape: Periphery and Plurality as Aspects of North Norwegian Regional Identity," Venke Åsheim Olsen discusses the complexity of regional identity and the role of landscape both in its constitution and its signification. North Norway has been described by various metaphors: "a diverse people in a rebellious land" reflects North Norwegian identity in terms of ethnic plurality and political periphery; "the meeting of three tribes" refers to the Norwegian, Saami, and Finnish components of the population; "the land of suspense" alludes to the challenges and potentialities of regional development. The emphasis on diversity and contradiction has led some to question whether a common North Norwegian identity can be said to exist. Yet a sense of belonging has been and continues to be culturally constructed through academic debate as well as regional identity projects such as the "Cultural History of North Norway," published in the 1990s. Historically, North Norway has constituted a distinct legal-administrative entity, known as Hålogaland, since the twelfth century. The region's location at the northern periphery of Europe is signified by distinctive features of both physical and human geography, while its location at the border between Western and Eastern Europe has through history made it a periphery with geopolitical significance. Ethnicity, cultural heritage, and origin stories are concepts that help us understand the link between identity and landscape in a multiethnic region. Cultural heritage expresses contrasts and sometimes conflicts between ethnic groups, while landscape and landscape elements provide sources of historical information about the use of resources and their users. Cultural information is transmitted through many types of sign. Heraldic arms provide an example of signs of medieval origin that have gained new popularity in recent years as expressions of local and regional identity. Heraldry conveys information about landscape and other features that characterize and hence symbolize local districts as well as wider regions. In the form of road signs and decorative elements on public buildings, heraldic arms become in turn part of the landscapes that they signify and demarcate.

KARMØY

Western Norway, with its dramatic mountains and fjords, has often been considered to derive its regional identity from the physical landscape. An alternative, historical approach that pays more attention to the role of human activity in shaping the area is advocated by Anders Lundberg in his chapter, "Changes in the Land and the Regional Identity of Western Norway: The Case of Sandhåland, Karmøy." Lundberg takes as his starting point recent regional divisions of the landscapes of Norway, one in a report for the Nordic Council of Ministers, where a regionalization based on biophysiographic criteria was made for purposes of nature conservation, and the other a division into vegetation regions in the *National Atlas of Norway*. One problem encountered in studying western Norway is how to deal with the long history of modification of the vegetation by human activity. While vegetation science, landscape ecology, and biogeography have contributed much to an understanding of processes of environmental change, the landscape is more than geomorphology, climate, and ecosystems. A full understanding of how the region and its vegetation have been shaped requires the analysis of landscapes as cultural entities that have undergone continuous change as a result of human activity through the course of history. In a detailed study of land-use change on the farm cluster of Sandhåland, on the island of Karmøy in southwestern Norway, Lundberg illustrates processes of change during the last two hundred and fifty years, some leading to habitat destruction and others to increased biodiversity. He identifies four periods: (1) before 1870, agriculture was characterized by the use of hand tools on fragmented holdings in hamlets; (2) land reorganization between 1870 and 1910 led to the disintegration of farm clusters and the consolidation of farm units into individual landholdings; (3) between 1910 and 1950, horse mechanization facilitated the process of bringing new land into cultivation; and (4) after 1950, tractors replaced horses, allowing further new cultivation and contributing to input-intensive farming, but also leading to the abandonment of areas less amenable to mechanized agriculture. These transformations are manifested in the ever-changing landscape of a region that, paradoxically, derives its identity from perceptions of natural stability.

INNER SOGN

Contrasting with the low-lying island of Karmøy is the vertical topography of the western Norwegian fjord landscape, presented by Ingvild Austad and Leif Hauge in their chapter, "The 'Fjordscape' of Inner Sogn, West Norway." Marked gradients of local climate and vegetation typify the fjord from the coast to the innermost settlements and from sea level to the mountains and glaciers. The cultural landscape of Sogn is characterized by small-scale agriculture. Animal husbandry, stretching back to 2500 B.C., has always been important. Fodder collection and grazing in the extensive mountain outfields have led to a variety of seminatural vegetation types, such as hay meadows and wooded pastures, pollarded and coppiced woodlands, and mountain pastures. The animals were moved annually to seasonal mountain farms at different altitudes. Trade was carried on by boat to Bergen and via the cattle paths across the mountains to southeastern Norway. Austad and Hauge describe the typical features of this landscape and the changes that have occurred through time. The land reorganizations of the nineteenth century led to the dispersal of many of the old clustered farms while facilitating the introduction of new farming methods and equipment. During the twentieth century, areas unsuitable for modern farming were abandoned and the number of seasonal mountain farms was greatly reduced. The authors also describe the landscape as experienced by artists: painters, poets, playwrights, and folk musicians have been inspired by the fjord landscape and through their work have contributed to a strong regional and national identity.

HJARTDAL-SVARTDAL IN TELEMARK

Landscape transformations over time are documented for the Hjartdal-Svartdal area in inner Telemark, southeastern Norway, by Ann Norderhaug in her chapter, "The Agropastoral Mountain Landscape in Southern Norway: Museum or Living Landscape?" From the time the area became permanently settled in the fifth century, land use has been dominated by animal husbandry. The small arable fields and the extensive outfields, the latter providing both summer grazing and winter fodder, were described in detail by the local pastor Hans Jacob Wille in 1786. The seminatural grasses and other vegetation used for fodder collection and grazing have a high biodiversity, but they began to disappear as a result of the intensification and specialization of agriculture during the second half of the twentieth century. The agropastoral economy has also left a rich cultural legacy in the form of wooden buildings, stone walls, and other built structures. This legacy, along with rich traditions of wood carving, rose painting, and folk music, have given the region a reputation as being a "typically" Norwegian landscape, reinforced by the depictions of landscape painters in the nineteenth and twentieth centuries. Norderhaug argues that the maintenance of the characteristic features of this cultural landscape" in the 1990s, requires sustainable agricultural management rather than turning the areas into a museum.

FINLAND

Introducing Finland, W. R. Mead begins his chapter, "Reflections on the Historical Landscapes of Finland," with a selection of impressions by the earliest British travelers to Finland. He then presents the historical sources available to the researcher reconstructing past landscapes in Finland. Accounts of the transformation of the natural landscape into a cultural landscape are followed by examples of how some present-day landscapes are conversely reverting from a cultivated state to wild landscapes, illustrated by depopulated islands of the west, such as the outer islands of Åland, and remote areas in the east. The former parts of Finland that were lost to the Soviet Union in 1944 remain in Finnish consciousness as landscapes of memory. In contrast, two distinctive features of the historical landscape have kept their place in the present-day physical landscape: the landscaped parks of country estates in southern Finland and the imprint of the Orthodox Church in eastern Finland. Pioneering research on landscape representation was undertaken by the Finnish geographer J. G. Granö, who in 1929 presented a unique method for mapping landscape features, not only the visual forms but also the smells and sounds of the landscape. Granö regarded the fundamental objective of geographical research to be the study of the environment as perceived by the senses. With the rise of the tourist industry, landscape has acquired new values. Today landscape is often regarded as a commodity to be sold to consumers, illustrated by the case of Finnish Lapland.

ÅLAND

Åland, an archipelago lying midway between mainland Sweden and Finland, is the only territorial political landscape in Norden that uses the term *landskap* to denote the administrative region. Since the beginning of the 1920s, Åland has been an internally autonomous, Swedish-speaking region under the sovereignty of Finland. In his chapter, "Landscape Territory, Autonomy, and Regional Identity: The Identity of the Åland Islands in a Cultural Perspective," Nils Storå defines landscape-regional identity as the part territory plays in the identity experienced by a group of people living in a particular region. Regional identity is given an intermediate position between national identity and local identity and is communicated by the region's inhabitants as well as attributed to them by outsiders. After autonomy, the government of Åland played an important role in shaping and communicating the cultural identity of the region. At the same time, Åland is a maritime society and thus mobile, and this has provided Ålanders with ample opportunities to communicate their identity. Like the rest of present-day Finland, Åland was part of Sweden until 1809. It was historically the smallest of Sweden's "landscapes," having in the Middle Ages its own law assembly or *ting* in addition to three local *tings*. Åland forms a cultural bridge between East and West. Until the nineteenth century, Ålanders had an obligation to transport mail between Finland and Sweden. The islands form a crossroads of the Baltic, and hence are open to outside influences from several directions and also subject to the ravages of war. The islands were occupied by Russian troops during the Great Northern Wars of the early eighteenth century. Then with Finland, they came under the rule of the Russian tsar in 1809. Suffering bombardment from British and French naval vessels in the Crimean War, the islands were demilitarized by the Treaty of Paris in 1856. After Finnish independence in 1918, Åland was claimed by both Finland and Sweden. The majority of Ålanders wanted reunion with Sweden. Intervention by the League of

Nations secured internal autonomy for Åland, with its own government and democratically elected parliament *(landsting)*, and the demilitarization and neutralization of the islands were confirmed in 1921. Physically and in the minds of the inhabitants, Åland comprises many landscapes: seascape, outer and inner archipelagoes surrounding the Ålandic mainland, winter icescapes, and summer recreational landscapes. The resource landscapes of fishermen in the outer archipelago, farmer-fishermen in the inner archipelago, and farmers on the main island have contributed to the island culture, although the predominance of shipping, the growth of the service economy, and depopulation of the more remote islands led to considerable changes in the twentieth century. Åland identity is officially recognized through the Swedish language, regional citizenship (the "right of domicile"), and the Åland flag. Images of island landscapes, reflected in the Åland anthem, the local newspaper, and school textbooks, have contributed further to the process of nation-building. The maypole, decorated at midsummer and evoking the summer landscape, has become the national symbol of Åland.

THE EASTERN FINNISH FOREST

In the process of nation-building, landscape has provided a framework for the construction of a local and national Finnish identity, according to Ari Aukusti Lehtinen in his chapter, "Landscapes of Domination: Living in and off the Forests in Eastern Finland." The landscape carries a double emphasis on aesthetic experience and territorial identity. For the Finns, landscape is the medium by which to evaluate, orientate, and even compete in the environing world. Finnish landscape is (1) a spatial framework by which the familiar and well known within the environing world is identified as a home region or arena and, hence, as a territory under domestic control. The landscape is also (2) a collective means to react against the pressures from the outside; it is a tool for categorizing what is unknown and, therefore, fearful. Landscape divides "us" and "them." Finally, (3) The Finnish landscapes can be identified at various levels of abstraction from short-lived "export landscapes" to archetypal landscapes that function as tools of spatial integration and separation. These landscapes play an especially important role in relating to the Finnish geopolitical position, within the circumboreal forest zone, between the East and the West. Identification with the eastern forest is balanced against Westernization and modernization. The landscape of fear refers to the shared collective experience of external threat, whereas the landscape of domination is an expression of the modernization and Westernization that promises to alleviate that threat, but at the expense of forms of landscape identity in which Finns have seen themselves as being at home in the forests of the eastern borderland.

VISUALIZING FINLAND

As a nation-state born in the twentieth century, Finland represented a modern national vision. The vision drew heavily on artists, who gave the Finnish landscape shape and form as place, as shown by Maunu Häyrynen in "A Kaleidoscopic Nation: The Finnish National Landscape Imagery." The landscape image of Finland acted as a scenic inventory of the nation's natural, economic, and aesthetic resources. It served to demarcate national borders and render the geographic makeup of the nation-state as both natural and God-given. The landscapes were "discovered" by individual authors or artists and became stock items only gradually. As a discursive model and signifying system, the national landscape imagery has preserved its symbolic force despite its recurrent reorganizations and reinterpretations. The structure and hierarchy of this imagery persists today in tourist advertising, popular geography, and in the values expressed in national landscape conservation policy.

LANDSCAPE AND REGION IN FINLAND

In "Finnish Landscape as Social Practice: Mapping Identity and Scale," Anssi Paasi discusses the relations between landscape, territory, and identity. He develops a conceptual framework to trace the meanings of landscape and region in different social and cultural practices and the power relations involved in the production of ideas of landscape and their identities on various spatial scales. He discusses landscape in terms of: (a) a nationalistic, ideological landscape; (b) a framework for administrative power; (c) an academic construct; (d) part of a regional image; (e) a form of regionalism and provincial promotion; and (f) an instrument of media discourse. Paasi suggests that landscape and identity are contested contextual categories. This, in turn, means that one must raise the question of just whose landscapes and identities one is dealing with, and for what purpose, when one addresses more general questions of landscape identity.

NORDEN

Two concluding chapters provide a brief presentation of some principal features of the geography of Norden: "The Nordic Countries: A Geographical Overview" by Michael Jones and Jens Christian Hansen and "Features of Nordic Physical Landscapes: Regional Characteristics" by Ulf Sporrong. These chapters provide a broad frame into which the previous chapters can be placed.

CONCLUSION

This book is the product of a cooperative effort by prominent Nordic scholars of landscape, region, and place. These chapters on the multiple meanings of landscape and region are conceptually situated within contemporary social and scientific theory and concretely linked to the broader history that relates Nordic experience to that of Western society. Some of these studies focus at the theoretical level on landscape and region as social constructions, whereas others dig deeply into the ecological interactions that have created environments that are now highly valued both ecologically and aesthetically. The chapters focus on the regional identity of landscape as place from a social and cultural perspective and from a natural science perspective. It is not our goal to create a holistic vision of landscape, but to show how different discourses meet and can speak to one another in the understanding of particular places.

This book should be useful in a number of contexts. Its many-faceted approach to the complexity of landscape and region provides a series of theoretically grounded case studies that should be of use to those concerned with region and landscape within fields such as geography, planning, landscape ecology, and landscape architecture. It is also intended to be of more general interest to those who are concerned with the role of landscape, region, and place in the environmental and cultural history of Western society. Finally, it should be of value to anyone whose geographic imagination is drawn to the northern edge of European civilization.

Acknowledgments

This book had its origins in the Nordic Seminar for Landscape Research, a series of meetings held in the 1990s by an interdisciplinary group of researchers with a common interest in landscape. The initial two meetings were held in Sweden, the first at Sigtuna in November 1993 and the second in Lund in June 1994. The third meeting was held at Sogndal in Norway in October 1996. The papers presented at the Lund and Sogndal meetings have been published as seminar proceedings.

At the Sogndal seminar an idea crystallized to collaborate on publishing a series of essays exploring different conceptions of landscape and their relevance for understanding regional identity, with case studies from the five Nordic countries of Denmark (including the Faeroe Islands and Greenland), Finland (including Åland), Iceland, Norway, and Sweden. The term "Nordscapes" was coined as a working title for the project, connoting the various ways in which the land in different regions of Norden has been shaped not only physically but also conceptually and institutionally. The suffix *-scape* has the same etymological derivation as the word "shape" in English; variants of the suffix are found in the various Germanic languages, including those of Norden (e.g., *-skap* in *landskap*). Our focus was on how human interaction with the environment is manifested and how it involves more than the ways human activity modifies the physical features of the surroundings and creates new features. It also reflects how humans think about their environment at different times and places and how that thinking is manifested in the institutions and polities that regulate their dealings concerning the environment.

A series of meetings was held for the express purpose of writing the book, allowing the various chapters to inform one another. At meetings held at Mariehamn in Åland in October 1997 and at Sørvágur in the Faeroes in May 1999, the contributors exchanged ideas and presented drafts for mutual criticism. Many contributions were also presented to a wider circle for criticism at a workshop held during the eighteenth session of the Permanent European Conference for the Study of the Rural Landscape in Trondheim, Norway, in September 1998. Additional contributors were invited to join the Nordscapes project along the way. This collection of essays is not intended to cover all regions in Norden; rather, it indicates the predilections of the twenty-one contributors.

As editors, we would like to thank especially Professor Emeritus Ulf Sporrong of the Department of Human Geography at Stockholm University for initiating the Nordic Seminar for Landscape Research and for encouraging the Nordscapes project. We express our thanks to the National Heritage Board in Sweden; the Royal Swedish Academy of Letters, History, and Antiquities; and the Bank of Sweden Tercentenary Foundation for financial support for the organization of our seminars and meetings. We are indebted to the Swedish Council for Planning and Coordination of Research for a financial grant to subsidize the final publication. We thank Catriona Turner for assistance in compiling the index.

Notes

1. Kenneth Olwig argues that the notion of landscape as a pictorial scenic surface developed in Britain as part of the process by which the state sought to consolidate Britain as a political unity (Olwig 2001, 2002). The contemporary British focus upon this scenic conception of landscape reflects the power of this British imperial tradition. There is also an older Germanic tradition, spanning from Norden to Germany, Switzerland and the Netherlands, and to the specific heritages of the separate nations of the British Isles, that has maintained an interest in the political landscape understood as a unity of polity and place. The original states of the United States once formed colonies within the British Empire, but they also formed independent regional polities, shaped in part by the different ethnicities (many of them Germanic) that settled there. The approach to landscape that sees it as the place and environs of a polity has had considerable influence in England and America through authors such as England's Walter G. Hoskins and America's John Brinckerhoff Jackson and Carl Sauer, both of whom were influenced by Germanic ideas of landscape (Sauer 1925; Hoskins 1955; Jackson 1984).

2. Finnish is not a Scandinavian language, but because of the long-standing historical ties between Finland and Sweden, the Finnish language contains words that convey the territorial sense of landscape.

References

Barnes, Trevor J., and James S. Duncan, eds. 1992. *Writing Worlds: Discourse, Text, and Metaphor in the Representation of Landscape.* London: Routledge.

Bender, Barbara. 1993. "Landscape: Meaning and Action." In *Landscape: Politics and Perspectives,* ed. Barbara Bender, 1–17. Providence, R.I., and Oxford: Berg.

Cloke, Paul, and Owain Jones. 2001. "Dwelling, Place, and Landscape: An Orchard in Somerset." *Environment and Planning A* 33: 649–66.

Conzen, Michael P. 1990. *The Making of the American Landscape.* Boston: Unwin Hyman.
Daniels, Stephen. 1989. "Marxism, Culture, and the Duplicity of Landscape." In *New Models in Geography,* vol. 2, ed. Richard Peet and Nigel Thrift, 196–220. London: Unwin and Hyman.
———, and Denis Cosgrove. 1988. "Introduction: Iconography and Landscape." In *The Iconography of Landscape,* ed. Denis Cosgrove and Stephen Daniels, 1–10. Cambridge: Cambridge University Press.
Duncan, James S. 1990. *The City as Text: The Politics of Landscape Interpretation in the Kandyan Kingdom.* Cambridge: Cambridge University Press.
Hirsch, Eric, and Michael O'Hanlon, ed. 1995. *The Anthropology of Landscape: Perspectives on Place and Space.* Oxford: Clarendon Press.
Hoskins, William G. 1955. *The Making of the English Landscape.* London: Hodder and Stoughton.
Ingold, Tim. 1993. "The Temporality of Landscape." *World Archaeology* 25, no. 2: 152–72.
———. 2000. *The Perception of the Environment: Essays on Livelihood, Dwelling, and Skill.* London: Routledge.
Jackson, John Brinckerhoff. 1984. *Discovering the Vernacular Landscape.* New Haven: Yale University Press.
Olwig, Kenneth R. 1992. "The European Nation's Nordic Nature." In *The Source of Liberty: The Nordic Contribution to Europe,* ed. Svenolof Karlsson, 158–82. Stockholm: The Nordic Council.
———. 1996a. "Nature: Mapping the 'Ghostly' Traces of a Concept." In *Concepts in Human Geography,* ed. Carville Earl, Kent Mathewson, and Martin S. Kenzer, 63–96. Savage, Md.: Rowman and Littlefield.
———. 1996b. "Recovering the Substantive Nature of Landscape." *Annals of the Association of American Geographers* 86, no. 4: 630–53.
———. 2001. "Landscape as a Contested Topos of Place, Community, and Self." In *Textures of Place: Exploring Humanist Geographies,* ed. Steven Hoelscher, Paul C. Adams, and Karen E. Till, 95–117. Minneapolis: University of Minnesota Press.
———. 2002. *Landscape, Nature, and the Body Politic: From Britain's Renaissance to America's New World.* Madison: University of Wisconsin Press.
Paasi, Anssi. 1986. "The Institutionalization of Regions: A Theoretical Framework for Understanding the Emergence of Regions and the Constitution of Regional Identity." *Fennia* 164, no. 1: 105–46.
Rose, Gillian. 1990. "Place and Identity: A Sense of Place." In *A Place in the World? Places, Cultures, and Globalization,* ed. Doreen Massey and Pat Jess, 87–118. The Shape of the World: Explorations in Human Geography 4. Oxford: The Open University and Oxford University Press.
Sauer, Carl O. 1925. "The Morphology of Landscape." *University of California Publications in Geography* 2, no. 2: 19–53.
Tilley, Christopher. 1994. *A Phenomenology of Landscape: Places, Paths, and Movements.* Oxford, Berg.

Denmark

1.
Danish Landscapes

KENNETH R. OLWIG

Any study of Denmark's landscapes must first wrestle with the problem of defining Denmark.[1] The first impulse is to go to the latest atlas and find the shaded area of color that is designated Denmark. But if the atlas was inherited from a grandparent, one would see that the shaded area is different from the area designated as Denmark in a more recent atlas. And if that grandparent inherited the atlas from a preceding grandparent, it would be different again. The one constant about Danish history is that the area colored Denmark keeps changing, and during the last four centuries it has, for the most part, become consistently smaller. In this book, we have organized the chapters on different landscapes according to their location within the spatial boundaries of present-day nation-states. The exceptions are the landscapes of the North Atlantic, all of which are, or were recently, under the suzerainty of Denmark but can hardly be said to lie within Denmark's spatial boundaries. Because we have not included these landscapes in the space we allocated to Denmark, the section on Denmark seems relatively short. If we took a more historical perspective, however, this book would be dominated by Denmark. At a certain point, as under much of the reign of Queen Margaret I and her son Erik in the late fourteenth and early fifteenth centuries, virtually the whole book could be placed under Denmark.

DENMARK'S SPLIT PERSONALITY

A first glance at the map will draw the eye to the 16,639 sq. miles (43,094 sq. km) of archipelago that forms the ancient kingdom of Denmark, located just south of the Scandinavian peninsula. It bridges the north/south gap between the Nordic region, Germany, and the rest of continental Europe. It also forms a kind of slash in the ocean, the peninsula of Jutland acting as a massive jetty, whose rugged flat and sandy western coast juts unbroken into the North Sea, dividing it from the Baltic, while Jutland's indented hilly eastern coast, rising to Denmark's highest natural elevation of 568 ft. (173 m), protects some 406 Danish islands (about 97 of which are inhabited) from Atlantic storms. This slash divides Britain and the smaller Atlantic islands in the west from the continental lands of Eastern Europe adjoining the Baltic. It is here that some five million Danes now live, summer and winter, amid the mediating vagaries of a damp, cool, and windy Atlantic climate.

The touristic image of Denmark, which is visible on the average tin of Danish butter cookies, with its cozy half-timbered, thatched farms, punctuated by stately manors in Renaissance red-brick or Enlightenment neoclassical white, is not far off the mark, even if the humble thatch has become a luxury item few working farmers can afford. Eastern Jutland and the islands, including Funen (Fyn) (1,152 sq. miles, 2,984 sq. km), Lolland (480 sq. miles, 1,234 sq. km), and Zealand (Sjælland) (2,709 sq. miles, 7,015 sq. km), the home island of Copenhagen, consist largely of gently undulating, fertile soils, with nary a rocky outcropping to spoil the softness of the waving pale blue-green barley or the bright yellow swath of rape. When the flinty geologic foundation of the islands finally does become visible, where cliffs rising from the sea mark the southeastern coast of the island of Møn, they are discreetly clad in soft white chalk. The morainic soils of Denmark, scraped by glaciers from the Scandinavian shield and dumped into the sea at its foot, have provided the foundation for agricultural communities so stable that it is not unreasonable to suppose that the farmer seen plowing the soil today is a distant relative of the people buried in the Iron Age barrow blocking his path. This, at least, is a favorite image of the national-romantic poets and painters, who created the image of the Danes as a peaceful population of egalitarian and democratically minded farmers.

A more circumspect look at a map will reveal that the areas shaded Denmark do not stop with the wrinkled, variegated coastline of the peninsula and isles that are stuffed, like a broken cork, into the mouth of the Baltic. Roughly 100 miles (160 km) to the east, in the Baltic, lies the isle of Bornholm with its dark granite outcroppings. In the opposite direction, roughly 700 miles (1130 km) out in the North Atlantic, there are myriad dramatically rocky specs called the Faeroe Islands (540 sq. miles, 1,399 sq. km), and to their north, about twice as far from Denmark, there is the enormous icy mass of Greenland (840,000 sq. miles, 1,833,900 sq. km), which dwarfs Denmark proper. These are the remnants of a vast empire that Denmark inherited when the Danish king became monarch of both Norway and Denmark in 1381 and king of a united realm in 1536. This empire included Iceland until Denmark's occupation by Germany during World War II, when Iceland became independent after a plebiscite in 1944. Another recent part of Denmark's empire is the Danish West Indies, which became the American Virgin Islands after their sale in 1917. (Further back in time, the Danish tropical empire included outposts in Africa and India.) Going back to Viking times, Denmark's empire included places as far-flung as Britain to the west and Tallinn (meaning "Danish castle") to the east in what is now Estonia.

Many British schoolchildren know the story of the Viking King Canute (or Knut) (c. 995–1035), whose power over Britain's shores was so great that he was thought to be able to stop the tide. Fewer have heard of Valdemar the Victorious (1170–1241), who laid the Baltic at his feet, and to whom the Danish flag, "Dannebrog," supposedly descended from heaven during a battle in Estonia. The eastern family connections of this potentate are suggested by the fact that his name is a Danified version of Vladamir. Norway to this day extends to a far northern border with Russia on the Barents Sea, and Danish sea kings, such as King Christian IV (1577–1648), sailed along this "northern way" and saw to the fortification of this coast under the Danish flag. Without this empire, it is hard to explain the anomaly of Copenhagen, which means "merchant's harbor" or "haven." What, one might otherwise ask, is a large cosmopolitan city, with its grand foreign ministry, its world-scale shipping companies, its girth of warlike fortifications and docks with names recalling distant exotic ports, doing amid a pacific terrestrial sea of butter, bacon, barley, and beer? Denmark is indeed the home of the homebody farmer, dwelling amid long-cultivated fields, but it is also the home of ruthless sea kings and naval captains, restless explorers, merchant seamen, mercantile merchants, and long-distance fishermen.

In part because the map encompasses all of Denmark within the same shaded contiguous space, one might think that the space within Denmark's borders is woven together in seamless web. A city is thus seen to be the nucleus and expression of its hinterland. But even if Copenhagen lies amid fertile agricultural land, it did not grow out of this land into a metropolis simply by trading agricultural products. Cities like Copenhagen were established by monarchs as fortifications against foreign foes and as nuclei for their military and economic power over a competing rural aristocracy. Here the modern centralized state took form. Copenhagen was a walled and enclosed bastion whose polyglot population of burghers (who were under the protection of the king's castle or "burg") as likely as not spoke a language different from the surrounding population of Danish farmers. The striking split in Denmark's personality between settled peaceful farmers and restless seafarers is related to the geography of the place. The area of Denmark is well suited to the stable bonds fostered by permanent agriculture, but at the same time it provides a location at the crossroads of east and west, north and south, that favors the overlapping interests of sea king and merchant trader. The two Denmarks of course coexist, but they have not always had the same shared interests, and they are difficult to encompass within a unified national identity, as the chapter on Jutland will show.

These different ways of thinking about Denmark's personality relate to two different ways of thinking about the way the land has been shaped as a landscape, and therefore to two different ways of thinking about land and the social and physical processes that shape it.

An Old, Yet New, Idea of Landscape

The definition of Denmark, and thereby the definition of what is meant by Danish landscape, depends to a certain extent upon what one means by "landscape." This is well illustrated by the example of the section on *Skåne,* or Scania in Latin and English (the root of *Scan*dinavia), which we have located in this book under Sweden. For most of its history, however, and until the mid-seventeenth century, Scania was part of Denmark. A modern map will place the capital and largest city of Denmark, Copenhagen, on the eastern edge of the country on the island of Zealand. The population of Denmark is disproportionately concentrated in the Copenhagen area. The population of Copenhagen proper today is about 500,000, but Greater Copenhagen is about three times that size, accounting for roughly a quarter of the population of Denmark. Capitals are administrative centers that, like Paris or Berlin, tend to be located somewhat centrally in a country, not balancing precariously on its edge—but Copenhagen's location is entirely asymmetrical. If, however, we take into account Scania, which lies across a narrow belt of water from Copenhagen, then the capital finds itself suddenly placed much more toward the middle of the country.

Much as Jutland is the land of the Jutes, Scania is the land of the *skåninger*. The land has been shaped by a polity, and hence is a "political landscape" in both the earliest and the most contemporary meaning of the word "landscape." At one time Scania, like Zealand or Jutland, had its own body of ancient law rooted in its customs. It is through the process of lawfully dwelling together that the Scanian habitus was generated as "a particular area of activity" (*Merriam-Webster's Collegiate Dictionary,* 11th edition, v. "landscape"; Olwig 1996). The writing down and formalization of unwritten law, according to principles heavily influenced by the written-law tradition of the Romans and the Roman Catholic Church, was part of the process by which Scanian law, and Scania itself, became unified with other Danish lands under the state of the Danish monarch. Scania nevertheless retained a separate identity as the place of a dwelling for a polity with its own characteristic mores and Scandinavian tongue, even if it gradually ceased to have its own formal legal institutions. Scania was shaped as a land, with the character of a land, and hence became known as a *"landskab"* or landscape with an ancient body of *landskabslov* or "landscape law."[2]

The Renaissance Meaning of Landscape

Scania was annexed by the Swedish state, together with two adjacent Danish territories, Blekinge and Halland, in the mid-seventeenth century after Sweden defeated Denmark in a war. This event had a major impact, of course, on the area of the map shaded as part of Denmark. It also was part and parcel of a change then occurring throughout Europe that brought with it a new concept of land and landscape. How, one might ask, could the Swedish state justify the policy of cultural suppression by which it sought, among other things, to force a population to learn a new language and transfer its loyalty to a new state? The justification lay in a new way of thinking about land and landscape. By annexing the areas of the southern Scandinavian peninsula that were designated on the map as Denmark, the Swedish state rounded out its borders along the *natural* land/water boundary of the peninsula. When Sweden later annexed Norway from Denmark in 1814, it completed that process, creating a great Nordic empire that was naturally bounded by water all around the coast to Norway's border with Russia. Looking at a map today, the border of Sweden seems to follow the coast in a most *natural* way, until it reaches Norway—which broke off from Sweden in 1905. Here, however, the spine of mountains down the back of the Scandinavian peninsula seems to naturalize the border to Norway as well. Looking at the map, it is easy to forget that there is nothing natural about these borders and that they were brought about by force and only naturalized later according to governmental policy.

The new way of thinking about the land as a naturally bordered area of soil, rather than as the land of a historically constituted polity, was to a certain extent the result of a form of cartographic imagination engendered by Sweden's pioneering effort to map, and thereby control, the extensive territories under its dominion. Sweden became one of Europe's great powers during the Renaissance. Sweden's hold over extensive territories rested in turn on the effectiveness of the land-based power of the Swedish army. Denmark, an archipelago, was held together by a water-borne infrastructure and defended by its formidable naval power. The shift in power from water to land was emphasized by a mid-seventeenth-century quirk of nature, an unusually cold winter that allowed the Swedes to march across the ice to Zealand and conquer most of the island except Copenhagen, whose staunch burghers refused to give way. From the perspective of military force, it was clear that it was difficult to defend Scania against a land-based army invading from the north if the strength of one's military power was seaborne.

The map provided a means of visualizing Sweden as an area demarcated by natural borders that bound Scania within the Swedish state. Cartography also provided the technique by which the area bound by the map could be visualized as three-dimensional. Basically, if the line of projection of a map is altered from a top-down position to a more oblique and horizontal view, what emerges is something that is midway between a map and a landscape picture (see Rantzau's map of Odense in chapter 2). Visualized this way, Scania is transformed from a land shaped by a polity to a dry area that forms the scenic backdrop upon which the people of a centrally regulated state demarcate their properties and *build* physical structures. The Swedish state, after 1660, actually did seek to alter the appearance of Scania so that it would come to resemble Swedish norms. One remnant of this effort, still visible today, involved the encrusting of local churches with a frosting of stucco in a contemporary Swedish style. To this day the human imprint of the Scanian landscape resembles neighboring Danish Zealand with its characteristic half-timbered whitewashed farm buildings, linked together around a courtyard in nucleated rural villages surrounded by open fields, rather than the landscapes dotted by colorful wooden structures surrounded by fields carved into the forest, which are characteristic of Sweden further to the north. It is as if a foreign power had invaded Zealand, replacing its red and white road signs with blue and yellow ones, and then went on to do funny things to its churches.

Now that a bridge links Zealand and Scania, there has been a remarkable proliferation of cooperative enterprises across the Sound involving everything from commerce to university teaching and research. Likewise, if on any given Sunday one visits a Danish cultural institution such as the Louisiana Art Museum, the number of people speaking with a Scanian dialect can be so great that it is difficult to tell whether one is on Zealand or in Scania. The Scanian dialect is probably as common on the streets of Copenhagen as at any time in history, yet it would be romantic folly to suggest, as some do, that Scania is on its way to becoming part of the Danish state. Cultural affinity between places and identification with a nation-state are two different things, even if they often overlap. Scanians are politically Swedes, even if they have cultural affinities with the Danes. This is why it can be difficult to demarcate just what is meant by Denmark, and Danish landscape, and why it is necessary to be clear about what one means by "landscape" when making such a demarcation.

Contemporary Danish Landscapes

Jutland is comparable to Scania as a landscape polity in that both have ancient roots upon which modern identity-building tends to feed. It is also, like Scania, a peripheral region, as distant from the capital in Copenhagen as one can get in modern Denmark, and the urge to create and re-create a distinctive Jutland identity likewise feeds upon a desire to counter its marginality. Jutland is both the landscape of an internal Jutish polity and an area of physical land, incorporated within a state centered outside its bounds. There are other ancient Danish landscape polities, but they do not maintain such a well-delineated historical identity as Jutland. As the seat of the capital, Zealand is so overwhelmed by the presence of this large center of world commerce and state administration that a distinctive, unified, historically grounded landscape identity is difficult to ascertain. Zealand has distinctive regional identities, supported by distinctive patterns of speech, but these do not lend themselves to the kind of analysis that we apply in this book to Jutland or Scania.

Natural boundaries, as noted above, are important to the constitution of the modern, scenic notion of landscape, and Denmark has its share of water-bound islands that make effective loci for insular identities. Some of these islands, such as Lolland or Funen, were landscape polities in the ancient past, but they fell under the dominant shadow of Zealand and Jutland long ago. The one exception would be the islands and coastal regions of Frisia, or North Friesland in Slesvig (or Schleswig in German), south of the present Danish border. This is an area colored Denmark on the map until 1864, when it was lost to the Germans after yet another military defeat. The myriad Frisian landscape territories resemble Scania and Jutland insofar as they are landscape polities founded originally upon ancient custom and law. The Frisian landscapes were a political reality as recently as 1864, when they were incorporated into a unifying German state. A distinctive grouping of languages still provides living material for the creation and re-creation of Frisian identity, even if these languages are under threat. The ancient Saxon territory of Ditmarsken (Dithmarschen in German), further south in Holsten (Holstein in German), another area of the map lost to Germany in 1864, was also a landscape polity *(Landschaften)* until that date. These areas, however, were never culturally Danish in the sense that Scania was. These landscape polities are thus beyond the scope of a book on Nordic landscape, though they are well worth a study of their own.

Landscapes of the Danish Imperium

One must distinguish between landscape polities that bond with culturally similar landscape polities to form larger lands (as was the case with Scania, Zealand, Funen, Lolland, and Jutland) and landscape polities that become incorporated within states with which they have relatively less legal and cultural affinity, as occurred when Sweden annexed Scania. Sweden, however, was not alone in expanding through the annexation of historically non-Swedish territories. The Danish state similarly absorbed other non-Danish landscape polities within the Nordic region. Queen Margaret (Margrete I) of Denmark became regent of the Norwegian kingdom in 1381, and Norway became part of a united Danish-Norwegian realm under Denmark in 1536. Thus when one seeks to define the extent of Denmark, one must consider the indelible mark that centuries of a Copenhagen-centered state administration had on the landscapes of Norway. Just as Scania never completely lost its identity as a polity, the Norwegian landscapes similarly retained a distinctive identity, which has provided the raw material for a process of national reconstruction that includes the invention of a new Norwegian language patterned on regional dialects. Norway, as noted earlier, was annexed by Sweden in 1814, but Sweden did not gain suzerainty over Iceland, Greenland, and the Faeroe Islands, though they had originally been settled from Norway and belonged to its realm.

Iceland became an independent nation during World War II, but the Faeroe Islands and Greenland remain under the Danish state. In a political sense it would thus seem logical to include the chapters on these places under the heading of Denmark, again raising the question of what it is we mean by Denmark and Danish landscape. We have opted to include these places within a separate Atlantic realm because they can hardly be said to be Danish landscape polities in the historical sense, even though they have a long history of ties with Denmark. Both Greenland and the Faeroes might be said to be moving toward some form of independence. Such touchy issues highlight the importance of reflecting on what it is one means by Denmark (or Sweden, Norway, or Finland, for that matter) and what one means by "landscape."

Old and New Meanings of Danish Landscape

The place-bound rural landscape polities that formed the historical core of Denmark also hosted an urban-centered monarchical state that provided the nucleus of an expansive sea-linked empire shaping a very different sort of Denmark. The two Denmarks coexisted in a form of tensive synergy that suggests it makes more sense to speak of a dual personality than to postulate that they formed two distinct entities. The rise of the centralized state in the Renaissance was linked to a new "Renaissance-modern" way of thinking about land and landscape. The focus shifted from the idea of land as a place or country shaped by a polity to land demarcated on a map as an area over which a state has propriety. It is upon this land, as demarcated and shaped by natural law and the law of the state, that the people play out their lives, as if on a stage (Olwig 2002). The backdrop of this stage is thus the landscape conceived primarily as scenery rather than as a polity shaping the land, as in the older sense of landscape.

Merriam-Webster's Collegiate Dictionary (11th edition) lists landscape in the Renaissance-modern sense of *vista* or *prospect* as being "obsolete"; the most recent meaning is "a particular area of activity" as in "political landscape." This brings us back full circle to the earliest meaning of landscape as a place shaped by a polity. These changing meanings reflect a historical process of development. In our globalized world, many of the functions formerly assumed by a territorially centralized state have now been taken over by such international bodies as the United Nations that need not have a bounded territorial identity or a single location or center. Likewise, as transnational bodies like the European Union gain in strength, the individual states are weakened at the same time as older landscape polities, whether Jutland, Scania, or the Faeroes, are increasingly able to act and shape themselves independently of the nation-state. Part of this shaping process is the reconstitution of historical memories in the construction of new/old place identities. The problematic question of defining Denmark is, thus, not likely to get any easier to answer in the future.

Notes

1. Useful English-language sources on Denmark are Fullerton and Williams 1972; Jones 1986; Mead 1981; Nordstrom 2000; Oakley 1972; and Thomas and Oakley 1998.
2. See chapter 2 on Jutland for a full discussion of ancient landscape law.

References

Fullerton, Brian, and Alan F. Williams. 1972. *Scandinavia*. London: Chatto and Windus.
Jones, W. Glyn. 1986. *Denmark: A Modern History*. London: Croom Helm.
Mead, William R. 1981. *An Historical Geography of Scandinavia*. London: Academic Press.
Nordstrom, B. 2000. *Scandinavia Since 1500*. Minneapolis: University of Minnesota Press.
Oakley, Stewart. 1972. *A Short History of Denmark*. New York: Praeger.
Olwig, Kenneth Robert. 1996. "Recovering the Substantive Nature of Landscape." *Annals of the Association of American Geographers* 86, no. 4: 630–53.
———. 2002. *Landscape, Nature, and the Body Politic: From Britain's Renaissance to America's New World*. Madison: University of Wisconsin Press.
Thomas, Alistair H., and Stewart P. Oakley. 1998. *Historical Dictionary of Denmark*. European Historical Dictionaries, no. 33. Lanham, Md.: Scarecrow Press.

2.

The Jutland Cipher: Unlocking the Meaning and Power of a Contested Landscape

Kenneth R. Olwig

Prologue: Landscape as Cipher

In 1859, Hans Christian Andersen left the cozy climes of eastern Denmark, where he was used to being wined and dined in the townhouses of Copenhagen's wealthy burghers and in the manors of the nobility on the islands of Zealand (Sjælland) and his native Funen (Fyn). He left this comfortable realm to make an excursion west into the wild heathlands of Jutland (Jylland). Andersen was famous in his day for his fairy tales and poetry, but he also wrote journals about his journeys to distant and exotic places (Mitchell 1957, 150). Jutland, to Andersen's readership, was just such an exotic place, and Andersen obliged his readers with a poem. The poem was put to music written a year later and became a well-known lyrical depiction of Jutland. In the poem, Andersen described the Danish peninsula as a rune stone lying between two seas, its runic ciphers foretelling the past and future. The poem provided a startlingly accurate runic reading of the peninsula's fortune by correctly predicting its pending transformation from heathlands to forest plantations and cultivated farms:

> Jutland between two seas
> as a rune-staff/stave[1] laid,
> The runes are great graves
> within the splendor of the forest
> and the great solemnity of the heath,
> here lives the desert's mirage.
>
> Jutland, you are the chief land,
> a highland with forest loneliness!
> Wild in the west, with sand-duned cliffs,

rising in place of mountains.
The Baltic and the North Sea's waters
Embrace over Skagen's sand.

The heath, yes, it's hard to believe—
but come yourself, look it over:
the heather is a splendid carpet,
flowers crowd for miles around.
Hurry, come! in a few years
the heath as a grainfield will stand.

Between wealthy peasant farms
soon steam dragons will fly;
where Loki now drives his herd,
forests will grow over the land.
The Briton will fly over the sea,
and visit prince Hamlet's grave.

Jutland between two seas
as a rune stone is laid,
the past is spoken by your graves,
the future unfolds your power;
the sea with all its breath
sings loud of Jutland's coast.

(Andersen 1964 [1860], my translation)

In this chapter, I will follow Andersen in interpreting the Jutland landscape metaphorically as a form of runic cipher. A cipher is a combination of symbolic letters, such as those "laid" on a rune stone, which require deciphering to be understood. The runes were the ancient Germanic (or Old Norse or Old English) alphabetical characters made up of angular lines whose name, "rune," could mean both mystery and poem. Andersen's poem can be interpreted as a deciphering of the mystery of the Jutland landscape perceived as a runic stave, or stone, whose prophetic ciphers give shape to the poem. In turn, I will use the poem as a key to deciphering the landscape as understood and shaped by Andersen and his contemporaries. This will involve the use of a form of conceptual archaeology to unearth not only the meaning of Jutland's landscape but also the meaning of the concept of landscape itself. Even as a physical entity, the "natural" landscape, as Andersen's poem illustrates, is subject to symbolic interpretation. It is not just the "runes," which have been carved by human hands into Jutland's surface, that give Jutland symbolic meaning, but it is also the spatial imprint of the peninsula itself, "laid" by physical forces between two seas. In interpreting the

Jutland landscape, we must be prepared to engage the many layers of meaning that emerge from such an approach.

DECIPHERING THE LANDSCAPE SCENE

Rune stones are characteristically graven with the mysterious, magical, and secretive interwoven symbolic ciphers of an ancient script (the futhark). The word "cipher" means, among other things, "code" and "a combination of [interwoven] symbolic letters" (*Merriam-Webster's Collegiate Dictionary*, 11th edition). It was, according to myth, the all-powerful Norse Aser god Odin—hero and god to kings, warriors, and skalds (bards)—who gained the power of the runes through his ritual, shamanistic hanging in a tree (Davidson 1964, 140–48). The runic stave engraved into a stone is a symbol of the presence of this power. The powerful act of gouging with hammer and chisel involves the use of constrained brute force against the blank surface of the stone, and this might help explain the power attributed to the runes. In the Germanic languages the words for political power, authority, rule, or governance and the word for force are closely related, or even identical (Langenscheidt 1967, s.v. "walten, Gewalt"; *ODS* 1931, s.v. "vælde, vold"), and the forceful permanence of the signs chiseled into the stone is continuous testimony to the power of those who had carved them. The blank surface of the stone, however, is a prerequisite to the making of a rune and provides the background for the image of its graven mark. For the exercise of power over the blank surface to have symbolic force, the surface itself must be invested with power. The Arabic word *sifr*, from which cipher derives, means empty or zero and as such had both religious and mathematical power as the mysterious nothing that, nevertheless, is something (Kaplan 2000). It is an absence, like the blank surface of the rune stone, that holds a place.

The binary language of the computer based on 0 and 1 illustrates the importance of the cipher today. Andersen's cipher, however, is not that of the absent presence of the "zero" but rather is the carved, graven, or graphic mark (the "one") that imbues the stone, or the landscape scene, with a powerful presence. The word "graphic" comes from a Greek word meaning to carve or write. In the first part of this chapter, I will discuss the notion of Jutland as a marked and graphically delineated landscape scene, which Andersen captures in his poem. In the second part, I will interpret the more invisible character of the landscape that might be identified, metaphorically, with the cipher in the original Arabic sense, the zero designating a place that gives meaning in a larger context. Whereas the first section takes its point of departure in a poem by an outside visitor, the second will take its point of departure in a short story by a native Jutland author, a contemporary of Andersen's, Steen Steensen Blicher.[2] The point of contrasting these two authors is not so much to distinguish the gaze of the outsider to the lived experience of the insider, but to bring the metapolitics of landscape transformation to the fore.

Jutland's Landscape Scene

Jutland is an area of 29,633 km² with a striking scenic surface, as noted by Andersen (Trap 1963, 4:3). It has forest highlands, wild western sand dunes, and heath. This heath, according to the British geographer Harry Thorpe, "formed a characteristic landscape covering approximately three million acres [1.2 million ha] in 1800, or about 40 per cent of the entire area of the peninsula" (Thorpe 1957, 87). The heath-covered plain of western Jutland owes to the division of Jutland rather neatly from north to south down the middle by a watershed, running south from the Limfjord. This watershed divide originated at the westward edge of the last glaciation. To the west, the melting waters of the glaciers created a vast, sandy outwash plain that was broken here and there by islands of hilly clay deposits from earlier periods of glaciation. To the east, the morainic deposits from the glacier left a hilly terrain with heavier, more fertile soils that supported both agriculture and forestry. The light soils of the outwash plain lent itself to clearing, pasturage, and cultivation with primitive equipment and, hence, to deforestation. Swidden agriculture, coupled with the effects of periods of cool, wet weather, as in the Iron Age, favored the development of leached, acidic soils and the consequent process of podzolization, by which a layer of hardpan, made up of mineral material leached from the soil, forms under the soil. All these factors favor the development of a low treeless vegetation complex known as heath, which is dominated, as the name suggests, by heather. The heath to some extent is the creation of such non-human physical geographical factors, but it is extended and maintained by human activities such as burning and grazing. Such unproductive soils are not conducive to large-scale manorial farming, and this left western Jutland in the hands of independent farmers, who learned to exploit the area's resources in myriad ways, ranging from cultivation and pastoralism to cattle driving and the widespread merchandising of wool products and, even, moonshine alcohol. At the time Andersen wrote his poem, the western portion of the peninsula, as well as large stretches of the northern and eastern part of the peninsula, were dominated by heathlands (Olwig 1984, xiv–xix).

The physical topography of the landscape can be experienced and interpreted as a form of scenery, such as we learn to experience and interpret the background scenery of a play in the theater. This is what Andersen does when he conjures up the image of the heathen Nordic god Loki. The image derives from the folk expression that Loki is grazing his flocks when the air shimmers, creating miragelike optical illusions on a hot summer's day (Bæksted 1963, 166). Loki is a duplicitous Prometheus figure who is always shifting form, sometimes parenting dragonlike beasts in the process (on Loki, see Davidson 1964, 176–82). The heath landscape, with its flat, treeless expanse of sandy soil, generates such optical effects, and the mind can easily imagine an ancient heathen setting because the dark flat surface of the heath gives added prominence to the elevated shapes of the ubiquitous prehistoric grave mounds. The very word "heath" conjures up (in both Danish and English) associations of the *heathen* (*ODS* 1931, "heden"; *Oxford English Dictionary*, 1989, s.v. "heathen"). This imagery is effective because, like Loki, the heath landscape has a complex, duplicitous quality. It can evoke passive romantic reveries stimulated by memories of the past that lie buried by the heath. It can also, as for Andersen, prompt visions of Promethean modern changes, worthy of the ancient heathen gods. The force of this ancient heritage is here symbolized by that fire-breathing offspring of Loki, the steam engine, now domesticated into a vehicle that transports the produce of prosperous farms to market while also conveying tourists to the site of Hamlet's grave.[3]

The idea of landscape as scene refers in one sense to a pictorial representation of a place perceived as visual scenery in (illusory) three-dimensional space and, in another, to a portion of territory defined by its being viewable, as if by someone taking a picture, "at one time and from one place" (*Merriam-Webster's Collegiate Dictionary*, 11th edition, s.v. "landscape"). In this sense, landscape involves a dual structure that the geographer Stephen Daniels terms "duplicitous." Daniels applies the term "landscape" both to a pictorial perception of a location as seen from a spatially distanced outside point of view and to the internal phenomena of a place that are given visual form in a picture (Daniels 1989). The landscape scene that Andersen, the consummate tourist, presents is viewed from the perspective of the outsider, and his poem is prescient in its prediction that he will be followed by foreign tourists, flying to Jutland to see its sights and heritage. Today, tourists indeed do come, in winged fire-belching dragons, to see the Jutland coastal strand, the remaining remnants of heath, and ancient archaeological sites like "Hamlet's grave." There is indeed a state-protected heritage site called "Hamlet's grave" to the east of Aarhus—though nobody can know whether Hamlet, if he ever existed, is actually buried there.[4] Embedded in this scenic landscape is the question of to what degree the landscape has been constructed through time by the imaginative representations of outsiders like Andersen, and, further, to what degree this idea of landscape has effected the transformation of that landscape.

Jutland, as Andersen describes, is a place whose destiny has been shaped by the fact that it is situated between two seas. To the west, where the sun travels across the North Sea, lies Britain, the site of modern progress, from which the British tourist will "come flying." To the east, where the sun rises, are the islands of Funen and Zealand, with the Danish state's capital of Copenhagen, the Baltic Sea, and further on, the ancient mythical homeland of Odin, god of the runes. Between the "two seas" lies the Jutland peninsular runic stave or staff, the "head" or chief land, which is about to have its ancient power unfolded at the expense of the heath's blooming carpet of tiny flowers.[5] To unlock the meaning of the Jutland landscape, the poem suggests, one must imagine Jutland both as a stretch of land with a strategic location between east and west, past and future, and as a location with a scenic surface expressing the hidden meaning of its history, as a key to understanding its future. Jutland also makes its mark as a boundary between north and south. South Jutland (Sønderjylland) is demarcated to the south by a fortified marchland (the Danes' mark) that separates Scandinavia from Holsten (Holstein in German) and continental Germany, the Holy Roman Empire of earlier times. South Jutland, which is demarcated from North Jutland by the King's River (Kongeåen), has had a somewhat separate, German-influenced political history, while North Jutland remained firmly within the Danish kingdom. In 1232, South Jutland was made a separate principality, which in the course of the Middle Ages became increasingly Germanized as the duchy of Slesvig (Schleswig in German). From 1459 the culturally mixed (Danish, German, Frisian) duchy of Slesvig and the culturally German duchy of Holsten to the south were linked politically under a largely German nobility that, in turn, owed allegiance to the Danish state (Trap 1963, 6:11, 10:5–34). The duchies were lost to Germany after Denmark was defeated in a war with Prussia in 1864.

JUTLAND AS A LANDSCAPE SCENE IN THE THEATER OF THE DANISH STATE

Andersen's poem envisions the landscape as if seen from the stars above, or as engraved on the space of a map, as it juts, like a rune stone, northward into the emptiness of the sea. The poem also requires that one imagine oneself positioned a bit lower, perhaps on a high hill, commanding the view of the panoramic scenes of heath and forest stretched out in front.

The use of map and landscape scenery to incorporate Jutland within the compass of the Danish imagination goes back to the origins of Denmark as a modern state. Henrik Rantzau (1526–98), a powerful court figure and enormously wealthy state governor *(statholder)* of Slesvig-Holsten, was the first person to present Jutland within the cartographic space of the Danish state. He became prominent throughout Renaissance Europe for his work in sponsoring the nascent disciplines of carto*graphy*, geo*graphy*, and the topo*graphic* depiction of localities and regions (choro*graphy*) illustrated by engraved scenes of the places under the domain of the Danish state.

Rantzau came from a prominent noble family whose regional power base was in the culturally German duchy of Holsten, at the southern root of Jutland. He breached the historical divide between his place of origin and the culturally and politically Danish regions to the north by casting his family as Cimbri, the Latin name for Jutland and its people. Identity derives hereby not from the historically constituted cultural and political identity of differing places, but from a geographic body, and the people living therein, as defined by the map and named by the Romans. His maps, engraved scenes, and chorographic description of Jutland (Ranzovii 1739 [1597]) helped to establish this identity. The maps provided the framework upon which his family tree, and the deeds of his family, was engraved. It was Rantzau who literally put Jutland, and with it Denmark, on the European map (Skovgaard 1915). He financed the earliest extant printed map of Jutland as part of a larger map of the Danish state (Figure 2.1).[6]

FIGURE 2.1. The oldest extant map of Denmark was prepared by Marcus Jordan for Henrik Rantzau in 1585. The map includes Latin inscriptions such as "here Johan Rantzau crushed the rebellious peasant army of 25,000 men and took Skipper Clement prisoner." The imprisoned Clement is depicted at the bottom right-hand corner. Copyright Kort & Matrikelstyrelsen (A. 156–01).

Many of Rantzau's maps, because of their angle of projection, are essentially landscape prospects created through cartographic transformations that he helped pioneer at a time when these techniques of perspective drawing were a recent discovery.[7]

Rantzau's family worked assiduously for generations to gain hegemony over Jutland and to promote the family's power as the Jutland ally of the Danish state. Henrik's father, field marshal Johan Rantzau (1492–1565), was literally a kingmaker, providing the military power to unite a Danish nation divided by civil war and enable a Holsten pretender to the throne to become King Christian III (1503–59, king from 1534–36 to 1559). Johan accomplished this by marching up the peninsula with his mercenary army in 1534 to suppress an extensively popular North Jutland insurrection led by a farmer's son and privateer, called Skipper Clement (c. 1485–1536). Johan and Christian were both Protestants, and by helping to reunite the country under the protestant Christian III, Johan helped pave the way for both the political and the Protestant religious unification of Denmark.

Henrik Rantzau, like Andersen, was fascinated by the runic monuments of ancient men of power. The most famous Jutland rune stones are no doubt those at the location of the two largest grave mounds in Denmark, at the site of the parish church in Jelling.[8] The earliest-known graphic depiction of Jelling and its stones (Figure 2.2) is reproduced in a chorographic book on Jutland by Rantzau (Ranzovii 1739 [1597]). One of the stones has been termed Denmark's birth and christening certificate (Glob 1967, 170–77). On this stone King Harald Bluetooth, c. 980, had cut the words: "King Harald ordered the placement of these stones in honor of Gorm, his father and Thyra, his mother, the Harald who won all of Denmark and Norway and made the Danes Christian" (Exner and Ebert 1968, 56–57). The stone, with its serpent-headed bands of runes and representation of Christ on the Cross (thought to be one of the oldest representations of Christ in Scandinavia), marks the site of the first Danish royal grave monument. An earlier rune stone at the same site, dedicated to Queen Thyra (d. c. 935) by Harald's

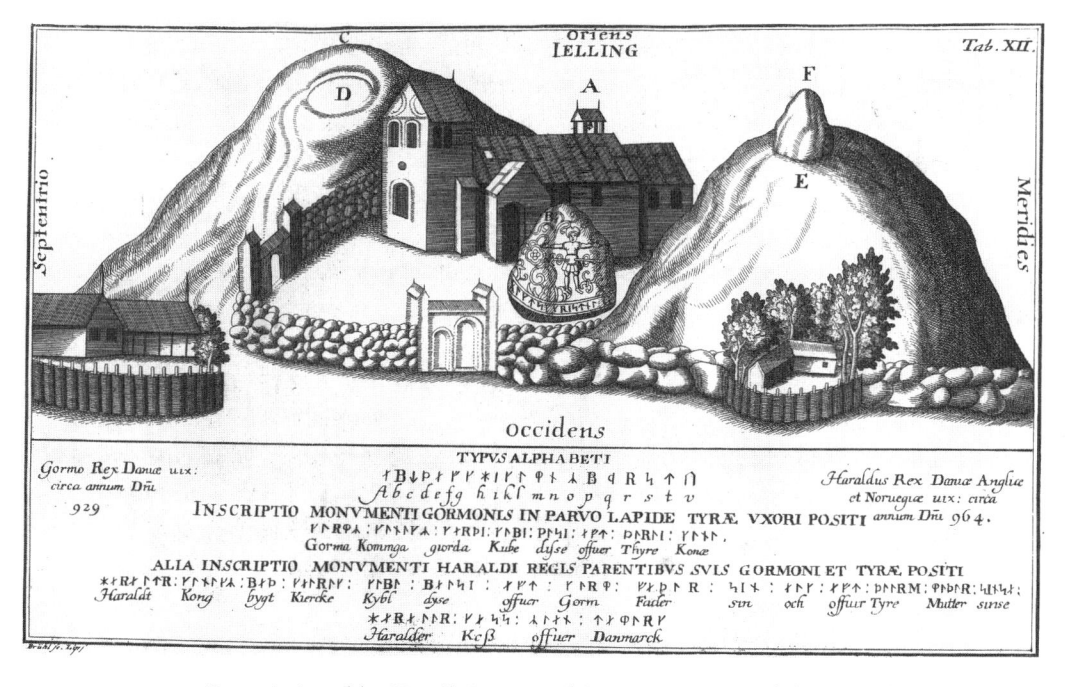

FIGURE 2.2. Commissioned by Henrik Rantzau, this 1597 engraving of the mounds and stones at Jelling is the first representation of this famous site and also was part of his project to establish Jutland as the home of the Rantzau dynasty. The Royal Danish Library, Photographic Atelier (negative no./Arkive no. 206.756).

father, Gorm, is presumed to mark the construction of the first mound. The second mound at Jelling is presumed to be the work of Harald, created as a monument to his parents and to himself. The Jelling site thus marks both a geographical crossroad, at the center of Harald's unified Danish state, and a historical crossroad between a loosely organized polytheistic Denmark and the unified Christian state proclaimed by Harald. Jelling was a pivotal point in a growing Danish nexus of power, which would soon, under Harald's son Svend and grandson Canute, include much of Britain to the west. Religion and political power meet at this point between the paganism still prevalent to the east and a western Christianity brought to Denmark by English and German missionaries. The parallels between Harald's efforts to centralize political and religious power in the hands of a nascent Danish state and the later efforts of the Rantzau family to do the same would hardly have been lost on Henrik.

MAPPING THE NATION-STATE'S PROGRESS

The chorographic work that Rantzau pioneered of combining map, landscape prospect (Figure 2.3), and regional depiction subsequently became part of a process by which a territorially defined Danish identity was mapped. This geographically defined identity was convenient for the many culturally German men of power, from the duchies, whom the Danish monarchs, at the expense of native elites, elevated to central positions of power. The Danish state was a multiethnic construction that included Germans, Frisians, Norwegians, Saami (formerly known as Lapps), Icelanders, Faeroese, Inuits (formerly known as Eskimos), and West Indians, as well as various ancient peoples, such as the Slavic Wends, to whose historical sovereignty the monarch laid claim. An important tool in this territorialization of identity was the monumental mid-eighteenth-century multivolume *Pontoppidan's Danish Atlas (Pontoppidans Danske Atlas)* (Pontoppidan 1763), which was initiated by bishop and university chancellor Erich Pontoppidan. This atlas of Denmark, which is as much a topography and regional geographical chorography as it is a collection of maps, was expanded from an earlier version that was published in German in Bremen in 1730. The title of the earlier work makes clear that the atlas was conceived as representing Denmark as a kind of theater, viewing place, or stage scene upon which the drama of history is played. The Latin/German title is *Theatrum Daniæ, oder Schaubühne des alten und jetzigen Dänemarks (The Danish Theater, or the Stage Scene of Ancient and Contemporary Denmark)* (Pontoppidan 1730). In this theater, the landscape scene sets the stage for the action and prefigures the action that will unfold upon it. The lengthy text was richly illustrated with maps, prospect drawings, pictures of the folk-costumed people, and depictions of ancient monuments memorializing the ancient glory of Denmark.[9] Pontoppidan, who was as enamored of monuments as Henrik Rantzau, also published a separate, similarly monumental two-volume work on Danish stone monuments titled *Marmora Danica* (Pontoppidan 1739–41). The exceedingly fine maps in the atlas, with inlaid

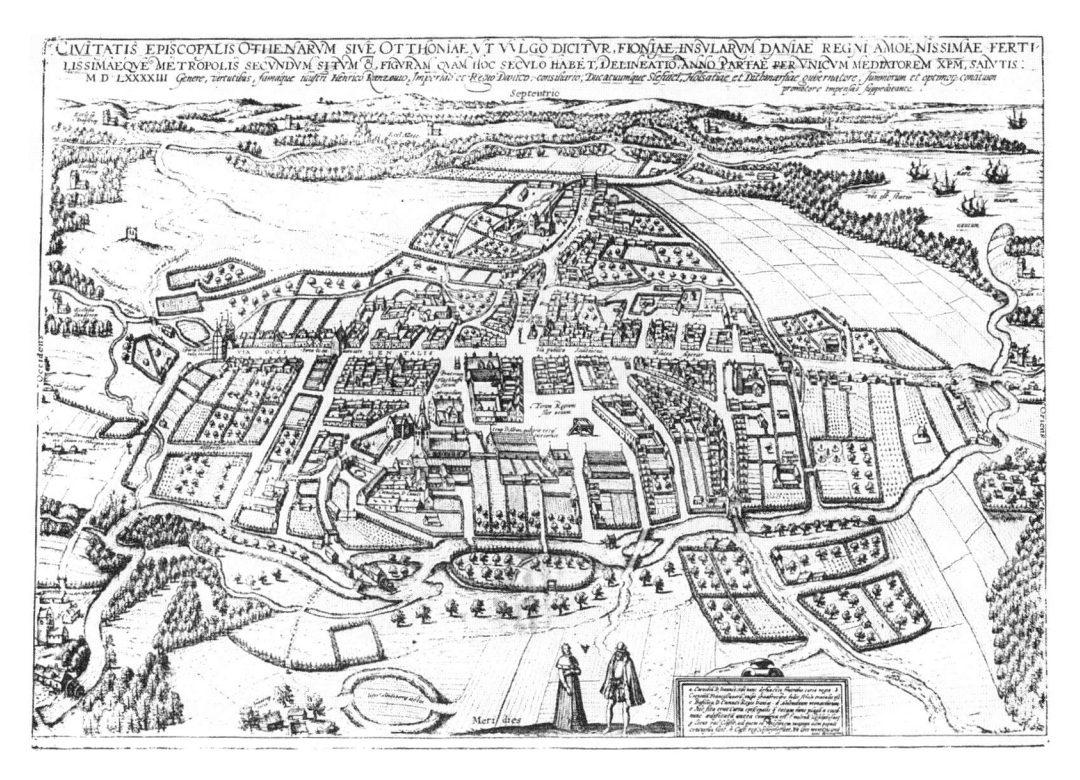

FIGURE 2.3. Henrik Rantzau commissioned this 1593 map and landscape prospect of Odense, Hans Christian Andersen's hometown (from Jørgensen 1981, n.p.).

landscape scenes, were the contemporary work of *Videnskabernes Selskab* (scientific society), of which Pontoppidan was a member. Mapping served not only a decorative and informative role at this time but also a military and increasingly important civil function as the organizational structure for what was to become an all-encompassing process of agrarian enclosure. This process would soon divide most of eastern Denmark into neatly mapped and demarcated parcels of arable land and forest under private or state ownership. This transformation of the variegated, historically generated places of eastern Denmark into the uniform geometric space of the map was envisioned, here as elsewhere, as a staging platform for the nation-state (Olwig 2002a). The trackless and open, unenclosed, and visually undemarcated grazing lands of the Jutland heaths were, by contrast, perceived as being unredeemably backward, as were the Jutlanders themselves (Henningsen 1995, 223–40; Matthiessen 1939).

JUTLAND WITHIN THE SPACE OF THE NATION-STATE

Andersen's interpretation of the Jutland landscape is very much the product of the national romantic era, when Denmark was in the process of redefining itself as a homogenous nation-state with a homogenous national landscape. National romanticism was a cultural movement identified with the arts and that paralleled the development of political nationalism more generally. Danish national romanticism was concerned, in great measure, to consolidate the Danish identity of the core area of the ancient Danish kingdom, midway between Germany and Norway. This consolidation tended to leave out most of the non-Danish cultural and ethnic groups, such as the Germans, whose territorial base was outside this area, even though they had long been an important presence within civil life throughout Denmark. This consolidation also had the effect of ironing out the cultural differences between the Danes living within the core area, thus blurring the differences between the Jutlanders and the Danes living on the eastern islands. Danish national romanticism was both inspired by and was a response to nationalist currents that were developing at the same time among Denmark's neighbors, particularly in Germany, where the former Holy Roman Empire was being reconsolidated along national and ethnic lines. This nationalism, in turn, threatened the identity of the Danish imperium, both to the south, where the German population was beginning to crave autonomous representation in government, and to the north, where Norway had moved in the direction of representative government just before its 1814 transferal, after the Napoleonic Wars, to Swedish sovereignty.

Andersen wrote his poem en route between the towns of Randers and Viborg. He was on the way to visit the ninety-three-year-old Danish Lutheran pastor Hans Bjerregaard, who had been a pioneer in encouraging the Jutland farmers to cultivate and afforest the heathlands (Andersen 1951 [1855, 1871], 451). Bjerregaard was the son of perhaps the most prominent peasant in eighteenth-century Denmark, Hans Jensen Bjerregaard, a man admired by agricultural improvers, besung by a famous German poet (Friedrich Gotlieb Klopstock), and honored by the monarch. The elder Hans was a Zealander who had risen from illiteracy and near serfdom to the ownership of his own enclosed farm and leadership in the emerging independent farmer estate. The younger Hans was just as ambitious as his father, on behalf of his rural Jutland parishioners, and he worked to bring Jutland into line with the productive scene of farms and landscape parks that he would have remembered from his Zealand childhood in eastern Denmark. His parsonage abutted a romantic landscape garden that he had created, complete with rune stones. The garden gave form to his vision of a lost national golden age (symbolized by the rune stones), which was on the verge of restoration (on the garden, see Molbech 1829, 154; Olwig 1984, 49). Andersen was thus not just on his way to visit a country parson, but to visit a man who promoted a particular landscaped vision of national progress, and that vision, including its rune stones, informs Andersen's poem.

The beginning of the artistic movement known as national romanticism is normally dated to 1803, when the geologist and natural philosopher Henrik Steffens held a famous series of lectures in Copenhagen (Steffens 1905). His audience included the intellectual elite that would shape the Danish national identity in the course of the nineteenth century, many of whom were close to Andersen. Among them were Steffens's cousin, N. F. S. Grundtvig, an educational and religious reformer; Adam Oehlenschläger, who became Denmark's most prominent romantic poet; and the internationally known electrophysicist, natural philosopher, and promoter of polytechnical education, Hans Christian Ørsted. In Steffens's hands, the landscape scene became a changing series of stages that created the framework within which the character of a particular people is molded:

> Through this interaction of the whole upon the individual, and the individual upon the whole, is generated an identical picture-history, which presupposes the entirety of nature as the foundation for all final existence, and all of humanity as the expression of this interaction itself. The expression of the coexistence of all these individuals' interaction in history and nature is *space*—eternity's continually *recumbent* picture. But the whole *is* only an eternal chain of changing events. Yes it *is* this constant alternating exchange, this eternal succession of transformations itself. The constant type of these changes is *time*—eternity's constant moving, flowing, and changing picture. (Steffens 1905, 91, my translation from Danish, italics in original)[10]

To those who attended his lectures, Steffens's ideas permeated the notion of national landscape identity. Hans Christian Ørsted later used Steffens's framework to present a powerful vision of a unified Danish national identity in a speech he titled "Danishness" (Ørsted 1836, 1843). Danishness is an expression of the evolution of the people within the geographical boundaries of the unified Danish state. King Frederik VI is proclaimed to be a genuinely Danish king who led the Danish folk to greater and greater "development" or "progress" in their Danishness. According to Ørsted, "The character of the Danish folk is in complete concord with the natural position of the land of their birth" (Ørsted 1836, 211–12).

Ørsted was the brother-in-law and friend of the poet Adam Oehlenschläger, who also attended Steffens's lectures. The two men spent long hours talking about Steffens's ideas, and it is from their meeting that Danish literary historians date the birth of romanticism in Denmark and the rise of Oehlenschläger as a prominent romantic poet (Albeck 1965, 23). The symmetry between the ideas of Steffens, Ørsted, and Oehlenschläger can be seen in the lyrics of the Danish national song, "There Is a Lovely Land" (c. 1819), which Oehlenschläger constructed according to the model of the landscape layer cake, each layer reflecting a new stage of cultural development:

> I know a lovely land,
> whose charming woods of beeches
> grow near the Baltic strand,

It waves from valley up to hill,
its name is olden Denmark,
And here dwells Freya still.

Here sat in time's past
the armor-clad warriors
rested from strife
they set out the enemy to harm
now they rest their legs
behind the barrow's monoliths.

This land is charming still
For blue are Belt and Ocean
And green are woods and hill
And noble women, lovely maids
Brave men and fearless boys
Inhabit Denmark's isles.

Hail king and fatherland!
Hail every Danish burgher,
Who does what he can!
Our ancient Denmark shall remain,
as long as the beech reflects
its top in the waves of blue.

(Folkehøjskoleforening 1966, 644–45, my translation)

The series of pictorial time frames that structures this poem contains many of the same elements that frame Andersen's poem about Jutland: the graves of the past, the land defined by its coast, the changing pictorial landscape scene, and the industrious modern Danes. Ancient gods like Loki and Freya (goddess of fertility and love) still dwell in Denmark, but their power has been tamed and put to the service of a progressive modern society. The difference is that Oehlenschläger describes an ideal eastern Danish landscape of the present, whereas Andersen calls forth a future transformation of the Jutland landscape into something approaching the eastern ideal.

NATIONAL ROMANTIC PROGRESS

In the work of Steffens, Ørsted, and Oehlenschläger, we see the emergent outline of an idea of national progress that, guided by a wise government inspired by philosophers and national romantic artists, gradually triumphs over, and harmonizes, the variegated physical and cultural terrain of the nation. Andersen's poem captures the progressive spirit of this national romanticism. With hindsight, his poem seems prophetic because it predicts a transformation of the Jutland landscape that soon was to become a reality. The success of its prophecy, however, owes to the way it so perfectly captures the spirit of the time. It was this spirit that would soon fuel a transformation that would attempt to reclaim the Jutland landscape by restoring it to an imagined ancient, golden-age state of peaceful forests and fields resembling the eastern Danish ideal depicted by Ørsted and Oehlenschläger. This transformation, as described by the geographer Harry Thorpe, was initiated when, "with the foundation of the Danish Heath Society in 1866, with its headquarters first in Aarhus and later at Viborg, a really determined attack on the heath began. By 1950 the extent of heath, dune and bog in Jutland had been reduced to 640,000 acres [259,000 ha] representing only 8.8 percent of the peninsular area" (Thorpe 1957, 87). A standard geography of Scandinavia notes how this occurred:

> Bravely and industriously, the local population waged war on the heather, encouraged after 1866 by the Danish Heath Society (Det danske Hedeselskab). How successful the reclamation has been is testified by the well laid out farms and plantations that now cover the outwash plains and the adjacent poorer moraines. After a century, over four fifths of the heathland have been converted to agriculture and forestry. . . . When the Heath Society was organised, the struggle against the heath was taken up in earnest. The road network was improved and railways laid. There was a needy population, altered conditions for agriculture and increased knowledge of management. Led by Enrico Dalgas, the founders of the Heath Society wrote about its local differences. . . . Only a few patches of heath remain as scar-like reminders of the former waste, but in several places, small areas of heath have been carefully preserved lest future generations forget the labours of the past. (Fullerton and Williams 1972, 111–13)

Enrico Dalgas, who founded the Heath Society (Figure 2.4), moved in social circles that included Andersen and many of the artists and scientists who had attended Steffens's lectures. He had an eye for landscape scenery, but also, like Ørsted, he believed in the power of science and technology to bring progress. He began his career as an army road engineer working in Jutland. He promoted the national cause of heathland development in a slim two-volume tract titled *Geographical Pictures of the Heath* (Dalgas 1867–68, my translation). To create these pictures he would stand on the top of an ancient grave barrow, from which he could survey the heath with spade, bore, and map in hand. With these tools he created a graphic word picture of a landscape scene that had gone through a series of stages, from a fertile, forested landscape during the heroic ancient ages when Denmark was an international power, to an era of decline under despotic rule, to the dawning progress in his own era, when the landscape would be returned to its original, natural state. The book concludes with the wish that "these Heath Pictures [may] bring you the conviction that the Heath Cause

FIGURE 2.4. Monument to Enrico Dalgas (1828–94), the "hero of the heaths." Photo by Kenneth Olwig.

is of very great importance to the fatherland, and that the difficulties which will be met can be overcome . . . and that it should be every Danish man's duty to join in removing the curse that seems to have lain upon West Jutland for half a millennium, a curse that, among other things, has resulted in the complete destruction of the forest" (Dalgas 1867, 125, my translation). Dalgas not only helped spark a massive transformation of the scene, but he became famous as the self-effacing "hero of the heaths," as the Danish-American photographer Jakob Riis called him (Riis 1910, 153–77).

Since the time of Dalgas, the drama played out upon, and within, the Jutland landscape has continued to develop from act to act. The heroic cultivation of the heaths was followed at the beginning of the twentieth century by an equally impassioned effort to preserve as "natural landscape" the remnants of heath that were left. Toward the end of the century, another change in landscape policy occurred and since then there has been a massive campaign to reforest the least productive agricultural lands in the name of nature and biodiversity. Whether as an area to be reclaimed for agriculture or as an area preserved as nature, the focus continues to be upon its scenic qualities. The vantage point from which the scene is surveyed today is no longer the elevated position of Dalgas's grave mound, but a Landsat satellite, or a computer simulation model, which can survey the entire Jutland scene by clicking the shutter or the mouse. The grave mounds have not lost their importance, however. As landscape heritage, they are under the watchful gaze of the same state agency that manages the forest and heathland scenes. Tourists flying to Denmark will have no trouble finding Hamlet's grave or the mounds and stones at Jelling. The scene changes, but the theme of progress through the application of science and landscape aesthetics is the same (Olwig 1984; 1986, 113–40). Another way to interpret Jutland as a landscape, as *a particular area of activity,* a "political landscape," is from the perspective of the "insider."

Landscape as the Place of a Jutland Polity

The Jutland-born author and rural pastor Steen Steensen Blicher was Andersen's contemporary, but his approach to landscape was quite different. The introduction, by the twentieth-century Danish author Martin A. Hansen, to an illustrated edition of one of Blicher's short stories, "The Three Holy Eves" (1827), set in the heathlands of mid-Jutland, brings out this contrast: "When one recollects this story one thinks one can remember how three seasons flow through a heavy Jutland landscape. There is absolutely no nature depiction in it. But it is so deeply sunken in the Jutlandic, that the Land in its entirety is present, unbound by words and so powerful that the pictorial artist, who illustrates Blicher, with sure instinct, does the opposite: all for him becomes Jutland landscape, where a people like this must grow, and where, between the high secretive heaven and the black soil, a story like this must be born" (Hansen 1950, my translation).

Whereas Andersen's poem is all landscape, with no people, this story is all people and no landscape, but the reader nevertheless experiences a landscape. In Andersen's poem, the landscape is the agent; in Blicher's story, the people are. The story suggests another way of thinking about the *land* in *land*scape. It begins with these lines: "If you, my reader! ever have been near 'Snabeshøy,' where the *Landsting* was held in olden times, then you could see from here, toward the south, a little, spread out, hamlet called *Uannet*. Here live, and probably never have lived, anything other than farmers" (Blicher 1930a [1841], 1, my translation).

Blicher follows the national romantic pattern of standing on a grave mound and looking out over the landscape, but then, as if to make a point, he does not describe the heathland scene (Hansen 1950). Instead, he introduces a dispersed collection of cottages where probably only farmers have ever lived. The hero of the story is just such a farmer, a man by the name of Stærke Sejr (Powerful Victory), who saves the day by single-handedly capturing the rovers who have been terrorizing the area in cahoots with a corrupt nobility (Figure 2.5). The name of the hamlet, Uannet, may contain a subliminal message. Spelled in standardized modern Danish with one *n*, the name means "unlooked for" or "unsuspected," and this is precisely the sort of unnotable place that Blicher describes the village as being.[11] Nevertheless, it is the home of a hero named "Powerful Victory"! Furthermore, the grave mound upon which Blicher stands is not just any grave mound; it was the supposed location of the ancient *Landsting,* the representative judicial body (called a *moot* or *thing* in English) that legally shaped the land of the Jutes, thereby generating the landscape polity of Jutland.

FIGURE 2.5. "Stærke Sejr hunting down the rovers," woodcut from F. E. Boisen's *Danish Reader (Danske Læsebog),* vol. 1, 1858.

Blicher believed, on the basis of local legend, that a barrow called Snabeshøj marked the site of the *Landsting* for the whole of Jutland, north of the King's River, which was commonly named the "Snapsting." He believed that the name was identical with that of this barrow (Blicher 1920 [1824], 161).[12] By drawing attention to the institution of the *Landsting*, Blicher was offering the possibility that it was the law of this ancient institution that had shaped Jutland as a polity. The idea that representative judicial bodies played a role in shaping the polity was controversial at a time when the state was ruled by an absolute monarch—whose prime minister and leading jurist was Anders Sandøe Ørsted, the brother of Hans Christian Ørsted. According to the ideology of the time, it was the state, under the absolute monarch, that naturally ruled on behalf of God and, by extension, Nature. The state thus ruled and molded the land by mobilizing law seen to derive ultimately from God and Nature. The idea that the law of the land could spring from the people as expressed through a parliament was particularly controversial because the Danish state had been pressured into allowing a form of advisory regional representative government, called an Assembly of the Estates, for the people of Slesvig and Holsten. Since the state wished to maintain a semblance of unity between its territories in the duchies and in the kingdom proper, it was thus forced, by extension, to grant similar representative bodies within the kingdom, one located in Viborg, Jutland, and the other in Roskilde, Zealand. By beginning the story of the powerful farmer, Stærke Sejr, with a reference to the ancient *Landsting*, Blicher was reminding the Danes that there was historical precedent for a representative Jutland government centered in the Viborg area. Since such parliamentary bodies throughout Europe were rooted in custom and historical precedent (Olwig 2002b), this was an important point. Although it only had advisory status, the Viborg gathering of the Assembly of the Estates set a precedent for separate Jutland representation that long frightened proponents of a centralized state located in Copenhagen (Frandsen 1996). The unsuspected and unlooked-for implications of this ancient site, coupled with the image of a heroic peasant who fought the corrupt nobility, was fraught with implications for Blicher's contemporary "political landscape."

THE POLITICAL LANDSCAPE

One definition of landscape is "a particular area of activity," as in "political landscape" (*Merriam-Webster's Collegiate Dictionary*, 11th edition, s.v. "landscape"). This definition shifts the emphasis from the landscape as a *vista* or *prospect*, which the dictionary now considers obsolete, to the landscape as an area defined by activity. What defines the landscape of the unsuspected village of Uannet is the action of the hero, and the actions of the people in times past in defining the laws by which they form their activity. This notion of landscape is commensurate with the earliest historical meaning of the term, which was tied to the activity at the *Landsting*. It is useful, therefore, to examine the praxis of the *Landsting* and the way it shaped a land as a landscape before returning to Blicher, who drew inspiration from it in shaping his own approach to the political landscape.

A land, as is stated in the preamble to the first written version of the Jutland "landscape" law from 1241, shall be "built" by law, or as it read in the original, *"Mæth logh scal land byggæs"* (Jansen and Mitchell 1971, 3–7).[13] The meaning of the word "built" at that time was related to such English words as "bower," meaning dwelling, as well as "abode," the place where one abides and dwells, the place of *being* home. A more correct translation might thus be: "The abode of the land is created by abiding by the law." This use of "to *build*" is still found in the English expression "nation-building" and, as the philosopher Martin Heidegger points out, in the word *neighbor*, the *near builders* or dwellers. At the existential level, this etymologically primary sense of building (*Bauen* in German) is, as the above suggests and as Heidegger argues, the root of the notion of *being* itself: "man *is* insofar as he *dwells*." Man is a dwelling being, and for Heidegger to dwell or not to dwell, that is the question. Dwelling, however, implies more than habitation and habituation; it also means "to cherish and protect, to preserve and to care for, specifically to till the soil" (Heidegger 1971a, 146–47). The Danish word for farmer, *bonde* (*Bauer* in German), is one who dwells (*ODS* 1931, s.v. "*bonde*"). A land is thus historically, according to the preamble to the Jutland landscape law, an area built and cared for according to precedence of habit and custom as formalized by the *Landsting* (thing or moot in English) (*ODS* 1931, s.v. "*land*" 3). The social life of the landscape polity, and thereby the character of the territory that it occupied, was shaped by the law formalized at the *Landsting*.[14] Whereas Heidegger focuses on the existential individual's mode of being through dwelling, tending toward a nostalgia for the life of the traditional farmer in Germany's Black Forest, the emphasis in this essay is on the land as a polity in which custom becomes formalized and is continually renewed as working common law. Heidegger's landscape is existential, concerned with being and dwelling, whereas the landscape here is a political landscape, concerned with the building of a polity.[15]

The *Landsting* meeting was a representative assembly that combined a form of legislative activity by which local custom was formalized and generalized as law and the activity of a kind of court where people were tried according to those laws. The *Landsting* was also the place where representatives of the polity met to make major decisions involving the land's external political situation. Denmark was an elective monarchy well into the Renaissance, and to become the monarch an eligible candidate of the right blood needed to be acclaimed at Viborg (civil war could result if the Jutlanders supported their own candidate). The last time a king was thus acclaimed was in 1655 (Trap 1963, 87).

The law, according to the preface to the Jutland law, shall be *"ærlic oc ræt, thollich, æftær landæns wanæ,"* "honorable, just, and tolerable, in accordance with the customs of the land" (Jansen and Mitchell 1971, 4–5). This customary law lay at the root of the historical Old Norse meaning of landscape, *landskapr,* which meant "conditions in a land, its character *[beskaffenhed]*, its traditions or customs." This meaning gives rise to the idea that the landscape is coequal with "the organization of things in a land" and finally the idea that the landscape is coequal with a "district" (Fritzner 1886–96, landskapr; Kalkar 1976 [1881–1918], landskab).[16] The land, as the place of a people, is *scaped/shaped* by, or is the creation of, its customs, organized as law.[17] Through the activity of this polity, the material forms of the landscape area were shaped in turn according to the principles of its law (Olwig 1996; Hoff 1998). Customary law, for the most part, is not formulated verbally but "lies upon the land," to borrow a phrase from the seventeenth-century English jurist Edward Coke. According to Coke, there are "two pillars" for custom: "common usage" and "time out of mind." It is on the basis of these pillars that customs "are defined as a law or right not written; which, being established by long use and the consent of our ancestors, hath been and is daily practised" (Thompson 1993, quotes 97–98, 128, 129). This law is not made visible through written texts or cadastral maps, but through customary community practice as manifested in its "moral economy" (185–351)[18] and in its "habitus."[19] The modern definition of landscape as "an area of activity" such as "the political landscape" thus retains something of this early meaning of landscape as a land, or polity, shaped by praxis and guided by custom.

The Jutland polity would have been centered in the area of the *Landsting,* which was located near the town of Viborg, further to the north of Jelling. But for all its historical importance, this site, when compared to the monumental constructions at Jelling, is notable for its nonpresence. Those in charge of heritage preservation have not attempted to mark a possible location for the site of the North Jutland *Landsting,* even though it is a significant place in the history of what is now Denmark. One reason for the contemporary nonvisibility of this place is that nobody is entirely sure of its exact location, particularly as Jutland is dotted with barrows bearing the name *Tinghøj* ("Thing hill") (Hald 1966, 37–38). It is not unlikely, however, that this place was found near one of the many ubiquitous grave mounds that mark the ancient lines of travel, which met at Viborg.[20] The *Ting* site near Viborg, like that at *Thingvellir,* near Reykjavik, was probably located at a spot accessible from a number of key roadways that met at Viborg, which is thought to have been an important heathen religious center.[21] The prefix *Vi* means "temple" or "high place" and *Viborg* is thought to mean "the holy hill" (Trap 1963, 7:87; Vinterberg and Bodelsen 1966, Vi).[22] The site Blicher identified was located at such a crossroads. The *Ting* site was likely to have been located on common land that at the same time belonged to both no one and everyone. *Snabeshøj* is located in a parish called *Almind,* meaning commons.[23] There was typically no building at a *Ting* site, and people met out in the open on a regular basis. This is remarkable given the characteristic inclemency of Danish weather, even in summer. Although various authorities on the historical topography of Jutland have given credence to

Blicher's identification of the *Landsting* site, little archaeological work has been done to corroborate it, and the exact location still belongs to the realm of speculation (Nielsen 1966).[24] The *Landsting* site remains, as it were, unlooked for and unsuspected.

THE UNWRITTEN LAW

The nonmonumental presence of the *Landsting* site corresponds to the non-thing-like quality of the *Landsting* institution itself, or to the landscape polity that it shaped. The word "landscape" did not refer, first and foremost, to a thing or object, but to the landscape polity as built upon the principles formalized at a *Landsting*. It was, as noted, at the *Ting* that *things* were ordered and the polity shaped. This, in turn, had an effect on the form of the physical environment as an expression of the law of the *Ting*. Things go better when you abide by customary law. The *Ting* gave meaning and form to the material world, transforming it into meaningful things—things that had become objects for discourse. The land is thus "built" in both a lawful and a material sense. In practice, as in the etymology of the word "thing" itself, the principles and values voiced at the *Ting* precede the designation of substantive meaning to things. This same principle applied to the law. It was not given the form and presence of an objective thing by writing it down; it was committed to memory, in particular by the law speaker, the head of the *Ting* assembly.[25] The law of the *Landsting* was made manifest through the praxis of dwelling and habitation, which concomitantly transformed the physical environment, not by being written down in books.

The limitation of the law to what could be remembered meant that a class of specialists, or a central authority, could not appropriate the law. It was constituted only at the *Ting*, and to engage with it one had to go to the *Ting*. Here the collective memory expressed by the law speaker and controlled by the memory of the other participants gained form. The basic principle behind the formalization of customary law is that of "time out of mind." A legal principle is hereby legitimized on the basis of a precedent that goes back as far as anyone can remember. This is practical because it also implies the correlate that what cannot be remembered ceases to have the force of law. Customary law was in effect always being reinterpreted and updated in light of the contemporary situation. The remembrance of custom is fundamentally flexible. "The human memory," according to the historian Marc Bloch, "is a marvelous instrument of elimination and transformation—especially what we call collective memory" (Bloch 1961 [1940], 114; see also Lowenthal 1985, 206–10).[26] Forgetting was thus as much a part of the nature of things as was remembering.

LANDSCAPE AS PLACE

The Arabic word *sifr*, from which *cipher* derives, means, as noted, empty or zero. A zero holds a place within the row of digits in a number and hence is vital to the total meaning of the number, even if the zero itself is essentially a circle marking the presence of the nothing that is within it. An absence that holds a place thus gives meaning in a larger context. The site of the *Landsting* was empty, but it marked a place where people could gather to constitute themselves as something more than a collection of individuals; the group became a polity. A polity is not a material thing, but it is nevertheless capable of concrete collective action. The site of the *Landsting* was a powerful place by virtue of its emptiness, waiting to be filled and given meaning when needed.

Historically in Indo-European culture, much ado has been made about the concept of zero because nothingness was attributed to the qualities of a deity, the unknowable nothing from which everything comes and of which no one must speak (Kaplan 2000, Seife 2000). This notion of nothing has been identified with the Greek concept variously spelled *choros* or *chora*, which Plato gave a prominent role in his cosmology as the vessel-like, feminine place from which the cosmos is born (Casey 1997) and hence of interest to some feminists (Olwig 2001). An extraordinary quality of *Ting* sites is their unmarked anonymity and the absence of monumental signs, and this is why they are often so difficult to locate today. Yet we know they were holy places, often the site of chambered graves, or tombs, marked by hallowed and hollowed, ubiquitous, and unimposing mounds of earth rounded in what some authors, including Blicher, have taken to be feminine shapes (Glob 1967).[27] At the *Ting*, law was committed to living memory, whereas the monument literally chisels memory into dead stone. The Jelling stone and the monstrous barrows marked a monumental royal presence, whereas the place of the Ting marked a womblike hollowness.

Choros not only meant place; it also meant land/country, and it provides the root of the geographical discipline of chorography (Olwig 2001, 2002b). The link between the idea of zero as an encircled nothing that holds a place and the idea of land as a polity may seem farfetched, but it is given perspective when seen in the light of Heidegger's analysis of the meaning of "the thing." Heidegger compares the meaning of *thing* to that of a jug as something that gathers a void: "The empty space, this nothing of the jug, is what the jug is as the holding vessel." "The thing," according to Heidegger, "things," and "thinging gathers" (Heidegger 1971b, 169, 174). The ancient institution *Ting* went under the name of *thing* or *moot*, in English, and a *moot* is the Old English word for a *meeting* or, as Heidegger would call it, "a gathering": "To be sure, the Old High German word *thing* means a gathering, and specifically a gathering to deliberate on a matter under discussion, a contested matter. In consequence, the Old German words *thing* and *dinc* become the names for an affair or matter of pertinence. They denote anything that in any way bears upon men, concerns them, and that accordingly is a matter for discourse" (ibid.). This connection to things in law, as established through discourse, is also true of the Latin word meaning "thing," *res*: "The Romans called a matter for discourse *res*. The Greek *eiro (rehetos, rhetra, rhema)* means to speak about something, to deliberate on it. *Res publica* means not the state, but that which, known to everyone, concerns everybody and is therefore deliberated in public" (ibid.).

One might then surmise that just as deliberating on things through discourse *(res)* generates a *res publica,* the deliberation on things at the *Ting* generates the *land* as a *res publica.* The land in this sense is also by extension something "real" because, as Heidegger put it, "The Roman word *res* denotes what pertains to man, concerns him and his interests in any way or manner. That which concerns man is what is real in *res*" (Heidegger 1971b, 176). The *land* in landscape is thus real in the substantive sense of being legally sanctioned by the people as constituted through the deliberation of the *Landsting.* It is thus through legal rights that the land is materialized as a physical thing for use. The social and judicial parallels, furthermore, between the Latin use of *res* and the Germanic concept of *thing* are striking:

> The Old German word *thing* or *dinc,* with its meaning of a gathering specifically for the purpose of dealing with a case or matter, is suited as no other word to translate properly the Roman word *res,* that which is pertinent, which has a bearing. From that word of the Roman language, which there corresponds to the word *res*—from the word *causa* in the sense of case, affair, matter of pertinence—there develop in turn the Romance *la cosa* and the French *la chose;* we say, "the thing." In English "thing" has still preserved the full semantic power of the Roman word: "He knows his things," he understands the matters that have a bearing on him; "He knows how to handle things," he knows how to go about dealing with affairs, that is, with what matters from case to case. (175)[28]

The discourse that arises through the gathering of the *Landsting* has no materiality, which makes it nothing in the physical sense of thing. It is not an objective thing, but it is nevertheless this substanceless discourse that gives real substance to material things, and, in this sense, to know your things is everything. The site of the *Landsting* was, like a jug, empty, because it was a gathering place, yet it was a powerful place because it was only here that the unwritten, memorized law upon which the land was built could fully be made manifest. This same mode of thinking could be applied to Blicher's authorship. His stories are simply nothing, in the sense that they have no substance beyond the printed words on a page, yet they participated in a discourse that helped create a place for the Jutlanders as the shapers of their land.

BLICHER'S LANDSCAPE

Blicher lived at a time when there was no representative government in Denmark to speak of, and the freedom of the press was a contested issue. The *Landsting,* in earlier times, had provided a means by which the land of the Jutes was represented and shaped. Blicher followed the time-honored custom of reshaping older forms of discursive praxis to new ends. He made manifest the everyday dignity and heroism of the ordinary people of Jutland and their language as a way of legitimizing their place in the political landscape. Whereas the Jutlanders, and the Jutland dialect, had often been looked down on and treated in literature as an object for comic relief, Blicher gave the dialect serious content. He wrote entire poems and stories in variants of the Jutland dialect, and by so doing gave voice to the language and the people who spoke it.

The *Landsting* was fundamentally a gathering, as Heidegger points out, and the principle of gathering is a central theme in Blicher's work. One of Blicher's contributions to Danish cultural history is his invention of an open-air folk festival. Like the *Landsting*, it was a seasonal gathering that attracted people for the oratory and poetry as well as for the song and entertainment. These festivals were, as one contemporary critic put it, Blicher's "very best poetry." Speeches took the place of parliamentary discourse, but in the absence of a parliament the festival did provide a forum for political discourse that made the implications of such a popular gathering readily apparent to Blicher's contemporaries, not the least to the wary watchdogs of the central absolutist state (Olwig 1984, 43–45).

The principle of "gathering" provided the subject for Blicher's famous and innovative story *The Knitting Bee* (*Æ Bindstouw*, literally "the knitting room"). Here men and women from Blicher's home district in mid-Jutland could gather to socialize, entertain one another with stories, verse, and song while doing handiwork. In fact, knitted goods made from the wool of the sheep that grazed the heather were a major Jutland export (Blicher 1930b [1842]). The landscape is thus made manifest in their language and discourse, not in terms of picturesque descriptions, but in terms of a language habitus (Bourdieu 1991) formed in relation to the activities through which the world is inhabited, and where the objects of the material world are given meaning as things for discourse. Blicher's stories and poems reveal the Jutland material and moral economy and its fauna and flora—so much so that a pioneering modern ecologist was able to make a botanical study of Blicher's poetry (Raunkiær 1938). This is why a reader can imagine a material landscape when reading Blicher, even when he does not actually describe it.

Blicher was not simply a pastor and one of the literati; he was also a kind of geographer, who was, like his father before him, the author of an extensive chorographical monograph on his home area. His stories on occasion might include depictions of landscape, but they were quite different from Andersen's grandiloquent style. His landscapes are a scene in the sense of being the place or occurrence of an action, an area of activity or praxis. Instead of Andersen's panoramic view of spectacular scenery seen from a fixed point on high, Blicher makes us bump along a deeply rutted road until we are enmeshed in the life of the people who dwell there, and are swallowed up by its impoverished but starkly beautiful environment. Instead of a fixed scene, the reader experiences the movement and process that gives character to a place. In the introduction to Blicher's "The Robbers' Den," he spoofs the romantic adventure stories of his day, with their florid picturesque landscapes, by contrasting their style to his own down-to-earth realism (Figure 2.6).

FIGURE 2.6. Lithograph of the *Alhede* heath published by Bærentzen's Lithographic Institute in 1855. It is prefaced with quotations from a poem by Steen Steensen Blicher and a passage from "The Robbers' Den." The editor Jens Peter Trap notes that "no-one has given a more truthful or living description of this [the heath's] nature than he in the overview with which he opens the well-known short story, 'The Robbers' Den'" (Bærentzen & Co., 1856, n.p.). Trap, presumably the owner of Bærentzen, would become Denmark's foremost topographer.

"The Robbers' Den" is a tongue-in-cheek romance in the Shakespearean mode: two star-crossed lovers from the upper echelons of society, with the help of nature and the common sense of common folk, find love (Olwig 1981). This flimsy story begins with a description of the landscape transformation that occurs as one moves west across Jutland into the heathlands:

> When one sails from the delightful land of Funen over to Jutland, one believes at first that one has just crossed a river, and one can hardly convince oneself that one is now on the mainland, so similar and nearly related to the islands is the peninsula's countenance. But the farther inland one gets, the more the country[29] changes: the valleys become deeper, the hills more precipitous; the forests look older and more decrepit; many a rush-grown bog, many a bit of ground covered with low heather, great rocks on the high

ridged fields—all testify to a lower state of cultivation and less population. Narrow roads with deep wheel ruts with high ridges in between indicate less travel and less intercourse between the inhabitants. The dwellings of the people become worse and worse, lower and lower, the farther we go, as if they were ducking the violent onslaught of the west wind. Just as the heaths become more frequent and larger, the churches and villages become fewer and farther apart. On the farms instead of the light frames for drying hay one sees instead stacks of black peat and instead of the orchards one sees cabbage plots. Great heather overgrown bogs, carelessly and wastefully treated, tell us: here are plenty of them. No hedges, no rows of willows make a boundary between man and man; one might think that all was still held in common. When at last we reach [the midway point of the north-south watershed demarcation of] the Jutland ridge,[30] the immense flat heaths are spread out before our eyes; at first they are strewn with grave-mounds, but gradually the number is lessened, which leads one to reasonably suppose that this region has never before been cultivated. . . . In the eastern side of this heather-grown plain we occasionally encounter groups of low, shrubby oaks, which serve the wayfarer as a compass, for the crowns of the trees are all bent toward the east. Otherwise we see but few touches of green on the great heather-clad slopes; an occasional patch of grass or a young quaking aspen causes one then to wonder: how did you come here? If a brook or a creek runs through the heath, no strip of meadow or bushy growth proclaims its presence; deep down between hollowed banks it winds secretly and with speed, as if it were hurrying to get out of the desert.

The introduction is critical to understanding the social and economic context of the story, helping to explain the situation of the two heroes: Sorte Mads (Black Mads), a poor farmer who dwells in this environment, and a traveling young nobleman in disguise (Holger), who is lost:

> Across such a brook, one fine autumn day, rode a young well-dressed man toward a little rye field which the distant owner had cultivated by burning its scraped off crust to ashes. The owner, himself, and his family were just engaged in reaping it, when the rider approached and asked the way to the manor of Aunsbjerg. After the farmer had repaid the question with another: namely where the traveler came from? He then told him what the traveler already knew, namely that he had lost his way. The farmer thereafter called a boy, who was tying the sheaves together, and told him to show the traveler the right road. (Blicher 1922 [1827], 53–54, my translation, with interpolations)

As the young nobleman and the farmer converse, "Black Mads" makes his appearance, a figure straight out of local folklore, unwillingly riding a deer—on the horns of a dilemma, as it were. The introduction provides the geographical background to understand how poverty leads an honest Jutlander, nicknamed Black Mads, to engage, like so many other Jutlanders, in a variety of activities that often were considered disreputable, such as poaching. Mads lives the difficult life of a poacher, but he is not unpopular with the farmers, whose fields are often damaged as a result of the nobility's passion for hunting. Mads, in fact, is forced to go underground because of the violent fury of the manor owner and the estate's game warden, who would deliver him to the state for punishment in Copenhagen. But Black Mads is on the side of the law when seen from the perspective of local custom. He also, like Stærke Sejr, becomes a hero because of his enormous knowledge of the land. He thus ends up playing a key role in bringing the young lovers together, who are also fleeing the ire of the manor owner, the girl's father. Just in the nick of time, he uses a local agricultural technique and sets a line of fire through the heath, which prevents the father and his minions from capturing the lovers. He then leads them to safety in his underground lodgings in an ancient heather-covered chambered barrow, which was thought to have once been a robbers' den (hence the story's title—there are no robbers in the story). The entrance tunnel is low and narrow, but there is room in the chamber for Mads's family and the young lovers. Although this is a popular adventure story, it also tells an enormous amount about the condition of the Jutlanders who dwell along the streams that thread their watery way through the heaths, and who exploit every niche in the physical and social environment. The story ends, as might be expected, with a gathering, a depiction of the spring folk festivities of earlier times (before the Pietists succeeded in putting a damper on them in the eighteenth century). These festivities provide the setting for an idealized, deliberately utopian scene of social harmony in which the nobility mingle with the common folk and the game warden and poacher, as good Jutlanders, sit down together for a friendly beer.

Blicher's landscape is a working polity, fundamentally a political landscape. The political depth of Blicher's writing was in fact recognized in an early review of his work by the budding philosopher Søren Kierkegaard. In the review, Kierkegaard favorably compares Blicher's ability to make manifest the voice of the individual as part of a collectivity, as opposed to Hans Christian Andersen's concern with the individual's private struggle to make peace with his own inner soul. Kierkegaard recognized the "autochthonous" originality of Blicher's ability to give new life to that which is rooted in place and the past. He concludes that in Blicher's work there is a "unity, which in its immediacy points meaningfully toward the future, and which necessarily must grasp the present much more than it has, and in that way have a happy influence on the prosaic way the politics have hitherto been treated" (Kierkegaard 1906, 60–61, my translation). Blicher created a place for the Jutlanders in the political landscape constituted by Danish literary and political discourse. If Andersen's poem reflected national concern with material things in Jutland, setting the stage for a heroic individual like Dalgas, Blicher's work helped create an interest in how things were for the Jutlanders as people.

Blicher's Landscape and Dalgas

It is true that Enrico Dalgas succeeded in mobilizing progressivist landscape imagery in order to attract Copenhagen political support and capital for projects, such as afforestation, that would create highly visible marks in the landscape scene (Henningsen 1995, 325–48). Dalgas would not have succeeded, however, were it not for the fact that he was able, concurrently, to work with the Jutlanders in the further development of less-visible cooperative projects, such as meadowland irrigation, which the local farmers had long since established on their own initiative. Heathland agricultural intensification was not based on radical changes in the structure of land use, but involved a gradual process of investment in the intensification of existent methods. The farms were largely located adjacent to watercourses, the vital waters of which were necessarily shared by many farmers up- and downstream. The ability to extend and improve the often sparse meadows along the streams was a key to agricultural survival on the sandy soils of the heaths. This is because they provided the fodder that fed the animals that produced the manure that fertilized the fields. When Dalgas arrived on the scene, farmers were already busy creating irrigation schemes that would improve the fertility of the meadows, by spreading nutrient-rich waters across the meadows. These schemes required collective action because use rights to the water belonged to a number of farmers along the stream and the sharing of these rights was regulated according to long-standing custom.

Blicher had helped open the eyes of the Danes to the potentiality nascent in the Jutland people, and Dalgas, who knew Blicher's work, saw that if the local population were given sufficient capital, technical advice, and infrastructure, the Jutlanders could take care of irrigation and heath cultivation themselves (Dalgas 1867, 39–41; 1868, 91; Skrubbeltrang 1966, 254–60). It is difficult to know to what degree Blicher might have contributed to the sea change that occurred from the characteristic eighteenth-century view that the Jutlanders were irredeemably backward to the idea, vital to Dalgas's project, that they themselves were capable of shaping and transforming their land.

Dalgas had begun his career as a road engineer, and he had seen how local efforts to improve the local agricultural economy were fostered by infrastructural improvements such as roads. This helps explain why Dalgas began his influential study, *Geographic Pictures from the Heath* (Dalgas 1867, 1868, my translation), with a long quotation from Blicher describing the changing view from an east-west road. Jutland was not, he argues, a place that was underdeveloped by nature; it had the human resources to reclaim soils impoverished by centuries of peripheralization under an absolutist government centered in Copenhagen. All that was needed now was for the nation to provide capital and infrastructure (Olwig 1984, 65). Blicher had made the Jutlanders a part of the political landscape, and Dalgas was putting them on the political agenda. Thus, when early in Dalgas's campaign the Royal Agricultural Society argued for the need to concentrate efforts at agricultural improvement on the rich soils to the east, Dalgas replied: "I must draw attention to the fact that Denmark is not made up of Zealand, Funen, and East Jutland alone. There is also something that is called West Jutland, where many people live; they also farm, and their agriculture needs to be intensified. This can best be achieved by the means suggested by the Heath Society: afforestation and the creation of artificial meadows. It is not only for the sake of the dead earth that we have come here, but also for the living people, who also have a claim to make" (quoted in Dessau 1866, 210).

THE INSIDE AND OUTSIDE OF LANDSCAPE

The story of the transformation of the Jutland landscape was told and retold at the beginning of this chapter from the perspective of outsiders like Hans Christian Andersen, who viewed Jutland in terms of landscape scenery. If one reexamines the story of Dalgas in the light of the second part of this chapter, from the insider perspective represented by Blicher, it becomes apparent that Dalgas's monumental transformation of the Jutland landscape was in fact something of a scenic veneer upon a process of ongoing change that the Jutlanders themselves were generating.

The landscape face of Jutland has a Janus countenance whose aspect depends on how it is interpreted and how this interpretation is represented. One aspect is that of a highly visible presence as physical thing, whereas the other is manifested by what might be termed the invisible things constituted through social practice and habitus that could be "built" into a polity. It is common for writers on landscape to emphasize the way in which the pictorial construction of landscape creates an "insider/outsider" dichotomy (Barrell 1972; Cosgrove 1984; Rose 1993). The outsider is the individual spectator whose gaze commands the scene, whereas the insiders are an anonymous collective whose labor shapes the landscape scene, following the patterns of an unreflected tradition. This vision of landscape encapsulates a tale of modernity, prefigured by the philosophy of a Steffens, as well as works like Goethe's *Faust,* in which progress belongs to the heroic individuals who conquer personal alienation while bringing a modern rational order to the scene (Berman 1982). For some, like the Welsh literary scholar Raymond Williams, this dichotomy can manifest itself as the existential dilemma of an insider who has become an outsider, who, like the Welshman ("Will") in the novel *Border Country,* returns home after spending years in London to discover that his country has become, on the one hand, a landscape scene and, on the other, a place from which he has become alienated:

> When Matthew got back from town, he walked slowly up the lane.... In Gwenton he had met nobody he knew, and the simple shopping had been difficult, after London: the conventions were different. He had felt empty and tired, but the familiar shape of the valley and the mountains held and replaced him. It was one thing to carry its image in

his mind, as he did, everywhere, never a day passing but he closed his eyes and saw it again, his only landscape. But it was different to stand and look at the reality. It was not less beautiful; every detail of the land came up with its old excitement. But it was not still, as the image had been. It was no longer a landscape or a view, but a valley that people were using. He realized as he watched, what had happened in going away. The valley as landscape had been taken, but its work forgotten. The visitor sees beauty: the inhabitant a place where he works and has friends. Far away, closing his eyes, he had been seeing this valley, but as a visitor sees it, as the guidebook sees it: this valley, in which he had lived more than half his life. (Williams 1960, 75)

No doubt the landscape duality that Williams captured is real and that its tensions reflect the existential position of place in modern society (Daniels 1989; Harvey 1996, 31–34). This landscape, however, is a phenomenon that represents only one aspect of a highly contested landscape terrain. The Janus character of the landscape is not simply the product of "a way of seeing" from the outside, which can be traced back to the Renaissance discovery of the science of perspective (Cosgrove 1984; Berger 1972; Edgerton 1975). The Jutland case suggests that the insides and outsides of landscape are as fundamentally political as they are existential and phenomenological. In Williams's case it is significant that the character in the novel who experiences these sentiments, like Williams himself, has moved to England, and the valley he writes about is located in the "border country" of his native Wales, a country that has maintained a separate place identity despite centuries of subjugation to England.

The landscape of the insiders is marked by the ever-changing principles of custom, not by the reified signs of tradition, where custom becomes costume, as in the works of Henrik Rantzau and Pontoppidan. The latent and tacit zero/cipherlike character of custom means that it must be made manifest in a larger context before it can have an effect in generating a polity. The *Landsting* at Viborg thus gave voice to local mores, thereby giving form to the principles upon which the land of the Jutes was "built." Jutland thereby took on the recognizable form of an established land. The early nineteenth-century revival of ideas of representative government in the Danish provinces was in turn linked to the revival of a Jutland identity fostered by those like Blicher, who argued that power, fundamentally, was in the hands of the people, as exemplified by Stærke Sejr and Black Mads. It was heroes like Sejr and Mads, who in turn worked together locally to make the transformation of Jutland's social, political, and economic landscape possible. This social transformation then made possible the landscape scene facilitated by Dalgas. When one thinks about landscape, it is important to consider its scenic imprint. This scenery should not, however, conceal the importance of the heritage of places like Uannet and Snabeshøj, which tell of the "unlooked for" and "unsuspected" powers as a force in land shaping and that lie in the invisible, historically rooted custom and culture of polity and place.

NOTES

1. Andersen's original reads *"Runestav,"* which the standard Danish dictionary defines as a wooden stave or stick upon which runes have been carved. Such sticks were used as magic wands and as calendars. The equivalent English word is *rune-staff.* The Danish suffix *stav* can also be translated to mean the lines (especially vertical) that make up a runic letter or symbol, which in English is spelled *rune-stave* (*ODS* 1931, s.v. "Runestav"; *Oxford English Dictionary,* 1989, s.vv. "Rune-staff, Rune-stave"). Either meaning can be used to describe the shape of the Jutland peninsula as it cuts its way between the North Sea and the Kattegat, which forms the entrance to the Baltic.

2. An English translation of many of Blicher's best-known works can be found in Blicher 1945.

3. Hamlet was a legendary Jutland prince whose story of feigned madness in a kingdom made rotten by deceit reached a wide European audience through its telling in the chronicle of Denmark's history, *Gesta Danorum,* by the Danish monk Saxo Grammaticus (1150–c. 1216) (Grammaticus 1998).

4. This is not the supposed "Hamlet's grave" visited by Andersen.

5. Andersen, for the sake of simplicity, describes Jutland as dividing the North Sea from the Baltic, or "Østersøen" (the Eastern Sea) as it is called in Danish. Actually, the waters to the east of Jutland include both the Baltic in the southeast and the Kattegat in the northeast (which lies between Jutland and Sweden, and which is technically north of the Baltic's mouth).

6. Henrik Rantzau's map (drawn upon by both the cartographers Abraham Ortelius and Gerhard Mercator) was made in 1585 by the Holsten-born University of Copenhagen mathematics professor and lecturer on Ptolomaen geography, Marcus Jordan (1521–95) (Jørgensen 1981, 15–18). This was a monument to the Rantzaus' territorial conception of the estate of the Danish king as encompassing the lands of Slesvig-Holsten and Denmark. It includes textual inserts memorializing the military victories of Henrik Rantzau's father over the enemies of the Danish state, as well as a pictorial insert where Skipper Clement (c. 1485–1536) is shown imprisoned while awaiting execution. Henrik used maps as a way of creating an image of Jutland as a geographical unit defined by physical features, according to chorographic principles based on those of the second-century astronomer and geographer Claudius Ptolemaeus (or Ptolemy), who named Jutland *Kimbriké Chersonesos* (The Cimbrian Peninsula) (Trap 1963, 7:11). Henrik saw himself in this spirit as *Cimbrian,* and Jutland, on his 1585 map, is called *Cimbricæ Chersone.* The maps were supplemented by Latin poems celebrating famous Cimbrians, beginning with the grandson of Noah and ending with Johan Rantzau. He also, as noted, published an illustrated regional geographic (chorographic) work on the "Cimbrian" peninsula, depicting scenes of the peninsula and its people in local costume.

7. Henrik Rantzau had particularly close ties to the prelate and cartography publisher George Braun, whom he supplied with important contributions to his *Civitates orbis terrarum* (including Jordan's map), which was first published in 1572. Braun, who was apparently of Dutch background, was a protégé of the great Antwerp humanist and cartographer Abraham Ortelius (1527–98), but he was educated in Germany and was based in Cologne (Jørgensen 1981, 7–18). Braun also published a Rantzau family tree for Henrik in which the tree is shown growing out of a map of the duchies and Denmark. The map is festooned with pictures and text illustrating the illustrious deeds of the Rantzaus, including a picture of the prostrate *Landschaft* representatives of Dithmarschen begging for mercy from their conqueror (Jørgensen 1981, 16–17).

It also, however, contains illustrations of more peaceful Rantzau contributions, such as the building of bridges and the planting of experimental orchards, which Henrik sponsored.

8. They are 60–70 meters across and 11 meters high.

9. The full title of this work in translation gives a good indication of its contents: "The Danish Atlas, or The Kingdom of Denmark, with its Natural Characteristics, Elements, Inhabitants, Vegetation, Animals and other Products, its ancient Occurrences and present Conditions in all Provinces, Cities, Churches, Castles and Manors. Presented through a thorough Description of the Land which is illustrated with Land maps of every Province which have been prepared for this purpose, with Urban Prospects, Ground Plans, and other notable Copper Engravings. At the Behest of the Royal Highness by Erich Pontoppidan, doctor of theology, professor and pro-chancellor of the University of Copenhagen and member of the Royal Copenhagen and Imperial Petersburg Scientific Societies" (Pontoppidan 1763, 126–53).

10. Steffens developed a pictorial conception of natural national development that appears to have inspired, and been inspired by, his contact with the Dresden school of painters, where landscape painting was being revolutionized by the work of Caspar David Friedrich and the theories of Carl Gustav Carus (1789–1869) (Boime 1990, 428–32).

11. It is probable that Blicher is referring to a place south of Snabeshøj, called Vandet (meaning "the water"). By spelling it according to the local dialect (in which *V* is pronounced like *W* [or *UA*] and an *N* is substituted for the *D*), he is able to transform the word into a close approximation of the Danish word for "unsuspected." Given the emphasis that Blicher places upon the anonymity of the place, it is not unreasonable to suppose that he intends a play on words suggesting that the place is "unlooked for," or "unsuspected." Blicher's father, Niels, engaged in similar, humorous plays on words in interpreting the etymology of nearby local place-names in his celebrated *Topography of Vium Parish* from 1795 (see Blicher 1978 [1795], 32).

12. There are those who assume that the word *Snapsting* derives from the vernacular Danish word for aquavite, or vodka, *snaps* (Kjersgaard and Hvidtfeldt 1963, 284). The restaurant at the Danish parliament, the *Folketing*, in Copenhagen is called the *Snapsting*, with humorous reference to this tradition.

13. This law became a foundation of subsequent Danish justice, though it was superseded (except in the duchies) by the codified body of law promulgated by the absolute monarchy in the seventeenth century (Kroman 1945; Benediktsson et al. 1981, 228–33).

14. Denmark was dotted by *Ting* sites that no doubt shared many of the features that can still be experienced at Thingvellir in Iceland. Not only did each *land*, such as Zealand, Lolland, and Scania have its *Ting*, but the lands themselves were divided up into *Herreder*, roughly equivalent to the English *Hundred*, which each had a *Ting* where local legal issues were adjudicated. Northern Jutland had approximately eighty to ninety *Herreder*, which dated back to prehistoric times, and many of which still have some administrative functions. At least as far back as Viking times, Jutland was also divided into approximately fourteen *sysler* (e.g., Vendsyssel, Thy syssel, Himmersyssel, Salling syssel, Jelling syssel, Almind syssel), which also had a *Ting* site. These were not for lawmaking and court cases, but for the discussion of larger political issues. These *sysler* lost their function at the beginning of the 1500s, and they exist primarily as loci of local identity, supporting historical societies and other forms of heritage-related activities. After 1100 the three *sysler* in southern Jutland held a common *Ting* at *Urnehoved*, the site of which, like that of the Viborg *Ting* (which was superior to that at *Urnehoved* in terms of lawmaking), is not known with certainty (Trap 1963, 11). These many *Ting* sites today are known primarily through the survival of place-names like the ubiquitous *Tinghøj* (Thing

barrow) or rarer names like *Tinghulen* (Thing hollow) or *Lovbjerg* (which might mean "law hill") (Hald 1966, 37–38). These preexisting features marked places that were not otherwise noted by the presence of permanent built structures. The *Ting* institution was gradually professionalized, with centrally appointed sheriffs and judges, in the fifteenth century, and the first attempts to legislate the construction of courthouses in the sixteenth century (Kjersgaard and Hvidtfeldt 1963, 283–86). Today, the *Ting* survives as the name for a court, and as the name for the Danish parliament, the *Folketing*. Finally, it should be noted, each village had a *Ting*-like institution, called a village (or neighbor) *stævne,* or meeting, held on the village green, through which the farmers regulated village affairs. This institution flourished well into the nineteenth century and still persists in some villages in the form of a nostalgic reminder of the past.

15. The question "To be or not to be" poses a conundrum, I would suggest, because the answer involves a struggle between individual existential needs and human existence as social beings.

16. I am indebted to Chris Sanders at the Arnemagnæanske Commission's Dictionary, Copenhagen, for his help in tracking the older Nordic meaning of this term. The word *beskaffenhed* (*Beschaffenheit* in German) has the root *skab* (or *schaft* in German) meaning *shape*. *Shape* can mean to create by shaping, but it can also be used to refer to the shape or form of that which has been shaped. The *Beschaffenheit* of something is thus literally the *shape* something is in. The term *Landschaft*, in this sense, literally refers to the shape the land is in with respect to its customs, the material forms generated by those customs, and the shape of the bodies that generate and formalize those customs as law.

17. The suffix *shaft* and the English *ship* are cognate, meaning essentially *"creation, creature, constitution, condition"* (*Oxford English Dictionary,* 1971, s.v. "-ship"). *Schaft* is related to the verb *schaffen,* to create or shape, so *ship* and *shape* are also etymologically linked (*Oxford English Dictionary,* 1971, s.v. "Shape").

18. Of the "moral economy," Thompson writes: "It is possible to detect in almost every eighteenth-century crowd action some legitimizing notion. By the notion of legitimation I mean that the men and women in the crowd were informed by the belief that they were defending traditional rights or customs; and, in general, that they were supported by the wider consensus of the community. . . . These grievances operated within a popular consensus as to what were legitimate and what were illegitimate practices in marketing, milling, baking, etc. This in its turn was grounded upon a consistent traditional view of social norms and obligations, of the proper economic functions of several parties within the community, which, taken together, can be said to constitute the moral economy of the poor. . . . While this moral economy cannot be described as 'political' in any advanced sense, nevertheless it cannot be described as unpolitical either, since it supposed definite, and passionately held, notions of their common weal—notions which, indeed, found some support in the paternalist tradition of the authorities; notions which the people re-echoed so loudly in their turn that the authorities were, in some measure, the prisoners of the people. Hence this moral economy impinged very generally upon eighteenth-century government and thought" (Thompson 1993, 188–89).

19. Thompson gives the following gloss of *habitus:* "Agrarian custom was never fact. It was ambience. It may be understood with the aid of Bourdieu's concept of 'habitus'—a lived environment comprised of practices, inherited expectations, rules which both determined limits to usages and disclosed possibilities, norms and sanctions both of law and neighbourhood pressures" (Thompson 1993, 102; see also Bourdieu 1977, 16–22).

20. The most important of these ancient thoroughfares was the road that followed the watershed divide separating eastern from western Jutland, bisecting the peninsula all the way south,

past Jelling, and past the ancient site of the lesser *ting* site for southern Jutland, at *Urnehoved*, just north of the border of what is now Germany, and on to continental Europe. This road has had many names; some of them, like the Old Viborg Road, hark back to Viborg's historical importance as a heathen religious center and as the location of the *Landsting* for northern Jutland, and as the continued site, during the Middle Ages, of the *Landsting* legal court, as well as the site of a cathedral that marked an important stopping point on a pilgrim route linking Iceland and Norway to the Continent and Rome. Another name for this road was *Studevejen* or *Oksevejen,* meaning the Road of the Oxen, and this name calls forth a memory of the peaceful occupations of the Jutlanders who raised livestock on the extensive pastures of the heath and then herded the stock south to the rich pastures of southern Jutland (such as those on the marshlands of the western coast) before they were sold to markets further south. A source from 1591 notes that 50,000 cattle were driven down this road in that year. As late as the 1860s, when the drives ebbed, equivalent numbers have been given. Finally, the road, particularly in the Middle Ages, came to be known as *Hærvejen,* The Military Road, or the King's Highway, marking yet another perception of the purpose of this road, which ran close by Jelling, linking together a number of fortifications with the *Dannevirke* fortification that crosses the narrow southern neck of Jutland and that is linked in origin to the era of Thyra, Gorm and Harald Bluetooth (Matthiessen 1962, 7–20; Trap 1963, 10:9).

21. Unlike the site at *Snabeshøj,* the Icelandic *Thingvellir* site is now a national monument for republican Iceland, visited by tourists from around the world because of its fame as the site of perhaps the world's oldest "parliament." Even though the place is now marked by an array of permanent buildings, a boardwalk, railing, and flag, what is striking to a visitor is the way the site seems to erase its own presence through its pronounced lack of monumentality. *Thingvellir,* located some distance outside Reykjavik, is essentially a large meadow formed within a broad valley through the diversion of the Öxará River by ancient Icelanders. It is thus the creation of human design, but one would hardly know it since the major features of the place, such as the *Lögberg (Law Rock),* from which the law speaker proclaimed the law, or the *Öxaráhólmi,* the little island upon which ritual combat was carried out, are preexisting features of the terrain. The meetings were held outdoors and they thus were, it might be said, "building" a land, not monuments (Thorsteinsson 1987; Olwig 1998).

22. Kristian Hald, however, points out that *Vi* can also be derived from the word *vidje,* meaning willow (Hald 1966, 57).

23. *Almind* is an abbreviated form of an earlier spelling, *Alminningh* (1499) (Trap 1963, 7:411), which means "commons."

24. Jeppe Aakjær, an author with a background in history who was an authority on the topography of Jutland, supported Blicher's claim (Blicher 1930a [1841], 275). Blicher himself published scholarly works on the topography of Jutland, and he was convinced, on the basis of local tradition, that this was the site of the ancient *Ting*. The story to *The Three Holy Eves* was based on a collection of Jutland rover stories published the same year (Blicher 1920 [1824], 161).

25. In 1241, King Valdemar promulgated the first written version of the Jutland law at a meeting with "the best men" of his kingdom at Vordingborg castle on Zealand, which marked a major transition in Danish history. This "landscape law," as it came to be known, hereby was frozen in time, becoming a material thing rather than an expression of an ongoing flow of things in law. The law had become something that the king claimed he had the power to "give," and that the "land" must necessarily "take" unto itself (presumably through approval at the *Landsting*), and, as such, it became the primary law not only of Jutland, but of all Denmark (Fenger 1992, 1993).

26. The historian Eric Hobsbawm has explained the logic of this mode of thought by comparing it to a motor: "'Custom' in traditional societies has the double function of motor and flywheel. It does not preclude innovation and change up to a point, though evidently the requirement that it must appear compatible or even identical with precedent imposes substantial limitations on it. What it does is to give any desired change (or resistance to innovation) the sanction of precedent, social continuity and natural law as expressed in history" (Hobsbawm 1983, 2–4).

27. Blicher thus compares a hollowed barrow to a womb in his short story "Røverstuen" ("The Robbers' Den") (Blicher 1922 [1827]).

28. My references to Heidegger should not be construed as an attempt to align this chapter with the phenomenology of the philosopher. Heidegger in the end is more interested in the phenomenon of the jug as a gatherer than in the gathering of a polity. I completed the first draft of this chapter before I discovered these highly perceptive remarks, and I include them for their insight, not for the author's philosophy, which is quite another thing.

29. An earlier translator, Hanne Aa. Larsen, translates the Danish word *egn*, meaning "area" or (part of a) country, as "landscape" (Blicher 1945 [1827], 79–80). In this way the subject of the passage is subtly changed from a place of dwelling to a visual scene.

30. The idea of a Jutland ridge, running north-south roughly down the middle of the peninsula from the Limfjord, is something of a myth because much of this divide is actually quite flat. The myth owes to the fact that at certain points along this line an apparent ridge of hills does create a rather dramatic contrast between the flat glacial outwash plains of West Jutland and the hillier terrain of East Jutland. These hills and the perception of a divide owe to the fact that the watershed of Jutland, because of glaciation, does divide along this line. The perception of a contiguous ridge derives from the days before cartographers were able to reliably measure altitude, showed a ridge of hills running along watershed divides. Maps from the eighteenth century thus showed a ridge of hills running down Jutland's back, and the myth of such a ridge persists to this day (Erslev 1886).

References

Albeck, Gustav. 1965. "Henrich Steffens: Stafetten med den tyske Romantikk." *Dansk Litteratur Historie,* vol. 2, ed. P. H. Traustedt, 18–23. Copenhagen: Politiken.

Andersen, Hans Christian. 1951 (1855, 1871). *Mit Livs Eventyr.* Copenhagen: Gyldendal.

———. 1964 (1860). "Jylland Mellem Tvende Have." *Folkehøjskolens Sangbog,* ed. Karl Bak et al., 714–15. Odense: Foreningen for Højskoler og Landbrugsskoler.

Bæksted, Anders. 1963. *Guder og Helte i Norden.* Copenhagen: Politiken.

Bærentzen & Co. 1856. *Danmark,* ed. Jens Peter Trap. Copenhagen: EM Bærentzen & Co. Lith. Inst.

Barrell, John. 1972. *The Idea of Landscape and the Sense of Place.* Cambridge: Cambridge University Press.

Benediktsson, Jakob, Finn Hødnebø, John Granlund, Allan Karker, Magnús Már Lárusson, and Helge Pohjolan-Pirhonen. 1981. *Kulturhistorisk Leksikon for Nordisk Middelalder fra Vikingtid til Reformationstid,* vol. 10. Copenhagen: Rosenkilde og Bagger.

Berger, John. 1972. *Ways of Seeing.* Harmondsworth: Penguin.

Berman, Marshall. 1982. *All That Is Solid Melts into Air: The Experience of Modernity.* New York: Simon and Schuster.

Blicher, Niels. 1978 (1795). *Topografphie over Vium Præstekald*. Herning: Blicher-Selskabet, Poul Kristensen.

Blicher, Steen Steensen. 1920 (1824). "Oldsagn på Alheden." *Samlede Skrifter* 6, ed. Georg Christensen and Jeppe Aakjær, 147–61. Copenhagen: Gyldendal.

———. 1922 (1827). "Røverstuen." *Samlede Skrifter* 10, ed. George Christensen and Jeppe Aakjær, 52–112. Copenhagen: Gyldendal.

———. 1930a (1841). "De tre Helligafterner." *Samlede Skrifter* 26, ed. Johs. Nørvig and Jeppe Aakjær, 1–15. Copenhagen: Gyldendal.

———. 1930b (1842). "Æ Bindstouw." *Samlede Skrifter* 26, ed. Johs. Nørvig and Jeppe Aakjær, 73–124. Copenhagen: Gyldendal.

———. 1945 (1827). "The Robbers' Den." In *Twelve Stories*, trans. Hanne Aa. Larsen, 79–121. Princeton: Princeton University Press.

———. 1945. *Twelve Stories*, trans. Hanne Aa. Larsen. Princeton: Princeton University Press.

Bloch, Marc. 1961 (1940). *Feudal Society*. Chicago: University of Chicago Press.

Boime, Albert. 1990. *Art in an Age of Bonapartism, 1800–1815*. Chicago: University of Chicago Press.

Boisen, F. E., ed. 1858. *Dansk Læsebog*. Copenhagen: Rittendorff and Aagaard.

Bourdieu, Pierre. 1977. *Outline of a Theory of Practice*. Cambridge: Cambridge University Press.

———. 1991. *Language and Symbolic Power*. Cambridge: Polity Press.

Bramsen, Bo. 1965. *Gamle Danmarkskort: En Historisk Oversigt med Bibliografiske Noter for Perioden, 1570–1770*. Copenhagen: Grønholt Pedersen.

Casey, Edward S. 1997. *The Fate of Place: A Philosophical History*. Berkeley and Los Angeles: University of California Press.

Cosgrove, Denis. 1984. *Social Formation and Symbolic Landscape*. London: Croom Helm.

Dalgas, Enrico. 1867–68. *Geographiske Billeder fra Heden*, 2 vols. Copenhagen: C. A. Reitzel.

Daniels, Stephen. 1989. "Marxism, Culture, and the Duplicity of Landscape." *New Models in Geography*, vol. 2, ed. Richard Peet and Nigel Thrift, 196–220. London: Unwin and Hyman.

Davidson, H. R. Ellis. 1964. *Gods and Myths of Northern Europe*. Baltimore: Penguin.

Dessau, D. 1866. *Den tiende danske Landmandsforsamling i Aarhus 25–29, Juni 1866*. Copenhagen: Triers Bogtrykkeri.

Edgerton, Samuel. 1975. *The Renaissance Rediscovery of Linear Perspective*. New York: Basic Books.

Erslev, Edvard. 1886. *Jylland: Studier og Skildringer til Danmarks Geografi*. Copenhagen: Jacob Erslevs Forlag.

Exner, Johan, and Jan Ebert. 1968. *400 danske Landsbykirker*. Copenhagen: Gyldendal.

Fenger, Ole. 1992. "Magten." In *Den nordiske verden*, vol. 2, ed. Kirsten Hastrup, 117–66. Copenhagen: Gyldendal.

———. 1993. *Fred og Ret i Middelalderen*. Århus: Bogformidlingens Forlag.

Folkehøjskoleforening. 1966. *Folkehøjskolens Sangbog*. Odense: Foreningen for Højskoler og Landbrugsskoler.

Frandsen, Steen Bo. 1996. *Opdagelsen af Jylland: Den regionale dimension i Danmarkshistorien, 1814–1864*. Aarhus: Aarhus Universitesforlag.

Fritzner, Johan. 1886–96. *Ordbog om Det Gamle Norske Sprog*. Kristiania: Ny Norske forlagsforening.

Fullerton, Brian, and Alan F. Williams. 1972. *Scandinavia*. London: Chatto and Windus.

Glob, P. V. 1967. *Danske Oldtidsminder*. Copenhagen: Gyldendal.

Grammaticus, Saxo. 1998. *The History of the Danes*, books 1–10, ed. Hilda Ellis Davidson, trans. Peter Fisher. London: Woodbridge Brewer.

Hald, Kristian. 1966. *Stednavne og Kulturhistorie*. Copenhagen: Dansk Historisk Fællesforening.
Hansen, Martin A. 1950. "Forord." *En Jydsk Røverhistorie: De Tre Helligaftener*, ed. Steen Steensen Blicher. Copenhagen: Det Berlingske Bogtrykkeri, u.p.
Harvey, David. 1996. *Justice, Nature, and the Geography of Difference*. Oxford: Blackwell.
Heidegger, Martin. 1971a. "Building Dwelling Thinking." *Poetry, Language, Thought*, ed. and trans. Albert Hofstadter, 145–61. New York: Harper and Row.
———. 1971b. "The Thing." *Poetry, Language, Thought*, ed. and trans. Albert Hofstadter, 165–82. New York: Harper and Row.
Henningsen, Peter. 1995. *Hedens Hemmeligheder: Livsvilkår i Vestjylland, 1750–1900*. Grindsted: Overgaard Bøger.
Hobsbawm, Eric. 1983. "Introduction." *The Invention of Tradition*, ed. Terrence Ranger and Eric Hobsbawm, 1–14. Cambridge: Cambridge University Press.
Hoff, Annette. 1998. *Lov og Landskab*. Aarhus: Aarhus Universitetsforslag.
Jansen, F. J. Billeskov, and P. M. Mitchell. 1971. *Anthology of Danish Literature, vol. 1: Middle Ages to Romanticism* (bilingual edition). Carbondale: Southern Illinois University Press.
Jørgensen, Ove. 1981. "Otonivm i Vrbivm Praecipvarvm Mvndi Theatrvm Qvintvm Avctore Georgio Bravnio Agrippinate, Odense 1593." In George Braun and Frans Hogenberg, *Civitates orbis terrarum*, V. del, Cologne, 1597, blad 30. Faksimileudgave af beskrivelsen og kortet med oversættelse til dansk. Odense: Odense University Press.
Kalkar, Otto. 1976 (1881–1918). *Ordbog over det ældre danske Sprog, 1300–1700*. Copenhagen: Akademisk Forlag.
Kaplan, Robert. 2000. *The Nothing That Is: A Natural History of Zero*. Oxford: Oxford University Press.
Kierkegaard, Søren. 1906. "Af en Endnu Levendes Papirer." *Samlede Værker*, vol. 13, ed. J. L. Heiberg, A. B. Drachmann, and H. O. Lange, 41–92. Copenhagen: Gyldendal.
Kjersgaard, Erik, and Johan Hvidtfeldt. 1963. *Danmarks Historie: De første Oldenborger, 1449–1553*. Copenhagen: Politiken.
Kroman, Erik. 1945. *Danmarks Gamle Love*, vol. 2. Copenhagen: Gad.
Langenscheidt. 1967. *Handwörterbuch Englisch, Teil II Deutsche-Englisch*, ed. Heinz Messinger. Berlin: Langenscheidt.
Lowenthal, David. 1985. *The Past Is a Foreign Country*. Cambridge: Cambridge University Press.
Matthiessen, Hugo. 1939. *Den Sorte Jyde*. Copenhagen: Gyldendal.
———. 1962. *Hærvejen: En tusindaarig vej fra Viborg til Danevirke. En historisk-topografisk studie*. Copenhagen: Gyldendal.
Mitchell, P. M. 1957. *A History of Danish Literature*. Copenhagen: Gyldendal.
Molbech, Christian. 1829. "Optegnelser paa en Udflugt i Jylland i Sommeren 1828." In *Nordisk Tidsskrift for Historie, Litteratur og Kunst*, vol. 3, 106–76.
Nielsen, Erik Levin. 1966. "Det ældste Viborg." *Fra Viborg Amt: Årbog Udgivet af Historisk Samfund for Viborg Amt* 32 (1965): 157–62.
ODS. 1931. *Ordbog over Det Danske Sprog*. Copenhagen: Gyldendal.
Olwig, Kenneth R. 1981. "Literature and 'Reality': The Transformation of the Jutland Heath." *Humanistic Geography and Literature*, ed. Douglas C. D. Pocock, 47–65. London: Croom Helm.
———. 1984. *Nature's Ideological Landscape: A Literary and Geographic Perspective on Its Development and Preservation on Denmark's Jutland Heath*. London: George Allen and Unwin.
———. 1986. *Hedens natur: Om natursyn og naturanvendelse gennem tiderne*. Copenhagen: Teknisk forlag.

———. 1996. "Recovering the Substantive Nature of Landscape." *Annals of the Association of American Geographers* 86, no. 4: 630–53.
———. 1998. "Community and Power." *Landscape Design: Journal of the Landscape Institute* (269), 39–42.
———. 2001. "Landscape as a Contested Topos of Place, Community, and Self." In *Place, Meaning, and Self: Geographies of Imagination and Experience*, ed. Steven Hoelscher, Paul Adams, and Karen Till, 95–117. Madison: University of Wisconsin Press.
———. 2002a. "Landscape, Place, and the State of Progress." In *Progress: Geographical Essays*, ed. Robert D. Sack, 22–60. Baltimore: Johns Hopkins University Press.
———. 2002b. *Landscape, Nature, and the Body Politic: From Britain's Renaissance to America's New World*. Madison: University of Wisconsin Press.
Ørsted, H. C. 1836. "Danskhed, en Tale." *Dansk Folkeblad* 1 (53/54): 209–16.
———. 1843. "Betragtninger over den danske Character." *Dansk Ugeskrift* 2 (85): 105–20.
Pontoppidan, Erik. 1730. *Theatrum Daniæ, oder Schaubühne des alten und jetzigen Dänemarks*. Bremen: Herman Jäger.
———. 1739–41. *Marmora Danica*, 2 vols. Copenhagen: E Typographéo S.R.M. privil.
———. 1763. *Den Danske Atlas eller Konge-Riget Dannemark*. Copenhagen: Godiche.
Ranzovii, Henrici. 1739 (1597). "Cimricæ Chersonesi, Ejusdemque Partium, Urbium, Insularum Et Fluminum, Nec Non Cimborum." *Monumenta Inedita Rerum, Germanicarum Præcipue Cimbricarum et Megapolensium*, vol. 1, ed. Ernestus Joachimus de Westphalen, 1–166. Lipsiæ (Leipzig): Christiani Martini.
Raunkiær, C. 1930. *Hjemstavnsfloraen hos Hedens Sangere, Blicher og Aakjær*. Copenhagen: Schultz.
Riis, Jakob A. 1910. *Hero Tales from the Far North*. New York: Macmillan.
Rose, Gillian. 1993. *Feminism and Geography: The Limits of Geographical Knowledge*. Cambridge: Polity Press.
Seife, Charles. 2000. *Zero: The Biography of a Dangerous Idea*. New York: Viking.
Skovgaard, Johanne. 1915. "Georg Braun og Henrik Rantzau." *Festskrift til Johs. C.H.R. Steenstrup paa Halvfjerdsaars-dagen*, ed. En Kreds af Gamle Elever, 189–211. Copenhagen: Erslev and Hasselbalch.
Skrubbeltrang, Fridlev. 1966. *Det Indvundne Danmark*. Copenhagen: Gyldendal.
Steffens, Henrik. 1905. *Indledning til Philosophiske Forelæsninger i København 1803*. Copenhagen: Gyldendal.
Thompson, E. P. 1993. *Customs in Common*. London: Penguin.
Thorpe, Harry. 1957. "A Special Case of Heathland Reclamation in the Alheden District of Jutland, 1700–1955." *Transactions of the Institute of British Geographers* 23: 87–121.
Thorsteinsson, Björn. 1987. *Thingvellir: Iceland's National Shrine*. Reykjavik: Örn og Örlygur.
Trap, J. P. 1863. *Danmark*. Copenhagen: Gad.
———. 1963. *Statistisk-topographisk Beskrivelse af Hertugdømmet Slesvig*. Copenhagen: Gad.
Vinterberg, Herman, and C. A. Bodelsen. 1966. *Dansk-Engelsk Ordbog*. Copenhagen: Gyldendal.
Williams, Raymond. 1960. *Border Country: A Novel*. London: Chatto and Windus.

The North Atlantic

3.
Icelandic Topography and the Sense of Identity

KIRSTEN HASTRUP

The subarctic Icelandic landscape has often been described as impressively vast and wild: volcanoes, glaciers, geysers, sulfurous pools, and colorful mountains tend to dwarf people and make culture disappear from view (Figure 3.1). In spite of a sparse population, in today's towns and villages one of course is constantly reminded of human habitation, but even there nature towers over society and penetrates the everyday. The point I want to make in this chapter is that in spite of the natural "wild" being so predominant, at least to the newcomer's eye, the Icelandic landscape is deeply marked by history and meaning for the Icelanders. Indeed, the landscape is a well-known history (to invert Lowenthal's [1985] phrase); it is a vital part of the local memory and hence of the sense of Icelandicness.[1]

It is often assumed that the relation between space and memory results from the capture of memories in architecture, monuments, or other visual landmarks, while the landscape itself is seen as an objective, fixed, and measurable surface that is unaffected by the processes of remembrance (Küchler 1993, 103–4). The baseline taken here is different. I see the Icelandic landscape as deeply historicized. It is not simply a surface, or a stage upon which people play their social roles; it is part of the social space. It infiltrates practice and makes history. There is, as it were, agency on both sides; the opposition between wilderness and culture dissolves.

The naturalistic view of the landscape as the backdrop of human activities is out of tune with anthropological insights into the culturally constructed environment that does not allow an external perspective of nature (Ingold 1992). "People and environment are constitutive components of the *same* world" (Tilley 1994, 23). As Gísli Pálsson has convincingly argued with special reference to Iceland, we have to get beyond the language of nature and reconcile discourse and ecology if we properly want to understand society's embeddedness in nature (Pálsson 1991).

In the arts of memory, studied so extensively by Frances Yates (1966), yet by her exclusively attached to human-made structures, the landscape stands out as the most generally applicable *aide-mémoire* of a society's knowledge of itself (Küchler 1993, 85). In that sense, landscape becomes a *topos* of identity. Identities relate to "sensory topics," moments or instances where a shared feeling for common circumstances is created and commonplaces are established (Hirsch 1995, 17). This is a key to my interest in the Icelandic landscape: the everyday spatial practice, imbued with feelings deriving from age-old notions and events of bygone centuries and sustained by a deep sense of tradition.

FIGURE 3.1. Nature dwarfs human habitation in Iceland, but the landscape acquires meaning from the routes walkers take and the names they give to its features, thus structuring the experience of those who follow. Photo by Kirsten Hastrup.

Primordial Land Takings

In the Icelandic tradition, the island was settled by Norsemen, who came either directly or by way of the British Isles, to where an earlier wave of emigration had taken them. The first true settler allegedly landed in A.D. 870, and by 930 the country was "fully inhabited" according to the first Icelandic chronicler Ari *inn fróði* ("the wise") in his *Íslendingabók* (The Book of the Icelanders), c. 1120.[2] The period of settlement, or *landnámsöld*, is extensively referred to even today. The First Times are celebrated as a time of discovery, independence, and heroism—and as the source of present-day Icelandic identity.

Viking Age Scandinavia was marked by unrest. Compared to the great migrations across Europe and parts of Asia in the fifth to seventh centuries, the Viking voyages may seem almost negligible. Qualitatively, however, they were to have remarkable repercussions. The voyage from various parts of the Nordic realm to the North Atlantic islands implied the birth of new peoples. In contrast to populations that exist by virtue of numbers, peoples exist by virtue of definition. Thus it took more than migrations to a far northern island to make the Icelanders.

As an island with a distinct name, Iceland literally appeared on the world map in the eleventh century; before that, it had appeared under the name of Thule or Ultima Thule, mentioned by the Irish monk Dicuil in 825, for instance, and generally referring to an ill-defined far northern island gradually being pushed further north as knowledge about the northern fringe expanded (see, e.g., Nørlund 1944). In local Scandinavian notions of geography, Thule has a parallel in Svalbarði ("the cold coast"), a name that is thought to have referred not only to present-day Svalbard but also to parts of eastern Greenland and other little-known coastlands in the far north (Holm 1925).

The Icelanders were more familiar with the North Atlantic than most other people. In the late Viking Age they traveled regularly between Scandinavia, the British Isles, and the North Atlantic islands. It was from Iceland that Greenland and later America were discovered, or, more correctly, discovered as already inhabited. The ancient Norse cosmology could fit in the new world remarkably well. The concentric view of the world as inhabited by gods and humans in the inner circle (Miðgarðr) and by monsters and giants in the outer circle (Útgarðr), separated by the world ocean (Úthaf), was replicated in the new territorial reality. In the process, a learned "European" world map was revised in the light of particular (local) experiences. Long before the Americas had become part of the European consciousness, the Icelanders knew it for a fact.

According to the Icelandic tradition of *Landnámabók* (The Book of Settlements, from the mid-twelfth century), the name Iceland is owed to one of the first would-be settlers, Flóki Vilgerðarson, who is described as *vikingr mikill*, "a great Viking."[3] Suffering the hardships of his first Icelandic winter in c. 870, and seeing the drift ice in the fjords, he awarded the land its present name. Whether historically correct in detail, the naming of Iceland is certainly part of an age-old myth of origin and a contemporary symptom of a process that definitively transformed a specific (and so far uninhabited) North Atlantic island from a natural to a cultural category. A "land" had been created.

The creation of a new land did not give birth to a new "people" at the same time, however. Defining the Icelanders as such took a little longer. Among the oldest sources on the early Icelandic world is the lawbook, *Grágás;* it dates back to 930, when the settlers decided to have their own law, more or less explicitly following the Norse dictum, *með lögum skal land byggja* ("with law shall a country be built"). One could say that the concepts of law and nation were conflated in the old Norse concept of *lög*.[4] Throughout *Grágás,* a distinction between the inhabitants of Iceland and foreigners is made. In the terminology of the law, foreigners were *útlendir menn* ("outlanders"), and the reference to the land is a clear reflection of the ancient Nordic practice of thinking in terms of spatio-legal "landscapes," while it gives us no clue to any specific cultural identity of the Icelanders. They would simply be those people who lived in Iceland and who were, as a matter of course, subject to Icelandic law. Anyone could become an Icelander by this criterion, well suited to a population of settlers.

Within the category of outlanders settling in Iceland, anyone speaking the "Danish tongue" was immediately eligible to hold political office in Iceland, while others had to have lived there for three years. In that sense, the early legal definitions do not refer to a people with a particular identity except for their being subject to a common law; yet there is a sense of distinction between Norsemen and others. A specific Icelandic identity was still to be formulated. This happened very soon after the laws had been written down with the advent of two major works in twelfth-century Icelandic scholarship, the *Íslendingabók* and the *First Grammatical Treatise*.[5]

Shortly after 1120, Ari *inn fróði* wrote his "The Book of the Icelanders," which endowed the Icelanders with a separate history. Although linking Iceland to Norway (in particular) in many ways, Ari established a distinct history of the Icelandic people, to whom the name "Icelanders" was now given with retrospective application from the settlements. The meaning of the Icelandic space expanded as a matter of course.

The next step in this process was taken by the First Grammarian, the anonymous writer of a grammatical treatise from the mid-twelfth century. In this most remarkable piece of linguistic scholarship, the notion of Icelanders was used as a means of conscious self-identification. The Icelanders were now definitively cast as a distinct people among other peoples. The First Grammarian writes about "us, the Icelanders," and he generally speaks of different peoples *(þjóðir)*[6] speaking different languages. The argument is that, because the Icelanders are (now) a distinct people with their own language, they also need a particular alphabet in which to accurately depict the sounds of Icelandic. In addition to the separate history written by Ari, and to which the First Grammarian refers admiringly, the Icelanders were now endowed with a distinct language (see Hastrup 1982). In this language, the famous saga literature was soon to be written and thenceforth to frame both the myth of origin and the identity of the Icelanders.

Thus, by the mid-twelfth century, the criteria of geography, law, history, and language had finally merged into a multiple definition of Icelandicness, which is still very much at play in present-day Iceland. Listen, for instance, to Jónas Kristjánsson, an eminent scholar of Icelandic literature, introducing an English translation of his work on early Icelandic literature:

> There is little change from the Icelandic version. As a consequence the book occasionally presupposes an Icelander's knowledge of the subject or gives a different emphasis than would have been the case if it had been originally intended for foreign readers. But if there is a slight disadvantage in the fact that the book was written *for Icelanders,* there ought to be a decided advantage in the fact that it was written *by an Icelander,* for none but Icelanders can fully participate in this unique national literature. Only we speak this ancient language and only we have the setting of the sagas in our daily view. Thus it ought to be of benefit for foreigners to be led through this landscape by an Icelandic guide. (Jónas Kristjánsson 1988, 7)

Having the language and landscape of the sagas in daily view offers a privileged understanding of matters Icelandic. Language and landscape are equal reference points for Icelandic identity.

The Poetics of Space

Like language, the landscape possesses a power to condense meaning; in this sense it is poetic (see Bachelard 1950; Tilley 1993b). The familiar landscape is the locus of dreams about unknown worlds, including the past, which for Icelanders has a peculiarly "Uchronic" quality (Hastrup 1992). By being attributed to the landscape, the past is cut loose from time. The landscape becomes a site of social memory; the society does not remember itself so much by rituals and bodily practices (Connerton 1989) as it does by simple everyday social practices.

Social spaces are always defined in actual practice. As Michel de Certeau (1984, 118) has said, "space is a practiced place." Spaces, in his terms, are always social, and the result of movements, orientations, and itineraries. Journeys and paths are never simply features of a map; they are routes of relevance, value, and meaning.

A significant part of the meaning of the present-day Icelandic landscape derives from the First Times, and hence from the settlers' original routes. From my fieldwork in Iceland in the 1980s, I know this to be a feature of prominence in any conversation about "the land," including both wild and the domesticated nature. On a number of occasions, the people living on the farm that was one of my field sites told me that the farm was situated at *landnámsland,* "settlement land." They would refer to *Landnámabók* and point out how a certain Hrollaugr had claimed land at the site around 900, coming from the sea and landing somewhere "down there" at the coast. The name of the first settler survives in the name of a group of small rocky islands off the coast, Hrollaugseyjar, visible from the farm in clear weather. According to the friend who first told me about it, Hrollaugr was the son of Rögnvaldr Jarl the Great of Norway, and a brother of Göngu-Hrólfr, who conquered the land of the Franks. In the original *Landnámabók,* the story is slightly more elaborate (see Benediktsson 1968, 316ff.), but the point is that the past is noted with veneration, and the fact that the actual farmstead of Hrollaugr probably was right here, at "our" place, is a matter of shared interest. There is no sense of regret or nostalgia; rather, there is a sense of interest and of historical depth.

All over Iceland, when engaged in conversations about the land, someone will refer to the *landnám.* It signifies the original domestication of the land and the roots of Icelandicness as something distinct from the shared Nordic past. The settlers were primordial Icelanders; they colonized virgin land and made a lawful society that before had been only wilderness. In one of the extant manuscripts of *Landnámabók,* it is claimed that it was written in order to establish the noble ancestry of the Icelanders, who allegedly had been accused of being descendants of slaves and robbers. In this aim, it certainly succeeded, even if the motive for writing was probably also a wish to keep track of landownership at a time when all available (inhabitable) land had been claimed and the population continued to grow (Rafnsson 1974). Today the reference to the *landnám* has another significance, which in its own way subordinates the other two. It is part of a comprehensive and continual discourse establishing the antiquity of Icelandicness. The Icelandic landscape is spoken of in terms of what happened during the *landnámsöld,* the age of settlement, or the First Times, in a way that is not totally unlike the way the indigenous Australian landscape is referred to as the Dreaming (see Layton 1995, 213). Antiquity and authenticity are conflated in the sense of Icelandicness.

Of course the landscape is also the object of subsistence activities and a discourse on subsistence, in which the Icelanders speak of fields and meadows, pastures and grazings, as well as fishing and other resources. In this sense, the landscape has a rather plain referential meaning, if not a stable one. No less important, however, is the "songline" created by the ancestral past and transmitted in words, the meaning of which transcends the present and the economic domain. It is a landscape of a shared knowledge of history.

In the renditions of the First Times, as marked in the landscape by the indications and names of settlement lands, there is a shared consciousness of the original journey to Iceland, by which Iceland was created and entered into history. It is this knowledge of the original journey that feeds into present-day sentiments about the landscape. In this way, the construction of the Icelandic landscape marks a collective identity, while the individual farmers and others may have completely different notions of their livelihood and relative fortune.

The original journey toward Iceland marked a discovery not unlike Columbus's discovery of the Americas some five centuries later. For the Vikings, too, the discovery of the New World (which included their discovery of Vinland) was based on a "knowing the unknown" (Paine 1995, 47ff.). From their oceanic practices, the Vikings knew the unknown shores before they left home, and the world they discovered was readily incorporated into previous understandings. Orkney and the Faeroe Islands, Iceland, Greenland, and Vinland did not have to be "invented"; they could be lived immediately in the Viking or Norse way. It was part of the Norsemen's canonical knowledge that new territories existed and could be colonized without profoundly affecting the Nordic cosmography. In a smaller way, the process was repeated when emigrants from Iceland established New Iceland in Canada (Winnipeg) in the 1870s, which has been recast as yet another *landnámssaga* (Jackson 1919). The New World is not ontologically new, just another place in the same Old World.

An important point, already foreshadowed but still worth emphasizing in connection with the constant reference to the settlements and the landscape's being redolent with memories, is the paradoxical fact that by fixing the ancestry of Icelandic society in the land, it becomes a timeless reference point. As in Australian Yolngu ontogeny (Morphy 1995, 188), place takes precedence over time in Icelandic notions of the Beginning. Without it actually being claimed that the Icelandic landscape is totemic, it does articulate a distinctive comment on society by way of historical poetics.

The poetics of history are found more readily at hand in the place-names of Iceland. All over Iceland pastness was evident in the naming practice. Place-names in Iceland are relatively transparent, precisely because the land was settled only on the verge of historical times. Consequently, there is a remarkable presence of the past tied to the landscape. History in Iceland is actually understood through a spatialization of time. Such understandings represent an active engagement in the landscape; individual biographies are formed in dialogue with particular places. On the surface, farmers have a more extensive relationship to the land than fishermen, but at closer inspection this is only true if we take the exploitation of the soil to be primary in the social relationship to the land. This is just part of it, however. The historical poetics not only extend to the sea, but the sea itself also provides a particular perspective upon the land. I did not realize this until I had gone fishing in the North Atlantic. After the gales at sea, approaching land gained new, even heightened significance because of the contrasting experience. If wildness reigned at sea, land was the ultimate source of safety and a manifestation of familiarity. The domesticated space ashore thus has no less totemic meaning for the fisherman gaining his livelihood from the untamed sea than it has for the farmer tilling the soil. What is different are individual experiences and memories, but the grand historical narratives are shared vehicles of a national identity, firmly rooted in nature (see also Brydon 1996).

Whether seen through the lens of primordial Icelandicness or lesser histories, one remarkable feature of the Icelandic landscape in all likelihood has deeply marked people's sensation of its historical magnitude, namely, its wide visibility. Unlike, for example, the ever-impeded vision of the world at large in the dense forest of New Guinea, in which the Umeda live (Gell 1995), the Icelanders live in an open space. The lines of the land are vast, and no trees impede the vision. There are mountains and valleys, ridges and canyons, but one is always close to a grander view. Around the corner the horizon expands infinitely. The vastness and the barrenness of Iceland combined with the clarity of the air create a sense of emptiness within which the observer has difficulty measuring distance and height. There seems to be no scale appertaining to the Icelandic space, except for time. The ethnographer arriving from lesser places, like myself arriving from Denmark, is constantly taken aback—and this evidently plays no small role in my tracking down the paths of Icelandic history. The very exposure to this landscape marked my own impressions vividly.

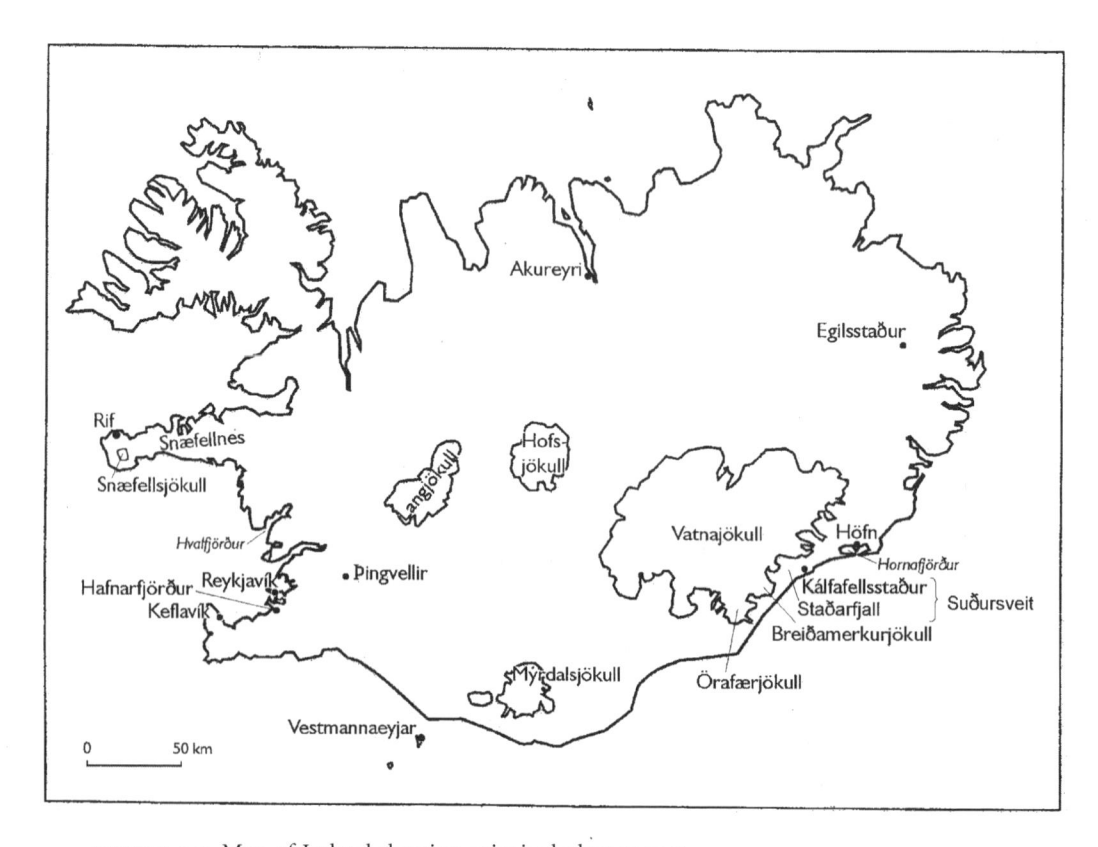

FIGURE 3.2. Map of Iceland showing principal place-names.

FIGURE 3.3. Sheep roundup in the mountains of Suðursveit. Photo by Kirsten Hastrup.

Walking in the rocky landscape of particular tracts is a journey in collective memory and narrative. Once in the mountains of Suðursveit (Figure 3.2) on an expedition to round up sheep (Figure 3.3), I ventured far afield with the farmers and virtually every top and turn, every rock and cave, had a name, and on my inquiry the names could all be explained. At Staðarfjall I was told that it had originally been Papýlisfjall, indicating that there had been *papar* (Irish monks) when the settlers first arrived. They had a church there, and when the newcomers arrived from Scandinavia they fled further into the mountains and dropped the church bell in a ravine about 300 meters deep that was still named Klukkugil, the "ravine of the bell." Another version of the name was that it was owed to the troll Klukki, since trolls had been known to reside there. The ravine was impressive, and one understood that it had caught the interest of the mountaineers. Close to Staðarfjall was Helghóll, and *"þar er huldufólk"*—"there are hidden people"—as I was told and for which I was referred also to an authority, Þorsteinn Jósephsson, whose work *Landið þitt* (Your Country, 1967) was part of the farm library and often consulted for such detail on particular places. In this view of the landscape, Irish monks, trolls, and hidden people belong to the same register of previous or other inhabitants; they have left their mark in legend and landscape alike.

At a more general level, "a journey along a path can be claimed to be a paradigmatic cultural act, since it is the following in the steps inscribed by others whose steps have worn a conduit for movement which becomes the correct or 'best way to go'" (Tilley 1994, 31). Paths create relationships and the more people have walked there, the greater the significance attached to the relation. The paths created by generations of people structure the experience of subsequent walkers, and the historical marks left by predecessors form the conceptual space of present-day travelers. One significant example is provided by the two-dimensionality of the terms of orientation almost everywhere in Iceland, and certainly also at my field sites. Chasing the stray sheep, or driving toward more distant goals from Suðursveit, was invariably conceived of as taking one of two possible directions: *suður* or *austur*, south or east. "South" covered a varied field of directions, largely toward the southwest, but due to the rocky and crumbled nature actually covering the entire compass. "East" was likewise, if generally heading toward the northeast. The point is that because of the topography and the ancient political geography of Iceland, most locations are en route—toward somewhere else. The route invariably consists in a line connecting two or more points on a (perceived) circular path, ultimately linking the four quarters of Iceland, established c. A.D. 965 as a judicial division. Thus, the recurrent reference to what appears to be a two-directional space is owed to the nature of Icelandic society having of necessity been built (more or less) along the coastline, and to the fact that the line thus created was punctuated by a political decision to subdivide the country into four smaller judicial units. They lasted only briefly, but they deeply marked the representations of the landscape in all quarters, where orientation is generally two-dimensional (Haugen 1957). In Suðursveit (and elsewhere), *suður* today refers to Reykjavík, capital and center of the ancient southern quarter, or to the direction toward it. *Austur* is the other way, toward Höfn in Hornafjörður, and farther (north, actually) toward Egilsstaðir, the capital of the Eastlands.

In addition to the original landtakings in Iceland, the domestication of the landscape consists in appropriation through a naming practice that leaves a lasting imprint on space, as do the thousands of cairns indicating old paths in the wilderness across vast and barren territories, which are thereby incorporated in the social space. Otherwise, the traces left by a thousand years of Icelandic history hardly amount to more than some moss-overgrown ruins of individual farms, originally built from stone and turf, some with wooden gables, but all of them very modest by continental European standards. There are no castles, churches, or feudal mansions standing from the Middle Ages. Every building in the country, except one or two eighteenth-century stone buildings in Reykjavík—since 1800 the administrative center of the country—was made from nature itself, as it were, and once abandoned was swallowed back into nature. Timber was extremely scarce, and recycled until left for the fire. Therefore, in addition to the landscape, *words* (including names) remain the most significant remains of the past; words carry the message of antiquity, as attached to little knolls, rocks, ruins, and cairns that are often barely visible. There is no immediate appearance of antiquity or age, only a sensation of a timeless nature. The remains of history are stubby, implying that the awareness of things past cannot readily be anchored in either visible antiquity or conspicuous decays (see Lowenthal 1985, 125). This probably is one reason that space takes precedence over time in the recollection of history in Iceland.

As for the space, the map is ever changing, not only because of the natural changes in the landscape but also because the meaning of the words and images of the past is emergent and changing. What at one point is a deified sign of nature, a holy wellspring, for example, may at another time be simply a source of water; landscapes are always contested (see Bender 1993). When moving about the landscape, people mark it deeply. They are by their own nature architectonic. In the landscape, boundaries are drawn between the cultivated and the wild, between inside and outside, and everywhere these boundaries are more than lines on a map. They are also social markers; people cannot be on both sides at the same time. Similarities and distinctions are created by way of borders that are projected as natural; territories are social spaces rather than geographical places. What gradually dawned on me during my own wanderings through the landscape is that nobody ever walks completely alone, not even the solitary wanderer like myself on my many excursions. By way of words and implicit knowledge, the landscape is always already populated—if sometimes by absent figures.

The landscape is something *seen,* but it cannot be seen from nowhere in particular. It is seen from the point of view of particular human agents, whose perspective is also historicized and directed by tradition. Thus, for example, in Iceland, petrified trolls are a part of the landscape in the shape of huge "troll-shaped" rocks. Even though they are referred to mostly in folktales, they have a certain material reality in the collective scheme of things. This also applies to other beings, such as hidden people, ghosts, and revenants, and it is not simply a feature of the rural areas. In Reykjavík, the town planning council a few years ago issued a map on which the dwellings of these beings were marked so that they could be respected by builders and entrepreneurs; this was in the 1980s. Rural or urban, the environment is a participatory field of many kinds of beings. In human territory, the consciousness of hidden and visible features inevitably becomes part of the landscape. There is no possibility of stepping out of it. Like the petrified trolls, people have no choice but to stay in place.

Collingwood (1945, 177) has argued that natural scientific facts are just one class of historical facts. In Iceland this has a peculiar truth in that the facts of nature are part and parcel of Icelandic history. The opposite also holds true when we take the point of view of the social agent, in Iceland as elsewhere. "Time and space are components of action rather than containers for it" (Tilley 1994, 19). The landscape is a total social fact. By its power of condensation, space is distinctly poetic.

Stones That Speak

In his chronicle of Suðursveit, Þórbergur Þórðarson (1981) uses the image of the speaking stones for his reminiscences of family and local history. It is an apt image in Iceland, where rocks and stones are named monuments of bygone life. They are not equal in this respect, because they are part of a political economy of the landscape, reflecting not only people's movements but also the structure of authority.

The First among stones in Iceland is the Lögberg, the Law Rock, standing majestically at Þingvellir, the site of the ancient Althing or people's assembly. The Althing, inaugurated in A.D. 930, met at the same site until 1800. Thus for close to nine hundred years, the Law Rock was the center of political attention and decision in Iceland. From it the law speaker would announce laws and verdicts, which for the first two hundred years had to be memorized and orally transmitted, whence the law *speaker*. Today Þingvellir is a site of veneration and festive celebration. It is also a camping site, giving anyone the possibility of resting in the vicinity of pastness—as I have also done with my family, hoping to inhale some of the political beauty of it all. The place is imbued with a sense of sacredness that derives both from history and nature, much in the way of other sacred sites, which are so classified because they are so *perceived* by humans (see Carmichael et al. 1994).

Magnús Einarsson has noted that Þingvellir "is a place where culture and landscape seem almost identical to Icelanders. This place contains simultaneously a sacred and a profane dimension, by indicating a sacred inspiration and a pleasurable recreation. Culture and history are engraved in the landscape" (Einarsson 1996, 224). The engraving is an engraving upon the consciousness of the Icelanders more than on the actual surface of the soil, where very little is to be seen.

Close to the place from where the law speaker would announce the law are a dozen or so remnants of booths where the traveling *þingmenn*, representatives to the Althing, would dwell. They are modest structures of low stone walls around a hollowing in the ground with room for two or three persons to sleep and over which the *þingmenn* had probably raised temporary hide roofs. This is all there is left in terms of human-made structures and they date not from the Saga Age but mostly from the seventeenth century. By comparison to Versailles or Persepolis, the scenery is austere and architecturally very modest; yet the whole place is tremendously impressive and symbolically loaded with Icelandicness. The Althing is a condensed symbol of the original society, of independence and of the will to order amid the demanding and rather wild nature of Iceland. The symbolic value was heightened during national romanticism and the incipient nationalist movement of the early nineteenth century.

The imprints are also of a more sinister kind, as testified by the name of the Drekkingahylur, "the drowning pool," where adulterous women and suchlike were drowned (men were beheaded in cases of capital offense) until the late eighteenth century. The pool may have changed shape and position over the centuries, as the roaring stream and waterfall down the rock are said to have done, but the name and wildness of the present pool leave nothing wanting in terms of its manifestation of horror inherent in past punishment.

In the Nordic countries of the early Middle Ages, laws were generally connected to "landscapes" (see Olwig 1993). The landscapes were localized sociopolitical units. Land and law were one, as evidenced by the proverbial entry to the earliest Nordic law-boks, *"með lögum skal land byggja, en eigi með ólögum eyða"*—"with law shall the country be built, and not with unlaw laid waste" (Hastrup 1985, 205ff.). Law was deeply rooted in the landscape, and, conversely, the landscape was deeply politicized from the beginning. The First Stone of Iceland is major evidence of this. The stone also provides testimony to the point Tilley made about Swedish megaliths—that their role in social reproduction resides in the authority structure that was able to raise them in the first place, even if the Law Rock was raised only metaphorically. However much the megaliths or the Law Rock seem to be nature-born, they reflect a society's view of itself, and "personal biographies are formed through encounters with particular places in the cultural landscape and the recognition and understanding of the panoply of codes constituting their meaning" (Tilley 1993a, 82). In Iceland, Þingvellir is encoded with Icelandicness as embedded in landscape and history, and to which individuals may relate differently.

Locally, other rocks and mountains tower over daily life, and, due to their magnitude, both in terms of sheer enormity and of impressive beauty, they are almost personified and seen more as social than natural features. The troll-rocks and other named and categorized stones are reminders that nature is always on the edge of the social, and impinges upon it. In Iceland, people are engaged in a never-ending conversation with nature, and it is in that sense that we may say that the mountains speak to people.

Lesser stones than the Law Rock may also speak of Icelandic history, and of the encounter between biography and place, such as Björnssteinn (Björn's stone) at Rif on Snæfellsnes. A local teacher first showed it to me when I had just arrived at the fishing village. It is a relatively small stone protruding from the earth some thirty meters from the seashore, and allegedly it is the stone where Björn *ríki*, "the rich," was killed by Englishmen in the fifteenth century (1467). Then, I was told, the stone had been right on the edge of the sea, and it would have been a dramatic sight with Björn standing on the top of it fighting fifteen Englishmen, the teacher said while enacting the drama himself. Afterward, "they cut him to goulasch upon the very stone," where the teacher pointed out a small hollow as the container of his blood.

The tale may have become more important than the stone itself, yet they belong together. When I asked a young colleague of mine (in the fish industry, which was another of my field sites) whether she had seen the stone, she said she had not heard of it, but she knew about Björn *bóndi*, "farmer," and his having been killed there. She then volunteered that "he had a wife who said that one should not cry over his death, one should revenge it. And so she did." Clearly impressed by this woman, whose name she did not know, the young woman's way of appropriating the tale was one of attributing to Björn's wife the strength that she herself did not have. The male teacher stressed the rich man's fight with the Englishmen, the holding off of the foreigners, while the woman laborer emphasized the social bond within the farmer's family and female power. The point is that the stone does not just speak; its tale is transformed and interpreted according to individual perspectives, all of which are, however, part of a larger social space, built not from stones but from human relationships.

While there is little doubt that the Icelandic landscape is deeply permeated by tradition, tradition itself does not mean the same to everybody, nor does it always make sense. Yet, through place-names and legends attached to almost every rock, knoll, and hillock, even the most arid and forbidding landscape is packed with latent meaning and memory. In that sense, the stones speak.

The Ways of Water

History not only connects to stone or other solid features of the landscape. In Iceland, as supposedly elsewhere, water is prominent. The island floats in the ocean, as it were, an ocean that both separates and connects Iceland to the rest of the world. Iceland was settled across the water, so the sea is not only a medium for present-day island life, it was also the original medium of history. As such, it features as the way toward whatever Iceland has become.

The sea also has been the way of the waterborne plague, the Black Death, which first reached the Icelandic shores from an English merchant vessel in 1402 and upset society for centuries to come. Between 1402 and 1404, about one-third to one-half of the population succumbed to the plague (see Hastrup 1990a). Firsthand evidence of the Great Plague *(plágan mikla)* is sparse, while more contemporary information on the Later Plague *(plágan síðari)*, which hit Iceland in 1494–95, is available. There is no doubt, however, that both of these plagues and the recurrent *drepsóttir*, "killer diseases," ravaging the island intermittently for centuries were thought to have come from overseas (which of course they did). In Fitjaannáll, it is told of the Later Plague: "The plague is said to have emerged from a blue cloth, which had come to Hvalfjörður (but some say Hafnarfjörður at the old booths)" (*Annálar Íslands* 1922–55, 2:27). The two best-known ports of Iceland are mentioned here as the passageway of the plague.

Epidemics of other kinds, notably smallpox and measles, came and went with ships from outside, adding disaster to poverty during the centuries of crisis. The Great Plague itself left an imprint on Icelandic memory for centuries; such was its effects that it marked the landscape by leaving fields to waste, by emptying farms, and destroying age-old patterns of family subsistence. Not only plague but also foreign merchants and rulers came across the sea, associating the waterways with ways of domination and exploitation. Even pirates from "Turkey" capturing slaves were all too well known on the Icelandic shores during the sixteenth and seventeenth centuries. Small wonder, then, that "overseas" was interpreted as impending danger by the Icelanders, who for good reason felt that they were vulnerable to others.

Adding to this danger, relating to the sea as a waterway for others, was the inherent uncertainty of the sea as a natural resource, which Icelanders had exploited since the earliest times. "Icelanders belonged to the land, but their fate was largely shaped by two kinds of uncertainties relating to the sea," namely, the uncertainty of the catch and the danger of the sea itself (Pálsson 1991, 97). Fishing implies the hunting of an invisible prey, yet the "blind date" with the sea was no random venture but relied upon the fishermen's *eftirtekt* (attentiveness) to natural signs. The appearance of particular birds, for instance, was an indicator of the arrival of fish. There was thought to be a special bond between different species, such as the seal and the black-beaked gull; the former provided the latter with food, such as the intestines of fish, while the latter would give cries of warning when the seal slept ashore. To seal hunters this bond was simply a matter of experience (L. Kristjánsson 1980, 449). The exploitation of the natural resources of the sea was not simply the individual hunter negotiating the uncertainties; fishing was deeply embedded in a social system of collective wisdom, including an elaborate folk meteorology. This wisdom was solidly based on local experience.

Today the fisheries are deeply embedded in a global capitalist economy (Figure 3.4), which has meant a shift in the folk model of fishing from the household producer engaged in a struggle against the elements to the model of the modern skipper competing with others (Pálsson 1991, 103ff.). Yet hunches and dreams still play an important part in decision-making (Pálsson and Durrenberger 1982); the sea becomes memorized in the subconscious, as it were. The official record of catches is everywhere complemented by a local recollection of relative *fiskni*, "fishiness," or the ability to get fish. Such recollections play an important part in the mapping of the sea and in the evaluation of particular fishing places. At another level, an inversion between humans and fish seems to have occurred, in that earlier the fish were seen as the active ones, those responsible for human life, while today, with the industrial (over-)fishing, humans become responsible for the maintenance of fish (Pálsson 1990). The main point is, however, that in the social history of the Icelanders the sea has always been both prey and partner. Either way, the sea is an active agent in Icelandic social life.

If the sea virtually frames island life, other waters permeate it from within. The spring torrents will stop people from moving about or continuing their journey; travelers being stuck at foreign farmsteads while waiting for possible passages to open is a recurrent theme in the literature. Unlike the great rivers of ancient civilizations, such as the Nile or the Tigris, the torrents of Iceland have rarely been conceptualized as

FIGURE 3.4. Although fishing is deeply embedded in the economy, the ability to catch fish is dependent on social memories of fishing places. Photo by Kirsten Hastrup.

arteries of fecundity. In modern times they provide energy for the welfare state, thus supplementing that which steams out of the volcanic interior.

The hot or thermal springs have reached mythical proportions in some narratives, but they have also had well-established historical use, such as the still-visible bath of Iceland's greatest medieval chronicler and saga-writer Snorri Sturluson. Visiting the bath, a slightly "edited" natural pothole with stones to sit upon in the warm water, one receives the impression of stones and waters speaking together of the past in a multi-vocal narrative of nature and history. Today, the thermal springs warm up the city of Reykjavík—giving a distinct smell of sulfur to every bath one should want, and blackening the silverware—and supply Icelanders with vegetables from distant hothouses. If not sources of fecundity, they are at least sources of much needed heat in the subarctic.

By contrast, one might claim that the area swept by the tide is an area of richness from water. The tide, relating the very concept of time *(tíð)* to the rising sea, meant changing access to the resources of the beach, and they were plenty. In medieval registers as well as in living memory, *fjörunytjar,* the usufruct of the beach, are extensively described. Seaweeds, strandings, driftwood, shellfish, and hosts of other material and edible items were important supplements to a marginal economy (see Lúðvík Kristiánsson 1980 for elaborate detail). Since here I am not primarily concerned with the economic assets of the beach but its position in the practical construction of space in Iceland, I shall just suggest from my own reading and direct experience that the beach constitutes a liminal space between land and sea, between private property and common. The ways of water make this space along with others and thus deeply affect the lives of people who cannot be separated from their environment.

The Times of Ice

The colder form of water, ice, also deeply marks life in Iceland, symbolically and practically. The ice defines time and history in many subtle ways. First of all, Iceland owes it name to the occurrence of drift ice in the fjords, as experienced by one of the first would-be settlers, Flóki Vilgerðarson, as mentioned above. Drift ice is a well-known phenomenon on the north coast of Iceland, and apart from the inconvenience caused to the traffic at sea, especially in earlier times, when boats were smaller, the drift ice has also caused peril ashore, when starving polar bears landed and went foraging on nearby farms. Although a rare event, bear hunts were a popular theme in tales of manliness and strength.

Ice, however, is not only a phenomenon of sea, it is also a remarkable feature of the landscape in Iceland, where glaciers take up a large proportion of the land. As it happened, both of my main field sites in Iceland were hovered over by glaciers, Breiðamerkurjökull (part of the huge Vatnajökull, Figure 3.5) and Snæfellsjökull, respectively, covering vast volcanic structures. Rain and fog, snow or sleet, would often block the view, but the more impressive the glaciers were when the view was clear. From their position in local conversation, I gathered that in both cases, the *jökull* was not an inanimate thing, it was a friend or, sometimes, an enemy. Whichever, it was a living force, and one that potentially infringed upon the social.

As Þórbergur Þórðarson (1981, 196–97) has it about the place under Breiðamerkurjökull: "Many hundred years ago, when there was more sunshine in Iceland and the weather was better and the country ready for feasting, when elves played in the woods and fairies drifted in the air, then there was a big farm under the mountain." This farm was called Fjall, and apparently it was a major place with a prominent history to its name. However, things deteriorated, and people had to fight poverty, famine, and all kinds of trouble, and, Þórbergur Þórðarson continues,

> the weather was destroyed, the sunshine became less, the downpour more, the cold harder, the woods disappeared and the elves fled into the rocks, and the fairies disappeared from the air, and Breiðamerkurjökull woke from its long sleep and began to slide down

on the plain between Fjall í Öræfa and Fell í Suðursveit, longer and longer, toward south, east and west, until Fjall fell waste and the glacier covered the mountain and everything around it, and nothing was left which reminded one about people living there, except a coldish place name. (ibid.)

The movement of the glacier will be noticeable in a long-term perspective, but at any point it threatens to wake up and to contest people's claims to space. This threat makes the Jökull an icy companion.

This is no less true for Snæfellsjökull. In the yearbook for Snæfellsnes, Haukur Kristjánsson (1982, 125) wrote about Neshreppur *utan* Ennis, the commune outside the mountain Enni: "above her tower Snæfellsjökull and Ennisfjall, dwelling places of good and bad landspirits." In Ennisfjall itself lives a troll woman, and the *jökull* is permeated by living forces. The point is that while few people in the 1980s (and later) would believe in trolls and land spirits, such beings still provide a vocabulary for speaking about the animate nature of mountains and rocks. Reference to beings of various kinds is not a matter of belief, let alone superstition; it is simply a symptom of an ongoing conversation with an animated environment.

FIGURE 3.5. Breiðamerkurjökull, part of Vatnajökull, Iceland's largest glacier. Photo by Kirsten Hastrup.

At Snæfellsnes, the life of the glacier is also conspicuous in another way of speaking; I was repeatedly told that the glacier evaporated *straumar* ("streams") of good or bad influence, much like the *vættir* (land spirits) would be good or bad. Individual life histories were thus deeply affected by this extremely beautiful, cone-shaped glacier that was nowhere to be hidden from in the area. In the literature, the entire area, which since the fourteenth century has been a prominent fishing site, is known as *undir jökli*, "under the glacier." According to one of my interlocutors at the place, the area *undir jökli* is "that place in Iceland to which most folktales and legends are attached." Whether statistically verifiable or not, the lore of the glacier and its people is preeminent in local discourse—as it is in the landscape.

The glaciers are "historical" in another sense as well. The most expansive of them, Vatnajökull, had always separated the northern farmers from the rich sea resources of the southern coast, yet they had known how to traverse it, if sometimes in great peril and with fatal losses, to be able to take part in the season's fishing (February–May). Many stories are still told about people who were lost in the ice; some had fallen into clefts and had been heard singing hymns from within the ice for decades after. With such stories we are in the field of reminiscing the (im-)possible, rather than in the field of actual memories. Yet even such stories, of which there are plenty, are prefaced with the old verbal entry into historical space: *svo er sagt*, "it is told thus." The stories need no other authority; once "said," they have their own. After this, the question of whether the Icelanders *believe* in these stories is as nonsensical as the question of "How wide is a Dreaming track?" within the discourse of the Dreaming (Layton 1995, 213).

One recurrent if unpredictable feature of the glaciers that has marked the Icelandic landscape as well as individual histories is that represented by the glacier bursts, *jökulhlaup*, where water that has been stemmed up beneath the glaciers bursts out and ravages vast areas of land. In 1996 the world was able to follow on television a glacier burst on the southeastern coast of Iceland, and thus to get an impression of the frightening power inherent in the glaciers should they choose to wake up. In nearby Suðursveit, neighbor to a *jökull* of prominence, glacier bursts would be an all-too-well-known phenomenon. In the sandy area close to "my" farm where some of the sheep strayed were the remains of an old mighty farm, Fell, which had been inhabited since *landnámsöld* and until one hundred years ago, when it had been devastated by a *jökulhlaup*. Allegedly it had been one of the biggest farms in the country, and it had housed *sýslumenn* (local officials), but even they did not have the power to subvert nature. They were powerful in other respects, and I was told that, when in 1720 a popular *sýslumaður* was to be buried at the church of Kálfafellsstaður and was carried eastward toward the church, the bells began to ring by themselves, "just like they did in the case of Jón Arason, Hólabishop." This is no insignificant comparison, since Jón Arason was the bishop of the North who resisted the Reformation for fifteen years until he was finally beheaded in 1550. In confrontation with the *jökull*, however, even the inhabitants of Fell came out losers.

The time of the ice, then, is both the beginning of Icelandic history and the long-term history of the slow withdrawal or progress of any single glacier, measuring small changes in average temperature over generations and leaving their mark on the conception of the landscape as well as the immediate results: giving or taking of possible grazing lands, blocking or opening passages between farms and friends, and bursting in upon people's lives in torrents.

Clues in Place

In this chapter, I have implicitly argued, following Yates (1966) and Radley (1990), among others, that remembering is something that takes place in a world of solid rock and other *things,* as well as in the world of *words* (see Hastrup 1998). When particular objects or sites are selected for preservation and awe, they turn into monuments of a shared past, which thenceforth govern the politics of remembering. The First Stone, and the ruin of a local farm, freeze the images and link remembering to ideology. Through the making of monuments, or the turning of particular objects into museum pieces, in short, by commemoration, other interpretations of the past are silenced (see Middleton and Edwards 1990, 8). This leads to the other side of remembering: forgetting. Like other areas of attention, memories are selective; something is forgotten, either by being silenced or repressed, or just by being deemed irrelevant. What makes the collective remembering so important is that it gives a shared sense to individual feelings—much in the way that tradition produces a set of shared images, which one can then attach to and interpret individually.

In this chapter, I have argued that remembering in Iceland is closely tied to the landscape, and that a sense of shared identity is to a large extent a function of living together in a mythical terrain. Any landscape can be *mapped* according to certain well-established codes of abstract notation because it is something substantial and measurable, but the actual territory of a people is not so easy to depict, whence the notion of mythical. The map is not the territory, as Bateson said (1972), because for one thing it strips away human movement, the points of remembrance, the articulations of space, and the bodily itineraries and routines.

The naming practices of particular topographical features such as mountains, rocks, bays, beaches, and settlements are crucial for the establishment and maintenance of their identity. "Through an act of naming and through the development of human and mythological associations such places become invested with meaning and significance" (Tilley 1994, 18). By recalling ancient times through movements in space, the Icelanders re-create local meanings and values as timeless features of their world.

The act of recalling, whether in spatial or some other social practice, constructs the past. Yet it cannot be constructed in just any way for it to make sense in this profound way; it must be grounded (see Shotter 1990, 133). In the Icelandic world, this grounding of the collective remembering must be taken very literally. Remembering Iceland, and thus to perceive it in the first place, is not to retrieve its history in accurate detail, but to move in the space of momentous pastness, along ancestral paths, and in a shared sensorial field of tactility, sound, smell, taste, and vision. Within this space the individual is constantly reminded of the collectivity; the past is not a foreign country in this sense (see Lowenthal 1985). Quite the contrary: the landscape is a well-known history.

The stress upon the landscape as the locus of history is also a stress upon unspoiled antiquity and purity in a world that is increasingly marred by modernity and noise of all kinds. "Icelanders like to present a romantic image by emphasizing the idea of purity in all spheres of the country—environmental, historic, linguistic, cultural and culinary" (Einarsson 1996, 228). Purity in Iceland is very much a function of antiquity, a perception that also permeates the Icelandic identity. In stressing the purity and the antiquity of Iceland, Icelanders may make a claim to Nordic aboriginality, not unlike the Greek claim to represent the cradle of European civilization (see Herzfeld 1987; Pálsson 1995, 20). An important feature of Icelandic topography is the power to sustain an image of a timeless and therefore pure history, to which the Icelanders may attach their sense of self. This poetic power of space is closely related to the "remoteness" of the island, or the feeling of vulnerability or marginality in relation to modern Europe (see Hastrup 1998, 185ff.; Ardener 1987).

What I am arguing here is that history is not visible simply as an imprint upon the landscape in the form of human-made material structures or other traces left by the ancestors. Far more important are the mental imageries and social practices linking modern Icelanders directly to their ancestors, who walked along the same paths. To comprehend the Icelandic space, we should not therefore aim at decoding the landscape. There is no code to be deciphered, no signs to read. There are clues to be understood, clues that connect people into a unified orientation in a space, where the social and the natural world are not seen as constituting a duality but a whole, of which people cannot have an external perspective.

Thus, memory and landscape, history and nature, or, even more generally, time and space are mutually implicated in this vision of the world. In Iceland, stones speak, ice grinds, and the water plays a symphony that the attuned ear may understand—not as a representation of history, but as a composition of practices, past, present, and future. This composition is what unites individual sensations of Icelandicness, and it is achieved not by way of reference but by way of resonance across time and space and between people. The landscape is part of a historical consciousness that is brought to bear on contemporary life by literally grounding it in space. Icelandic topography may *look* historically empty, but it certainly *feels* packed with meaning and memory.

NOTES

1. This chapter draws on more detailed works of mine on the Icelandic world; see Hastrup 1985, 1990a, 1990b, 1998. For a comprehensive account of my view of social and practical experience as the basis of knowledge, see Hastrup 1995.

2. *Íslendingabók* was published by Jakob Benediktsson in 1968. It was written sometime between 1120 and 1130 with a keen eye to an authentication of the sources.

3. *Landnámabók*, written in the mid-twelfth century and in part based on older genealogical material, was known to be among the earliest writings in Iceland. It has been transmitted in various manuscripts; in Benediktsson's 1968 edition, the differences between the manuscripts are noted. Rafnsson's 1974 study is also highly illuminating.

4. *Grágás* is a compilation of early laws dating from A.D. 930 to the mid-thirteenth century. In the nature of ancient Nordic lawmaking, new laws were simply added on top of old ones. The totality of laws, deriving from distinct periods and more or less visible in the layers of the *Grágás*, was rendered obsolete by the introduction of Norwegian rule in 1262–64.

5. The *First Grammatical Treatise* was published by Hreinn Benediktsson in 1972. In it the First Grammarian, as the author is called, refers to Ari's history of the Icelanders and to the allegedly extensive production of genealogical material.

6. The letter þ ("thorn") in lowercase, Þ in uppercase is an unvoiced consonant, pronounced as *th* in English "thorn"; the letter ð ("eth") in lowercase, Ð in uppercase is a voiced consonant, pronounced as *th* in English "then."

REFERENCES

Annálar Íslands. 1922–55. Annales Islandici. Posteriorum saeculorum. Annálar 1400–1800, 4 vols. Reykjavík: Hið íslenzka bókmenntafélag.

Ardener, Edwin. 1987. "Remote Areas: Some Theoretical Considerations." In *Anthropology at Home*, ed. Anthony Jackson, 38–54. ASA Monographs 25. London: Routledge.

Bachelard, Gaston. 1950. *The Poetics of Space*, ed. John Stilgoe. Boston: Beacon Press.

Bateson, Gregory. 1972. *Steps to an Ecology of Mind*. New York: Ballantine Books.

Bender, Barbara. 1993. "Stonehenge: Contested Landscapes (Medieval to Present-Day)." In *Landscape: Politics and Perspectives*, ed. Barbara Bender, 245–79. Providence and Oxford: Berg.

Benediktsson, Hreinn, ed. 1972. *The First Grammatical Treatise*. Publications in Linguistics, vol. 1. Reykjavík: University of Iceland.

Benediktsson, Jakob, ed. 1968. *Íslendingabók, Landnámabók*. Íslenzk fornrit, vol 1. Reykjavík: Hið íslenzka fornritafélag.

Brydon, Anne. 1996. "Whale-Siting: Spatiality in Icelandic Nationalism." In *Images of Contemporary Iceland*, ed. Gísli Pálsson and E. Paul Durrenberger, 25–45. Iowa City: University of Iowa Press.

Carmichael, David L., Jane Hubert, Brian Reeves, and Audhild Schanche, eds. 1994. *Sacred Sites, Sacred Places*. One World Archaeology Series. London: Routledge.

Certeau, Michel de. 1984. *The Practice of Everyday Life*. Berkeley and Los Angeles: University of California Press.

Collingwood, R. G. 1945. *The Idea of Nature*. Oxford: Clarendon.

Connerton, Paul. 1989. *How Societies Remember*. Cambridge: Cambridge University Press.

Einarsson, Magnús. 1996. "The Wandering Semioticians: Tourism and the Image of Modern

Iceland." In *Images of Contemporary Iceland*, ed. Gísli Pálsson and E. Paul Durrenberger, 215–35. Iowa City: University of Iowa Press.

Gell, Alfred. 1995. "The Language of the Forest: Landscape and Phonological Iconism in Umeda." In *The Anthropology of Landscape: Perspectives on Place and Space*, ed. Eric Hirsch and Michael O'Hanlon, 232–54. Oxford: Clarendon Press.

Hastrup, Kirsten. 1982. "Establishing an Ethnicity: The Emergence of the 'Icelanders' in the Early Middle Ages." In *Semantic Anthropology*, ed. David Parkin, 145–60. ASA Monographs 22. London: Academic Press.

———. 1985. *Culture and History in Medieval Iceland: An Anthropological Analysis of Structure and Change*. Oxford: Clarendon Press.

———. 1990a. *Nature and Policy in Iceland, 1400–1800: An Anthropological Analysis of History and Mentality*. Oxford: Clarendon Press.

———. 1990b. *Island of Anthropology: Studies in Past and Present Iceland*. Odense: Odense Universitetsforlag.

———. 1992. "Uchronia and the Two Histories of Iceland." In *Other Histories*, ed. K. Hastrup, 102–20. London: Routledge.

———. 1995. *A Passage to Anthropology: Between Experience and Theory*. London: Routledge.

———. 1998. *A Place Apart: An Anthropological Study of the Icelandic World*. Oxford: Clarendon Press.

Haugen, Einar. 1957. "The Semantics of Icelandic Orientation." *Word* 13 (3).

Herzfeld, Michael. 1987. *Anthropology through the Looking-Glass: Critical Ethnography in the Margins of Europe*. Cambridge: Cambridge University Press.

Hirsch, Eric. 1995. "Landscape: Between Place and Space." In *The Anthropology of Landscape: Perspectives on Place and Space*, ed. Eric Hirsch and Michael O'Hanlon, 1–30. Oxford: Clarendon Press.

Holm, Gustav. 1925. "De islandske Kursforskrifters Svalbarde." *Meddelelser om Grønland*, vol. 59. Copenhagen: Nyt Nordisk Forlag Arnold Busck.

Ingold, Tim. 1992. "Culture and the Perception of the Environment." In *Bush Base, Forest Farm: Culture, Environment, and Development*, ed. Elisabeth Croll and David Parkin, 39–56. London: Routledge.

Jackson, Þorleifur Jóakimsson. 1919. *Landnámssaga Nýja Íslands í Canada*. Winnipeg: Columbia Press.

Jósephsson, Þorsteinn. 1967. *Landið þitt*. Reykjavík: Þjóðminnafélagið.

Kristjánsson, Haukur. 1982. *Lýsing Snæfellsnes frá Löngufjörum að Ólafsvíkurenni, Árbók 1982*. Reykjavík: Ferðafélag Íslands.

Kristjánsson, Jónas. 1988. *Eddas and Sagas: Iceland's Medieval Literature*, trans. Peter Foote. Reykjavík: Hið íslenska bókmenntafélag.

Kristjánsson, Lúðvík. 1980. *Íslenzkir sjávarhættir*, vol. 1. Reykjavík: Menningarsjóður.

Küchler, Susanne. 1993. "Landscape as Memory: The Mapping of Process and Its Representation in Melanesian Society." In *Landscape: Politics and Perspectives*, ed. Barbara Bender, 85–106.

Layton, Robert. 1995. "Relating to the Country in the Western Desert." In *The Anthropology of Landscape: Perspectives on Place and Space*, ed. Eric Hirsch and Michael O'Hanlon, 210–31. Oxford: Clarendon Press.

Lowenthal, David. 1985. *The Past Is a Foreign Country*. Cambridge: Cambridge University Press.

Middleton, David, and Derek Edwards, eds. 1990. "Introduction." *Collective Remembering*, 1–22. London: Sage.

Morphy, Howard. 1995. "Landscape and the Reproduction of the Ancestral Past." In *The Anthropology of Landscape: Perspectives on Place and Space,* ed. Eric Hirsch and Michael O'Hanlon, 184–209. Oxford: Clarendon Press.

Nørlund, N. E. 1944. *Islands Kortlægning: En historisk Fremstilling.* Copenhagen: Munksgård.

Olwig, Kenneth R. 1993. "Sexual Cosmology: Nation and Landscape at the Conceptual Interstices of Nature and Culture; or What Does Landscape Really Mean?" In *Landscape: Politics and Perspectives,* ed. Barbara Bender, 307–43.

Paine, Robert. 1995. "Columbus and the Anthropology of the Unknown." *Journal of the Royal Anthropological Institute (Incorporating MAN)* 1: 47–65.

Pálsson, Gísli. 1990. "The Idea of Fish: Land and Sea in the Icelandic World-View." In *Signifying Animals: Human Meaning in the Natural World,* ed. Roy Willis, 119–33. London: Unwin Hyman.

———. 1991. *Coastal Economies, Cultural Accounts: Human Ecology and Icelandic Discourse.* Manchester: Manchester University Press.

———. 1995. *The Textual Life of Savants.* New York: Harwood Academic Press.

Pálsson, Gísli, and E. Paul Durrenberger. 1982. "To Dream of Fish: The Causes of Icelandic Skippers' Fishing Success." *Journal of Anthropological Research* 38: 227–42.

Þórðarson, Þórbergur. 1981. *Suðursveit.* Reykjavík: Mál og menning.

Radley, Alan. 1990. "Artefacts, Memory, and a Sense of the Past." In *Collective Remembering,* ed. David Middleton and Derek Edwards, 46–59. London: Sage.

Rafnsson, Sveinbjörn. 1974. *Studier in Landnámabók. Kritiska bidrag til den isländska fristatstidens historia.* Bibliotheca Historica Lundensis, 31. Lund: Gleerup.

Shotter, John. 1990. "The Social Construction of Remembering and Forgetting." In *Collective Remembering,* ed. David Middleton and Derek Edwards, 120–38. London: Sage.

Tilley, Christopher. 1993a. "Art, Architecture, Landscape [Neolithic Sweden]." In *Landscape: Politics and Perspectives,* ed. Barbara Bender, 49–84.

———. 1993b. "Introduction: Interpretation and a Poetics of the Past." In *Interpretative Archaeology,* ed. Christopher Tilley, 1–27. Providence and Oxford: Berg.

———. 1994. *A Phenomenology of Landscape: Places, Paths, and Monuments.* Oxford: Berg.

Yates, France. 1966. *The Art of Memory.* London: Routledge and Kegan Paul.

4.
Land Divisions, Land Rights, and Landownership in the Faeroe Islands

Arne Thorsteinsson

The Sheep Islands

The Faeroes (*Føroya* in Faeroese, *Færøerne* in Danish) are the "Sheep Islands," in name as in fact (Figure 4.1). The sheep grazing on the extensive hill pastures immediately strike the visitor. Much of the landscape bears the imprint of sheep, and sheep breeding is deeply embedded in Faeroese identity. Sheep have always been important for the economy of the Faeroes, giving meat, wool, skins, and tallow, and for centuries they provided the most important exports from the islands. The law code for the Faeroe Islands from 1298 was called the *Seyðabrævið*, the Sheep Letter, which was an amendment to the law for regulating sheep grazing in the hill pastures. Many of its provisions passed into new legislation in 1866, upon which management of common pastures is still based. The Faeroese language contains a multiplicity of terms related to the keeping of sheep. Rights of use to the hill pastures, as well as to other resources characteristic of the Faeroes, such as the yields of whaling and fowling, are regulated by a complex set of institutions and a terminology based on historical assessments of the value of the different resources. This terminology has shaped the landscape conceptually and in the way that it has been used and physically transformed. Yet despite the dominant role that sheep breeding has had in Faeroese history, this terminology appears to have its fundamental starting point not in sheep breeding but in cultivation. This paradox can only be explained by going back to the cultural background of the original Norse settlers in Viking times and their response on meeting the previously unsettled islands.

West Norwegian Cultural Background, Influence, and Adaptation

The extensive emigration from Norway in the Viking Age led to the establishment of Norse farming communities in areas inhabited by the Celts on the islands and coasts

FIGURE 4.1. The Faeroes, or "Sheep Islands." View from Sørvágur toward the island of Mykines with the jagged form of Tindhólmur on the left. Photo by Michael Jones.

of north and west Scotland and on the coast of Ireland, as well as on islands in the North Atlantic that were previously unsettled. Those who settled in the Faeroes brought their culture with them. They had been born and raised on Norwegian farms and had intimate knowledge of the natural surroundings as well as their domestic livelihoods, including working methods, tools, and building techniques. They brought with them their customary values, lifestyle, beliefs, language, house construction, and community structure—everything that we associate with the concept of culture.

Excavations of Viking dwelling places in the Faeroes have revealed Norwegian artifacts and building types. The language now known as Faeroese is of Norwegian origin. The farmsteads were built on delimited dwelling areas surrounded by cultivated land—the homefields (Faeroese: *bøur*) within the infield *(innangarðs)*—through which a lane connected the settlement with the outlying grazing lands, or outfield *(uttangarðs);* this too was a Norwegian practice. Without further enumeration, such comparisons suffice to demonstrate that Faeroese culture is essentially of Norwegian origin.

One might ask which elements of their culture the emigrants would have brought with them to a new country. The answer seems obvious: their complex cultural background, ranging from their customary social structure, settlement patterns, and livelihoods to their tools and implements, eating habits, religious practices, and folktales.

They would have attempted subconsciously to transfer the West Norwegian farming communities in all their aspects to the Faeroes, where, in spite of the geographical differences, they could have exploited the land with the knowledge and skills they had acquired at home. If West Norwegian farming families considered the Faeroes as having potential for settlement, it was because they found opportunities for continuing their existence there on the basis of their previous customs and knowledge. In the adaptation process that followed, it would not be surprising to find preserved some of their original cultural elements that seem inexplicable seen against the background of the new natural surroundings; nor would it be surprising to find elements of culture preserved that have subsequently disappeared from their homeland.

The Viking period was a period of exodus. People were on the move, encountering and influencing one another. Further south the new settlers came into contact with the existing Celtic population, with inevitable, reciprocal cultural impacts as a consequence. The contacts between the different Norse settlement areas led to the spread of Celtic cultural influences further northward in the North Atlantic. Thus, in the Faeroese and Icelandic languages, one can detect words of Celtic origin, along with other cultural traits originating in Scotland or Ireland. The Celtic influence in the Faeroes and Iceland is not direct but rather is the result of people moving from the Norse colonies in the Celtic areas. However, intensive research into these relationships over several generations has resulted in a strong exaggeration of the significance of Celtic cultural traits brought from the Norse settlements in the south. Yet, if linguistic influence is used as a yardstick, no more than half a percent of the Faeroese vocabulary shows Celtic influence. The significance of these occurrences lies mainly in their documentation of the contacts between the various Norse settlements during and after the colonization period.

Natural Surroundings

The Faeroes are the fissured and eroded remains of an extensive basalt plateau that was formed by a series of volcanic eruptions 55–60 million years ago. The basalt layers are about 6,000 m thick, of which about half lies below the present sea level. The eroded glacial valleys have become sounds and fjords, which are carved into the islands. Their grass-covered sides, broken by high crags, extend up the mountains on both sides, reaching as much as 900 meters above sea level. To the west and north, the islands have been eroded by the sea to form steep cliffs that are rich in bird life.

The Faeroes lie at 62°N and 7°W and consist of eighteen main islands (Figure 4.2), of which seventeen are inhabited. The total land area is 1,399 square kilometers. The climate is Atlantic, with much precipitation. Because of the Gulf Stream, it is relatively mild, with average temperatures of c. 3°C in February and 12–13°C in July.

The natural environment of the Faeroes is unlike West Norway in many ways, with

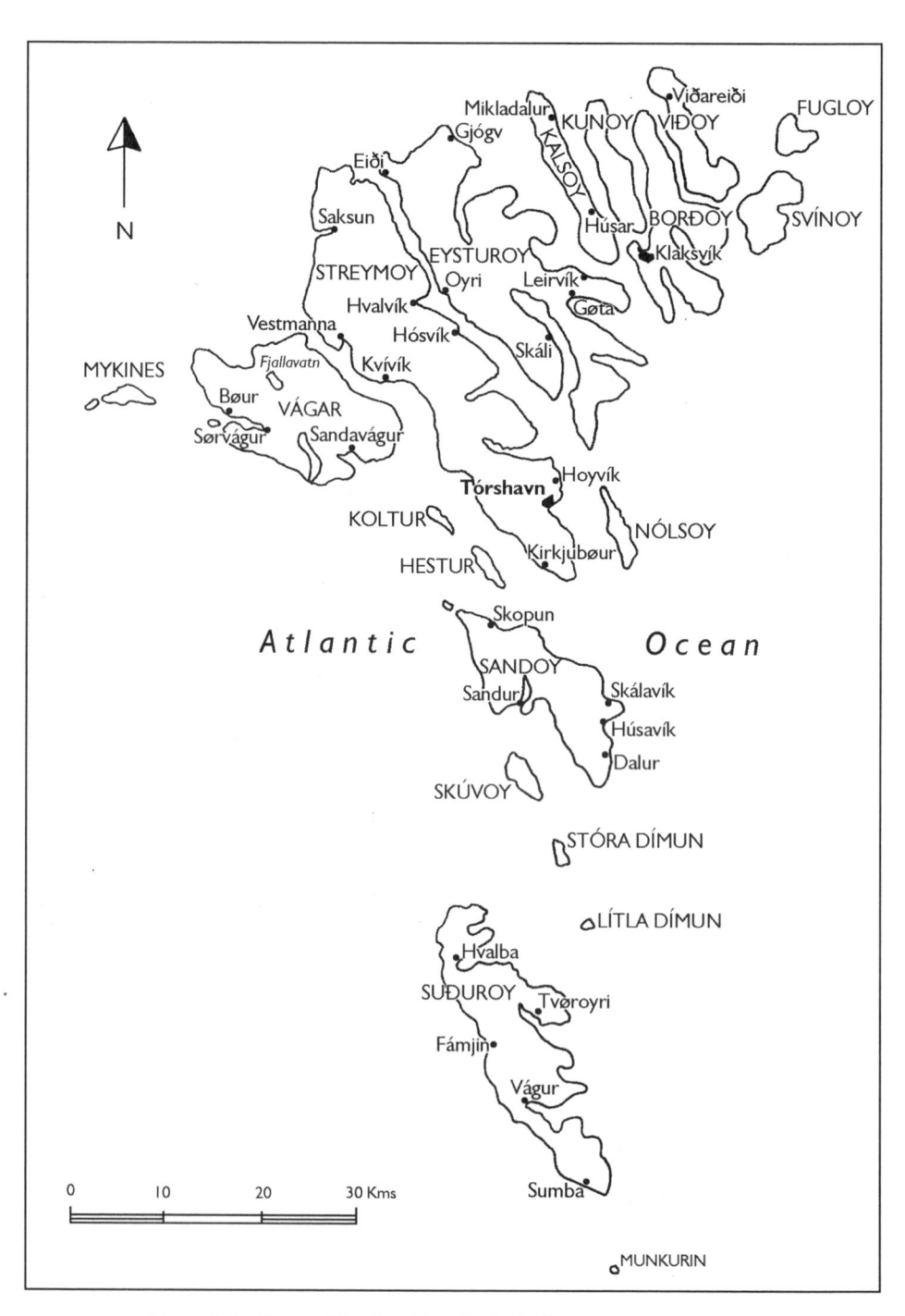

FIGURE 4.2. Map of the Faeroe Islands with principal place-names.

geological, climatic, zoological, and botanical differences. Most conspicuous is the lack of trees and bushes, while grasses and ling *(Calluna vulgaris)* dominate. Nevertheless, the landscape of the Faeroes, with its mountains, hill slopes, outcrops, and small fjords, is still somewhat reminiscent of West Norway.

Production and Trade

Until 100–150 years ago, the Faeroes were an agrarian society. Production, however, was not derived only from farming and livestock. Bird catching, seal hunting, whale hunting, and especially fishing all played a significant part in the economy. The farming community integrated all exploited resources into a single production system based on landownership. The farmer was cultivator, cattle breeder, sheep farmer, fisher, whaler, fowler, and gatherer and adapted his work according to the source that gave the best returns at any given time. With the exception of fish and sea mammals harvested at sea, the rights to all produce belonged to landowners. The yield was divided among the landowners according to the size of the owners' property following given rules.

Farm produce in the Faeroes was largely for home consumption. However, the Norse population was not self-sufficient. Important consumer goods, such as timber, metals, soapstone and slate, at times also corn, and luxury goods, had to be imported. This in turn necessitated the production of goods for export. In the Middle Ages and later, production also had to cover the payment of land dues *(landskuld),* tithes, and other duties. Trading abroad was always a necessary condition to maintain the Norse population in the islands. The history of the Faeroes has never been one of total isolation.

I will discuss the agrarian production of the Faeroes in detail below. This description will of necessity suffer from the weakness that it is to a great extent based on post-Reformation evidence, to a lesser extent on evidence from the Middle Ages, and to an even lesser extent from the Viking period.

Corn Growing

The Faeroes lie at the climatic limit for growing grain. Barley and some oats have been cultivated in the past. Only some settlements had the right growing conditions for corn. Corn was grown for straw, thatching, and fodder as well as for food; corn cultivation also formed part of the tillage system. The corn itself was exclusively for home use. The average corn yield at the end of the eighteenth century was about 2,000 *tunnur* (278,800 liters or 8,000 bushels) a year. This yield was far from sufficient to meet the needs of a population of around 4,500, and about the same amount of corn had to be imported. The average yield was seven- to eightfold. Only in the best areas did the yield reach twentyfold, and for some farmers thirtyfold. The best corn-growing areas were Suðuroy, Sandoy, and Svínoy, where it was usually possible to grow sufficient crops for home use.

In such a marginal corn-growing area as the Faeroes, the crops did not ripen every year and crop failures were not unusual. The prudent household would have in reserve at least a year's supply of corn. The corn dried poorly in the field and the drying process had to be done over a fire or in a special drying and threshing building.

There can be little doubt that corn production in the Middle Ages was far greater than in the post-Reformation period. Place-names that refer to corn production, as well as hay harvesting, are found throughout the outfield areas. On Suðuroy, Sandoy, and Mykines, one can find on hill slopes exposed to the sun in the outfield extensive areas with ancient clearings, where the cleared stones were used to build parallel walls at intervals of 10–15 m, stretching upslope, occasionally with dividing walls across the slope. Research into some of these systems has revealed evidence of barley and wheat. On Suðuroy, these grain-growing areas in the outfield are termed *deildir* (boundary markers), which can also be found in the infield. One of the ancient outlying fields was referred to in a contract dated 1412 from the settlement of Sandur on Sandoy as *ruddstaðir* (cleared place) in a way that indicated the special economic significance of the area.

The decline in corn production thought to have taken place in the late Middle Ages is not necessarily explained by local circumstances. The explanation could equally have been the market economy. A considerable demand for fish, together with low prices for imported corn, may well have led to greater emphasis on fishing and the consequent abandonment of the highly labor-intensive corn cultivation in the outfield. Although there are many local legends of depopulation as a result of the Black Death, which had serious economic consequences on the Norwegian mainland in the fourteenth century, there is no definite scientific evidence that it reached the Faeroes.

LIVESTOCK BREEDING

On Suðuroy and Sandoy, cattle breeding was as important as sheep breeding. The cows were kept in stalls during the winter and fed on hay harvested from the infield. In summer they grazed on the lower-lying parts of the outfield. In the early summer and early autumn, the cows were herded into the stalls overnight, though in midsummer they could remain in the outfield, often corralled in a fenced-off area *(tippi)* for the night. At the end of the eighteenth century, there were an estimated 2,824 cows in the Faeroes. They were small and probably did not give more than four *pottar* of milk a day (1 *pottur* = 0.968 liter or c. 1 quart). Milk, dairy products, meat, and skins were normally for domestic use and apparently were not exported. Nevertheless, butter was mentioned among the payments of land dues. In some places oxen and fatstock were also kept in the outfield. In the early stages of settlement, and for several hundred years into the Christian Middle Ages, the people also had summer farms in the hills where cows were milked. The term for summer farm was *ærgi,* a word of Celtic origin, which is reflected in some present-day place-names in the Faeroes. The *ærgi* system may have been inspired in some way by the Gaelic shieling (summer farm) system found in Scotland (Mahler 1998, 56).

Sheep breeding has always played a dominant role in the economy of the Faeroes. According to assessments in the nineteenth century, the value of sheep breeding was on average almost twice that of field cultivation and cow breeding combined. Sheep farming also yielded the most important exports: wool, wool products, skins, and tallow.

The Faeroese generally kept about sixty sheep or fewer, which remained outdoors year-round. In summer they grazed in the higher mountain areas or on the lower slopes of the outfield; between October 24 and May 14, they were kept in the infield. Thus, sheep farmers differentiated between mountain (summer) pastures and lower (winter) pastures. Selected sheep, such as castrated sheep and rams, were fattened on special areas of the outfield known as *feitilendi*, which could be found in fairly inaccessible clefts in the mountains, as well as on small islands, stacks (freestanding rocks off the coast), and screes.

Foddering was practically unknown, just as sheep houses were rarely found. Nevertheless, horseshoe-shaped shelters are found in the winter pastures, where the sheep could shelter against snow. Poor care frequently caused a catastrophic fall in the numbers of sheep in hard winters, and the average yield of meat was low. At the end of the eighteenth century, the number of sheep was recorded as being 75,539 ewes, although formerly the count had been 96,549 ewes. The decline was attributed to a reduction in the quality of the grazing and to the introduction of larger animals to replace an older, smaller species of sheep. The number of sheep for slaughter was estimated at 36,175 in an average year; in other words, two ewes had to be kept for every sheep slaughtered. The value of sheep for their wool, however, was at least as much as the value for their slaughter. Meat and viscera were used for domestic purposes, while wool and wool products were always the agrarian society's most important export.

In the Middle Ages, wool was made into homespun cloth *(vaðmal)* and even until 1600 it was an important export. However, by the mid-sixteenth century at the latest, the knitting of hosiery had begun, and by the mid-seventeenth century it had become the major export, the main purchasers being the Danish-Norwegian and Dutch navies. The knitting of hosiery expanded as other occupations declined, including even the fishing industry.

In the early phase of settlement, sheep were milked, but this practice disappeared from the Faeroes in the thirteenth century or earlier. Nevertheless, it is reflected in the occurrence of *lambhagar* (lambing enclosures) throughout the islands (Thorsteinsson 1982). An apparent decline in milk production could have contributed to drying becoming almost the only method used for preserving food.

Place-name evidence shows that pig farming was not entirely insignificant in the Middle Ages. In the royal cadastres, records have been found of individual farms with a single pig listed on its inventory and on which rentals were payable. Similarly, place-names bear witness to the keeping of goats.

FOWLING

Bird catching could have considerable importance in the settlements that had nesting cliffs and other fowling places in their area. The most important birds caught were the puffin and the common guillemot. Puffins were found in holes on the grass-covered slopes down toward the edge of the cliffs, and they were caught either by pulling them out of the holes or capturing them in flight using a long stick with a net. The guillemots were caught directly from the cliff ledges. The catchers were either lowered down on a rope from the top of the cliff or they climbed up from the base of the cliff. The guillemot's eggs were taken in the same way. The eggs and bird meat were kept for domestic use, while the feathers and down were exported. Fowling rights on the nesting cliffs belonged to the land. When the catch was divided, a greater or lesser share, customarily half, was paid to the landowner; or, in other words, the catchers received a greater or lesser share of the catch from the landowner as payment for their work. Another method consisted of catching the birds in a net from a boat at the foot of the cliffs as the birds flew from the ledges. In this case the landowner did not receive any part of the catch so long as the catchers did not come into contact with the land.

WHALING AND SEAL HUNTING

Whaling has been of major economic importance throughout the Faeroes. Possibly large whale species were hunted in the Middle Ages, but in the post-Reformation period the pilot whale was caught almost exclusively. While there were only a few places where pilot whales could be killed, men from a wide area took part in the hunt, and the amount of whale meat and blubber that resulted was valued for domestic purposes. The boat that found a school of whales at sea hoisted a flag on the mast as a signal to the other boats and, similarly, a message or signal was sent around the whole district. The whales were driven in toward the most convenient inlet and then driven up the beach as far as possible, where they were killed. The greater part of the catch was for domestic use, but the whale oil, from the heads and blubber, was a valuable export. Another small whale species, the bottle-nosed whale, most often visited Suðuroy and was used exclusively for oil production. The Faeroese showed little interest in the large whale species, which, understandably, they were apprehensive about encountering at sea. The latter were taken only when they were found dead in the sea. The practice of whaling, which almost certainly was for pilot whales, was mentioned in the *Seyðabrævið* of 1298, but whaling has existed in the Faeroes for as long as it has been populated. When the whales were driven up onto the beach, the catch belonged to the owner of the adjoining land and the landowner paid half of the yield from the catch to those who took part in the hunt; or, from the hunter's point of view, a half share of the catch was paid as rental to the landowner. For whales caught at sea, however, the hunters could keep the whole whale and the landowner had no claim to a share. The economic significance of whale hunting is best indicated by the fact that when land was leased out the whaling rights were not included. Instead, the landowner, for example, the Crown, kept the whales. Similarly, a seller could retain the whaling rights upon the sale of his land. In principle, according to existing laws, whales could be freely hunted at sea. In spite of this there was, for example, a case in 1617 in which a court made a judgment based on illegal regulations that prohibited the harpooning of pilot whales before the whalers made an attempt to drive the whole school ashore.

After a school of pilot whales had been captured and killed, each whale was assessed and its value expressed monetarily in guilders *(gyllin)* and skins *(skinn)*. The valuation was used as the basis for dividing up the catch. The value of pilot whales was not documented before the beginning of the eighteenth century (Thorsteinsson 1996) but nevertheless must be considerably older.

Seal hunting had been carried out in the past. The seals bred either in deep caves or on open beaches. They were caught at their breeding sites, or on the rocks where they rested, by hitting them with a wooden club. The meat was eaten and the skin used to make bags and shoes. The blubber could also be eaten but was especially used for the production of oil. The seal catch belonged to the land and usually the landowners and tenant farmers supplied the labor in proportion to their landholdings. Otherwise the hunters' share of the catch was as a rule a third and in some places half of the catch, depending on how difficult it was to reach the breeding site and carry the catch home.

FISHING

Fishing was the most important of the agrarian community's supplementary sources of income. According to ancient Norse law, the sea was free and everyone was free to find food there. Unlike fowling and whaling, fishing was not subject to landowner rights. A large number of different fish species have been caught in the sea around the Faeroes, but cod is the most important. The catch was partly for domestic use, but dried cod was an important export. Cod fishing was carried out using a hand line from a rowboat that could also be used with sails. Large boats with four to six oars could be used close to land and in the fjords, and in summer they could go further out to sea. In winter, large boats with six to eight oars were used. The typical fishing boat was the eight-oared *áttamannafar*. The catch usually was divided equally among the crew, with a share for the boat owner. One might think then that fishing was the only occupation in the old agrarian community that could be practiced freely and independently of landownership. However, the right to keep a boat was usually linked to landownership, a customary law that the Faeroese parliament apparently respected. Usually it was the largest landholders in a settlement who owned boats and in cases when they could not man them with their own men they could require other people in the settlement to serve as crew. The latter, in return, also had a right to a place in the boat. In other cases, especially where the land was more evenly divided in the settlement, several landowners could share the ownership of a boat. The working fellowship that keeping a boat required appears to have been linked with specific land in the settlement, land that in an earlier period had been worked as a single farm. A regulation from the thirteenth century concerned whales caught out at sea by a farmer's housecarl: at this early point one could scarcely imagine any other crew manning a boat.

DRIFTWOOD AND SEAWEED

Driftwood that washed ashore belonged to the land. In places it was a significant source of income for the landowner, assuming that the driftwood consisted of unworked tree trunks. Worked timber was considered wreckage and belonged to the Crown, assumingly on behalf of the unknown owner. There appears always to have been considerable use of timber in the Faeroes, as witnessed by the predominant building types. It is not known how much was covered by driftwood, but the import of timber has been quite high in recent times and presumably this was also the case in the Middle Ages.

Seaweed, an important product from the beach, was an important fertilizer. As with other products from the beach, such as whales and driftwood, seaweed that drove ashore was divided in proportion to the shares of land belonging to each owner in the settlement. Seaweed that did not loosen and drift ashore by itself was cut offshore. In such cases, the coast was divided into delimited stretches where certain owners or groups of owners could harvest the seaweed.

PEAT CUTTING

Peat was a primary agricultural product and almost met the population's need for fuel. There is never any mention of fuel shortages, nor have they had any influence on buildings or construction styles in the Faeroes, as they had, for instance, in certain parts of Shetland and Iceland. Peat was usually cut in one's own outfield, but a farmer also had the right to cut peat from a neighbor's outfield, in other settlements, and even on other islands. Sometimes the farmer paid for the peat in cash or instead granted grazing rights for a certain number of a neighbor's sheep. Usually, though, no one paid for peat because the old natural rights were established at a time when no rights of ownership to land in the outfield existed.

Administrative Divisions in the Faeroes

The Faeroes were divided into six main administrative districts *(várting)*, with a central parliament *(løgting)* for the islands as a whole. Originally there was an Althing

(Alting), where all free men had the right to meet, but after the Magnus Code *(Magnus Lagabøters Landslov)* was adopted by the Althing between 1274 and 1298 it became the *løgting*, which was a representative parliament. Until 1400, a governor *(sýslumaður)* was responsible for the Crown administration, after which the islands were divided into six districts, identical with the *várting*, and with a bailiff as the king's senior administrator. In the Middle Ages, bishops were responsible for ecclesiastical administration; after the Reformation, a rural dean *(prostur)* was in charge. The ecclesiastical division consisted of seven parish districts with various numbers of parishes in each, totaling thirty-nine parishes altogether. The parish districts coincided with the *várting*, with the exception of Streymoy, which was divided into two parishes.

With regard to land use, the islands are divided into eighty-five settlements *(bygdir)*, with a high degree of common land use by the holdings in each settlement. Today, this number is the officially recognized number, although over time some settlements were subdivided into smaller units with regard to sharing land rights and thus could be interpreted as independent settlements. Further, at times, individual settlements were merged and later separated by new boundaries.

Settlement Patterns and Rights of Use

A settlement *(bygd)* will always be divided into a number of functional areas. The broad distinction between the infield and the outfield—*innangarðs* and *uttangarðs* (*inside* and *outside* the enclosing fence) was mentioned in the *Seyðabrævið* of 1298. However, a more exact description of the settlement's different areas provides a more nuanced picture.

HEIMRUST

Heimrust is the area where the buildings are located. In addition, the *heimrust* provides a place for walled enclosures, where angelica and cabbages are grown, and for *heimabeiti* (home grazing). However, *heimabeiti* can also be found at some distance from the *heimrust*, thus presumably indicating a settlement that has disappeared. From the *heimrust*, a cattle path *(geil)* runs through the cultivated infield to the outfield and, where needed, there is a path to the beach. The *heimrust* and cattle path are always fenced off from the infield. About half of the settlements have only one *heimrust*, while the remaining settlements have from two to thirteen different *heimrustir*, in which the settlement's *býlingar* are located. *Býling* refers to the group of houses. The area between the houses was known as *tún*, the same word as is used in Norwegian for a clustered farmstead or for a farm courtyard. Where only one *heimrust* is found it belongs to all the settlement's farms. The same is the case in some settlements with several *heimrustir*, but elsewhere each *heimrust* has particular lands around it, which allows reconstruction of the extent and size of the original farms.

BØUR

The term *bøur* (Figure 4.3) for the cultivated infield was a term also used in West Norway. The infield is entirely fenced in against the outfield, cattle path, *heimrust,* and often also the shore. Otherwise the limit of vegetation constitutes the boundary between the *bøur* and the shore. Sheep were excluded from the infield between May 14 and October 24. However, the old fences, constructed of stone, grass turfs, and earth, were so low and broad that they kept the cattle out of the infield, but not the sheep. In the summer, a watchman and his dog would keep the sheep out of the infield. From the late eighteenth century and throughout the nineteenth century, however, higher walls of stone were constructed in most of the settlements. Only during the winter were the sheep allowed to graze in the infield.

The infield is divided among the owners into named lands or *jarðir,* and again into smaller parcels. Each person cultivated his or her own holding, but in some places the farmers cooperated on the hay harvest. They harvested and dried the grass in common, then divided the hay according to each landowner's share.

FIGURE 4.3. The settlement of Funningur on Eysturoy. The boundary between the *bøur* or *innangarðs* (infield) and the *uttangarðs* is clearly marked. In the nineteenth century, the *bøur* was extended to accommodate population growth. Funningur consists of three *býlingar* (the original groups of houses). Photo by Michael Jones.

The infield was used for the rotation of crops and hay in such a way that the same piece of land was dug and manured about every seventh year, after which it was left to grow grass again. This was the only fertilization the land received. Owing to lack of manure, the land could be left for longer periods before it was worked again, which could lead to the spread of moss and unwanted plants and weeds that choked the grass. The land was worked by spade. The plow was never introduced in the old Faeroese farming communities. Fertilization consisted mainly of cow manure, but seaweed was also important in settlements where large amounts of seaweed floated onto the beaches.

Crop rotation was not universal. According to information from the eighteenth century (Svabo 1959) and from the existence of the place-name elements *fori* and *ong* found throughout the Faeroe Islands, it is possible to discern an older practice, one in which the field was cultivated year after year, while grass was harvested from meadows.

The most frequently occurring field forms are *teigar,* which are long strip fields, 5–6 *alin* wide (1 *alin* = 0.6277 meter or 24.7 inches), which run along the hillside, inclining sideways toward a drainage ditch.

Over the last two centuries, the cultivated area has doubled in size, partly due to an extension of the existing infield areas and partly due to the establishment of new farms. However, these newly cultivated areas, called *trøð, gerði,* and *byrgi,* do not have a share in other land rights. The size of the infield areas also increased in earlier times. These terms constituted elements in a number of earlier place-names. Among the old lands listed in the royal cadastres preserved since 1584 were a number of holdings, presumed to have been independent farms, which consisted of *hagaleys* (hill-less) land, that is, without outfield or other rights. Presumably these were areas cultivated in the Middle Ages at a time when all resources had already been divided among existing lands.

In the *Seyðabrævið* of 1298, restrictions were placed on the clearance of new land. No one could establish a farm with fewer than three cows, and, if anyone was found supplying someone with less *land,* both would be fined. The term *land* rather than *jørð* (land in the infield) was used when outfield areas were brought into cultivation. Presumably it was not uncommon for farmers to allow others to cultivate land in their outfield. Those who cleared the land appear to have been tenants without a share in the use of the outfield.

UTTANGARÐS

The outfield can be divided into *hagi, húshagi,* and *feitilendi.* To the old settlements belonged one or more *hagar* with a fixed number of sheep, which belonged either to particular owners or groups of owners in common. The separate *hagar* belonged to certain lands in the infield. Sometimes the lands to which a particular *hagi* belonged were grouped around a *heimrust,* but often they were dispersed throughout the whole settlement, in which case one can assume that a division of the hills *(hagar)* occurred at a later point, after the original farm on the settlement's *heimrustir* had been divided up and new patterns of ownership established.

The division of the *hagi* into mountain hills *(fjellhagi)*, where the sheep graze during the summer, and lower hills *(undirhagi)*, where the cows graze during the summer and the sheep in winter, is a practical division without fixed boundaries. In practice, the snowline determines the limit of the sheep's grazing during winter. In the lower hills, various constructions are found: sheepfolds *(ból)*, where the sheep can seek shelter; fences to keep the cows away from dangerous areas; and sometimes *tippi*, enclosures where the cows are kept at night during the summer.

The boundaries between settlements, and between *hagar* within the settlement, are often natural boundaries. In some cases they are marked by buried peat ashes, stone cairns, or fences. Occasionally fences are placed between *hagar* in an attempt to prevent the spread of contagious sheep diseases.

In a number of settlements, one or more areas called *húshagi* are found in the part of the outfield closest to the cultivated infield. All landowners in the settlement have a share in a *húshagi* or have *húshagi* rights in particular *hagar*. Separate *húshagar* belong to specific lands in the outfield and are often fenced in. The *húshagi* is used for grazing cattle in the summer, for collecting stone and grass turfs used as building materials, and for grazing small flocks of sheep.

On more or less inaccessible areas of grass in the mountains, on islets and stacks, and in other remote outlying areas are the pastures known as *feitilendi*. In general, these have natural boundaries that the sheep were not able to cross (Figure 4.4). The right to keep sheep in such areas does not belong to the *hagi* where the *feitilendi* is found, but to the entire settlement or to particular lands in the settlement. Usually, castrated and old rams and barren ewes are placed in the *feitilendi* for fattening up for slaughter. A large *feitilendi* that can hold larger flocks may be used as a normal *hagi*.

Often a *feitilendi* is also a place for catching birds. The right to catch birds on land belongs to the landowner, but it is never the same group of owners that has the right to keep sheep and catch birds in the same area. One might ask what the rights consist of: An ownership right to the land, or only the right of use of the land?

The outfield system is regarded as an ownership right to the outfield area. The *Seyðabrævið* of 1298, for instance, mentions of one or more men owning *hagi* together. It is evident, however, that it was not the "men" but the lands of the infield that owned the *hagi*. Thus, it was only a manner of expression, one that is also found today, to talk in terms of one person or another owning a particular *hagi*. Yet this view has led to numerous court cases, not only concerning settlement and *hagi* boundaries but also about the right of ownership to *húshagar, feitilendi,* and fowling sites in the *hagi*. It is likely that the original right of ownership was attached only to the cultivated land and that the remainder of the land was common land *(almenningur)*. The outfield boundaries have arisen from the delimitation of rights of use for sheep grazing, cow grazing, *ærgi* use, *lambhagar,* peat cutting, hay harvesting, *feitilendi,* and fowling sites. A number of boundaries formed in this way appeared and disappeared according to use. Others continue to overlap one another for different uses. Since the predominant use was for sheep, the sheep-breeding boundaries took form as ownership boundaries.

The Seashore and the Sea

The seashore forms a special area in terms of rights (Figure 4.5). The resources from the shore consist of the landowners' share of whale catches, driftwood, seaweed, and mussels, together with stones, sand, and gravel. The rights to the shore usually belong

FIGURE 4.4 Kleggjaberg, two-hundred-meter-high cliffs on the southern side of Vágar, forming part of the hill pastures belonging to the settlement of Sørvágur. Halfway down the cliffs are grassy slopes, fertilized by guano from seabirds and used to graze rams or lambs. Such pastures, or *feitilendi,* are difficult to reach and difficult to escape from; the sheep are lowered down the cliffs on ropes. Such areas are also used for fowling. Photo by Michael Jones.

to the settlement's landed property as a whole. In a few cases, the rights to particular stretches of shore belong to particular properties, and in the past this could have been the case in many settlements. It is possible to imagine landowners combining shore rights in order to avoid conflicts, especially if whaling rights transgressed the settlement's limits. There are several records of whaling rights being combined where settlements border the same fjord or inlet. For example, the whale rights in Vestmanna and Kvívík at one point were combined between 1781 and 1870.

The inner limits of the shore are undoubtedly the outer limit of vegetation growth. The outer limits of shore rights are more diffuse. Among definitions of the boundary between shore and sea are the lowest ebb; the *marbakki* (the point where the sea bottom drops to deeper water), where such exists; as far out as a horse can wade at the

FIGURE 4.5. The harbor at Gjógv, located in a long, narrow inlet eroded by the sea along a zone of weakness in the rocks. The landowners of the settlement have joint rights to resources on the shore (sea mammals, driftwood, seaweed, shellfish), whereas hunting and fishing at sea can be practiced freely. Photo by Michael Jones.

lowest ebb; and as far out as a boat can float. Regarding rights to whales and driftwood, undoubtedly stranding is the deciding factor. A stranded whale or log belongs to the landowner. If it is floating, it belongs to the finder. In one court case, a whale that had floated in and touched a precipitous cliff was awarded to the finder because it was not considered stranded. Since the borderline between land and sea is of such great significance for rights belonging to a landed property, it could be expected that there was a well-defined, commonly accepted boundary. Instead, this boundary is surprisingly poorly defined in Faeroese customary law.

Shore rights are the landowners' rights to the resources from the shore. This does not mean that the landowners have ownership rights to the shore, especially in cases where harbors have been established and when compensation has been paid not only for the loss of the shore areas but also for any loss of user rights that might occur.

While all rights on land are appurtenances to the land *(jørð)*, this is not so in the case of the sea. Catching and fishing were practiced freely, and, unlike catches on land, it was not necessary to pay half a share of the catch to the landowners. Nevertheless, fishing was linked to the landowner through boat ownership. However, it was customary that boats from particular settlements favored particular fishing grounds. Although no one had a right to them, some tried to prevent nonlocal boats from finding these grounds by withholding information about them from strangers. In certain cases, individual settlements felt they had exclusive rights to "their" fishing grounds, but the courts have dismissed these cases when there has been a conflict. Today, quotas and other regulations have broken with the old principles of freedom to fish and represent a large step toward expanding private ownership rights to include the resources of the sea.

The *Grannastevna*

Grazing and other rights of use in the outfield are regulated by the neighborhood council or *grannastevna*. Such a body was mentioned in the Magnus Code of 1274 and was required to convene annually, and again in Christian IV's Norwegian Law of 1604 and Christian V's Norwegian Law of 1687. Village councils were similarly mentioned in the Danish provincial "landscape" laws of the thirteenth century, in the Funen (Fyn) Decree of 1492 and in Christian V's Danish Law of 1683. However, although Norwegian law applied to the Faeroes in the seventeenth century, the age of the *grannastevna* in the Faeroes remains uncertain. It has often been assumed in literature from the end of the nineteenth century and later that such meetings were held on a regular basis in Faeroese settlements from early times. Yet they were not mentioned in the seventeenth-century descriptions of the Faeroes by Thomas Jacobsen Tarnovius (1950 [1669]) and Lucas Debes (1963 [1673]). These descriptions were not comprehensive, and their silence on this does not provide conclusive evidence that the *grannastevna* did not exist in the

Faeroes at this time. But it is also possible that regulation occurred through the practice of unwritten customary rules, without regular meetings of neighborhood councils.

J. C. Svabo, traveling to the Faeroes in 1781–82, refers to a set of regulations for a *grannastevna* on Sandoy dated 1692, but the reference may have been just to Christian V's recently issued laws since there is no evidence that meetings were ever held (West 1975, 48–50). According to John West, the term *grannastevna* does not appear to have come into regular use before the 1840s. Earlier, occasional Faeroese references to *grannastevna* can have referred to irregular meetings for a variety of purposes. The term is found in conveyancing documents of 1708–9 from Sumba and of 1709 from Vágar. These were penned by Danish officials, who were familiar with the Danish village meetings, and do not refer to rights of resource use in the outfield. Although a document mentioning the *grannastevna* in the settlement of a dispute on Suðuroy in 1803 involved such resource use, there is no evidence of regular meetings. Nor was this the case in an official letter written in 1836 requesting the calling of a *grannastevna* to obtain permission from the joint landholders for the building of a house on their land (50–51).

The oldest printed reference to *grannastevna* in the Faeroes is a communication of 1836 from the Danish government stating that the rules requiring village councils to meet regularly should be applied to the Faeroese *grannastevna*, this apparently not having been the case earlier (Bjørk 1956, 1:20). West (1975, 52–55) argues from the context of this reference that meetings were held to conduct police business and arose out of correspondence between the Faeroese governor and the Danish government concerning police meetings on Vágur. Such police meetings had their origin in government ordinances of 1757 and 1775. Police records from Sandoy for the period from 1777 to 1838 document similar police meetings. According to West (51), a government decree of 1698 and the ordinance of 1775 proclaimed that local decisions were otherwise to be made by the principal resident landholders. Disputes were settled through the intervention of the Danish officials on the Faeroes, as occurred in the case of a dispute at Gjógv in 1825 and five times on Nólsoy between 1814 and 1841.

Whether or not the *grannastevna* is of great antiquity on the Faeroes, its importance as a regular meeting increased after 1836. In 1842 an order was issued that the minutes of the meetings were to be recorded in the official police records, and after this archive references to the *grannastevna* become frequent (West 1975, 55). However, decisions had to be unanimous until the Hunting Act of 1854 made the regulation of sea-fowling and seal hunting possible by a majority of the owners, provided that together they controlled at least half of the rights in question. In 1847 the Land Reorganization Act *(Lov om Udskiftning af Fælleshauger)* gave powers to the *grannastevna* to act by majority rule in matters concerning the enclosure of *hagi*. The Hills Act *(Haugelov)* of 1866 contained a long list of matters to be dealt with at the *grannastevna*, including regulations concerning the keeping of various types of animal, fencing, grazing, sheep marks, and the inspection of bridges. A specific act defined the legal position and functions of the *grannastevna* in 1891. The provisions were broadly similar in a new act that replaced this in 1937. The *grannastevna* deals with matters concerning the whole *bygd*. At the same time, in 1937, separate councils and elected boards of management were introduced for the *hagi* to regulate internal matters, such as grazing, peat cutting, and fishing. West (57) argues that it was the pressure on resources resulting from the large increase in the number of private landholders and the fragmentation of land that led the *grannastevna* to become an institution of importance in the nineteenth century because it provided a decentralized system for regulating common resources.

Land Inheritance

In the Viking period the farm belonged to the farmer. Inheriting land was firmly established in the Faeroese ownership system, as we also know from recent times. The same can be read from legislation in the Middle Ages, by which division of land on inheritance was entirely legitimate. Until 1273, when the *Gulating* Law of West Norway was enforced, the Faeroese had their own laws. The Faeroese Law Book for that period has since been lost, but stipulations from it appear to have been incorporated in Duke Hákun's law amendment for the Faeroes, the *Seyðabrævið* of 1298. This was an amendment to the Magnus Code, which had been adopted at the Faeroese Althing sometime after 1274. Inheritance of *óðalsjørð* (allodial or freehold land) was not a specifically Faeroese phenomenon. On the contrary, it was found in all Norse legislation. Evidently, from the restrictions introduced in the Magnus Code, problems existed as early as the thirteenth century. This legislation stipulated the right of the first-born male to the home farm, a right that was conditional on the other heirs receiving other, equally good *óðalsjørð*. Limitations on free inheritance became more stringent under Christian V's Norwegian Law of 1687, but the law did not seem to have any influence on the inheritance system as it was practiced in the Faeroes. The unrestricted subdivision of the farm on inheritance was sometimes carried out with such consistency that each parcel of land owned in different places was divided equally among all inheritors. The result was that *óðalsjørð* later became greatly fragmented.

Upon inheriting, one could build a new farm on the outskirts of the infield or create a new *heimrust* with new farm buildings, or one could use the existing buildings of the original farm included in the inheritance. Terms for outbuildings are frequently found in the names of dwelling houses.

During the Middle Ages, the amount of *óðalsjørð* declined and by the end of the Middle Ages it comprised only one-third of all land owned. Ownership rights to the remaining two-thirds passed into the hands of ecclesiastical institutions, the Norwegian nobility, and the king.

Landholdings

Land holdings in the Faeroes are assessed in relation to one another in *mørk,* which functioned as a unit of land value by which payment of rents and calculation of shares in an inheritance were determined. Its main function in later times has been as a unit for determining relative shares in outfield and other common property.

The total number of *merkur* in the Faeroes is almost 2,400. Approximately 1,300 *merkur* are crown land and 1,100 are *óðalsjørð*. Until the mid-sixteenth century, around 300 *merkur* belonged to the nobility, such as the Norwegian families of Benckestock and Rosenkrantz. Almost all the land owned by the nobility was later sold privately and is now included in the amount of *óðalsjørð*. The greater part of the crown land is accounted for by medieval ecclesiastical land that was confiscated by the Crown after the Reformation. Only a small part is derived from medieval crown land. The distribution of landholdings at the end of the Middle Ages was approximately 1,100 *merkur* of ecclesiastical land, 800 *merkur* of *óðalsjørð*, 300 *merkur* of land belonging to the nobility, and 200 *merkur* of crown land.

At this point, the *óðalsjørð* had already been extensively fragmented for hundreds of years as a result of the terms of the laws of inheritance. Of the two ways of dealing with inherited land, either operating the farm as a whole and paying land dues to the other owners or physically subdividing the land, it appears that only the latter has been practiced in the Faeroes. Joint ownership occurred in certain cases, but the land was soon divided.

The greater part of the crown land is almost just as fragmented as *óðalsjørð*. It lies dispersed throughout the Faeroes in large and small concentrations. The greatest concentration is on Suðurstreymoy, the bishop's seat. Here all land belonged to the Crown. In the southernmost settlements of Norðstreymoy, too, are found comparatively large crown holdings. The size of the leaseholds varies from half a *mørk* to 32 *merkur,* while the smallest tenancies are independent holdings within a *býling*.

On Suðurstreymoy in the late Middle Ages, there were 25 *býlingar* of various sizes: 4–10, 12, 13, 15, 16, 20, and 32 *merkur*, totaling 289 *merkur*. The largest was the bishop's estate. In the bishop's outfield there were two small farms of four and two *merkur* of *hagaleys* land, known, respectively, as á Argjum and í Sandagerði. They did not have any share in the outfield. The farm of á Argjum is the only known example of a farm established on an abandoned *ærgi*. The farm of í Sandagerði was probably a medieval clearing of a type that was forbidden in the *Seyðabrævið* if it did not support three cows. In the rest of the Faeroes, both crown land and *óðalsjørð* is fragmented, apart from a few limited areas and individual large estates.

Where crown land is fragmented or intermixed with fragmented *óðalsjørð,* it seems reasonable to imagine that the land was acquired by the Church or Crown in the late Middle Ages. On the other hand, where a whole *býlingar* forms a consolidated holding, it is a matter of an estate that was acquired in the early Middle Ages and not been divided. Landownership, land distribution, and settlement on Suðurstreymoy thus to a great extent reflect the situation of the early Middle Ages.

Land Evaluation

The size of an estate is expressed in terms of the number of *merkur*. In order to give an impression of the size referred to by a *mørk*, it can be mentioned that, according to nineteenth-century sources, a man with one *mørk* of *jørð* on average could grow c. one *tunna* (139.4 liters, or four bushels) of corn and also keep one cow and thirty sheep. Average values are rarely found in practice, and in any individual case the average *mørk* can be multiplied or divided by two or three. However, one cow per *mørk* of *jørð* appears to have been an old method of reckoning.

Unfortunately, *merkur* are seldom mentioned in medieval sources. The first time *mørk* was mentioned in connection with land was in 1412, but it appears to have been alluded to in the *Seyðabrævið* of 1298. Inadequate documentation cannot be given any decisive significance in estimating its age. The medieval source material is sparse and fragmentary, and one is occasionally surprised by what is not mentioned when one might have expected otherwise. In six letters, dated 1403–5, concerning a dispute about an inheritance from Guðrun Sjúrðardóttir in Húsavík, land in every settlement on Sandoy is referred to, yet the word *mørk* is not mentioned. In my view, these letters cannot be considered as evidence that the *mørk* was introduced after 1405 and before 1412, when it is encountered in a similar situation. Later medieval sources mention no other type of land evaluation. Such a phenomenon does not arise at the time it is first recorded in writing. On the contrary, *mørk* could have existed long before this. I am unable to envisage a medieval agrarian society without any form of land evaluation. The existence of a phenomenon is documented by its mention in the source material, but one should be cautious about regarding its lack of mention as evidence of its absence altogether.

The *Mørk*

The nature and origins of the *mørk* have been the subject of much discussion. What kind of land measurement is it? Is it primarily a measure for land dues and taxation, or a measure of production, or a measure of the purchase value of the land? Is it a measure of surface area or a measure of value? What is it that is measured or valued? Is it the infield or the outfield, or both, or the entire production potential of the farm or settlement? When the main emphasis in Faeroese production has always been based on sheep breeding, it is not so remarkable that many writers have interpreted the number of *merkur* as an expression of the value of the outfield. Was the value of the *mørk* calculated within the farming community according to their own need, or was it determined by a higher, royal authority based on the needs of the king? All possibilities have their advocates. Seduced by the fact that the measure of length called *alin* (1 *alin* = c. 63 cm or 25 inches) forms a subdivision of the *mørk*, a number of earlier writers have claimed that one *mørk* of land is an area of either infield or outfield. Today, there is general agreement that the *mørk* is an indication of value.

A *mørk* is an expression of the land's commercial value or capital value at the time its price was first settled. A *mørk* is as much land as a farmer could purchase for one *mørk* in money. The history of the *mørk* can be described in three stages:

1. A value in burnt (unminted) silver. A subdivision into *øre* and *øretog*, found elsewhere in the Nordic countries, does not seem to have existed in the Faeroes, or it disappeared during the Middle Ages.
2. At some point during the Middle Ages, 1 *mørk* of burnt silver became equivalent to 320 *alin* of homespun cloth *(vaðmal)*, allowing land values of less than 1 *mørk* to be expressed. In the Middle Ages homespun was universal as a unit of measurement in the Faeroes, except in the case of fines. The division of land into units of homespun (1 *mørk* = 320 *alin* of land) is not, however, documented before 1584, by which time this value had outlived itself and the real price of land had fallen to 160 *alin* per *mørk*.
3. The third stage in the history of the *mørk* began in the late Middle Ages, when the international currency of guilders *(gyllin)* was introduced to the Faeroes. This became popular and the Faeroese guilder was the main unit in the accounting system up until 1790—and for the assessment of land dues right up until 1902. The Faeroese guilder apparently originated from the Rhenish guilder and corresponded in value to an old Danish *daler* or two Bergen guilders.

About 1600 a Faeroese guilder had a fixed value for a range of export goods and corresponded to 1 *vág* (an old unit of weight equivalent to c. 18.5 kg or 41 lb.) of fish, feathers, or tallow, 10 *alin* of homespun, and 20 lambskins. Thus 1 guilder became divided into 20 skins.

In the late sixteenth and early seventeenth centuries, 1 *mørk* of land cost about 16 guilders. Accordingly, 20 *alin* of land (equivalent to one-sixteenth of a *mørk*) was worth 1 guilder of land and 1 *alin* of land was worth 1 *skinn*. Thus, 1 *mørk* of land was divided into 16 guilders, and 1 guilder of land into twenty skins of land. The valuation of land in homespun disappeared at the same time, though its memory lives on both in the names of areas of land and among landowners. The new land assessment can be found documented in court cases, the first entered in the oldest preserved court register, from 1617, as well as in contemporary writings. This did not appear in the Crown's land register before the beginning of the eighteenth century, by which time the land evaluation units no longer bore any relation to the actual price levels. This seems to confirm that this land evaluation was established and developed in the farming community itself and over time gradually became accepted by the authorities, not the other way round, as is often claimed.

THE LAND

The number of *merkur* thus expresses the value of the land, but what is included in the concept of land, that is, *jørð*? Since the eighteenth century, the concept *jørð* is linked to both *innangarðs* and *uttangarðs,* that is, the land both within and outside the fence around the infield. Thus a settlement's total number of *merkur* is attached to both the infield and the outfield. The concept "outer farm"—*uttangarðs*—seems to be exceedingly old. It is mentioned in the *Seyðabrævið* of 1298. Outfield is also mentioned in 1412 in such a way that indicates it is not considered itself to be *jørð* but an appurtenance of the *jørð*: "*hagan allan, som till for nefndar jordh ligur*" ("the common grazing that belongs to the land below"). The concept of *jørð* thus relates to the infield only, and it is the value of the infield alone that the number of *merkur* expresses. Similarly, land purchases in the seventeenth century, as mentioned in court records, included the standard phrase "*jord med tilliggende haug*" ("farm with adjoining hill"). The same phrase can be found in the works of seventeenth-century writers, such as Arent Berntsen (1971 [1656]) and Lucas Debes (1963 [1673]).

This interpretation is supported further by a number of other circumstances. First, the land rent for *hagaleys* land is the same as for land with adjoining *hagi*. Second, the island of Lítla Dímun, which has never been inhabited or cultivated but has been used only for sheep grazing and fowling, has never had any *mørk*. Third, the designation *jarðir* (lands) is used only to refer to infield areas. Fourth, the suffix *jørð* is frequently found in place-names in the infield, but never in the outfield.

Now it has become common to refer to a *hagi* that belongs to, for example, 8 *merkur* of *jørð* as 8 *merkur* of *haga*, thus giving the *hagi* its own number of *merkur* corresponding to that of the infield.

If we return to the *Seyðabrævið* of 1298, we find the concept *jørð* used in connection with corn cultivation—and only in this connection. One can therefore envisage that the concept *jørð* originally only related to the cornfields, not to infield used for harvesting hay—nor to other meadowland.

The number of *merkur* that a *jørð* had was used for fixing land prices; subdividing infield plots among individual owners; stating the individual owner's share in the appurtenances of the settlement outside the fence, such as grazing, peat cutting, fowling, whaling, driftwood, and seaweed; and fixing land rents and annual taxes. However, the number of *merkur* was not used to assess the total land dues or taxes collected.

LAND DUES AND TAXES

The land dues *(landskuld)* in the Faeroes consist of a number of rents: land rent, sheep rent, cow rent, and, in rarer cases, horse, pig, and bull rent. The livestock rents are often summarized by the term *innstøða* and the rent from this is termed *innstøðuleiga*. *Innstøða* never includes a farm's total stock, but only the share of it that belongs to the landowner, while the farmer's own animals are not included.

The land rent is calculated according to the number of *merkur*, but since 1584 it has varied in different parts of the Faeroes. On Suðuroy, Sandoy, and Mykines, the rate is normally 1 guilder = 20 skins per *mørk* of *jørð*, while on Norðuroy the amount is half this, namely, 10 skins per *mørk* of *jørð*. In the remainder of the Faeroes, the rent varies from 10.5 skins on Skúvoy to 12, 15, and 16 skins in other places. If we assume that the land rent had originally been a fixed share of the land value, then this would indicate that rents have become differentiated in different places. Unfortunately, there is only one source on the amount of rents prior to the Reformation and that is a record in Munkeliv Monastery's ledger from the second half of the fifteenth century. From this we can ascertain that *jørð* in the Northern Isles of the Faeroes—which later gave 10 skins per *mørk*—at this point gave twice that amount, or 10 *alin* of homespun.

The annual tax was the same over the whole of the Faeroes: 10 skins per *mørk* of *jørð* every three years. For cows, horses, and pigs, the rate was normally 5 skins per head. The sheep rent was calculated according to the number of sheep *merkur*. In 1584, the sheep rent per *mørk* was uniform in the Faeroes: 2 guilders per *mørk* of sheep—twice the highest rent per *mørk* of *jørð*.

The different rents have different weights in a farm's total land dues, dependent on the number of *merkur* and on the size of rented stock. A small farm with a large *innstøða* might thus have to bear higher land dues than a large farm with low *innstøða*. The number of *merkur* and amount of land dues are therefore not comparable.

CROWN TAX

The only tax that the Crown levied in the Middle Ages was the crown tax *(kongsskattur)*, a poll tax, that was, as in Iceland, 10 *alin* homespun per taxed farmer. The land as such was not taxed before the introduction of the cadastral tax *(Matrikulskatten)* in 1657, at 3 skins per *mørk* of crown land and 5 skins per *mørk* of *óðalsjørð*. This tax is not related to the land dues.

SHEEP *MERKUR*

The number of sheep *merkur* is first mentioned in the *Seyðabrævið* of 1298—*sauða mørk*—and can be explained as a statement of value for sheep parallel to the number of *merkur* stated for *jørð*. The term was used throughout the Middle Ages and until the eighteenth century, but it has since disappeared. A flock of sheep was thus not described in terms of number of sheep, but in terms of *merkur* of sheep. The size of the sheep *mørk* differed between the north and south of the Faeroes. The size of one sheep *mørk* in the north was reckoned at forty ewes compared with forty-eight in the south. Even today, the sheep of the south are considered to be worth less than those of the north. The great importance of sheep breeding is evident from the question that farmers invariably pose to each other when they meet: "How many are there to the *mørk*?" meaning: "How many ewes per *mørk* can the outfield support?" Each individual *hagi* has a fixed stock that cannot be exceeded. This is termed *skipan* and can, for example, be 20, 30, 40, 60, or 80 ewes per *mørk* of *jørð*. In earlier times, the expression *skipan* referred directly to the relation between the number of *merkur* of *jørð* and the number of *merkur* of sheep. A *hagi* could have a full *skipan* when it could hold 1 *mørk* of sheep per 1 *mørk* of land. A double *skipan* had 2 *merkur* of sheep per *mørk* of *jørð*. A half or three-quarters *skipan* had, respectively, a half or three-quarters of a *mørk* of sheep per *mørk* of *jørð*.

The Origin of the System of *Merkur*

Because of the dominant economic role of sheep breeding in Faeroese society, several writers, including the former archivist Anthon Degn (1930), have considered the outfield as the basis for land evaluation. If this had been the case, however, it would logically have been the number of *merkur* of sheep, not *merkur* of *jørð*, that all other rights were related to. In all likelihood, this would have been the case if the system of *merkur* had developed in the Middle Ages on the basis of Faeroese circumstances alone. That sheep breeding did not assume the primary role must presumably be because the system was introduced to the Faeroes from outside. An evaluation of the cornfield as the basis for all other production rights appears too exotic to have been developed in the Faeroes. It has also been claimed that the king introduced the system of *merkur* to the Faeroes as the basis of taxation or as some other kind of official valuation. Various named kings have been presented as the instigators and at different times in the eleventh, twelfth, and thirteenth centuries. However, it appears extremely unlikely that the king had anything to do with the development of the Faeroese system of *merkur*. First, it is difficult to imagine his interest in such matters. The number of *merkur* did not form the basis of any taxation before 1657, and, as a landowner, the king had very little interest in the Faeroes in the Middle Ages. Second, a land assessment similar to the Norwegian system might have been expected if it had been developed by the king in the Middle Ages. The system of *merkur* is unlikely to have its origin in official evaluation, but was more likely to have been developed in the farming community itself as a result of its own need to express relative value in sales, inheritance, and rentals.

In the Old Norse territories, three types of land valuation expressing monetary values were found. These were essentially different from one another. The first expressed the capital value of the land and is found in the Faeroes as well as in Iceland, Shetland, and Orkney. In Iceland, it is expressed as *hundrað,* homespun, corresponding to the Faeroese division of the *merkur* into units of homespun. In the Faeroes, it is expressed as *merkur* of *jørð,* and in Orkney and Shetland it was similarly expressed in marks of land or merks of cleared land. The second type of evaluation expresses the farm's total land dues *(landskuld)* and in the Norse territories is found only in Norway, where apparently it was universal in the Middle Ages and also formed the basis for tax assessment. In Norway, the land dues were an evaluation in terms of *merkrból, ertogból, mánaðarmatarból,* and *laupsból.* The third type of expression was found only in Orkney and Shetland, where the taxation value of land was expressed in terms of "pennylands." Thus, two parallel systems existed. In Norway, it is thought that the land assessment based on land dues developed in the early Middle Ages in connection with a tax giving exemption from naval defense obligations *(ledang).* If that was the case, it is difficult to imagine that throughout the Iron Age and Viking period the Norwegian farming community did not have or need any unit of measurement for land by which land sales, land inheritance, and land rentals could be determined. Since this type of land assessment is found throughout the North Atlantic, there can only be a common origin in Norway.

It is not entirely clear how widespread land assessment based on the land's capital value was in the Nordic countries. It was at least found in Denmark, where it was known as the gold evaluation. The assessment of capital in gold is also known from Shetland. In Denmark, land assessment based on land dues is a more recent system, which replaced the older capital value system. It may be possible to detect a pattern: in the old agrarian system of the Nordic countries, land was indicated in everyday dealings by its once-fixed value for the purpose of trade and inheritance, and the land rental was calculated on the basis of this value. The value was fixed in the community without the assistance of the official authorities, that is, it was fixed *"ejerne imellem"* ("among the owners themselves"), as it was expressed in Øm Monastery's chronicle in 1168. This land assessment survived in many places. Only in some places, such as in Norway and Denmark, was it replaced by official land assessments based on the land dues or tax. Only the land's own value was included in the assessment of land value, whereas land dues would have been based not only on the land rent but also on all that was taxable, which is why these systems were not directly comparable.

The Danish gold evaluation, as well as the assessment of land dues, is regarded as having been introduced in the early Middle Ages. The same applies to the Danish *bol* assessment. All are said to have been introduced as official assessments. However, there is no actual dated evidence from that period to confirm that all were introduced by official initiative. Thus one should be skeptical of linking together such diversified systems. One cannot avoid an impression that the chronology is based on a belief that there is a fundamental difference between the documented, historical period and the foggy, undocumented prehistoric period. This assumed difference may not reflect anything other than a difference in the nature of the source material.

The transition between the prehistoric and historical periods is characterized by widespread upheaval in the Nordic countries. The unification of nations and the introduction of Christianity resulted in the production of written source material in a previously unknown quantity. Thus, at a given point, historical phenomena begin to be recorded in writing. Yet history did not begin with the written sources. On the contrary, both the monarchy and written sources are products of a preexisting society. The history of agriculture is several thousand years old. The history of land cultivation in the Nordic countries begins five thousand years before written sources appear. In prehistoric agrarian society there existed norms, rules, and measures that are difficult to document. The appearance and the study of written documents can itself be seen as creating a barrier, leading to assumptions that the first written documentation of a phenomenon is from the same time as its establishment or introduction. Yet there is no reason to believe that land assessment could not have had its origins in prehistoric society, rather than being created by higher authorities in the Middle Ages.

Rights of Landownership and Use

The cultivated infield has a value that is expressed in *merkur* of *jørð*. It is owned and can be inherited, rented out, and sold. The sheep that graze in the outfield have a value that is expressed in *merkur* of sheep. They are the property of the people and can be used as items in independent economic transactions. The same is the case for cows, horses, pigs, and buildings. These are values created by people: the land through clearing and cultivation, the animals through domestication and their introduction to the Faeroes, and the buildings through their construction.

When one looks at the natural landscape, or on nature as a whole, we get another picture. The hill pastures are certainly discussed in terms of belonging to people, but, as has been shown, they are really appurtenances of the lands, or *jarðir*, shaped by people, and they cannot be disposed of independently from these lands. The same applies to *heimrustir* and *húshagar*. Concerning the *feitilendi* and fowling sites, it is doubtful whether they can be regarded as belonging to the *jarðir*, but rather the owners of the *jarðir* exercise the rights of use in these areas. Similarly, the *jarðir* bring with them rights of use to shore areas even though these are not considered to belong to anyone. Whales, birds, and fish are not owned by anyone, but on land the right to catch them belongs to the owner of the land and they do not have any independent value before they are caught. Only then, for example, is the whale assessed and assigned a value, expressed in guilders and skins.

The right of ownership in the Faeroese agrarian society was thus originally limited to what was created by humans. Nature, on the other hand, could be used but not owned by humans. Possession, as seen in the relationship between the *jarðir* and areas such as the *heimrust* and *hagi*, can be understood as a gradual development from rights of use to a situation of ownership.

When the Faeroes were settled in the Viking period, they were regarded as a commons, or as a no-man's-land. Farmsteads were built, land was cultivated, peat was cut, and sheep and other animals were grazed in the outfield without the idea that anything more was owned than the cultivated land and the domestic animals.

In research on Faeroese settlement history, there has been a perception that (as found in the Icelandic literature from the Middle Ages) individuals took possession of large areas of land and through the subsequent division of the land developed the settlement pattern that we recognize today. If my discussion here has any validity, then it is time for a reevaluation of our understanding of the organization and development of the settlement of the Faeroes.

Note

The principal sources for this chapter include Aakjær 1936; Bjørk 1956–63; *Kulturhistorisk leksikon for nordisk middelalder* 1980–82 (1956–78); Matras 1932; Myhre 1975, 1990; Thorsteinsson 1977, 1978, 1979, 1981a, 1981b, 1990, 1991, 1993, 1996; and Zachariasen 1959–61. Catriona Turner translated the chapter from Danish and Michael Jones edited it and wrote the introductory section as well as the section on *grannastevna*. Símun V. Arge advised on Faeroese terminology.

References

Aakjær, Svend, ed. 1936. *Nordisk kultur; Maal og vægt*. Stockholm: Bonniers; Oslo: Aschehoug; Copenhagen: J. H. Schultz.
Berntsen, Arent. 1971 (1656). *Danmarckis oc Norgis fructbar herlighed*. Copenhagen: Selskabet for udgivelse af kilder til dansk historie.
Bjørk, E. A. 1956–63. *Færøsk bygderet*, 3 vols. Tórshavn: Matrikulstovan.
Debes, Lucas. 1963 (1673). *Færoæ & Færoa reserata*. Tórshavn: Einars prent og forlag.
Degn, Anthon. 1930. "Marken som værdienhet på Færøerne." *Historisk Tidsskrift* 10: 68–78. Copenhagen.
Kulturhistorisk leksikon for nordisk middelalder fra vikingetid til reformationstid. 1980–82 (1956–78). Copenhagen: Rosenkilde og Bagger.
Mahler, Ditlev L. 1998. "The Stratigraphical Cultural Landscape." In *Outland Use in Preindustrial Europe*, ed. Hans Andersson, Lars Ersgård, and Eva Svensson, 51–62. Lund Studies in Medieval Archaeology 20. Lund: Institute of Archaeology, Lund University.
Matras, Christian. 1932. *Stednavne paa de færøske Norðuroyar*. Aarbøger for Nordisk Oldkyndighed og Historie III–22. Copenhagen: Gyldendal.
Myhre, Bjørn. 1975. "Gårdshusenes konstruksjon opg funksjon i jernalderen." *Arkeologiske Skrifter fra Historisk Museum, Universitetet i Bergen* 2: 73–105. Bergen.

———. 1990. "Hvor gamle er gårdsgrensene?" *Namn og eldre busetnad. Rapport frå NORNAs femtande symposium på Hamar 9.–11. juni 1988*, ed. Tom Schmidt, 125–39. NORNA-rapporter 43. Uppsala: NORNA-förlaget.
Svabo, J. Chr. 1959 (1781–82). *Indberetninger fra en reise i Færøe 1781 og 1782*, ed. N. Djurhuus. Copenhagen: Selskabet til udgivelse af færøske kildeskrifter og studier.
Tarnovius, Thomas Jacobsen. 1950 (1669). *Ferøers beskrivelser*, ed. Håkon Hamre. Færoensia 2. Copenhagen.
Thorsteinsson, Arne. 1977. "Heimildir um seyðamjólking í Føroyum." *Fróðskaparrit* 25: 84–94. Tórshavn.
———. 1978. "Forn búseting í Føroyum." *Fróðskaparrit* 26: 54–80. Tórshavn.
———. 1979. "Ruddstaðir í Brekkum—ein muturgøla frá 1412." *Mondul* 5–1: 14–21. Tórshavn.
———. 1981a. "On the Development of Faroese Settlements." *Proceedings of the Eighth Viking Congress, Århus, 24–31 August 1977*, ed. Hans Bekker-Nielsen, Peter Foote, and Olaf Olsen, 189–202. Odense: Odense University Press.
———. 1981b. "Jordforhold i det gamle landbrugssamfund." *Landinspektøren* 30: 10 (90): 664–78. Copenhagen.
———. 1982. "Lambhagar." *Mondul* 8–3: 13–14. Tórshavn.
———. 1990. "Bebyggelse og bebyggelsesnavne på Færøerne. En idéskitse." *Namn og eldre busetnad. Rapport frå NORNAs femtande symposium på Hamar 9.–11. juni 1988*, ed. Tom Schmidt, 115–24. NORNA-rapporter 43. Uppsala: NORNA-förlaget.
———. 1991. "Landbrugstradition og jordetal på Færøerne." *Nordatlantiske foredrag. Seminar om nordatlantisk kulturforskning i Nordens Hus på Færøerne 27.–30. august 1990*, 20–26. Tórshavn: Norðurlandahúsið í Føroyum.
———. 1993. "Merkur, alin og gyllin—gomul føroysk virðismeting." *Frøði* 7: 4–9. Tórshavn.
———. 1996. "Grindaprísir og skinnatal 1584–1638." *Fróðskaparrit* 44: 57–77. Tórshavn.
West, John F. 1975. "How Old Is the Faroese grannastevna?" *Fróðskaparrit* 23: 48–59. Tórshavn.
Zachariasen, Louis. 1959–61. *Føroyar sum rættarsamfelag, 1535–1655*, 3 vols. Annales Societatis Scientiarum Færoensis, Supplementum 4. Tórshavn: Føroya Fróðskaparfelag.

5.
Perceiving Landscapes in Greenland

Bo Wagner Sørensen

Landscape seems to be routinely invoked in ethnopolitical and nationalist discourse in the sense that a particular territory is identified with a particular people, emphasizing rootedness and distinctiveness as well as epitomizing the very soul of the people (Löfgren 1989; Malkki 1992; Alonso 1994). This idea of a correspondence is also reflected in Greenlandic discourses on culture and identity, which usually refer to the impact of nature and special living conditions in an Arctic environment. A particular "national landscape" is thus invoked that is believed not only to fit in with its natural inhabitants but also to mold them in a certain way, the result being that different landscapes produce different kinds of people.

The creation of a national landscape, however, is fraught with inconsistencies. Although all the inhabitants of Greenland are surrounded by an essentially Greenlandic landscape according to a national logic, they nevertheless have different national origins, besides living in different places that are subdivided along a Greenlandic/Danish binary opposition (see Figure 5.1 for place-names). This means that not all places are considered equally Greenlandic, even if the degree of cultural authenticity accorded specific places is inherently contested. While local people tend to differentiate between true or ideal Greenlandic places, on the one hand, and culturally ambivalent places, on the other, it is also important to see their perceptions and characterizations in a processual perspective. Ambivalent places may be culturally retrieved and consequently turned into more truly Greenlandic ones.

My special focus will be Nuuk, the capital of Greenland, which stands out as an example of a modern urban landscape (Figure 5.2). The town is thus widely contrasted with other places, of which the settlements especially epitomize a more authentic cultural landscape.[1] I lived in Nuuk for nine months in 1988–89, when I did fieldwork, and again in 1992–95, when I worked as a curator at the Greenland National Museum

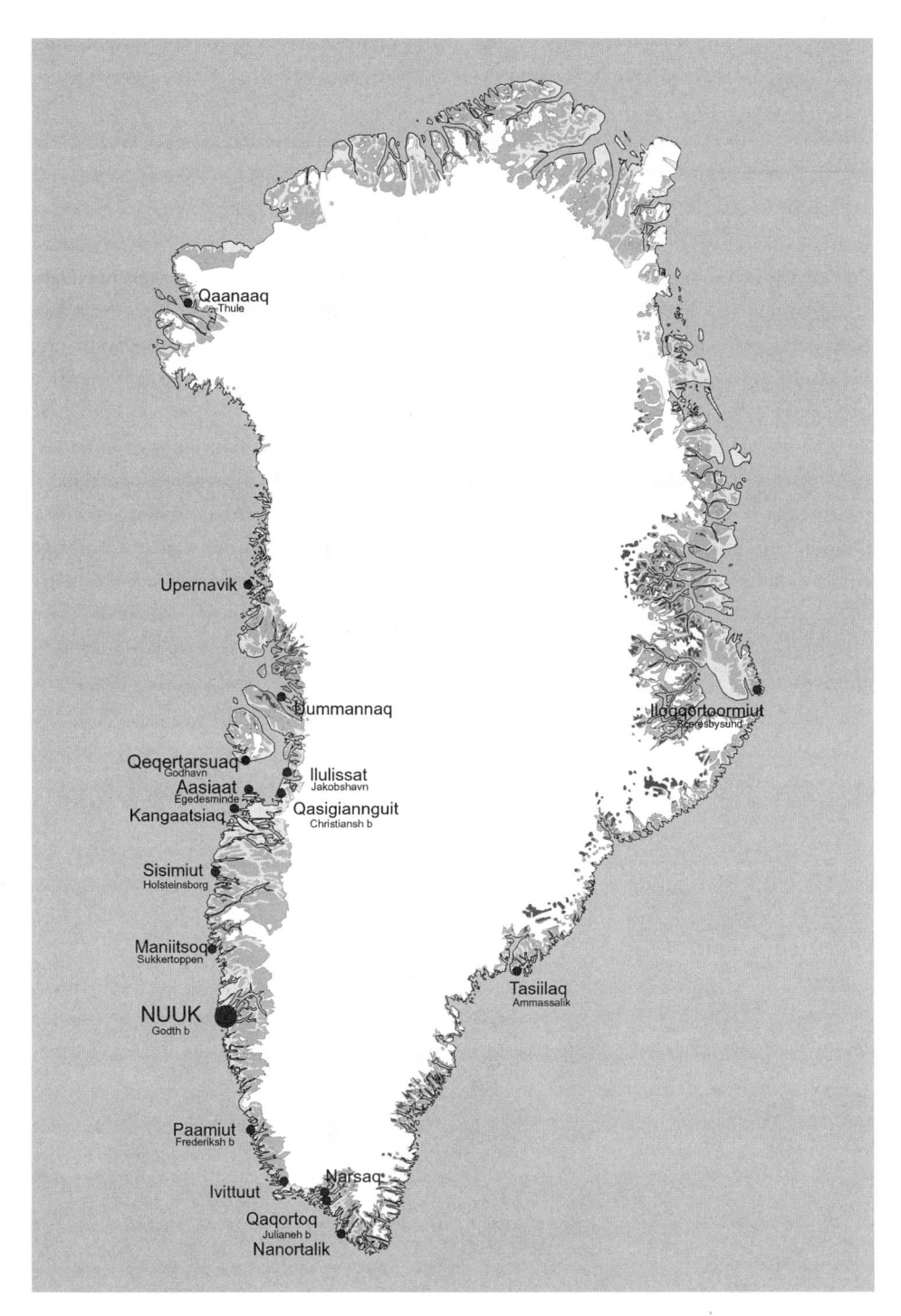

FIGURE 5.1. Map of Greenland with principal place-names. Copyright Kort & Matrikelstyrelsen (A. 163–01). Published with permission.

and Achives. Some of the data presented in this chapter stem from my original fieldwork (Sørensen 1994a), while the interview citations are part of a minor research project that was carried out in 1994.[2]

Landscape in Anthropology

In his exploration of the concept of landscape in anthropology, Hirsch (1995) comments that it has been largely unproblematized.[3] Yet it has had a submerged presence and significance in anthropological accounts in two related ways. First, landscape has been deployed as a framing convention that informs the way the anthropologist brings his study into "view." Second, it has been used to refer to the meaning imputed by local people to their cultural and physical surroundings.

The first of these uses of landscape is well known from classic monographs, where the scene(ry) is set initially in order to portray people in a recognizable landscape or picturesque view: the "objective" landscape of a particular people. The arrival scenes that establish an authorial presence within a text are illustrative examples of the framing convention, emphasizing that the author has truly "been there" (Geertz 1988; Pratt 1986).

FIGURE 5.2. The main street running through the center of Nuuk. Photo by Bo Wagner Sørensen.

As Hirsch points out, the two uses are nevertheless related, and the "objective" outsider's perspective is soon left behind in order to capture the native's point of view: how a particular landscape "looks" to its inhabitants. There is thus the landscape we initially see and a second landscape that is produced through local practice. This second landscape remains hidden until we manage to "unpack" it, recognize and understand it, through fieldwork and ethnographic description and interpretation. Basso writes in a similar vein that "an unfamiliar landscape, like an unfamiliar language, is always a little daunting, and when the two are encountered together . . . the combination may be downright unsettling" (Basso 1992, 220). What Basso refers to is the initial predicament of the stranger or ethnographer who just sees the physical landscape "out there," but who has no idea about the local meanings invested in the very same landscape.

According to Hirsch, the two related ways of considering landscape need not be understood as alternatives. Rather, the Western convention of landscape representation may be used as a productive point of departure from which to explore analogous local ideas, in that the Western representation is a particular expression of a more general foreground/background relationship that is found cross-culturally (Hirsch 1995, 2–3). Landscape painting, which has its origin in late sixteenth-century painting, can, for instance, be seen as a framed miniature representation of ideal life. Its ultimate goal is to achieve a correspondence between the pictorial ideal and the countryside itself (see Okely 1997 for a fine illustration) or bridging the gap between town and country.

Although Hirsch does not engage the issues of ethnicity and nationalism, his point about a general foreground/background relationship—the relationship between an everyday, workaday life and an ideal, imagined existence—seems useful in trying to make sense of the dubious image of Nuuk versus the idealization of the settlements. While the contrast between town and settlement, which I suggest can be framed as a contrast between town and country, is still emphasized in contemporary Greenland, there are also other processes at work. "Bridging the gap between town and country" or "the marrying of town and country" are phrases that aptly summarize some of those processes. The perception of Nuuk is contested among insiders and outsiders in addition to being susceptible to historical change that relates to both large-scale politics and discourses on urban life.

The Framing Convention

Literature on the Arctic tends to focus on the special natural environment and its effects on human life. In monographs on traditional Inuit life, the opening chapters or arrival scenes, in the case of field-based representations, contain descriptions of frostbound Arctic landscapes that spell harsher than harsh and thus, on the face of it, are not fit for human existence. Yet, against all odds, these vast Arctic expanses *are* inhabited, which means, in turn, that the Inuit are usually depicted as heroic survivors with a fascinating ability to adapt. They overcome the fight against the cold and the quest for food by technological ingenuity and social flexibility (Sørensen 1994a, 128–32).

It seems likely that the "force of tradition in Eskimology" (Riches 1990) has had an impact on both external representations and self-representations. Fienup-Riordan (1990) notes that Western observers have naturalized Eskimos as paragons of simplicity and virtue. She also shows how the highly diversified groups that occupy a variety of environments across the Arctic have melted into a single Eskimo stereotype—the peaceful, original ecologist and fur-clad survivor of the High Arctic—and how contemporary Eskimos' representation of themselves often is as much a product of nonnative expectation as of their own history. To some extent they have become what others have made them. Appadurai (1988, 37) likewise argues that anthropologists have contributed to the incarceration of natives in certain places. Native peoples are usually represented as naturally rooted and ideally adapted to a specific environment.[4]

A recent book on Greenland (Gynther and Møller 1999), written mostly by Greenlanders, illustrates the framing convention. The first chapter, "The Arctic World," includes descriptions of Greenland's regional placing, geology, climatic conditions, sea and ice, and flora and fauna. After the frame has been set, the Inuit are brought alive and immediately merge into the Arctic scenery. The book conveys the message that life in Greenland is permeated by nature and by adaptations to climatic and environmental conditions. The generalized Greenlander appears to engage in a special, and respectful, relationship with nature. The focus is not on a fight *against* nature but rather on a recognition that nature is powerful and that human life should be lived with due respect according to nature's whims. Greenlanders stand out as part of a particularly harsh and impressive landscape to a much higher degree than Europeans, even if Europeans are inscribed into specific national landscapes as well. The metaphorical fusion of Greenlanders, sea, ice, and mountains tends to be much stronger and more contemporary than that of Danes, oat fields, and beech trees.

A NATION IN THE MAKING

The West Greenlanders did not regard themselves as a unit at the time of the colonization in 1721, but they have been increasingly unified by colonial administration (Kleivan 1984, 595). According to Thuesen (1988), the emergent Greenlandic national consciousness became pronounced in the early twentieth century, when Greenland was still a Danish colony.[5] Thuesen identifies the founding in 1908 of the first Greenlandic religious association, *peqatiginniat,* as a turning point toward a national revival on a Christian basis. Berthelsen (1988, 134) likewise comments: "The years at the turn of the century were a time of upheaval. . . . The period was characterized by a spiritual and national awakening." The pioneers were Greenlanders who were trained as teachers and catechists at the teacher's college in Nuuk. This cultural institution, which was originally established in 1845 but reemerged in 1907 with the opening of a new, large, and flamboyant building, has played a central role in the recent history of Greenland. The catechists were diligent debaters and writers of poetry and prose whose works and ideas were disseminated through the printed press and thus contributed to make it possible to "think" the nation (cf. Anderson 1991).[6]

The upsurge in literary and artistic production had a decidedly national bent. Berthelsen (1983) comments on the patriotic depictions of nature and landscape of this period, and, according to Petersen (1984, 645): "The first patriotic songs published after 1900 were abstractions of the authors' native areas, but they developed a love of Greenland in which the 'spirit of the people,' expressed in the community of language and culture, was formed and in which the history of the people gradually became a synonym for Greenlandic identity." In her book on the art of Greenland, Kaalund (1983, 191) refers to "the period of revival," when artists' depictions of the nation's scenery and people raised a national awareness and strengthened Greenlandic identity. Presenting some of the most influential painters of this national-romantic genre, she writes: "The extreme precise rendering of people and scenery characteristic of such painters . . . was nurtured by profound love for Greenland's mountains and fjords, weather and changing light, and for her people. Though all were deeply religious . . . these painting pastors elected not to paint biblical themes. Instead, nature became the grand subject of their art" (Kaalund 1983, 192).

A polemic debate on Greenlandic identity, *kalaaliussuseq*,[7] appeared in the newspaper *Atuagagdliutit* around 1911 and continued for almost ten years (Berthelsen 1988; Thuesen 1988, 143, 160). While seal hunting was still the main occupation, sheep breeding and fishing were being developed. More Greenlanders were also being employed as store assistants and as catechists and teachers in the church and the schools. The debate, which engaged both hunters and employees, revolved around the relationship between occupation and identity. The hunters equated being a hunter with being a Greenlander, whereas the catechists argued that the command of Greenlandic language was the main criterion of identity (Berthelsen 1988, 140). Acknowledging the importance of seal hunting, which was considered the national trade, the catechists nevertheless saw themselves as pioneers who should serve as models and work in the service of their country (Thuesen 1988, 1995).

The early twentieth century saw an increasing social differentiation and stratification while, at the same time, a national identity emerged, based on the Greenlandic language, *kalaallisut*, the land, *nuna*, and a common history (Thuesen 1988, 160). Localized identities remained important, and still are, but the overarching identity as *kalaallit*—Greenlanders—brought diverse groups together in a common destiny that pointed simultaneously back and forth in time. The category *kalaaleq* (pl. *kalaallit*) is surely a product of colonialism; before then, individuals and groups were identified according to geographical location.[8]

Greenland Home Rule

Malkki (1992, 26) shows how an identity linking people and territory is expressed and naturalized in everyday language, but also through the visual device of the map, which represents the world of nations as a discrete spatial partitioning of territory with no vague nor fuzzy spaces and no permeable boundaries. Each nation is fixed in space; it is sovereign and limited in its membership. Hylland Eriksen (1993, 106) likewise points out: "Although maps existed before nationalism, the map can be a very concise and potent symbol of the nation. Country maps, present in classrooms all over the world, depict the nation simultaneously as a bounded, observable thing and as an abstraction of something which has a physical reality."

Greenland is the world's largest island. It is thus clearly demarcated geographically and on the map.[9] During the debate before the introduction of Greenland Home Rule in 1979, both the special Arctic landscape and the geographical demarcation were often highlighted. The argument that contrasted the Greenlandic landscape in all its grandeur and wildness with the domesticated Danish landscape espoused that Greenlanders and Danes were separate peoples of a quite different nature. The change in emphasis from Arctic to *Greenlandic* landscape is not accidental. The landscape has, so to speak, been culturally appropriated.

Since 1979, Greenland has been known officially as Kalaallit Nunaat—the Greenlanders' Land. Nuttall (1992; 1994) speaks of the emergence of a "homeland." The nation-building process, which intensified under Home Rule, has involved the creation of national symbols such as the national anthem, the flag, a national day on June 21, and a coat of arms (Kleivan 1988; 1991).[10] The term *kalaallit* emphasizes the ethnic roots of this micro-nation, or nation in the making. People of Danish origin who are born in Greenland are still referred to as Danes, not as *kalaallit*, which so far remains an ethnically exclusive term (Sørensen 1993a). Some people, however, have argued in favor of a broader, nonethnic definition of "citizenship" in Greenland that is more in tune with reality.[11] Thus Robert Petersen (1994; 1995) argues against an ethnic dichotomous model of society, suggesting instead that all people living and working in Greenland should be regarded as citizens of Greenland. Such a perspective is not common, though. Danes are often conceived of as transitory and not really belonging (Nuttall 1992, 102–4; Sørensen 1993a).

The National Landscape

"Everywhere in Greenland the landscape has an unmistakable 'Greenlandic' character," writes Rosing (1999, 45; my translation). Although he puts "Greenlandic" in quotation marks, he subsequently remarks that while the Greenlandic character of the landscape may sound obvious, it is nevertheless remarkable considering the vast size of Greenland.

In any case, the seemingly Greenlandic character of the landscape *in* Greenland is telling. At the same time, his remark also suggests that there is such a thing as a typical or true Greenlandic landscape. The positions of geologist and Greenlander are conflated.

The identity linking people and landscape is often brought about by the use of botanical metaphors that "suggest that each nation is a grand genealogical tree, rooted in the soil that nourishes it" (Malkki 1992, 28). Trees are largely absent in most parts of Greenland, yet the lack of trees does not rule out the use of botanical metaphors and ideas of rootedness and belonging. Tracing family genealogies is a popular occupation in contemporary Greenland. Greenlandic family trees tend to intersect to a high degree, making the Greenlanders one big, potential family. Whether family ties are actualized is, however, a matter of choice based on a sense of closeness (cf. Nuttall 1992, 81). A young woman states (in Nilsson 1984, 107; my translation): "A Dane can never have the same strong feelings that I have for my country. We Greenlanders often say *nunarput*—our country—with broad gestures and strong feelings; it means so much to us. We are part of it; part of the mountains, nature, the air, the hunting game, the people—it is all inside our bodies, inside our hearts."

In a newspaper article, Thue Christiansen (1994), an ardent debater and former politician, reflects on the concept of culture, emphasizing the relationship between physical environment—the ground we step on each day, its placing on the map of the world, its shape, and the surrounding sea—and culture. According to Christiansen, people are shaped in harmony with their environment. The ideal presented seems to be based on congruence: a Greenlandic landscape in harmony with Greenlandic culture: "Our Greenlandic culture will always be different. . . . As long as our country continues to exist and has an ice cap 'hat,' the inhabitants will maintain their peculiar way of life, have a culture of their own—no doubt about that" (Christiansen 1994, 5; my translation).

Some Greenlanders also take a radical view when it comes to ethnic identification. Ono Fleischer, who is a teacher and adventurer, is illustrative in this respect. Since the early 1990s, he has followed in Knud Rasmussen's footsteps on several sledge trips in the Canadian and Alaskan Arctic with the main purpose of building bridges between the Inuit peoples. In an interview he states: "The Inuit are one people. We look alike. We smell alike and speak the same language. . . . We think alike, too. . . . When we are far away in Canada's deserted snow-filled wilderness, we always end up meeting our own kind, and feel at home immediately. We can sense that the people there are just like us and we are absorbed naturally into the community. . . . We can see who is Inuit. Or perhaps feel it. The Inuit are one people" (Brønden 1999, 20). Fleischer's message is clearly essentialist and pan-nationalistic, invoking common ancestry and a common feel; the Inuit are one body.

Mere exposure to the Arctic landscape on the part of migrant workers and other "aliens" in Greenland is usually not believed to make radical changes in their basic anchoring in national landscapes elsewhere. A case in point is Jess Berthelsen, chairman of the Greenland Labor Union, who used an interesting botanical metaphor during his public speech on Labor Day in 1998, comparing "people from outside" with "artificial flowers" and "plastic flowers" (*Sermitsiaq* 1998, 17). His point was that Greenland is controlled by people from outside—a euphemism for Danes living and working in Greenland—who are not naturally rooted in the Greenlandic soil and therefore, inevitably, must influence the development in a wrong, if not unnatural, direction. A "true development," according to Berthelsen, is based on the native inhabitants' "brains, muscles and hands." It springs from the national soil, so to speak. Plastic flowers do not perish easily, and at the same time they are replaceable, referring to the fact that individual Danes come and go, yet the Danes stay on. The lack of roots also keeps them out of touch with real life and natural, continual growth.[12]

Landscape serves as a metaphor for cultural distinctiveness and bounding practices. Although the Greenlandic landscape is accorded agency in its capacity to mold a special ethnic identity or national character, local people are, not surprisingly, the true agents in this national game, even if they often tend to naturalize it. The introduction of Greenland Home Rule can be seen as a temporary culmination of an ethnopolitical or micronationalist movement toward self-government. Interestingly, the political struggle has been widely understood and framed as a struggle between two cultures or mentalities with a special focus on the preservation and retrieval of the peculiar features of Greenlandic culture (Sørensen 1997a, 1997b). In turn, the contrast between Greenlandic and Danish cultures has been played out within the Greenlandic society, and even among individuals.

As the self/other contrast is a necessary condition of one's own definition, it follows that a cultural self-definition will imply a cultural "other" that serves as a contrast. According to Hastrup (1988, 126–27), cultures exaggerate themselves to stand out more clearly vis-à-vis other cultures. Cultural exaggeration is necessary in order for cultures to become distinct. Moore talks about "a passion for difference," which implies processes of identification and differentiation: "These processes are engaged for all of us, in different ways, with the desire to belong, to be part of some community, however provisional. Belonging invokes desire, and it is in this desire that much of the passion for difference resides" (Moore 1994, 2).

Although the national-geographic in Malkki's (1992) sense tends to rule out vague or fuzzy spaces and boundaries, it can be argued that some internal spaces, or landscapes in Greenland, are more culturally ambivalent than others are. This leads us to Nuuk.

Perceiving Nuuk

When I came to Nuuk for the first time, I had no idea that it was widely considered a Danish-like town. Some of the Greenlanders I spoke with readily characterized it as "a piece of Denmark on rocks," "miniature Denmark," or "a northern counterpart to a Copenhagen suburb, only smaller." I did not think of it then, but in retrospect these statements can be seen as oral, yet framed, picturesque, and condensed representations of a particular Arctic landscape, which is contrasted with the symbolic Arctic landscape of the rest of Greenland. This other landscape seems to epitomize a more symbolic point of identification. In short, the statements referred to life as it is in a particular setting versus life as it can be, could be, or even should be.

During fieldwork, many people also requested that I go to other places than Nuuk if I were to study Greenlandic culture. Seemingly, culture could be "found" primarily outside Nuuk, most likely in the small places in the peripheries where it had survived so far (Sørensen 1993b). Alternatively, it would be located in certain cultural institutions such as the Greenland National Museum and Archives, which can be seen as the repository for the national heritage, and the culture center, *Katuaq* (Figure 5.3), which was opened in 1997. Most anthropological studies in Greenland have in fact been carried out in the settlements or in the "outlying districts"—that is, northwestern and eastern Greenland.

Initially, I was puzzled by the cultural status of Nuuk, not least considering the fact that Nuuk has the largest Greenlandic population of all Greenlandic towns. I wondered

FIGURE 5.3. *Katuaq*, Nuuk's culture center. Photo by Bo Wagner Sørensen.

how a town inhabited by some 8,000 Greenlanders out of a total of about 46,000 could be considered non-Greenlandic or, at least, less Greenlandic than other places. The image of Nuuk was also puzzling seen in a comparative perspective where capitals tend to stand out as symbols of the nation. The reflections of local people, however, showed clearly that it had to do with the high concentration of Danes, who made up a little less than one-third of the total population of Nuuk. Danes are thus more visible in Nuuk than in other parts of Greenland and their influence is more readily experienced and felt in everyday life. Danish is widely spoken both in private and in public, not least because many Danes are employed in the Home Rule and municipal administrations. The concentration of non-Greenlandic-speaking Greenlanders is also much higher in Nuuk than elsewhere on the coast. Such "halfies" often present a dilemma when it comes to ethnic categorization and cultural order in general (Sørensen 1997a, 1997b).

Spoken Greenlandic is generally used as an indicator of Greenlandicness among local people. During an interview, a woman stated: "The most important thing is that I speak Greenlandic. That is why I can call myself a Greenlander." Confronted with the issue of non-Greenlandic-speaking "halfies," however, the woman modified her statement somewhat: "Many children of mixed marriages do not speak Greenlandic, yet perceive themselves as Greenlanders. How they identify themselves as Greenlanders then, I do not know. Even so, I think of them as Greenlanders because I cannot really think of them as Danes. . . . A Dane is somebody who comes from Denmark. Well, usually they come from Denmark—that is, some of them are born in Greenland and have lived here always, but that does not mean they speak Greenlandic."

It appears from the interview that real life is often more ambiguous than people's categorizations. The "halfies" are nevertheless encompassed by the term "Greenlander"; they are just somewhat removed from the prototype. Nuuk can be seen as a less prototypical Greenlandic town as well, according to the same informant:

> Well, the way of life and the way people think are very close to a Danish provincial town. You only experience the real Greenland if you get outside of Nuuk. If you go to Sisimiut, for instance, your first impression will be that people speak Greenlandic. If you go to Nuuk, you are almost one hundred percent sure to come across a Danish cab driver and be taken to the hotel where they speak Danish too. So Nuuk is not a decidedly Greenlandic town. If you go to *brættet* [a stall where Greenlandic food is sold], however, you will hear Greenlandic spoken. *Then* Nuuk is a Greenlandic town.

The characterization of Nuuk as Danish-like also has to do with its relatively high pace and general townlike appearance. Nuuk has sometimes been singled out as the ultimate symbol of cultural alienation, not least exemplified by the big apartment buildings in the central part of town, Narsarsuaq (Figure 5.4). They were built in the late 1960s in

FIGURE 5.4. Apartment buildings in Narsarsuaq, a central area of Nuuk. Photo by Bo Wagner Sørensen.

response to the acute housing shortage due to heavy migration to Nuuk and the general population explosion during that period (Rosendahl and Laage 1978). The area may look rather desolate in dull weather, but the critical voices raised against these ten four-story apartment buildings, and one five-story building, basically questioned whether they were in accordance with the true Greenlandic lifestyle and worldview (cf. M.-L. D. Petersen 1986). In her description of modern history, a Greenlandic woman puts a strong focus on alienation:

> The secure and familiar world of the villages was replaced by the completely foreign one of larger towns, and many people suddenly found themselves living in huge apartment blocks filled with total strangers. At the same time, the new world brought new temptations. Modern stores appeared with all the amenities of the European lifestyle, including alcohol. It did not take long for people to discover drinking as an effective way to escape the stresses of the modern world.... Virtually overnight, the Greenlandic people lost their culture and their identity and became spectators in their own land. (T. S. Petersen 1994, 139)

Nuuk has a quite extensive system of roads, lots of cars and taxis, a well-developed system of buses that run from every fifteen minutes to once an hour, a few traffic lights, and even three traffic circles. With its many supermarkets and specialized stores, Nuuk may also appear to be a consumer's paradise to people visiting from smaller places, some of which have just one store for basic necessities. The town, moreover, offers a wide selection of restaurants, including a Thai restaurant, cafés, bars, and discotheques. A cinema has opened in the culture center, *Katuaq*, and a bowling alley was opened in 1999 as an extension to *Hotel Godthåb*, which is a complex of bars and restaurants.

Part of the reason why Nuuk is contrasted with other Greenlandic towns is also due to the fact that Nuuk is an administrative and educational center par excellence. This means that hunting and fishing, which are closely associated with Greenlandic culture, are not as conspicuous as elsewhere. Occupational hunters and fishermen make up a tiny part of the population, whereas most people are employed in white-collar and blue-collar occupations within Nuuk Municipality, or Greenland Home Rule, which is Greenland's largest employing agency. Hunting and fishing do take place in Nuuk, though, not least as a part-time or spare-time activity, and many people own boats that are moored primarily in either the marina in suburban Nuussuaq or in the commercial harbor in Nuuk. In the smaller places, the harbor is quite central and what goes on there can be looked over from most houses. Nuuk, in contrast, is a relatively large and spread-out town consisting of Nuuk proper and Nuussuaq, which began to be developed in the late 1970s. If one does not happen to work in the fishing industry down at the commercial harbor or live close to it, one can easily forget that fishing is the prime industry of Greenland.

One of the reminders that hunting and fishing play a role in the modern life and economy of Nuuk is *brættet* or *kalaaliaraq* in Greenlandic. It is a roofed stall situated close to the Colonial Harbor (Figure 5.5) where local hunters and fishermen or their wives sell fresh seal and whale meat and different kinds of birds, and fish. Frozen Greenlandic food is nevertheless sold in all supermarkets, being part of the usual assortment, even if they are usually kept in separate frozen-food counters.

The fact that Greeenlandic food is kept separate from imported goods is interesting and may indicate an underlying cultural order. Food seems to be intimately related to cultural identity, and thus also to nationalism. In her article on Greenlandic food, Kleivan (1996, 155) concludes: "One confirms and maintains one's Greenlandic identity, among other ways, by eating and liking Greenlandic food in a world where there are many other options. As the process of internationalization has become more striking in Greenland, and in the area of food too, Greenlandic food as an ethnic symbol has taken on new dimensions."

Eating Greenlandic food may be interpreted as a way of connecting oneself historically to a specific cultural landscape inhabited by one's forefathers as well as by different species of "indigenous" land and sea mammals, fish, and birds. It is a way of

FIGURE 5.5. The Colonial Harbor in the old part of Nuuk. Photo by Bo Wagner Sørensen.

connecting with a hunting way of life that is generally seen as the essence of Greenlandic culture. In her article on a so-called reinvented Icelandic tradition, Hastrup (1991) likewise suggests that the Icelanders are "eating the past" when they indulge in a yearly communal consumption of traditional peasant food. Being interested in how the nation is felt and experienced by people in their everyday life, Palmer (1998) also points to food as one of three "flags of identity"—the others being the body and the landscape—that provide a system of reference within which aspects of the material world are used, consumed, and experienced.

Town Versus Settlement

There is a certain truth to comments by some of the local people that everyday life in Nuuk may not be very different from life in any odd provincial town in Denmark. Most adults go to work, children go to school, people shop in the supermarkets, prepare dinner, have friends or family over for coffee, and watch television or videos before going to bed. In contrast to the rest of the week, weekends are often set off for relaxation and fun that may involve dinner parties, going out, pub-crawling, or family outings and boat trips. The immediate familiarity with life in Nuuk should not, however, mislead one into thinking that cultural difference is not highlighted in everyday life. On the contrary: living in Nuuk means living in a highly politicized cultural space.

Especially the younger generation seems to reflect a good deal about culture and identity in their everyday life. This in turn is reflected in their readiness and ability to engage in conversations about these issues. A man said during an interview:

> Well, I call myself a Greenlander and also speak Greenlandic, but I live as a European, which makes me a sort of mixed Greenlander. If we are to talk about a scale of Greenlandicness, there are still some aboriginal Greenlanders living primarily from hunting and fishing in the outlying districts and settlements. They speak a very pure Greenlandic that is hardly influenced by Danish or other languages. I would call them true Greenlanders—that is, they are more Greenlandic than others that are more or less mixed, including myself.

During another part of the interview session, the same man nevertheless seemed to fend off potential challenges to the legitimate basis of his own ethnic identity:

> I will have to admit that I am not one hundred percent Greenlandic, yet I call myself a Greenlander because I was brought up in an all-Greenlandic environment. My mother tongue is Greenlandic, and during the first fifteen years of my life I was part of the decidedly Greenlandic hunting and fishing culture. I was born in a small settlement and lived there for two years. After moving away, however, we spent our holidays there. All in all, this is why I think of myself as a one hundred percent Greenlander in spite of the fact that I am biologically mixed, but others may question that.

It appears from the interview that not all Greenlanders are equally Greenlandic, and also that some places are considered more Greenlandic. Such marked places in turn can be used as part of individual narrative strategies that direct themselves toward producing a story of an unmistakable anchoring in a decidedly Greenlandic landscape in spite of certain obvious flaws such as being of mixed origin. Cultural purity seems important as a basis for evaluation. Put in another terminology, the semantic density (Ardener 1982) revolves around a complex of a pure spoken Greenlandic, hunting activities and products, and settlements. This complex, exemplified by the settlement (Figure 5.6), can be referred to as an ideological cultural landscape.

The relationship between town and settlement is also central in Nilsson's book (1984). Her interviews with fourteen young Greenlanders show clearly how they tend to contrast life in town with life in the settlements. The town comes to stand for alienation and all sorts of trouble, whereas the settlement stands for the healthy way of life, including traditional values, warmth, social cohesion, cooperation and solidarity, and not least, a close relation with nature. In my own interviews, conducted in Nuuk in

1988–89, many interviewees made use of a before/now opposition, where "before" represented a settlement way of life that was simpler and easier to grasp in addition to being harmonious and spiritually rich, compared with "now." People from settlements who had recently moved to Nuuk also commented that the townspeople were highly reserved toward newcomers. The latter were used to people greeting one another when they met and often found it hard to fit in.[13]

In his monograph on Kangersuatsiaq, a settlement in northwestern Greenland, Nuttall (1992) also deals with the contrast between town and settlement life. One of the locals, who had lived in Nuuk for ten years, explained his decision to leave town life, commenting that on the one hand life is better in Nuuk, and it is easier to earn a lot of money. But on the other hand, there is a lot of drinking and fighting, and he himself also drank and smoked a lot of hash. Being back in his native settlement, he now lives as a hunter, earns little money from selling sealskins, and has to live in his father's house, all of which he describes as hard. Life in the settlement is thus not glorified, and yet he says: "But I can trust people here . . . in Kangersuatsiaq you can be a *kalak* [a real Greenlander]. Here, I can eat Greenlandic food every day. In Nuuk I longed for seal meat all the time because you only eat Danish food. You have Denmark in your mind and in your stomach" (151). Nuttall comments that whatever this

FIGURE 5.6. Qoornoq, a settlement in Nuuk Municipality. Photo by Bo Wagner Sørensen.

informant's own personal ideas of being a "genuine" Greenlander, he nevertheless seemed to have learned new ways during his absence. Many of his co-villagers thus felt that he acted and thought "like a Dane." According to Nuttall, sharing is a metaphor for village life: "To give is to be a *kalak,* a 'real' or 'genuine' Greenlander. When people say somebody is 'like a Dane' *('soorlu qallunaaq'),* it is used as a symbol of the values perceived to be pervasive in the dominant society, values that are beginning to take a hold in the community. The attitude is quite clear: Greenlanders give freely, Danes sell. However, today only real Greenlanders give freely, while Greenlanders who act like Danes sell" (148).

The Greenlanders' "Wild"

During my first stay in Nuuk, it soon appeared that people who characterized Nuuk squarely as Danish-like were mostly newcomers from other parts of Greenland. The true local, the *nuummioq,* may share the migrants' view on Danish influence and dominance but tends to perceive his or her hometown as merely different from other Greenlandic towns. Local persons also sometimes ridicule the migrants by questioning why they tend to stay on in Nuuk considering they find it so dreadful. Cultural authenticity is thus inherently contested as people with different local identities, experiences, and outlooks claim to be Greenlanders in their own right. It also seemed that people of mixed origin in general were more sympathetic to cultural ambiguity in the sense that they did not find it threatening. Some spoke of themselves as "hybrids" (*bastarder* in Danish) and seemed to welcome any signs of hybridization, which may not be surprising given their own position (see Sørensen 1994b). One such self-declared "hybrid" said about the contested cultural status of Nuuk: "When people say that it is not a Greenlandic town, it may be because there are so many Danes here and also more Danish things than there are in the other towns, but even so, it *is* a town in Greenland, only somewhat different. It may also be due to the fact that there are so many Greenlanders who speak Danish and dress more European—that is, they are more concerned about what to wear, and also interested in different kinds of food. Nuuk also has more to offer as regards European goods, for instance, and it is much less isolated than the other small towns."

The same woman spoke about three different "mentalities" in today's Greenland: Greenlanders of Greenlandic origin, Danish-Greenlanders of mixed origin, and Danes proper. Other interviewees brought up the importance of education in relation to the social differentiation of the Greenlandic population, exemplified by the hunters versus the highly educated as the extremes on a continuum, which likewise undermined the idea of a common Greenlandic mentality vis-à-vis a Danish mentality. A man mentioned that Nuuk was much less homogeneous than the other Greenlandic places. Nuuk leaves room for difference, which was the reason he liked it and wanted to stay on.

Once he went back to his place of birth, where he was received with shouts of "What are you doing here? Your are not a Greenlander anymore."

New arrivals or visitors to Nuuk especially seemed more inclined than the *nuummioq* proper to try to separate Greenlandic culture from Danish culture, arguing sometimes that Nuuk is not the real Greenland or even stating that it is more Danish than Greenlandic. Yet they always ended up saying that there is an element of Greenland after all. "The Greenlandic mentality makes itself felt in many ways after all," as one woman put it. It is only wrapped up in a general Danish appearance. What the interviewees pointed at may be summed up in the expression "the inner life of Greenland" (Lynge 1981, 1988). This expression sometimes appears in local discourse, in which connection it serves as a metaphor for an esoteric cultural space from which aliens are largely left out. It also seems that some Greenlanders are more securely placed within this cultural space. The "inner life" seems to stand for a particular way of life and values that are anchored in people's ideas of life as it used to be, but at the same time it holds a potential for life as it could be.

In any case, "the inner life" has certain parallels to Ardener's concept of muted groups and their "wild." This he explains as a definitorial space created by the lack of congruence between the bounding models of the politically dominant group and the bounding models that are set up by the subordinate group (Ardener 1989). The subordinates are thus not totally encompassed by the dominants' models. In spite of the fact that Greenland has Home Rule, many Greenlanders still think and feel in terms of Danish dominance. However, the very idea that the Greenlanders' "wild" or "inner life" is beyond the reach of the Danes may be important. Thus an element of virgin and immaculate culture is even present in the hybrid urban landscape of Nuuk. It just does not show much in public or on the surface. The settlements, on the other hand, tend to be seen as the roots of life in town and as lifelines for the town-dwellers, representing a kind of safeguard against total sociocultural disruption and alienation. They are the epitome of a Greenlandic cultural landscape, a landscape rather untainted by Danish influence.

Cultural "Treasure Islands"

After World War II, Denmark ended its isolationist policy toward Greenland, which had been a Danish colony since 1721. Colonial status was abolished in 1953 and Greenland became an integral part of the Kingdom of Denmark, thus giving the Greenlanders equal status to Danes. Emphasis was now placed on social welfare and on infrastructural change as part of a process of modernization that was explicitly aimed at raising the general standard of living and narrowing the gap between living conditions in Greenland and Denmark. The ending of colonial rule marked the beginning of an era characterized by extensive changes: "A massive construction programme was undertaken, to cater for the health, educational and housing needs of a growing population, and the increased rationalization of the fishing industry was placed at the centre of economic development. During the 1960s, as part of a policy of centralization and urbanization, people were 'encouraged' to migrate from outlying areas to west coast towns by the closing down of what the Danish authorities regarded as 'unprofitable' settlements" (Nuttall 1994, 7). During the same period, the numbers of Danes in Greenland increased significantly because of the need for workers in construction and in services such as health and education.

In his dissertation on the development policy debate in Greenland, Forchhammer (1997) argues against the present, widespread idea that the centralization policy was an entirely Danish enterprise. The policy was launched in the 1950s, yet had its most pronounced effects in the 1960s, until it was discontinued in the early 1970s. Reading through the minutes of the meetings of central political bodies, he shows that Greenlandic politicians and representatives have generally been in favor of such a policy. However, they have been divided as to how it should be implemented in practice, in which connection their own local "roots" and loyalties conflicted with their general stand.

In any case, the yearly migration from settlements to towns amounted to 300 persons during the 1950s, while it was twice as high during the 1960s. The settlement population of the so-called fishing region—that is, the west-coast area from Nanortalik in the south to Ilulissat in the north—dropped by 20 percent between 1960 and 1968. Furthermore, 19 settlements out of a total of about 64 had been depopulated in this region, while the population of the so-called open-water towns had almost doubled (Forchhammer 1997, 59). Nuuk, for instance, had a population of about 1,000 in 1950. Ten years later, the population had grown to about 2,500, and around 1970 it numbered some 7,000. At present, Nuuk has about 13,000 inhabitants. There are only two proper settlements left in the municipality of Nuuk: Qeqertarsuatsiaat and Kapisillit, with a population of 261 and 117, respectively (Statistics Greenland 2000, 263). In 1960, the settlement population in the Nuuk area numbered about 1,000 people, distributed among nine places of varying size (MfG 1974, Table 5).

Forchhammer writes how the tendency to romanticize the past and settlement life became widespread during the 1970s, and still is the case. He maintains that this tendency is not grounded in settlement life as it really was and is, but rather in an ideological construction that operates with two conflicting cultures: "On the one hand, there is the traditional or Greenlandic culture prevalent in the settlements, which is among other things characterised by collectivity, solidarity, small scale societies, ecological way of life, hunting sea mammals, and being close to nature. On the other hand, there is the Danish (European) culture—prevalent in the towns—which is thought to be diametrically contrary to the Greenlandic one" (Forchhammer 1997, 115).

Even so, Forchhammer argues that the general focus of the early centralization policy debate was on social problems presumably arising from the settlement migrants' problems of adapting to life in town, in which connection the "conflicting cultures" would seem to provide a meaningful explanatory framework. It was only later that nationalist ideas became fused with settlement policy. Settlements eventually came to be viewed as the roots of Greenlandic culture, as retreats or "Treasure Islands" (Forchhammer 1997, 139), harboring the original and pristine Greenlandic culture that had vanished in the more modern parts of the country as a consequence of colonialism. When the settlements were accorded this new status as symbols of true Greenlandic culture, both politicians and ordinary people at large took a general stance against the former centralization policy. This policy, by implication, was seen as an entirely Danish enterprise that threatened to undermine and wipe out the foundation of Greenland's cultural distinctiveness. Consequently, when Greenland Home Rule was introduced, a pro-settlement policy was part of the program.

"Treasure Islands" Revisited

The public debate on settlements in the 1990s was largely informed by the idea that the settlements are places where Greenland's cultural heritage is both preserved and kept alive. At a conference on settlements that took place in Ilulissat in 1992, Marianne Jensen, a member of the Home Rule Government, said: "From old times, the settlements have been of great importance for the preservation of our culture. They have coped spiritually and physically by adapting to the natural environment. They contribute to the preservation and continuation of our customs and traditions, and our inner life. They have indeed a lot to pass on to the townsmen, and so they secure our cultural heritage in the future" (A/G 1992, 10; my translation).

Another participant, the renowned writer and pastor emeritus Otto Sandgreen reminded of the importance of the settlements being rather untainted by Danish influence and thus all the more representative of a true Greenlandic cultural landscape worth protecting: "In the time of dawn . . . people have settled in places which they found most beautiful and lovely, and where they could make a living. In the settlements, Danishness has hardly penetrated. It is up to you to secure that Greenlandicness will live on" (A/G 1992, 10; my translation).

Both citations appear in a newspaper article covering the conference. Interestingly, the journalist remarks in conclusion that it is townspeople especially who represent the settlements as the spiritual strongholds of Greenland. They are the ones who romanticize life in the settlements, whereas people living in the settlements are down-to-earth pragmatics who are occupied with making ends meet. What they are interested in is local development, which, again, means grants. The article ends with a citation from an interview with another Home Rule Government representative, Emil Abelsen, who is also a townsman, but nevertheless says: "I cannot bear these romantics who praise the small settlement they have had to leave a long time ago, missing it so much.

However, they would never dream of going back because there is no flush lavatory, running water and other kinds of luxury" (A/G 1992, 10; my translation).

Forchhammer (1997; 1998) likewise shows how the settlements are culturally glorified, being used as symbols in a nationalist discourse. At the same time, however, people in the settlements often live miserable lives compared with the townspeople because the settlements are economically starved.[14] The cultural glorification is also somewhat at odds with the way the settlement population is often portrayed and treated by town-dwellers in everyday life (Forchhammer 1997, 148).

Even so, many national politicians and other prominent townspeople tend to represent the settlements as the authentic Greenland. The settlements are depicted as repositories of a cultural capital that transgresses local boundaries, being ultimately common property. It is stated, one way or the other, that townspeople have a lot to learn from settlement people, who as keepers of tradition are the last cultural stronghold in a world on the move.

In another newspaper interview, Josef Motzfeldt, a Home Rule Government member, says: "It is in the settlements that we shall find our roots and the original community of solidarity" (*Sermitsiaq* 1999a, 17; my translation). He also comments that the settlements especially suffer from migration by people with initiative. Yet he draws a gloomy picture of the urban destiny of these people: "They leave their houses in the settlement, and at the same time they face the almost insuperable problem of getting a place to live and a job in one of the magnetic towns—very often Nuuk. The settlements, however, cannot do without these fiery souls. They are often leading figures in the settlements, but in the towns they risk ending up in the same state of powerlessness and hopelessness as the settlements they left behind" (ibid.). To counter this general tendency, he suggests that the living conditions of the settlements must be improved through a targeted effort in order that even well-educated people would like to return or stay put and enjoy the special quality of life in the settlements.

In an election pamphlet, Prime Minister Jonathan Motzfeldt (1999, 9) points out that the settlement population is part and parcel of society. He also emphasizes how the calm surroundings leave their mark on the settlement population, and how a Greenlandic spirit of independence has existed in the settlements almost always. He concludes his appraisal by underlining that people in the outlying districts, the settlements, and the sheep-farming areas live a rooted life and are among the pillars of society. The emphasis on a Greenlandic spirit of independence originating from the settlements is repeated by Motzfeldt in another newspaper article, only this time he speaks of human freedom as a driving force: "For thousands of years the strong Arctic nature has ruled over our land, Kalaallit Nunaat. The Inuit have survived due to the experiences of their ancestors and respect towards the enormous forces that ruled over life and death. In our culture, human freedom and survival have always been a driving force" (Motzfeldt 2000, 8; my translation).

A tempting interpretation could be that neither the Greenlandic landscape nor its true inhabitants are possible to control and domesticate by foreign powers, no matter how hard they try. Such an interpretation seems not totally unfounded in light of the following quotation from Søvndahl Petersen, who likewise speaks of forces that are impossible to control: "Greenland has been under the Danish crown since the 1700's. Unlike so many stories of colonization around the world, the Danish colonization of Greenland was virtually bloodless. Yet it was typical of attempts by world powers to acquire a land whose geography and climate made it impossible to control" (T. S. Petersen 1994, 136).

Siting Culture

It seems that Greenlandic culture is not a straightforward matter in today's Greenland, where many people are involved in processes of finding a place for culture. The concepts of "siting culture" and "cultural sites" (Olwig 1997) may be useful in this connection. Based on empirical data from a West Indian community, Karen Fog Olwig shows how a strong propensity to migrate is counterbalanced by an equally strong notion of attachment to place. Most important, she challenges the idea that a sense of uprootedness or displacement fuels only imaginary worlds of ordered and stable lives that are long lost and gone. Her move is to substantiate the migrants' longings and attachment to place by showing how the global network of social relations is grounded in particular places through cultural institutions such as family land and the family house, which represent "cultural sites" in her terminology—that is, focal points of identification.

On the face of it, many of the often fragmented Greenlandic statements of cultural loss and longings for a more whole, simple, ordered, and stable life, represented by the Greenlandic settlement landscape, could be interpreted as sheer nostalgia. I believe that Olwig is right, however, in suggesting that we try to find out how such longings are part of the processes of actual social practice.

The Greenlandic settlements are not all long gone but still exist. Many town-dwellers have friends or family who live there temporarily or on a more permanent basis. The townspeople thus go visiting and the other way around, depending on people's economic situation. Most families also tend to be identified with, and most individuals identify themselves with, a particular place—town or settlement—that represents their ancestral point of origin. Family names are usually geographically anchored, even if individuals move around in contemporary Greenland or even leave Greenland to stay in Denmark or elsewhere. Special events, for instance, town or settlement jubilees, therefore tend to be major attractions, not just to former inhabitants but also to their children and grandchildren. They often turn into public events with local politicians and journalists attending as well. Major jubilees are also covered in the national newspapers, *Atuagagdliutit* (A/G) and *Sermitsiaq,* both of which tend to insert a full coverage of several pages.

An unveiling ceremony I attended in 1993 in a settlement in southern Greenland had attracted a good deal of former inhabitants and their offspring, who mostly lived in two nearby towns, the nearest of which can be reached by boat within about an hour. The event started out in a highly solemn spirit when people gathered in the church and afterward in the graveyard to commemorate one of the founding fathers of the settlement. Local hymns were sung and speeches were given at the site, where not only the principal speaker but also part of the audience burst in tears. After this emotional event, people went off in family-based groups before most met again in the evening to eat, drink, and dance in the community center. By this time the atmosphere had changed completely and spirits ran high. Since the reunion of 1993, it has turned into a tradition with people getting together in the settlement once a year.

Some people living in Nuuk have weekend cottages in the settlement of Kapisillit, which has about 120 permanent inhabitants, or in some of the old deserted settlements in the municipality of Nuuk. Qoornoq, which is one such old settlement, had about 80 inhabitants in 1970, but only two registered in 2000 (Statistics Greenland 2000, 263). Most houses have been kept intact; some are privately owned, while schools and other institutions in Nuuk have others at their disposal. The old school, for instance, belongs to *Meeqqat Illuat,* the children's home in Nuuk. In addition, new private houses are built in moderate numbers. In the summer, Qoornoq is a very lively place with people coming and going by boat. Thus people's ideas of a simple life close to nature are kept alive in practice. A woman in Nuuk often spoke about her plans to seek early retirement and go to live in her cottage in Kapisillit with her grandchild in order to give the child the opportunity of experiencing what she referred to as a more genuine Greenlandic way of life.

Weekend cottages are still for the relatively few. However, boat trips out in the fjords and countryside with family and friends also accord with people's ideas of Greenlandic sociability and culture. Such trips are often combined with hunting, fishing, and berry-picking in season. Eating Greenlandic food, such as newly caught fish, for instance, prepared on an open fire and served directly on a rock, forms an integral part of the trip. On the national day, many people in Nuuk have a family picnic in combination with the many public activities that take place.

The last example shows that it may not even be necessary to leave Nuuk in order to find a place for culture. The late Greenlandic painter and writer Hans Lynge is a case in point. In his large number of books about Nuuk, he never seems to deplore the fact that his native town has changed so dramatically over the years (Lynge 1978a, 1978b, 1978c, 1981, 1988). Although his narratives, which are based on collective memory and his own memoirs, concentrate on topography and life as it used to be, he nevertheless evokes a cultural landscape where the past is still part of the present to himself and other true locals. Detailed descriptions of nature, Nuuk, and the surrounding area, place-names, human activities, and stories and gossip about named people are all intertwined to produce a fascinating local universe (cf. Nuttall 1991 on the sense of locality).

Presenting himself as a *nuummioq*, Lynge (1978c, 6) explains that it means that, although he lives in Denmark for the time being, he is still in Nuuk because "that is how we feel always." Being a *nuummioq* is thus a condition rather than a mere description of place of birth or place of living. He opens the book with a tribute to his native town: "There is no place on earth as delightful as Nuuk. I praise the destiny that I have been born in Nuuk." His description of the landscape of Nuuk likewise reads as a tribute to a specific place of belonging: "From early childhood, an overwhelming impression of nature's grandeur, an impression of beauty, height, wide expanse and grandeur, remains deeply stamped on one's mind. In spite of the relentless climate, the abundant precipitation and long periods of stormy weather, the most pronounced impression remains one of indescribable beauty which is primarily due to the presence of the characteristic mountains" (Lynge 1978c, 6; my translation). Each of the mountains is then described in turn. *Sermitsiaq,* for example, which is the prime landmark of Nuuk, "towers in the background like an ally on guard" (ibid.).

The material presented seems to indicate that the issue is both about attachment to specific places and about identification of a true Greenlandic cultural landscape. In Lynge's case, the two are congruent as Nuuk is truly home, but others may have to search outside Nuuk for their focal points of identification.

The Pilgrimage: Coming Full Circle

A film from 1997, entitled *Heart of Light,* engages the issue of cultural identity in Greenland, focusing on the importance of recovering a true Greenlandic identity.[15] The film illustrates the gap between a culturally ambiguous town(scape) and an essentially Greenlandic landscape, which are finally brought together as a result of a hunting trip that serves the purpose of a pilgrimage.[16]

Heart of Light was launched as the first real Greenlandic feature production because it is based on a manuscript by a Greenlandic writer, Hans Anthon Lynge, in cooperation with the director, Jacob Grønlykke (Lynge and Grønlykke 1997). It was shot on location in Ilulissat and hired Greenlandic actors who spoke Greenlandic. The film therefore got a lot of publicity. According to both Greenlandic and Danish newspapers, the premiere audience in *Katuaq,* the culture center in Nuuk, was deeply touched by the way it broke the silence and engaged the problems of contemporary Greenland (see Pedersen and Sørensen 1999 for a review).

The main character is a middle-aged man, Rasmus, who has an understanding wife and two adolescent sons. Yet he seems troubled, is out of work, and tends to drink too much, which makes him a pathetic character who fails as a provider and father figure, and he is sometimes ridiculed in public. It soon turns out that his predicament is not due to an ordinary midlife crisis, but to a cultural identity crisis that is ultimately rooted in an unsolved childhood experience. Rasmus's deceased father was a politician who was collaborating with the Danish authorities and sympathetic toward the policy of "progress" during the late years of colonial rule. Progress meant cultural assimilation, and Rasmus remembers how his father brought him up to be Danish. This in itself may account for Rasmus's present dilemma, but in addition he recollects how his father made him sing a Danish children's song in the presence of a visiting Danish delegation. Allegedly, this early event made the boy sell out his Greenlandicness and sacrifice his true identity. He thus appears to have a "split cultural identity" that has plagued him ever since. And he tends to react strongly toward expressions of "mixed culture" in his everyday life, precisely because he himself does not have a firm and stable cultural identity.

Rasmus's crisis reaches a peak point in the wake of a family squabble that leads in turn to a tragic event: his eldest son drinks heavily and runs amok, shoots two young people at a party, and subsequently takes his own life. More than ever after this tragedy, Rasmus needs to be healed, and he seems to have known the cultural remedy for healing all along: a hunting trip on a dog sledge out in the Greenlandic landscape. His many previous failed efforts in this respect recur in the film. Although he is not really fit for this traditional activity and his gear is inadequate, he manages to do the right thing and even survive. Being out there in the uncontaminated Greenlandic landscape, he meets a *qivittoq*,[17] among others, who guides him on his therapeutic quest. Perhaps not surprisingly, his quest leads him to a small settlement that appears to be out of time. People live in small wooden houses and traces of hunting activities can be seen everywhere. Entering one of the houses, he sees an old dying man lying on a plank bed. In his feeble condition, this old man, who used to be a political adversary of Rasmus's father and a strong opponent of cultural assimilation, guides Rasmus on with small hints, making him finally unravel his traumatic childhood experience and realize that he did not obey his father back then. In fact, he stood up to him, singing a Greenlandic children's song instead in front of the Danish delegation. With this realization, Rasmus's search for his inner Greenlander (Sørensen 1994c) comes to an end; he has been healed. He even manages to catch seals before embarking on the trip back to town.

It seems that town and country are initially set apart in the film. The town represents an ambiguous, hybrid culture—exemplified by a Danish-speaking Greenlandic girl, and scenes from a discotheque and a bingo hall where the youngest son works—whereas the country, including the prototypical settlement out of time, represents a pristine Greenlandic space inhabited by spiritual beings, hunters, and animals. Cultural healing seems to involve a (de)tour into an unambiguously Greenlandic cultural space whereby the self is discovered through confrontation with the cultural "other," including the "other" in Rasmus himself (cf. Peacock and Holland 1993, 372). Once the cultural self has been (re)discovered, life in town is tolerable. The place has been transformed into a cultural space. Town and country have come together, so to speak. It is tempting to think of this movement in the cultural landscape as a Greenlandic version of the pilgrimage.[18] In any case, the main character returns as a new man, mature and serene, and he becomes reconciled with his remaining son and the community as such.

Retrieving Dormant Cultural Landscapes

So far I have shown how town, exemplified by Nuuk, and country, exemplified by the settlements as representing the true Greenlandic cultural landscapes, tend to be separated in conventional thinking. At the same time, however, I have also shown that the *nuummioq* may perceive things differently, and I have touched on processes that work against this general separation, pointing rather to the marrying of town and country. It seems that the pessimistic antiurban tales, which have been quite pronounced since the 1950s, are slowly giving away to a more positive view of life in town and its possibilities. This change of perception may be a reflection of the fact that Greenland has had Home Rule for more than twenty years. Town planning and development are domestic affairs, which makes it still harder to uphold a close association between towns and Danes or the simple idea that towns are alien entities that are inherently incompatible with a Greenlandic way of life. It also seems that the antiurban discourse that has been dominant among scholars and intellectuals, not just in a Greenlandic context but globally (cf. Finnegan 1998), has increasingly been replaced by more positive and varied perspectives of urban life.

In a recent New Year speech, Prime Minister Jonathan Motzfeldt stated: "Many of us have left our beloved settlements, but in most cases it has been a choice of our own. From old times we follow the game—break up and seek new opportunities. In modern society of today we migrate according to job opportunities. This is part of our living conditions" (Motzfeldt 2000, 9; my translation).

The emphasis now is on individual choice, which in turn is backed up by cultural tradition: we follow the game. In 2000 a political proposal for the establishment of four regional centers of power, involving the municipalities of Ilulissat, Sisimiut, Nuuk, and Qaqortoq, caused public debate. While some people saw the proposal as an onslaught on the settlements and were concerned that it would imply a stagnation of the development of the settlements and outlying districts, others believed that it was an economic necessity that need not have such dire implications (*Sermitsiaq* 2000, 13). In any case, the proposal may be seen as a tendency toward centralization.

The fact that Greenlandic politicians are in control of development and town planning means that neither is necessarily identified with colonial masters or Danes in general. Nuuk is not irretrievably Danish-like. Since Home Rule, Nuuk has, for instance, seen initiatives toward a Greenlandization of street- and place-names. The new blocks are kept low and tend to be in bright colors, which is widely assumed to be in line with Greenlandic tradition. In addition, the numbers of Danes living in Nuuk has decreased by about 750 since 1990.[19] A case in point when it comes to the issue of cultural retrieval of the towns is once again Jonathan Motzfeldt. He is clearly proud to have played a central role in building up a big, modern town, consisting of Nuuk and Nuussuaq, where people go to work in the morning: "When I look out of my window in the morning and see people on their way to work in Nuussuaq, I am proud. We have built a modern town as big as Ilulissat. When I see the towns along the coast, I note the big improvements that have taken place.... No one shall say that nothing has happened during the twenty years of Home Rule" (*Sermitsiaq* 1999b, 22; my translation).

The days are surely over when town planning and large-scale construction work could readily be identified with the Danes. Likewise, the pessimistic antiurban tales are on the wane. Motzfeldt speaks of Nuuk as a modern town that is the product of local effort. Nuuk may in fact be on its way to claim a status as a national symbol on a par with capitals in most other places.[20] Nuuk is often spoken of as the place where things happen and one is close to people in power and decision-making. This all makes Nuuk more vibrant and alive than most other places in Greenland.[21] The cultural status of places is susceptible to historical change, and the possible transformation of Nuuk is a case in point. Having usually been depicted in a rather unflattering light in the scattered references in the literature as well as among Greenlanders in general,[22] except for the *nuummioq,* it seems that Nuuk is in the process of cultural retrieval. The gap between the foreground and background of social life may still be present, but it is closing up.

Notes

1. "Town" and "settlement" are basically administrative terms. Greenland is divided into the subprovinces of West Greenland, North Greenland, and East Greenland. West Greenland is further divided into fifteen municipalities, while there is only one municipality in North Greenland and two in East Greenland (Statistics Greenland 2000, 12). Each municipality has a center or town and a number of settlements. Both towns and settlements are thus of varying size. In local speech, however, the concept of "settlement" entails notions of traditional culture and a particular way of life. From an outsider's point of view, Nuuk may seem like a settlement, although it is the capital of Greenland. On a local scale, however, Nuuk stands out vis-à-vis other towns and settlements given its large population and general townlike appearance.

2. My original fieldwork was part of a Ph.D. project that was financed by the Danish Social Science Research Council. The other project received support from the Danish Research Council for the Humanities. It dealt with Greenlandic discourses on culture and identity and was carried out in Nuuk and the Copenhagen area. Of the thirty-two interviews conducted, twenty

took place in Nuuk. Local students did part of the interviewing and transcribed them: Hulda Holm, Anders U. Jensen, Kistaaraq Vahl, and Tittus Grønvold. The citations that appear in this chapter have been translated from Danish by the author.

3. Several anthropological books on landscape and related issues were published in the 1990s. See, for example, Bender, ed., 1993; Olwig and Hastrup, eds., 1997; Gupta and Ferguson, eds., 1997; and Lovell, ed., 1998.

4. The spatial confinement of natives means that deviations tend to be seen as pathological (Malkki 1992, 31). The same tendency is often reflected in the case of Greenlanders in Denmark (Sørensen 1993a). They are not just regarded as being "out of place," but they are expected to *feel* "out of place," which shows in both textual and film representations. A recurrent theme is that Greenlanders, in addition to their general coping problems, have trouble breathing once they are placed in such a foreign cultural landscape. They suffocate and have fits of claustrophobia.

5. Greenland was a Danish colony from 1721 to 1953, when colonial status was abolished and Greenland became an integral part of the Kingdom of Denmark. Greenland Home Rule was introduced in 1979 in the wake of a decade of ethnopolitical mobilization (see Larsen 1994).

6. A printing house was established in Nuuk in the mid-1800s (Berthelsen 1983, 37–38). Before that time, reading material consisted mainly of the Bible and hymnals (Berthelsen 1988). One of the most significant developments was the publication of the newspaper *Atuagagdliutit* in 1861. According to Berthelsen (1988, 133–34): "It was in the columns of this newspaper that the ordinary Greenlander started to express himself in writing." *Atuagagdliutit* has been issued ever since. In 1952, it merged with the ten-year-old *Grønlandsposten* and soon was published in Danish as well (Kleivan 1970, 241). Consequently, the name was changed to *Atuagagdliutit/Grønlandsposten* or *A/G* in everyday speech. *Sermitsiaq,* the other national newspaper, started as a Greenlandic-language community newspaper in Nuuk in 1958, but later that year it had already become bilingual. It was named after a local mountain that is the prime landmark of Nuuk. In about 1980 it became a national newspaper. The editorial office kindly supplied this information. See also http://www.sermitsiaq.gl.

7. *Kalaaliussuseq,* which stems from *kalaaleq,* "Greenlander," translates as "being a Greenlander" or "the Greenlandic identity."

8. The Greenlandic self-designation *kalaaleq,* pl. *kalaallit,* seems to derive from Old Norse *skrælingr.* While *kalaallit* was used only in southern Greenland in the early twentieth century, it has since come to be used as the general word for Greenlanders in standard Greenlandic (Kleivan 1984, 620). Today's Greenland is officially known as Kalaallit Nunaat, "the land of the Greenlanders." *Kalaallisut,* that is, Greenlandic language, means "in the way of a Greenlander."

9. For the same reason, it can be easily appropriated by the handicraft industry, as demonstrated by the large-scale production of jewels depicting the map of Greenland, such as brooches, necklaces, charms, and earrings made of bone or tooth. The map of Greenland also appears on T-shirts, lighters, and a whole range of other souvenirs.

10. Jørgen Fleischer's recent article in *Sermitsiaq* has shown that there is some uncertainty about which of two songs should be regarded as the national anthem: *Nunarput,* which became famous nationwide in the 1930s, or *Nuna asiilasooq,* which became widespread during World War II. The first song is influenced by the turn-of-the-century thoughts about national awakening and progress, while the other is more decidedly a tribute to a particular landscape and its people (Fleischer 2000, 19).

11. Greenlanders have Danish citizenship. I use the term *kalaallit* in a broader sense than the strictly legal sense. Perhaps "(fellow) countryman" is more adequate. My point about reality

relates to the number of Danes living and working in Greenland, many of whom are married to or living with Greenlanders. See note 19 on recent statistics.

12. The article mentioned that Berthelsen's speech met with enthusiastic applause.

13. An interview with the Greenlandic actress Nukâka Coster-Waldau, who grew up in Greenland but now lives in Denmark, also touches upon her own childhood experience moving to Nuuk from a small town in Northwest Greenland (Honoré 1999). In addition to the small town–big town issue, it engages the contrast between the urban landscape of Copenhagen and the Greenlandic landscape: tower blocks, freeways, trains, and occasional claustrophobia are set against mountains and a wide view. The actress emphasizes what living in, and being at one with, nature has meant, and still means, to her personally.

14. Criticism of the actual settlement policy during Home Rule is occasionally raised in the newspapers. A case in point is an interview with Samuel Knudsen (Madsen 1994, 12), who accuses the national politicians of electioneering promises and compares them with colonial masters, only now they live in Nuuk. His point is that everything gets centralized in practice, in spite of the official pro-settlement policy.

15. The Greenlandic/Danish title is *Qaamarngup uummataa/Lysets hjerte*. My short presentation may not do justice to the film, and even less so because I use it in a certain context.

16. The term "pilgrimage" draws on Anderson (1991, 53, 55), who in turn is inspired by Victor Turner. Anderson speaks about both religious and secular pilgrimages. The term refers to the journey, between times, statuses and places, as a meaningful experience. Initially, I had used the term "Grand Tour" as a synonym for the Danish term "dannelsesrejse" ("Bildungsreise" in German), but Kenneth Olwig suggested "pilgrimage" instead.

17. Nuttall's description seems precise: "A *qivittoq* is a mysterious, supernatural wanderer, a person who has left the warmth and security of his home community to live alone in the mountains. In the past this was said to be a response to personal pressures and problems with others. The most common expression is that *qivittut* are mainly men who have been unlucky in love" (Nuttall 1992, 112).

18. The pilgrimage is not just an ideological construct that is confined to literary and artistic production. In the late 1970s, many young politically active Greenlanders went to live in a settlement for about a year with the explicit purpose of connecting with their cultural roots and learning "pure" Greenlandic. They were often employed as teachers in the local school (cf. Forchhammer 1997, 146–47).

19. The figures show that 3,898 persons "born outside Greenland" lived in Nuuk in 1990, compared with 3,153 in 1999, equivalent to, respectively, 31.9 and 23.9 percent of Nuuk's total population. Of the total population of Greenland, the number of persons "born outside Greenland" were 9,416 out of 55,558 in 1990 and 6,806 out of 56,087 in 1999 (Grønlands Statistik 1999, 431, 433). The statistics operate with the categories "born in Greenland," "born outside Grenland," or just "total population." The first category is mainly but not exclusively synonymous with Greenlanders. The second is likewise mainly but not exclusively synonymous with Danes, not least considering that about 12,000 Greenlanders live in Denmark on a permanent or temporary basis.

20. In a newspaper interview in 1994, the mayor of Nuuk, Agnethe Davidsen, revealed her ambitious plans for the future development of the town. She established definitively that Nuuk is the capital of Greenland, and that she expects it increasingly to make its mark on the international scene (*A/G* 1994, 11).

21. Teenagers from the settlements often complain about being bored because there are few activities and possibilities (see, for instance, *Paarisa* 1999, 4–5).

22. An early description from the 1840s by the Danish missionary Carl Emil Janssen (1913, 191–92) is illustrative. The Greenlanders in Nuuk are held in contempt, being depicted as half-civilized people who lack the distinctive *national* character of true Greenlanders. They go to work in the morning, dress in semi-Danish clothes, and live in semi-Danish houses. Altogether these cultural halfbreeds are contrasted with proud and independent Greenlanders in other places (cf. Sørensen 1994a, 48). Another example is a Danish civil servant's description of how he felt depressed when he got to Nuuk in the late 1960s and saw the "madness concentrated in this pseudo-town," which he characterized as a product of "abortive optimism" (Jacobi 1971, 109). He also commented on the concrete building blocks and cultural decline among the Greenlanders in general.

References

A/G. 1992. "En egentlig debat om bygdekultur udeblev." *Atuagagdliutit/Grønlandsposten* 67: 10.
———. 1994. "Nuuk er Grønlands ansigt udadtil." *Atuagagdliutit/Grønlandsposten* 82: 11.
Alonso, Ana María. 1994. "The Politics of Space, Time, and Substance: State Formation, Nationalism, and Ethnicity." *Annual Review of Anthropology* 23: 379–405.
Anderson, Benedict. 1991. *Imagined Communities: Reflections on the Origin and Spread of Nationalism,* rev. ed. London: Verso.
Appadurai, Arjun. 1988. "Putting Hierarchy in Its Place." *Cultural Anthropology* 3: 36–49.
Ardener, Edwin. 1982. "Social Anthropology, Language, and Reality." In *Semantic Anthropology,* ed. D. Parkin, 1–14. London: Academic Press.
———. 1989. "Belief and the Problem of Women." In E. Ardener, *The Voice of Prophecy,* ed. M. Chapman, 72–85. Oxford: Basil Blackwell.
Basso, Keith H. 1992. "'Speaking with Names': Language and Landscape among the Western Apache." In *Rereading Cultural Anthropology,* ed. G. E. Marcus, 220–51. Durham: Duke University Press.
Bender, Barbara, ed. 1993. *Landscape: Politics and Perspectives.* Oxford: Berg.
Berthelsen, Christian. 1983. *Grønlandsk litteratur: Kommenteret antologi.* Copenhagen: Centrum.
———. 1988. "Main Themes in Greenlandic Literature." *Folk* 30: 133–48.
Brønden, Jens. 1999. "Travelling Is Life Itself: Ono Fleischer Will Build Bridges Between the Inuit Peoples." *Suluk* 2: 20–23.
Christiansen, Thue. 1994. "Kulturen: Et fremmed element?" *Atuagagdliutit/Grønlandsposten* 87: 5.
Eriksen, Thomas Hylland. 1993. *Ethnicity and Nationalism: Anthropological Perspectives.* London: Pluto Press.
Fienup-Riordan, Ann. 1990. *Eskimo Essays: Yup'ik Lives and How We See Them.* New Brunswick: Rutgers University Press.
Finnegan, Ruth. 1998. *Tales of the City: A Study of Narrative and Urban Life.* Cambridge: Cambridge University Press.
Fleischer, Jørgen. 2000. "Hvilken nationalsang?" *Sermitsiaq* 11: 19.
Forchhammer, Søren. 1997. "Gathered or Dispersed? Four Decades of Development Policy Debate in Greenland." Ph.D. diss., University of Copenhagen.
———. 1998. "Kulturelle skatteøer eller marginaliserede områder? Bygder og bygdepolitik i Grønland." *Jordens Folk* 4: 11–19.
Geertz, Clifford. 1988. *Works and Lives: The Anthropologist as Author.* Cambridge: Polity Press.
Grønlands Statistik. 1999. *Grønland 1999. Statistisk Årbog.* Nuuk: Atuagkat.

Gupta, Akhil, and James Ferguson, eds. 1997. *Anthropological Locations: Boundaries and Grounds of a Field Science*. Berkeley and Los Angeles: University of California Press.

Gynther, Bent, and Aqigssiaq Møller, eds. 1999. *Kalaallit Nunaat: Gyldendals bog om Grønland*. Copenhagen: Gyldendal.

Hastrup, Kirsten. 1988. "Kultur som analytisk begreb." In *Kulturbegrebets kulturhistorie*, ed. H. Hauge and H. Horstbøll, 120–39. Kulturstudier 1. Århus: Århus Universitetsforlag.

———. 1991. "Eating the Past: Some Notes on an Icelandic Food Ritual." *Folk* 33: 229–43.

Hirsch, Eric. 1995. "Introduction. Landscape: Between Place and Space." In *The Anthropology of Landscape: Perspectives on Place and Space*, ed. E. Hirsch and M. O'Hanlon, 1–30. Oxford: Clarendon Press.

Honoré, Pernille. 1999. "Jeg er blevet et lettere menneske." *Ud & Se* (DSB) 4: 17–24.

Jacobi, Hans. 1971. "Triste tanker om Grønland." *Tidsskriftet Grønland* 4: 109–16.

Janssen, Carl Emil. 1913. *En Grønlandspræsts Optegnelser, 1844–1849*. Copenhagen: Gyldendalske Boghandel, Nordisk Forlag.

Kaalund, Bodil. 1983. *The Art of Greenland*. Berkeley and Los Angeles: University of California Press.

Kleivan, Inge. 1970. "Language and Ethnic Identity: Language Policy and Debate in Greenland." *Folk* 11–12: 235–85.

———. 1984. "West Greenland Before 1950." In *Arctic: Handbook of North American Indians*, vol. 5, ed. D. Damas, 595–621. Washington, D.C.: Smithsonian Institution Press.

———. 1988. "The Creation of Greenland's New National Symbol: The Flag." *Folk* 30: 33–56.

———. 1991. "Greenland's National Symbols." *North Atlantic Studies* 1, no. 2: 4–16.

———. 1996. "An Ethnic Perspective on Greenlandic Food." *Cultural and Social Research in Greenland 95/96*, ed. B. Jacobsen et al., 146–57. Nuuk: Ilisimatusarfik/Atuakkiorfik.

Larsen, Finn Breinholt. 1994. "The Quiet Life of a Revolution: Greenlandic Home Rule, 1979–1992." In *Sustainability in the Arctic: Proceedings from Nordic Arctic Research Forum Symposium 1993*, ed. T. Greiffenberg, 169–206. Aalborg: Aalborg University Press/NARF.

Lovell, Nadia, ed. 1998. *Locality and Belonging*. London: Routledge.

Lynge, Hans. 1978a. "Godthåb gennem tiderne." In *Godthåb—Nûk 1728–1978 i tekst og billeder*, ed. U. Kristiansen et al., 9–34. Nuuk: Sydgrønlands Bogtrykkeri.

———. 1978b. "Erindringer fra Godthåb." In *Godthåb—Nûk 1728–1978 i tekst og billeder*, ed. U. Kristiansen et al., 124–42. Nuuk: Sydgrønlands Bogtrykkeri.

———. 1978c. *Nûk: Hvad der i vore dage huskes fra Godthåbs fortid*. Nuuk: Det Grønlandske Forlag.

———. 1981. *Grønlands indre liv: Erfaringer fra barndomsårene*. Nuuk: Det Grønlandske Forlag.

———. 1988. *Grønlands indre liv II: Erindringer fra seminarietiden*. Nuuk: Atuakkiorfik.

Lynge, Hans Anthon, and Jacob Grønlykke. 1997. *Qaamarngup uummataa/Lysets hjerte. Filmmanuskript*. Nuuk: Atuakkiorfik.

Löfgren, Orvar. 1989. "Landscapes and Mindscapes." *Folk* 31: 183–208.

Madsen, Lone. 1994. "Tilskuere til udviklingen." *Atuagagdliutit/Grønlandsposten* 35: 12.

Malkki, Liisa. 1992. "National Geographic: The Rooting of Peoples and the Territorialization of National Identity among Scholars and Refugees." *Cultural Anthropology* 7, no. 1: 24–44.

MfG (Ministeriet for Grønland). 1974. *Grønland 1973: Årsberetning*. Copenhagen: Schultz.

Moore, Henrietta L. 1994. *A Passion for Difference: Essays in Anthropology and Gender*. Cambridge: Polity Press.

Motzfeldt, Jonathan. 1999. "Bygdebefolkningen er en integreret del af det grønlandske samfund." *Siumut: Inatsisartunut qinersineq/Landstingsvalget 1999*, 9.

———. 2000. "Mennesket det vigtigste ressource og råstof." *Atuagagdliutit/Grønlandsposten* 1: 8–9.
Nilsson, Eja. 1984. *Menneskenes land: 14 grønlændere fortæller*. Copenhagen: Københavns Bogforlag.
Nuttall, Mark. 1991. "Memoryscape: A Sense of Locality in Northwest Greenland." *North Atlantic Studies* 1, no. 2: 39–50.
———. 1992. *Arctic Homeland: Kinship, Community, and Development in Northwest Greenland*. London: Belhaven Press.
———. 1994. "Greenland: Emergence of an Inuit Homeland." In *Polar Peoples: Self-Determination and Development*, ed. Minority Rights Group, 1–28. London: Minority Rights Publications.
Okely, Judith. 1997. "Picturing and Placing Constable Country." In *Siting Culture: The Shifting Anthropological Object*, ed. K. F. Olwig and K. Hastrup, 193–220.
Olwig, Karen Fog. 1997. "Cultural Sites: Sustaining a Home in a Deterritorialized World." In *Siting Culture: The Shifting Anthropological Object*, ed. K. F. Olwig and K. Hastrup, 17–38.
Olwig, Karen Fog, and Kirsten Hastrup, eds. 1997. *Siting Culture: The Shifting Anthropological Object*. London: Routledge.
Paarisa. 1999. "Meeqqat takorluugaat/Børnenes drømme." *Paarisa* 13: 4–5.
Palmer, Catherine. 1998. "From Theory to Practice: Experiencing the Nation in Everyday Life." *Journal of Material Culture* 3, no. 2: 175–99.
Peacock, James L., and Dorothy C. Holland. 1993. "The Narrated Self: Life Stories in Process." *Ethos* 21, no. 4: 367–83.
Pedersen, Kennet, and Bo Wagner Sørensen. 1999. "Grønlandsk sønneoffer." *Tidsskriftet Antropologi* 39: 233–40.
Petersen, Marie-Louise Deth. 1986. "The Impact of Public Planning on Ethnic Culture: Aspects of Danish Resettlement Policies in Greenland after World War II." *Arctic Anthropology* 23, nos. 1–2: 271–80.
Petersen, Robert. 1984. "Greenlandic Written Literature." In *Arctic: Handbook of North American Indians*, vol. 5, ed. D. Damas, 640–45. Washington, D.C.: Smithsonian Institution Press.
———. 1994. "Grønlandsk etnisk identitet og voksende fællesskab." *Atuagagdliutit/Grønlandsposten* 37: 5.
———. 1995. "Lidt om grønlandsk etnicitet og identitet." *Atuagagdliutit/Grønlandsposten* 49: 2.
Petersen, Tove Søvndahl. 1994. "Superwoman and the Troubled Man." In *The Nordic Countries: A Paradise for Women?* ed. B. Fougner and M. Larsen-Asp, 136–41. Nord 1994, 16. Copenhagen: Nordic Council of Ministers.
Pratt, Mary Louise. 1986. "Fieldwork in Common Places." In *Writing Culture: The Poetics and Politics of Ethnography*, ed. J. Clifford and G. E. Marcus, 27–50. Berkeley and Los Angeles: University of California Press.
Riches, David. 1990. "The Force of Tradition in Eskimology." In *Localizing Strategies: Regional Traditions of Ethnographic Writing*, ed. R. Fardon, 71–89. Edinburgh: Scottish Academic Press; Washington, D.C.: Smithsonian Institution Press.
Rosendahl, Gunnar P., and Vagn Laage. 1978. "Byudviklingen siden 1950." In *Godthåb—Nûk 1728–1978 i tekst og billeder*, ed. U. Kristiansen et al., 143–69. Nuuk: Sydgrønlands Bogtrykkeri.
Rosing, Minik. 1999. "Landet." In *Kalaallit Nunaat: Gyldendals bog om Grønland*, ed. B. Gynther and A. Møller, 45–53. Copenhagen: Gyldendal.
Sermitsiaq. 1998. "1. maj præget af skolekrisen." *Sermitsiaq* 19: 17.
———. 1999a. "Sejrherren." *Sermitsiaq* 8: 17.

———. 1999b. "Alle skal tage en uddannelse." *Sermitsiaq* 51: 22.

———. 2000. "Bygderne siger nej til de fire storkommuner." *Sermitsiaq* 27: 13.

Sørensen, Bo Wagner. 1993a. "Bevægelser mellem Grønland og Danmark." *Tidsskriftet Antropologi* 28: 31–46.

———. 1993b. "Kapløb med tiden: Museale ideer om kultur og etnografi." In *Grønlandsk kultur- og samfundsforskning 93*, ed. C. Andreasen et al., 185–97. Nuuk: Ilisimatusarfik/Atuakkiorfik.

———. 1994a. *Magt eller afmagt? Køn, følelser og vold i Grønland*. Copenhagen: Akademisk Forlag.

———. 1994b. "Culture as Politics: Experiences from Greenland." In *Sustainability in the Arctic. Proceedings from Nordic Arctic Research Forum Symposium 1993*, ed. T. Greiffenberg, 207–16. Aalborg: Aalborg University Press/NARF.

———. 1994c. "Jagten på den indre grønlænder." *Kvinder, Køn & Forskning* 2: 53–68.

———. 1997a. "Når kulturen går i kroppen: 'halve grønlændere' som begreb og fænomen." *Tidsskriftet Antropologi* 35–36: 243–59.

———. 1997b. "Between Two Cultures? Notes on 'Halfies,' Culture and Identity." *Yumtzilob* 9, no. 4: 349–70.

Statistics Greenland. 2000. *Greenland 2000–2001 Kalaallit Nunaat: Statistical Yearbook*. Nuuk: Atuagkat.

Thuesen, Søren. 1988. *Fremad, opad. Kampen for en moderne grønlandsk identitet*. Copenhagen: Rhodos.

———. 1995. "Eksempel til efterfølgelse: Seminarieuddannede kateketer i Grønland i det 19. og 20. århundrede." In *Ilinniarfissuaq ukiuni 150-ini. Festskrift i anledning af Ilinniarfissuaqs 150-års jubilæum i 1995*, ed. D. Thorleifsen, 88–137. Nuuk: Ilinniarfissuaq/Atuakkiorfik.

Sweden

6.
The Swedish Landscape: The Regional Identity of Historical Sweden

ULF SPORRONG

Although the Nordic countries are often regarded as a uniform region, not least from the political point of view, there are considerable variations in the landscape, as we can see if we travel through the physical landscape or meet the landscape in literature, art, or music. Political stability has meant that few adjustments to national frontiers have been made in the past two hundred years. On the other hand, as distinctive regional features were allowed to develop in their own way, they often changed, with the result that these features were strengthened rather than weakened in the latter part of the twentieth century. One of the aims of this book about Nordic landscapes is to highlight the differences that exist in the Nordic countries as to how the landscape is perceived. The object of the introduction to the section on Sweden is to describe features that are characteristic of Sweden (Figure 6.1) and the Swedish landscape.[1]

If we begin with the physical landscape, we find one or two features that are almost unique to Sweden and Finland. One is the bedrock consisting largely of gneiss and granite; another is the fact that the country was covered in ice until about ten thousand years ago. Morphological research on glaciers has become of prime importance for knowledge of our land formations, which have largely been created by the action of the inland ice. Furthermore, the land was pressed down by kilometer-thick ice and is still in the process of rising again out of the sea. The coastline is therefore constantly changing, especially in the northern part of the country, where shore displacement in the course of one generation is clearly discernible. Here the land uplift is almost one meter in a hundred years and new land is constantly emerging. Charts must therefore be revised continually and the depth of shipping lanes checked. In the far south of Sweden, on the other hand, the land is sinking into the sea.

The most conspicuous effects of the inland ice are the soils it gave rise to. The bedrock, which is acid and resistant to weathering, eventually breaks down into poor till

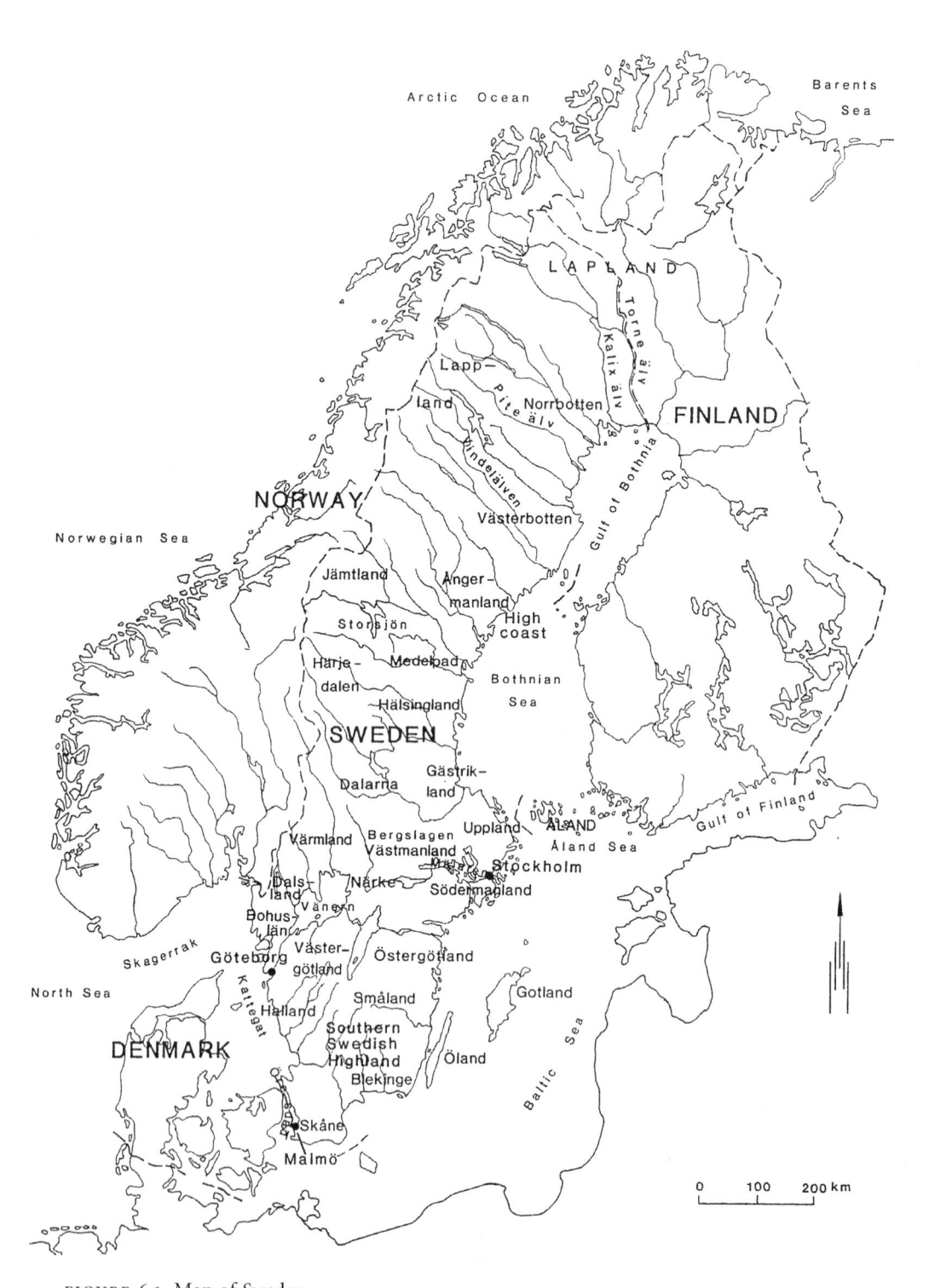

FIGURE 6.1. Map of Sweden.

with big boulders and large deposits of gravel and sand. The percentage of land under cultivation is consequently low in the whole country, about 7–8 percent, although it varies considerably from region to region. However, the land is eminently suitable for forests, mainly coniferous, although there are natural broad-leaf forests in the southernmost parts of the country. As much as 55 percent of the country is covered by forests and 35 percent consists of unusable land such as coarse-grained till, mires, lakes, and watercourses. Only a few percent is covered by settlements.

The bedrock in a few places, particularly in the south, is of a different nature, often calcareous and easily eroded. In these places the soil consists of clay till, and it is precisely there we find cultivated regions. One of the effects of the land uplift is that cultivable land is to be found along the coast. What was once sea floor is now dry land and consists of stiff clays. The level the sea reached when the land was glaciated—termed the highest shoreline—forms an important dividing line between coarse till and finer sediment. This line has been of great importance in the history of colonization: the landscape above the highest shoreline was generally not colonized until the Middle Ages or later.

If we look at the land as a whole, we see cultivation along the coasts and in those regions that are below the highest shoreline, particularly in a belt across the country from east to west. Nonetheless, settlements have long been found above the highest shoreline, too, especially where the soil consists of clay till or alluvial sediment. In the south lie the southern Swedish highlands, where cultivated areas alternate with extensive forests to form a "smiling" and varied landscape whose red-painted timber cottages with white corners are considered typical of Sweden. In the north are forest-covered highlands, which, as they approach Norway, become mountain ranges that are alpine in character. It is in Norrland that Sweden is richest in forests and it is here that we find undisturbed forest and mountain regions that are unique in Western Europe. Here, too, we have extensive mires and swampy land as well as lakes and waterways that give this region its untamed feeling. Several of Sweden's mightiest rivers flow from northeast to southwest through this wasteland. Most of them have been regulated to provide hydroelectric power, but four of the greatest are still undammed: from north to south, they are the Torne Älv, Kalix Älv (Figure 6.2), Pite Älv, and Vindelälven. Several smaller rivers are still untouched and have been given protection from exploitation.

To sum up, we can say that the Swedish natural landscape has a character of its own in several regions, as the conservation laws have recognized. Special legislation regulates the exploitation of the mountain ranges in the west, the untouched rivers in northern Sweden, and, in the south, the archipelagoes both in the Baltic and on the west coast toward the Skagerrak. This also applies to the two large islands, Öland and Gotland, in the southern Baltic. Some of the most valuable types of landscape have recently been included in the World Heritage List: the heath landscape (Sw. *alvar*) of southern Öland (Figure 6.3), the Laponia area in Lapland, and the High Coast. The latter is the area by the Gulf of Bothnia, where the postglacial land uplift has been greatest—in total more than 800 meters, which means that the highest shoreline or upper marine limit is now situated 286 meters above present sea level. In the same area, present land uplift is approximately 9 cm per 100 years. A remarkable coastal region has developed at sea level surrounded by 300-meter-high mountains that give the landscape a fjordlike character.

A Historical View of Swedish Cultural Landscapes

Although the frontiers of the Kingdom of Sweden have remained unchanged for almost two hundred years, adjustments to the frontiers with other Nordic countries were earlier fairly common. The geographical extent of Sweden was greatest around 1660, after it had gained provinces in the south from Denmark—Scania (Skåne), Halland, and Blekinge—as well as Bohuslän and Jämtland, which formerly belonged to Norway. Finland was also part of the realm, as were Estonia, Latvia, and certain areas round the Gulf of Finland in what is now Russia. When it was a great power, Sweden also ruled

FIGURE 6.2. The Kalix River, still undammed, in northern Sweden. The long island in the middle is former grazing ground and meadowland. Photo by Ulf Sporrong.

over areas in northern Germany. The country subsequently decreased in size and today only the former Danish and Norwegian provinces remain within its frontiers. However, they still have many of their original characteristics. Scania, in particular, resembles Denmark in many respects, for example, in how the natural landscape is used and in the history of its settlements. The Norwegian origins of Bohuslän and Jämtland are obvious to anyone with a knowledge of history.

The history of the colonization of Sweden is not yet fully known, but the main part of the population came from the south as the inland ice relaxed its grip on northwestern Europe. In recent years, however, it has become clear that there was an early immigration of people from northeastern Europe to the northern part of Scandinavia. They settled in the northernmost part of the country, around the Gulf of Bothnia. Because of the land uplift, all traces of the earliest settlements are high above sea level.

The original immigrants from the south settled along the coasts and in regions that were easy to cultivate and build on, where the soil and vegetation were suitable. The agricultural population early settled down on what is still the best farmland in Sweden: in Scania, Halland, central Västergötland, Östergötland, Öland, Gotland, and in the Mälar Valley, as well as in a few places in the north of Sweden. These early settlers were

FIGURE 6.3. The heath landscape of southern Öland, now a World Heritage Site known as Ölands Stora Alvar. Photo by Ulf Sporrong.

also interested in exploiting the country's natural resources such as forests and iron ore. Forest districts were therefore colonized early in the southern Swedish highlands and in Bergslagen, an early industrialized forest region in central Sweden, which produced iron, copper, and silver. These were minerals that the rest of Europe was interested in. It was not until after World War II that this core industrial region lost some of its economic importance.

The population is still concentrated in the regions that were first colonized. It is there that towns have grown up and there that industry has its strongest foothold. There are many lines of communication between modern centers situated on ancient farm land. However, this does not mean that different parts of the country are similar in appearance. The natural landscape and cultural traditions ensure that there are noticeable differences between one region and another—differences that the regional policies of the European Union (EU) in many respects reinforce.

We can trace the earliest division of the land into administrative regions to soon after A.D. 1000. This enabled several provinces or "landscapes" (Sw. *landskap*) to develop a system of self-government, each with its own laws and regulations. Several of these medieval provincial lawbooks have been preserved, and gradually, toward the end of the Middle Ages, they were codified into a common corpus when Sweden began to emerge as a united country. Until that time, each province had had an identity of its own, which, from the administrative point of view, was documented in its lawbook and so preserved for the future—we can at least say what their normative values were. These provincial lawbooks reflect a society in which regional differences could be quite considerable as far as administration was concerned. When Sweden became a centrally governed country in the sixteenth century, the independence of the provinces was reduced, except in one interesting respect. Even today, most Swedes associate their origins and identity with one of these old administrative provinces when they say who they are and where they come from. A person would say, "I am a Västgöte," rather than name one of the modern administrative regions into which Västergötland is included. The folklore movement and weather forecasts on radio and television also use the old names of these provinces, which are still in common use in everyday speech. It is almost impossible to describe the history of the Swedish landscape without using the old names of the provinces, thereby awakening associations with the old heartlands, which still have the power to move Swedes. Three provinces that, each in its own way, have been of great importance to the development of the country will be discussed in separate chapters. Over the centuries, they have preserved a distinctive culture, not least where their dialects are concerned: Scania has its eyes fixed on Denmark and the Continent, Värmland is a mining district with industrial and literary traditions, while Dalecarlia (Dalarna) has a folk culture that has gained recognition in the outside world and is not infrequently seen as a symbol for everything that is genuinely Swedish.

Any attempt to portray the history of the Swedish landscape must therefore be based on a knowledge of the old Swedish provinces. Folk culture, the development of industrial life, dialects, and patterns of settlement have fused together into something that helps to create identity. For many Swedes it is still a living force and is easy to recognize in paintings and literary works through the centuries. However, this may mean less to the new generation of immigrants to Sweden who come from a background with quite different cultural traditions.

The history of the Swedish landscape is deeply rooted in the man in the street. Sweden was a nation of farmers well into the twentieth century and, by European standards, industry developed late. Let us now look at some phenomena that, in an international perspective, lend luster to the country's historical development and that are still an important part of the cultural heritage.

Because the country is so sparsely populated, Sweden is unusually rich in prehistoric remains. Natural conditions have played an important part here; burial grounds and other remains from the Iron Age were placed on infertile land and other uncultivated spots. There is a burial ground in practically every village in southern and central Sweden, often adjacent to the settlement itself. The medieval lawbooks show a clear relationship between a person's right to own land and farm it and his ability to prove that he is related to "those in the tumulus." As a rule, it is only in the southernmost provinces that continuous cultivation and new layers of soil have covered most of the remains. The prehistoric remains that have managed to survive are all the more impressive. Descriptions of the open landscape in Scania have long been based on clearly visible Stone Age constructions or Bronze Age tumuli on high ground. In the northern part of the west coast, there are unusually rich and extensive finds of rock carvings—in the surroundings of Tanum more than fifteen hundred carvings have been placed in the World Heritage List.

Another phenomenon that is only to be found in the Nordic countries is the multitude of rune stones that were erected in the latter part of the Iron Age to commemorate the dead and often to record their right to land. This was the only way a society without any other written sources could document such things. These rune stones also tell of great voyages, from England and Ireland in the west to the Caspian Sea in the east.

On the whole there is no such documentation of how the prehistoric population in the north of Sweden lived, with the exception of those who lived in the coastal regions and who were in close touch with the southern part of the country. Instead, we find traces of a hunting and trapping culture with roots as deep as those in the south. The visible traces in the interior of Norrland are slight, often merely the foundations of huts, trapping pits, and faint indications of burials.

If we turn to the Middle Ages, it is buildings that come to mind as representing what is typically Swedish. As in the rest of Europe, this was a time when many impressive buildings were erected, not least churches, first of wood and then stone. In the

Mälar Valley there are still many well-preserved granite churches that look the same today as they did at the end of the Middle Ages. They are of great symbolic value. Another medieval phenomenon found widely in the Nordic countries is the shieling (Sw. *fäbod*, No. *seter*)—the summer grazing farm—which may also be of prehistoric origin. In the forest regions of northern Sweden, the inhabitants were compelled to make use of all available resources in every possible way according to the season. In summer they could use the forests for grazing and make dairy products. Few things are more closely associated with the romantic conception of country life in the olden days than shielings. Today, they are generally well-tended summer cottages.

Finally, let us look at another factor that is firmly rooted in history. Swedish ironworks are of great significance (Figure 6.4). In the seventeenth and eighteenth centuries, the iron industry underwent a rapid expansion and in many places proved to be economically viable, to a large extent because a skilled labor force immigrated early. Swedish iron became famous in Europe, above all in England. Ironworks and blast furnaces sprang up in a belt known as Bergslagen, which stretches across the country.

FIGURE 6.4. Gysinge ironworks on the Dalälven River in the province of Gästrikland, built in the mid-seventeenth century. Photo by Ulf Sporrong.

Here there were rich finds of iron ore, plenty of water power, and any amount of wood for charcoal-burning. In many places small manufacturing centers emerged in which the ironmasters achieved an economic standard unprecedented in Swedish industry. The favorable economic situation of the ironworks made them not only economic and technical hubs but also enabled them to attract a cultural elite. This was true of many of the finest ironworks in Bergslagen. One of those ironworks, Engelsberg's Ironworks near Fagersta, has been included in the World Heritage List.

Sources Used in Studying the Historical Landscape

Important historical sources that describe landscape are literature, art, and music. To illustrate this, we can take a few examples of landscape as portrayed by three Swedish authors whose fame has spread far beyond Swedish borders. They lived in different eras and had different conceptions of the Swedish landscape: Carl von Linné (Linnaeus), born in 1707, scientist and traveler; Selma Lagerlöf, who wrote in the late nineteenth century; and our contemporary, Astrid Lindgren, who wrote children's books. Lindgren's descriptions of the landscape are undoubtedly linked to what she experienced early in life, particularly in childhood.

The sharp-eyed explorer and scientist Linné was born in Stenbrohult, in the southern part of the province of Småland. He undertook many journeys and his diaries are available to us in print. Like most scientists of his day, his interests were wide and his knowledge was deep. Even though his energies were mainly devoted to collecting plants and natural-history specimens, many descriptions of people and the landscape are to be found in his writings. The following is his discription of the landscape between Malmö near Öresund in western Scania and Trelleborg, situated on the south coast:

> The countryside between Malmö and Trelleborg was the most delightful in the world, mostly resembling Flanders, for it was a plain, without hills, brooks, stones, rivers, trees and bushes; it gradually merged into even convexities in the length of a quarter of a mile that were barely perceptible; everything at that time was covered with the finest crop of winter rye and barley as well as some fallow fields left for grazing. It was remarkable to see that so many cows and horses, so many sheep and white geese together with pigs could find sustenance on those fields, which bore so little grass and where grain had grown the previous year.... Here and there amidst the fields there were small damp parks which bore some grass, and in them there were generally small pools where earlier inhabitants had dug peat.... The villages generally lay in these narrow vales, particularly if there was a spring, and in the middle of the village a shallow pond had generally been dug to which cattle, geese and ducks might go to drink. Most of the farms had gardens and white willows had been planted on the banks between fields. (Linné 1874, 156 [1759]; translated by Michael Stevens)

Selma Lagerlöf, who was a teacher as well as an author, was born at Mårbacka in Värmland. At the beginning of the century she was given the task of describing the geography and landscape of Sweden for schoolchildren. She chose to write a bird's-eye view of the country without ever having been in any sort of aircraft. The result was the book *Nils Holgerssons underbara resa genom Sverige* (Lagerlöf 1906–7), translated into English as *The Wonderful Adventures of Nils* (Lagerlöf 1947), about the boy who flew over Sweden, from south to north and back again to the south coast, on the back of a goose. However, Lagerlöf drew inspiration for most of her works from the people in the mining districts of Värmland and the forests of Dalecarlia. She was awarded the Nobel Prize for Literature in 1909.

The following quotation illustrates how people left their home villages in the province of Hälsingland in central Sweden to go to the shielings in the mountainous woods for summer grazing:

> The following day the boy travelled over Hälsingland. It spread beneath him with new, pale-green shoots on the pine trees, new birch leaves in the groves, new green grass in the meadows, and sprouting grain in the field. It was a mountainous country, but directly through it ran a broad, light valley from either side of which branched other valleys—some short and narrow, some broad and long.
>
> "This land resembles a leaf," thought the boy, "for it's as green as a leaf, and the valleys subdivide it in about the same way as the veins of a leaf are foliated." The branch valleys, like the main one, were filled with lakes, rivers, farms, and villages. They snuggled, light and smiling, between the dark mountains until they were gradually squeezed together by hills. There they were so narrow that they could not hold more than a little brook.
>
> On the high land between the valleys there were pine forests which had no even ground to grow upon. There were mountains standing all about, and the forest covered the whole, like a woolly hide streched over a bony body.
>
> A little later in the morning there was life and movement on every farm. The doors of the cattle sheds were thrown wide open and the cows were let out. They were prettily coloured, small, supple and sprightly, and so sure-footed that they made the most comic leaps and bounds. After them came the calves and sheep, and it was plainly to be seen that they, too, were in the best of spirits.
>
> It grew livelier every moment in the farm yards. A couple of young girls with knapsacks on their backs walked among the cattle; a boy with a long switch kept the sheep together, and a little dog ran in and out among the cows, barking at the ones that tried to gore him. The farmer hitched a horse to a cart loaded with tubs of butter, boxes of cheese, and all kind of eatables. The people laughed and chattered. They and the beasts were alike merry—as if looking forward to a day of real pleasure.
>
> A moment later all were on their way to the forest. One of the girls walked in the lead and coaxed the cattle with pretty, musical calls. The animals followed in a long line.

The shepherd boy and the sheep-dog ran hither and thither, to see that no creature turned from the right course; and last came the farmer and his hired man. They walked beside the cart to prevent its being upset, for the road they followed was a narrow, stony forest path.

The boy saw how processions of happy people and cattle wandered out from every valley and every farm and rushed into the lonely forest, filling it with life. From the depth of the dense woods the boy heard the shepherd maidens' songs and the tinkle of the cow bells. Many of the processions had long and difficult roads to travel; and the boy saw how they tramped through marshes, how they had to take roundabout ways to get past windfalls, and how, time and again, the carts bumped against stones and turned over with all their contents. But the people met all the obstacles with jokes and laughter. (Lagerlöf 1947, 414)

In many of her books for children, Astrid Lindgren has described the countryside in which she grew up, the province of Småland. An entire theme park based on her stories has been built in her hometown of Vimmerby. Her rural surroundings play a central part in many of her books. The action of her stories takes place against the background of buildings and farms around the villages, for example, in *Alla vi barn i Bullerbyn* (1948) or *Emil i Lönneberga* (1970), the latter translated into English as *Emil and His Clever Pig:* "If you have ever been in a forest in Småland on an early summer morning in June, you know just what it is like. You hear the cockoo call and the black birds whistle, you feel the softness of the path strewn with pine needles under your bare feet and the warmth of the sun on your neck. You walk along smelling the resin from the fir trees and the pine trees, and in the glades you see the white blossoms of the wild strawberries" (Lindgren 1973, 44).

There are three major sources for obtaining information about the history of the Swedish landscape. In some provinces, detailed information about individuals can be found in parish records from the end of the seventeenth century onward. For the country as a whole, official statistics exist from 1750 onward. These records show where each individual lived and all of his or her moves from cradle to grave. Records of the taxation of property go back even further. From the mid-sixteenth century onward, these taxation registers record landowners and all land transactions.

Some of the most valuable sources from the historical point of view are the old, large-scale cadastral maps (Figure 6.5). Farms and cultivated land were mapped on the scale of 1:5,000 beginning as far back as the first half of the seventeenth century. Not all parts of the country were included in the first mapping program, but, later, a map was made of every hamlet roughly every fifty years between 1700 and 1850, when the scale was even larger, 1:4,000. The reason for this long period of intensive mapping was the introduction of enclosure, which has thus provided us with a good knowledge of what changes in the cultural landscape took place between those years. These maps can give us information about continuity and discontinuity in the older use of land—information that is also of great value for research into the history of vegetation.

Modern Sweden

The history of the Swedish landscape explains features that are still essential to its character region by region. As in many other Western countries, however, a process of change began with the industrial revolution in the nineteenth century. Agricultural Sweden became a modern state, a development that, although it began relatively late, has progressed all the faster in modern times. The extensive, and often radical, land reforms of the eighteenth and nineteenth centuries set the ball rolling. Traditional structures of ownership and patterns of farming that were common throughout Europe were replaced by farms that made effective use of the physical layout of arable land and new methods of cultivation. The urge to mechanize agriculture led to the rise of a new industry that was quite different from the traditional ironworks—the engineering industry. It was situated in large agricultural districts, although it was still dependent on water power, as were other industries such as textile factories and sawmills.

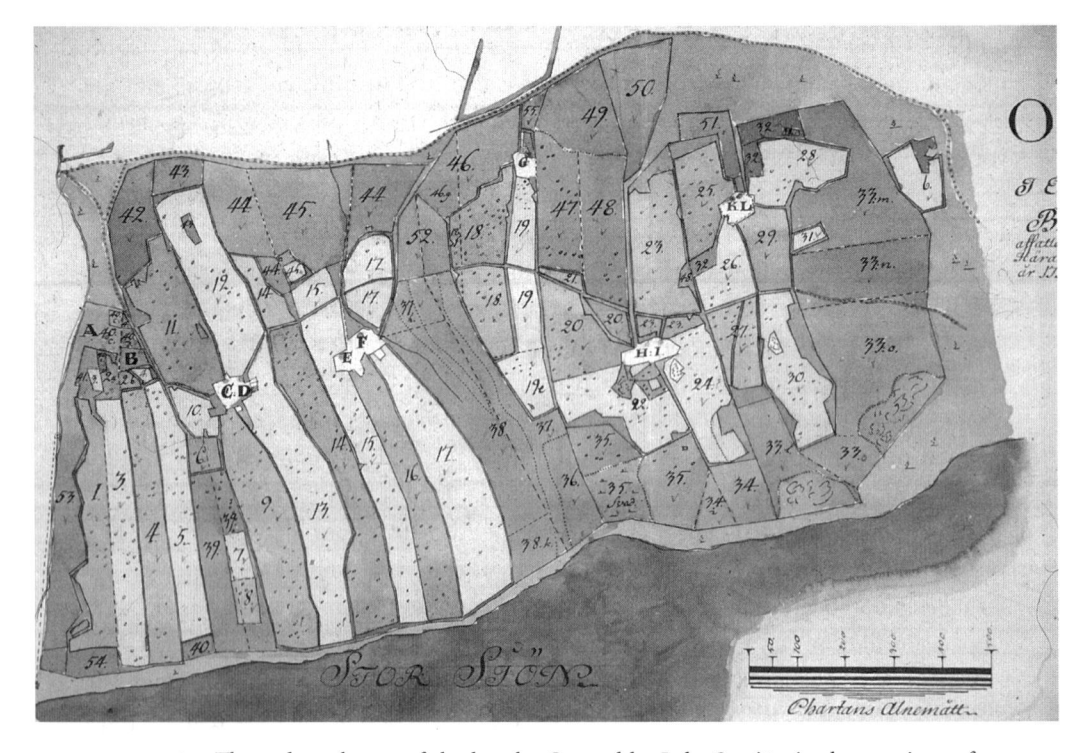

FIGURE 6.5. The cadastral map of the hamlet Optand by Lake Storjön in the province of Jämtland, c. A.D. 1750.

These land reforms allowed the creation of a number of new farms. The land-division reforms of the nineteenth century reshuffled the rural landscape, particularly the pattern of settlements, as farm buildings were moved out of the villages, in some areas as many as half. Abandoning the old farming methods also led to changes in land use; for example, when hay was grown on arable land, the extensive meadow landscape disappeared.

The reforms did not go far enough, however, in providing new opportunities in farming for the growing population. Much of the population increase was swallowed up by the wave of emigration, mainly to North America. Hardest hit were the forest regions in the west, which lost more than 10 percent of their population between 1800 and the 1890s. A second wave of emigration took place about 1920, with Dalsland and Värmland losing a large number of inhabitants. These provinces did not have the means to support them, largely because rationalization and mergers in the traditional ironwork industry reduced the need for labor. The eastern and northern parts of the country got off more lightly, mainly because the increase in population was not so great since large-scale agriculture was linked to primogeniture. This difference between eastern and western Sweden has deep roots and can still be observed in modern society.

A successive decrease in the number of farmers after 1920 resulted in a large-scale redistribution of jobs. Today, less than one percent of the population is engaged in agriculture. The number of people living in built-up areas, on the other hand, has grown. By about 1940, the urban population outnumbered the rural one. This tendency accelerated after World War II, when migration to towns and other built-up areas increased dramatically. The housing problem thus created was solved by launching what was, by Swedish standards, a gigantic building program in the suburbs of many towns. The architectural style was functional. The process of industrialization continued and, as early as the 1960s, labor began to immigrate, first from southern Europe and later from other parts of the world. These immigrants were to have a noticeable effect on Sweden's cultural life. In the Mälar Valley, around Göteborg and Malmö, the number of people born abroad often exceeds 20 percent; indeed in certain suburbs it can be as high as 40 percent or more.

The concentration of people in large towns and cities means that 85 percent of the population is tied to an urban environment and that the rural population must fight for survival. This is especially true in the north, where whole districts have lost their inhabitants as agriculture ceased and farms were abandoned. As forests replace farmland, the face of Sweden changes. Perhaps the traditional character of the Swedish landscape will soon be viewed by people with other values; the new urban generations will lose their links with the countryside, while new immigrants will bring a different cultural background and values.

The result of these processes, which have now been at work for about 150 years, has been the extreme concentration of the population in a few expansive regions, mainly the Mälar Valley and Östergötland, but also the west coast and southwestern Scania. Traditional industries emigrate and are replaced with high-tech firms engaged in education, research, or information technology. In other parts of the country, particularly in forest regions, the population is decreasing, the average age is rising, services are deteriorating, and opportunities for earning a living are steadily growing fewer. The government is attempting to mitigate these effects by granting large subsidies to sparsely populated districts and by redirecting income from wealthy municipalities to poorer districts. The EU's geographical target regions have developed special forms of support for the Nordic countries. In Sweden, more than 75 percent of the country receives subsidies in one form or another.

Is nothing dynamic happening outside the expansive regions? Yes indeed. First and foremost, forestry is being allowed more and more elbow room, at least where transport facilities are favorable. Swedish forests have an annual regrowth rate that exceeds the requirements of raw material for saw and paper mills. This cannot be regarded as purely positive since international competition is increasing and the surplus that is not felled is clogging the forests. Another factor that affects sparsely populated districts is the large-scale recreational facilities that are being opened, particularly in the southern mountain region. Furthermore, many Swedes own a summer residence. The demand for these houses is greatest where the population pressure is greatest, that is, along the coast, as well as in Swedish traditional heartlands, where more and more farm buildings are being converted into summer houses. Their owners often have a link to the district where the house is situated.

A third aspect of the heavy concentration of the population is that the sparsely populated districts are increasingly in demand for those seeking a physically active outdoor life. Forest districts, which were colonized late, were always sparsely populated. Large parts of inland Norrland have a density of fewer than four people per square kilometer. This means that much of Sweden can be regarded as a wilderness, at least from the urban-dweller's point of view. Particularly in the southern parts of the Swedish highlands and forest districts, for instance, in Värmland and Dalsland, people are attracted to an outdoor, open-air life. Large parts of the mountain ranges are also valuable for tourism. There are large areas in Lapland that cannot be reached at all by train or road, and the Swedish authorities are determined that it shall remain that way.

In Lapland, historically the Saami sphere of influence, the Saami have practiced reindeer husbandry for hundreds of years. Conflicts of interest arise as increasing numbers of people engage in outdoor activities. There have already been conflicts, and the Saami minority of about twenty thousand has reacted against the loss of some of the important rights that have their origin in a nomadic culture. Due to their cross-country vehicles, the Saami have become more or less domiciled. One serious threat to Saami culture is the danger of losing the Saami language. Saami has therefore been given an official status as a minority language within the Swedish realm, as has Finnish and *meänkieli* ("our language"), a variety of Finnish that is spoken in the Swedish-Finnish frontier district along the Torne Älv in the north.

Developments in the twentieth century led to great changes in living conditions as well as in the pattern of settlement. In spite of all action taken to support it, what remains of the old agricultural society will disappear except in places where it can be preserved as a kind of open-air museum. This will inevitably change the Swedish landscape. Agriculture will be concentrated on soils that give the best yield, forests will again spread out over the landscape, and most of the population will drift to a few relatively large urban regions. Young Swedes are still aware of the values that pertain to the historical landscape, but future generations with different cultural values may lose touch with the past. The next three chapters on Sweden will examine what lies behind the traditional views of the various Swedish provinces. We might say that we need the past to explain the present, or, to quote the Swedish author Jan Fridegård: "It is often the backward-looking eye that takes in the whole view" (Fridegård 1964, 3).

Note

1. An excellent source for detailed information is the seventeen-volume *National Atlas of Sweden,* particularly the volumes: *Agriculture* (Clason and Granström 1992), *Climate, Lakes, and Rivers* (Raab and Vedin 1995), *Cultural Heritage and Preservation* (Selinge 1994), *Geography of Plants and Animals* (Gustafsson and Ahlén 1996), *Geology* (Fredén 1994), *Landscape and Settlements* (Helmfrid 1994), *Maps and Mapping* (Sporrong and Weenström 1990), *Sea and Coast* (Sjöberg 1992), *The Environment* (Bernes and Grundsten 1992), *The Forests* (Nilsson 1990), and *The Geography of Sweden* (Helmfrid 1996). A regional description can be found in *Swedish Landscapes* (Sporrong, Ekstam, and Samuelsson 1995).

References

Bernes, Claes, and Claes Grundsten, eds. 1992. *The Environment. National Atlas of Sweden.* Stockholm: SNA Publishing.

Clason, Åke, and Birger Granström, eds. 1992. *Agriculture. National Atlas of Sweden.* Stockholm: SNA Publishing.

Fredén, Curt, ed. 1994. *Geology. National Atlas of Sweden.* Stockholm: SNA Publishing.

Fridegård, Jan. 1964. "Det karga Uppland." In *Uppland,* ed. Richard Holmström and S. Artur Svensson. Malmö: Allhems förlag.

Gustafsson, Lena, and Ingemar Ahlén, eds. 1996. *Geography of Plants and Animals. National Atlas of Sweden.* Stockholm: SNA Publishing.

Helmfrid, Staffan, ed. 1994. *Landscape and Settlements. National Atlas of Sweden.* Stockholm: SNA Publishing.

———, ed. 1994. *The Geography of Sweden. National Atlas of Sweden.* Stockholm: SNA Publishing.

Lagerlöf, Selma. 1906–7. *Nils Holgerssons underbara resa genom Sverige,* 2 vols. Stockholm: Albert Bonnier.

———. 1947. *The Wonderful Adventures of Nils,* translated from the Swedish by Velma Swantson Howard. Cambridge, Mass.

Lindgren, Astrid. 1948. *Alla vi barn i Bullerbyn.* Stockholm.

———. 1970. *Emil i Lönneberga.* Stockholm.

———. 1973. *Emil and His Clever Pig,* translated from the Swedish by Michael Heron. Stockholm.

Linné, Carl von. 1874 (1759). *Carl Linnaei skånska resa: på höga öfverhetens befallning förrättad år 1749: med rön och anmärkningar uti oeconomien, naturalier, antiqviteter, seder, lefnadssätt.* De skånska landskapens historiska och arkeologiska förening, 1. Lund: Gleerup.

Nilsson, Nils-Erik, ed. 1990. *The Forests. National Atlas of Sweden.* Stockholm: SNA Publishing.

Raab, Birgitta, and Haldo Vedin, eds. 1995. *Climate, Lakes, and Rivers. National Atlas of Sweden.* Stockholm: SNA Publishing.

Selinge, Klas-Göran, 1994. *Cultural Heritage and Preservation. National Atlas of Sweden.* Stockholm: SNA Publishing.

Sjöberg, Björn, ed. 1992. *Sea and Coast. National Atlas of Sweden.* Stockholm: SNA Publishing.

Sporrong, Ulf, and Hans-Fredrik Wennström, eds. 1990. *Maps and Mapping. National Atlas of Sweden.* Stockholm: SNA Publishing.

Sporrong, Ulf, Urban Ekstam, and Kjell Samuelsson. 1995. *Swedish Landscapes.* Stockholm: Swedish Environmental Protection Agency.

7.
The South of the North: Images of an (Un)Swedish Landscape

Tomas Germundsson

The Swedish author August Strindberg once described how a person from central Sweden, on his way home from the European continent after several years abroad, would apprehend the southernmost parts of his home country. Based on his own experience, Strindberg argued that leaving the lowlands of northern Germany, crossing the Baltic Sea, and eventually coming ashore on the Swedish mainland would, to the traveler's surprise, not result in a feeling of coming home: "From the shore, which gives the impression of a French rather than a German coast, the land stretches inwards in open plains, now smooth as a floor, now sweeping in lovely wave-lines, where the fields spread out like well-combed skin rugs.... It is a completely foreign landscape for a stay-at-home Swede from up-country, but for a widely traveled person it is a northern French landscape" (Strindberg 1985 [1896]).[1]

Travelers both before and after Strindberg have made similar reflections. This southernmost part of the country, the province of Skåne,[2] is not Sweden proper. An often-cited passage is by estate owner R. H. Stjernsvärd (1838): "Attached to Sweden is a small patch of land which is called Skåne, to show Sweden what the rest of Europe looks like."[3] The expression "attached to" is rather adequate because what on a map might look as a "natural" part of Sweden is historically and culturally something else. As a consequence of great-power politics nearly 350 years ago, the Danish province of Skåne (together with the provinces of Halland and Blekinge) was attached to Sweden by force of arms. This shift in nationhood has been of great significance for the identity and image of the province, and a Danish heritage is very palpable.

Skåne is one of Sweden's twenty-five *landskap*, which are historical provinces that often have their roots in prehistoric times and thus predate the Nordic nations. It is also one of Sweden's twenty-one *län*, which are the current regional administrative districts (counties), most of them originating from the seventeenth century. In some cases (as with Skåne), the *landskap* and *län* cover the same area. While *landskap* has a more cultural and "personal" tone, *län* has a more official and neutral one. If you ask a Swedish person which part of the country he or she comes from, a *landskap* will often be the answer. Yet if you look in the same person's passport, it will say in which *län* he or she was born.

Skåne is an undisputed and wholly integrated part of Sweden, but because of its turbulent history it is one of the provinces with the most pronounced regional identity (Tägil 1995). Dialects, cultural habits, and building traditions are examples of formative elements in the construction of this identity—and so is the shape of the regional landscape. A dominant theme here is the open landscape with its intensive farming. Skåne's agriculture is described in every schoolbook as prominent, productive, and important for the national food supply. Not surprisingly, the fertile plains of Skåne stand out as one of the standard images of the region, and the whitewashed, half-timbered farmhouse, square-built around a yard, preferably embedded in billowing grain fields, has turned out to be one of the most popular motifs representing the province (Figure 7.1).

FIGURE 7.1. "Greetings from Skåne," a typical postcard. Photo by P. Andersson. Published with permission.

Skåne also houses many of the country's biggest landed estates, with imposing castles and manor houses, often built in a manner not found elsewhere in the country. In pictures, paintings, maps, and literature with themes from Skåne, these landed estates are seldom far away. Also, the cities in Skåne have an image that differs from the rest of Sweden. Stone and red brick are more common building materials than elsewhere, and again the Danish and north German influence is evident. Some parts of Skåne, especially in the west, belong to the most urbanized areas in the country.

The picture of the province as a slightly exotic and un-Swedish part of Sweden is primarily applicable to the southern and western parts of Skåne, but it should be remembered that the province consists of several physical landscape types. In the north, coniferous forests dominate and give the landscape a character that does not differ much from other forested areas in southern Sweden. Thus, a traveler going from south to north through Skåne leaves what is generally apprehended as the typical regional landscape long before he or she leaves the formal region of Skåne, penetrating a former border district to the north. This is because of the geopolitical situation of the Swedish and Danish kingdoms when they began to appear as centralized nations in medieval times. At the same time as uninhabited or sparsely populated areas to the south, west, and north surrounded the core of Sweden in the area controlled from Stockholm, the Danish realm was taking shape on islands and coastal lands. The desolate areas that bordered Sweden to the south formed a barrier to Danish expansion to the northeast. From the perspective of the early Nordic nation-builders, this area constituted a peripheral no-man's-land. Gradually, however, the geopolitical spheres came into closer contact with one another in this border region. The exploitation of this area's resources of timber, bog-ore, wood and iron products, and taxes were of growing interest to the central power both to the south and to the north. Therefore, between the earliest boundary stones placed in the eleventh century, a borderline of national status began to wind around villages, through woods and bogs, and along rivers and roads (Karlsjö 1995).

While the construction of the border was in accordance with the logic of nation-building, it was hardly meaningful to the people living in the region. This was reflected in the "peasant peace-treaties" (Sw. *bondefreder*), concluded between many villages when the inhabitants suddenly found themselves living in two different countries (Johannesson 1984, 140). For centuries these local agreements were to confuse Swedish and Danish officers fighting for their respective armies.

Skåne's gradient of material landscape types is often noticed in descriptions of the province. One example is the map in Figure 7.2, found in a recent information folder presenting environments of great value according to the preservation authorities (*Skånska bygder* 1997). The areal differentiation observable in this often-used map is based on an ethnographic study where farming practices and material cultural traits from the mid-eighteenth century were identified and mapped. On the original map about twenty cultural districts were identified in Skåne, and these districts were generalized into three generic landscape types, called the Plain district, the Intermediate district, and the Forest district (Campbell 1928). In their main outlines these three districts coincide with Skåne's geological conditions, and they are also quite evident in today's landscape.

In the south, a flat and open agricultural landscape has expanded on lands where the bedrock consists of limestone, shale, or sandstone and where the melting inland ice left a layer of fertile, clayey till. In northern Skåne, the forest dominates, mainly consisting of coniferous trees. It grows on rather poor soils overlying bedrock consisting

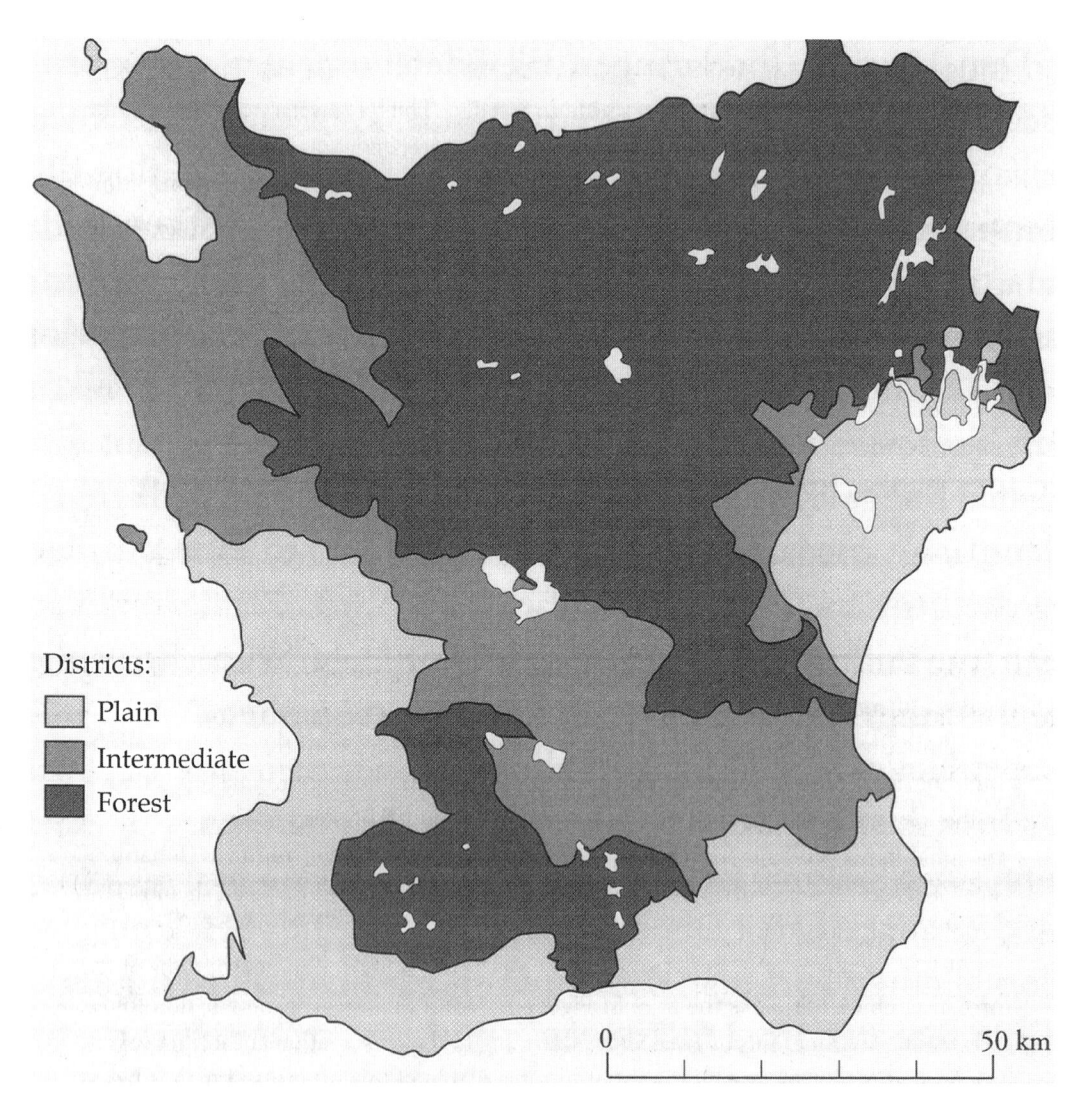

FIGURE 7.2. The three main cultural landscape districts in Skåne. Generalization of a map by ethnologist Åke Campbell (1928, 279), often used to describe regional differences in Skåne.

of gneiss and granite. The intermediate district is characterized by a topographically more varied landscape, where an old Pre-Cambrian basement alternates with other geological formations. The flatter areas have a sandier till than in the plain district, and in the intermediate district grazing lands and meadows have been much more widespread than in the plains. Historically, low and thin deciduous woods have dominated the vegetation in this district. Here many of the landed estates are situated and on their domains beech forest is a characteristic feature.

The way of representing the regional variation of the province's landscape shown in Figure 7.2 is grounded in field observations and surveys of artifacts and other culturally determined physical and visible imprints in the landscape. It expresses a discrete classification principle that has resulted in a static image of the province. This gives the map an impression of objectivity and neutrality, which I think is why it has been used in many different contexts. I shall come back to the question of how this representation of Skåne highlights or hides different aspects of landscape and land-life relations, but I present it here because I believe its frequent use has been of importance in forming the image of Skåne. This picture is not false, but there are also other dimensions.

This type of mapping is a common way of representing the differentiation of an area's landscape. On a national or Nordic scale, various examples can be found focusing on the natural preconditions and the morphology of the cultural landscape (Carlsson 1987; *Natur- og kulturlandskapet i arealplanleggingen,* 1987, vol. 3; *Kulturlandskapet och bebyggelsen* 1994, 66–77; Sporrong et al. 1995, 40–43). In such maps the landscape of Skåne is intersected by delimitations assigning different part of the province to different regions based on the physical landscape. This leaves us with a picture of Skåne as split up between different geographical realms, yet it is still a cultural area of its own. Carl Sauer esteemed the art of analyzing how areas are made up of distinct associations of both physical and cultural forms, thus creating a regional personality (see Crang 1998, 18). Sauer wrote that the cultural region would almost inevitably not map neatly onto the physical because the boundaries, rather than centers, of physical regions are likely to be centers of culture areas in order to use a varied resource base (Sauer 1963, 363 [1941]). Much of this is true for Skåne, but my aim here is not to try to make a refined map of such aspects. I will instead try to grasp some processes and moments in the region's history that have been of vital importance in forming its cultural landscape and its image.

This effort to give a historical background to what Skåne is in a Nordic context will involve two basic, interrelated meanings of the concept of landscape. The first is landscape as a historical province, territory, or area—in the case of Skåne evolving into a formal region (both *landskap* and *län*). The second is landscape as the expression of our interaction with land and environment, seen both as a medium for land-life relations, including such aspects as natural conditions, economy, social classes, and culture, and as the visible result of these relations: landscape as scenery, involving aspects such as symbolic meaning and landscape politics. These different meanings of landscape have a long and intricate history, and it is clear that the concept, not least in a Nordic context, is complex (Jones 1991; Olwig 1993, 1996b, this volume). While the double meaning of the word "landscape" may be unfamiliar to some English-speaking people, it has an ordinary and uncomplicated function to Scandinavians. This can be demonstrated by the common use of landscape pictures in Swedish elementary schools from the mid-nineteenth century. To learn more about distant areas and places, children should not only be able to locate them on a map but also comprehend them as homes of other people. Schools achieved this by using pictures. "Each picture was chosen to show a local landscape (meaning scenery), representing a wider landscape (meaning territory). In this way the two meanings merged" (Hägerstrand 1991 [1984], 4).

The discussion so far indicates that landscape can never be a neutral or objective entity. Questions concerning landscape are always situated in social and cultural contexts that have implications for how they are perceived, represented, and transformed (Mels 1999, 35–67). The materially real dimension of landscape "can only ever be experienced from within our perspective as embodied, socially shaped human beings using various modes of representation and performance" (67). With these insights in mind, I will try to give some images of Skåne as landscape.

The Earliest Landscapes

The name *Skåne* is thought to derive from *Skathinawjô*, an old Nordic word meaning approximately "the dangerous peninsula." The peninsula referred to is the one in the extreme southwest of Skåne. The name is believed to have been given by seafarers, who often encountered great difficulties here, since the sandy bottom constantly changes and new reefs are created. It seems as if the word describing the conditions near the coast eventually became a designation also for the land beyond the seashore. In a regional context, the old Nordic name was shortened to *Skaane* (or Sw. *Skåne*), but in its Latinized form it was changed to *Scandinavia* (Hallberg 1999). The common root of these two names, and the fact that Skåne gave its name to the whole of Scandinavia, reflects the contact between Norden and the rest of Europe.

The coast of Skåne became settled early on. Hunters and gatherers lived here and, when the inhabitants developed agriculture, fertile soils were abundant. As the Roman Empire expanded, an elite culture appears to have developed in southern Scandinavia. Roman products were imported to Skåne, both through trade and as gifts and tributes in regional contacts with Sjælland (Zealand) and Bornholm in present-day Denmark. Archaeological finds show that Skåne during the Roman era was at times well integrated in a stable network in the southern Baltic, while contacts at other times were more hostile (Larsson 1999).

In the fifth century, a marked change occurred as Europe entered a turbulent period on the dissolution of the Roman Empire. In Skåne, conflicts between different groups were common, and one reflection of the changes was that some habitations grew to a size that sharply distinguished them from other settlement forms. In Uppåkra, a few kilometers southeast of Lund, one such large settlement site has been recently found. It covers an area of about 50 hectares, and it was probably constantly inhabited up to the Viking Age. This is very rare, and stands in stark contrast to the mobile character of the typical settlement during these times. Uppåkra functioned as a center for what today is western Skåne, and it may have been a forerunner of the city of Lund, founded in the tenth century (Larsson 1999).

Much speaks in favor of the possibility of identifying a "Primordial Skåne," an ancient territory in the southwestern parts of present-day Skåne, into which more or less independent cultural areas—small "lands"—to the northeast gradually were integrated. An expression of this is that the "free men" of these small lands had to come to the assembly or "thing" (Sw. *ting*) that gathered somewhere in the vicinity of present-day Lund. Here disputes concerning various matters—not least questions about land rights—were settled (Bolin 1933, 69–84). Society became increasingly "territorialized," and some time during the ninth century it seems that Skåne could be identified as a "region" of roughly the same extent as today. It must of course be remembered that this region was of a very embryonic character and had no definite boundaries. We know, for instance, of no internal leadership for the whole of Skåne, but we can still distinguish it as a part of an early Danish kingdom at the beginning of the ninth century. A time of disintegration followed, but during the tenth century the kingdom was strengthened again. This was manifested on the famous runic stone at Jelling in Jutland, where it is stated that King Harald conquered both Denmark and Norway, and that he Christianized the Danes (105–12).

Parallel to this political development, a gradual a change in agriculture into a more permanent use of arable land and meadows resulted in a fixation of settlement in Skåne. The pattern of nucleated rural settlement started to take shape. The settlement was concentrated in low-lying areas, more specifically in dry parts near areas that were periodically flooded. Meadows were a renewable resource thanks to the natural flooding bringing in fertilizing nutrients. The settlement and the arable fields were situated on the dryer spots, but it should be remembered that the fields were small and that the most important localization factor was access to productive fodder-land (Fries 1963, 142–74). What in later times has been a labor-intensive rationalization, namely, the drainage of wetlands, would in those days have been a disastrous devaluation of land. Abundance of water was of the utmost importance. Today we find the villages in Skåne situated in the flat valleys surrounded by vast arable fields, and it is not unusual that the village centers and the oldest churches lie close to a small stream. Such a small rill of water might today appear very insignificant, but its potential as a water supply in a damming system should not be underestimated.

Although overlaid with centuries of changes, especially profound during the last two hundred years, the historical villages in Skåne are still one of the landscape's main features. The Swedish landscape writer Carl Fries gives this picture: "Such was the Skåne of the peasant and the grazing animals, gently and compliantly incorporated in the terrain, everywhere close to the water-rich meadow, the good and naturally dry arable soil, and the free outlying lands which gave the animals summer pasture. The Southern Plain itself was just as much meadow as arable; the low-lying green meadows along all the watercourses played a part in the picture which we today can hardly imagine" (Fries 1963, 153). Today the two original localization factors of nearness to meadowlands and water are invisible or diminished. Because of modern changes, a landscape character that once predominated in the province has become rare and exotic.

A Regional Landscape in Denmark

Beside the villages and agrarian settlements, there were constructions representing different forms of power. These included religious buildings such as village churches, monasteries, and cathedrals, as well as buildings representing secular powers, such as castles and state demesnes. King's farms (Sw. *kungsgårdar*) were large farms strategically dispersed, constantly prepared to accommodate the king and his escort whenever the political situation demanded his presence (Anglert 1995, 43–45).

These processes took place within a discourse of nation-building—not in the modern sense of the word, but in a more original one where nation could be connected to the meaning of "birth" and "lineage"; a group of people with reference to their geographical descent. Skåne was gradually integrated in the Danish kingdom from the tenth century onward. A new elite of powerful men closely connected to the king emerged. They built strongholds of power in castles, initially along the coasts of southwestern Skåne, and then representing the king in the emerging towns. The earliest towns in Skåne were inland, which shows that they had a more agrarian background than the later urbanization, which was more coast-bound. The most important city was Lund, soon becoming the "capital" of Skåne, and Denmark's most important religious center (Anglert 1999).

The societal changes expressed in the fixation of villages and the emergence of market places, churches, and cities made the struggle for power and control into an increasingly spatial and territorial question. Control of areas and land became more and more important and clearly affected the regional integration of Skåne (Schmidt Sabo 1997). Around the year 1000, under the leadership of King Knut (Canute) the Great, a Danish empire based on territorial consolidation and strengthened by successful Viking expeditions and trading became the leading power in Norden. In this empire

Skåne was now a well-integrated region. Fertile soils, relatively high population density, and a strategic position on the important trade routes in the southern Baltic area helped to consolidate this position. It is difficult to say when Skåne acquired a regional identity of its own, in the sense that the inhabitants identified themselves as *skåningar* ("Scanians") and felt that they belonged to a region outside the local community. Much suggests that this started to take shape in the period A.D. 1000–1100. At that time there existed a juridical authority for all of Skåne and the administrative hundreds were brought together in one Scanian unit. Beside the Scanian assembly we now also find a special official for Skåne appointed by the king (Sw. *gälkarlen*) and *Skånemarknaden* ("the Skåne Fair"), an important regional market (Skansjö 1997, 13–14).

The political project of making Skåne a recognized region was made possible by several circumstances. One of these was that the region could quite easily be comprehended as a "natural" territorial unit. The delimitation of this unit was the coastlines to the west, south, and east, and the vast forest area in the north. This geopolitical entity, in medieval times labeled *terra Scaniae*, seems to have been conceptually formed about a thousand years ago, even if it would take more than half a millennium before it was represented on a map in any recognizable form (Richter 1929). The forming of a regional identity in Skåne during medieval times was reflected in different ways. There occurred strong protests from both peasants and noblemen when Bishop Absalon, in cooperation with the Danish king, installed commissioners from Sjælland in Skåne in the twelfth century. The protests, which concerned the collection of tithes, were not only against an exploiting power as such. There were also clear signs of discontent because the commissioners came from another *land*, Sjælland (Skansjö 1997).

At the same time as the region of Skåne was consolidated during medieval times, more and more pronounced subregions reflecting land-life relations and the use of resources were emerging. In the south, an open and flat landscape took shape, while the forest was predominant in the north. Agriculture was organized in an open-field system of the same type as in other similar regions in northwestern Europe. The use of the landscape during this time could be seen as a variation on a theme that was common in Nordic landscapes. It could be expressed as a trinity of land use: good, naturally drained arable land; rich pasture used during summer; and productive meadows for feeding the animals during winter. The division of land for these purposes is reflected in *Skånelagen*—the Law of Skåne—written down in the fourteenth century (and also valid in the provinces of Halland and Blekinge): "Fences are set around arable fields and meadows to keep the livestock out and to protect the hay and grain, and therefore no one has the right to remove his own or anyone else's fences from the enclosure around the fields or meadows until the hay or grain is harvested and all the sheaves have been carried away from the fields" (*Svenska landskapslagar* 1979, xxvii). Such rules, here regulating land use, had grown out of practice and were codified in the Law of Skåne. As the law was valid for a certain region, this process could symbolize how negotiations and rules concerning social life, emanating from local practices, turned into general rules and an expression of power within a certain area.

A Classic Landscape?

The development of a village landscape through the process of settlement fixation started earlier than the spread of Christianity in Skåne, but soon these processes became parallel. Thus church-building and the administrative division into parishes became material manifestations of the changes in land-life relations, and as such they can be followed in time and on the land up to the present. During the eleventh and twelfth centuries, hundreds of parish churches were erected in Skåne, and for nearly a thousand years they have been one of the most characteristic landmarks in the region. It would be hard to produce a picture of a Scanian landscape without one.

The eleventh and twelfth centuries were also the time when the city of Lund and other towns started to take shape and urban life entered the region. Such a shift in practice, beliefs, and social life affected the relationship between humans and their material environment. One way of reading this could be with Olwig and his use of the writings of Virgil. With agriculture—the active cultivation of the soil—an earlier stage of pastoral economy was continued in parallel and eventually was abandoned as the main form of livelihood. In the first stage, the "character of the 'pastoral' environment, with 'no fence or boundary-stone' (Virgil, *Georgics*), was expressive of the cultural breeding of the community which created it" (Olwig 1993, 313). In the following phase, when agriculture became increasingly dominant, the use of land was different, at the same time as the community values of the pastoral predecessors to a great extent were preserved. It is perhaps from this classical view of the development of society that the original meaning of "culture" comes. Culture, then, is not seen as the opposite of nature. Instead it denotes a range of practices such as "inhabit," "cultivate," "protect," or "honor with worship." Culture became society's way of participating, via care (e.g., of the land), in a cyclical natural process in which the natural, inborn potential of society and its environment was made manifest. The next historical stage was the development of the city, which enabled a highly regulated exchange of products between areas producing different things.

This evolutionary story can also be interpreted in landscape characteristics, stretching from the wilderness to the garden (the agricultural landscape) and into the city. In this cosmology, both nature as perceived by humans and the cultivated land, the garden, are symbolized as female: Mother Earth as the source of new life through her periodical fertility. The male counterpart in this picture would be the sky (Father Air in Virgil's poem), but also the city. This is because the practice of the city represents the same kind of rational, mathematical, and geometrical knowledge as the knowledge of the nature of the heavens (Olwig 1993).

Even if the rational, geometrical aspect was more pronounced in the classical world and later during the Renaissance, I think this interpretation of the cosmological order—here without adopting its gendered framework—offers an interesting way of seeing Skåne's landscape in the eleventh century. During this time, when the garden, the agricultural land, and the city, and the developing towns with their commerce and centrality of power, occupied more and more land and developed a sort of infrastructure, there emerged an increasingly pronounced distinction between different land-use areas. Arable fields, meadows, and coppice woods became distinct cultivated areas at the same time as the forest, used as grazing grounds and a resource base, was increasingly involved in the spatially stable and permanent agriculture. The interdependence among arable land, meadows, pasture, and forest lands, which was a result of the production system's emphasis on cattle breeding, was typical of northern Europe. It distinguished this "barbarian" part of the world from the Roman system, where the limit between arable land *(ager)* and wilderness *(saltus)* was much sharper (Duby 1973, 1981, 30ff.).

One reflection of how the landscape in Skåne was a mixed landscape when it comes to land use and scenery, yet losing the wilderness, can be found in the church elite's efforts to re-create pristine nature and pastoral idyll in Skåne in the eleventh century. The blurred limits between wild nature and cultural lands, emanating from everyday practices and societal changes, gave rise to different symbolic landscapes and also a wish to create the lost or missing ones. An example reflecting this could be found near Dalby in southern Skåne, where a church and a monastery were founded in the eleventh century. Ten minutes' walking distance from the church and the adjacent monastery, a small forest, called Dalby Hage, is situated in the otherwise open agricultural landscape. It is known that the area has been used for grazing for a long time, and it has also been shown that it has had a very old connection with the monastery and church in Dalby. One plausible explanation of the origin of this area is that it once was enclosed and cared for as a representation of wilderness, or as the wilderness transformed into a pastoral landscape (Andrén 1997). In this way Dalby Hage could be understood as an Arcadian landscape, and as such was an ideal birthplace for lyric songs, poetry, and love. In this construction of a paradise lost beyond the city and the garden, wild animals and natural plants were allowed to live free (within fences), and humans (the monks and priests) could enter to get a foretaste of the heavenly paradise. As such Dalby Hage would be a Danish example of a medieval deer park, or *Tiergarten,* which were such "wilderness parks" that can be traced in many places in Europe. Finds supporting this theory are both human structures and bones from animals such as fallow deer, that is, animals that did not live wild in the area at the time.

Landscape changes and the example of Dalby Hage reflect some themes in the relationship between man, power, and landscape that can be followed up to the present. Although predominately agrarian, the province of Skåne (like Sjælland) has been comparatively highly urbanized since the time the cities started to emerge. The division between town and countryside has long since been manifest in both social classes and landscape, and the domination of the towns has meant that many expressions in the landscape outside towns and cities are based on an urban culture. The classic "Arcadia," for instance, both in its idyllic and its wild forms, "are landscapes of the urban imagination, though clearly answering to different needs" (Schama 1995, 525). Later this vision of the landscape can be seen both in the cities' use of their rural hinterlands for consumption and also in the wish to create landscapes of pleasure and recreation. By looking at both the historical landscape's real and symbolic dimensions, the dichotomies of urban-rural and culture-nature become situated in time and space and thus conceptually dissolved. One example could be the establishment of national parks in Skåne, where a Swedish, national(istic) discourse of preserving untouched nature and wilderness tends to overlook the fact that areas seen as "natural" have long been cultural landscapes (Mels 1999, 170–74). Dalby Hage, for instance, was in the early twentieth century thought to be the remnant of a natural southern Scandinavian forest, and it was proclaimed a national park in 1918. Two other national parks exist in Skåne: Stenshuvud on the east coast, established in 1986, and Söderåsen in central Skåne, established in 2001. As Mels has shown, these landscapes have a long history of human utilization. Yet in official descriptions and presentations, they are not only referred to as "nature," but they have also been "made wild" with the help of pictures, texts, maps, and practice (175–84).

Landscape with Cities

Developments during the medieval period were characterized, on the one hand, by increasing trade and stronger integration within different regions in northern Europe, and, on the other hand, by disruption and disintegration on other levels. The national attire in which the northern European countries were dressed during the preceding centuries turned out to be a misfit, as territories and nodes became increasingly important categories of power. Strategically important areas and centers crystallized. The German Hansa towns are examples of how power and influence were expressed in nodal geopolitical forms. Within this system the tradesmen in the northern German cities were entirely dependent on free passage for their ships through Öresund (the Sound, between Skåne and Sjælland) and the Belts (farther west). From their point of view, it was a disadvantage that these waters were situated within the Danish kingdom because it meant that one single nation could regulate commerce and shipping—and perhaps block it. A national border in Öresund was desirable for many actors in the region, and after a political agreement with the Swedes in the 1360s the dukes of Mecklenburg led an attack against fortifications in Öresund and on the southern coast of Skåne. The operation succeeded and the Danish hegemony was temporarily broken. However, it was soon reestablished and Skåne remained within the Danish nation.

Despite this, and despite the fact that the Nordic countries united in the Kalmar Union at the end of the fourteenth century, Skåne clearly stood out as a pawn in the geopolitical game in the Baltic region. This was to be the situation for centuries and a circumstance that was to torment the population in the province for ages (Åberg 1994).

Skåne's geopolitical position, the upswing of trade, and the increasing importance of the cities were factors that affected the landscape. The images of Malmö and Landskrona in the copperplates from the end of the sixteenth century could illustrate this (Figure 7.3). Here the cities lie as walled islands in the surrounding landscape. The wall and the moat are obviously motivated by military defense but also by other factors. Trade was a privilege of the cities and was allowed to take place only within the city walls. Craftsmen also belonged to the city and crafts could be made only outside the walls. The obligations of the cities to the surrounding countryside were expressed in a royal decree, stating that the city should hold "good tradesmen and that everyone can obtain, at a reasonable price, hops, salt, steel, cloth, boots, shoes, and other things that the nobleman and the peasant need and which should be on sale in the market

FIGURE 7.3. View of the city of Landskrona ("Landeskron"), c. 1580. Copper engraving in Georg Braun's and Franz Hogenberg's *Liber quartus urbium praecipuarum totius mundi.* Reproduced by permission of Lund University Library.

towns" (Johanesson 1984, 134). Outside the city, on the strait of Öresund, ships are sailing. The jetty extending out over the water is a connection with distant markets. On land-cultivated areas, both arable fields and gardens surround the city. The printed prospects were published in 1594 in a famous book by Georg Braun and Franz Hogenberg, and they are the first known pictures showing a total view of the cities.[4]

Seen together, the prospects give an image of a strong system of cities in the province, reflecting ambitions of influence in the city-based economy of northern Europe. In the individual pictures, the supremacy and power connected to the cities are reflected in different ways. The wall and the moat have already been mentioned, and it may also be noted that important buildings like the churches are emphasized (and sometimes "moved" a bit from their real position so that they could be better seen). Outside the city, windmills are dominating elements and may symbolize how the products from the countryside were refined and sent to the city. The gallows-hills were also important elements, including the gibbit, where the cut-up body of the hanged criminal was placed.

On the whole, the prospects give an image of a controlled and designed environment. Artists place themselves on an imagined height in the landscape and from this position they depict the city and its surroundings with great accuracy, as viewed by the individual eye. In such central perspective, figures could be palpably presented in space, which gives the picture a high illusion of reality. One could argue that the pictures are examples of how the relationship between measurement and vision was elaborated during the Renaissance era. The pictures could then be seen as a chorography, which, during this period, and as a heritage from Ptolemy, had the aim of creating recognizable images of the visible features of single parts of the oecumene (Nuti 1999, 90; see Olwig 1996a). This was in a way in opposition to the geographical aim, which sought mathematical abstraction and global knowledge. As this latter view in general was the one accepted as superior, the town prospects can be seen as an effort to ground the representation in both absolute measurement (the street layout) and in the representation of the visible world (a total vision of the town "as it looked"). Lucia Nuti writes: "It is indeed through the combination of the two 'pingendi rationes'—'geometrica' and 'perspectiva'—that Georg Braun claimed the superiority of the town images in his books" (Nuti 1990, 94).

The type of town pictures that are shown in Figure 7.3 became popular during the sixteenth century, and long afterward, and I think this reflects how the artists managed to elaborate a novel way of representing a reality that was of vital importance to the spectators; the emergent towns were becoming places of power, energy, and domination. Interpreted in this way, the apparently realistic and accurate scenes in the copperplates also stand out as a way of seeing the landscape that reflects how the cities during this era were integrated in the Danish nation's efforts to create an image of its power and strength.

The Landscape of the Estates

The great landed estates of Skåne mostly have a long history, and their origin can in many cases be traced back to the spatial expression of power and use of resources in early modern times. During this era the nobility in Denmark had a strong position, controlling, among other things, the regional administration. This was an important position at a time when the nation was extending its influence in Europe. Taxes were collected by region, and the army was also organized on this territorial basis. At the same time as the nobility had strong political and military positions, the development of the market gave them an opportunity to enhance their economic position. During the sixteenth and early seventeenth centuries, the prices of agricultural products rose, partly as an effect of increasing urbanization. This meant that the nobility could use their traditional, more or less feudal, privileges in a new era. While the estates were exempt from taxes, their owners had access to free labor. This was due to a system whereby the peasants on the farms within the estate domains had to work at the manor a certain number of days per year. This gave the nobility the opportunity both to produce food for a growing market and to create milieus and landscapes that manifested their position in the society. Traditionally many manor houses were situated in or near the villages, but during this time there are many examples of how new imposing buildings were erected in more secluded positions. Improved fortification and access to pasture and hunting grounds, together with a wish to mark a new social status, were some of the motives. The development meant that an "estate landscape" began to take shape, especially in the central parts of Skåne (Skansjö 1987).

One example of how the emergence of such a landscape was actively promoted by conscious and strategic actions of the estate owner may be found at the estate of Svenstorp near Lund. On this domain, known in records since the fifteenth century, the owner in the 1590s built a magnificent manor house with an impressive garden. At this time the land belonging to the estate was spread out, mainly in the two parishes of Igelösa and Odarslöv, but within a few decades the estate was very purposefully collected into a single territorial unit:

> Tenant farms in the vicinity of the estate were acquired by exchanges with the Lund Cathedral Chapter. . . . The cultivable lands of the manor farm had been organized in a way that must be described as cunning tax planning for those days: two of the manor's three fields lay within the bounds of Igelösa Parish and one in Odarslöv. In this way it had two parishes with peasants obliged to do day-labor, about 45 farms in all. Around 1650 the estate-owner at Svenstorp acquired the patronage of the two churches and thereby obtained the right to the use of a couple of the church's homesteads in the parishes. Around 1650, thus, Svenstorp was in complete control of both parishes. It was a closed estate unit, virtually a state within the state. (Skansjö 1997, 161)

During the capitalist transition, the status of land was uncertain: "Its redefinition, from use value to exchange value, was a long and hard-fought process.... For a long period land was *the* arena for social struggle: it was both status and property" (Cosgrove 1984, 63), and it thus had a certain artistic and moral force. In his studies of the English landscape in the seventeenth century, Cosgrove finds different aspects of this force. One is the analogy between the rural estate and the state of the realm—just as in the case of Svenstorp: "A well-managed country house and its land form a self-sufficient world, a microcosm of the mercantilist state." Here "harmony rests ultimately on its subordination to the care and authority of one all-powerful lord" (196). This power was expressed in the physical landscape and also as a way of seeing, for instance in estate paintings typical of the times, where "painting, language and nature become transferable one with another in their subordination to the 'lordship' of the eye" (194). Besides the pictorial technique of realist representation, the perspective in such a painting carries another sense, namely, the one "in which we use the word when we tell someone to get a problem 'in perspective.' In this second sense, the perspective meant the correct way of seeing the social, moral and political order of the world" (ibid.). This is illustrated in Figure 7.4, showing a painting of the estate of Maltesholm in eastern

FIGURE 7.4. The Maltesholm estate in Skåne (artist unknown). The date of the poem on the print, written by a physician visiting the estate, is 1638. Photo courtesy of Elvy Engelbrektsson, Malmö.

Skåne in 1638. Here the artist has represented the estate owner's efforts to arrange nature into a harmonious and fruitful order. Included in the picture is also a poem in which the estate itself speaks to its guests *("Malthisholm al hospites")*. After telling how this place—wild, gloomy, and difficult to access—had formerly been an awkward residence for three peasants, the estate pays homage to its new owner: "He scrutinized my location more closely, considered what I could become by and by, found pleasure in the crystal-clear water which I pour out with a pleasant and constant ripple—the only advantage I possessed by nature—, he began to make roads, fields, pastures, dams, and the buildings you now see on the hillside, surrounded by water, and finally he named me after himself. How shall I repay this gentleman who has shown me such honor?" (Cited in Wåhlin 1931, 52).

In a broader context, the shift in the way of seeing the landscape may be said to be a part of the elite culture that emerged in Europe during the Renaissance. As indicated above, new discoveries of how the outer world ("nature") was constructed were an important ingredient in this change. Progress in science led, among other things, to the identification of a higher ("divine") order in nature, which in turn meant that the landscape could be understood as a reflection of this order. In paintings and poetry, the landscape was often represented as an orderly arrangement of objects (Barrell 1972). As a consequence, a new meaning of the concept of landscape was gaining ground, denoting a view over land. This scenic concept of landscape was different from the traditional meaning, which was more connected to custom, territory, and community (Olwig 1996b). The poets, for instance, using words like *view, prospect,* or *scene* interchangeably with the word *landscape* made "the land something out there, something to be looked at from a distance, and in one direction only" (23). This attitude of "looking over" the landscape had the result that the artist "manipulated the objects in them, simply according to the rules and structures sanctioned by a pure and abstract vision, and without any reference to what the function of those objects might be, what their use might be to the people who lived among them" (59). When the ruling nobility adopted the new concept, it could be used as a way of exercising power. It was linked to the capacity to regard society in an objective and rational way. "It also supported the 'natural' authority of upper-class men over other men and women in general," writes Susanne Seymour on Britain, and concludes: "The ability to see landscape in particular ways became part of the construction of the position of the aristocracy and newly developing middle class, within the countryside, within state politics and thus also the position of the labouring poor" (Seymour 2000, 200). I would argue that this applies to southern Scandinavia as well. In the landscapes of the landed estates, these circumstances and changes had quite pregnant expressions in Skåne. Their landscapes are reflections of a hierarchical society where almost everything was put in place by the estate owner. This is also true for different representations of the landscape, and has a bearing in modern times, for instance, in estate paintings, pictures, and maps, so often giving the "prospect" view of the estate. The "community" experience of the landscape is mostly hidden or tacit, and it is only in later times that landscapes and histories of laborers, crofters, and other similar groups have been made visible, for instance, in literature, art, museums, and planning. Yet still the estate landscape, and representations of it, is a socially stratified landscape. To understand its roots, it is important to remember Olwig's call for a substantive understanding of landscape's tensive dual meaning. It is not enough to study landscape as a scenic text; one must also recognize the historical and contemporary importance of community, culture, law, and custom (Olwig 1996b, 645). I will return to the landscape of a particular estate in Skåne in connection with some questions concerning "modernity," but first I should describe another formative moment in the province's history, namely, its shift of nationhood.

From Danish to Swedish

Skåne continued to strengthen its regional position and identity during the fifteenth and sixteenth centuries. The combination of a city-based power structure and a hierarchical national administration helped to shape Skåne into a more pronounced province of Denmark. However, during the early seventeenth century, an imperialistic Swedish policy of expansion was about to change the power balance in the Baltic area. The direct background was that the Danish king declared war on Sweden in 1657, seeking revenge for provinces lost to Sweden in an earlier war. At this moment most of the Swedish troops were involved in operations in Poland, and the Danes hoped to take advantage of this situation. However, from their position in Poland, the Swedish troops turned west, then north, and marched through southern Jutland. Due to a harsh winter, the troops were able to cross the straits of the Little Belt and the Great Belt on the ice, and in December 1657 Swedish troops attacked Sjælland—the heart of the Danish realm—from the west. The hazardous operation succeeded. The Danish troops were forced to surrender and, by the peace treaty in February 1658, Skåne, together with the provinces of Halland and Blekinge, became Swedish. That same autumn the Swedish king tried to conquer the rest of Denmark in order to establish a Nordic union. However, the Danish resistance was hard, and a fierce Swedish attack on Copenhagen was fought back. A new peace treaty in 1660 did not change much of the situation in 1658.

The process of implementing Swedish supremacy in Skåne, and thereby making the province "Swedish" after it was conquered in 1658, was at first quite militant. Skåne was a valuable prey for the Swedish state. It was rich country with prosperous agriculture (at least potentially). It was strategically an important piece of land, as its incorporation meant that the monopolistic Danish control of the entrance to the Baltic Sea was broken. Therefore several measures were taken to deliberately erase the "Danishness" of Skåne. In schools children were forced to learn Swedish, and in the churches the services had to be held in Swedish. Instead of singing their traditional hymns, people had to mumble along in a new language they did not quite understand (Åberg 1994).

Toward the nobility, the tactics were somewhat different. The Swedish government let this class keep many of its earlier Danish privileges. This measure facilitated the acceptance of Swedish supremacy because it occurred at a time when the Danish king Frederik III was strengthening his position by becoming an absolute ruler, thus diminishing the nobility's influence in Denmark. The Danish-Scanian nobility was also blended with native Swedes from the upper social strata. This was made possible by a maneuver in which the island of Bornholm was given back to Denmark in exchange for land in Skåne owned by the Danish nobility. On these lands a "new" Swedish nobility was installed. Another important step in this process was taken in 1668, when the University of Lund was established. Now the military system, the civil administration, part of the nobility, the leaders of the church, and the education system were Swedish, but not yet the mass of people—and far from all their leaders. There were many protests, and in combination with Danish efforts to recapture the province, Skåne was put in a state of war, which was to last with varying intensity for about forty years. Several bloody battles were fought, and in some parts of Skåne guerrilla warfare was waged against the Swedes. The punishments for such actions were hard and many of the rebels were killed. The last Danish attempt to invade Skåne was in 1709–10.

In a historical perspective, the effectiveness of Swedish supremacy, combined with a wish by a large proportion of the population to live in peace after a long period of wars, made the "Swedification" of Skåne a relatively short and definitive process. This was largely a political process in that for most common people the shift in nationhood was probably not a deep personal question. Rather it was seen as an authoritative command to change loyalties (Linde-Laursen 1995, 52–65). The fact that the struggles in Skåne in the decades after 1658 were extremely bitter and ruthless should not be hidden. During the change of nationality, thousands of people had their lives and their world totally destroyed. Homes and farms were burned and plundered by Swedish soldiers, as well as by gangs of robbers—and sometimes also by Danish soldiers, when the logic of war demanded it. The last decades of the seventeenth century were a terrible time in the history of the province.

The strongholds of the resistance were the forests in northern Skåne. The groups that put up armed resistance were called *snapphanar*. Most of them were peasants and crofters from the forests, but burghers and priests also joined in. As the wars went on, many of these people lost contact with their civilian lives and some of them formed gangs of robbers. One explanation for the word *snapphane* is that it is an old Swedish term of abuse originating from the German *schnappen*, meaning "steal" or "rob" (Skansjö 1997, 183). Another explanation is that it comes from a Dutch word meaning "pecking bird," which is a description of the cock on the type of guns that the guerrillas used (Lauring 1964, 120). In its Scanian context the latter meaning of *snapphane* is probably a construction from later times, with the aim of reevaluating its negative tone.[5] In any case, the occurrence of both explanations reflects different views of the resistance.

Up to the present, stories and myths of the *snapphanar* are vivid, and they are often connected to the physical landscape where the typical *snapphane* belonged—the forestlands. Life here was different from life on the plains in the south. Work for a living followed a different rhythm than that of the intensive agriculture on the plains. Boundaries were less obvious, and most people moved daily or periodically over longer distances. There was a different way of adapting to nature in order to collect wood, berries, and mushrooms and to catch animals. Many of these stories about *snapphanar* describe how the landscape creates and forms the people: the wildness and the uncertainty in the forests become an attribute of the people living there. The picture of the *snapphane* as an armed woodsman suddenly emerging out of nowhere is often repeated and at times contrasted to the more earthbound and slow peasant from a village on the plains.

Many of the historical traits in Skåne's landscape are Danish. The old churches, castles, official buildings, and national monuments from the mid-seventeenth century are Danish, with inscriptions in Danish and so on. Farmhouses and other constructions are also Danish in the sense that they reflect a cultural realm that had its center to the south and west—not to the north.

Different Landscapes

When the political and military struggle was over at the beginning of the eighteenth century, Skåne in many respects was an impoverished region. It slowly recovered, though, and from about 1750 onward the population increased. Peace, a growing population, a growing market, and improved means of transportation stimulated intensification and areal specialization. A "regionalization" in the use of resources became increasingly palpable in Skåne. The historical sources—such as tax records, tithe records, and maps—testify that the three different districts referred to in the introduction took a more definite form (see Fig. 7.2). As mentioned, an open and flat landscape took shape in the south, and agriculture was organized in an open-field system of the same type as in other, similar regions in northwestern Europe (Frandsen 1983, 11–38). The villages were large and could have more than forty farms. An ecologically important division of land in the village was between infields and outfields (Dahl 1942, 104–14). The infields were arable and meadow, fenced in to protect them from the cattle. The manure from the cattle was essential. The arable land was utilized in a three-year rotation system. Two-thirds of the area were sowed each year, while one-third lay fallow. Each farm's arable land was found in narrow strips in all parts of the village's fields. The outfields were woods and pasture.

This basic division had been maintained for ages, but the proportions had changed. With the aim of increasing grain production, the infield areas were extended to the detriment of the outfields. During the eighteenth century, the outfields almost disappeared in the plain district. Instead the cattle grazed on the fallow, but it is obvious that the system was under heavy negative stress in many areas in the eighteenth century because of lack of manure. Here woods and trees became a rarity—almost all land was open (Emanuelsson et al. 1985, 98).

In the northern, forested areas, the villages were small and outspread, and there were also many single farms. Arable land was sparse and lay in isolated openings in the forest. The forest was the basis for the economy, producing timber, charcoal, tar, and potash. In central Skåne, the intermediate district, also called the "brushwood district" (Sw. *risbygd*), took shape. The name probably alludes to the appearance of the important grazing and meadowlands when they were periodically cleared to increase productivity. The denomination is common in tax records from the eighteenth century. When a village was assigned to this region, it meant that it could not fully pay its taxes in either grain or forest products. Cattle breeding was important and the sale of livestock was a common source of income for the villages. Records show that common people called this area the "forest rim" (Sw. *skogskanten*) or "the land between the forest and the plough" (Campbell 1928, 14).

In Skåne, these districts were complementary and the reason they could exist was the exchange between them. Each of the main products—grain, cattle, and wood—could be produced with a surplus in each region, while there was a shortage of the

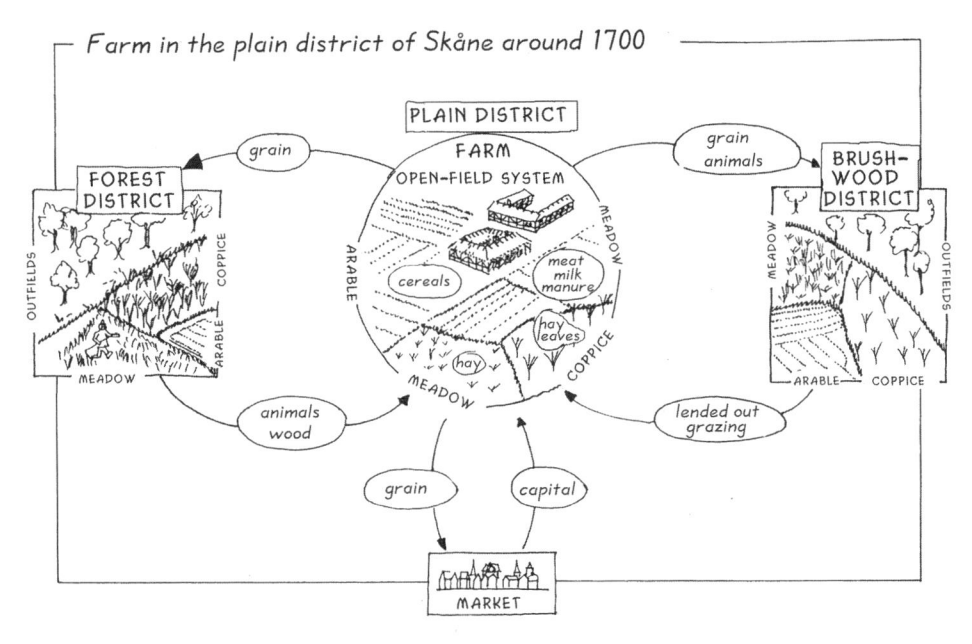

FIGURE 7.5. Drawing by Ulrich Lange showing the interdependence among the traditional districts in Skåne (cf. Fig. 7.2) (Strömberg 1993, 66). Used by permission.

other two products in the same region (Sjöbeck 1936, 49–51). Figure 7.5 illustrates the flows of products in the case of a farm in the plain district. Against this, it is obvious that the map in Figure 7.2 gives a static image of the Scanian landscape in the eighteenth century, void of all the movements of products and people that were the prerequisites for the existence of the diversified Scanian cultural landscape and its role as a human habitat. The generalized map from 1928, so typical of statistical thematic mapping during the twentieth century, thus distances the map reader from the changing landscape and its inhabitants in a way that hides the fact that the areas represented are cultural products, shaped by people's daily lives. As Denis Wood puts it: "It is this *subsumed and amassed cultural capital* that mapmaking societies bring to the task of making maps; not the patiently acquired mastery of this or that individual more or less carefully passed on—often in secret—through speech or gesture or inculcated habit" (Wood 1993, 48, italics in original). It is therefore crucial to complement the often-used map of the Scanian landscape districts with more place-bound experiences and studies, for instance, in preservation planning. Otherwise the preservation of cultural landscapes will be grounded only in visual aspects, fitting into an idealized pattern, and it will fail to communicate the landscape's role as lived environment. I find the latter aspect very important in present-day Skåne, where modern agriculture and forestry have completely changed the way the landscape is managed, and where, for instance, the "intermediate district" has almost disappeared.

A Modern Landscape?

The image of the Scanian landscape illustrated above was the result of a long and slow process. Even if important technical improvements within agriculture had been made, the spatial and ecological practices within farming and forestry in the mid-eighteenth century were not essentially different from those of several hundred years earlier. This situation, however, changed dramatically during the late eighteenth and the nineteenth centuries. The changes were part of the transformation of society during this dynamic period. From a political and national perspective, they took place at a time when Sweden was healing its wounds after losing its position as one of Europe's great powers. As a consequence, efforts to enhance the national strength were turned inward and a will to survey and use the country's resources rationally became evident (Guteland et al. 1974, 27–53). The extraction of Swedish iron ore and the production of iron and other metals intensified, for example, but there was a particular focus on agriculture as the foundation of the wealth of the nation. In this process a perspective of "utility," mirrored in the progress of science and a mercantile policy, was increasingly established. With "utility" as an argument, a great many changes occurred and new activities were realized. In a broader perspective, these ideas and practices were aspects of the great scientific revolution and the Enlightenment in Europe.

A prominent figure in this context was Carl von Linné (Carolus Linnaeus) (1707–78). His scientific work, his view of nature, and his career came to be integrated in the Swedish state administration in different ways during the eighteenth century. He undertook a number of journeys in different Swedish provinces in order to report to the government on natural and economic resources, on work and activities, and on landscape and culture. To Linné, the concept of utility was central, as it was to many philosophers and scientists at the time. This is obvious in his travel reports, and here it is also clear that Linné included an aesthetic dimension in the concept of utility. What was beautiful was also good because it reminded one of God's goodness: there is not only an earthly economy but also a heavenly one, and Linné's way of visualizing the latter was to investigate and describe the order that existed in the world and in Creation (Liedman 1997). This philosophy is most elaborated in his famous sexual system of classification, but it is also reflected in his travel reports.

In 1749, Linné traveled in Skåne and wrote a voluminous report on his observations (Linné 1975 [1759]). It has become a classic description of the province and its landscape and is a beautiful piece of literature. The sharp eye of the author not only catches the state of things but also sees the underlying processes. Although Linné expresses respect for traditional knowledge and practices, he often criticizes the peasants' methods of treating arable lands and woods. In this way Linné's report on Skåne is representative of the Enlightenment. It has a rational and scientific basis coupled with an aim not only of explaining but also improving the phenomena it describes.

If Linné was the person who presented the first modern view of Skåne, there is another person, almost contemporary with Linné, who may be said to have given the province its first modern landscape.[6] Rutger Macklean (1742–1816) was baron on the estate of Svaneholm in southern Skåne. In the late eighteenth century, he introduced a strict form of enclosure called *enskifte* ("one shift") in Sweden. The principle of the *enskifte* was that all land belonging to a farm should be collected into one piece. The aftermath of Macklean's actions was a set of official rules and regulations that promoted an implementation of the *enskifte* enclosure in the open plains in Sweden. The reform completely altered landscape and settlement, and it may be seen as the Swedish equivalent of the British "Parliamentary enclosure." There are also many parallels to the development in Denmark and the *udskiftning* ("out-shift") of the villages in the late eighteenth and nineteenth centuries.

Instead of the old system with its vast number of strips in the open-field system, where every farm had a share in all of the village's fields, a strict geometrical principle was to be the guiding star: one farm, one plot (with the possible exception of some of the outfields). As the farmsteads were to be situated on these new plots, the *enskifte* implied a breakup of the old nucleated villages. The process of enclosure spread fast on the Swedish plains during the nineteenth century, and, not least in Skåne, Macklean's spirit rests over the modern agricultural landscape with its spread-out farms

(a)

(b)

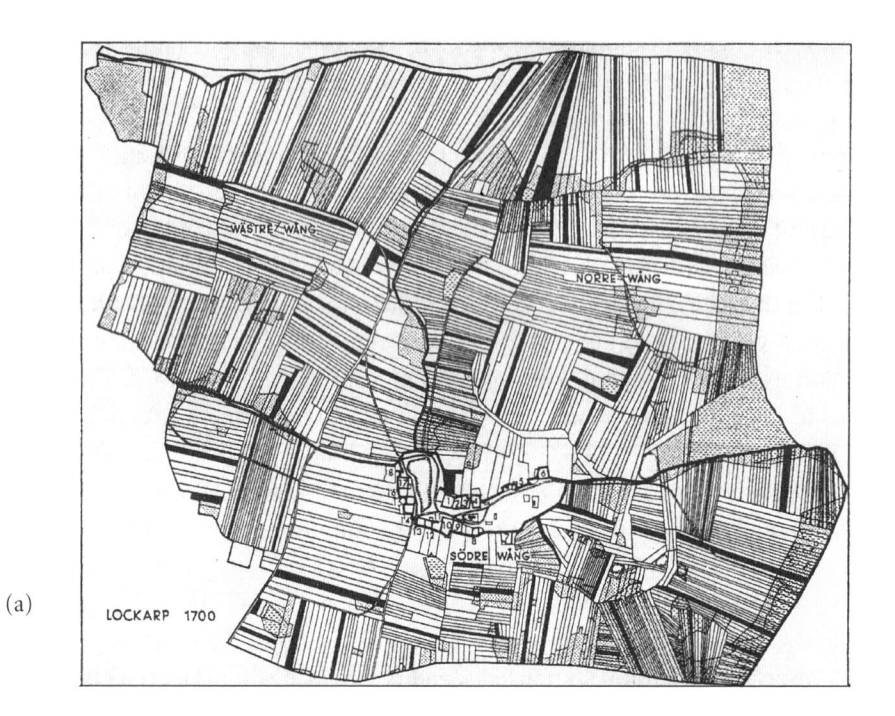

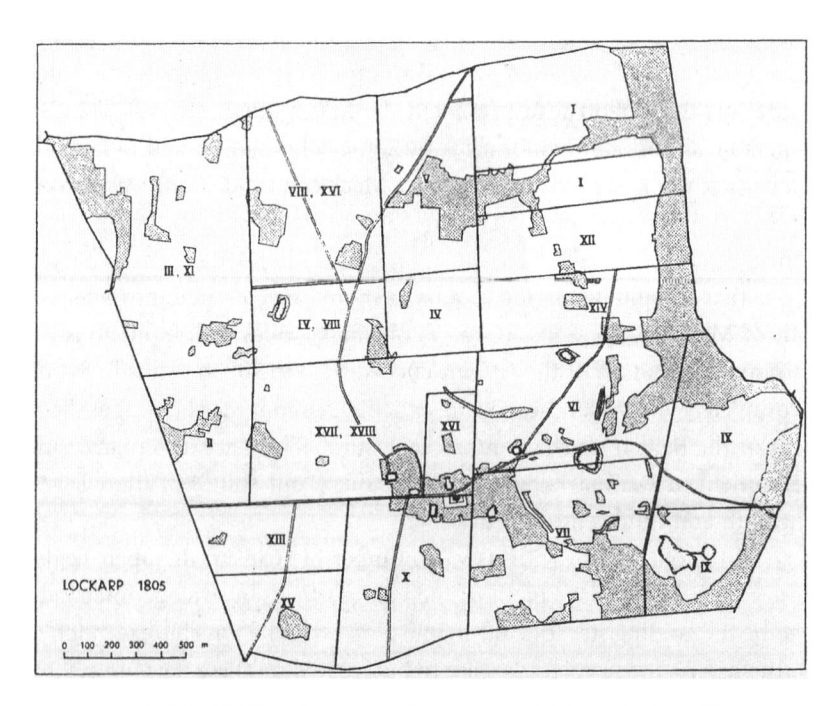

FIGURE 7.6. *"Enskifte"* enclosure on the southern plains of Skåne. The village of Lockarp (a) in 1700 and (b) according to the enclosure plan in 1805 (Wester 1960).

and rationalized lands. Figure 7.6 captures cartographically the essential change of the village lands as the result of an *enskifte* enclosure.

We do not know when Macklean visited the estate of Svaneholm for the first time. It was owned by his mother's family, which belonged to the nobility in southern Sweden and were proprietors of several estates and big farms. Macklean's father was also an estate owner and member of the nobility, but his main occupation was that of a military man.[7] He served as an officer in Karl XII's army, was badly injured in a battle, and taken as prisoner of war by the Russians at Poltava. When he returned to Sweden, he married at the age of over fifty the twenty-year-old Vilhelmina Eleonora Coyet, Rutger Macklean's mother. In 1782, at the age of forty, Rutger Macklean inherited the Svaneholm estate, where he then lived until his death in 1816. During his first years as an estate owner, he was also an active member of the Swedish parliament in Stockholm.

Like his father, Rutger Macklean started his professional career as an officer in the army, but his prime commitment was the political career he made parallel to the military one. The basis for this was his position as chamberlain, first to a princess, and from 1780 to the queen. These positions at the Swedish court, given to him by his birth, introduced him to a court life with a strong touch of the ancien régime under the guidance of King Gustav III, a great admirer of France and Louis XIV. During this period we can observe Macklean as one of the young noblemen playfully taking part in court spectacles, dances, masquerades, and intrigues (Mårtensson 1997).

During this time, Macklean soon lent his ear to other voices in the political life of Stockholm. They came from abroad and were most clearly expressed in prerevolutionary France, where a middle-class bourgeoisie of tradesmen, bureaucrats, and artisans together with radical elements from the nobility and the church criticized the old societal order with arguments formulated within the Enlightenment. The natural legal system and human rationality were founding concepts, and many saw the British parliament and its form of political representation as an ideal. These thoughts had a strong impact on Macklean and soon led him into direct opposition to the king. On several occasions he criticized the royal authoritarian regime, both within and outside the parliament. He was even arrested in the revolutionary year of 1789, when the Swedish king managed to strengthen his position. This ended Macklean's political career in Stockholm, and when he was released he retired to Svaneholm. From here he followed political developments both in Sweden and abroad very closely. In his library he collected the acts of the French Revolution, and he also worked on a form of government for Sweden that was similar to the French constitution of 1791 (Waller 1953).

Even before his retirement from active political life, it is obvious that Macklean became deeply involved in running and developing the Svaneholm estate. The political and agricultural commitments were parallel, and one more look in his library shows that he systematically collected the latest literature on agriculture. He had works on how to raise productivity through working the land more efficiently, on new machines, on the importance and handling of manure, on how to construct buildings, how to improve grazing, and how to grow sugar beets and potatoes (Germundsson 1999).

When Macklean inherited the estate in 1782, day-laboring peasants from four underlying villages ran the manor farm and paid their rent in grain and animal products. The total area of the villages was a little less than 2,000 hectares. Around 600 hectares were pasture and the rest was arable land. Agriculture was organized in a traditional open-field system typical of the plains in Skåne. The arable land was very scattered, and a farm could have around seventy strips of land spread out in the village's domains. Macklean noted that almost one-third of the strips in the villages were uncultivated, since they were lying too far from the farmsteads in the village center. The conditions on the manor farm were also bad, and Macklean to a great extent blamed the day-laboring system. At the same time as it forced the peasants to work on the manor land, to which they obviously did not feel any strong commitment, it kept them from properly running their own farms. Many of the peasants were also in debt to the estate owner. Macklean characterized the situation on the estate with the words "disorder, powerlessness, and decay" (Åberg 1953). His solutions to the problems were radical—and geometrical.

In 1785 he asked the office of the ordnance survey to send a surveyor to Svaneholm. When he arrived, he started his work by thoroughly measuring the estate domains and the underlying villages. On Macklean's orders he then depicted the area on a map and divided it into seventy-three homesteads. The allotments were arranged in squares. A farmstead was placed in the middle of each square and was surrounded by one hectare of garden. Each allotment had about 20 hectares of arable land and 5 hectares of meadowland (Figure 7.7) (Andersson 1991). Macklean invented a plan that totally dissolved the traditional village society and the traditional way of farming. Instead of the nucleated villages with the farmsteads situated less than a minute's walk from one another, they were optimally spread out in the landscape. Each farm had its land in one piece, and the old organization of strip cultivation, which required that a good deal of work be carried out collectively, was abandoned. The farmers and other subordinates had to implement the plans within a few years. The villages were physically deconstructed and the farmsteads were torn down and rebuilt on the new plots. To the villagers, the changes were fundamental because their homes, work, and connection to the land had been completely altered. Many families fled from the experiment, and, in the years following the redistribution of land and settlement at Svaneholm, a large number of farmsteads were empty. However, on many of the farms that were in operation, the new system started to bear fruit. The trend was reversed and eventually all the farms were inhabited.

The changes implemented at Svaneholm during the late eighteenth century became a model for a law in 1803 that regulated how *enskifte* reforms in Skåne should be carried out. It was the landowner who decided whether to introduce enclosures. In villages controlled by the estates, the estate owner made the decision. In villages where the peasants were freeholders, the peasants themselves made the decision. The rules said that if just one farm owner wanted an enclosure, a plan had to be made up and the majority of the farmers had to accept the plan (Holmberg 1939). After 1803 the *enskifte* enclosure spread fastest in areas dominated by villages inhabited by freeholders (Fridlizius 1979). This can partly be explained by the fact that the class of estate owners

FIGURE 7.7. Baron Rutger Macklean's map of the enclosure in the four villages belonging to the estate of Svaneholm in southern Skåne, 1785. Courtesy Swedish National Land Survey.

was far from homogeneous. Some of its members shared the views and values of Macklean, but others stuck to more traditional ideals. The freeholding peasants also had varying views. While some saw advantages and possibilities in the new system, others feared that the changes would be negative for them. Both arguments could, from different point of views, be on good grounds.

Many of the ideas behind the changes at Svaneholm were inspired by developments and experiences elsewhere. Similar enclosures had been established in England and Denmark. It was not only the enclosure technique and new methods that were implemented in the landscape. Other changes Macklean introduced were the payment of rents from the farms in cash, instead of labor, and the termination of small tenanted crofter holdings. Wage laborers and smallholders, not the peasant farmers, were to carry out the work on the manor farm (Weibull 1923, 130–31). In the long run, a new relationship developed between people and their land. In a broad socioeconomic perspective, the changes Rutger Macklean set in motion at Svaneholm can be seen as one step in the transition from a feudal system to a capitalistic one (see Hoppe 1997, 264). Analogically it can be argued that it was a step toward modernity and that Macklean himself was (becoming) something of a modern figure. By understanding the rational relationships between man and land in the kind of society he wanted to shape, he was also able to formulate a new order (Nyström 1942). Here it is tempting to see a close association between his political and philosophical thoughts and his efforts to change the landscape at Svaneholm. Obviously he wished to ground his life and work in rational thought, and he was a man who made rational decisions. It can be noticed in the way he examined the political system within which he was fostered and how, with the help of new ideas, he dismissed it as inefficient and worked to replace it with a new one. Likewise, he wanted to understand the principles of agriculture in order to improve them. This relation between what people know and what they do is fundamental for understanding the modern history of ideas, writes Liedman: "The Enlightenment project is fundamentally nothing but the idea and hope that people, by knowing more, will also act in a more conscious, better way" (Liedman 1997, 53). This was the way Macklean wanted to live, and it was also the way he wanted the peasants at Svaneholm to live. His project was a fascinating link between a new world of ideas gaining ground in Europe and a local rural society. As the situation in France was evolving into a revolutionary stage in 1789, a development that he closely followed, Macklean was, on the one hand, working to reform the political system in Sweden and, on the other, performing a cultural revolution on the domains of his estate. Needless to say, it was not the democratic dimensions of the new ideas that Macklean brought to Svaneholm. "Enlightened absolutism" would be a more proper metaphor.

The main point of sketching a picture of the enclosure in Skåne and showing a close-up of Rutger Macklean is not to emphasize the role of a single person or event. I have tried to put the *enskifte* process in a broader historical perspective and demonstrate how ideas and knowledge within the project of Enlightenment were concretely applied to the landscape. I might add that the process was similar to the English version of the Enlightenment, which is said to have been "less oppositional than its continental counterpart, more comprehensive, practical rather than speculative," and "less a shining path through a dark forest of reaction than an integral part of the cultural landscape, materially so . . . in the transformation of town and countryside, such as street lighting and enclosure" (Daniels et al. 1999, 345).

The interplay between theory and practice at Svaneholm during the late eighteenth century is only one example of these processes, but history has shown that it became very formative. The *enskifte* was paralleled in village after village in Skåne, and thus the experiment at Svaneholm became a model for a reform that completely altered the landscape and turned it into a more or less modern one. Today the expression of the *enskifte* is one of the most pregnant and identity-shaping features of the open landscape in Skåne. The landscape of the *enskifte*, however, should not be seen simply as a result of political and socioeconomic factors, such as the rise or fall of rents, the relationship between landlord and tenant, the rise of the market, and so on. Matthew Johnson argues that there are alternative understandings. One is that landscape and material culture can "be seen as active creations and as such take on a power and meaning of their own, one not necessarily intended by any one social actor" (Johnson 1996, 66). Allan Pred too sees the landscape of Skåne not just as a reflection or result of societal and cultural processes but also as a structure deeply embedded in the life of individuals, groups, and classes (Pred 1986). With this in mind, "active creations" inevitably lead to real persons performing activities in the landscape; in the case of the *enskifte*, the peasants, smallholders, and farmhands carried out the changes, sometimes in accordance with their ideals and wishes, sometimes in opposition to them.

The modern era in Skåne saw its breakthrough with the *enskifte*. Not only were the forms of the dominant occupation, agriculture, changed, but a new order was also formed. Conceptions of time, changing cultural habits, and the spread of urban influences underwent dynamic changes during the nineteenth century. The countryside in Skåne became "a training field for modernity" during the *enskifte* era, and even more so when the agricultural revolution continued in the form of drainage works, the introduction of new crops and crop rotation, advanced animal husbandry, mechanization, and artificial fertilizers (Hägerstrand 1961).

In the late nineteenth century, when the Swedish industrialization process really took off and industries, railways, and built-up areas started their triumphal progress, the characteristics of modern Skåne were further chiseled toward their present appearance. This was taking place in a regional setting where one of the significant historical happenings was the profound change within agriculture during the eighteenth and nineteenth centuries. Gradually these changes were integrated in the industrialization process, materially expressed in, for instance, sugar refineries, distilleries, and mills, as well as in the dwellings of a rural proletariat. The result is that Skåne today is a region where modernity is part of the landscape history, also outside the cities, in a way that is atypical of Sweden. The ground for this was laid in the late eighteenth century, when "the effect of *enskifte* on the landscape of the plains was similar to that of Taylorism in the landscape of industry" (Svensson 1997, 26). In most other places, historical agrarian landscapes—and especially the *idea* of a historical landscape—are of a more "idyllic" style (cf. Crang 1999). Because of this prevalent national norm of a pastoral landscape as the ideal for preservation, the Scanian agricultural landscape's primary qualities—its openness and the constant work on vast arable fields—have only recently been understood as an important cultural heritage.

A Bridge in the North

Western Skåne is one of the most urbanized parts of Sweden, with a considerable area built up and a dense and intensive communication network. At the same time, this is an agricultural area that contains the country's most productive arable land. Other parts of Skåne have a different character. The southeastern parts are also agricultural, but urbanization is on a more modest scale and several problems typical of sparsely populated rural areas are found, such as a declining and aging population and increasing unemployment. The northern areas, dominated by forest, have similar problems. As we have seen, its historical development has meant that Skåne is a long-established region in a formal administrative sense as well as in regard to its cultural history. Thus problems of a regional dimension are generally incorporated in a Scanian discourse: they are problems that concern Scanians, and they are expected to be handled by the county administration in Skåne. In a broader geographical perspective, it may be a matter of safeguarding Scanian interests in the national sphere. Skåne is furthermore the Swedish part of the Öresund region—to some, one of the dynamic new regions in Europe (Törnqvist 1998, 151–72). In the tourist literature, Skåne is marketed as "a country in Sweden"—where "country" (Sw. *land*) has about the same connotations as in English: both as countryside and as a country in terms of a national territory.

Skåne is a societal and cultural construction that has been created and re-created throughout history. For more than a thousand years, the concept of Skåne has been associated with home, belonging, supremacy, politics, power, and visions. In other words, there is a Scanian regional identity that one can relate to and discuss. The landscape is an integral part of this process, and the Scanian landscape is also a construction—a "production of space" constructed by being lived in conveyed through emotions, experiences, and visions, in maps, texts, and pictures.

The latest spectacular addition to the Scanian landscape is the bridge that now links Skåne with Sjælland. On the open lands south of Malmö, sweeping away one of the most characteristic *enskifte* landscapes, and on the old common land of Amager south of Copenhagen, huge road and tunnel constructions channel traffic to the bridge. The fixed link has a powerful symbolic meaning. In official planning and for regional business, the bridge is a manifestation of the regional integration of western Skåne and eastern Sjælland, a project that has been described as an "invocation of a transnational metropolis" (cf. Berg et al. 2000). For others the bridge is a link to the Continent, a symbol of a northern yearning to be connected to the center of Europe. Yet there are fears as well as hopes about the new link; the bridge will bring an influx of Danes, Germans, and other foreigners, who will (over)exploit the leisure landscape and the Swedish right of public access, buy up abandoned farms in the forested parts of Skåne and adorn them with their own national flags. For many who have protested against the building of the bridge, it stands as a symbol of a technocratic vision of the future; it is regarded as an affirmation of the motorcar and an environmentally nonchalant industry that is slowly poisoning us. For many in the green movement, the landscape of the bridge is also a devastated landscape, achieved in terms of efficiency and speed, without the slightest regard for the disappearance of the world's best arable soil under asphalt.

Now the bridge is a part of the landscape of Skåne and Sjælland and it cannot be left out of the regional identity. Like the medieval deer park outside Dalby, or the landscape of the first towns, or Svaneholm after the enclosures, the landscape of the Öresund Bridge is not just a stage or an arena for people to observe or perform on. These places make up the "real" landscape, which acquires meaning through constant human activity, whether in action or in thought and communication. In this respect, landscape may be seen as a social reinvention of space, stemming from the dialectic between performance and representation and emphasizing a concern with what "landscapes do rather than simply are" (Mels 1999, 177–78).

Speaking of bridges: in a technical respect it might be said that the span of steel and concrete across the water connects Skåne to Denmark, Sweden to Europe, Scandinavia to the Continent; but looking at the cultural meaning of movements through these regional spaces, could it not be said that it is the landscape of Skåne that constitutes the substantive bridge?

Notes

1. The passage is from Strindberg's essay *Svensk natur* ("Swedish nature") (Strindberg 1985, 245 [1896]). All citations from Swedish literature in this chapter were translated by Alan Crozier.

2. I will use the Swedish/Danish name *Skåne* throughout (I think it is a dimension of the region's identity), even if the Latinized *Scania* is a more or less official translation into English (in German Skåne is called *Schonen*). But when it comes to the adjective—*skånsk*—I use the word "Scanian."

3. Quoted in Hägerstrand 1961, 33.

4. Braun and Hogenberg produced six volumes of views of European cities, published in Cologne between 1572 and 1618 with different titles (Wåhlin 1931). Volume 4 depicts the Nordic countries: *Liber quartus urbium praecipuarum titius mundi* (Braun and Hogenberg 1588).

5. The word also exists in English: snaphance (or snaphaunce), with the meaning of (1) "an armed robber or marauder; a freebooter or highwayman; a desperate fellow or thief" (obsolete); (2) "an early form of flint-lock used in muskets and pistols; also the hammer of this"; (3) "a musket, gun, etc., fitted with a lock of this kind, in use in the 16–17th centuries" (*Oxford English Dictionary* 1989, s.v. "snaphance").

6. In a broad sense I understand "modernity" and "the modern period" as, on the one hand, the development and expansion of capitalist economies and their new modes and relations of production, distribution, and consumption during the eighteenth, nineteenth, and twentieth centuries. On the other hand, it also comprises new concepts of rational knowing and, as Catherine Nash puts it, "discourses of 'newness' which articulated a sense of a shift from traditional society governed by communal custom and religion to a secular state of autonomous individuals" (Nash 2000, 13).

7. The name Macklean, which is quite un-Swedish, was taken by Rutger and his brother Gustav when they were appointed lords in 1783 and introduced to the House of the Nobility (Sw. *riddarhuset*). Before that, they and other members of the family bore the name Macklier (sometimes Makelier), which is of Scottish origin. The Swedish branch of the family immigrated around 1600. The reason Rutger and Gustav changed their name to Macklean was probably that their father—falsely—got the information that the family was a branch of the Scottish noble family Maclean (S.B.L. 1982, Macklean).

REFERENCES

Åberg, Alf. 1953. *När byarna sprängdes*. Stockholm: Natur och kultur.

———. 1994. *Kampen om Skåne under försvenskningen*. Stockholm: Natur och Kultur.

Andersson, P. G. 1991. "Carl Gideon Wadman—kartans bortglömde general." *Svaneholm Årsskrift 1991*: 6–12.

Andrén, Anders. 1997. "Paradise Lost: Looking for Deer Parks in Medieval Denmark and Sweden." In *Visions of the Past: Trends and Traditions in Swedish Medieval Archaeology*, ed. H. Andersson, P. Carelli, and L. Ersgård, 469–90. Stockholm: Almqvist & Wiksell International.

Anglert, Mats. 1995. *Kyrkor och herravälde*. Lund: Almqvist & Wiksell International.

———. 1999. "Det förhistoriska och medeltida Skåne." In *Atlas över Skåne. Sveriges Nationalatlas*, ed. T. Germundsson and P. Schlyter, 64–70. Gävle: Kartförlaget.

Barrell, John. 1972. *The Idea of Landscape and the Sense of Place*. Cambridge: Cambridge University Press.

Berg, Per Olof, Anders Linde-Laursen, and Orvar Löfgren, eds. 2000. *Invoking a Transnational Metropolis: The Making of the Øresund Region*. Lund: Studentlitteratur.

Bolin, Sture. 1939. *Skånelands historia—skildringar från tiden före försvenskningen. Andra delen*. Lund: Borelius.

Braun, Georg, and Franz Hogenberg. 1588. *Liber quartus urbium praecipuarum titius mundi*. Cologne.

Campbell, Åke. 1928. *Skånska bygder under förra hälften av 1700-talet: Etnografisk studie över den skånska allmogens äldre odlingar, hägnader och byggnader*. Uppsala: A.-B. Lundequistska Bokhandeln.

Carlsson, Dan. 1987. "Regionindelning av det agrara kulturlandskapet i Sverige." In *Natur- og kulturlandskapet i arealplanleggingen, 1: Regioninndelning av landskap,* 353–89. Copenhagen: Nordisk Ministerråd.

Cosgrove, Denis. 1984. *Social Formation and Symbolic Landscape.* London: Croom Helm.

Crang, Mike. 1998. *Cultural Geography.* London: Routledge.

———. 1999. "Nation, Region, and Homeland: History and Tradition in Dalarna, Sweden." *Ecumene* 6, no. 4: 447–70.

Dahl, Sven. 1942. *Torna och Bara: Studier i Skånes bebyggelse och näringsgeografi före 1860.* Lund: Carl Bloms boktryckeri.

Daniels, Stephen, Susan Seymore, and Charles Watkins. 1999. "Enlightenment, Improvement, and the Geographies of Horticulture in Later Georgian England." In *Geography and Enlightenment,* ed. D. N. Livingstone and C. W. J. Withers, 345–71. Chicago: University of Chicago Press.

Duby, Georges. 1981. *Krigare och bönder—den europeiska ekonomins första uppsving 600–1200.* Stockholm: P. A. Norstedt & söners förlag.

———. 1973. *Guerriers et paysans VII–XII siècle: premier essor de l'économie européenne.* Paris: Gallimard.

Emanuelsson, Urban, Clas Bergendorff, Bengt Carlsson, Nils Lewan, and Olle Nordell. 1985. *Det skånska kulturlandskapet.* Lund: Signum.

Frandsen, Karl-Erik. 1983. *Vang og tægt.* Esbjerg: Bygd.

Fridlizius, Gunnar. 1979. "Population, Economy, and Property Rights." *Economy and History* 22: 3–37.

Fries, Carl. 1963. *Den svenska södern.* Stockholm: Wahlström & Widstrand.

Germundsson, Tomas. 1999. "Mackleans bokhylla." In *Geografi i Lund: Essäer tillägnade Gunnar Törnqvist,* ed. K. Cederlund, T. Friberg, and M. Wikhall, 56–63. Lund: Lunds universitet.

Guteland, Gösta, Ingvar Holmberg, Torsten Hägerstrand, Anders Karlqvist, and Bengt Rundblad. 1974. *The Biography of a People: Past and Future Population Changes in Sweden.* Stockholm: Allmänna förlaget.

Hägerstrand, Torsten. 1961. "Utsikt från Svaneholm." In *Svenska turistföreningens årsskrift 1961,* 33–64. Stockholm: STF Förlag.

———. 1991 (1984). "The Landscape as Overlapping Neighbourhoods. Carl Sauer Memorial Lecture." In *Om tidens vidd och tingens ordning: Texter av Torsten Hägerstrand,* ed. G. Carlestam and B. Sollbe, 47–55. Stockholm: Byggforskningsrådet.

Hallberg, Göran. 1999. "Ortnamn." In *Atlas över Skåne: Sveriges Nationalatlas,* ed. T. Germundsson and P. Schlyter, 80–81. Gävle: Kartförlaget.

Holmberg, Nils. 1939. *Enskiftet i Malmöhus län—förebilder och resultat.* Lund: Gleerups förlag.

Hoppe, Göran. 1997. "Jordskiftena och den agrara utvecklingen." In *Agrarhistoria,* ed. B. M. P. Larsson, M. Morell, and J. Myrdal, 254–70. Stockholm: STF förlag.

Johannesson, Gösta. 1984. *Skåne, Halland och Blekinge: Om skånelandskapens historia.* Stockholm: Norstedts.

Johnson, Matthew. 1996. *An Archaeology of Capitalism.* Oxford: Blackwell.

Jones, Michael. 1991. "The Elusive Reality of Landscape: Concepts and Approaches in Landscape Research." *Norsk Geografisk Tidskrift* 45: 229–44.

Karlsjö, Bertil. 1995. *Skånes och Blekinges riksgräns: Dokumentation av de två danska landskapens gräns mot svenska Småland.* Lund: Dialekt- och ortnamnsarkivet.

Kulturlandskapet och bebyggelsen. Sveriges Nationalatlas. 1994. Höganäs: Bokförlaget Bra Böcker.

Larsson, Lars. 1999. "Det förhistoriska och medeltida Skåne." In *Atlas över Skåne: Sveriges Nationalatlas*, ed. T. Germundsson and P. Schlyter, 54–59. Gävle: Kartförlaget.

Lauring, Palle. 1964. *Danmark i Skåne*. Lund: Berghs.

Liedman, Sven-Erik. 1997. *I skuggan av framtiden: Modernitetens idéhistoria*. Stockholm: Bonnier Alba.

Linde-Laursen, Anders. 1995. *Det nationales natur: Studier i dansk-svenske relationer*. Copenhagen: Nordisk Ministerråd.

von Linné, Carl. 1975 (1759). *Carl Linnaei Skånska resa förrättad år 1749*. Reprint, ed. C.-O. von Sydow. Stockholm: Wahlström & Widstrand.

Mårtensson, Eric. 1997. "Rutger Macklean—om den stora jordrevolutionen och den nya skolan." *Svaneholm Årsskrift 1997*: 16–24.

Mels, Tom. 1999. *Wild Landscapes: The Cultural Nature of Swedish National Parks*. Lund: Lund University Press.

Nash, Catherine. 2000. "Historical Geographies of Modernity." In *Modern Historical Geographies*, ed. B. Graham and C. Nash, 13–40. Harlow: Longman/Prentice-Hall.

Natur- og kulturlandskapet i arealplanleggingen. 1987. Miljørapport 1987: 3, 3 vols. Copenhagen: Nordisk Ministerråd.

Nuti, Lucia. 1999. "Mapping Places." In *Mappings*, ed. D. Cosgrove, 90–108. London: Reaktion Books.

Nyström, P. 1942. "En dag på Svaneholm." *Ord och Bild 1942*: 537–47.

Olwig, Kenneth R. 1993. "Sexual Cosmology: Nation and Landscape at the Conceptual Interstices of Nature and Culture, Or: What Does Landscape Really Mean?" In *Landscape: Politics and Perspectives*, ed. B. Barber, 307–43. Oxford: Berg.

———. 1996a. "Nature—Mapping the Ghostly Traces of a Concept." In *Concepts in Human Geography*, ed. C. Earle, K. Mathewson, and M. Kenzer, 63–96. Lanham, Md.: Rowman and Littlefield.

———. 1996b. "Recovering the Substantive Nature of Landscape." *Annals of the Association of American Geographers* 86, no. 4: 630–53.

Pred, Allan. 1986. *Place, Practice, and Structure: Social and Spatial Transformations in Southern Sweden, 1750–1850*. Oxford: Polity Press.

Richter, Herman. 1929. *Skånes karta från mitten av 1500-talet till omkring 1700*. Lund: Lunds universitet.

Sauer, Carl O. 1963 (1941). "Foreword to Historical Geography." *Land and Life: A Selection from the Writings of Carl Ortwin Sauer*, ed. J. Leighly, 351–79. Berkeley and Los Angeles: University of California Press.

S.B.L. 1982. *Svenskt Biografiskt Lexikon*, vol. 24. Stockholm: Bonnier.

Schama, Simon. 1995. *Landscape and Memory*. New York: Alfred A. Knopf.

Schmidt Sabo, Katalin. 1997. "'Now the peasants want to build a village . . . ': Social Changes during the Period of Village Formation in Skåne." In *Visions of the Past: Trends and Traditions in Swedish Medieval Archaeology*, ed. H. Andersson, P. Carelli, and L. Ersgård, 671–96. Stockholm: Almqvist & Wiksell International.

Seymour, Susan. 2000. "Historical Geographies of Landscape." In *Modern Historical Geographies*, ed. B. Graham and C. Nash, 193–217. Longman/Prentice-Hall.

Sjöbeck, Mårten. 1936. *Skåne: Färdvägar och vandringsstigar utgående från stambanorna*. Stockholm: Statens Järnvägar.

Skansjö, Sten. 1987. "Estate Building and Settlement Changes in Southern Scania c. 1500–1650 in a European Perspective." In *The Medieval and Early Modern Rural Landscape of Europe under the Impact of the Commercial Economy*, ed. H.-J. Nitz, 105–14. Göttingen: Department of Geography, University of Göttingen.

———. 1997. *Skånes historia*. Lund: Historiska Media.

Sjöbeck, Mårten. 1936. *Skåne: Färdvägar och vandringsstigar utgående från stambanorna*. Stockholm: Statens Järnvägar.

Skånska bygder—vårt odlingslandskap genom tiderna. 1997. Malmö: Länsstyrelsen i Skåne län.

Sporrong, Ulf, Urban Ekstam, and Kjell Samuelsson. 1995. *Svenska landskap*. Stockholm: Naturvårdsverket.

Strindberg, August. 1985 (1896). *Svensk natur*. Reprinted in *August Strindbergs samlade verk* 29: 245–72. Stockholm: Almqvist & Wiksell.

Strömberg, Lars G. 1993. "Kulturmiljön och lokala näringssystem." In *Vad berättar en by? Om äldre kulturmiljösystem i odlingslandskapet*, 31–88. Stockholm: Riksantikvarieämbetet.

Svenska Landskapslagar. 1979. *Fjärde serien: Skånelagen och Gutalagen*. Tolkade och förklarade av Å. Holmbäck och E. Wesén. Stockholm: AWE/Gebers.

Svensson, Birgitta. 1997. "Vardagsmiljöer och söndagskulisser." In *Moderna landskap*, ed. K. Saltzman and B. Svensson, 21–44. Stockholm: Natur och kultur.

Tägil, Sven. 1995. "Skåne: En region i Sverige och Europa." In *Skåne: Svenska turistföreningens årsbok 1996*, ed. H. Bauer, 8–23. Stockholm: Svenska Turistföreningen.

Törnqvist, Gunnar. 1998. *Renässans för regioner*. Stockholm: SNS Förlag.

Wåhlin, Hans. 1931. *Scania Antiqua: Bilder från det forna Skåne*. Malmö: John Kroon.

Waller, Sture. 1953. *Rutger Macklean och 1809–1810 års riksdag*. Skrifter utgivna av Fahlbeckska Stiftelsen 39. Lund: Gleerup.

Weibull, Carl Gustaf. 1923. *Skånska jordbrukets historia intill 1800-talets början*. Lund: Gleerup.

Wester, Ethel. 1960. "Några skånska byar enligt lantmäterikartorna." In *Svensk Geografisk Årsbok*, 162–80. Lund: Gleerup.

Wood, Denis. 1993: *The Power of Maps*. London: Routledge.

8.
The Province of Dalecarlia (Dalarna): Heartland or Anomaly?

ULF SPORRONG

Dalecarlia (*Dalarna* in Swedish) is one of the best-known provinces of Sweden, both nationally and internationally, and is a popular tourist destination along with the seaside resorts on the coast, the mountain regions, the northern wilderness, and parts of the densely populated south. Dalecarlia has received much publicity, and it is no exaggeration to say that the province has meant and still means a great deal to Sweden's image. Indeed, Dalecarlia has time and again been represented as a Swedish ideal.

What, then, are people looking for in Dalecarlia? The Dala mountains in winter are a skier's paradise. Only the mountain regions in Jämtland can boast a larger number of visitors. Summer tourists are drawn primarily to the region of Lake Siljan. Here the folklore tradition is important alongside the panorama of lakes and distant blue mountains in the beautiful Dala countryside This province has a distinctive character that is not to be found in any other part of Sweden (Figure 8.1).

For centuries, Dalecarlia has been the center of attention, first for the part it played in the creation of the national state, and then as a tourist destination. Various scientific disciplines have contributed to draw attention to the region. Other factors are the countryside itself, the way business life is run, and the economic and above all the social life that has developed (Rehnberg 1972, 142). The starting point for this chapter will be the historical and cultural background of the province. I shall then examine the physical landscape to see what it can tell us about Dalecarlia's special character. I will compare it with other regions in Scandinavia and to some extent with the rest of Europe. Dalecarlia here is limited to the physical landscape north of Djurmo, that is, the river valleys of the Österdalälven and the Västerdalälven, particularly the region around Lake Siljan.

Dalecarlia as a Cultural Region

To answer the question how Dalecarlia achieved its position among the provinces of Sweden, I will start with the view of Dalecarlia as a Swedish ideal, a view that in many respects is based on historical national propaganda of unclear origin. When King Gustav III, faced with the threat of a Danish invasion in 1778, traveled to Dalecarlia, he clearly did so to seize Swedish patriotism by the roots (Klein 1926, 103f.), for it had been the Dala people who had rallied around Gustav Vasa when he made his reckless attempt to break up the Nordic Union at the beginning of the sixteenth century. Generations of Swedish schoolchildren have been brought up on the oral tradition that

FIGURE 8.1. The province of Dalecarlia (Dalarna) in central Sweden.

surrounds this "heroic deed," and Gustav Vasa's victory later led to his being elected king of Sweden in 1524, making his triumphal entry into Stockholm on June 6, now the Swedish National Day.

It was perfectly natural for Gustav Vasa to turn to the people of Dalecarlia for help. Political figures earlier had appealed to Dalecarlia for help in carrying out Swedish national projects. Best known is the fight against the oppression of Danish sheriffs during the period of the Nordic Union, which was begun by Engelbrekt at the beginning of the fourteenth century and continued by the Sture family in the fifteenth century. Gustav Vasa evidently felt that he and the Stures shared a common cause; both the latter and Engelbrekt had recruited picked troops in Dalecarlia. It is difficult to say why the people of Dalecarlia took part in national politics, though some reasons were certainly economic. It is sufficient to say that in the later part of the Middle Ages the Dala peasantry must have been mobile so that they could quickly be called to arms, much as the Vikings had been elsewhere. The conclusion must be that a strong sense of Dala patriotism survived in the province well into the sixteenth century. The tales and myths about the unification of the country lived on in Swedish national patriotism. King Gustav was regarded as a Swedish William Tell from the "Ur-Canton" of Dalecarlia (Hård af Segerstad 1972, 16).

Gustaf Näsström, author and cultural historian, has devoted a great deal of attention to the idealized picture of Dalecarlia by studying the Romantic nineteenth-century literature on the province. Among other authors, he quotes from Fredrika Bremer's novel *I Dalarna* ("In Dalecarlia") from the mid-nineteenth century:

> You must go to Dalecarlia if you wish to see a landscape so magnificent in its innocence, a people still in the patriarchal way of living that is gradually disappearing from the earth but which has features of such great and moving beauty. As the Dalälven River runs through Dalecarlia—a great and bright thought through a grave and arduous life—so runs religion through the laborious life of the Dala people, and the centuries have passed over this people without leaving any dross. In appearance, customs, dress, and temperament, they are still the same as they were in Engelbrekt's and Vasa's days. Work and prayer have preserved their health and their industriousness. Low are their dwellings. They bow their heads at the doors of their cottages, but they have never yet bowed them to the yoke of an oppressor. Historic events have hallowed this ground—the native soil of Swedish freedom. (Näsström 1937, 51)

Näsström sums up his views:

> Swedish feelings of patriotism flowed out from Dalecarlia first over the small medieval kingdoms and peasant republics, then over the kingdom welded together by Gustav Vasa. Memories of the part the province played in Engelbrekt's revolt and the war of liberation

in 1520–21, as well as of the important part the Falu Mine played in financing the wars Sweden fought when it was a great power, have echoed through the Dala soul for centuries and finally created an awareness—sometimes gratefully acknowledged, sometimes hotly denied—that Dalecarlia and its population have become a kind of measure of Swedishness. The tenacious conservatism of its economy and customs made the province stand out in the nineteenth century as a symbol of an old-fashioned strict and aesthetically attractive way of life, and while industrialized civilization rolled victoriously forward over other parts of the country, bringing feelings of unease and loss in its wake, Dalecarlia shone like an Arcadian refuge for tired city souls and all kinds of longing for firm roots and a country idyll in a picturesque old background. (Näsström 1937, 11–12)

Näsström also hints at the conditions under which modern tourism was established: people simply tend to seek out these "untouched" country surroundings. Of course there are many descriptions of the same theme, but common to all of them is that

the Dalecarlian himself appears in a historical perspective as the austere, independent udaller [owner of an ancestral farm] who, free and without ties, has always tilled the soil he inherited from his forefathers, been jealous of local traditions and, on occasions, stepped out of his anonymity impudently to stand up for his own opinion before the lords of the realm. In his capacity as a udaller, however, he has suffered a constant series of serious worries, and the Dalecarlian who today struggles against the poor returns his farm gives him has taken over the inheritance of his forefathers in this respect, too. The problems of sparsely populated districts are nothing new to him; they are persistent historical facts. (Hård af Segerstad 1972, 14)

In addition to the national Romantic picture of Dalecarlia, it is also a province in which one has to struggle for one's existence, a theme I shall return to.

Another factor that continually turns up in descriptions of Dalecarlia's special character is the "attractive countryside," representative of both central Sweden and the northern provinces:

Anyone coming from the gently rolling or flat lands of southern Sweden to the northern Swedish countryside and settlements can hardly escape being struck by how the character of the landscape changes abruptly when he enters southern Dalecarlia, irrespective of whether he does so at Krylbo or at Västanfors. The same experience meets him on all roads. Distant blue heights appear briefly in the northwest and soon surround the road. The countryside grows larger and distances and heights are greater. One is in a region that is hilly in a quite different way from the country around Lake Mälar. Put briefly, it is the Bergslag or, if you like, the Norrland countryside that has begun. (De Geer 1926, 22)

Similarly, the ethnologist Mats Rehnberg considers the southern part of the Norrland countryside an important ingredient in Dalecarlia: "It is here that the distant blue mountains, like forestwaves, roll away into the distance. . . . The distant blue forests that cover the greater part of the country districts play an important part" (Rehnberg 1972, 140–43).

The landscape of Dalecarlia has both grandiose qualities and intimate features suitable for agriculture in the valley floors. From the mountains, one looks out over a cultural landscape created by a socioeconomic system peculiar to the province. Its cultivable assets lie on loose boulder clay and sedimentary soil along the two arms of the Dalälven river. To the casual observer, the cultivated areas resemble light, open windows in the predominantly wooded landscape. In addition, the large lake, Siljan, is another element of great importance to the beauty of the landscape: "We drove by farms and cottages, mounted a hill, and . . . paused to admire lake Siljan in all its beauty—calm and serene—with blue hills in the distance, a pavé of turquoise mixed with amethyst" (Marryat 1862, 2:217).

The writer Johan Nordling depicted the beauty of the landscape in his novel *Siljan: A Book about the Heart of Sweden*:

> The outlook tower rises up before them, and Karl-Olof gives his companion strict orders not to look around until they reach the top. They clamber higher and higher up through the stuffy twilight of the wooden staircase toward the vista and the light. When they finally reach the top, they are first blinded for a few moments by the flood of sunlight, then pressed backward by the force of the wind until they find a good handhold on the railing. [Then] they turn their gaze on the green waves and discover that there are forests, mile after mile of billowing pine forests rolling out of the depths on the horizon over height after height in a rising rhythm until they reach their goal at his feet. They turn to one side [and look at a fresh segment of their surroundings]. Here, too, there is a sea of forest, but over there—halfway—it is broken by something that glimmers blue, something metallic with a shining reflection. It is Siljan looking at its reflection among the ancient mountains. Gradually more details become clear. Villages burst forth like blossoming tussocks in the light glades, on the hem of the ridges, along the winding ribbons of water. There, the buildings of the village of Leksand flock together like a herd around its shepherd, there Siljansnäs Church kneels before its reflection in the waters of Byviken. And like a precious bejeweled frame joining and holding the vast panorama together, the grave, high mountains, more and more remote, bluer and bluer, vault across the horizon. And what a play of light, what variations of living spirit over this world! Each cloud that moves across the heavens chases its shadow before it across the billows of the forest sea. You can sense the shivers of cold through the earth's limbs, hear the sigh that is forced from her distressed breast. But the next moment, when the dark wing of the storm bird has passed over, how the region flashes into life again with

the abandoned smile of a mistress drunk with love under the fiery lash of the young sun god. These are some of the scenes from the eternal drama of the great open spaces, a few poor strokes of the pen about what Sten and Karl-Olof felt when, early one morning in May, they embraced Dalecarlia from the summit of Mount Käringberget. (Nordling 1907, 38ff.) (Figure 8.2)

Nordling's purple passage may not be the finest description written by a Swedish author, but it is typical of the times. His aim was to entice people to the countryside and experience its beauty for themselves. The printrun for Nordling's novel was thirty-five thousand copies, a unique figure for Swedish literature of the day!

FIGURE 8.2. A famous outlook tower in Leksand parish, Käringberget. Postcard, c. 1900. Municipality Archives, Leksand.

The historical aspects of Dalecarlia's economic life also attracted interest outside the province. The government was concerned about the low returns on agriculture because land in this region had become so fragmented and the small properties yielded little in the way of taxes. The government tried to consolidate agriculture and force the people of Dalecarlia to treat the ownership of land in the same way as it was treated in other parts of the country. However, the customs regulating inheritance and the acquisition of land were quite different from those practiced in other parts of Sweden and the government was unsuccessful.

In general, the small amount of arable land (3–4 hectares, or 7.5–10 acres) per farm was only *one* side of a larger, coherent use of resources or system of diversified occupations typical of the region, as it was in other forest and mountainous areas of Europe. Furthermore, the diversity of occupations was highly seasonal. Agriculture, however, was the kernel of a farm's economic life. Various social rights and obligations in society were linked to ownership of land. Owning land gave the farmer social legitimacy.

Wetlands and forests were used alongside fields for purely agricultural functions. Hayfields were extensive along the shores of lakes and the banks of rivers, on mires, and also on mountain slopes. The forest was grazed, not least on shielings (summer farms). Some shielings were situated near villages, while others were farther away. The youngest and oldest women moved to the more distant shielings in summer to make use of the grazing and to do dairy work (Brorson 1985). The men worked in the forest, which gave wood, charcoal, and tar.

Winter was spent on home arts and handicrafts, an occupation that in places was extensive. In some cases the sale of handicrafts could assume almost protoindustrial dimensions. In summer, large numbers of people left Dalecarlia, mainly for the Lake Mälar district, where the men did laborers' work such as digging ditches, building, and cartage, and the women also did heavy jobs such as working for masons or in breweries and rowing boats. This type of work was known as *herrarbete* (people working for a landlord). Some women sold handicrafts until late in the autumn. These migrant workers were well known in central Sweden and certainly helped to spread knowledge about conditions in Dalecarlia among ordinary folk. The Dala people preferred to keep together in groups; they had their own dialect and frequently wore the folk costumes of their parish. Those who found work in country districts also came into contact with the landless in Swedish agrarian society, which must have strengthened their self-esteem. In Dalecarlia there was really no agrarian proletariat—almost everyone owned some land. Landless people in the Mälar Valley must have thought this quite fantastic.

One of the original strongholds of the Swedish folklore movement is in Dalecarlia, where Jones Mats Persson opened a folklore museum in Leksand in 1899. The journeys Artur Hazelius made to Dalecarlia in the mid-nineteenth century led to the foundation of the National Museum of Folklore and the open-air museum at Skansen in Stockholm

(Klein 1926, 107; Crang 1999). Folklore centers and museums were established throughout Dalecarlia. Dalecarlia's Archaeological Association (1863) was one of the first associations of this kind to be founded in the country. Yet it was not until well-known artists and authors began to discover the peasant culture of Dalecarlia that the folklore movement gained strength. At that time the church congregation still wore the parish folk costumes, which were admired and described by tourists, ethnologists, and littérateurs (Figure 8.3). Women wore these costumes until well into the twentieth century. Another two important elements in folk culture should be mentioned: the local building tradition (Andersson 1989) and Dalecarlia's peasant painting and folk music.

FIGURE 8.3. People in folk dress leaving a church service. Postcard, c. 1900. Municipality Archives, Leksand.

In 1896, the artist Anders Zorn moved a *härbre* (a log cabin raised from the ground) to his farm in Mora and converted it into a studio, thereby linking painting and folk culture. Zorn was interested in more than painting; in 1906 he organized the first rally of folk musicians in Dalecarlia (Rosander 1976, 52). Another portal figure was the painter Gustaf Ankarcrona, who founded the Leksand Art and Crafts Association at a handicraft exhibition in 1904 and Dalecarlia's Local Folklore Society in 1915. He also established a folklore museum on his farm at Holen in Tällberg, Leksand parish.

The first monuments in Dalecarlia, such as the Falu Copper Mine and Ornässtugan, a building associated with stories about Gustav Vasa, had already attracted attention in the eighteenth century. By the second half of the nineteenth century, Dalecarlia was attracting tourists in greater numbers. Tourist guides publicized the midsummer festival in Leksand (Figure 8.4). Tourists were transported to the Siljan district by boat,

FIGURE 8.4. Raising the maypole, Noret village, Leksand parish, c. 1925. Municipality Archives, Leksand.

and later, after 1890, people could travel by railway. Many tourist hotels were built and accommodation in private homes was arranged (Rosander 1976, 50ff.).

Tourism began to flourish when artists and authors moved to Dalecarlia and inspired the people to embrace their own culture (Rosander 1976, 50). The effects of such publicity cannot be exaggerated—Dalecarlia became a name to conjure with! Dala romanticism flourished! Among the new spokesmen after Zorn moved to Mora in 1896 were K. E. Forslund, who settled in Ludvika in 1898, Carl Larsson in Sundborn in 1901, E. A. Karlfeldt in Sjugare village in Leksand in 1920, and Hugo Alfvén, who moved to Tibble in 1908 and later to Tällberg (Rosander 1976, 51). Alfvén composed the music to Bishop Thomas's *Frihetssång* ("Song of Freedom"), which every schoolchild once had to learn. At the beginning of the twentieth century, he also composed symphonies based on Dalecarlia's folk music; the best known are *Midsommarvaka* ("Midsummer Watch") and *Dalarapsodi* ("Swedish Rhapsody"). Already in the nineteenth century, many of the Nordic cultural élite had shown their appreciation for old Dala culture, for example, the Danes Hans Christian Andersen and artist Wilhelm Marstrand. Naturally the enormous publicity these celebrities gave Dalecarlia created curiosity about the province and helped to attract tourists to what they imagined to be the well-preserved remains of a European peasant culture. This image of Dalecarlia captured the imagination of tourists well into the twentieth century.

The growth of mass tourism coincided with the passing of the holiday laws. A legal minimum of two weeks' holiday was introduced in 1938, allowing more people time to travel. In winter, the center of the industry shifted from the area around Lake Siljan to the mountain regions. In summer, tourism still has a firm base in the central districts of Dalecarlia.

Special property laws in the upper part of the province have made many children and grandchildren part-owners of properties, although they no longer live in the region. *Härbren,* barns, and shieling buildings have been converted to weekend or summerhouses. Large numbers of weekend or summer houseguests come from outside the province. This has greatly increased the size of the summer population and helps to explain the care people take to preserve their traditions.

In Dalecarlia there is a clear connection between tourism and the cultural life of the province. The attention paid to Dalecarlia as early as the Middle Ages led to a kind of cultural fixation. The people began to see their history in a positive light and to hold on to old ideals. Another factor was the great increase in population, especially in the eighteenth century. Nevertheless, the old customs governing landownership were retained, which meant that most of the population belonged to the land-owning class. The code of behavior that evolved to maintain this aim required strict social control within the village and parish. This was further emphasized by the fact that people clung to visible, outer attributes such as peasant costumes, their dialect, folk music, and building style. The poverty of the province is probably attributable to these phenomena.

When the Dala people migrated to find summer work, they did not wear their folk costumes to show them off. They wore them because that was what they always wore: "The women wear red bodices and white sleeves, petticoats of a dingy blue, yellow aprons, red caps with yellow hoods; the men, a long dark coat, yellow leather vest and breeches, the latter cut down very low in front; long lambswool stockings with the fluff outside; buckles on their shoes, of which the heel is placed in the center; blue ribbons with red bobs dangle jauntily from their knees" (Marryat 1862, 2:216).

A considerable amount of ethnographic and anthropological research exists on Dalecarlia as a cultural region, but there has been little research on the development of the cultural landscape. Much has been written about local building traditions, but hardly anyone has addressed how the special form of the buildings and details of the landscape are related to the social code, particularly to landownership. This has been the essential factor in determining land use and, consequently, the development of the landscape. In what follows, I shall look closer at the physical landscape, its natural resources and its anthropogenic transformation into a cultural region, seen from an agrarian historic point of view. What can an analysis of the cultural landscape tell us about the development of what has come to be known as the heartland of Sweden? Is it really so different, and, if so, what explanation can we find?

The Physical Landscape

The Natural Landscape

It is the "intrinsic" beauty of the physical landscape that has attracted attention, especially in the tourist context. Here we think first of the hilly terrain characteristic of the northern Swedish landscape in a transitional zone between coast and inland. If we regard the landscape as a stage, we can compare the heights to backdrops placed at increasing distances from the observer. The distant blue hills often disappear into a haze the farther away they are. Lake Siljan, an essential ingredient in Dalecarlia's beauty, is 290 square kilometers, the seventh largest lake in Sweden. Other important physical features are the two branches of the Dalälven River, Österdalälven and Västerdalälven. The Swedish name Dalarna, meaning "The Valleys," was probably derived from the two main river valleys north of Djurmo.

If one looks a little closer at the details of the landscape, a good many features appear that not only bring out what is special about the landscape of Dalecarlia but also explain the use of its resources at different times of the year. Siljan is 160 meters above sea level, just below the highest postglacial shoreline, which in this part of Sweden is about 200 meters above sea level. This means that a deep bay once existed where the river now flows. Running water deposited material in this bay, and later, when the land uplift began, the river cut through its own sediment, bringing with it further deposits. The soil that covers the lower parts of the river valleys thus consists of fine sand and silt and, as a result, there are a considerable number of ravines. These conditions have produced the Siljan region's cultivated central districts with villages that are unusually large for Sweden, for the landscape has a high degree of cultivation. On the mountain slopes, where the till is rough, the cultivated land comes to an abrupt end and the landscape becomes covered with forest. The forests belong to the undulating, hilly terrain of southern Norrland, with a difference in height between the valley floor and the summits that frequently exceeds 400 meters. Geologically, the Cambro-Silurian rocks (remnants of old coral reefs) are the most remarkable feature in the region, found in a ring north of Lake Siljan.

Conditions are different above the highest shoreline. As the sea never reached this height, the moraines here have never been abraded and there are ridges of sandy till here and there. Cultivation has made these ridges look like light, open windows in the forests, often giving a wide field of view to the settlements that have grown up here.

THE CULTURAL LANDSCAPE

If we look at resource use from a historical point of view, we find mainly arable land and meadows in the valley floors. Grazing begins higher up, toward the forest, which grows on till. It was the "natural beauty" of just this sort of countryside that attracted tourists (Figure 8.5). *Säters sköna dal* ("Säters Beautiful Valley") in the southern part of the landscape was famous, and the hills near the village of Bergsäng in the Siljan district attracted attention as early as the nineteenth century.

Shielings were situated above the highest shoreline. Whether or not the meadows and grazing grounds were linked to the shielings depended on the distance between

FIGURE 8.5. Postcard of Sjugare village in Leksand, c. 1900. Collection of Gerda Söderløund, Municipality Archives, Leksand.

them. Both situations, village and shieling, were part of the same economic system in which the seasons determined where the people and animals lived. We are fairly certain that the villages in the valleys and the shielings are about the same age (Emanuelsson 1977, Lange 1994, Emanuelsson and Segerström 1998). Thus we find an example of what is common in forest and mountainous terrain: people have learned to use various niches in the landscape depending on its physical features and the changes in seasons. Here lie the foundations of multiple occupations—no part of the landscape predominated in the use of resources. Infields in the valley villages were the basis of the rights of ownership that gave social legitimacy. As a landowner, one could have a voice in social decisions, even if the amount of land owned was small. Infields were the basis on which state tax was levied and also taxes to the community. Other assets were regarded as more flexible since there was less control over them.

THE GROWTH OF THE RURAL LANDSCAPE

Historical agrarian research on Dalecarlia began early in the twentieth century. No doubt researchers were enticed by rumors that here was an unspoiled population of the kind that no longer existed elsewhere in Western Europe. However, they were also interested in agriculture and how it was practiced. Ever since Gustav Vasa's days, attention had been drawn to the disadvantageous physical structure of the arable landscape and the fragmentation of arable land, which meant farmers would pay fewer taxes. Reforms were introduced to bring agriculture in Dalecarlia into line with other parts of the country. The most important reform was probably "enclosure," which was not introduced until the nineteenth century, about one hundred years after the rest of Sweden.

Reearchers soon discovered the reasons for this fragmentation of land. In Dalecarlia, the system of partible inheritance was strict. The German agrarian historian Wilhelm Aabel produced a map of inheritance customs in Europe in the middle of the twentieth century. On this map, Dalecarlia is completely isolated in northern Europe with regard to its inheritance system (Thirsk 1978). Even though this first attempt at charting inheritance was couched in general terms, researchers clearly were aware that Dalecarlia was different from the rest of northern and central Europe.

The rules governing inheritance in Dalecarlia specified that all siblings were to receive part of the assets, although before 1845 boys received twice as much as girls. What was unique was that girls were allowed to inherit (arable) land in central positions. They thus became landowners and in some respects were the social equals of men. The land they brought with them when they married was documented and in a way remained within the woman's family. This meant that both man and wife were landowners and, at the same time as one property was divided among siblings on the death of their parents, new properties were built up from the land they had inherited and corresponding land inherited from other farms (Figure 8.6). One might say that

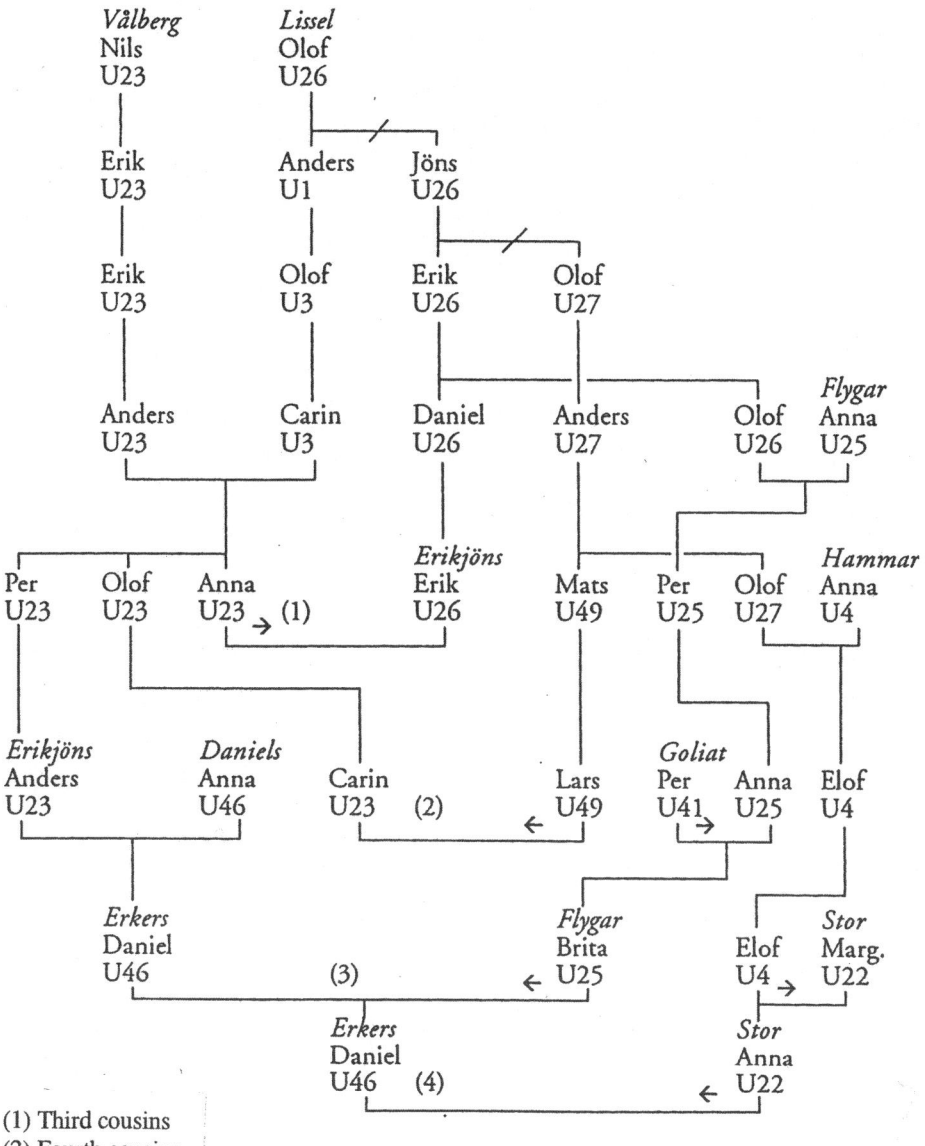

(1) Third cousins
(2) Fourth cousins
(3) Fifth cousins
(4) Sixth cousins

FIGURE 8.6. Marriage and kinship register at Danielsgården in Ullvi, Leksand parish, from c. 1600 to c. 1830. The letter U with a number identifies individual farms in the village (based on Wennersten 2001, 239).

the fragmentation of land was immediately compensated for by the creation of new properties. However, these transactions were not ratified in the same way as they were in other parts of Sweden: the agreement between the contracting parties was verbal (Sw. *sämjodelning*) or was simply summarized in documents relating to inheritance. The system made it impossible for the state to exercise any control over these transactions, which were not always legally registered in court. This was probably why the state was suspicious of a system that in itself seems to have worked perfectly: it did not produce social dropouts, and Dalecarlia is the one province in Sweden that has neither an agrarian upper nor lower class. The system strengthened the position of farmers and families were able to keep their grip on the land, among other ways by encouraging strategic marriages (Sporrong 1998).

In 1827, Anders Grafström, rural dean and author, described the situation:

As, with few exceptions, every farmer in the region himself owns the farm he cultivates, it gives him, however small it may be and however little his gain, a feeling of independence and being his own master, which is reflected in his disposition, habits, and actions; this is the spirit that even the least part of his native soil infuses in its owner's breast. (Sandberg et al. 1827; Holmström and Svensson 1971, 1:9)

In 1862, the English author Horace Marryat noted:

In no province are lands so subdivided as in Dalarne; the peasant never sells, even to a member of his own family, the heritage derived from his forefathers. Hence the traveller beholds long narrow strips of land from six to ten ells in length, and ofttimes but one ell wide, divided off by a slight wooden fence; so infatuated is the peasant, that, should a man at his death leave six plots and six children, instead of each taking one piece whole and entire, they carefully divide them into six lozenges, share and share alike. A bonde [farmer] seeking a wife prefers a helpmate with a plot adjoining his own; so the villagers marry and are given in marriage among their own plots and people. Woe betide the venturesome gallant who comes over the border a-courting! Indignantly expelled with clubs and stones by the young men of the district, he is requested to be off and pay his attentions elsewhere. (Marryat 1862, 2:208)

These descriptions are relevant only to the time *after* enclosure in the nineteenth century. Judging by estate inventories, it was not common to divide up the small strips between interested parties before enclosure. Instead, where what actually happened *before* enclosure has been documented, small parcels were allotted intact to the heirs so that the final sum was fair. One might compare the procedure with a game of cards. The parents had a number of cards to divide among the children so that each child received a certain number of cards to use as the basis on which to build a new property, that is, a nucleus, a certain area of land. With this nucleus in hand he or she looked for—or in most cases was allotted—a partner in marriage, who was of course also a landowner. The effects of enclosure are important because the "modern" idea of Dalecarlia's nonproductive agriculture is based precisely on the effects caused by enclosure (Sporrong 1998).

The disadvantage of the system before enclosure was that the location of parcels in the new property built up by the marriage could be unfavorable geographically. The new owners thus immediately had to start a process of exchange so that their land could be concentrated in the vicinity of their farmstead. To achieve this, owners could also buy up land. This had one very important consequence for the appearance of the landscape. Although each parcel might change hands, its physical boundaries were fairly stable. The checkerboard of small plots that was characteristic of Dalecarlia in the eighteenth and nineteenth centuries was thus a part of the system, so to speak. Small areas were eminently suitable for the kind of transaction that was involved in partible inheritance.

From the practical point of view, two tallies were assigned to each parcel in order to keep a check on this multitude of parcels. On each tally was noted the surface area, the type of land (arable, meadow, etc.), and its yield. The owner kept one tally and the other was kept in a storehouse (usually a barn) by a person who kept all the tallies. When a transaction was made, tallies had to be transferred between owners both at their homes and in the parish storehouse. The system was common in many parts of Europe, even though the token itself, in this case a wooden tally, might vary. I should also mention that these very small parcels—100–200 square meters was common—required special measures. Dalecarlia had small square measures suitable for the system used in the province in which parcels of land were constantly changing hands—the two most basic units were *bandland* and *spannland*, which were only 12 and 120 square meters, respectively (Sporrong and Wennersten 1995, 55ff.).

This system seems to have worked fairly well both economically and socially. However, the central government was unable to exercise any fiscal supervision. Thus in the nineteenth century, enclosure was introduced and the division of land was made uniform throughout the country. The intent of the model, which came from England and Denmark, was to consolidate properties. In Dalecarlia, farms were therefore allotted continuous blocks of land stretching from low-lying lakes and watercourses, across good arable land, up to meadows and pastures, and finally ending in forest. This was definitely something new, for the forest had until now been common land.

The result of this land reform was that every property had a long strip of land, which could be as narrow as 20–30 meters and as long as several kilometers. After the reform, the landscape was "striped" instead of "chequered." Each property now had fixed economic and juridical boundaries, which was something quite new. The problem was that landowners kept to the old system of *sämjodelning*, with the result that the long parcels created by enclosure were constantly being divided lengthways with almost grotesque consequences (Sporrong 1998) (Figure 8.7). Parcels that were 5–10 meters broad might be several kilometers long. The boundaries between joint owners in one and the same mother unit (property after enclosure) were often poorly marked. The system fell into decay when owners lost the flexibility to exchange parcels. The ever-more-fragmented properties could no longer support their owners and dropouts and emigration were the result (Ostergren 1982). Dalecarlia became poor. This is the situation that ethnologists, artists, and authors found when they came to the province in the late nineteenth century—a landscape with agriculture in decay. This course of events certainly contributed to the reputation of Dalecarlia as a poor, conservative province where the people held on to traditions that seemed out of date, but at the same time made their special mark on the physical features of the landscape. In this situation, a government commission—the Dala Commission (SOU 1931, 19)—was appointed to investigate how land was partitioned in the province.

Not until 1962 was *sämjodelning* prohibited in the Swedish constitution. Then all partible inheritance ceased. After that, a commission on ownership was set up to ascertain exactly who owned what in this tangle of narrow strips, but now it was more

FIGURE 8.7. The view from Tibble mountain, c. 1890, Leksand parish. Municipality Archives, Leksand.

complicated because many people had emigrated or abandoned agriculture for other occupations in Sweden. Two phenomena remained: a fragmented landscape, and a large number of people who considered themselves landowners even if they did not farm the land. People who had their roots in Dalecarlia kept their land and returned to their native districts in summer. As landowners, they were welcomed and even played a part in making decisions on matters affecting, for example, the economic conditions of the landscape.

After the Dala Commission's report on ownership, the government began to look into the unfavorable physical distribution of land. Large forest areas were needed to carry out conservation and lumbering and thus a new phase of development began. Enclosure was to be implemented now that *sämjodelning* was prohibited and it met with stubborn resistance because this new system might not take into account private ownership. Smallholdings were to be amalgamated into a form of common land again, and this aroused protest in many places. This system of ownership can be understood only when the code of social behavior is considered, much of it having to do with the ownership of land in past generations—the time before enclosure.

THE PEOPLE

The customs of the province were distinctive and essentially based on landownership. The ethnographer Mats Rehnberg wrote in the Swedish Tourist Club's Yearbook: "It may be seen as a contradiction in terms that the century-old custom of dividing up the land into small parcels as though each man wishes to be his own master still goes on at the same time as the Dala farmers are those who cooperate more than any others in the Swedish agricultural society" (Rehnberg 1972, 147f.). He continued: "By owning land, a person acquired rights on common land and other benefits and fitted into the many-sided cooperation that was a necessary condition in this type of society. The far-reaching pattern of cooperation seems to have contributed to maintaining the forms of social life that have given Dala culture an old-fashioned character. The numerous forms of ownership and cooperation helped to preserve a system of cooperation and equality in which there was little scope for anyone to break away, to run his farm and his life in a way that was not dependent on other people" (148).

No one could express the idea better than this. Here it is a question of strategies for survival in a physical landscape that could not offer a surplus of resources in proportion to the size of the population. Over the centuries, a social code was built up based on landownership, the family, and kinship. As a landowner, a person participated in social life and had social legitimacy—what might be called a "vote." This, in turn, guaranteed that no one dropped out of the community. Yet at the same time the landowner was obliged to work with and cooperate with others. Now questions of inheritance and the division of joint property come into the picture. It is apparent that the people of Dalecarlia followed strict rules when dealing with real estate. The rules are encoded in the building and marriage Acts in the medieval Dala provincial or "landscape" laws, which in their turn were influenced by other laws that served as models, primarily the Uppland laws (Holmbäck and Wessén 1936). However, the landownership and the rights of inheritance were at times at odds with the letter of the written law. The question then arises: Did these deviations occur *after* the laws had been promulgated or are we seeing remnants of older customs that have survived in spite of the law?

The diocese of Västerås has collected population details from parish registers since the seventeenth century, and a state register of farms and their owners has been kept since the mid-sixteenth century, as well as records since the end of the seventeenth century of local land measurements undertaken by the government to make taxation assessments. We thus have a good opportunity to study the development of the population in detail over a long period. In favorable cases we can reconstruct the history of families and farms from the sixteenth century. Furthermore, estate inventories and title deeds can help us with this reconstruction. Concentrating on partible inheritance in Dalecarlia and taking judicial sources and personal histories as a starting point, we can see how partible inheritance was applied.

Marriage patterns show that the choice of a partner meant a great deal if a kinsman were to retain some control over the land. Since most marriages were between partners from the same parish, it seems that partners must have followed clear rules in order to avoid problems of close consanguinity. My opinion is that the women's task was to apply these rules so that marriages that were "too close" did not take place. Here one can draw parallels with other societies where relations are based on kin, as in Africa. The relationship in Dalecarlia between the two contracting parties in a marriage is far too "calculated" to be mere chance. Rules must have stipulated who might marry whom in certain families. These rules refer to inherited ancestral land. Parallels with the medieval Norwegian provincial or "landscape" law codes (*Gulatingsloven* and *Frostatingsloven*) seem particularly clear (Sporrong and Wennersten 1995, 25ff.)

Let us follow developments on two "original" farms from the later part of the sixteenth century (see Figure 8.6). If the heirs were two sons (the second generaration—Anders and Jöns—in the right part of the diagram), this family will divide into two branches. If we chart what happens from generation to generation, the principles of inheritance and marriage seem at first to be about the same as in the rest of the country. However, after four or five generations, marriages clearly become arranged so that contact is established between the two branches emanating from the original farm (marked with an arrow). Thus marriage was permissible between fifth cousins or more distant relatives, exactly as it was in the medieval Norwegian laws. Closer relatives were not allowed to marry without a special dispensation from the Crown. In the archives of Leksand church, for example, records can be found of persons who applied for permission to marry their cousins.

What practical steps could be taken to retain control of the land without confronting problems of consanguinity? One common method was an exchange of siblings. Both the person and the land were transferred from one farm to the other and vice versa. Another way was marriage outside the family, but in that case the land inherited by the partner who left his or her native village (usually the woman) was carefully registered. This land was then farmed by the family into which the woman had married. One or several children of the marriage (once again, usually the women) moved back to their mother's village and became landowners there. In this way they brought new blood into the family. If a married couple was childless, their land returned to the original farms: the wife's land went to her family and the husband's to his. The couple could also take a foster child from either of the two families, in which case the child inherited the whole farm. If the parents died before the children had grown up, the boys could "move back" to their father's family, taking their father's land with them, and the girls likewise could move back to their mother's family together with their mother's land.

These rules of inheritance are not unusual in Europe (Siddle 1986; Sabean 1990, 96–111). What makes Dalecarlia unique is that the rules were applied with a surprising firmness and consistency. This means the rural landscape must be regarded as clearly fenced in by rules governing ownership and cultivation. The small parcels that were such a thorn in the side of authorities were obviously part of a stable society whose ambition it was to survive both socially and ecologically, to use a modern term. The successive fragmentation of land that the authorities believed they could observe never actually took place. On the contrary, between 1734 and 1820 (before enclosure, that is), the average area of a farm actually *increased*. This was true not only in the physical sense but also in the yield. The land was rated parcel by parcel; the total value of the yield on each farm rose on average. It is interesting to note that the increase in area was greatest when both husband and wife came from the same village, which also indicates that people planned strategically to ensure that a marriage was the "right" one. When enclosure was introduced in the nineteenth century, this positive trend came to an end. The new fixed boundaries made it impossible to remain flexible in handling the system of parcels that previously had been characteristic of Dalecarlia. Enclosure was thus an important turning point in the history of Dalecarlia's agriculture: a socioeconomic system collapsed and no one was able to replace it with something radically new, unlike what happened in the rest of the country as a result of the second round of enclosure (Sw. *laga skifte*) in the nineteenth century, when properties were consolidated. In Dalecarlia, however, *sämjodelning* continued (Sporrong 1998).

Finally, if we imagine we are looking out over the rural landscape around Lake Siljan just before enclosure began, without the opportunity to use any form of maps, we could describe the scene. A settlement lies concentrated in kinsman-related clusters, around several centers in the same village domain. Arable land is divided into small square parcels, on average about 300 square meters. The name of each parcel tells us where it is situated. Where the arable land is extensive, arable land enclosed within the same fence is given a name ending in *-åker,* for instance, *Våtåkern* (the wet field) and *Krokåkern* (the hook field). In outlying areas the word *gärde* is used, for example, *Trangärdet* (the crane field) and *Råggärdet* (the rye field). Smaller fenced-in fields have their own names, such as *Erkresåkern* (Erker's field).

The arable landscape looked something like a patchwork quilt, with the smallest fields centrally placed in the landscape. Areas of meadowland were often larger. The outfields nearest the village served as grazing ground; thus the land here was cropped bare but offered a broad view. It is typical of this landscape that everything is on a small scale and open to view. Although there was great mobility in ownership, the boundaries between parcels of land changed little because of the effects of partible inheritance. Landowners did not willingly divide up such small parcels; instead they were part of the system. So these parcels were seldom partitioned when they were passed on to the next generation; heirs were allotted whole parcels. According to contemporary observations, agricultural production in Dalecarlia functioned well. In his 1822 report to the provincial governor, Hans Järta certified that the yield was above average compared with Sweden as a whole (Bortas 1992).

Settlements

Settlements are an important part of Dalecarlia's cultural identity. Nowhere in Sweden do we find such large conglomerations of agrarian settlements. One can almost speak of agro-towns in the continental sense. It is not unusual to see seventy farms in the same village. In addition, there is a comprehensive system of shielings in the forest regions around the mother village.

Partible inheritance undoubtedly was important to the growth of these villages. Anyone holding a certain minimum area of land after the distribution of his parents' property had the right to build on that land. Conversely, if the holding was split up and no longer met the required minimum, the buildings were dismantled and moved to the family's main farm. The population grew greatly in size in the eighteenth century, which meant that new farms were constantly being established. New farmsteads were built on suitable sites around the core of the old village and were linked together with a network of roads, which could be exceedingly crooked.

Futhermore, strategic marriages, which partible inheritance demanded, have left clear imprints on settlements. Closer study shows that adjacent farms were owned by people who were distant relatives. There are two reasons for this: certain families preferred to marry into certain others, which to some extent was connected with economic status; and when a property was distributed among the heirs, they received some centrally situated parcels of arable land and built on these parcels because they were near the settlement. As a result, clusters of farms belonging to related families can be seen as forming a kind of secondary center in large Dala villages, which is part of the reason villages spread out over such large areas and also why valuable arable land was used for building.

THE TOFT AND THE SURROUNDING LAND

Another important detail regarding the cultural landscape and settlements in Dalecarlia is how farm centers were organized. Having a toft was a practical way of showing landownership in the old agrarian society. One might call it a sort of certificate showing that the title to the land had been registered. The organization and function of the toft and the fields immediately surrounding it, as well as the adjacent arable land, are important to our understanding of how society worked from the perspective of landownership. They are also important in the planning of the cultural landscape. Since there are no cadastral maps of Dalecarlia before the enclosure movement of about 1800, we have little information about what the toft looked like and how it functioned under partible inheritance. The parishes of Floda and Nås, however, are exceptions, for in both parishes around the mid-seventeenth century arable land as well as meadowland was mapped.

In her doctoral thesis, Brita Pallin (1977) used these unique cadastral maps on a scale of 1:1,000 to elucidate how land had previously been divided. These maps show in detail what the field system looked like in the seventeenth century. Here was a distinctive fenced-in open-field system with a fully developed pattern of fragmented holdings. The geographic relationship between farm and field undoubtedly suggests that this was originally a single-farm system that was successively built up into villages due in part to partible inheritance. Indeed, using the oldest maps as a starting point, one can say that Pallin has charted in-depth historical phenomena that, to the best of our knowledge, no other researcher has done, and that her findings may be relevant to the interpretation of the partitioning of land in European countries when it was based on partible inheritance.

Pallin shows that in the practical partitioning of land there is a relationship between a farm's toft and the immediately surrounding arable land (Sw. *tomtåker*), which in turn defines the other rights that pertained to the farm. The relation between the area of the toft and that of the *tomtåker* was 1:6, and the relation between the *tomtåker* and other arable land was generally the same (Pallin 1977, 40; 1986). Thus the system of evaluating land based on the toft concept probably governed not only the amount of arable land a farm had, but other rights as well. As far as we know, these observations have never been described before and they are not mentioned in the medieval Dala law. However, if we look outside Dalecarlia, we can wonder whether these conditions also existed in other regions where the principles of partible inheritance were applied, such as areas where Norwegian law was valid.

The *tomtåker* obviously played a special part in the land-partitioning system, but what kind of legal functions did it have? As in Western Europe, specific rules were connected with the concepts of toft and *tomtåker,* mainly of a traditional nature, not only regarding crime and punishment but specifically concerning inheritance, for instance, who should inherit the main farm. The toft and *tomtåker* frequently fell to the eldest son. Pallin suggests that the concept of *bördsjord* (inherited ancestral land, "udal right") might be linked to the *tomtåker* concept insofar as the area surrounding the toft could not be sold without the consent of *bördsmän* (all kinsmen with the right to inherit ancestral land [Pallin 1977, 42]). The "Thing" (*ting,* the representative council or court) had to be notified three times that land was to be sold before the family could actually sell it. Where the toft and *tomtåker* were concerned, there were many obstacles to selling it. The sale had to be approved by all male members of the family. Anyone who owned part of the *tomtåker* also had the right to take part in village discussions (Pallin 1986, 102). There are many parallels to an early situation in Norway and in the Atlantic realms—not least in many details (Jones 1996).

These examples lead to some interesting observations. It is true that there are no special rules in the Dala law concerning the toft and its functions other than those in central Sweden that deal with the medieval form of land division known as *solskifte* ("sun division"). Yet, as far as we know, *solskifte* was never practiced in the region around Lake Siljan. However, the relationship that Pallin shows between the area of the toft and the size of other assets points to parallels in many directions. For example, the toft was often enclosed (fixed) as soon as the property had increased to ideal proportions in relation to the fiscal evaluation of the arable land—a newcomer could not establish a farm within that specific cultivated area. The original farm was thus defined by this physical means of expression. It was possible to return to the toft and its actual size to decide matters connected with the farm itself, such as rights and responsibilities, which might be necessary if the estate were to be divided and the entire property reallocated. It is also possible that the toft had other functions related to social life in the village. In a way, it was the toft that symbolized ownership and participation in making decisions in village affairs. Members of a certain family could also move here if they found themselves in difficult financial straits. Thus, the toft played an important role in the old agrarian society in this part of Europe.

If we take a final look at the layout of settlements, arable land, and meadows, we can say that the patchwork of the arable landscape functioned on firm principles regarding both agriculture and inheritance. The amount of land available was small, but a landowner had the right to build on his own property. Over the years, the landscape became densely populated with highly built-up villages that are still the region's most important feature.

Dalecarlia: Heartland or Anomaly?

Dalecarlia must now be placed in a larger regional perspective. It does in fact differ from other provinces in Sweden and the rest of Scandinavia. If we look at the landscape, it is primarily the fragmentation of land and the dense and extensive settlements that are characteristic features, even today. A stringent social code led to its distinctive physical features. Dalecarlia enjoyed social homogeneity and was egalitarian in its rules. Another characteristic is the special method used to mark ownership of both arable land and toft. A comparison of Dalecarlia with the surrounding provinces is instructive. North of Dalecarlia lies Härjedalen, which was colonized relatively late as an agrarian region. Seventeenth-century maps show how villages evolved from a single-farm structure, and there was little intermingling of parcels. Probably the settlements in Dalecarlia were similar in nature. There was no organized division of settlements and cultivated land in Härjedalen. Inheritance customs regarding the allocation of land were clearly different in the two provinces, and Härjedalen never had the same degree of land fragmentation as Dalecarlia.

To the east of Dalecarlia is Hälsingland, where the parish of Järvsö has been studied fairly closely. Here there are many parallels with Dalecarlia, such as the large villages, and arable land was fragmented at the beginning of the eighteenth century. A kind of partible inheritance system similar to that in Dalecarlia was practiced, but the natural resources of the Ljusnan Valley were different, which meant the distribution of assets between heirs was different as well. As in Dalecarlia, the eldest son inherited the farm. The other siblings, however, did not always obtain land in central positions, but rather in *bodland*—summer farms as they might be termed—on which the heirs could settle permanently after the property had been divided. The historical difference between the two provinces is, however, that inheritance customs were not followed as systematically in Hälsingland. The laws and customs were followed less strictly than in Dalecarlia (Wennersten 1999). Furthermore, far-reaching fragmentation of ownership was permitted, which in turn often led to social mobility—that is, an heir could no longer keep his status as a farmer but was obliged to settle on a croft, sometimes on a cottage. This, however, did not alter the fact that a suitable marriage might lead to upward mobility later in life. Here there is a fundamental difference between the two provinces. Another is that the marriage pattern in the province of Hälsingland had a larger geographical base.

To the south lie the Mälar provinces of Södermanland, Uppland, and Västmanland, where settlements and cultivation were quite different. The most important difference was the law of inheritance. Here primogeniture was the rule, the farm being bequeathed to the eldest son. The other siblings had to be compensated in another way, either economically or, as in Hälsingland, by receiving land in an outlying area. None of the flexibility of landownership that was typical of Dalecarlia was true in the

Mälar region. Settlements and cultivation were organized and regulated according to the medieval reform known as *solskifte* ("sun division"). Arable land was in the framework of an open-field system like that found on the Continent and in England. Settlements were planned on the same strict rules. Geographically, the sequence between farms was the same on the toft as on arable land. One of the objects of the reform was that ownership of the land should not be split up by the division of land among the heirs. Feudal ideas from the Continent may have led to an order that suited the Swedish political environment. Otherwise, development during the Middle Ages seems to have gone different ways in other parts of Scandinavia. The Mälar landscape, however, was linked to similar systems and ways of thought in the southwest of Finland (then a part of Sweden), the southern Swedish provinces of Östergötland and Öland, and large parts of Denmark (Helmfrid 1962; Göransson 1971, 1979; Hannerberg 1977; Sporrong 1985; Roeck Hansen 1999). The southeastern part of Dalecarlia really belongs to these regions, which may be why the medieval Dala laws contain sections regulating the toft that were more like those in the Mälar region than in upper Dalecarlia, according to the earliest maps from the mid-seventeenth century.

To the west of Dalecarlia is Norwegian territory, and here perhaps we can make the most interesting comparisons. We have earlier mentioned the regulations in Norwegian customary "landscape" or provincial law about marriage between relatives and inheriting udal land or ancestral land. Studies of western Norway and the Atlantic islands, especially Orkney, Shetland, and the Faeroes, show clear parallels in the registration of landownership. In Dalecarlia, the toft and the surrounding fields correspond to specially marked land for settlements and udal land (that is, land that controlled the right to use resources in that settlement). This law applied not only to arable and meadowland but also to grazing, fishing, and hunting. Obviously this was an older way of arranging settlements and cultivation and it may be a Scandinavian predecessor to the *solskifte*, which clearly has an "external" origin and may have replaced an older system.

Interesting parallels in the "western Nordic" sphere are worth noting. The differences between this western Nordic region and the strictly planned and regulated pattern of settlement and cultivation in eastern and southern Scandinavia are apparent in many ways. In the latter, primogeniture was the rule, and there was a fixed and consistent farm and ownership structure manifest in an ingeniously constructed land division. Dalecarlia and adjacent regions, by contrast, had a clear farm- and kin-related system of ownership that was exceedingly flexible when parcels of land were exchanged, bought, or pledged. Conditions in Norden may once have been considerably more homogeneous, that is, before the time of the provincial lawbooks and regular strip fields. In that case the transformation of the cultural landscape in Denmark and Scania (Skåne), eastern Sweden, and southwestern Finland may have meant that these regions abandoned a rule of landownership that lived on in western and northern Scandinavia (Karlqvist 1977; Sawyer 1983).

This opens the exciting prospect that opposition to the intermingling of different owners' strip fields grew strong in feudal parts of Europe because power there was connected to the possession of land, often in the form of big estates that practiced primogeniture. New laws were written in Europe, including southern and eastern Sweden, to prevent the weakening of this power. Perhaps an older, more egalitarian influence over landownership thus disappeared in parts of Scandinavia. Instead, principles of landownership were introduced that have lasted to the present—a farm is inherited by only one of the heirs.

Systems of inheritance that were valid in feudal Denmark, including what is now southern Sweden, might indeed have spread to nearby regions, and perhaps to areas with traces of large-scale agrarian production. The principle was that only one heir took over the farm and had the right to farm it. In more outlying areas of northern Europe, feudal ideas had a weaker foothold, perhaps because arable land played a smaller part in the economy. The coastal, forest, and mountainous districts were characterized by diversified livelihoods, working on a small scale and using multiple resources, and they often had an egalitarian structure of ownership.

The people in these marginal regions had learned to make use of *all* the ecological niches in the landscape and based their livelihood on the multitude of natural resources. Thus the pattern of settlements spread out seasonally over a wide area. Here the study of Dalecarlia comes into the picture. Because the people of Dalecarlia have held on to their traditions of land inheritance and rights to the use of land, and also have a well-developed sense of cooperation, we can discover clues to deep roots in Nordic society and how its cultural landscape has been formed. We thus return to the question of regional identity and Dalecarlia's distinctive character from both the historical and the geographical view: Is Dalecarlia the heartland of Sweden or an anomaly? Seen through the eyes of a researcher into landscape, the answer must almost certainly be both. Some features in Dalecarlia's cultural landscape are deeply rooted in Nordic history. If the hypothesis that the province embraced the customs sanctioned in law during the earlier part of the Scandinavian Middle Ages holds, it conjures up a picture of Dalecarlia as a true heartland. Other parts of Sweden developed in different ways, especially in regard to the social code that applied to landownership. We also think we can discern an affinity with western Scandinavia that seems to have been there from the beginning. On the other hand, Dalecarlia's cultural image differs from that in the other parts of Sweden, not least in how the cultural landscape was formed, because the people of Dalecarlia followed the rules of medieval laws and later legislation more literally than elsewhere, and they observed these laws from generation to generation. Although the people dealt flexibly with the practical questions of landownership, in most other matters conservatism has grown into a kind of cultural fixation (Rosander 1987).

Note

Direct quotations from sources in Swedish have been translated by Michael Stevens.

References

Andersson, R. 1991. Byggnadskultur i övre Dalarna. *Bygga och bo i Dalarna,* ed. G. Ternhag, 9–51. Hedemora: Gidlunds, Stockholm.

Bortas, M. 1992. *Storskifte i Dalarna 1803–04 med exempel från Leksands och Åls socknar.* Leksand: Leksands kulturnämnd.

Bremer, F. 1845. *I Dalarna.* Stockholm.

Brorson K. 1985. *Sing the Cows Home: The Remarkable Herdswomen of Sweden.* Mount Vernon, Wash.: Welcome Press.

Crang, M. 1999. "Nation, Region, and Homeland: History and Tradition in Dalarna, Sweden." *Ecumene* 6: 447–70.

De Geer, S. 1926. "Dalanatur och Dalabygder." *Svenska Turisföreningens Årsskrift 1926*: 21–48. Stockholm.

Emanuelsson, M. 1977. *Bosättning, agrarkris och fäbodväsende. Vegetations- och markanvändningshistoria i Läde, Dalarna.* Falun: Dalarnas Forskningsråd.

———, and U. Segerström. 1998. "Forest Grazing and Outland Exploitation during the Middle Ages in Dalarna, Central Sweden: A Study Based on Pollen Analysis." In *Outland Use in Preindustrial Europe,* ed. H. Andersson, L. Ersgård, and E. Svensson, 80–94. Lund Studies in Medieval Archaeology 20. Lund: Institute of Archaeology, Lund University.

Göransson, S. 1971. *Village Planning Patterns and Territorial Organization: Studies in the Development of the Rural Landscape of Eastern Sweden (Öland).* Acta Universitatis Upsaliensis, Faculty of Social Sciences 4. Uppsala.

———. 1979. "Regular Settlement in Scandinavia: The Metrological Approach." *Landscape History* 1: 76–82.

Hannerberg, D. 1977. *Gård, by och territoriell indelning i den äldre kumlabygden. Kumlabygden IV.* Kumla.

Hård af Segerstad, U. 1972. "Svenskt ideal?" *Svenska Turistföreningens Årsskrift 1972*: 7–20. Stockholm.

Helmfrid, S. 1962. "Östergötland Västanstång: Studien über die ältere Agrarlandschaft und ihre Genese." *Geografiska Annaler* 1962: 1–2.

Holmbäck, Å., and E. Wessén. 1936. *Svenska landskapslagar tolkade och förklarade för nutidens svenskar. Andra serien: Dalalagen och Västmannalagen.* Stockholm: Geber.

Holmström, R., and S. A. Svensson, eds. 1971. *Dalarna.* Malmö: Allhems förlag.

Jones, M. 1996. "Perceptions of Udal Law in Orkney and Shetland." In *Shetland's Northern Links: Language and History,* ed. D. Waugh, 186–204. Edinburgh: Scottish Society for Northern Studies.

Karlqvist, 1977. "Vad säger runstenarna?" *Meddelanden: Arkivet för folkets historia 1977*: 4.

Klein, E. 1926. "Hembygdsrörelsen i Dalarna." *Svenska Turistföreningens Årsskrift 1926*: 103–28. Stockholm.

Lange, U. 1994. "Dammskog och Skallskog." *Bebyggelsehistorisk Tidskrift* 25: 7–24.

Marryat, H. 1862. *One Year in Sweden,* 2 vols. London.

Näsström, G. 1937. *Dalarna som svenskt ideal.* Stockholm: Wahlström & Widstrand.

Nordling, J. 1907. *Siljan: En bok om Sveriges hjärta,* 2 vols. Stockholm: Fritzes.

Ostergren, R. 1982. "Kinship Networks and Migration." *Social Science History* 6, no. 3: 293–320.

Pallin, B. 1968. "The 'Bytomt' (Village Toft): Its Significance and Function." *Geografiska Annaler* 50 B: 52–61.

———. 1977. *Bälg och Bondelag. Några drag i jordfördelning och uppodling i Västerdalarnas södra del 1539–1670*. Meddelanden serie B 34, Kulturgeografiska institutionen, Stockholms Universitet. Stockholm.

———. 1986. "Medieval Hamlets in Northern Western Dalarna: Development of Physical, Economic, and Social Organisation." *Geografiska Annaler* 68 B: 95–104.

Rehnberg, M. 1972. "Folket i de stora byarna." *Svenska Turistföreningens Årsskrift 1972*: 140–55. Stockholm.

Roech Hansen, B. "Land Organisation in Finnish Villages Around 1700." *Fennia* 177, no. 2: 137–56.

Rosander, G. 1976. "Dalturismen under hundra år." *Dalarnas hembygdsbok 1976*. Falun: Dalarnas hembygdsförbund.

———. 1987. "Från herrskapsturism till turistindustri." In *Turisternas Leksand: Turismen i Leksand, Siljansnäs och Ål genom tiderna*, ed. G. Rosander, 13ff. Leksands sockenbeskrivning IX. Leksand: Leksand kommun.

Sabean, D. W. 1990. *Property, Production, and Inheritance in Neckarhausen, 1800–1870*. Cambridge Studies in Social Anthropology 68. Cambridge: Cambridge University Press.

Sandberg, J. G., A. Grafström, and C. Forssell. 1827. *Ett år i Sverige: Taflor af svenska almogens klädedrägt, lefnadssätt och hemseder*. Stockholm: J. Hörberg.

Sawyer, B. 1988. *Property and Inheritance in Viking Scandinavia: The Runic Evidence*. Occasional Papers on Medieval Topics 2. Alingsås: Viktoria bokförlag.

Siddle, D. 1986. "Inheritance Strategies and Lineage Development in Peasant Society." *Continuity and Change* 1: 333–61.

SOU 1931:19. *"Dalautredningen." Betänkande med förslag till lag med särskilda bestämmelser om delning av jord inom vissa delar av Kopparbergs län mm*. Statens Offentliga Utredningar 1931: 19. Stockholm.

Sporrong, U. 1985. *Mälarbygd: Agrar bebyggelse och odling ur ett historiskt-geografiskt perspektiv*. Meddelanden serie B 61, Kulturgeografiska institutionen, Stockholms Universitet. Stockholm.

———. 1998. "Dalecarlia in Central Sweden Before 1800: A Society of Social Stability and Ecological Resilience." In *Linking Social and Ecological Systems: Management Practices and Social Mechanisms for Building Resilience*, ed. F. Berkes and C. Folke, 67–94. Cambridge: Cambridge Univerity Press.

———, and E. Wennersten. 1995. *Marken, gården, släkten och arvet. Om jordägandet och dess konsekvenser för människor, landskap och bebyggelse i Tibble och Ullvi byar, Leksands socken, 1734–1820*. Leksands sockenbeskrivning X. Meddelanden serie B. 91, Kulturgeografiska institutionen, Stockholms universitet. Stockholm.

Thirsk, J. 1978. "The European Debate on Customs of Inheritance, 1500–1700." In *Family Inheritance, Rural Society in Western Europe, 1200–1800*, ed. J. Goody, J. Thirsk, and E. P. Thompson, 177–91. Cambridge: Cambridge University Press.

Wennersten, E. 2001. *Gården och Familjen. Om jordägandet och dess konsekvenser för människor, landskap och bebyggelse i Säljesta by, Järvsö socken 1734–1826*. Meddelanden nr 108, Kulturgeografiska institutionen, Stockholms Universitet. Stockholm.

9.
Selma Lagerlöf's Värmland: A Swedish *Landskap* in Thought and Practice

Gabriel Bladh

> I must now describe this long lake and the rich fields and the blue mountains around it, since they were the scene where Gösta Berling and the gay cavaliers of Ekeby spent their joyful life.
> —Selma Lagerlöf, *Gösta Berling's Saga*

The epigraph presents the initial lines of the Swedish author Selma Lagerlöf's description of the landscape around the long Lake Löven in *Gösta Berling's Saga*. When she published the novel in 1891, she made a lasting contribution to the characteristic image of *landskapet* Värmland (here both the province and the landscape of Värmland). She connected the (hi)stories of the province with the geographical landscape and cast them into a romantic light that still shapes the images of Värmland and Värmlanders. An important issue here is the pattern of connotations between the multilayered concept of *landskap*.

Fryksdalen (Lake Löven of the Saga) is a typical part of Värmland, but not every part of Värmland was in the center of the setting. In contrast to the rich and lively *bygd* (countryside) around Fryksdalen, the main setting of Lagerlöf's story, she also depicts the everlasting forests of Finnskogen in her story. In the travel diaries of the Finnish writer Carl Axel Gottlund, we will discover another Värmland in those forests. Thus Selma Lagerlöf's Värmland can also be seen as a territory with a center and periphery and to some degree as an area for inclusion and exclusion.

In this chapter I will describe and analyze some characteristics of the landscape region Värmland in a historical perspective and problematize the way concepts and categories are used. Starting with the different meanings of *landskap* and landscape, I will explore the links between territory, landscape, and regional identity. How can we think about the relations between *landskap* as a historical province and as scenery?

Through a reading of *Gösta Berling's Saga*, I will discuss some aspects of the historical geography of Fryksdalen and Finnskogen, starting with the landscapes of the 1820s and comparing them with the classical image of Värmland.

Landskap, Landschaft, and Landscape: More Than a Question of Translation?

The term "landscape" has a long and ambiguous history. Much thought has gone into trying to find an essential definition of the concept of landscape, especially in the wake of its scientific use in geography during the interwar period.[1] When German geographers adopted the concept of *Landschaft* in the nineteenth century, it came to represent a portion of the tangible, physical earth studied as a physical or cultural landscape. However, in practice, different conceptions were mixed. In northern Europe from at least the Middle Ages, *landskap* has stood for a historical province, which has generally been culturally and geographically homogenous. With the influence of the Dutch *landschap* painters, the term "landscape" came to represent a view of inland scenery. Primarily, the object of study was a visible part of the Earth's surface in accordance with the predominant morphological view, in which the role of people was often unclear. In the background, those invisible relations in and to the landscape as an object of study were often unspoken. With the scientific use of the term followed connotations that affected the term's relations, often taken for granted, to concepts such as region, nature, culture, and nation. Denis Cosgrove (1998 [1984]) emphasized landscape as a historically specific way of seeing the world. He linked the idea of landscape with changes that have occurred in the relations between humans and their land with the emerging capitalist economy of southern and northwestern Europe. Thus connecting changing social relations of production with changes in human consciousness, the idea of landscape indicated a new vision, which signified a specific view of a subject. However, Kenneth Olwig (1992, 1996) has stressed the history of the Nordic concept of *landskap* as related to ideas and practices that dealt with law, justice, and culture. It is important to grasp the Nordic and German linking of landscapes, territories, and cultural habits when discussing the *landskap* of Värmland.[2] The Scandinavian term *bygd* has similar connotations found in connections between a lived practice, a territory, and a cultural landscape (Nelson 1913; Enequist 1941).

The concept of *landskap* as a lived territory or province has a long history in Sweden, at least as far back as the early Middle Ages. Although a strong Swedish nobility came into being during the twelfth and thirteenth centuries, the peasants retained their influence in local matters and kept old jurisdictions. Until the early sixteenth century, the *landsting* (the court of the province) was the place where the men of the *landskap* territory gave, or withheld, their support for going to war, often in a direct meeting with the king. However, this was not an egalitarian society. Communalism meant a balance of power between nobility and peasantry, and it also had clear geographical differences (Blickle, Ellis, and Österberg 1997). The tradition of communalism and provincial autonomy was strong, and a properly representative Swedish diet was established for the first time in the sixteenth century. In the seventeenth century, government became increasingly formalized and centralized, with the introduction of absolutism in the 1680s as an intense period of change in the local administration. Government made crucial decisions in standardizing justice and controlling regional and local administration. At the regional level, a new division of the country into *län* (counties), with provincial governors, replaced the older division already of the 1630s. Yet the regiments first established during the reign of Gustav II Adolf were still known as *landskapsregementen*, a way to apply an old legitimacy and social memory to a new order.

The concept of *landskap* as scenery spread to Sweden during the period of Romanticism and was practiced especially by the upper class in the style of *"voyage pittoresque"* (looking for the sublime). Nevertheless, the concept of *landskap* understood as province obtained new recognition in the second half of the nineteenth century, during the period of national awakening and "the discovery of the people." Because of Sweden's political position during this period, there was little need for self-assertion toward other nations and the country's ethnic character was not threatened.[3] Folkloristic interest and other processes of identity-formation were directed to the provinces *(landskap)* instead of to the national level, called by the ethnologist Mats Rehnberg (1980, 29) *"landskapens nationsbyggande"* (the national construction of the provinces). Descriptions of provinces were published and historical societies looking back to the myths and memories of old times were founded in the different provinces. The historical legitimacy gained from keeping the old *landskap* unit was important in a rapidly changing Sweden, whose ongoing process of modernization was triggered by migration, industrialization, and urbanization. These ideas were often initiated by student organizations at universities and organized according to the students' home *landskap*.

Artists and writers also played a part, often connecting the aspects of *landskap* as an inhabited and lived province and *landskap* as scenery. The "naturalistic romanticism" that characterized much Nordic art in the 1890s was, in Sweden, connected to an aesthetic nationalistic revival, which departed from the old militaristic symbols of "Swedish punch patriotism" (Björck 1946). The sense of national distinctiveness at this time called for a pictorial language open for place-bound identification and reflection, which was found in the genre of landscape painting (Brummer 1997). Here artists depicted typical Swedish nature and historical culture in different geographical areas and stressed their organic connections. Thus a kind of cultural regionalism was developed where the unit of *landskap* had a central position as both an arena of practices and a symbolic territory.[4] Selma Lagerlöf's book *Nils Holgerssons underbara resa* (1906–7) (*The Wonderful Adventure of Nils,* 1947), used as a geography reader in elementary schools, is a principal example of this connection between landscape as lived province and landscape as scenery.[5] In a similar way, the open-air museum of Skansen, which opened in Stockholm in 1891, the same year that *Gösta Berling's Saga* was published, provided a material "landscaping" of the idea of the national construction of the provinces *(landskapens nationsbyggande)*. At Skansen, farmsteads and buildings, moved from different parts of the Swedish countryside, were rebuilt to represent and celebrate the various facets of the nation. Regional cultures were enframed and pictorialized in a combined process of naturalization, exoticization, and romanticization (Crang 2000). This appeal to "Swedishness" would transcend established social boundaries and promote social harmony in a changing society (Facos 1998; Mels 1999).

In scientific geography in Sweden, literally a child of its time, we can identify this blurring of categories concerning the concept of *landskap* as province or scenery. Sometimes the concept of "landscape picture" *(landskapsbild)* was used explicitly in an attempt to clarify the meaning (Kristoffersson 1924). Interestingly, the leading Swedish geographical concept up to the 1950s was *bygd,* which also was connected to multiple layers of meaning in a similar way to the concept of *landskap.* It was often used as a synonym for the meaning of cultural landscape as a settled agricultural area. Both *bygd* and *landskap* as province very much belonged to the disappearing peasant Sweden. *Bygd* quickly became outdated in the highly urbanized Swedish society of the 1960s and 1970s, though traces of it remained in terms such as *landsbygd* (countryside) or the new *glesbygd* (sparsely populated area). Nonetheless, at least outside urban regions, *landskap* is still a locus of regional identity in Sweden, now also promoted by the tourist industry and other image-producers.

In contemporary Anglo-American cultural studies, including cultural geography, the popular concept of landscape has mainly been seen as a form of scenery. Landscape seen as text or as a symbolic arena has promoted significant research (see the introduction in Cosgrove 1998 [1984]). By contrast, the Nordic use of the concept still has connotations of the making of landscape through social practices in a territory. The differences between the terms *landskap* and "landscape" are more problematic than simply a question of translation. The concepts reflect related, but not identical, "ways of seeing." In the evolving English-speaking global academic network, we have to understand how we decontextualize geographically situated cultural connotations and social practices when we try to develop a common geographical discourse. A historical comprehension of how the concepts and categories have been shaped and reshaped in different cultural contexts through time and space, both in thought and practice, thus makes up an important field of research.

The search for an essential definition of the concept of landscape is a futile point of departure. Consequently, the term "landscape" (which I use as an inclusive concept here despite the problematic language connotations) has different meanings in different historical-geographical contexts, and the construction and use of the different meanings have shaped different scientific traditions (Jones 1991). Thus we can detect tension between different discourses using the term "landscape." Landscape can be seen as the appearance of an area, the assemblage of objects used to produce that appearance, or the area itself. It can also be seen as a way of seeing, and as a part of different cultural and social processes and practices.[6] The term "landscape" can be described as a "chaotic concept" (Sayer 1992) and in that sense it could be argued that it should be abandoned. From my point of view, it should instead be seen as a fruitful concept for studying the tensions and relations between different aspects of the Earth and the world and how these change. Here I will use landscape as a reciprocally related way of seeing, representing, and making land or as enacting territory. This is always bound to situated, embedded, and embodied practices, and it puts the perspective of the actor or observer in the center. Thus landscape has to be seen as a relational concept. This landscape can be looked at as an arena, where natural processes belong along with human-land relations in the form of social practices, "ways of seeing," and power relations. A (physical) landscape in the traditional sense is not something given "out there" as a preconstituted entity, and it cannot be seen independently of meaning, social relations, and human actions. This is certainly not to deny that there are physical and organic processes, but representations of these processes are always interpreted in an embedded context of human practices and signification. I use the concepts "landscape of signification," "institutional landscape," and "action landscape" as analytical tools alongside the term (physical) "landscape" to stress the relational character of landscape as a meeting place between "society" and "nature" (Bladh 1995). People "scape land," but they do not do it under circumstances they have chosen themselves. Every generation is born into an existing landscape, where the given (but changing) material landscape and spatial structure, its social structure and cultural and economic relations, specific "ways of seeing," and rights of disposition provide opportunities or impose restrictions for action in the landscape. We have to turn to specific areas at different historical periods to put thought and practice in context. Here I will shift the perspective to the geo-historical setting of the *landskap* of Värmland, and some situated perspectives, where the landscape is seen as a transformative and contextual arena or enacted territory.

Värmland: Territory, Landscape, and Regional Identity

Värmland (Figure 9.1) can in many ways be looked upon as a border region in terms of its physical and cultural geography (Furuskog 1924; Widgren 1988; Bladh 1990, 1995). The topography changes from low plains around Lake Vänern in the south to hilly country toward the north. The land is incised by a number of north-south-running rift valleys, such as Klarälvsdalen, Fryksdalen, and Glafsfjorden. The bedrock consists mainly of Precambrian rocks (granites and gneisses). Climatic conditions also vary: the northern part, for example, is snow-covered nearly one hundred days more than the south. The transition belt from mixed forests in middle Sweden to coniferous forests in the north also runs through Värmland.

Glaciation, melting water, and changing sea levels in the postglacial period deposited different landforms, which later guided the zoning of vegetation, settlements, and cultivation. The maximum limit of the sea after the last ice age is an important geographical borderline, known as the "highest coastline." Above this borderline, moraine predominates, while sedimentary deposits are found mainly below it. Traces of agricultural settlement, located in the sedimentary areas to the south, date from Neolithic times. The political border with Norway to the west has in different periods worked as an obstructing barrier and provided a possibility for exchange and innovations.

Although a few large estates near Lake Vänern had been established by the late Middle Ages, the valleys around the lakes of Glafsfjorden, Värmeln, and Fryken and

FIGURE 9.1. The province of Värmland.

the Klarälven River were dominated by peasant freeholders. The farms were solitary with fields mostly held in severalty. Often just small fractions of the arable land were left as fallow (Jansson 1998). Cattle raising and milk production were important means of livelihood, especially in the north and west. Here cooperation between farms existed, which was regulated by customary law regarding land use. As in other Swedish *landskap*, there was a special *landskapslag (law of the province)* during the Middle Ages. The *ting* (a combined communal assembly and court) was held at Tingvalla (today's Karlstad). In specific *bygder*, such as Fryksdalen, the legal district was called *härad* (hundreds). At the *häradsting* (the court of the hundred), the freeholders were also a prominent group during the early modern period. This enabled the farmers to uphold a strong position vis-à-vis the central state in questions regarding the use of land. During the sixteenth and seventeenth centuries, for example, the district court in Fryksdalen was able to oppose the claims by the Crown on the farmers' outlying land (Bladh 1995). The courts were important places for interaction connected to how the institutional landscape was enacted, both in relation to formal legal practices and regarding use of customary law. These institutional and informal practices were important in the formation of social memory and a feeling of identification and belonging to the *bygd* and to the larger *landskap* territory. Yet the institutional continuity of a communalistic system should not be understood as reflecting a static or egalitarian society. For example, periods of colonization as well as desertion can be traced to this period. Such processes also effected changes in the social structure of the society.

Since the early modern period, Värmland has been characterized by dynamic change. Strong economic dependence on the trade conditions for iron and wood has given preconditions for both improvements and setbacks. The use of the forest has been a dominant project in the landscape. Eastern Värmland, a part of the Bergslagen region, was colonized during the late Middle Ages. There iron mining opened up the region to outside influences. The expansion of mining activities also meant that the peasantry was involved, for example, through cattle transport, cereal production, charcoal production, or different forms of transport in connection with iron production in Bergslagen. This expansion also influenced the agrarian settlements of Värmland. From the seventeenth to the early nineteenth century, the population increased rapidly and most solitary farms were divided and villages formed. The geographer Mats Widgren (1988) has characterized Värmland as an area of "immature cultural landscape" because of its continuous transformation.

In the 1630s, a geographical division of labor was formed between mines and smelting works and the ironworks and forges, with the result that a large number of ironworks were gradually established in west and central Värmland (Furuskog 1924). The ironworks became an important part of the landscape of Värmland, except in the plains around Lake Vänern. Until the end of the seventeenth century, areas such as Fryksdalen formed a homogenous farming society, but this changed when the iron industry expanded. The industry's golden era in Värmland occurred during the eighteenth and the first half of the nineteenth century. The province was one of the most important iron-producing areas in Europe. The farmers in the valleys became connected to the smiths and ironmasters of the ironworks and manors. After the restructuring of the iron industry at the end of the nineteenth century, many old ironworks were changed to pulp mills. Thus the geographical structure of the old industry was also important for the later industrialization period of Värmland, when forestry and the forest industry became the most important forms of production.

At the beginning of the seventeenth century, large parts of the previously uninhabited moraine areas of Värmland were colonized by Forest Finns (Bladh 1995). During the sixteenth and the early seventeenth century, a rapid population expansion occurred in all directions from the Savo area in Finland. In the Norwegian-Swedish border areas in Värmland-Solør, a continuous area of settlement known as Finnskogen ("the Finn Forest") was established. Initially, during the colonization period of the seventeenth century, slash-and-burn cultivation was an important livelihood, and later, cattle raising and milk production became dominant. From the end of the eighteenth century, timber production was an important livelihood. The population in Finnskogen made up a distinct ethnic group until the twentieth century.

During the early modern period, a number of new towns were founded in Sweden, including three in Värmland: Karlstad (1584), Filipstad (1611), and Kristinehamn (1642). These towns were situated on existing trading routes. From the beginning, the towns were linked to the iron trade and had strong connections to Nya Lödöse and its successor, Gothenburg (Göteborg). Karlstad became the seat of a bishopric in 1647, but not a seat of county government until 1779, when the county of Närke-Värmland was split into two new counties, Värmland and Örebro *län*.

I have presented the geo-historical structure of Värmland before the changes brought about by industrialization and urbanization. Geographically, the province *(landskap)* of Värmland appears to form a heterogeneous area.[7] In spite of this, the image of Värmland is to a large extent homogenous: the popular image of the physical landscape of Värmland, as seen in tourist brochures or picture books of Värmland, seems to display unifying characteristics. What is the reason for this? The explanation, I would argue, is that regional identity is a social construction, shaped by situated practices, where specific forms of collective and individual consciousness have been produced and reproduced (Paasi 1986).[8]

In the meeting between the old farming society of Värmland and the new ironwork sector, we find a developing culture that produced the breeding ground for a changed landscape of signification and new forms of regional identity. Such famous Swedish writers as Erik Gustaf Geijer, Esias Tegnér, Gustaf Fröding, and Selma Lagerlöf were rooted in this landscape of ironworks and mansion houses. Clear differences exist between Geijer and Tegnér, living at the beginning of the nineteenth century and still inspired by the Gothic heritage, and Fröding and Lagerlöf, living at the end of the century and imparting a backward-looking and romantic picture of Värmland. Yet in different ways they are connected to what is known as the classical period of Värmland (c. 1790–1840). This period has solidified the image of Värmland as a joyful and adventurous province. Important symbolic cultural values are also heard in the regional anthem, "The Värmland Song" (A. Fryxell), and the popular burlesque comedy known as the "Värmländers" (F. A. Dahlgren), both written in the mid-nineteenth century (Olsson 1950). This image of classical Värmland has been promoted most effectively in the writings of Selma Lagerlöf, the Nobel Prize–winning author, and especially through *Gösta Berling's Saga*.[9] Whereas this image is still promoted by the tourist industry, and perhaps has lost some of its romantic appeal, it has shaped the ideas of several generations about Värmland and Värmländers. *Gösta Berling's Saga* is of special interest because Lagerlöf shaped her plot around the legends and stories of her childhood in Fryksdalen. Those stories came into being through the meeting between people from the farming area, the ironworks, and the manors. While the romantic spirit often conceals existing power relations and material conditions, the latter are still present. Although the stories could be interpreted as being old-fashioned, Lagerlöf has composed a significant lasting landscape in which she mixes realism and fiction. Here the multilayered unit of *landskap* has a central position as both an arena of practices and a symbolic territory, connecting "nature" and "culture." Through reading *Gösta Berling's Saga*, I will compare the image of Classic Värmland with a historical-geographical narrative about Fryksdalen and Finnskogen, two neighboring districts in upper Värmland. I will also compare Lövsjö hundred, which Lagerlöf constructed in her novel, with the idea of Finnskogen hundred proposed by the Finnish student C. A. Gottlund in the 1820s. I will elucidate the counterpoint between Värmland's center and its periphery, its culture and its "nature," by comparing two imagined communities.

Selma Lagerlöf and Värmland

Gösta Berling's Saga takes place in Lövsjö hundred (Fryksdals Härad) in Värmland in the 1820s. On Christmas Eve, twelve aging, debauched, and homeless men known as "the cavaliers" are given one year's possession of the estate of Ekeby (Rottneros, Figure 9.2) and its ironworks through a contract with the devil, provided they remain true to their cavalier code to do nothing that is useful or sensible. During the Christmas dinner party at Ekeby the following day, the Lady of Ekeby, famous as the owner of seven foundries and one of the most powerful persons in Värmland, is thrown out of her home when the secret of an old love affair is revealed. The prophecies of the Evil One are thus fulfilled. The lady's husband, the Major of Ekeby, gives the huge property inherited from her wife's lover into the hands of the cavaliers, thinking their bad management would do the greatest harm. The leader of the group of men is the

FIGURE 9.2. The Rottneros House, named "Ekeby" in *Gösta Berling's Saga,* is a main tourist attraction in Värmland. Photo by Bertil Ludvigsson, reproduced with permission.

defrocked and ruined pastor Gösta Berling. He is also known as a devoted womanizer, but at the end of the story he finds the woman who becomes his savior. The saga is a series of connected adventures featuring different members of the cavaliers. The cavaliers send "a wave of adventure" up and down the shores of Lake Löven, which are played out from ballrooms to crofters' houses during the year of affliction that follows. The novel addresses the perennial struggles and joys of human existence, but it also presents the landscape and people of the area in a lively way. The novel presents Lake Löven, the fertile plain and the blue hills around it, as well as the peasants' huts, the mansion houses, and the foundries, fields, and forests. Often Lagerlöf includes the entire *landskap* of Värmland in her story. She portrays the mountains, lakes, and fields as dynamic living actors, but also as related to the people and their life in the countryside of Lövsjö hundred.

For Gösta Berling, the smiling Lövsjö hundred is contrasted to Finnskogen, the great forests of the north. When this defrocked priest came to the parish of Bo (Sunne) as a miserable beggar during his wanderings, he imagined these forests as he awaits Death in a snowdrift. From Lövsjö hundred, "in this land of wealth and joy, where the estates lay side by side, and the great iron foundries adjoined one another" (Lagerlöf 1997 [1918], 19), he longs for the quiet of the great forest. In Lövsjö, "he heard the thunder of the flails on every threshing floor, as if the grain were unfailing; here loads of timber and charcoal came in endless succession from the inexhaustible forests, and the heavy wagons of the ore obliterated the deep ruts cut into the roads by the preceding carts. Here, sledgefuls of guests drove from one estate to another and it seemed to him as if Joy held the reins, and Youth and Beauty stood on the runners" (20). Yet Gösta Berling instead longs for the everlasting forests in Finnskogen, "where the trees rise straight and column-like from the level, snow-covered ground."

Lagerlöf presents the areas Fryksdalen and Finnskogen (Figure 9.3) as a contrast of center and periphery, from both a metaphorical and geographical perspective. These neighboring regions can be seen as contrasted ethnic landscapes but also as related to different histories dividing the area into *bygd* and forest. How can we understand this contrasting effect in the framework of the image of classical Värmland that Lagerlöf so successfully depicted in this novel? We can fruitfully interpret it in the light of a review of the book in 1891 by Gustaf Fröding: "This Värmland she [Lagerlöf] delineates resembles the real, but it is a fantastic enlarged and concentrated Värmland. You have all the characteristics of the *landskap* assembled in one place, lakes and rivers,

FIGURE 9.3. Winter landscape in Finnskogen. Photo by Gabriel Bladh.

high mountains and deep valleys, wilderness life and mansion life, everything brought together and intensified to work as a symphony where all strains join in. And if you sometimes think that exaggerations have manifested themselves too much, nevertheless you feel that the keynote is genuine" (cited in Olsson 1950, 30f.). We can compare this with the well-known recitation in the novel by one of the cavaliers, Squire Julius, from the summit of the Dunder Cliff (Gettjärnsklätten): "Ah, Värmland, my beautiful, my glorious Värmland! Often when I have seen thee before me on a map, I have wondered what thou didst represent, but now I know what thou art" (Lagerlöf 1997 [1918], 231). While viewing the wide landscape lying below the mountain, he continues: "Wide forests are thy dress. Long bands of blue waters and chains of blue hills border it. Thou art so simple that the stranger sees not how lovely thou art. Thou art poor, as the devout desire to be. Thou sittest still, while Vänern's waves wash thy feet and crossed legs. To the left thou hast thy mines and thy fields of ore; there is thy beating heart. To the north thou hast the dark, lonely regions of wilderness, of mastery, and there rests thy dreaming head" (232). Although much could be said about this way of representing Värmland, it would be wrong to charge Lagerlöf with exclusion. Fröding points to the way Lagerlöf places Fryksdalen in the center of her centrifugal representation of Värmland. The resulting effect is that the *bygd* of Fryksdalen is seen as *landskapet* Värmland. Moreover, nature and landscape are seen together. Everything is included—the mines in Bergslagen, the shores of Lake Vänern, the big forests in the north—yet Fryksdalen is in the middle of this image of classical Värmland.

When Lagerlöf was writing her story in the 1890s, the old society of Värmland was rapidly disappearing. She has described how she experienced the death of the ironworks in the 1870s as a disaster for her native *bygd*. Her uncle was afraid "the whole of Fryksdalen would collapse." The story of Gösta Berling was written in Landskrona in Scania, where Lagerlöf was a schoolteacher, and she completed it in Sörmland. The book was clearly triggered by the loss of the family home Mårbacka, which had to be sold in 1890 (Edström 1991). Her wish was "to write an epos for a whole *bygd*" and "to preserve the old stories and myths" from the period of classical Värmland (Ahlström 1959). Although the classical image of Värmland is surely an image of the elite in Värmland, Lagerlöf also describes the stratification in the society, and there is an obvious foreboding in the novel of a coming upheaval.

Lagerlöf mixes historical characters and places with fiction when she constructs her area of Lövsjö hundred. She connects everyday practices to ideals and utopias, where the landscape is the enacted territory, and deliberately uses both a geographical and metaphorical perspective. The metaphorical perspective weaves together her realistic descriptions of landscape and people with sentiments and affection. The anthropologist Eric Hirsch (1995) discusses something similar when he defines landscape as entailing a relationship between the "foreground" and "background" of social life. The foreground can be seen as the concrete actuality of everyday social life, while the background suggests the perceived potentiality thrown into relief by our foregrounded existence. Here landscape becomes a cultural process existing between the two poles of "the way we are now" and "the way we might be." I suggest that this ongoing process between everyday practice and normative ideas is also at stake in the double meaning of *landskap* that can be followed in Lagerlöf's writing: *landskap* as lived territory and scenery.

What, then, is the relation between the image of classical Värmland and the images from other historical sources? How can the contrasting histories of Fryksdalen and Finnskogen be thought of as practices, power relations, and ways of seeing situated in the landscape? In the 1820s, ongoing economic and social processes occurred that increasingly interrelated and integrated the *landskap* of Värmland in a material sense, as seen in Gösta Berling's description of Fryksdalen above. But this adds to the differences between areas such as Finnskogen and Fryksdalen in other ways. Here the history of Finnskogen points to the question of inclusion and exclusion, which also play an important part in the forming of a regional identity.

Fryksdalen in the 1820s

In Fryksdalen in the 1820s, farming (Figure 9.4) and iron production were the dominant occupations. Tradition and a small-scale pattern characterized farming activities. In the small, scattered fields of the sedimentary areas, oats were the most important crop, but the region was not self-supporting in cereals. Contributory forms of livelihood provided the possibility of obtaining an income that enabled the freehold farmers to pay taxes and buy cereals. Cattle raising and milk production provided an important share, as did harvesting fodder and moving the cattle to the summer farms. An increase in population started to widen the gap between the ordinary farmers and the people without their own land, but workers were still needed. In the eighteenth and nineteenth centuries, northwest Värmland was characterized in Swedish terms by an enormous growth in population and settlements. In the second half of the nineteenth century, the population began to decrease as a result of migration to Norrland, emigration to the United States, and eventually migration to towns and cities. This process coincided with the emergence of industrial society.

In the 1820s, iron production created opportunities for wealth that was concentrated with the ironmasters and mansion houses of Fryksdalen. The transport of iron ore from Bergslagen, of charcoal to the ironworks, and the ready-made bar iron toward Karlstad on its way to export was an important source of income for the farmers. Areas where little or no charcoal was produced for the ironworks became possible sources of raw material for the timber trade, and these included the forest areas of upper Fryksdalen and Finnskogen. The different drainage basins formed distinct production areas because of the importance of timber floating as a means of transport. The iron and

FIGURE 9.4. Fryksdalen in the mid-nineteenth century, in "Värmland in Drawings" (Schéele 1858–67). Such drawings promoted the idea of *landskapet* Värmland.

timber trade had led to the formation of concrete links between the landscape and the people of northern Värmland and a production, distribution, and consumption system, in which the trading houses in Gothenburg acted as middlemen, mostly for English buyers. Wholesalers and owners of ironworks such as John Hall and Niklas Björnberg, the richest men in Gothenburg, had direct connections to Fryksdalen. Many of the trading houses were represented by male members of the families living on the spot, who became ironmasters, or by their wives, like Björnberg's daughter, who lived in Rottneros. The representatives of the trading houses at the Fryksdal mills were also able to purchase timber from upper Fryksdalen for their sawmills along the River Göta, which had been promoted by the freeing of the timber trade in Värmland in 1784. Their mansion houses, as described in *Gösta Berling's Saga*, were rural, idyllic spots of intense social activity. Persons of rank in Värmland were seen and treated as members of a community connecting the class of ironmasters and other members of the upper class. Retired military officers and priests were often guests in this vibrant social scene. It was important to maintain one's position, as revealed by estate inventories, but the mansions were also places for production and innovation (Lindahl 1988). It was here that a large-scale nineteenth-century reclamation of land began, and new cultivation techniques and forms of crop rotation were introduced and spread (Figure 9.5). It was

FIGURE 9.5. View from Mårbacka, Selma Lagerlöf's family home. "For many, many generations, the plain[s] have been cultivated, and great things have been done there" (Lagerlöf 1918, 30). Photo by Bertil Ludvigsson, reproduced with permission.

the ironmasters who predominated, for example, when the first Swedish agricultural society was founded in Värmland in 1801.

A narrative based on historical source material can thus verify many of the components of the image of the classical Värmland. Mansion houses like Ekeby (Rottneros) in Lagerlöf's novel were in the center of ongoing activities in Fryksdalen. A good description of life in the mansion houses can also be found in the diaries of a Finnish student in Uppsala, Carl Axel Gottlund, who traveled in Värmland in the 1820s (Gottlund 1986). Gottlund described the social life, the romantic playing of music and picturesque evening excursions, and the hunting at several mansion houses where he stayed as a guest. In his description we can feel the atmosphere of *Gösta Berling's Saga*. In fact Gottlund himself can be seen as a personification of that atmosphere.

Finnskogen: The Other Värmland

The aim of Gottlund's travels was to study the Forest Finns and the areas of Finnish settlement in Värmland and Norway (Gottlund 1986; Tarkiainen 1990, 1993). What about the lonely everlasting forests that Gösta Berling longed for? The Forest Finns who settled in Värmland came to make up an important ethnic group for several centuries (Bladh 1995). In some places in Finnskogen, Finnish-speaking networks existed until the early twentieth century. At the time of Gottlund's visit in 1821, Finnskogen was witnessing a growth in population as well as rising poverty and increasing social differentiation. The effect was an increased density of settlement as a result of the division of farms and the building of new crofters' houses. Gottlund estimated the number of Finns to be around eight thousand, so it was in no way an empty forest as suggested by Lagerlöf. Especially among the increasing population of crofters, the ethnic borderline was starting to loosen. Agriculture was underdeveloped and backward, and the landless people had difficulty gaining a livelihood. The most important occupations were cattle raising, milk production, and forestry. Since the second half of the eighteenth century, timber trading had evolved into an important source of income, thus linking Finnskogen to the production network previously described. The people still had an awareness of the colonizers that arrived in the seventeenth century, and stories were told about the rich men who grew huge amounts of rye through their slash-and-burn cultivation (Bladh 1998). Other stories Gottlund tells deal with injustice and the history of the Swedish repression of the Finns. Such stories suggest how the relations between Swedes and Finns were experienced in the 1820s, although research reveals that few such historical events actually took place (Broberg 1981).

Gottlund's diaries also describe the different aspects of Finnskogen's landscape. During his walking tour in Finnskogen, he could see "the Finnish farms and crofters houses, big and small, one after the other. Their small arable fields, their green slash-and-burn plots, their old *ahos* [old slash-and-burn plots], everywhere the forest has clearings" (Gottlund 1986, 177). He asked about the history of the Finnish farms (Figure 9.6), about their special Savo family names, and he collected Finnish place-names, magic readings, and stories. He proceeded as a skilled ethnographer.

Gottlund described a landscape of production, where different practices and resources gave coherence and significance to the landscape. To be acquainted with forests and land involved knowing where the necessary resources could be found. It was a question of practical knowledge, sanctioned by use, which was tied to being in the landscape. A successful realization of a slash-and-burn clearing meant a complex "reading" of the landscape to be able to chose the right type of soil, suitable local climate, adequate dimension for the slash-and-burn area, and so on (Bladh 1998). In addition to functional knowledge, people also had to know about different events and persons who had used the landscape. These landscapes of signification are reflected in place-names or in tales and stories connected to the landscape. We can find many examples of Finnish family names denoting localities in the landscape, for example, Häckfallet (the slash-and-burn area of Häkkinen). Such naming then provides a clue to the origin of the locality. In this way the landscape received a historical depth grounded in

FIGURE 9.6. View of Finnskogen in the border region between Sweden and Norway. The old Finnish farm "Purala" appears almost as an island in the forest landscape. On the opposite side of Lake Rögden is the old customhouse. Photo by Gabriel Bladh.

stories handed down about relatives or the activities of previous generations. This is certainly not part of just the Finnish heritage. Reading Selma Lagerlöf's novel about Mårbacka, and the stories of the old housekeeper and her grandmother, brings out the same historical depth in a landscape full of events and actions (Lagerlöf 1922). Spirits and other creatures also populated the landscape. Such supernatural beings were all part of the personification of the courses of events in nature. Yet there are different myths circulating in the *bygd* and in the forest. Glimpses of the Forest Finns' landscape of signification can be found in fragments of Finnish folk poetry, which Gottlund collected in Finnskogen.

In mid-September 1821, Gottlund arrived near Lake Kymmen in Gräsmarks Finnskog. At a farm, Honkamakk, north of the lake, he met the Finnish wife Anna Hakkarainen: "She was, under the name of Pasu-Anni, known as a famous sorceress. She was young, and knew considerable incantations. She had an excellent memory, so when I read some of my poems for her, she could immediately recite a whole part of it by heart" (Gottlund 1986, 110f.).

In the old Finnish culture, the seer or shaman *(tietäja)* had a central position as an intermediary of the traditional knowledge that was preserved in the form of epic poetry, stories, incantations, or other forms of word-magic (Talve 1997). With its specific metrical form, the word-magic has been passed on through oral tradition. The poems are often expressed in the form of cosmic metaphorical language with a mythical origin (Kuusi and Honko 1983).[10] The oldest poems often deal with birth and origin, like the origin of fire, read by Anna in 1821.

Anna's incantations are good examples of the worldview and view of nature preserved through epic folk poetry. In such stories there is a trace of a pantheistic attitude to the world. Even though the influence of Christianity had been felt for several centuries, the landscape of signification for people in Finnskogen was still affected by such thoughts. Gottlund had brought Christfrid Ganander's *Mythologica Fennica* (1789) to Finnskogen and spoke with some Finns about the traditions of Finnskogen. From the answers we can see that those beliefs, myths, and ritual practices were part of the why and how things were done in the landscape.

What were Selma Lagerlöf's relations to this seemingly archaic world? One important character in *Gösta Berling's Saga* is the Dovre witch. Representing the incarnation of evil by putting a curse on the valley of Lövsjö, she is also described as the daughter of the ancient Finnish sorcerers, the proud follower of her forefathers' mighty wisdom and god-inspired magic. Although named after the Norwegian mountain, she clearly represents the Finnish population. Wandering around as a poor beggar wearing birch-bark shoes, she also impersonates the less romantic image of the Finnish population during the nineteenth century. The people in Fryksdalen had often seen Finns begging during the famine years, especially in the 1830s. Lagerlöf knew of the Dovre witch and in a typical way she intervened in her text: "And she is still alive today. For I, who write this have seen her" (1997 [1918], 212). The romantic and somehow fantastic myths of the Dovre witch can certainly be deconstructed as a fantasy fiction and as a construction of the "other" through our (post)modern gaze, but the links to the archaic world we have glimpsed in Finnskogen are also clear. The past is a foreign country, as David Lowenthal (1985) reminds us.

In 1821, some days before the meeting with Pasu-Anni, Gottlund had visited the manor house of Kymsberg on the eastern side of Lake Kymmen, just a few kilometers from Honkamakk. There he walked through the remnants of an English park, laid out by the first ironmaster about fifty years earlier. Gottlund (1986, 102) wrote: "The fruit-trees and the caves, among other things, show that the English taste has been here." The manor house had been built in the 1770s on the same site as the cottage where the first Finns settled in the 1620s. Gottlund could still see the remains of the typical Finnish smoke cottage.[11] Here we can literally follow how the bourgeois landscape of signification overlaid that of the Forest Finns. Daniel Lagerlöf, Selma's grandfather, worked for several years as the inspector of this manor house at Kymsberg around 1800, when he married the young Lisa-Maja Wennervik, the priest's daughter, who lived at Mårbacka. The meeting between Forest Finns and ironmasters was surely also part of Lagerlöf's heritage.

A conclusion from this reading of Gottlund's diaries and *Gösta Berling's Saga* would be that social memory and myth play an important part in the construction of a history of the enacted territory or landscape. Another is that Finnskogen was very much used as a contrasting area to complete the classical picture of Värmland. While it was a part of the material practices connected to the ironworks and timber production, it was in other ways seen as an "other" Värmland.

Constructing a Region I: Finnskogen Hundred

From reading *Gösta Berling's Saga*, we can see how Lagerlöf constructed her own *bygd* through her writings on Lövsjö hundred. Carl Axel Gottlund actually became involved in his own project of constructing a region. These are different projects in many ways, but they can give us some idea about how a territory is constructed and how regional identity is formed. Paasi (1986) has analyzed the links between regional transformation and the long process of the institutionalization of a region and distinguishes four stages: (1) the constitution of territorial shape, (2) symbolic shape, (3) institutions, and (4) the establishment of the territorial unit in the regional structure and social consciousness. The Finnish hundred that Gottlund envisioned can be seen as a type of "imagined region" in the form of a nearly utopian landscape. It can thus be studied as an example of the attempted construction of a region that failed.

Gottlund had grown up in Juva in Savo and had taken an early interest in Finnish culture and language in the spirit of the romantics. As a teenager, he began collecting folk poems in Savo, and was won over to the cause of promoting Finnish culture. As a student in Uppsala, he learned about the Forest Finns living in Scandinavia and in 1817 made a journey to the area of Finnish settlement in southern Norrland and Bergslagen, where he collected folk poems and described the settlement history.[12] During his journey to Värmland in 1821, he became interested in the situation of the Forest Finns. His goal after a while was to establish a Finnish hundred *(härad)* and three Finnish parishes in both the Norwegian and Swedish parts of the Finnskogen area. He intended to establish a Finnish-speaking area, a new institutional landscape, with a program along radical ethnic lines, where only Finnish-speaking persons could hold official positions and only Finns could own land.

On the Norwegian side, Gottlund's activities led to a revival of Finnish identity, followed by political mobilization (Godø 1975). The Finns in Norway lived under more or less feudal conditions in relation to the timber masters in Kristiania, who owned the land. On the Swedish side, the enthusiasm for Gottlund's work was not particularly evident, but most Finns agreed with his ideas by signing a petition to the king. Through intensive work in Uppsala and Stockholm after his return from Värmland,

Gottlund was able to introduce the idea of the Finnish hundred to some influential persons, including Crown Prince Oscar. The subject was also raised in the Swedish parliament in 1823. In March, a deputation of twelve Forest Finns (six from Värmland and six from the Norwegian side, Norway then being under the sovereignty of the Swedish king) arrived in Uppsala and later in April obtained an audience with Crown Prince Oscar and King Karl Johan in Stockholm. Gottlund's petition for the Forest Finns' hundred was nearly five hundred pages long. It contained a close description of the areas settled by the Finns, and was signed by six hundred Finns from Finnskogen. On a map (Figure 9.7), Gottlund had marked the names of the three new parishes: one was Juvaniemi, named after Gottlund's birthplace. A seal for the hundred was also made. Thus, in accordance with Paasi (1986), the first two stages of constructing a region had been completed.

Initially the action was favorably received, but soon Gottlund's ideas met with resistance, mostly from priests and owners of ironworks from Värmland. Gottlund was prohibited from interfering in politics and was sent back to Uppsala. The matter was submitted to the county governor of Värmland, Johan af Wingård, who was positive about some aspects but wrote: "He is sorry to deprive Gottlund and his friends the

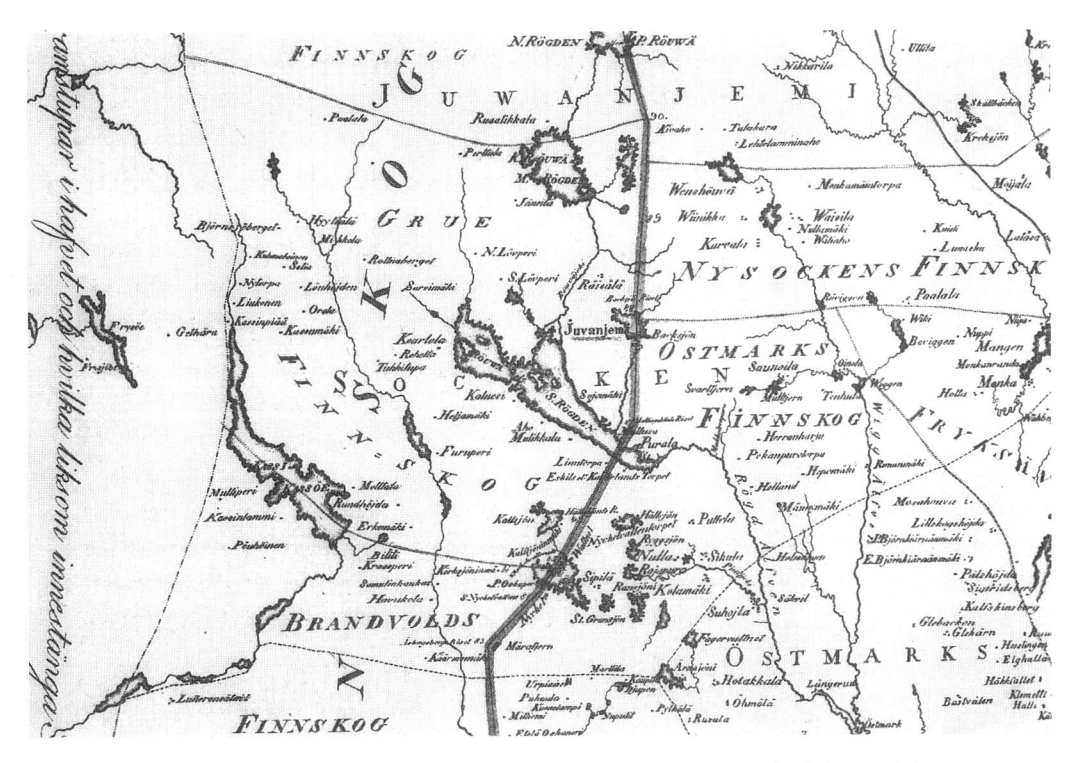

FIGURE 9.7. A partial view of Gottlund's map over Finnskogen hundred (Suomalaisen Kirjallaisuuden Seura, Helsinki). Photo by Gabriel Bladh.

Finns of their Eldorado in the borderland" (cited in Tarkiainen 1993, 144). The romantic utopia was crushed by a return to reality and the existing power relations. Gottlund's idea that all the land from the ironworks should be handed over to the Finnish hundred was too revolutionary. A joint Swedish-Norwegian hundred was seen as incompatible with the constitution and was rejected by the Norwegian and Swedish parliaments. However, a measure of success was achieved when the Norwegian parliament allowed the estate of the timber master Anker to be split and sold at auction. This decision led to a land reform whereby most of the Finns in Norway were allowed to buy their own farms. The Swedish parliament investigated the situation in the Swedish part of Finnskogen. At first nothing came of this and all claims promoting Finnishness were rejected.

Eventually, as a result of the investigation, a few new parishes were established, but the church was maintained as an important Swedish-speaking institution. Paradoxically, the unintentional result of Gottlund's activities was to speed up the process of making the Finns Swedish or Norwegian. During the second half of the nineteenth century, the Forest Finns became increasingly assimilated with the Swedish population (Bladh 1995). Changes in the Finns' livelihood were a central factor, but other institutional factors, such as the establishment of new parishes and churches and the introduction of elementary schools, also had a marked effect on the complex process of assimilation. From the 1870s until World War I, the timber trade in Finnskogen changed considerably from a farmer-dominated industry to a company-dominated one at all stages of production. Large sections of the farmers' lands were sold to Norwegian and Swedish companies or timber dealers. These sales enabled many Finns to emigrate to the United States. The population now started to decrease rapidly. The beginning of the twentieth century also saw the disappearance of an increasing number of small farms, and many smallholders became tenants or seasonally employed woodsmen.

Along with changes in the material culture of the Finns—smoke cottages (Figure 9.8), for example, ceased to be built—there were also great changes in the Forest Finns' landscape of signification. For most Finns, there was no longer anything *pyhä* (sacred) in nature, as one of the last wise men in Östmark put it at the beginning of the twentieth century. By the mid-nineteenth century, most Finns were involved in several different types of both Swedish- and Finnish-speaking networks and institutional projects and the status of the Finnish language was clearly oppressed. As an example, the verb *finska* (to talk Finnish) was synonymous in the Värmland dialect with "talking rubbish." Many of the practices connected with Finnish culture were significantly affected when the language started to disappear, including communication within the household. Yet the process of becoming Swedish and modern, and the changes in the landscape of signification, was complex, even at the level of the individual. With the breakup of the Swedish-Norwegian union in 1905, most of the Finnish population sided with their home nation, rather than with Finns across the border.

The Swedish-language magazine *Finnbygden* played an important role in making a new landscape of signification in the 1920s. Here an idealistic group of Finnish descendants, mostly smallholders and forestry workers, found a platform. The magazine included articles on Esperanto and veganism alongside poems about the golden days of the Finnish settlement area in the seventeenth century. The group also wanted to preserve the cultural heritage in the form of buildings and landscapes. The Finnish language was not promoted, even if Finnish was spoken until the 1970s.

Gottlund's vision of seeing Finnskogen as a unit has attracted renewed attention. A heritage movement to revive Finnish culture has sprung up in recent decades, and Gottlund's vision has been used in promoting local development and tourism. Now it is clearly the celebration of the cultural heritage of a long-gone ethnic "other" that creates new interest in the area, and also a more positive regional identity. Yet it is still a question among developers whether Finnskogen should be described as a wilderness area in a sparsely populated area or as a historical cultural landscape. Finnskogen thus remains the contrasting "other" to classical Värmland.

FIGURE 9.8. Ritamäki, Lekvattnet parish, Finnskogen. This smoke-cottage was in use until 1964; today it is owned by the local folklore society. Photo by Gabriel Bladh.

Constructing a Region II: Lövsjö Hundred and Värmland

Compared to Gottlund's project to construct Finnskogen hundred, Selma Lagerlöf's Lövsjö hundred was a literary construction that had much more success. It was shaped on an existing historical territory and can be seen more as part of a representation and reconstruction of Fryksdalen and Värmland. *Gösta Berling's Saga* is an excellent example of how, through narrative, it is possible to give a landscape and a *bygd* an experience of related wholeness, thus constructing an imagined community. Lagerlöf depicts a landscape and a region with a powerful expressiveness and luminosity. The novel's literary success led to a large volume of literature of the same genre. Many of the seemingly exotic circumstances described in the novel can be placed within the context of the dynamic history of Värmland. It is easy to argue that the identity of Värmland and the myth Lagerlöf constructed simply define each other. Yet she succeeded in presenting "an enlarged and concentrated Värmland" centered on Fryksdalen, which was strongly attached to the image of classical Värmland. Her writing plays an ongoing role in the formation both by Värmländers and tourists of a regional identity.

Even before Lagerlöf rebought her family home, Mårbacka, in 1907, it had begun to attract visitors. As Gunnar Ahlström (1959) noted, Lagerlöf was a popular writer who enjoyed huge sales of her books. She wrote a famous reader for schoolchildren and attracted new social groups to books around the turn of the century. Lagerlöf worked with all the new media of the new century. *Gösta Berling's Saga* was initially published in a weekly magazine. She continued to publish and soon was among those celebrities interviewed and covered by the weekly press. Her novels later inspired Swedish filmmakers. She once reported to her faithful film producer, Victor Sjöström, the new experience "that film also promotes the books." Selma Lagerlöf was also a successful radio broadcaster.

Many of Lagerlöf's themes in *Gösta Berling's Saga* were constructed out of oral storytelling in the context of face-to-face interaction, and in that way they were grounded in tradition. Lagerlöf's own career is thus a good example of how media came to transform the role and character of tradition. This content was now shared by people and "stretched" across time and space. While orally transmitted traditions continue to play an important role for most people, traditions have been gradually and fundamentally transformed by the mass media. John B. Thompson (1996) stressed three consequences of this transformation. First, traditions have been partly deritualized, thus the fixation of symbolical content into a book means that they have to some extent been separated from a need of continual reenactment. Second, traditions have been partly depersonalized. They have been detached from the individuals with whom one interacts in day-to-day life. Instead, media have created new links between the authority of tradition and those individuals who transmit it. Third, traditions have been gradually and partially delocalized and uprooted, and the bonds that tied traditions to specific localities of face-to-face interaction have been weakened. However, tradition was not de-terriorialized but, importantly, was refashioned in ways that enabled it to be reembedded in a multiplicity of localities and reconnected to territorial units, such as regions or nations, that exceeded the limits of face-to-face interaction. The schoolteacher Lagerlöf, sitting in Landskrona in Scania, writing the stories about old Värmland for her first book, exemplifies this delocalization. Yet it is possible to argue that one explanation for her success was her ability to personalize, territorialize, and make new constructions out of traditional material, a kind of reflexive engagement with that tradition, and in that way bridge the gap between traditional and mediated forms of symbolic content. Here it is important not to contrast a static tradition with a degenerative change into modernity. Innovation and performance has always been an important feature of folk tradition (Crang 2000).

In the early 1920s, Lagerlöf rebuilt Mårbacka (Figure 9.9) as the neoclassical manor house that it is today. While preserving the old Mårbacka, she turned its facade into a suitable environment for a world celebrity. As is typical for Lagerlöf, even the building is a mix of realism and fiction, and has become part of her own image as well as the image and heritage of Värmland. Other practices were more directly connected with the promotion of tourism in the area of Fryksdalen. In Värmland, organizations such as the Swedish Tourist Federation published travel guides about the "real" context of *Gösta Berling's Saga,* and schoolchildren, not least, took part in the materialization of the stories through excursions around Fryksdalen as part of their education.[13] Lagerlöf lived to see places like Mårbacka and Rottneros become celebrated tourist destinations. In this way Fryksdalen has also come to represent a typical image of the physical landscape of Värmland. Later interpreters of the image of classical Värmland contrast marginalized areas such as Finnskogen. As we have seen from Gottlund's narratives and practices, the place of Finnskogen in this image can be said to be ambivalent. In the new geographical division of labor, Finnskogen was the producer of raw material for the forest industry and was once again seen as a wilderness area. Here it was not possible to use the double meaning of *landskap* as scenery and lived territory, nor the connotations made between *bygd* and *landskap,* so successfully developed by Lagerlöf linking Fryksdalen and Värmland.

Representing the Landscape

To read Selma Lagerlöf's epos about people and landscape in Fryksdalen as a geographer poses several questions about distance and nearness, representation and text strategies. It also provides an opportunity to reflect about descriptions of land and life in a historically specific region, a problematic belonging to the tradition of regional geography as well as the study of landscape.

In an article on Lagerlöf's fiction, Lagerroth (1998) points to the difference between dealing with the landscape in a narrative epic composition compared with describing the aesthetics and symbols of a landscape through art and iconography.[14] He argues that, in the latter case, the way of looking at the landscape tends to be visual, static, and analytical, while narrative epic primarily stresses relations, processes, and totalities. The landscape can attain meanings, content, and attraction through its relations to people and events. Lagerroth differentiates between epic relations in the form of factual and casual relations and lyrical relations in the form of similarity and feeling. He stresses that Lagerlöf is notable for her impressive manner of weaving together epic and lyrical relations between people and landscape. The descriptions of events, visual landscapes, and people strengthen and elucidate one another, but they preserve a contrapuntal independence between the categories. The result is that, unlike conventional

FIGURE 9.9. Mårbacka. Photo by Bertil Ludvigsson, reproduced with permission.

metaphorical language, these contrapuntal compositions open the border between picture and epic reality.

The work of Selma Lagerlöf shows that landscape identity can be created in a forceful way by a conscious textual strategy. It also demonstrates how nature/landscape and territory are connected. Lagerlöf's aim was to give her readers a literary experience through her stories about landscape and people in Fryksdalen. The nearness to actual forms sought by such a description is of course very different from that sought by a traditional geographical landscape analysis (Meinig 1983).[15] Poststructuralist accounts of culture celebrating difference and situatedness have given new challenges to universalistic versions of knowledge. The fact that many of the most successful regional geographical work has been produced by authors writing about a region as insiders suggests the way in which such knowledge is situated. Perhaps it is possible, as the writer Bengt-Emil Johnson (1997) noted, that intimate knowledge of a landscape results in respect for its integrity and its untranslatability. He believes that this kind of concrete experience is contrary to self-righteousness and isolationism, which gives the opportunity to use landscapes not as havens of refuge, but as viewpoints.

It has been seen from the examples of Lagerlöf's Värmland how media has come to transform the role and character of tradition, and how new practices have come about to organize the representation of symbolic content. Orvar Löfgren (1989) has noted how the nationalization of shared cultural understanding and knowledge has been established through different practices shaping a common frame of reference. New national arenas are thereby created, as through the shared knowledge of schoolbooks in the national school curriculum. Selma Lagerlöf was an important actor in developing a shared understanding of what constituted Sweden's geography and landscape through her book *The Wonderful Adventure of Nils*. This book was one of three readers that were produced through an initiative by the association of Sweden's elementary schoolteachers as a means of reforming the writing of school texts (Edström 1991). The schoolteacher Lagerlöf took the challenge, but she insisted on writing a book that was both realism and fiction. The bewitched boy, Nils, flying around the country with the wild geese, has, perhaps more than any other book, given an integrated picture of Sweden and at the same time put Sweden on the international map. In a letter to her friend Sophie Elkan, also a writer, in April 1906, Lagerlöf wrote:

> Do you remember my description of *landskap* in Gösta Berling? It is my intention, for each *landskap*, to do a short and energetic description with the help of images and inventions, and also fables, which some *landskap* have composed about themselves. This little introduction cannot be longer than a couple of pages, but they give a feeling of homelike atmosphere, you know where you have come, if you have to deal with a plain or a mountainland, a coast or woodland. I think that those short descriptions will be very beneficial. (Toijer-Nilsson 1996)[16]

In that way each Swedish *landskap* received its own representative description, where territory and scenery, province and nature, were seen together. Generations of Swedes have later learned to see Sweden through those glasses. Through Nils's goose-eye view, the nation shrinks into perceivable proportions. This vertical top-down axis is complemented by the north-south axis connecting Nils's home in Scania with the home of the wild geese in Lappland. In this way Lagerlöf successfully communicates "the unity of differences" inside Sweden's territory, which can be seen as one of the defining elements in the concept of nation (Thorup Thomsen 1998). Like Hazelius's open-air museum in Skansen, Lagerlöf's representation of regional differences within the frame of the nation-state can be seen as an important part of shaping a common national frame of reference. Here landscape is enacted territory, both in thought and practice, where Swedish nature and scenery are connected through "a way of seeing" to the lived territory binding together the provinces and the nation. Here the dimension in which landscape is constituted through the relationship between the foregrounded (the way we are now) and the backgrounded (the way we might be), as Hirsch (1995) suggests, helps form the nation as "we."

Edström (1986) characterizes Lagerlöf as our great epic interpreter of space and movement. Lagerroth (1998) pointed out how Lagerlöf, through her narrative epic, first of all emphasizes relations, processes, and totalities in landscape and nature. She formulates her aim with the book about Nils Holgersson: "To bring the map to life" (cited in Edström 1991, 71). Here it is possible to find points in common between Lagerlöf and the Swedish geographer Torsten Hägerstrand (1984, 1993), who developed the idea of "landscaping" or the "process-landscape" *(förloppslandskap)*. His "way of seeing" is characterized by "the never-ending flow of interrelated presences and absences in the world," and in that way he looks at the landscape as "a fine-grained configuration of meeting places" (1984, 378).[17] His representations are more formal than Lagerlöf, but they share a common worldview of seeing the landscape as process. Other applications of landscape as processes have already been incorporated in environmental planning and in GIS systems. Such applications are once again challenging and questioning the meaning and the use of the concept of landscape.

Conclusions

My experience from teaching geography at Karlstad University is that many students, in an unreflected situation, have difficulty discriminating between the aspects of Värmland as a province (Värmland as *landskap*) and the characteristics of what is seen to be a typical landscape of Värmland. I would argue that such blurring is not accidental. It belongs to a collective socialization process whereby one learns to see representations of a region and its landscape together. People thus learn to conceive of territorial identities in terms of landscape scenery. Although perhaps the idea of *landskap* as province is used infrequently in everyday language, the multilayered connotations of the term still make it easy to connect the concept with territory and area. Every nine-year-old child in Sweden still learns his or her national geography by interpreting Swedish provinces as *landskap*.

In this chapter I have tried to show how Selma Lagerlöf's writings a century ago have been part of a cultural regionalism, where the unit of *landskap* had a central position as both an arena of practices and a symbolic territory. Thus, the narrative of Värmland can be linked to the making of a national Swedish landscape. Here landscape has been understood as a reciprocally connected way of seeing, representing, and making land or enacting territory. The relations between people, territory, landscape, and regional identity in regions like Värmland are thus continually reproduced or transformed through changing practices and discourses. As the case of Finnskogen shows, these are also processes where people and places are included or excluded. The discourses connected with classical Värmland are mainly promoted by the tourist industry, and Selma Lagerlöf has become part of a cultural heritage that is acknowledged in different ways by European tourists, Värmländers, or international businesspeople on tour in Fryksdalen. How such recent changes in practices, as well as representations of the national Swedish landscape, or Lagerlöf's Värmland, have affected the links between landscape and territory as seen by Swedes or Värmländers should be a fruitful area for further research.

Notes

All translations from Swedish are by the author, except quotations from Selma Lagerlöf's *Gösta Berling's Saga*, which are from the translation by Lillie Tudeer (Lagerlöf 1997 [1918]).

1. Among others, Hartshorne (1939) made an issue of the problems of finding an unambiguous definition regarding the content of landscape.

2. The Nazi *"Blut und Boden"* ideology has certainly cast a shadow on connections between culture and land seen as organic communities, but it is also possible to find links to communalism and regionalism connected to other (hi)stories.

3. After the loss of Finland in 1809, the elite in Sweden first redirected their search for identity to Scandinavia. This was politically appropriate after the union with Norway in 1814, but also later the idea of Scandinavism was an influential alternative discourse.

4. It is symptomatic of the primacy of the national-state paradigm that studies of regional identity and regionalism are still few in the Swedish context; see Aronsson (1995).

5. On the double meaning of landscape, see also Hägerstrand (1991).

6. See entries "cultural landscape," "landscape," and *"Landschaft"* in *The Dictionary of Human Geography* (ed. Johnston, Gregory, and Smith 1994).

7. Compare the regional descriptions and categorizations in the *National Atlas of Sweden* (Helmfrid 1994).

8. See also articles in Blomberg and Lindquist 1994.

9. Selma Lagerlöf (1858–1940) has exerted great formative power in Scandinavian literature. She was the first woman to win the Nobel Prize for literature, in 1909. *Gösta Berling's Saga*

(1891, English translation 1918) and her children's geography book of Sweden, *Nils Holgersson underbara resa* (1906–7) (*The Wonderful Adventures of Nils* [1947]), have been especially important in shaping images of the landscapes of Värmland and of Sweden.

10. Becoming part of the national epic of Finland as *Kalevala*, folk poetry has been part of building a Finnish national identity and thus has been part of romanticized discourse of "the golden days."

11. The simple smoke cottage, *pirtti*, usually had one room and only a wooden chimney. It had a big stove made of natural stones piled one on top of the other and vaulted with the application of clay (Talve 1997).

12. In a review in 1817, Gottlund suggested the possibility of combining Finnish folk poetry and literature. This was a formula for the birth of the Finnish national epic, *Kalevala*, compiled from folklore by Elias Lönnroth (Honko 1980).

13. See, for example, Petré (1958), itself part of this materialization.

14. The literary historian Erland Lagerroth has treated this theme in his doctoral dissertation (1958), "Landscape and Nature in the Story of Gösta Berling and Nils Holgersson."

15. For discussion about geography and literature, see Pocock (1981) and surveys in Pocock (1988) and Lando (1996).

16. Letter to Sophie Elkan from Selma Lagerlöf, April 18, 1906 (Toijer-Nilsson 1996).

17. For a more detailed discussion of Torsten Hägerstrand's and Allan Pred's different applications of the time-geographical approach, see Bladh (1995).

REFERENCES

Ahlström, G. 1959. *Kring Gösta Berlings saga*. Stockholm: Natur och kultur.

Aronsson, P. 1995. *Regionernas roll i Sveriges historia*. Expertgruppen för forskning om regional utveckling, rapport 91. Stockholm.

Björck, S. 1946. *Heidenstam och sekelskiftets Sverige: Studier i hans nationella och sociala författarskap*. Stockholm: Natur och kultur.

Bladh, G. 1990. "Några nedslag i det värmländska kulturlandskapet." In *Geografisk utbildning och forskning—rapport från 20 års erfarenheter och utblickar inför 90-talet*, ed. G. Gustafsson, 31–44. Forskningsrapport 90:11, Högskolan i Karlstad. Karlstad.

———. 1995. *Finnskogens landskap och människor under fyra sekler—en studie av natur och samhälle i förändring*. Meddelanden från Göteborgs universitets geografiska institutioner, Serie B nr 87. Göteborg.

———. 1998. "Colonization and the Second Stage: Changing Land Use in Finnskogen 1650–1750." In *Outland Use in Preindustrial Europe*, ed. H. Andersson, L. Ersgård, and E. Svensson, 184–94. Lund Studies in Medieval Archaeology 20. Lund: Institute of Archaeology, Lund University.

Blickle, P., S. Ellis, and E. Österberg. 1997. "The Commons and the State: Representation, Influence, and the Legislative Process." In *Resistance, Representation, and Community*, ed. P. Blickle, 115–53. Oxford: Clarendon Press.

Blomberg, B., and S.-O. Lindquist, eds. 1994. *Den regionala särarten*. Lund: Studentlitteratur.

Broberg, R. 1973. *Språk-och kulturgränser i Värmland*. Stockholm: Norstedt.

———. 1981. "Äldre invandringar från Finland i historia och tradition." *Fataburen* 1981: 32–65.

Brummer, H. H. 1997. "Prins Eugen och det nordiska stämningslandskapet." In *Naturen som livsrum*, ed. H. H. Brummer and A. Ellenius, 102–24. Stockholm: Natur och kultur.

Crang, M. 2000. "Between Academy and Popular Geographies: Cartographic Imaginations and the Cultural Landscape of Sweden. In *Cultural Turns/Geographical Turns,* ed. I. Cook, D. Crouch, S. Naylor, and J. Ryan, 88–108. Harlow: Prentice-Hall.

Cosgrove, D. 1998 (1984). *Social Formation and Symbolic Landscape,* 2nd ed. London: Croom Helm.

Edström, V. 1986. *Selma Lagerlöf's litterära profil.* Stockholm: Rabén och Sjögren.

———. 1991. *Selma Lagerlöf.* Stockholm: Natur och kultur.

Enequist, G. 1941. "Bygd som geografisk term." *Svensk Geografisk Årsbok* 7: 7–21.

Facos, M. 1998. *Nationalism and the Nordic Imagination: Swedish Art of the 1890s.* Berkeley and Los Angeles: University of California Press.

Furuskog, J. 1924. *De värmländska järnbruken.* Filipstad: Bronellska bokhandeln.

Ganander, Christfrid. 1789. *Mythologia fennica, eller Førklaring øfver de nomina propria deastrorum, idolorum, locorum, virorum &c.* . . . Åbo: Frenckellska Boktryckeriet.

Godø, H. 1975. *Grue Finnskog: Etniske aspekter ved sosio-økonomisk utvikling.* Hovedoppgave i etnografi, Universitetet i Oslo (Master's thesis in ethnography, University of Oslo).

Gottlund, C. A. 1986. *Dagbok över mina vandringar på Wermelands och Solörs finnskogar 1821.* Kirkenær: Gruetunets Museum.

Hägerstrand, T. 1984. "Presence and Absence: A Look at Conceptual Choices and Bodily Necessities." *Regional Studies* 18, no. 5: 373–79.

———. 1991. "The Landscape as Overlapping Neighborhoods." In *Om tidens vidd och tingens ordning: Texter av Torsten Hägerstrand,* ed. G. Carlestam and B. Sollbe, 47–55. Stockholm: Byggforskningsrådet.

———. 1993. "Samhälle och natur." *NordREFO* 1993: 1: 14–59.

Hartshorne, R. 1939. "The Nature of Geography." *Annals of the Association of American Geographers* 29: 3–4.

Helmfrid, S., ed. 1994. *Landscape and Settlements. National Atlas of Sweden.* Stockholm: SNA Publications.

Hirsch, E. 1995. "Landscape: Between Place and Space." In *The Anthropology of Landscape: Perspectives on Place and Space,* ed. E. Hirsch and M. O'Hanlon, 1–30. Oxford: Clarendon Press.

Honko, L. 1980. "Upptäckten av folkdiktning och nationell identitet i Finland." *Tradisjon* 10: 33–51.

Jansson, U. 1998. *Odlingssystem i Vänerområdet.* Meddelanden nr 103, Kulturgeografiska institutionen, Stockholms Universitet, Stockholm.

Johnson, B.-E. 1997. "Naturen som livsrum." In *Naturen som livsrum,* ed. H. H. Brummer and A. Ellenius, 46–63. Stockholm: Natur och kultur.

Johnston, R. J., D. Gregory, and D. M. Smith, eds. 1994. *The Dictionary of Human Geography,* 3rd ed. Oxford: Basil Blackwell.

Jones, M. 1991. "The Elusive Reality of Landscape: Concepts and Approaches in Landscape Research." *Norsk geografisk Tidsskrift* 45: 229–44.

Kristoffersson, A. 1924. *Landskapsbildens förändringar i norra och östra delen av Färs härad under de senaste tvåhundra åren: en kulturgeografisk studie.* Lund: Lunds Universitet.

Kuusi, M., and L. Honko. 1983. *Sejd och saga.* Stockholm: Raben & Sjögren.

Lagerlöf, S. 1891. *Gösta Berlings saga.* Stockholm: Hellberg.

———. 1906–7. *Nils Holgerssons underbara resa genom Sverige.* Stockholm: Bonnier.

———. 1922. *Mårbacka.* Stockholm: Bonnier.

———. 1947. *The Wonderful Adventures of Nils,* translated from the Swedish by Velma Swantson Howard. Cambridge, Mass.

———. 1997 (1918). *Gösta Berling's Saga*, translated from the Swedish by Lillie Tudeer. Iowa City: Penfield.

Lagerroth, E. 1958. *Landskap och natur i Gösta Berlings saga och Nils Holgersson*. Stockholm Studies in History of Literature 4. Stockholm.

———. 1998. "Landskapet hos Selma Lagerlöf." In *Miljöhistoria idag och imorgon*, ed. M. Johansson, 63–84. Högskolan I Karlstad, Forskningsrapport 98: 5. Karlstad.

Lando, F. 1996. "Fact and Fiction: Geography and Literature." *GeoJournal* 38, no. 1: 3–18.

Lindahl, G. 1988. "Byggnadskonst i Värmland." In *Ditt Värmland. Kulturmiljöprogram för Värmland och värmlänningar, Första delen*, 132–61. Karlstad: Länsstyrelsen i Värmland.

Löfgren, O. 1989. "The Nationalization of Culture." *Ethnologia Europaea* 19: 5–23.

Lowenthal, D. 1985. *The Past Is a Foreign Country*. Cambridge: Cambridge University Press.

Meinig, D. W. 1983. "Geography as an Art." *Transactions, Institute of British Geographers*, new series 8: 314–28.

Mels, T. 1999. *Wild Landscapes: The Cultural Nature of Swedish National Parks*. Lund: Lund University Press.

Nelson, H. 1913. "En Bergslagsbygd." *Ymer* 33: 278–352.

Olsson, H. 1950. *Fröding: Ett diktarporträtt*. Stockholm: Norstedt.

Olwig, K. 1992. "Sexual Cosmology: Nation and Landscape at the Conceptual Interstices of Nature and Culture, Or: What Does Landscape Really Mean?" *Man & Nature. Working paper* 5. Odense: Humanities Research Center, Odense University.

———. 1996. "Recovering the Substantive Nature of Landscape." *Annals of the Association of American Geographers* 86: 630–53.

Paasi, A. 1986. "The Institutionalization of Regions: A Theoretical Framework for Understanding the Emergence of Regions and the Constitution of Regional Identity." *Fennia* 164, no. 1: 105–46.

Petré, M., ed. 1958. *Selma Lagerlöf och Mårbacka: En bildkrönika*. Stockholm: Bonnier.

Pocock, D. C., ed. 1981. *Humanistic Geography and Literature*. Beckenham: Croom Helm.

———. 1988. "Geography and Literature." *Progress in Human Geography* 12: 87–102.

Rehnberg, M. 1980. "Folkloristiska inslag i olika tidevarvs idéströmningar kring det egna landet." *Tradisjon* 10: 17–32.

Sayer, A. 1992. *Method in Social Science*, 2nd ed. London: Routledge.

Schéele, F. von. 1858–67. *Vermland i teckningar*. Stockholm.

Talve, I. 1997. "Finnish Folk Culture." *Studia Fennica Ethnologica* 4. Helsinki.

Tarkiainen, K. 1990. *Finnarnas historia i Sverige* 1. Stockholm: Nordiska Museet.

———. 1993. *Finnarnas historia i Sverige* 2. Stockholm: Nordiska Museet.

Thompson, J. B. 1996. "Tradition and Self in a Mediated World." In *Detraditionalization*, ed. P. Heelas, S. Lash, and P. Morris, 89–108. Oxford: Blackwell.

Thorup Thomsen, B. 1998. "Terra (In)cognita: Reflection of the Search for the Sacred Place in Selma Lagerlöf's *Jerusalem* and *Nils Holgerssons underbara resa genom Sverige*." In *Selma Lagerlöf Seen from Abroad—Selma Lagerlöf i utlandsperspektiv*, ed. L. Vinge, 131–41. Kungl. Vitterhets Historie och Antikvitets Akademien, Konferenser 44. Stockholm: Almqvist & Wiksell International.

Toijer-Nilsson, Y. 1996. *Du lär mig att bli fri. Selma Lagerlöf skriver till Sophie Elkan*. Stockholm: Bonniers.

Widgren, M. 1988. "Det omogna kulturlandskapet." In *Ditt Värmland: Kulturmiljöprogram för Värmland och värmlänningar, Första delen*, 110–31. Karlstad: Länsstyrelsen i Värmland.

10.

The Swedish Agropastoral *Hagmark* Landscape: An Approach to Integrated Landscape Analysis

Margareta Ihse *and* Helle Skånes

The *hagmark* landscape is a scientifically defined functional landscape region of high significance for both nature conservation and cultural heritage. It is a semi-open agropastoral moraine landscape, forming a small-scale and often complex pattern of forests, fields, and grasslands. The most characteristic components in this landscape are grasslands and deciduous woods with a wide variation in tree and shrub cover and species. *Hagmark* is wooded grassland, the long-term outcome of grazing, mowing, and occasional cultivation. It is a botanically valuable rural landscape shaped by a multifunctional agriculture based on animal husbandry over centuries. We characterize the *hagmark* landscape mainly by its physical, geographical, and ecological conditions. In the Nordic countries, this landscape type provides the closest equivalent of the Mediterranean pastoral landscape described by Farina (1997). Many of the pastures and heterogeneous deciduous forests have high biodiversity and make good examples of sustainable land use.

The *hagmark* landscape is one of the typical Swedish landscape scenes that contribute to perception of Sweden. Carl Fries, a famous writer on the Swedish landscape, expresses this in his book: "The most immense efforts of cultivation in the past was not the establishment of the first arable fields, but the transformation of the natural landscape for the sake of animal husbandry, the creation of the deciduous rich, amiable peasant landscape, which has for us all, through a thousand years, been the essence of Swedish nature" (Fries 1965, 36, our translation).

This chapter will focus on a description of landscape regions, based on natural sciences, landscape as the local and historical expression of the use of natural resources, and natural and cultural values with specific regard to biodiversity. We propose an approach to landscape analysis that integrates natural and cultural properties reflecting the interrelations between humans and nature.

The Need for an Integrated Landscape Approach

In Sweden today, there is a rapidly increasing call for comprehensive methods integrating natural and cultural values in landscape monitoring, in nature conservation as well as in cultural heritage management, and especially for the maintenance of biodiversity (Barr et al. 1993; Bunce et al. 1996; Ihse and Blom 2000). When we discuss natural and cultural values in this chapter, our point of departure is the established set of values described and applied over the years in general advice for nature conservation by the Swedish government (e.g., Environmental Protection Agency 1975; Bernes 1993, 1994). Therefore, we will not enter a general philosophical discussion of valuation. The global significance of biodiversity, and indirectly also of cultural heritage values, are further expressed in the international agenda of environmental issues formulated at the United Nations Conference on Environment and Development (UNCED) in Rio de Janeiro in 1992 and at the Convention on Biological Diversity and in Agenda 21 (UNEP 1992). The environmental questions are also dealt with according to the global understanding of the concept of sustainable land use, as described by the Brundtland Commission (World Commission on Environment and Development 1987). As a step toward managing a landscape with sustainable development and high biodiversity according to these international goals, we shall discuss the necessity of an integrated landscape analysis, using sources and methods from physical geography, ecological geography, and historical geography.

The demand for integration is common to the Swedish Environmental Protection Agency, the Central Board for Antiquities and National Heritage, and the Boards of Agriculture and Forestry. These authorities are responsible for environmental monitoring and the conservation of nature, cultural heritage and landscapes, and for agricultural and forest production. They are also responsible for the maintenance of biodiversity in the landscape. By tradition, these sectors have handled most of the research and management of the landscape. Each sector has suggested its own solutions and applied them to the same landscape. Often these solutions have been quite disparate and the results sometimes unexpected and contradictory and not feasible for the sustainability and diversity of the dynamic and complex landscapes.

To achieve true integration and overcome the traditional differences in methods and the intrinsic ambiguity in terminology applied between these research fields, intercommunication is necessary between these authorities (Skånes 1997) and must be based on a holistic view of the landscape. By holism we mean the original concept developed by Smuts (1926)—that the whole is more than the sum of the individual parts. Sectorial knowledge must be supplemented with holistic knowledge. It is not sufficient simply to add existing individual items of knowledge to each other. This knowledge should not only satisfy the interest of the individual sector but also should contribute to understanding the whole. Without a holistic approach, different authorities may, unintentionally and unknowingly, offer advice and support that threaten landscape values instead of preserving them. Holistic views must be applied to enable planning for nature conservation and cultural heritage while concurrently meeting the new demands for sustainable development (Ihse and Norderhaug 1995).

Biological diversity (biodiversity) is a matter of global importance. It implies a rich variation in genes, species, and ecosystems. The ecosystem comprises biotopes and landscape. Biodiversity is often referred to as a list of threatened species, but it implies much more than that. It is essential for the life-giving systems on earth and is a resource with actual and potential value for humans. An ecosystem or a landscape with high biodiversity is more resilient and thus provides a buffer against future environmental threats and changes. It also has a potential for future human utilization. The concept of maintaining the biodiversity of ecosystems, landscapes, and biotopes includes the necessity for species to move and disperse in the landscape. The landscape structure and patterns, providing isolation or connection to plants and animals, must be considered.

Biodiversity ranges across biological scales from genes to species and ecosystems, and in this way it also influences the socioeconomic, cultural, and aesthetic dimensions of the landscape. Consequently, the biodiversity and biological properties of the landscape are not simply a sectorial issue for nature conservationists but a question of utmost importance for all the sectors of society.

Discussions on biodiversity often concentrate on tropical rain forests, but the Nordic countries are also rich in biodiversity (Bernes 1994). Although the species and ecosystem are completely different, in the global perspective they are as important as tropical forests. We have a responsibility to maintain the biodiversity found in natural boreal forests and in the ancient parts of the traditional agricultural landscape. Ihse and Norderhaug (1995) stress the necessity to elaborate a strategy and develop methods that include biodiversity in planning and management of the Nordic cultural landscapes.

The implementation of the new environmental laws and international agreements requires methods and models for landscape analysis for the Nordic countries that integrate natural and cultural aspects of the landscape. Integrated landscape analysis, based on geographical and ecological research and linked to environmental monitoring and landscape planning, could thus play an important role in the work of maintaining biodiversity (Sporrong 1993a; Berglund 1991; Skånes and Tollin 1991). This work must embrace the whole concept of biodiversity, not only threatened and rare species but also diversity and patterns of biotopes and ecosystems on a landscape level.

TRADITIONAL SWEDISH LANDSCAPE RESEARCH

Landscape research in Sweden has been undertaken in different disciplines and using different approaches, with different temporal and spatial scales. Elements of these approaches in physical and human geography, landscape ecology, Quaternary geology, and archaeology are used in the proposed integrated landscape approach.

Landscape research in physical geography (geomorphology, glaciology, hydrology, climatology) has specialized in describing landforms and their processes in a long-term perspective, one thousand to ten thousand years or longer, thus addressing the natural landscape and its formation. Remote sensing techniques provide information on the present-day landscape, for the last sixty to seventy years using aerial photographs and the last thirty years using satellite imagery. Recent ecological-geographical research within physical geography focuses on the last 150 to 250 years. This time period allows us to describe and understand the ecological situation and also describe the potential for future landscapes (Ihse et al. 1991; Skånes 1991; Ihse 1994a, 1995; Skånes 1996).

Landscape research in human geography has focused on historical geography, including analysis of old cadastral maps, which are a major tool in landscape analysis (Kristoffersson 1924; Sporrong 1990; Tollin 1991). These maps, with a good coverage in time and space, are unique to Sweden and a few other countries, such as Finland and the Baltic countries.

From ecology, we use knowledge about ecosystems and biotopes, their functions, distribution, and processes linked to landscape patterns mainly derived from research within landscape ecology and plant ecology. Major research in landscape ecology has specialized in describing the flora and fauna in a landscape perspective and the distribution and processes linked to the patterns (Naveh 1982; Forman and Godron 1986; Zonnerveld and Forman 1990; Naveh and Lieberman 1994; Forman 1995). Biodiversity studies have been performed mainly at the species and population level on selected plants and animals.

A Conceptual Model for the Dynamic Landscape

Landscape research and management require a new conceptual model to meet the demands for integration and sustainability. Fundamental to this model is how the landscape itself is defined. We define the landscape as "the tangible and characteristic result of the interactions between a specific society, the given physical conditions, and biotic as well as abiotic processes" (Sporrong 1993b). Since the integrated landscape concept includes dynamic phenomena, the processes shaping the landscape are especially important to address. The focus is placed on the ecological and physical conditions of this environment and how the natural prerequisites have governed the use of landscape resources over time, which in turn have resulted in a complex, dynamic landscape.

Sporrong (1993a) has established five fundamental prerequisites for an integrated landscape approach: the physical landscape, the history of the landscape, the dynamics of landscapes, regional characteristics, and interaction and scientific relevance. The physical landscape includes bedrock, soils, geomorphology, topography, climate, and plant life. With regard to plant life, the current vegetation can be seen as the long-term response to all the physical conditions and the human impact that has occurred over time. In the history of the landscape, the most important factors are the evolution of landownership and property rights, social strategies, decision making, and ideas and knowledge of landscape management. The present-day landscape can be comprehended only if we know its history. The history of land use, which partly explains vegetation conditions, provides a basis for understanding consequences of and threats from the ongoing changes today and the possible effects on biodiversity and sustainable production. To understand the dynamics of landscapes, including rate, succession, and exchange of landscape features, we must address the drivers behind the changes, such as demographic, economic, ecological, and technical factors.

Time geography, developed by Hägerstrand (1991), has been one method in social science to study phenomena in space, their individual positions and movements during the studied period as well as connections to other phenomena. These longitudinal studies describe flows out from as well as into the landscape. The landscape can thus be seen as a result of sedimentation of earlier happenings. Skånes (1997) has further developed this concept and describes the landscape as a result of sedimentation of earlier happenings, a mosaic where some parts are eradicated and where others remain whole or fragmented (Figure 10.1).

To understand the concept of landscape and biodiversity, one needs to have a background not only in the natural sciences, such as physical geography and ecology, but also needs to know how to consider and interpret human actions, which can create, maintain, or threaten biodiversity. The contribution from human and historical geography accounts for how different landscape types have developed through varied landownership and management regimes.

SOURCES OF DETAILED LANDSCAPE INFORMATION

Methods used in integrated landscape analysis include descriptions of changes in seminatural vegetation and small biotopes, description and measurement of the spatial pattern of landscape in ecological terms, and the monitoring of the dynamics and biodiversity at biotope and landscape levels (Skånes and Tollin 1991; Sporrong 1993a; Skånes 1997).

To fulfill Sporrong's five fundamental prerequisites for an integrated landscape approach, we propose the combined use of a variety of sources and materials that help us understand a dynamic landscape. The main sources, providing important information from different time periods, are aerial photographs and historical maps (Figure 10.2). Ancillary data complement the main sources in the analysis. A common denominator for these data sources is that they all represent geo-referenced spatial data. The focal point of data collection is the physical landscape and its ecological properties, its history, and its dynamics.

The present-day total landscape
A composite of physical and intangible structures and elements from recent and earlier time periods.

After AD 1940—The industrial farm landscape
Disappearing structures. Abandonment, fragmentation, and isolation of agricultural land.

1900–1940 (transition period)

AD 1750–1900—The scattered farm landscape
Reorganization of holdings, new fencing systems. Large-scale arable cultivation, mainly for arable land.

1100–AD 1750—The village landscape
Spatial organization of settlements. Permanent *infield–outland system* including vast areas of fodder-producing ecosystems.

1500 BC–AD 1100—The prehistoric landscape
Prehistoric stone walls, clearance cairns.

The physical landscape
A prerequisite for human utilization.

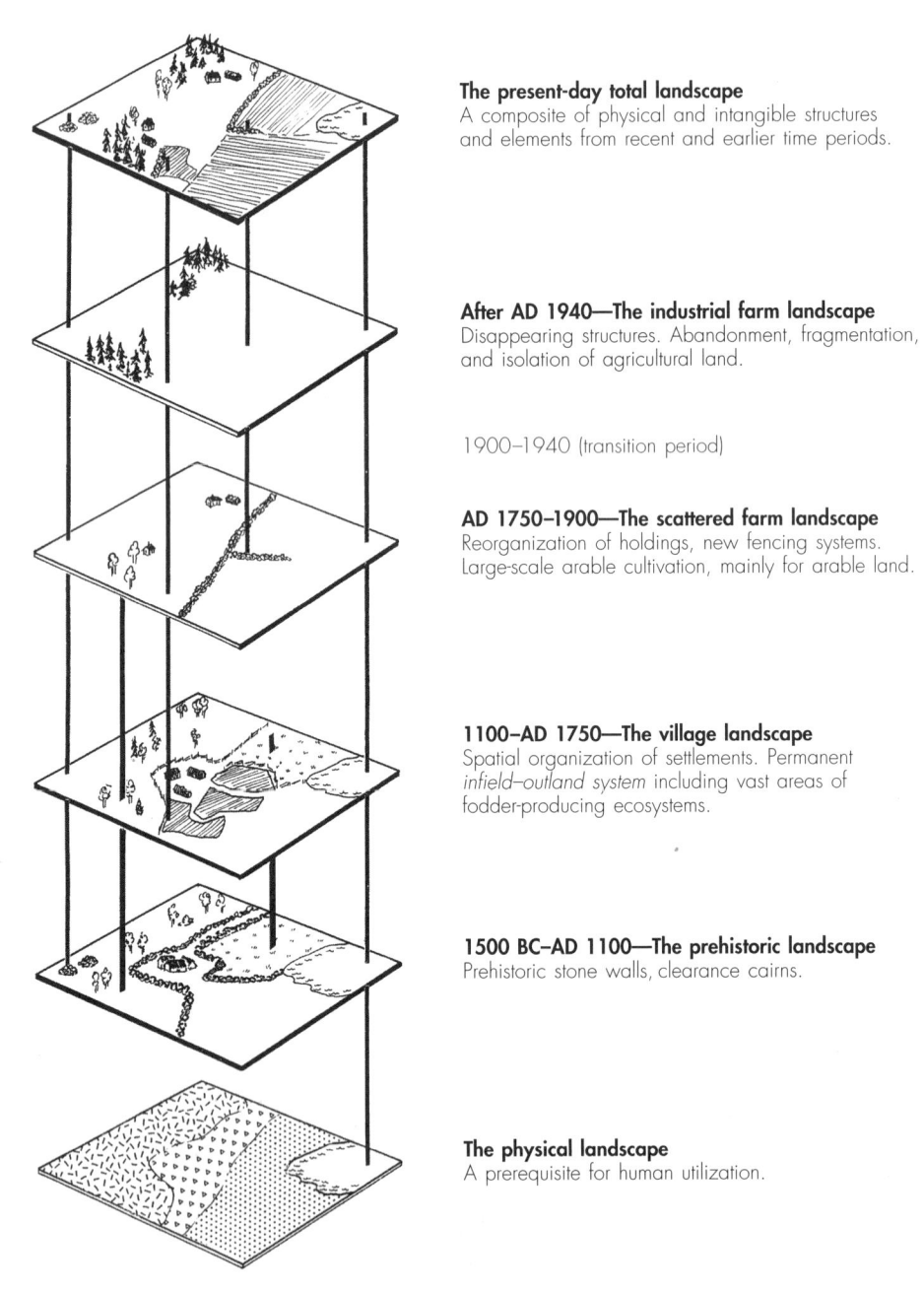

FIGURE 10.1. Conceptual model of the integrated ecological-geographical approach. The principal periods are shown as a system of opaque overlays indicating the origin of elements in the present-day landscape mosaic. Thus it is possible to use the present as a means of understanding the past, but, equally, knowledge of the past is needed to understand topological and chronological relationships within the contemporary landscape (modified after Skånes 1997).

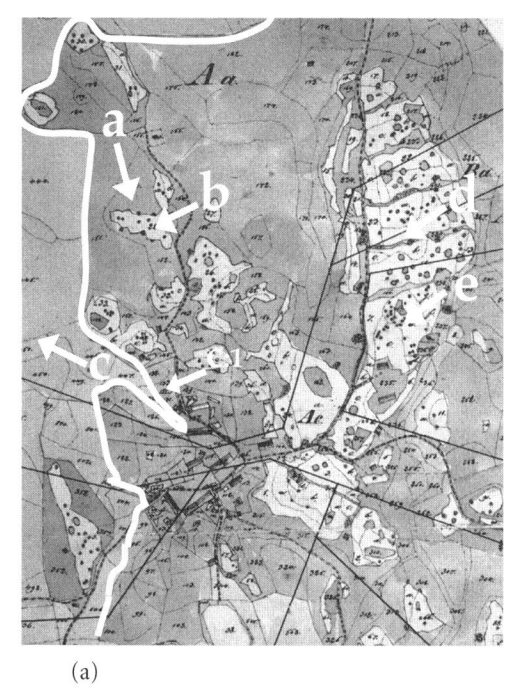

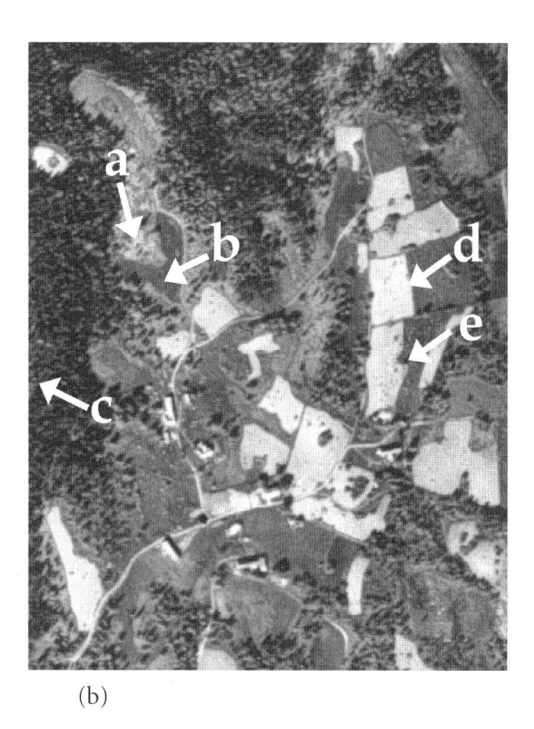

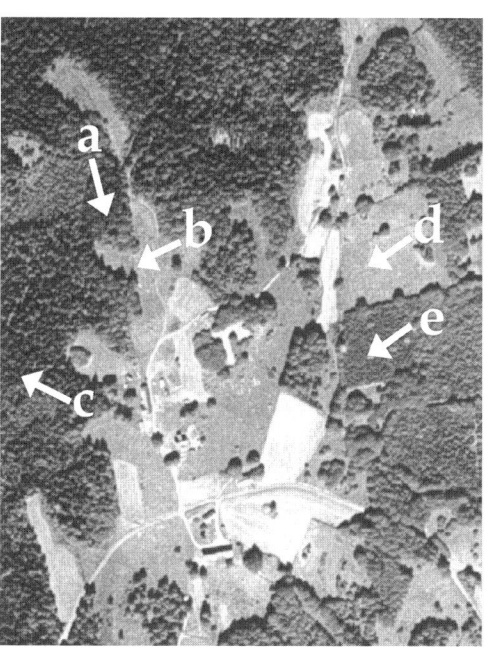

FIGURE 10.2. The *hagmark* landscape on a local level in Virestad parish: (a) cadastral map of 1851. (b) black-and-white aerial photograph, 1946. (c) color infrared (CIR) aerial photograph, 1986. In all three illustrations, the arrows (a) through (e) represent: (a) Hay meadow in the transition zone between arable land and outland forest. In 1946, the grassland is still open; by 1986 it had become dense mixed forest. (b) Arable field no longer in arable use by 1986, though its heart-shaped form can still be identified. (c) Outland pasture separated from the infield area by a fencing system (marked in Figure 10.2a). This area was wooded in 1946 and remained so by 1986. (d) Arable field in 1851 and 1946 and open cultivated pasture in 1986. (e) Arable field in 1851 and 1946; by 1986, the arable field was completely covered by a young planted conifer forest. Arrows (d) and (e) point to parts of the landscape with similar land-use history, but where a marked difference occurs in potential present-day and future biodiversity. Courtesy National Land Survey of Sweden, 1994.

In Sweden, the earliest black-and-white aerial photographs date to the end of the 1930s (Figure 10.2a). Since then there is repeated photography with an interval of five to ten years (scale 1:20,000–1:60,000). These photographs provide good information on land use and key aspects of vegetation. Since the 1970s, color infrared sensitive (CIR) aerial photographs have become available (1:30,000–1:60,000). These provide excellent detailed information of ecological properties and land use in the present-day landscape (Figure 10.2b) and are invaluable for the integrated approach. The period covered by aerial photographs is short but crucial and characterized by major changes within the Swedish economy clearly seen in the landscape. The oldest aerial photographs often show a small-scale rural landscape before mechanization with remains of the nineteenth-century enclosure landscape. The most recent aerial photographs represent the present-day landscape. Aerial photographs between these periods provide information on the direction and rate of changes, illustrating ongoing mechanization and specialization. The rate of change is so high that even the most recent photographs rapidly become out of date.

The photographs are stereographically interpreted using zoom stereoscopes providing a three-dimensional view of the landscape (Ihse 1978; Ihse and Nordberg 1984; Ihse 1987). The additional information this view provides is crucial for correctly interpreting and contextually understanding landscape features. The zoom of the instrument enables the combination between overview and detailed studies ranging from 20 km^2 to a few square meters. This zoom facilitates the interpretation and aggregation of details below one meter in the most common scale aerial photographs (1:30,000). These details are important in characterizing key landscape elements in the Swedish *hagmark* landscape. They are also important in comparison with detailed historical maps.

To increase the historical depth and visualize successions over time, researchers must interpret historical maps and their written descriptions. This enables the interactive detection of continuity in land use, especially in key elements like arable fields and grasslands, highlighting vegetation successions. Geometric and cadastral maps (1:4,000–1:8,000) describe the historical agricultural and enclosure landscape from the 1630s onward (Tollin 1991). They document the spatial organization of many farms, hamlets, and villages in the settled parts of the country before and during the enclosures of the eighteenth and nineteenth centuries. They also indirectly provide information on land use, ecological conditions, and past biodiversity (Figure 10.2c). This information, however, must be aggregated and interpreted (e.g., Aronsson 1979a, 1980; Emanuelsson and Bergendorff 1986; Tollin 1991; Skånes 1996). This interpretation is similar to that done for aerial photographs. The enclosure landscape is also described in old hundred maps (*häradskartor*, 1:20,000–1:50,000).

Ancillary data on the physical landscape are found in thematic maps of bedrock and soils (1:50,000–1:200,000), providing the prerequisites for potential vegetation and indicating land use. Topographical maps (1:50,000) give, through the depiction of relief, the basic features for the landscape scenery. The topography also indirectly provides information on soil distribution and hydrological conditions, important factors for vegetation pattern and potential land use. The contemporary economic maps (1:10,000–20,000) show landownership boundaries and distribution of arable production in combination with raw data in the form of ortho photos printed on the maps. Additional information comes from Statistics Sweden, with its regular Census of Land Use and Population since the 1920s. Field data from ecological inventories are available from the Swedish Environmental Protection Agency—for example, national inventories of wetlands, meadows, and pastures—and from nature conservation plans by local authorities. The Central Board of Antiquities provides descriptive and cartographic information on ancient monuments, rural buildings, and their function. Written historical sources and oral traditions provide a variety of names for meadows and pastures, which offer additional information on their properties (Ihse 1997). These ancillary data, useful to the analysis, are mostly descriptive but are not always spatial and completely geo-referenced.

The Concept of the Swedish *Hagmark* Landscape

The Swedish concept of the *mellanbygd,* the agropastoral zone between forest and agricultural regions, has been used principally to describe the production capacity in the agricultural landscape. *Mellanbygd* is by tradition a narrow definition of a semi-open mixed landscape, and only a few regions in Sweden are officially classified as such (Figure 10.5a). We want to extend the concept of semi-open mixed landscape, here called the *hagmark* landscape, to become a functional unit for the landscape type as a whole, taking into account the structures and ecological components of the landscape, on different scale levels, and the connections between them.

We define the *hagmark* landscape as a semi-open landscape, where the component grasslands, arable fields, and heterogeneous deciduous forests often create a complex mosaic pattern in the surrounding matrix of coniferous forests, mires, and lakes. Grasslands are the key components in terms of both nature conservation and cultural heritage values. We define grassland as the highest hierarchical level of all land-cover types currently influenced by or still showing evidence of grazing or mowing (Skånes 1996). They are characterized and mostly dominated by light-demanding herbaceous vegetation in the field layer, but they also might be dominated by heath vegetation (e.g., *Calluna vulgaris*). These grasslands can be open or wooded, with a wide variation in moisture regime, tree, shrub, and field-layer cover and species. Today, grassland ecosystems dominate the small biotopes, which are remnants of the former widely distributed seminatural hay meadows and pastures. Examples of these small biotopes are linear elements, such as ditches and road verges, or point elements, such as midfield "islets" (Ihse 1994b). The deciduous forests are also important because most of them were formerly used as pasture. The content and distribution of these components is not only dependent on physical conditions such as bedrock, soil, topography, and climate but also to a high degree on past and present land-use regimes, intensity, and ownership history. Grassland studies of special interest for the Swedish *hagmark* landscape include aspects of successions and dynamics of vegetation and management methods (Bengtsson-Lindsjö et al. 1991; Skånes 1991, 1996; Ihse 1995).

SCALE DEPENDENCY AND DISTRIBUTION

A mixed landscape can, in a general perspective, be perceived as a transitional zone between arable land and forests on a local village scale (Figure 10.3) or between forest regions and agricultural plains (Figure 10.4). Toward the forest landscape it consists of open glades in the forest, and toward the agricultural plains it consists of forest "islets" surrounded by open ground. Depending on how the landscape is perceived,

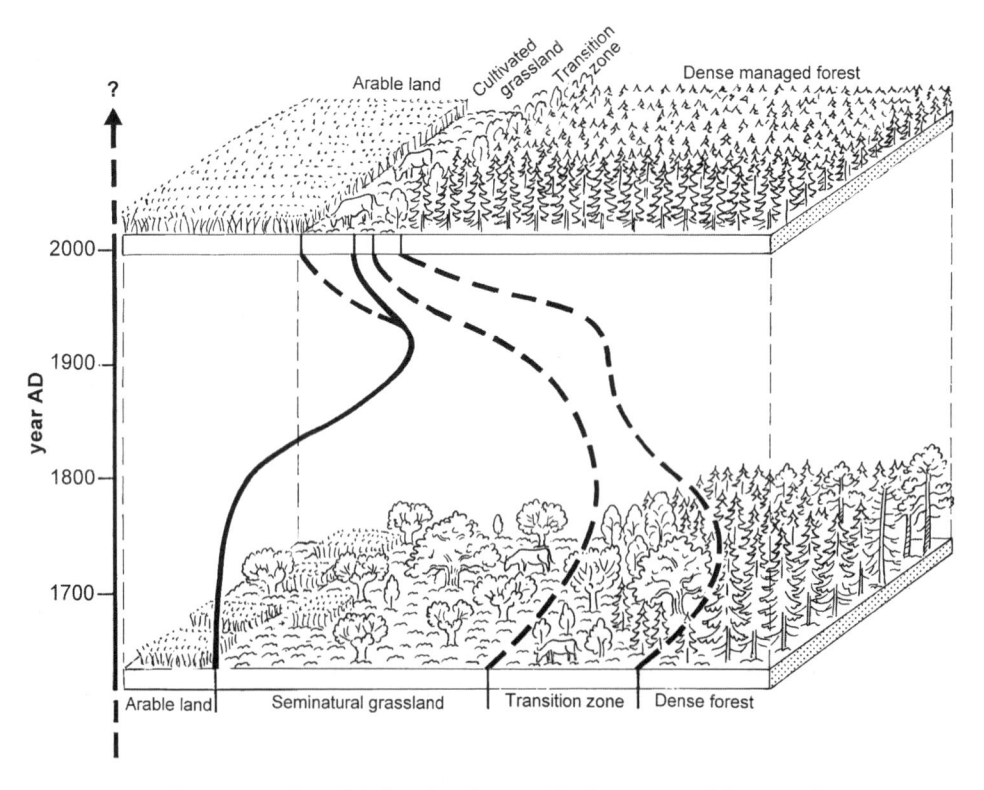

FIGURE 10.3. A conceptual model showing the grasslands as a transition zone between open, arable land and forest on a village or farm level. It is also a schematic visualization of the spatial distribution of the *hagmark* landscape as a transition zone from forest to open landscape (see Figure 10.4). Today, the arable land adjoins managed forest to a greater extent, and the heterogeneity provided by the grasslands is lost (from Skånes 1996).

the concept of semi-open mixed landscapes is scale-dependent and in general could be difficult to demarcate. Yet the landscape can also be seen as constituting a region of its own, with a common historical development from land that previously was more open than today. In general, this landscape type is distinguished neither in agricultural statistics nor in physical and environmental planning. This is problematic since it harbors a major part of the landscape biodiversity.

The ratio between forest and open ground is used as a key factor to landscape qualities. It is important for the visual perception of the landscape as well as the ecological functional connections. The mixed landscape type we describe here generally consists

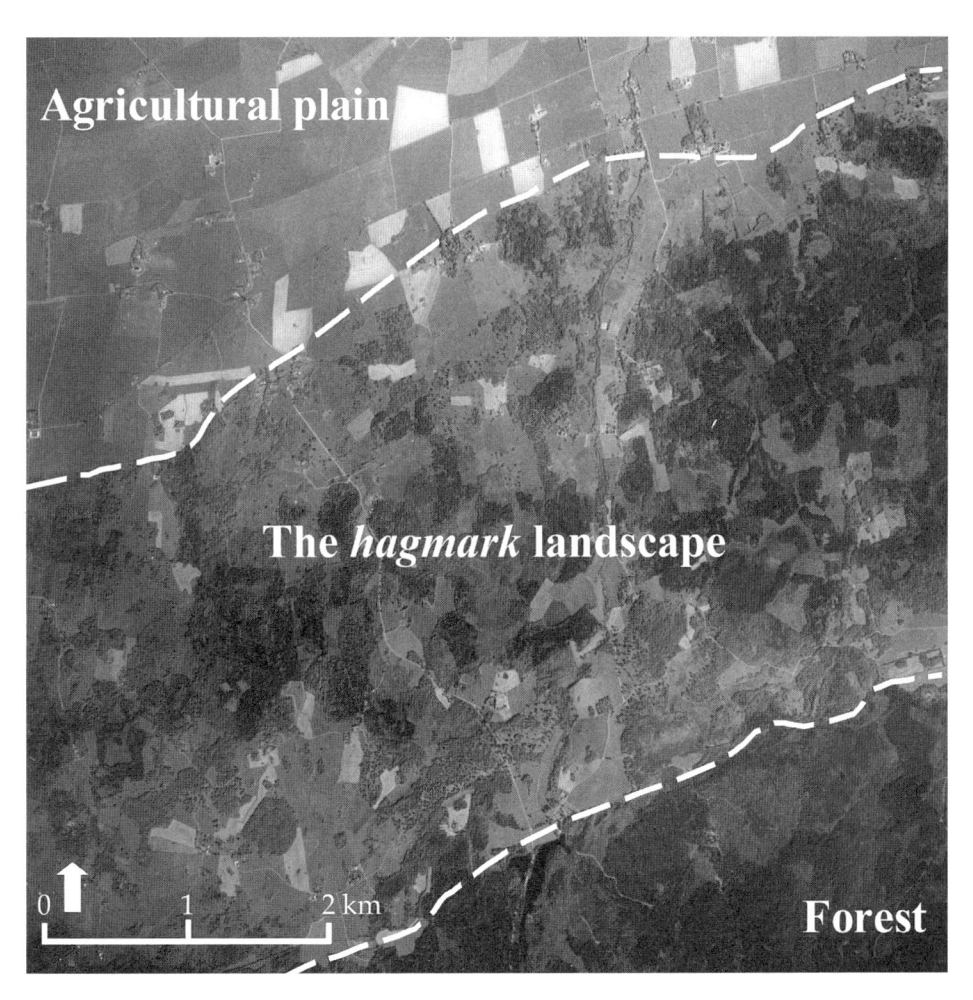

FIGURE 10.4. The functional *hagmark* landscape on a regional level can be seen as a transition zone (see Figure 10.3). This CIR photograph from 1983 shows the zone, 3–5 km wide, situated between the northern agricultural plains and the forest region of Götaland. Courtesy National Land Survey of Sweden, 1994.

of 25 percent to 75 percent forest. A higher share of forest is not unusual due to the recent increase in land abandonment and active afforestation. However, most important for the biodiversity is also the variation and connection within the open land between the tilled fields, the small biotopes, and the different grassland types. It is not feasible to put a percentage figure on the shares of different components today, since this in the highest degree depends on the scale in which the landscape is viewed. The *hagmark* landscape may occupy large landscape segments and characterize a wide area, or it may exist only as "islets" on the local level.

To illustrate the effect of scale-dependence, we will give some examples of how the *hagmark* landscape can appear on different scales.

At the national level (maps on scales 1:1 million or smaller), only a few areas in Sweden are classified as semi-open mixed landscape or *mellanbygd* according to the

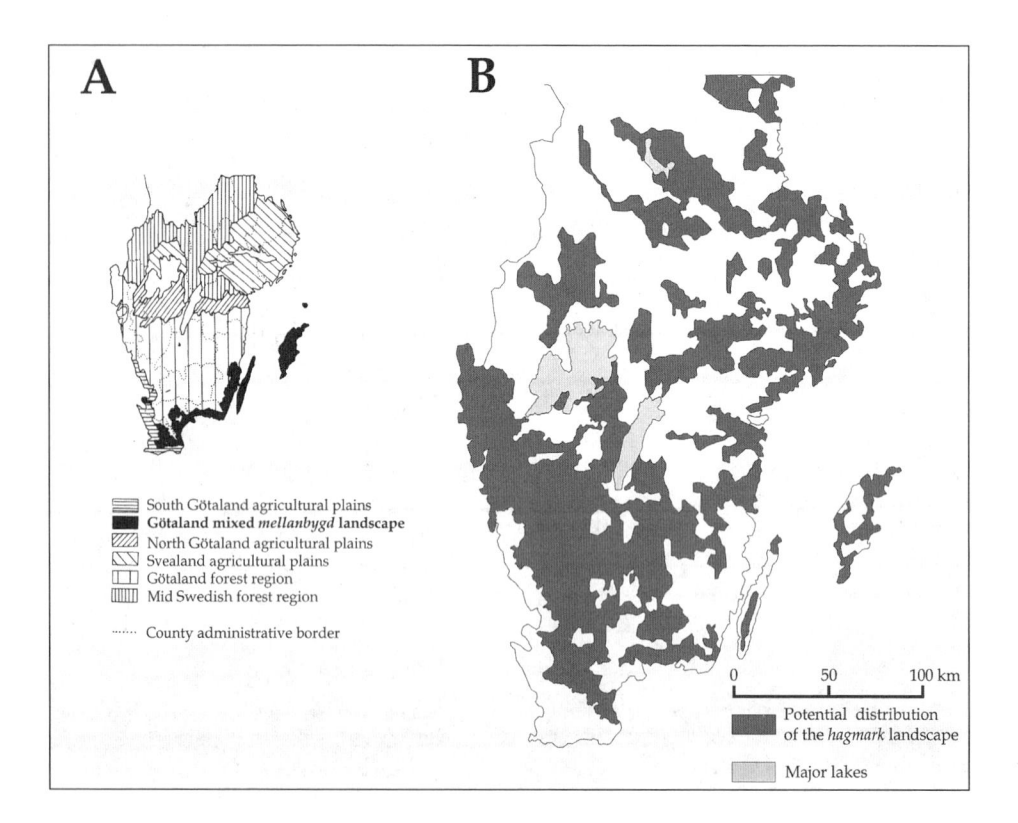

FIGURE 10.5. (A) The distribution of the mixed landscape (Sw. *mellanbygd*) (Statistics Sweden 1987). Only a few areas (in black) in southern Sweden are officially classified at this national level. (B) Preliminary map of the distribution of the Swedish *hagmark* landscape based on four *mellanbygd* classes (from Grundberg 1966). The distribution (dark gray) indicates the regions where the *hagmark* landscape is an important component in the agricultural landscape. Note the difference between the two classifications.

existing classification for agricultural production (Statistics Sweden 1998). This *mellanbygd* includes parts of the Scanian ridges, parts of the coastline of the province of Blekinge, and the coastal regions of the province of Kalmar along with the islands in the Baltic Sea (Figure 10.5a). We argue that it is not restricted to one single place or geographical region, but is a functional unit that can be found in several places in southern Sweden, most frequently in the south Swedish highlands but also in parts of Östergötland, the Lake Mälaren region, and northern Uppland. Grundberg (1966) formulated a regional classification for agricultural production describing landscape openness. Based on this classification, we present a preliminary compilation of the potential distribution of the Swedish *hagmark* landscape (Figure 10.5b).

At the regional county level (maps on scales 1:250,000 or 1:100,000), the difference between forest landscape, semi-open mixed landscape, and agricultural plains is no longer clear. On this scale, "local plains" connect open rural areas in the forested landscape. The map of Kronoberg county clearly shows the characteristic pattern of the open areas on the regional scale (Figure 10.6a). They follow the physical conditions closely, here as elongated and narrow strips following the streams in a north-south direction and more irregular shapes around the lakes. The transition zone, on a regional level, can be seen in aerial photographs, where it clearly diverges from the agricultural plains and the forest. In some cases it is a well-defined and distinct zone as in Östergötland (Figure 10.4).

The local parish level can be subdivided into village level (maps on scales 1:50,000 or 1:20,000) and farm level (maps on scales of 1:10,000 or larger), showing the details in the landscape. The map of Virestad parish in 1993 (Figure 10.6c) illustrates how the present-day landscape is composed of small "local plains" in a forested landscape, while in 1946 (Figure 10.6b) it was a mosaic of open land consisting of small fields and vast grasslands mixed with forest in a complex pattern. The *hagmark* landscape can also appear as local, sometimes isolated, objects in the landscape (Figure 10.7a). In many cases stag-headed deciduous trees are scattered in the grassland, a structure that is still visible despite ongoing abandonment and subsequent spontaneous encroachment. In the field, wooded grasslands are often perceived in positive terms, partly as an effect of their diversity in field layer, bushes, and tree cover (Figure 10.7b).

THE *HAGMARK* LANDSCAPE IN RELATION TO EXISTING SWEDISH LANDSCAPE CLASSIFICATIONS

Different types of landscape classifications have been used for various purposes at different times. Most planning today still uses the traditional administrative division of Sweden into three hierarchical levels, counties, local authority districts, and parishes (Statistics Sweden 1985). These levels are important in describing the *hagmark* landscape.

The classification for agricultural production (Grundberg 1965) is the most common system for official statistics on agricultural production. In this classification Sweden

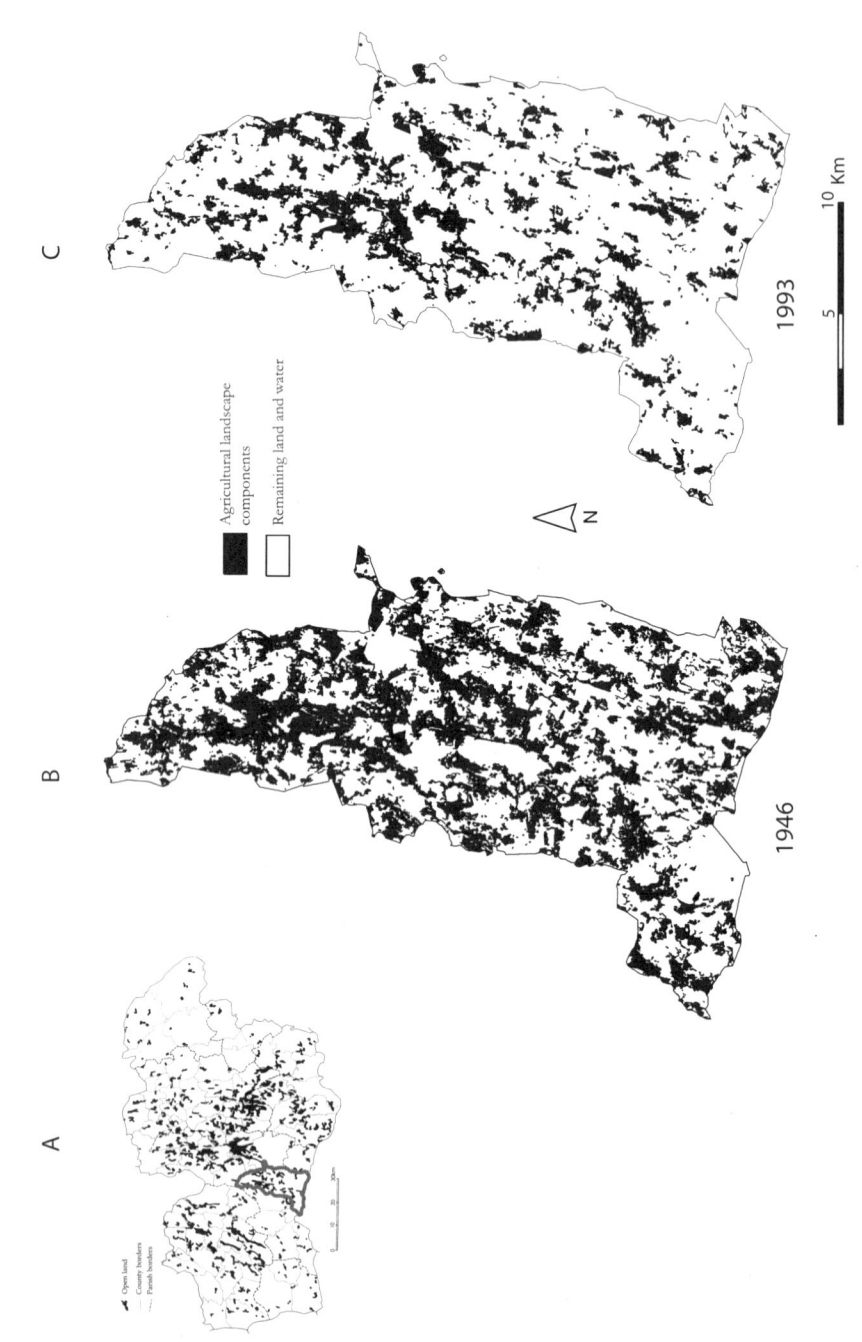

FIGURE 10.6. (A) The regional perspective, here the county of Kronoberg, shows the distribution between open land (in black) and remaining land (in white). The connection between the soil type, where the fine sediments occur mainly around the lakes and along the small rivers and valleys, and the plains and their pattern and distribution is clearly seen. (B and C) The local perspective, Virestad parish in 1946 and 1993, shows the change in distribution of agricultural areas (in black), including arable land and all types of grasslands, and remaining land, including forests, lakes, and mires (in white). Major changes over the last fifty years have resulted in a great loss of agricultural land (compiled from Skånes 1996).

FIGURE 10.7. (A) CIR aerial photograph from the oak-dominated *hagmark* landscape along the river Stångån. The stag-headed oaks appear as round dots (Courtesy National Land Survey of Sweden, 1978). (B) Detail of the *hagmark* landscape in the field: wooded pasture with a mixture of stag-headed trees and scattered shrubs over an open dry-mesic field layer. Photo by Helle Skånes.

is divided into three major landscape types: plains, mixed areas, and forested areas. These are in turn divided into eight agricultural regions according to geographical location, and then further divided into eighteen subregions on the basis of dominating land-use type and administrative boundaries. The system was originally constructed during the 1920s and is based on the assumption that land use is a result of physical and economic conditions only. It does not, however, adequately describe the present state of the landscape because it was constructed at the time when the Swedish area of arable land was near its maximum. According to this classification, the *hagmark* landscape is restricted to the zones of Götaland forest and mixed landscape.

The Nordic physical geographical classification often used in nature conservation was developed during the 1980s as a joint effort by all the Nordic countries (Nordic Council of Ministers 1977). Approximately forty classes based on potential vegetation zones concern Sweden (Sjörs 1956). They are subdivided according to physical conditions, such as bedrock, soils, relief, and climate. In this classification, the actual land use is not taken into account, nor are the cultural aspects of the landscape. The *hagmark* landscape belongs mainly to the boreo-nemoral zone (Sjörs 1963).

The *National Atlas of Sweden* includes a recent classification of integrated cultural landscape regions on a national level based on physical, ecological, and cultural features. This classification, highly relevant to the *hagmark* landscape, consists of approximately fifty regions (Helmfrid et al. 1994). These regions, based on natural features, vegetation, settlement, and present and historical land use, have been further aggregated into eleven major regions (Sporrong et al. 1995). The different types of ownership are important factors influencing the entire landscape, not only the land use but also the scenery and biodiversity. As a result of the landscape's complexity, these regions are not discrete but rather are separated by fuzzy transition zones (Helmfrid et al. 1994).

Natural and Cultural Values of the Swedish *Hagmark* Landscape

The description of natural and cultural values below should be viewed in the light of the narrow and sectorial definition of values. Criteria used to determine nature conservation values differ greatly from the criteria used to determine cultural historical values attached to grasslands, meadows, and pastures. From a biological viewpoint, the focus is on the composition of the plant communities and species, where nutrient levels and ground moisture constitute the most important factors. Species and biotopes are generally considered valuable for nature conservation when they are rare, unique, characteristic, well developed, or interesting in other aspects without a specific cultural-historical context. Cultural values relate to ancient monuments, buildings, and traces from agricultural history without a specific ecological context.

NATURE CONSERVATION AND BIODIVERSITY VALUES

Natural values embrace the whole concept of biodiversity, not only threatened and rare species but also diversity and patterns of biotopes and ecosystems on a landscape level. These values are often classified into botanical, zoological, and geological properties with criteria such as species-richness, rarity, and diversity. Only in recent decades has the significance of management regimes and land-use history gained recognition. Grasslands, including hay meadows and pastures, often have high botanical value. The importance of these grasslands, as key biotopes in the Nordic landscapes, cannot be overestimated in terms of biodiversity (Aronsson 1979b; Haeggström 1987a, 1987b; Ihse and Norderhaug 1995; Norderhaug 1996). They are species-rich and many of the species are now threatened. Of about four hundred threatened vascular plant species on the Swedish national list of endangered species, nearly three hundred belong to the agricultural landscape (Aronsson 1999). There are no comprehensive statistics that show how many of these species belong to the *hagmark* landscape. As a high proportion belongs to the grassland vegetation, and the grasslands are the most important feature in the *hagmark* landscape, we assume that many of these species will be found in this landscape. This argument is strengthened by the survey of ancient meadows and pastures in Sweden that revealed a high proportion of the most valuable (class I) meadows and pastures, in some cases over 50 percent of the total surveyed areas in the counties, including the *hagmark* landscape (Lindahl 1997).

The grasslands consist of many different plant communities. The concept of "plant community" is important for the botanical classification of plants. A plant community is the collective of jointly growing species, united by ecological bonds. They are all viable within the environment that makes up the location. Physical (abiotic) factors traditionally have been considered as the most important explanation factors, while land use has played a subordinate role until the last few decades. We know that the natural values of the grasslands are mostly related to traditional land use and continuity, and that factors like disturbance and stress (as in grazing or mowing) are most important for the existence of species. Plant communities are constructions that allow vegetation to be described and more easily compared in different regions. Several classification systems exist, based on different principles (Arnberg and Ihse 1996), but there is no universal classification. In the Nordic countries, a system describing the vegetation types is commonly used (Nordic Council of Ministers 1994).

The distribution, type, and number of plant species and plant communities are often a prerequisite for the existence of many animal species and hence for zoological values. Especially important groups are birds, insects, frogs, and bats. Of more than the 140 threatened higher animal species, more than 50 have their main habitat in the agricultural landscape. The grassland plants are mostly important for butterflies and dragonflies, while many birds and beetles depend on old deciduous trees and open, sunlit ground.

The biodiversity and function of the grasslands not only depend on the total area and species content but also on the distribution and thus the pattern and structure, together with the dynamics and successions. Thus the grasslands must also be described in landscape ecological terms, such as fragmentation, isolation, and connectivity.

For the extreme botanical conservationist, the natural vegetation, unspoiled by human influence, represents the greatest value and gives the greatest emotional charge, while what is seminatural, affected by human impact, accordingly has lesser value. This is evident in terms such as "secondary" plant communities.

CULTURAL HERITAGE VALUES

The cultural heritage is described in relation to the properties of settlements, archaeological sites, and buildings and in relation to physical and biological resources. Values associated with the cultural landscape also include immaterial aspects that we do not address in this chapter. In the present context, cultural values are mostly related to the agricultural history and traces of the agricultural practices over time (Table 10.1). There are also major immaterial and aesthetic values. The old, traditionally managed agricultural landscape and its grasslands have inspired poets, writers, artists, and composers. Meadows and pastures in this semi-open landscape are described in folklore songs and tales, creating a national image of red cottages, white birches, and pastures rich in herbs and flowers.

Cultural heritage values are viewed in a wide perspective, where humans, production, and management are the most important factors. Productive grasslands were crucial to agriculture in ancient times, but they are often neglected in terms of cultural

TABLE 10.1.
The classification of cultural historical elements in the *hagmark* landscape (modified from Ihse and Lindahl 2000).

Elements from grazing practice and fodder production	Elements from agricultural practices and archaeological monuments
Cattle path	Old field pattern with small and lobated fields
Stone walls and earthwork	Clearance cairns and cleared areas
Wood fences	Levelled terraces
Hedges	Old farm types and ruins
Coppiced and pollarded trees	Rune stones
Ditches and dam constructions	Burial mounds
Hay barn	Old curved roads or sunken tracks

heritage. Cultural valuation criteria also concern the landscape's openness and configuration, ancient character, relics of the past, style of architecture, and settlement location. "Culture" is often a positively charged term when it is connected with something that humans have made or done long ago, leaving traces in the form of ancient remains and old buildings.

INTEGRATED LANDSCAPE VALUES

The *hagmark* landscape is highly dynamic in time as well as in space. It has historical depth with specific natural and cultural features showing some of Sweden's agricultural history. Both natural and cultural values must be integrated in descriptions and analysis of the landscape in a holistic perspective (Ihse and Norderhaug 1995). It is a challenge to find mutual and positively charged concepts. The living heritage represented by meadows and pastures must be appreciated both from a natural and cultural point of view. To strengthen our approach, we have made efforts to bridge disciplines and bring clarity to the concepts (Skånes 1991; Skånes and Tollin 1991; Skånes 1997).

The integrated values are mostly connected to the wooded or open grasslands and patches of heterogeneous deciduous woods. In these biotopes we find rare and typical plant species, depending on continuous management. Old and sometimes pollarded trees, frequent in these grasslands, contain many rare insects as well as mosses and lichens. The grasslands frequently contain clearance cairns, stone walls of different ages, and old terraces from ancient agricultural fields. The presence of ancient settlement places and the old building style in the remaining houses also contribute to the overall values.

We argue that the natural and cultural landscapes are not mutually exclusive. Instead, the cultural landscape is seen as superimposed on the natural landscape in layers of varying transparency and degree of amalgamation, hence sharing the same space without cancelling out one another (see Figure 10.1). One important overall value is the sustainability of production and the ecosystem's goods and services in this landscape.

The necessary integration of cultural and natural values in a landscape context is just beginning to be understood (Ihse and Blom 2000; Ihse and Lindahl 2000). Many of these values are connected to the grassland ecosystems, with the ancient hay meadows and unfertilized pastures. This landscape also has high aesthetic and recreational value (Emanuelsson and Johansson 1989).

DESCRIPTIVE CHARACTERIZATION OF THE SWEDISH *HAGMARK* LANDSCAPE

Our description of the Swedish *hagmark* landscape is based on criteria in accordance with the integrated landscape analysis proposed by Skånes and Tollin (1992):

- Physical conditions: geology (bedrock, soils), geomorphology (landforms, mainly glacial), morphometry (relief, topography), and climate (temperature, precipitation).
- Vegetation and landscape ecological patterns (patches of key biotopes, point elements, as stepping stones, and linear elements like connecting corridors and barriers).
- Land use, historical and present, land organization and ownership (agriculture fields, pastures, forestry, semi-open mixed forestry/farming, commons, company-owned, private large estates, or private small farmers).
- Administrative boundaries and restrictions.
- Values: natural (botany, zoology, geology) and cultural (archaeological sites, arable systems, building sites and building styles, and specific characters).
- Dynamics and changes.

In the description of the landscape and its ecological content, grasslands are a key component regarding ecological significance and, indirectly, cultural historical significance. Grasslands, such as meadows and pastures, constitute an important part of the open agricultural landscape. In the agricultural plains, most of the grasslands are fertilized and hence species-poor. The unfertilized seminatural grasslands are more or less wooded. Former hay meadows are now exclusively used for grazing. Only a few open or wooded hay meadows remain in management. The amount of meadows and pastures is still relatively high in the Swedish *hagmark* landscape, in some cases up to 30 percent of the total cultivated land, a high figure compared to less than 6 percent in other parts of Sweden (Lindahl 1997).

PHYSICAL CONDITIONS: PREREQUISITES FOR UTILIZATION

The bedrock consists mainly of Precambrian acid rocks such as granites and gneisses, slowly weathering and producing nutrient-poor soils (spodosols) unsuitable for agricultural cultivation. Basic bedrock, like gabbro, diabase, and limestone, which weather more easily and produce fertile soils, is rare and important only on a local level.

The soils are mostly till, often rich in boulders. In some places, the bedrock protrudes through the soil layer. Organic soils like peat occur in all the small depressions and also in large wetlands, such as fens, mire complexes, and raised bogs. Glacial fine sediments are found in areas below the highest shoreline (HS) during the last deglaciation or in local lake sediments, and postglacial fine sediments are found in river valleys and along shorelines. As landforms and soils differ above and below HS, this is a critical boundary for differentiation between soils suitable for cultivation or not, and an important factor in understanding the landscape and the land-use pattern over time. Fine sediments are suitable for agriculture. The large Swedish plains and central agricultural districts of today are situated below HS, while most of the *hagmark* landscape is situated above it. Above HS are many lakes, mires, and raised bogs and midsize terrain forms from deglaciation, often eskers and drumlins. Here we find old agricultural fields on top of the drumlins and moraine hills, with light and dry soils, cultivated long ago. In other areas, we find newer fields in low-lying parts, such as drained wetlands on organic soils and on the bottoms of former drained lakes. Preferences for cultivation have shifted over time in accordance with technical developments.

The topography is moderate and gently undulating with low relief but with great variation. The differences in altitude are seldom dramatic. The area is partly situated on the south Swedish highland. In the northern and northeastern part, there are fissure valleys and faults, running in a west-east or northwest-southeast direction, with rows of lakes or fens on the valley floors. Other parts are flat or gently undulating, with clay and good soils, giving rise to small productive agricultural areas. In some parts the plains are fragmented by bedrock outcrops covered with till, emerging from the shallow clays.

The climate is cold temperate, mostly maritime, moist, with frequent precipitation during the whole year, cool summers with rainstorms, and front rain during spring and autumn. The annual precipitation is between 500 mm and 600 mm. The precipitation during the winter is partly snow. The average temperature for January is −1° to −3°C, and the average temperature for July is +16°C. The topographic conditions create a small-scale mosaic landscape, with a variation in local and microclimate, giving a mix of favorable and unfavorable areas for cultivation.

The vegetation belongs to the boreo-nemoral zone, with potential vegetation of mixed coniferous and deciduous forests (Sjörs 1963). The dominant forest is coniferous and consists mostly of plantations of spruce *(Picea abies)*. In the fissure valley landscape, there are fragments of old-growth pine forest *(Pinus sylvestris)*, natural or naturalized, on thin till soils near bedrock outcrops. Solitary hardwood deciduous trees are frequent, and more seldom are large blocks of forest dominated by pedunculate oak *(Quercus robur)* and mountain ash *(Fraxinius excelsior)* on fertile soils, or on sites with milder local climate along the lakes. Most of the hardwood deciduous woods today were formerly open or semiopen grasslands (Ekstam and Forshed 1992). The dominating deciduous trees are silver birch *(Betula pendula)*, downy birch *(B. pubescens)*, and aspen *(Populus tremula)* and occur on poor soils, mainly on till or peat. Mires, bogs, and fens are frequent and vary in size.

CULTURAL CONDITIONS: TRACES OF HUMAN ACTIVITIES IN HISTORICAL TIMES

The cultural landscape we see today is like a mosaic, composed of elements from different periods (Figure 10.1). It has been affected by a complex combination of factors during prehistoric and historical times through human use of natural resources. The different factors that have modified the landscape are related to population growth, different landownership forms, different management regimes and techniques, and various types of economy and politics. The time of settlement, historical buildings, archaeological sites, and the location of settlements and buildings describe the cultural features.

The early settlements were located in areas with a certain combination of natural features, and there is a good correlation between physical conditions and land use, as Kristoffersson pointed out (1924). This relationship is still detectable, although weakened, in the landscape (Figure 10.8a). The spatial forms of the fields were also closely adapted to the terrain, often resulting in irregular and complex forms. Three different types of village settlement location, according to the physical conditions, are easily found (see Figure 10.8b). Below HS, in the local plains, the most favorable farmstead location was in the valleys, close to the wet or mesic meadows or along small river valleys. The most typical location was higher in the terrain, on dry till, with bedrock outcrops, close to water and areas with fine sediments. Above HS, the typical location of settlements was on poorer soils on top of hills or drumlins with a good groundwater supply.

The historical development of the landscape is not a continuous process but rather a series of expansion and regression periods of new settlement or abandonment. We can still distinguish some elements of the natural landscape, the prehistoric landscape, the Viking Age landscape, a historical agricultural landscape, the enclosure landscape, and the rationalized agricultural landscape of today (Figure 10.1). The Viking Age landscape can be described as a village landscape, while the enclosure landscape is a scattered farm landscape and the modern rationalized agricultural landscape is an industrial farm landscape (see Figure 10.8a).

The village landscape that evolved from prehistoric times to the enclosures of the eighteenth and nineteenth centuries consisted of small clusters of farms. The plains and banks of watercourses were settled early, while the more marginal forested areas were settled during medieval times, or in the sixteenth century. Many archaeological traces exist from the early and late Iron Age in this region, such as settlements and burial fields, as well as frequent remnants from the Viking Age. The historical buildings are churches, many from medieval times, and some convents from the twelfth century. The villages were often single farms or small clusters of farms (Sporrong 1993c). The land around the village was organized as infield (Sw. *inäga*) and outland (Sw. *utmark*). The infields were dominated by hay meadows, with small tilled fields in a mosaic between. The outland was characterized by common pastures and outfields and multiple uses of forests and wetlands (Skånes 1997). Most of the forests were grazed, resulting in grass-dominated field layers and an open structure with a variety of coniferous and deciduous tree species. The wetlands were used for hay production, but also could be used for grazing and peat mining. Changes in this village landscape occurred within given frameworks provided by physical conditions, ownership, and management practices. The tilled fields covered a small area compared to the grasslands.

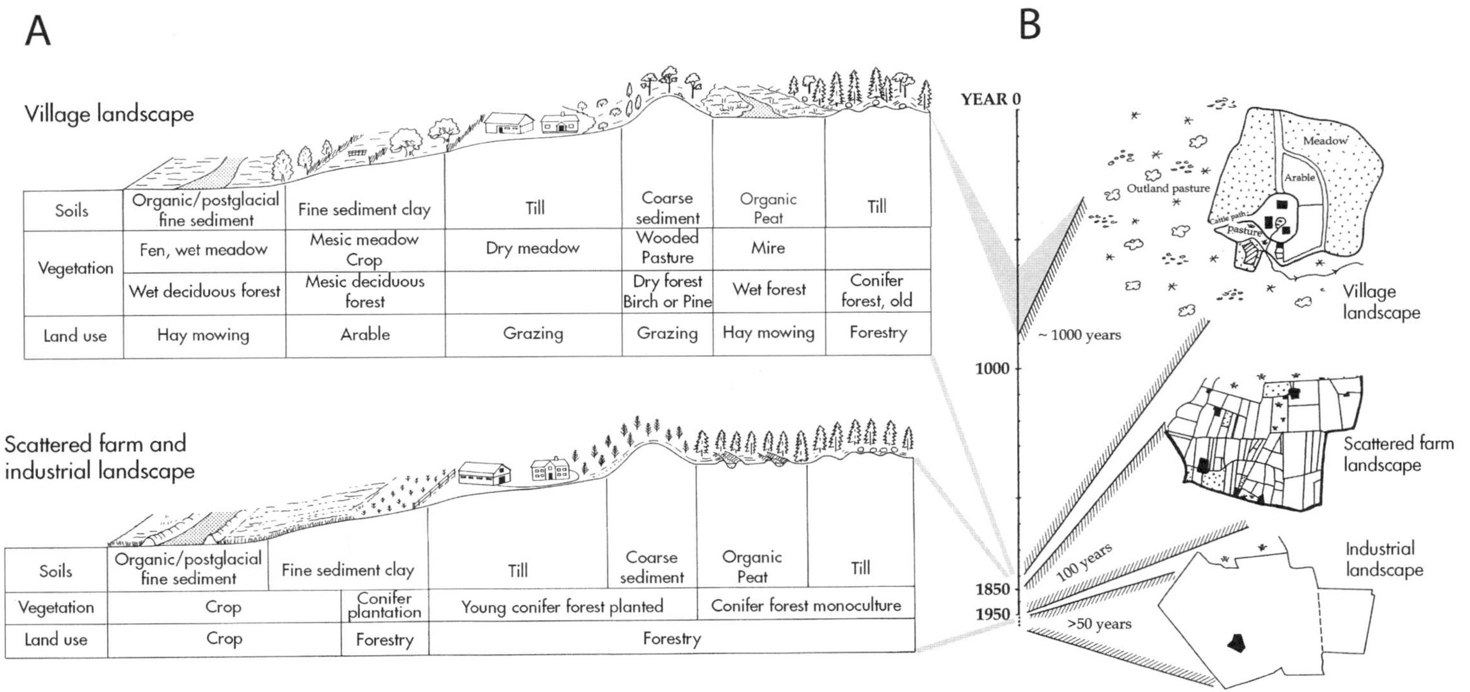

FIGURE 10.8. The development from a village landscape to a rationalized industrial farm landscape. (A) As the landscape structure shifts, the relationship between soil, land use, and vegetation is lost (modified from Ihse 1994). (B) In the village landscape, nested farms are surrounded by small arable fields, large meadows, and vast outland pastures. The scattered farm landscape is a mix of arable fields and grasslands. In the industrial landscape, only a few specialized farms remain and the size of the arable fields increases (modified from Ihse et al. 1991).

At the beginning of the nineteenth century, the total area of agricultural fields was about 1.5 million hectares.

Land use was organized according to the same principal pattern that existed over a period of more than one thousand years. This long period of continuity allowed vegetation types to develop and adjust to changes. Over recent centuries, however, the duration of stable periods has decreased and the rate of change has increased. The village landscape existed for more than a thousand years, the scattered farm landscapes for a hundred years, and the industrial farm landscape has developed since World War II and thus so far has lasted only sixty years (Figure 10.8b).

Cultural traces from farming activities are still found intact in the *hagmark* landscape. Many features are clearly connected to different types of boundary, for example, administrative, ownership, and land-use boundaries. The richness of boulders has given this landscape a specific character through the abundance of stone wall systems. Many derive from the enclosure period of the second half of the nineteenth century, but there are also relict stone walls from prehistoric and medieval times. The old tilled fields can be recognized by the many clearance cairns on boulder-rich till. The ancient fields can also be seen as flattened cleared areas, with shallow ditches and terraced areas in undulating terrain. The intensive removal of stones in recent decades has partly changed the natural conditions and eradicated these cultural historical traces. Here are many old, curved, and winding gravel roads with broad road verges holding a species-rich flora. The old roads often follow the terrain closely, on drumlins or eskers, the high and dry landscape elements. Many beautiful old stone bridges can be found along these roads. Many roads, some of which originated as cattle tracks, are bordered by stone walls. Fossil and relict roads can be traced as sunken tracks in the soil.

The scattered farm landscape was created by a series of laws, the last and most efficient of which was from 1827, when the land was reallotted and the villages became fragmented and replaced by solitary farms scattered in the landscape. Each farmer now owned one or a few blocks of land often marked by stone walls. Many stone walls, still characterizing this part of Sweden, are from this period. The large meadows were fragmented when large areas of former meadows were plowed for crops that were necessary to feed the growing population. The meadows and grasslands were dispersed over the landscape in small or large patches, connected by wide linear elements. Each farm had all types of land, meadows, pastures, and fields. Meadows and wetlands still had high economic value in the nineteenth century, but they successively lost their importance for hay production as ley production started.

Farming after 1827 was still a combination of animal husbandry and a wide variety of crops in rotation in small fields, normally less than one hectare. Tilled fields expanded at the expense of the former meadows, and the frequent lowering, or total draining, of small lakes and wetlands contributed to new arable fields as well as new hay meadows. The amount of open agricultural land was highest during the 1920s, around 3.7 million hectares, but by 1988 had decreased to 2.9 million hectares. The influence of the village landscape can be seen in the flora composition of the grasslands. Many plant species, with specific requirements regarding management, still occur in these grasslands and are keys to the landscape history. Such areas with relatively long continuity are richer in species and plant composition than more recently cleared grasslands.

PRESENT LAND USE

The present land use is a combination of animal husbandry, especially dairy or meat production, and forestry. Cereal crops are grown only in a few places. The farm economies are seldom based solely on farming, but this is usually combined with forestry and sideline work outside the farm. Small-scale mining and the glass industry were features of the most forested parts for several hundred years but now are of minor importance. The semi-open mixed landscape, like most rural areas in Sweden, suffers from an ongoing demographic impoverishment that started in the 1950s. Many small farms in the former semi-open mixed landscape have been abandoned. Marginalization has led to afforestation of large areas, transforming the open land of fields and pastures into closed monocultures of coniferous plantations. The agricultural practices have become increasingly intensified and specialized and, at the same time, concentrated in the central parts of the region. Some animal husbandry has moved to the more productive plains in southern Götaland, where the animals are now grazing on former arable land. Where the semi-open landscape becomes "closed" by forest, its history becomes less visible (Ihse 1995). The changes contribute to a decrease in biotope diversity and thus a loss of species on a local scale.

Since World War II, the coniferous forests have grown at the expense of low-production arable fields, pastures, and heterogeneous deciduous forests. The general trend between 1944 and 1988 in all the Swedish agricultural landscapes has been a decrease in open cultivated land, including fields and grasslands. The decrease by 40 percent to 45 percent in the south Swedish mixed forested landscapes has in many areas transformed the landscape into a densely forested landscape (Statistics Sweden 1990a). The counties of Jönköping, Kronoberg, and Kalmar together with Östergötland still have the highest share of seminatural grasslands, although less than 20 percent of the farmed land comprises pastures (Statistics Sweden 1990b; Lindahl 1997). Although the village landscape and the scattered farm landscape have ceased to function, its imprints are still visible in the *hagmark* landscape.

There has been a rapid decrease in seminatural vegetation, grasslands, and deciduous woodlands. Much of the ecological content and biodiversity as well as the land-use history are connected to these elements. Many seminatural grasslands have been fertilized and thus transformed into cultivated grassland with low conservation values.

Seen in the six-thousand-year-old perspective of cultivation history in Sweden, the accelerating changes of the last fifty years are even more pronounced.

Major Threats against the Swedish *Hagmark* Landscape

A great part of the *hagmark* landscape is facing an immediate threat of disappearance. There is no comprehensive planning for the management of this landscape, which, in the long run, may have negative consequences on the regional identity of the landscape and its biodiversity. Three major threats are affecting the biological and cultural values of the *hagmark* landscape.

The first threat comes from the land-use changes in agricultural areas. The marginalization processes cause a decrease in mosaic patterns necessary for high biodiversity. Changed proportions of land use between agriculture and forestry lead to a homogenization and "darkening" of the landscape increasing the contrast between open land and forest. The previously strong correlation between land use and soil and topographic parameters is weakened or lost (Figure 10.8a). The landscape character and scenery are altered toward lower nature conservation and heritage values. This trend is similar for the entire countryside of Sweden (Ihse and Blom 2000).

The second threat comes from the management methods in forestry. Monoculture results in the loss of old trees and dead wood as well as of deciduous trees and spontaneous successions on former grasslands. The modern forestry practices create clear cuttings, succeeded by even blocks of single-age monoculture plantations almost entirely lacking a field layer during half or more of the forest's lifetime.

The third threat comes from urban sprawl and the expanding communication infrastructure. Motorways lead to fragmentation and isolation of seminatural vegetation types such as grasslands and wetlands, which are important for biodiversity. The transformation from former small farms to summerhouses often leaves the landscape without the necessary management.

Conclusions

To understand and describe scientifically a cultural landscape, one must adopt a holistic perspective. The physical conditions provide a framework and essential basis for natural resource utilization. A landscape characterization should include historical depth, cultural heritage, and natural values. Historical depth means the development of the landscape from the village era that can be studied in maps from the early seventeenth century onward.

We argue that landscape analysis, integrating natural and cultural properties, described by human geography, physical geography, and ecology, is crucial for characterization of the landscape and for understanding the potential for biodiversity.

Integrated landscape analysis is useful in planning for future sustainable landscape management. We believe that the *hagmark* landscape needs to be taken into special consideration in planning and management if its characteristics and natural and cultural values are to be maintained.

The *hagmark* landscape is an ecological functional unit unique to the Nordic countries and is mainly found in Sweden. This means that in order to delineate the functional *hagmark* landscape, historical depth also has to be assessed. To conclude, grasslands are of major importance because they represent a living part of the landscape memory and cultural heritage.

This presentation of the Swedish *hagmark* landscape should be viewed together with the descriptions of the Norwegian landscapes presented in this book, (1) the fjord landscapes in Sognefjord (Austad and Hauge, this volume, chapter 14) and (2) the mountain agricultural pasture landscape in Hjartdal in Telemark (Norderhaug, this volume, chapter 15). All these landscapes have, despite their different scenery and physical conditions, a common context in natural and cultural values, in ecology and biodiversity. In all these areas, grassland is the most important characteristic feature, together with old deciduous trees and woods. Farming and animal husbandry, the most important historical land use in the Nordic countries, shaped them all for several hundred to thousands of years. Specific valuable features in Sognefjord are the meadows with pollarded trees; in Hjartdal, the mowed hay meadows; and in the Swedish *hagmark* landscape, the grazed pastures.

References

Arnberg, W., and M. Ihse. 1996. "Environmental Planning with a GIS Decision Support System Based on Vegetation Information in Colour Infrared Air Photos CIR." In *Remote Sensing and Computer Technology for Natural Resource Assessment. Proceedings from IUFRO World Conference, Tampere, Finland—University of Joensuu, Faculty of Forestry*, ed. J. Saramäki et al., 149–70. Research notes 48.

Aronsson, M. 1979a. "Det relikta odlingslandskapet i mellersta Kalmar län." *Svensk Botanisk Tidskrift* 73: 97–114.

———. 1979b. "Slåtter och beteslandskapet i det äldre odlingslandskapet." *Bygd och Natur, årsbok 1979*: 72–96.

———. 1980. "Markanvändning och kulturlandskapsutveckling i södra skogsbygden." In *Människan, kulturlandskapet och framtiden*, ed. A. Renting, 221–31. KVHAA konferens 5. Stockholm: Kungl. vitterhets historie och antikvitets akademien.

———, ed. 1999. *Rödlistade kärlväxter i Sverige—artfakta* (Swedish Red Data Book of Vascular Plants), 2 vols. Uppsala: ArtDatabanken, SLU.

Barr, C. J., R. H. G. Bunce, R. T. Clark, R. M. Fuller, M. K. Gillespie, G. B. Groom, C. J. Hallam, M. Hornung, D. C. Howard, and M. J. Ness. 1993. *Countryside Survey 1990: Main Report*. London: Department of the Environment.

Bengtsson-Lindsjö, S., M. Ihse, and G. Olsson. 1991. "Changes in Landscape Pattern and Grassland Plant Species Diversity in the Ystad Area during the Twentieth Century." In *The Cultural*

Landscape during 6000 Years in Southern Sweden: The Ystad Project, ed. B. E. Berglund, 388–96. Ecological Bulletins 41. Copenhagen: Munksgaard.

Berglund, B., ed. 1991. *The Cultural Landscape During 6000 Years in Southern Sweden: The Ystad Project.* Ecological Bulletins 41. Copenhagen: Munksgaard.

Bernes, C., ed. 1993. *Nordens Miljö: Tillstånd, utveckling och hot.* Monitor 13. Stockholm: Swedish Environmental Protection Agency.

———, ed. 1994. *Biological Diversity in Sweden. A Country Study.* Monitor 14. Stockholm: Swedish Environmental Protection Agency.

Bunce, R. G. H., C. J. Barr, R. T. Clarke, D. C. Howard, and A. M. J. Lane. 1996. "Land Classification for Strategic Ecological Survey." *Journal of Environmental Management* 47: 37–60.

Ekstam, U., and N. Forshed. 1992. *Om hävden upphör.* Värnamo: Naturvårdsverket.

Emanuelsson, U., and C. Bergendorff. 1986. "History as a Guideline to Nature Conservation and Urban Park Management in Scania, Southern Sweden. In *Ecology and Design in Landscape. Proceedings from the 24th Symposium of the British Ecological Society, Manchester 1983,* ed. A. D. Bradshaw, D. A. Goode, and E. H. P. Thorp, 237–55. Oxford: Blackwell Scientific Publications.

———, and C.-E. Johansson, eds. 1989. *Biotopvern i Norden: Rekommendationer för kulturlandskapet.* Nordiska Ministerrådet Miljörapport 1989: 5.

Environmental Protection Agency. 1975. *Översiktlig naturinventering och markvårdsplanering. Råd och anvisningar.* Stockholm.

———. 1987. *Inventering av ängs- och hagmarker. Handbok.* Stockholm.

Farina, A. 1997. "Landscape Structure and Breeding Bird Distribution in a Sub-Mediterranean Agro-Ecosystem." *Landscape Ecology* 12: 365–78.

Forman, R. T. T. 1995. *Land Mosaics: The Ecology of Landscapes.* Cambridge: Cambridge University Press.

———, and M. Godron, M. 1986. *Landscape Ecology.* New York: John Wiley & Sons.

Fries, C. 1965. *Gammalsverige.* Stockholm: Wahlström & Widstrand.

Grundberg, O. 1966. *Områdesindelningar för jordbruksstatistiskt bruk.* Kulturgeografiska institutionen, Uppsala universitet, Forskningsrapport nr 4. Uppsala.

Hägerstrand, T. 1991. *Tidens vidd och tingens ordning: Texter av Torsten Hägerstrand.,* ed. G. Carlestam and B. Sollbe. Stockholm: Byggforskningsrådet.

Haeggström, C.-A. 1987a. "Löväng." In *Biotopvern i Norden: Biotoper i det nordiska kulturlandskapet,* ed. U. Emanuelsson and C.-E. Johansson, 69–88. Nordiska Ministerrådet Miljörapport 1987: 6.

———. 1987b. "Hage." In *Biotopvern i Norden. Biotoper i det nordiska kulturlandskapet,* ed. U. Emanuelsson and C.-E. Johansson, 89–100. Nordiska Ministerrådet Miljörapport 1987: 6.

Helmfrid, S., U. Sporrong, C. Tollin, and M. Widgren. 1994. "Sweden's Cultural Landscape: A Regional Description." In *Landscape and Settlement,* ed. S. Helmfrid, 60–77. National Atlas of Sweden. Stockholm: SNA förlag.

Ihse, M. 1978. *Flygbildstolkning av vegetation i syd- och mellansvensk terräng.* SNV rapport PM 1083. Solna: Statens naturvårdsverk.

———. 1987. "Air Photo Interpretation and Computer Cartography: Tools for Studying the Changes in the Cultural Landscape." In *The Cultural Landscape: Past, Present, and Future,* ed. J. Birks, 153–63. Cambridge: Cambridge University Press.

———. 1994a. "Landskapets förändring under 1900-talet." In *Miljöforskare berättar,* ed.

F. Eberhardt, 105–27. Stockholm: Centrum för Naturresurs- och Miljöforskning, Stockholm University.

———. 1994b. "Fragmentation of Grasslands during the Last Two Hundred Years in Southern Sweden." In *Fragmentation in Agricultural Landscapes. Proceedings of the 3rd Annual IALE (UK) Conference in Myerscough College, Preston, England, September 1994*, ed. J. W. Dover, 194–96.

———. 1995. "Swedish Agricultural Landscapes: Patterns and Change during the Last Fifty Years, Studied by Aerial Photos." *Landscape and Urban Planning* 41: 11–36.

———. 1997. "Kan ängen vara en hed? Några reflektioner kring begreppet äng från kulturhistorikerns och botanistens synpunkter." *Svensk Botanisk Tidskrift* 91: 211–21.

———, and G. Blom. 2000. "A Swedish Countryside Survey for Monitoring of Landscape Features, Biodiversity, and Cultural Heritage: The LiM-Project. In *Consequences of Land-Use Changes*, ed. Ü. Mander and R. G. H. Jongman, 39–74. Advances in Ecology. Southampton and Boston: Wit Press.

———, and C. Lindahl. 2000. "A Holistic Model for Landscape Ecology in Practice: The Swedish Survey of Ancient Meadows and Pastures." *Landscape and Urban Planning* 50: 59–84.

———, and M.-L. Nordberg. 1984. "Landsbygdens förvandling—studerad med flygbilder och datorteknik." *Ymer* 104: 53–71. Stockholm.

———, and A. Norderhaug. 1995. "Biological Values of the Nordic Cultural Landscape: Different Perspectives." *International Journal of Heritage* 1 (3): 156–70.

———, B. Justusson, and H. Skånes. 1991. "Slättbygden i Skåne och Halland—landskap i förändring." *Naturvårdsverkets rapport* 3887. Solna: Statens naturvårdsverk.

Kristoffersson, A. 1924. *Landskapsbildens förändringar i norra och östra delen av Färs härad under de senaste tvåhundra åren. En kulturgeografisk studie.* Akademisk avhandling. Lund: Lunds Universitet.

Lindahl, C., ed. 1997. "Ängs och hagmarker i Sverige." *Naturvårdsverket rapport* 4819. Solna: Statens naturvårdsverk.

Naveh, Z. 1982. "Landscape Ecology as an Emerging Branch of Human Ecosystem Science." In *Advances in Ecological Research* 12: 189–237. London: Academic Press.

———, and A. S. Lieberman. 1994. *Landscape Ecology: Theory and Applications*, 2nd ed. New York: Springer Verlag.

Norderhaug, A. 1996. *Hay Meadows: Biodiversity and Conservation.* Doctoral thesis. Report series from the Department of Botany, Gothenburg University, paper VI: 32. Gothenburg.

Nordic Council of Ministers. 1977. *Naturgeografisk regionindelning av Norden.* NU-rapport 1977: 34. Stockholm.

———. 1994: "Vegetationstyper i Norden." *Tema Nord* 1994: 665.

Sjörs, H. 1956. *Nordisk växtgeografi.* Stockholm.

———. 1963. "Amphi-Atlantic Zonation, Nemoral to Arctic." In *North Atlantic Biota and Their History*, ed. A. Löve and D. Löve. New York: Pergamon Press, 109–25.

Skånes, H. 1991. *Förändringar i odlingslandskapet och dess konsekvenser för gräsmarksfloran. En studie från södra Halland.* Forskningsrapport nr 86, Naturgeografiska institutionen, Stockholms Universitet. Stockholm.

———. 1996. *Landscape Change and Grassland Dynamics: Retrospective Studies Based on Aerial Photographs and Old Cadastral Maps during Two Hundred Years in South Sweden.* Department of Physical Geography, Stockholm University, Dissertation series no. 8. Stockholm.

———. 1997. "Toward an Integrated Ecological-Geographical Landscape Perspective: A Review of Principal Concepts and Methods." *Norsk geografisk Tidsskrift* 51: 146–71.

Skånes, H., and C. Tollin. 1991. "Integrated Landscape Analysis." In *Proceedings of the European IALE-Seminars on Practical Landscape Ecology*, ed. J. Brandt, 2: 21–31. Roskilde: Roskilde Universitetsforlag.

———. 1992. *Landskapet på Krapperups gods—en samordnad analys. Krapperups odlingslandskap, historia, nutid, framtid*. Föredrag hållna vid seminarium på Krapperups borg 20–21 augusti 1991. Gyllenstiernska Krapperups stiftelsen.

Smuts, J. C. 1926. *Holism and Evolution*. New York: Viking Press.

Sporrong, U. 1990. "Land Survey Maps as Historical Sources." In *Maps and Mapping*, ed. U. Sporrong and H. F. Wennerström, 136–45. *National Atlas of Sweden*. Stockholm: SNA förlag.

———. 1993a. "Agrarian Landscapes in Sweden That Are of Particular Scientific Interest." *Bebyggelsehistorisk tidskrift* 26: 71–90.

———. 1993b. "Landskap." In *Encyclopedia Suesica*, vol. 12, 95–96. Bokförlaget Bra Böcker.

———. 1993c. "By." In *Encyclopedia Suesica*, vol. 3, 467–70. Bokförlaget Bra Böcker.

———, U. Ekstam, and K. Samuelsson. 1995. *Swedish Landscapes*. Stockholm: Naturvårdsverket.

Statistics Sweden. 1985. *Atlas över rikets indelningar i län, kommuner och församlingar*. Stockholm: Lantmäteriet kartförlaget.

———. 1990a. "Betesmarker—historiska data." *Statistiska meddelanden* Na 36 SM 9001. Stockholm.

———. 1990b. "Betesmarkernas omfattning och användning 1989—specialstudie." *Statistiska meddelanden* J13SM 9003. Stockholm.

———. 1998. "Områdesindelningar i lantbruksstatistiken 1998." *MIS* 1998: 1.

Tollin, C. 1991. *Ättebackar och ödegärden. De äldre lantmäterikartorna i kulturmiljövården*. Uppsala: Riksantikvarieämbetet.

UNEP. 1992. *Convention on Biological Diversity. June 1992*. United Nations Conference on Environment and Development, Rio de Janeiro, Brazil, June 1992. New York: Department of Public Information, United Nations.

World Commission on Environment and Development. 1987. *Our Common Future*. Oxford: Oxford University Press.

Zonneveld, I. S., and R. T. T. Forman. 1990. *Changing Landscapes: An Ecological Perspective*. New York: Springer Verlag.

Norway

11.
The "Two Landscapes" of North Norway and the "Cultural Landscape" of the South

MICHAEL JONES

"Two landscapes" have been considered to characterize North Norway. One is the landscape of coastal fishermen. The other is that of Saami (Lapp) reindeer herders, who move seasonally between grazings on the coast and grazings in the inland mountain and plateau regions. Some writers have presented these "two landscapes" with their distinctive resource uses as the essence of North Norway, giving the region its economic, social, and cultural distinctiveness, and thus implicitly constituting the basis of North Norway's identity.

The term "Two Landscapes of North Norway" was launched by the northern Norwegian philosopher Jakob Meløe at the International Wittgenstein Seminar arranged by the University of Tromsø at Skibotn in the autumn of 1986 (Meløe 1988, 1990). In the summer of the same year, another international symposium was held in West Norway to celebrate the one hundredth anniversary of the Botanical Museum in Bergen. The theme was "The Cultural Landscape: Past, Present, and Future," focusing on the human effects on vegetation, especially in agricultural landscapes (Birks et al. 1988). Although not new, the concept of "cultural landscape" has since 1986 been the subject of an academic debate in Norway concerning its definition and meaning (Jones 1998b). Some have associated "cultural landscape" with cultivated land, particularly in the context of the environmental management of agricultural landscapes. This emphasis favors a small part of Norway's area, with an obvious South Norwegian bias. Cultivated land accounts for less then 3 percent of Norway's land area, and more than 90 percent is located in the southern part of the country (Figure 11.1). This focus on what later was termed more precisely "the cultural landscape of agriculture" has arguably tended to draw attention away from environmental issues in other types of landscape: forests, mountains, coastal areas, and urban areas (Jones 1993a). Among these comparatively neglected landscapes are the "two landscapes" of fishing and reindeer herding in North Norway.

In this chapter I shall examine these two concepts, the "two landscapes of North Norway" and "cultural landscape," in relation to each other. My point of departure will be the Norwegian cultural landscape debate. Researchers concerned with the Saami minority people of northern and central Norway have criticized the use of the "cultural landscape" concept as overly focused on physical, mainly agricultural, landscapes, rather than on mental landscapes of cultural knowledge concerning the resource potential for highland activities such as reindeer herding, hunting, and fishing. Similarly, the sea bottom has been presented by other researchers as a mental landscape of knowledge related to fishing banks and navigation channels. Are the "two landscapes" of sea and highland adequate representations of North Norwegian distinctiveness and identity, in contrast to "the cultural landscape of agriculture," which might be seen as reflecting South Norwegian political domination?

Resources and Representations of Identity

Theoretically, I am interested in ways in which different resource uses are made into representations of a nation's or a region's identity through the concept of "landscape." In Norway, as elsewhere, the term "landscape" has evolved during the last twenty years from being something neutral and amenable to unbiased academic description to find itself at the center of the rhetoric of environmental management policies (Jones 1988, 1991; Jones and Daugstad 1997). When particular landscapes are seen as the physical product of particular forms of resource use and are made to represent the identity of a nation or region, then perceived environmental threats to these landscapes implicitly come to be presented as threats to national or regional identity. In this way, the values and interests of particular groups in society are presented as general values and interests of society as a whole. While regional representations may be used to counter or contest national representations, it should not be forgotten that a representation is no more than a partial truth; representations are susceptible, for example, to gender-blindness as well as to social and cultural discrimination (Duncan and Ley 1993, 5–7).

The Cultural Landscape Debate: Agriculture and Environment

In Norway, the concept "cultural landscape" has come particularly into focus among agricultural and environmental policymakers. As elsewhere in Western Europe, government subsidies for agricultural production met increasing criticism in Norway in the mid-1980s as being both economically and ecologically unsustainable. Agricultural surpluses, international demands for a reduction in production subsidies, and criticism

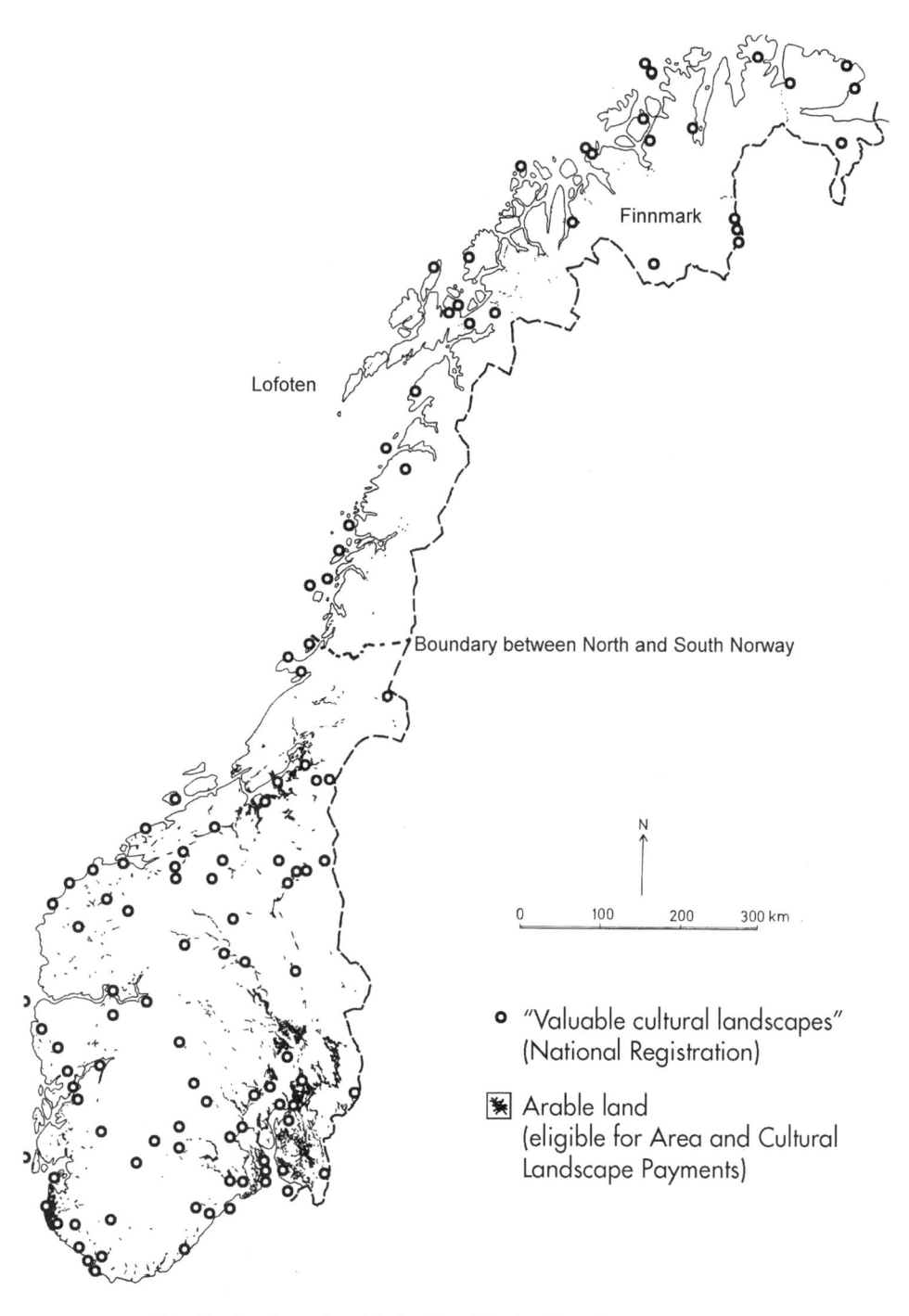

FIGURE 11.1. The distribution of arable land and "valuable cultural landscapes" in North and South Norway (*Nasjonalatlas for Norge* 1983; Jones 1993b, 59; *Verdifulle kulturlandskap i Norge,* 1994, State of the Environment Norway. http://www.mistin.dep.no/).

of the negative environmental impacts of intensive agriculture led agricultural policymakers to become increasingly interested in a more environmentally aware agriculture (Jones 1993b, 1998a). As surplus production undermined Norway's strict policy of protecting agricultural land against urban development, the emphasis shifted toward arguments for the protection of the cultural landscape, understood in this context as synonymous with the agricultural landscape (Jones 1992). Production payments were from 1989 partly replaced by hectarage payments, termed Area and Cultural Landscape Payments, aimed at keeping arable land in production without encouraging further intensification. All farmers are eligible for these payments. From 1991 the payments were made conditional upon farmers not engaging in specified practices considered detrimental to the landscape (Jones 1993b). By 1996, they accounted for more than 25 percent of the government's agricultural budget (Jones 1998a). A smaller amount was made available for active landscape management, based on management agreements. In 1993, the Ministry of Agriculture formulated a specific "cultural landscape policy" that had the objective of reconciling considerations of economy, ecology, and cultural history. The "cultural landscape of agriculture" was seen as an "environmental good" for which farmers were to receive payment, and it was defined as compromising—besides productive farmland—adjacent natural or seminatural biotopes, features of cultural-historical interest, and areas no longer in agricultural production (Jones 1993b, 1998a).

As agricultural policymakers showed increasing interest in the environment, environmental policymakers became more interested in agricultural landscapes. Between 1991 and 1994, the Ministry of the Environment initiated a "National Registration of Valuable Cultural Landscapes." This program initially identified 104 agricultural areas considered nationally important, primarily for their biological diversity and to a lesser extent for their cultural heritage. Of these, 27 were in North Norway (*Verdifulle kulturlandskap i Norge* 1994), a surprisingly high proportion in view of the small amount of agricultural land there. Later, another eight "valuable cultural landscapes" were identified in central Norway, bringing the total number for the whole country to 112 (Figure 11.1). Nevertheless, the 27 sites in North Norway still constitute 24 percent of the total. This may be explained by the historical predominance of seminatural grassland in North Norwegian agriculture. Much of this land is economically marginal, and large areas are on their way out of production. The botanical interest of such land is thus threatened by scrub growth. The geographer Bjørg Lien Hanssen (1998, 187, 269) has shown in her doctoral thesis how the Registration emphasized that such landscapes, termed "traditional cultural landscapes," were to be protected as a matter of national concern. Legitimized in terms of national identity, their value was thus taken for granted. She demonstrates how professionals made decisions on what were to be considered "valuable cultural landscapes" in a top-down process in which public participation was minimal. Local people in the selected localities were not involved in the process of definition and selection.

In 1991, the Norwegian Agricultural Research Council and the National Committee for Environmental Research initiated a five-year applied-research program on the "cultural landscape of agriculture." The program focused on the management of biodiversity in intensively cultivated land and the management of seminatural vegetation on economically marginal agricultural land; it also studied landscape perceptions and values and management policies (Framstad et al. 1998). Zoology and botany dominated the funded projects, but disciplines such as agricultural economics, landscape planning, and geography were also involved. Of the thirty projects funded, two-thirds were at institutions in the Oslo area (including the Agricultural University of Norway, which alone had twelve projects), while only one was from North Norway.

Agrarian interests have had a strong position in Norway, lasting throughout the nineteenth century and long into the twentieth. Agriculture was considered more important than fishing among the landowning civil servants that ruled the country until the securing of parliamentarianism in 1884. To carry on agriculture was regarded as a patriotic duty. Farmers were considered the most important tradition-bearers of Norwegian history. Other activities, such as fishing in North Norway, were looked down on as leading to the neglect of agriculture (Tveite 1959; Jones 1985a, 73–75). Farmers' interests continued to have a strong position in parliament long after 1886. The national emphasis given in recent years to the "cultural landscape of agriculture" thus has a solid historical anchoring, but one with a strong South Norwegian bias.

The Cultural Landscape Debate: The Saami Landscape

The agenda set by agricultural and environmental policymakers focused especially on agricultural landscapes, with a strong emphasis on biodiversity and cultural heritage in the physical sense. However, researchers on Saami culture in central and northern Norway criticized the conception of "cultural landscape" as a physical category (Grønningsæter 1997; Jones 1998b). Some argued that the Saami cultural landscape represents more than the physical traces of Saami cultural activity and includes natural features with cultural meaning such as old Saami offering stones and sacred mountains. They further emphasized that the cultural meaning of the landscape is related to the knowledge of the terrain that the Saami have acquired through generations of seasonal movements with their reindeer and when hunting and fishing. Immaterial cultural heritage such as place-names and local traditions regarding places are an important part of the Saami cultural landscape (Schancke [Schanche] 1987; Schanche 1990a, 1990b; Fjellheim, 1989, 1991, 1995; Nystø 1991, 1992; Gaski 2000). The archaeologist Audhild Schanche (1995) summarizes the landscape's importance for Saami identity as forming:

1. A historical landscape through Saami settlement and land use.
2. A magical landscape through memories of the pre-Christian Saami religion, which gave meaning to landscape elements in the form of cult places, offering stones, graves, and other sacred places.
3. A mythical landscape through narratives on the origins of landforms.
4. A political landscape through the strategic use of landscape symbols in the field of cultural politics. The establishment of institutions to manage Saami cultural heritage in the mid-1980s recognized that the Saami people have their own historical past.

The Saami contribution to the cultural landscape debate helped put on the agenda the notion that landscape is not a given, objective physical entity but rather is a selection of elements conditioned by the social and cultural perspective of the observer. The debate demonstrates that different groups in society have different criteria of what is important in a landscape when they describe or characterize it.

The Cultural Landscape Debate: The Sea as a Cultural Landscape

The discussion of the Saami landscape led the ethnologist Asbjørn Klepp (1991, 89) to pose the question: "Is the sea bottom a cultural landscape?" He argued that if the main emphasis is put on the landscape's cultural significance, then most, if not all, landscapes are cultural landscapes. This broad definition is not peculiar to Saami culture. Norwegian coastal culture involves a system of knowledge related to the natural features of the shore. Although invisible, the topography of the seabed, important for navigation and fishing, is remembered in relation to navigation markers and landmarks, presupposing a system of knowledge based on experience. Klepp argued that the focus here is on culture rather than landscape, and suggested that terms such as "landscape's cultural meaning" or "mental landscape" are more appropriate than "cultural landscape."

Inspired by the work of the anthropologist Tim Ingold (1992, 1993a, 1993b, 1998), Mayvi B. Johansen wrote a dissertation in social anthropology in 1999 on the maritime landscape around the island of Røst, in Lofoten, and examined the mutual relationship between coastal fishermen and the sea. She found that the maritime landscape is structured through the fishermen's active engagement with the sea in their orientation and navigation when searching for fishing places and when setting and taking up fishing lines and nets. In responding to the resource possibilities afforded by the environment, they give significance and meaning to particular features of the environment. Their knowledge of what is under the sea is based on experience, and is remembered and communicated by bearings taken from the features of the coastal topography. Names are given not only to coastal features but also to the crossing places of important bearings (Figure 11.2). Such naming is used to describe the position of fishing gear in communication between fishermen. This knowledge, based on experience, is extended by the use of modern electronic equipment such as echo sounders, map plotters, and satellite navigation systems. The maritime landscape is also formed by the meanings given to different sea areas among fishermen using different types of equipment. As new methods and types of equipment entered the Lofoten fisheries at the end of the nineteenth century, the need arose to organize relationships between fishermen in a new way. Conflicts were resolved by dividing the sea into zones reserved for the use of different types of equipment: nets, lines, and seines. Figure 11.3 shows the division of the sea around Røst for the winter of 1996. This division is again described by bearings. It is regulated by cooperative management institutions in which responsibility is divided between state fishing authorities and local fishermen's committees. Local regulation is based on the fishermen's acquired knowledge of both ecological and social factors.

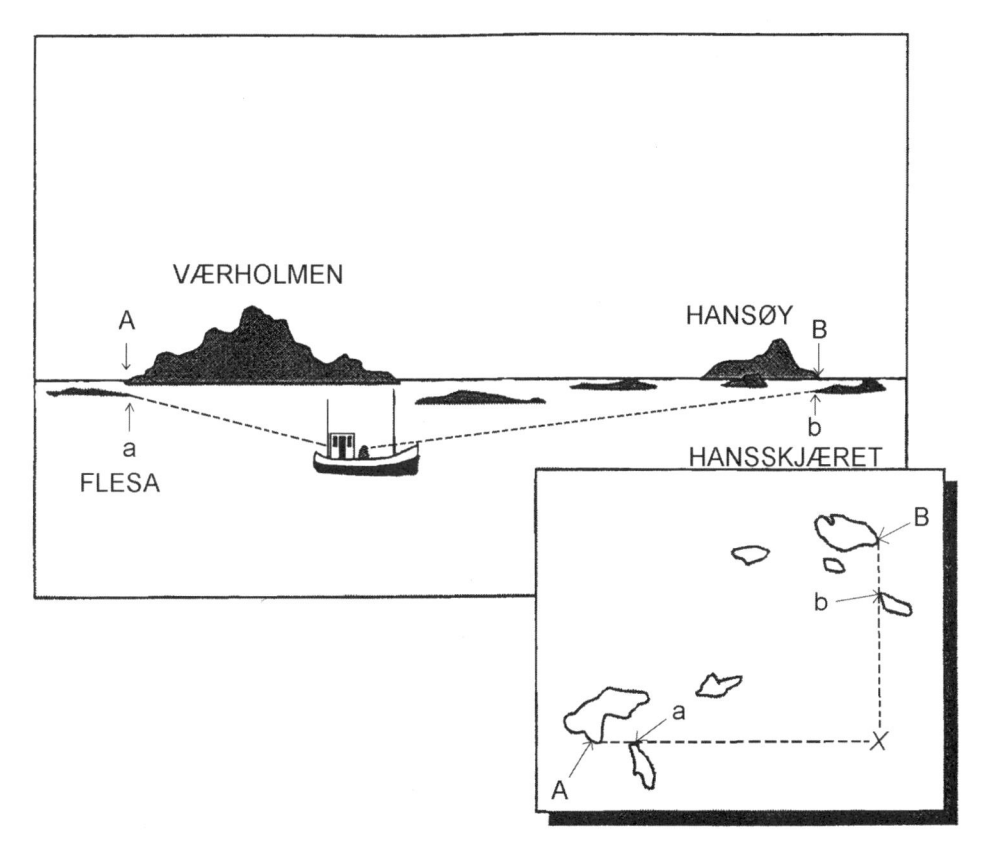

FIGURE 11.2. The crossing place of the bearings to "Flesa on the edge of Værholmen" and "Hansøya in Hansskjæret" (Johansen 1999, 31; reproduced with permission). Drawing by Mayvi B. Johansen.

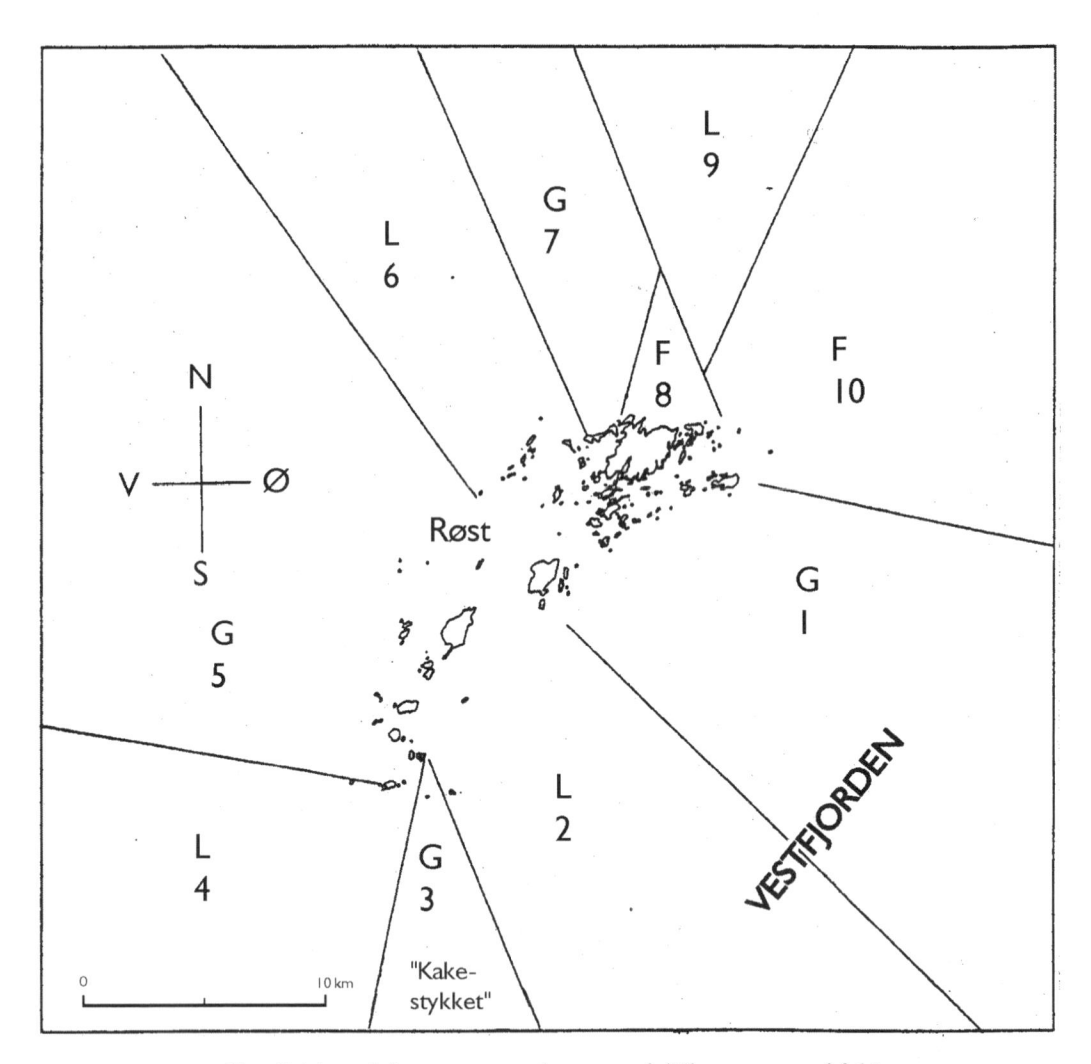

FIGURE 11.3. The division of the sea among the users of different types of fishing gear around the islands of Røst for the winter of 1996. G = nets (Norwegian *garn*), L = lines *(liner)*, F = common areas *(felles)* (Johansen 1999, 64; reproduced with permission). Map by Mayvi B. Johansen.

This example illustrates how the sea becomes meaningful for fishermen as a "mental landscape." Natural resources and topographical features, including those invisible to the naked eye on the seabed, are given shape by the activities and experience of the fishermen. The methods and terminology the fishermen used to refer to this landscape in communication among themselves also provide the basis for the system of cooperative management that regulates the relationships between different groups of fishermen in the use of resources.

NORTH NORWAY AND ITS "TWO LANDSCAPES"

North Norway (Figure 11.1), comprising the three northernmost counties, accounts for 35 percent of Norway's land area but only 10 percent of the population (Statistics Norway 2000). Extending from 65° to 71°N (excluding Svalbard), it lies largely north of the Arctic Circle. Because of the tempering effect of the Gulf Stream, farming, predominantly animal husbandry, can be carried on in the far north, but commercial agriculture is economically marginal in much of the region. Substantial areas of agricultural land have gone out of production; nonetheless, 9 percent of Norway's agricultural area lies in North Norway. Until the mid-twentieth century, the prevailing pattern consisted of fisher-farmer holdings, with the men predominantly fishermen and the women farmers (cf. Bratrein 1976; Jones 1985b). Norway's richest fishing grounds, principally off Lofoten but also in the Barents Sea, are in the north and attract fishermen during the winter season from other parts of the coast. The majority of Saami reindeer herders are similarly found in North Norway. However, as in the rest of Norway, the population of North Norway has become highly urbanized in the last fifty years. Most now live in small towns and urban agglomerations and work in secondary and tertiary occupations; less than 9 percent of the employed population work in primary occupations (Nordic Statistical Secretariat 1996).

In 1994, a two-volume cultural history of North Norway (Drivenes et al. 1994a; cf. Olsen, 1999, 476–77) was published, with chapters written mostly by North Norwegian academics. The introductory chapter discusses the foundations of North Norwegian identity. "Stubbornness" and "diversity" are the two words used to sum up North Norwegian culture. The region's inhabitants have shown a historical stubbornness in the struggle against a harsh nature of wild mountains, vast plateaus, naked shores, and enormous expanses of sea. Marine resources have been a major source of livelihood, but not an easy one. It is a landscape where the struggle for existence has been a struggle against storms, winter darkness, and despair. At the same time, the nature of North Norway has become a symbol of belonging and identity. North Norwegians have also shown stubbornness in their resistance to the perceived impositions of the south. Tension exists between North Norway as a periphery and the central authorities in the south of the country, which throughout history have attempted to exert their control over the resources of the north. While consciousness of being different from the south brings North Norwegians together, the region is characterized internally by diversity. There are considerable climatic and topographical variations from south to north, from coast to inland, and between sea and mountains. In addition to reindeer herding and fishing, a varied resource basis has supported both agriculture and industry. Three languages—Norwegian, Saami, and Finnish—are spoken (Drivenes et al. 1994b, 11–13).

Jakob Meløe's notion of the "two landscapes" of North Norway was not intended as a criticism of the South Norwegian domination of the meaning of "cultural landscape" in the agricultural and environmental management sectors. Nonetheless, it can be seen in the context of the strong feeling of cultural distinctiveness that many North Norwegians feel (and indeed that many in South Norway assign to North Norwegians). Meløe uses the term "landscape" to conceptualize the world of activities. In describing the two worlds of Norwegian fishermen and Saami reindeer herders, Meløe (1988, 388) states: "These are the two landscapes of Northern Norway. There are others, but not as illustrious." He mentions Saami fishermen and Norwegian women farmers but does not go into detail.

The first landscape is a perspective on the world viewed from the fishing boat. It is a "terrain" of waves and currents, shallow and deep waters, rocks and skerries, fjords and islands, fishing grounds, landing places, and harbors. Meløe uses the concept of the harbor to illustrate a landscape element that has no meaning without the boat. The second landscape is the terrain seen from the perspective of reindeer herding. This landscape is illustrated by the example of a patch of snow that survives the melting of the snow in spring and remains through the summer, termed a *jassa* in the Saami language. This is where a flock of reindeer in their summer grazing land can go to cool themselves and escape mosquitoes. It is this function that gives it meaning for reindeer herders. For a Saami fisherman, a patch of snow on a mountainside might have meaning as a landmark for taking bearings when fishing. For a Saami who is neither a reindeer herder nor a fisherman, it is simply a patch of snow. Important for both fishermen and reindeer herders is the need to use the landscape continually to take their bearings. The landscape is seen in terms of what one does in it and is learned through lore and experience.

Meløe's concept of "two landscapes" was engaged by the sociologist Svein Jentoft (1998a, 1; 1998b, 27) in two books in which he presented the results of a research program on the sustainable management of biological common resources, financed by the Research Council of Norway in the 1990s. The focus of the program was on environmental problems arising from the overexploitation of reindeer pasture and marine resources in North Norway. Jentoft (1998b, 27, my translation) states: "Fishery and reindeer herding are not the only livelihoods of great importance for this region, but they are two basic livelihoods that have always given North Norway its economic, social, and cultural character." He points out that while reindeer herding is a specifically Saami livelihood, fishing is carried on by both Norwegians and Coast Saami. He describes reindeer herding and fishing as "the material basis" of Saami culture and identity. Fisheries and reindeer pastoralism are similar in that both are dependent on common resources, in the meaning of open-access resources that are not privately owned. Both have suffered overexploitation, partly because of the use of increasingly efficient modern technology. The Barents Sea fisheries have suffered periodic resource crises because of overfishing, while the overgrazing of reindeer lichen in winter has led to conflicts in part of the Finnmark plateau. Government subsidies have tended to make the situation worse by encouraging expansion. Historically, for both fishermen and reindeer herders, local rules and institutions have encouraged cooperation and have reduced conflicts and regulated the use of the common resources. Both groups often face problems adjusting to increasing state intervention through national and international systems of regulation, based on scientific registration and terminology (Jentoft 1998b, 26–33, 81–132).

Jentoft (1998a, b) discusses the significance of local knowledge systems for resource management. He quotes Einar Eythórsson and Stein Mathisen (1998, 208) on the Coast Saami: "As in similar hunting- and fishing-based cultures, the economic adaptation of the coastal Saami is dependent upon an extensive knowledge of the natural environment which consists of identification of natural resources and how these resources can be utilized, knowledge of ecosystem functions, relations between species and carrying capacities of different resources. Ecological folk-knowledge is usually situated in a local harvesting area and concerns annual cycles of harvesting seasons. Social norms about access and management forms are integrated in the local knowledge system." This knowledge is acquired through informal socialization and participation in practical work. Fishermen form "mental maps" of the sea, fishing places, and the topography of the sea bottom. Knowledge of resources depends in part on the type of fishing gear used (Eikeland 1998). According to Jentoft (1998a, 9), what fishermen "need to know is determined by what kind of fishery they are participating in, where they usually fish, the particular gear they are using, and what sorts of fish finding devices they are using." Further, "local knowledge is more than ecological knowledge, it is also social knowledge highly relevant for management" (10; cf. Eythórsson 1998). Local understanding of customary rights to fishing places is a significant part of this knowledge (Meløe 1995, 12–13). Similarly, reindeer herders have numerous definitions of snow and names for eight seasons,[1] and describe the landscape in relation to their own activities (Meløe 1988). The resources of the sea and the highlands are structured as a "cultural landscape" in which both ecological and social limitations exist. Local knowledge is often "silent," expressed by what people do rather than what they say. Such knowledge forms a part of cultural identity. It is not, however, unchanging but adapts to new technology, new scientific knowledge, and new institutions (Jentoft 1998b, 116–26).

Meløe's (1998; 1990) idea of "the two landscapes of North Norway" as expressing relationships between people's activities and their concepts of the world has been taken up by the archaeologist Reidar Bertelsen (1999). Bertelsen notes that Meløe's example of "observing the world from a boat" focuses on the fisherman's landscape with little reference to activities on land and the interdependence of land and sea. He stresses that the fisherman's landscape changes as technology changes, and notes particularly the importance of the transition from sailing to motorized boats. Sailing required different skills and knowledge, and different types of landing places, than diesel-engine boats. Bertelsen's interest is the study of boathouses and landing places as a key to understanding the North Norwegian economy in the Iron Age and Middle Ages. Noting that men fished and women farmed, he emphasizes that proximity to the sea has been the main factor in the location of farms in North Norway. Bertelsen views the growth of commercial fishing in North Norway in the Middle Ages not primarily as a strategy for coping with the marginality of farming in northerly latitudes, but as a logical development from the maritime lifestyle of fishing and the hunting of marine animals. Nonetheless, he makes the point that landing places and other structures on the borderline between sea and land are neither exclusively related to life at sea nor are merely secondary to agrarian sites on land.

Conclusion

I have argued that the concept of "the cultural landscape of agriculture" is a planners' and environmental managers' perspective on landscape, and that it is primarily a South Norwegian perspective. The idea of "two landscapes of North Norway" was not formulated expressly as a criticism of the "cultural landscape" concept, but emphasizes other characteristics of North Norway than the agricultural landscape, namely, fishing and reindeer herding. As an expression of North Norwegian identity, it can indirectly be seen as an expression of North Norwegian opposition to South Norwegian political and cultural dominance.

However, I would further argue that, as a representation of North Norwegian distinctiveness, it is only a partial representation. Examined more closely, it appears to be a representation that is gender-blind, emphasizing male-dominated activities while underplaying female activities in the landscape. Moreover, it neglects a significant part of the cultural-historical diversity of North Norwegian landscapes. I would argue that North Norwegian distinctiveness and identity are bound up not just with two but with many landscapes. I will summarize some of them:

- The "cultural landscape of agriculture" is defined primarily in terms of government payments for keeping agricultural land in use, as well as its importance for biodiversity and to a lesser extent cultural heritage, and its recreational potential for the urban population. This is a landscape defined and structured by administrators and scientists who are mostly based in South Norway. North Norway has a disproportionately large number of designated "valuable cultural landscapes," which have been designated for their botanical values by experts rather than by local people.
- This contrasts with the idea of the "two landscapes of North Norway," which throws into relief the role of local knowledge systems in structuring the landscape. The "two landscapes" of reindeer herding and fishing are presented as representing the essence of North Norwegian distinctiveness and identity. They are

distinguished by common property regimes, seasonal movements between different resource niches, and specific cultural knowledge concerning the mountain terrain and the sea bottom.
- This focus, however, neglects the historical significance of North Norway's "third" landscape, that of subsistence agriculture. Historically, the fisher-farmer holding was the mainstay of North Norwegian coastal communities and involved a complementary division of labor between men and women. Farming was the main responsibility of the women, assisted by their children; the men were away fishing for a significant part of the year but contributed to the heavy work on the farm during the summer (cf. Bratrein 1992, 102–15). The farm can thus be seen in large part as a woman's landscape, in contrast to the maritime landscape of fishermen, which in large degree is a man's landscape. Furthermore, fisher-farmer holdings were typical not only of Norwegian culture in the north, but they have also been equally characteristic of the Coast Saami and Finnish cultures of North Norway. Nor are these different groups functionally isolated from one another. Through trade and other interactions, there has been a degree of symbiosis between farmers and reindeer-herding Saami.
- Finally, there are the modern landscapes of North Norway. Agriculture has become marginalized in the modern commercial economy, except in a few places where it has tended to become a specialized, male occupation. Fishing and reindeer herding have similarly become increasingly specialized livelihoods, in which market orientation and profit maximization have led to periodic resource crises as a result of overexploitation. Fish-farming has been one response to this. Although women take part in both fishing and reindeer herding, these remain male-dominated livelihoods (Jentoft 1998b, 97–100; Kvendseth 1998). Other landscapes that remain a local reality in parts of North Norway include forestry landscapes and military landscapes, both of them generally more familiar to men than women. In the twentieth century, North Norway also had localities with significant resource-based landscapes of mining and heavy industry, although these are now in decline. Meanwhile, the great majority of Norwegians, in the North as in the South, today work in service occupations and live in modern, urbanized landscapes.

It is further obvious that the 65 percent of Norway's area that constitutes South Norway comprises far more than the "cultural landscape of agriculture." As in North Norway, the combined fisher-farmer holdings that once characterized the western and southern coasts have largely disappeared. Norwegian agriculture, while still small-scale by North American and European standards, has undergone a process of modernization and economic rationalization since World War II, resulting in declining numbers of farm units and increasingly specialized, mechanized production. Besides agriculture, South Norway has its fishing landscapes (now highly industrialized), mainly in the west, and fish-farming, mainly in the coastal regions of western and central Norway. The South Saami reindeer-herding landscapes extend 350 km into South Norway along the Swedish border. Extensive forestry landscapes are commercially important in central and southeastern Norway, while the main urban and industrial landscapes are in South Norway. The exploitation of oil from Norway's continental shelf has led to new offshore landscapes of oil production, as well as making a direct physical impact locally on the mainland and an economic impact nationwide. Large areas of the south are characterized by mountains, which are increasingly in demand as recreational landscapes. Recreation also sets its mark on the coasts of the south, not least in the far south and along the shores of the Oslo fjord.

Concepts such as "cultural landscape" and "the two landscapes of North Norway" are often used as standard concepts in academic and political discourse. Used rhetorically, they can have a powerful effect. However, such concepts are not neutral but can be understood only in the context within which they are used. Our interests and activities structure our perceptions of what is significant in the environment when we engage with it, describe it, and constitute it as landscape.

A more complete view of the complexity of the reality that lies behind our experience of landscape requires comprehensive study. Ways in which concepts of landscape are used require critical examination. Different representations of landscape should be compared with one another. There is a need for empirical studies of resource use and livelihoods, including ways in which people engage with and give form to landscape through their activities. Finally, the social relations, institutions, and customs regulating such activities in the landscape need to be investigated. Such studies would allow more inclusive representations of people's feelings of regional identity to be established.

Notes

This chapter is a revised version of a paper presented at the nineteenth Session of the Permanent European Conference for the Study of the Rural Landscape (PECSRL) in Aberystwyth, Wales (UK), in September 2000 (Jones 2003).

1. The eight Saami seasons in the North Saami language are: *dálvi* (winter, January–February), *giððadálvi* (spring–winter, March), *giðða* (spring, April–May), *giððageassi* (spring–summer, June–July), *geassi* (summer, July–August), *čakčageassi* (autumn–summer, August–September), *čakča* (autumn, September–October), and *čakčadálvi* or *skábma* (autumn–winter or twilight period, November–December).

References

Bertelsen, Reidar. 1999. "Settlement on the Divide between Land and Ocean: From Iron Age to Medieval Period along the Coast of Northern Norway." In *Settlement and Landscape.*

Proceedings of a Conference in Århus, Denmark, May 4–7, 1998, ed. Charlotte Fabech and Jytte Ringtved, 261–67. Højbjerg: Jutland Archaeological Society.

Birks, Hilary H., H. J. B. Birks, Peter Emil Kaland, and Dagfinn Moe, eds. 1988. *The Cultural Landscape: Past, Present, and Future*. Cambridge: Cambridge University Press.

Bratrein, Håvard Dahl. 1976. "Det tradisjonelle kjønnsrollemønster i Nord-Norge." In *Drivandes kvinnfolk. Om kvinner, lønn og arbeid*, 21–38. Oslo: Universitetsforlaget.

———. 1992. *Karlsøy og Helgøy bygdebok. Folkeliv—næringsliv—samfunnsliv*, vol. 3, *Fra år 1860 til 1925*. Hansnes: Karløy kommune.

Drivenes, Einar-Arne, Marit Anne Hauan, and Helge A. Wold, eds. 1994a. *Nordnorsk kulturhistorie*, vol. 1, *Det gjenstridige landet*; vol. 2, *Det mangfoldige folket*. Oslo: Gyldendal Norsk Forlag.

Drivenes, Einar-Arne, Marit Anne Hauan, Einar Niemi, and Helge A. Wold. 1994b. "Den besværlige identiteten." In *Nordnorsk kulturhistorie*, vol. 1, *Det gjenstridige landet*, ed. Einar-Arne Drivenes, Marit Anne Hauan, and Helge A. Wold, 8–17. Oslo: Gyldendal Norsk Forlag.

Duncan, James, and David Ley. 1993. "Introduction: Representing the Place of Culture." In *Place/Culture/Representation*, ed. James Duncan and David Ley, 1–21. London: Routledge.

Eikeland, Sveinung. 1998. "Flexibility in the Fishing Commons." In *Commons in a Cold Climate*, ed. S. Jentoft, 97–114.

Eythórsson, Einar. 1998. "Voices of the Weak: Relational Aspects of Local Ecological Knowledge in the Fisheries." In *Commons in a Cold Climate*, ed. S. Jentoft, 185–204.

Eythórsson, Einar, and Stein Mathisen. 1998. "Ethnicity and Epistemology: Changing Understandings of Coastal Saami Knowledge." In *Commons in a Cold Climate*, ed. S. Jentoft, 205–20.

Fjellheim, Sverre. 1989. "Det sørsamiske kulturlandskapet." *Spor* 4, no. 2: 26–28, 48.

———. 1991. *Kulturell kompetanse og områdetilhørighet. Metoder, prinsipper og prosesser i samisk kulturminnevernarbeid*. Snåsa: Saemien Sijte.

———. 1995. "Det samiske kulturlandskapet." In *Fragment av samisk historie. Foredrag Saemien Våhkoe, Røros 1994*, ed. Sverre Fjellheim, 58–81. Røros: Sør-Trøndelag og Hedmark Reinsamelag.

Framstad, Erik, Ingunn B. Lid, Asbjørn Moen, Rolf A. Ims, and Michael Jones, eds. 1998. *Jordbrukets kulturlandskap. Forvaltning av miljøverdier*. Oslo: Universitetsforlaget.

Gaski, Lina. 2000. "Landskap og identitet." *Fortidsvern* 26, no. 2: 18–20.

Grønningsæter, Toril. 1997. *Landskap, kulturminner og identitet. Om det samiske kulturminnevernets betydning for diskusjonen om kulturlandskapsbegrepet og for samisk identitet*. Hovedoppgave i geografi. Trondheim: Geografisk institutt, Norges teknisk-naturvitenskapelige universitet.

Hanssen, Bjørg Lien. 1998. *Values, Ideology, and Power Relations in Cultural Landscape Evaluations*. Ph.D. diss., Bergen: Department of Geography, University of Bergen.

Ingold, Tim. 1992. "Culture and the Perception of the Environment." In *Bush Base: Forest Farm. Culture, Environment, and Development*, ed. Elisabeth Croll and David Parkin, 39–56. London: Routledge.

———. 1993a. "The Temporality of the Landscape." *World Archaeology* 25: 151–74.

———. 1993b. "Tool-Use, Sociality, and Intelligence." In *Tools, Language, and Cognition in Human Evolution*, ed. Kathleen R. Gibson and Tim Ingold, 429–45. Cambridge: Cambridge University Press.

———. 1998. "Culture, Nature, Environment: Steps to an Ecology of Life." In *Mind, Brain, and the Environment: The Linacre Lectures, 1995–1996*, ed. Bryan Cartledge, 158–80. Oxford: Oxford University Press.

Jentoft, S., ed. 1998a. *Commons in a Cold Climate: Coastal Fisheries and Reindeer Pastoralism in North Norway: The Co-management Approach*. Man and the Biosphere Series, vol. 22. New York: Parthenon.

Jentoft, Svein. 1998b. *Allmenningens komedie. Medforvaltning i fiskeri og reindrift*. Oslo: Ad Notam Gyldendal.

Johansen, Mayvi B. 1999. *Det maritime landskapet under lofotfisket. En kulturøkologisk studie av relasjonene mellom kystfiskere ved Røst og deres omgivelser*. Hovedoppgave i sosialantropologi. Tromsø: Institutt for sosialantropologi, Universitetet i Tromsø.

Jones, Michael. 1985a. "Datakilder, datainnsamling og verdisyn." In *Metode på tvers. Samfunnsvitenskapelige forskningsstrategier som kombinerer metoder og analysenivåer*, ed. Britt Dale, Michael Jones, and Willy Martinussen, 57–84. Trondheim: Tapir.

———. 1985b. "Cultural Landscape Change on the Outer Coast of Central Norway." *Northern Studies* 22: 1–27.

———. 1988. "Progress in Norwegian Cultural Landscape Studies." *Norsk Geografisk Tidsskrift* 42: 153–69.

———. 1991. "The Elusive Reality of Landscape: Concepts and Approaches in Landscape Research." *Norsk Geografisk Tidsskrift* 45: 229–44.

———. 1992. "Byvekst og jordbruk i perspektiv." In *Levekår og planlegging. Festskrift til Asbjørn Aase*, ed. Michael Jones and Wolfgang Cramer, 83–105. Trondheim: Tapir.

———. 1993a. "Om tverrfaglig samarbeid—kommunikasjon, verdisyn og begrepsdrøfting." In *Jordbrukets kulturlandskap. Forskerkonferansen 1992, 26.–27. oktober—Sundvollen Hotell*, 84–86. Ås: Forskningsprogram om kulturlandskap, Norges Forskningsråd.

———. 1993b. "Economy versus Ecology: Challenges for Agriculture in Norway in the Light of Some West European Experiences." In *The Future of Rural Landscapes*, ed. Ulf Sporrong. *Bebyggelsehistorisk Tidsskrift* 26: 43–64.

———. 1998a. "På vei mot en mer miljøvennlig jordbrukspolitikk?" In *Jordbrukets kulturlandskap. Forvaltning av miljøverdier*, ed. Erik Framstad, Ingunn B. Lid, Asbjørn Moen, Rolf A. Ims, and Michael Jones, 192–99. Oslo: Universitetsforlaget.

———. 1998b. "Kulturlandskapsdebatten i Norge 1987–1996." In *Nordisk landskapsseminar Sogndal 1996. Foredrag*, ed. Ann Norderhaug, 47–64. Rapport no. 7/98. Sogndal: Avdeling for Naturfag, Høgskulen i Sogn og Fjordane.

———. 2003. "Resources and Representations: Sea and Highland as the 'Two Landscapes' of North Norway." In *European Landscapes from Mountain to Sea. Proceedings of the 19th Session of the Permanent European Conference for the Study of the Rural Landscape (PECSRL) at London and Aberystwyth (UK), 10–17 September 2000*, ed. Tim Unwin and Theo Spek, 146–55. Tallinn: Huma Publishers.

Jones, Michael, and Karoline Daugstad. 1997. "Usages of the 'Cultural Landscape' Concept in Norwegian and Nordic Landscape Administration." *Landscape Research* 22: 267–81.

Klepp, Asbjørn. 1991. "Hvorfor havbunnen ikke er et kulturlandskap. Kulturlandskapsbegrepet drøftet med eksempler fra kystkulturen." *Norveg* 34: 81–94.

Kvendseth, Chris Helen. 1998. *Reindrift—næring og livsstil. Kvinners identitet i Kautokeino/Guovdageaidnu*. Hovedfagsoppgave i geografi. Trondheim: Geografisk institutt, Norges teknisk-naturvitenskapelige universitet.

Meløe, Jakob. 1988. "The Two Landscapes of Northern Norway." *Inquiry* 31: 387–401.

———. 1990. "The Two Landscapes of Northern Norway." *Acta Borealia* 7: 68–80.

———. 1995. "Steder." *Hammarn* 3: 6–13.

Nasjonalatlas for Norge. 1983. "Kartblad 8.2.1. Skog og jordbruksområder. Forest and Agricultural Areas," by R. Aaheim. Oslo: Landbruksdepartementet/Norges geografiske oppmåling.

Nordic Statistical Secretariat, ed. 1996. *Yearbook of Nordic Statistics 1996. Nordisk statistisk årsbok 1996*. Nord 1994: 1. Copenhagen: Nordic Council of Ministers.

Nystø, Nils Jørgen. 1991. "Samisk kulturlandskap." In *Vern og virke. Årsberetning fra Riksantikvaren 1991*, 13–15. Oslo: Riksantikvaren.

———. 1992. *Samisk kulturlandskap*. Riksantikvarens rapporter, 21. Oslo: Riksantikvaren.

Olsen, Venke Åsheim. 1999. "North Norway: A Multi-Ethnic Landscape." In *Shaping the Land, Vol. 2: The Role of Landscape in the Constitution of National and Regional Identity. Proceedings of the Permanent European Conference for the Study of the Rural Landscape, 18th Session in Røros and Trondheim, Norway, September 7th–11th 1998*, ed. Gunhild Setten, Terje Semb, and Randi Torvik, 475–88. Papers from the Department of Geography, University of Trondheim, New series A 27. Trondheim.

Schanche, Audhild. 1990a. "Samfunnsteori og forvaltningsideologi i kulturminnevernet." In *Kulturminnevernets teori og metode. Seminarrapport fra Ulstein kloster 8.–11. mai 1989*, 79–97. Oslo: FOK, Norges allmennvitenskapelige forskningsråd.

———. 1990b. "Samisk kulturlandskap." In *Samisk kulturminnevernforskning. Rapport fra seminar i Guovdaggeaidnu/Kautokeino 22.–24. november 1989*, 31–38. Oslo: FOK, Norges allmennvitenskapelige forskningsråd.

———. 1995. "Det symbolske landskapet—landskapet og identitet i samisk kultur." *Ottar* 207: 38–47.

Schancke [Schanche], Audhild. 1987. "Det samiske landskap." *Fortidsvern* 3: 17–19.

Statistics Norway. 2000. *Statistisk årbok 2000, 119. årgang*, Norges offisielle statistikk—Official Statistics of Norway, C 600. Oslo and Kongsvinger: Statistisk sentralbyrå.

Tveite, Stein. 1959. *Jord og gjerning. Trekk av norsk landbruk i 50 år. Det Kongelige Selskap for Norges Vel, 1809–1959*. Kristiansand: Bøndenes Forlag.

Verdifulle kulturlandskap i Norge. Mer enn bare landskap! 1994. "Del 4—Sluttrapport fra det sentrale Utvalget. Verdifulle kulturlandskap i Norge. Vurdering og virkemiddel. Tilråding." Trondheim: Det sentrale utvalget for nasjonal registrering av verdifulle kulturlandskap/Direktoratet for naturforvaltning.

12.

The Landscape in the Sign, the Sign in the Landscape: Periphery and Plurality as Aspects of North Norwegian Regional Identity

VENKE ÅSHEIM OLSEN

> On the one hand we *are* modernized and integrated [in the nation]. We enjoy in general terms the same welfare goods as people in other places in the country. . . . On the other hand we are still anchored, at least mentally, in social structures, economic adaption, and cultural traits that in some way open up for another mental world, other world views, other value preferences. . . . The close connection to the Arctic, the rest of the North Calotte and Northern Russia, with other cultures and languages, have stamped their marks on identity and world view.
>
> —Håvard Dahl Bratrein and Einar Niemi,
> "Inn i riket: Politisk og økonomisk integrasjon gjennom tusen år"

The focus of this chapter is local and regional heraldic arms in North Norway.[1] "All local societies are striving to find their 'special traits'—amongst others through the choice of heraldic commune arms" (Drivenes et al. 1994, 11). Heraldic arms have figures that are signs of nature or culture and can be interpreted as conveyors of information about cultural contexts, ecological diversity, and physiographic peripheries. They symbolize significant qualities of a local or regional political-administrative area and the population within this area. Heraldic arms are visual signs for the identification of territory and communicative signs in the cultural-heritage discourse.[2]

That the sense of belonging to a given landscape or region is culturally constituted does not mean that the properties or qualities of a given landscape are shared only by those from this specific landscape. A landscape is a physical reality a person relates to emotionally from experiences there and associations to earlier experiences. Identification with something involves the articulation of signs, whether verbal or visual, or whether constructed or represented by nature elements. The signs express and represent what a person feels he or she is, or what he or she feels other persons are. The epigraph above expresses an insider view of the region of North Norway in the early 1990s. There is an ongoing debate on images of North Norway, related both to crisis and possibilities, to ecological sustainability and economic development. Pessimistic scenarios represent a general anxiety for the future development of the region, with depopulation, and for the effects on the environment, with increasing maritime pollution, constant overexploitation of fish resources, scarcity of grazing land for reindeer, and global market liberalism (cf. Jentoft 1998). Optimistic visions are the technological development of fish farming, oil production from rich resources in the Barents region, computer technology, telecommunications, increased transactions with the vast Russian market, and the potential of international tourism with nature as a tourist magnet.

Opposition to and mobilization against certain central political decisions and international market forces are found both within and outside the region, and so are counter-opposition and mobilization for advanced technological and industrial production for the global market (Arbo 1997, 1998; Brox 1998). Whose picture is true? From a geographical point of view, "all territories have their problems and the perception of them differs according to whether one is born to them or is viewing them clinically from outside" (Mead 1974, 45). In a cultural-communicative or semiotic perspective, there is competition over images of North Norway and the identity of the northerner. In this discourse on landscape, understood as social and physical environment, male researchers are central producers of theoretical analyses and scenarios on the national arena, and the image of the "Northlander" *(nordlending)* is a male figure, as most national, regional, and ethnic stereotypes traditionally are (Edvardsen 1997; Arbo 1997, 1998; Brox 1998; Fulsås 1999; cf. Olsen 1993a). Researchers on gender in North Norway have in recent decades made coastal women more visible in an international forum and described them as the "cornerstone sex," the "flexible sex," "farmers of the coast," and "ground crew of the fishing industry" (Balsvik and Gerrard 1999, 1). They have become aware of how tied small fishing communities in the North Atlantic were to wider global system both in terms of historical processes and contemporary events (Davis 1999, 151). In the academic discourse on strategies for a sustainable future in North Norway, researchers are also conscious of their subjectivity and political engagement and the impact of language: "The ruling pictures of reality have power over minds. They take part in forming our visions of future and our self-understanding. They give our actions and strivings direction" (Arbo 1997, 322). There is agreement that "*all* attempts to describe and analyze North Norwegian conditions" are also ways of taking part in current regional conflicts. The verbal representations given in the discourse are "weapons in the fight over resources, legitimacy, and political support" (Brox 1998, 93).

The multicultural dimension of North Norway is subjectively experienced and articulated through the cultural heritage of the region and the nation, and it includes the cultural landscape. Multiethnic diversity is not a major theme in cultural, historical, or geographical research in Norway, even though some researchers have focused on the Saami dimension of the landscape since the late 1980s (Schancke [Schancke] 1987; Olsen 1996, 1999; Jones 1999; Bratrein 2000). So much of the human landscape was damaged during World War II in the trilingual part of North Norway that research data in the field are almost nonexistent. Hence pictures and photographs have an important function as representations of the historical landscape and a special value as cultural heritage for the inhabitants themselves (Olsen 1984, 1989).

My source material consists of authoritative heraldic texts used as official reference works on commune (local authority) and county levels, such as the jubilee book, *Norske kommunevåpen* (Norwegian Commune Arms), issued on the 150th anniversary of the Local Government Act of 1837 (NKV 1987). Supplementary articles are found in the Scandinavian heraldry journal (Nissen 1990, 1991, 1993, 1998).

The Production of Regional Identity

A Diverse People in a Rebellious Land

The two-volume "Cultural History of North Norway" *(Nordnorsk kulturhistorie)* (Drivenes et al. 1994a) was politically initiated and financed by the three county councils of North Norway. Their idea was to strengthen the self-image of the region (Olsen 1999, 2000). Nearly fifty researchers and other authors contributed to the richly illustrated work. Identity as a relation between periphery and plurality is reflected in the titles of the two volumes: *The Rebellious Land* and *The Diverse People* (*Det gjenstridige landet* and *Det mangfoldige folket*). The first is a metaphor for a population that does not always conform to national patterns of thought or action; the second alludes to the multicultural society. "The diverse people" fits with a much-used literary metaphor of North Norway as "the meeting of three tribes," originally a book title from 1918, *Tre stammers møte* (Schøyen 1918). It is a metaphor for multiethnicity and connotes the interaction between people of Saami, Norwegian, and Finnish cultural backgrounds (cf. Olsen 1986) (Figures 12.1, 12.2, and 12.3).

After World War II, the Finnish language, as well as Saami, disappeared from dispersed rural communities, from the Tromsø area in the west to Varanger in the east. This was partly an effect of the strict Norwegianization policies before the war and partly an effect of the wartime evacuation of local communities, causing the breakup of social and lingual networks (in addition to macrostructural changes common to all countries affected directly by the war and its social, economic, and technological consequences). From the 1960s the regional Finnish language tradition has been inspired by a new wave of immigration from Finland, especially into the trilingual region, but also to other parts of Norway (Olsen 1992a, 1995; Saressalo 1996; Anttonen 1999). By Amendment 110A to the Constitution in 1988, the Saami population received special recognition by the Norwegian state. As an indigenous people, the Saamis are also protected by the Indigenous and Tribal Peoples Convention of 1989.[3] In 1999, Norway integrated several human rights conventions into national law, and the North Norwegian Finnish minority (*kvener*, or Quains) gained the legal status of national minority, as did the Forest Finns (*skogfinner*) in southeast Norway, the Jews (*jøder*), the Romani People (*romanifolket, tatere*), and the Roma or Gypsies (*roma, sigøynere*).

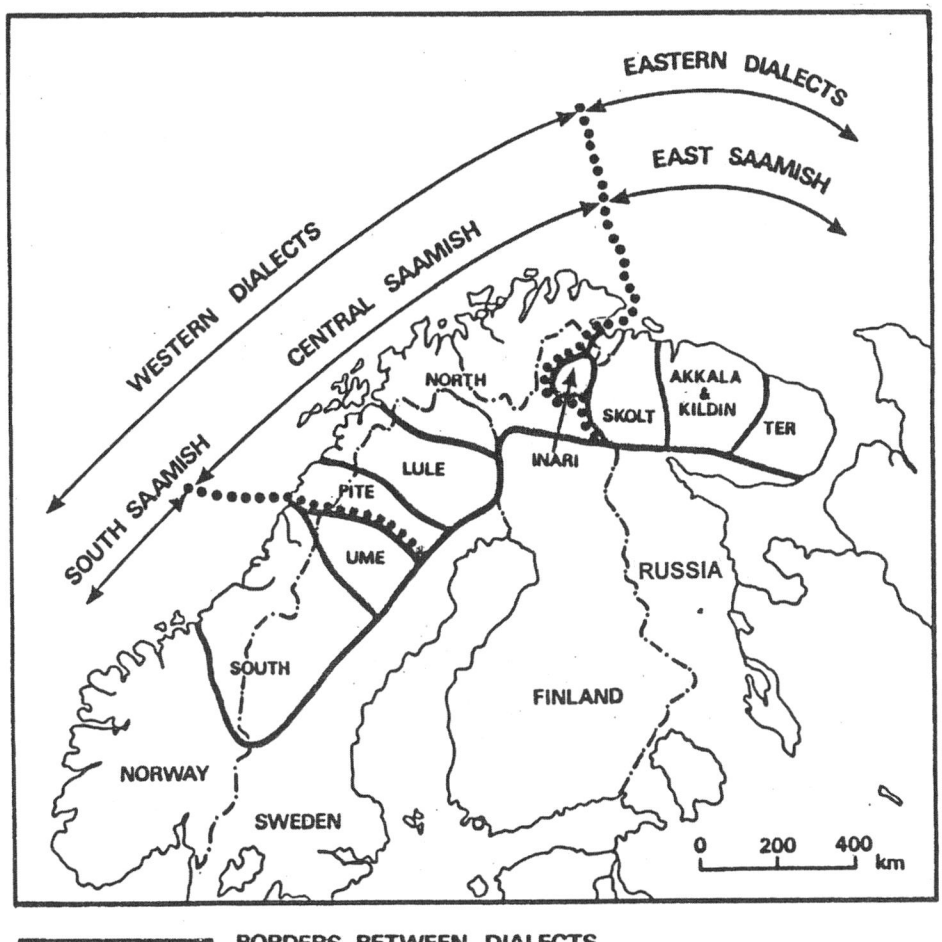

FIGURE 12.1. Saami dialect regions (based on NOU 1984, 97–99; Olsen 1986, 54; and Olsen 1999, 479).

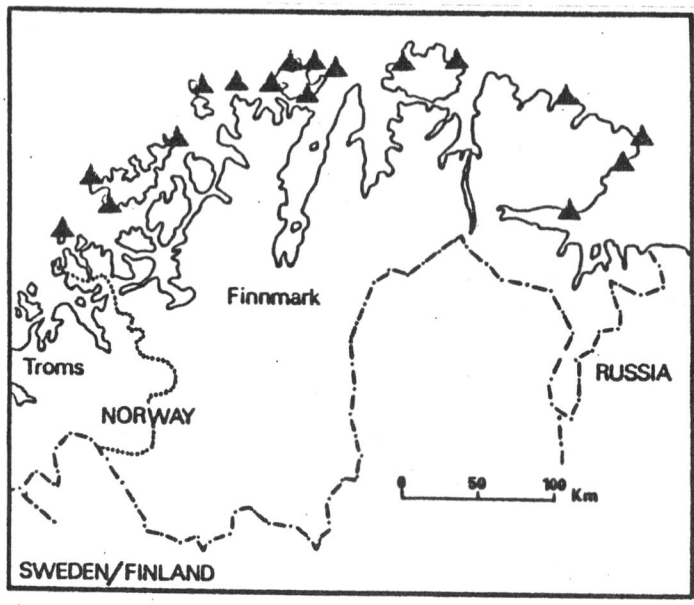

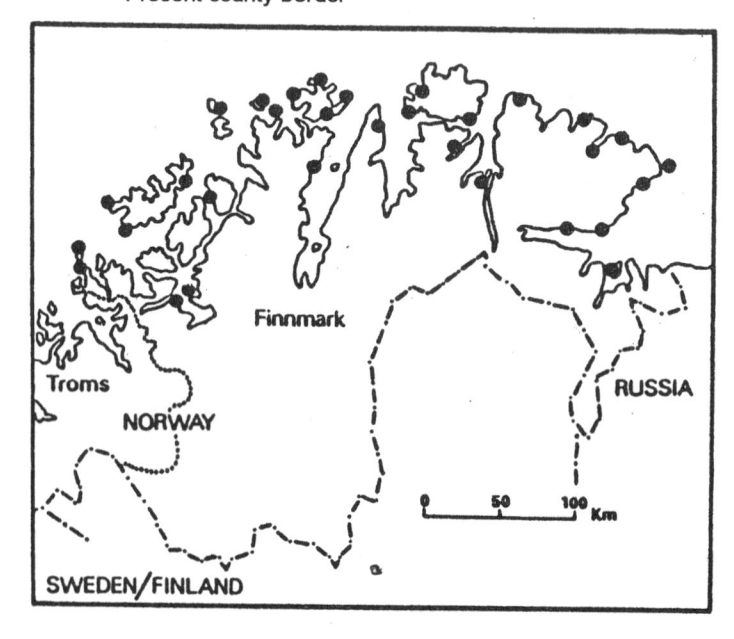

FIGURE 12.2. Sixteenth-century Norwegian settlement in Finnmark (based on Niemi et al. 1976, 32, 34; Olsen 1986, 37).

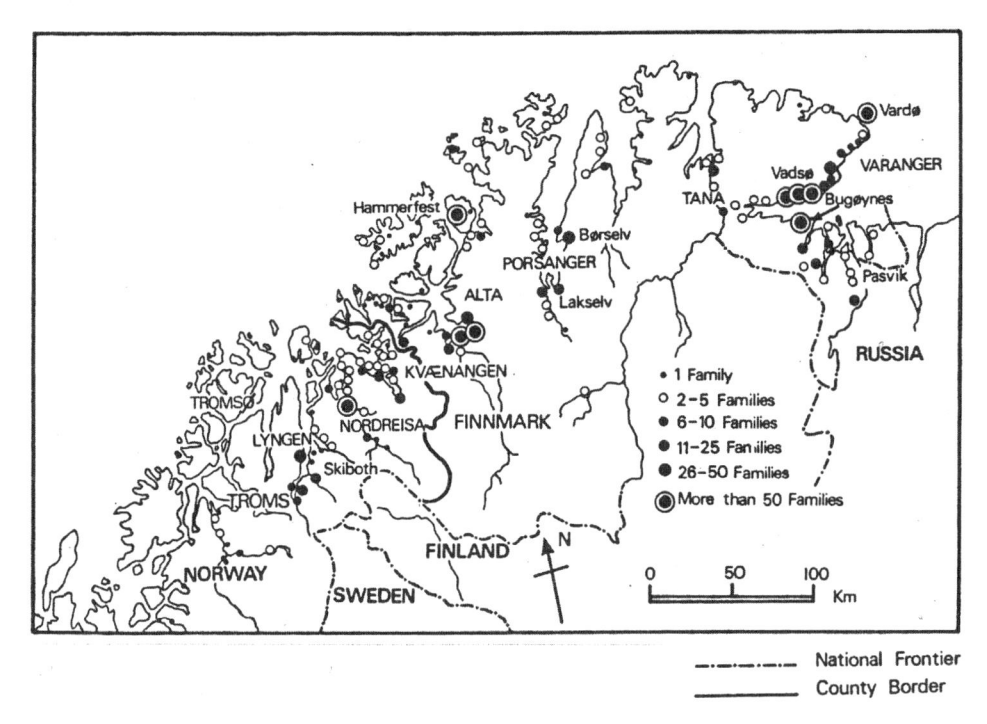

FIGURE 12.3. Finnish settlement in North Norway in the 1880s, interpreted from the ethnological maps by the linguist J. A. Friis (based on Jokipii 1982, 45; Olsen 1986, and Olsen 1999, 480.)

The "meeting of three tribes" as a metaphor of multiethnicity is still transmitted from text to text, context to context, and discourse to discourse (cf. Drivenes et al. 1994b, 11; Bratrein 2000). It points to ethnic diversity as a symbol of complementarity and symbiosis, not competition and conflict. "The land of suspense," *Spenningens land*, is another well-known book title, translated into many languages (Berggrav 1937), and a much-used positive metaphor about North Norway. The metaphors of "meeting" and "suspense" were used in an academic report in 1990 on future regional research policy for economic development in North Norway. According to the report, "nature and culture" gave "richness" to the region, based on "the diverse cultural variations, created by the meeting of the three tribes"; "suspense" was equated with possibilities, and contrasted to "crisis" (Bjørgo 1990, 9).

The editors of the "Cultural History of North Norway" announced: "The relation to the center, to the South, and to the one sitting with the upper hand, is an old North Norwegian trauma" (Drivenes et al. 1994b, 9–10). Besides, "those who had to pay the highest price for the integration into state and nation—with the loss of language, culture and self-esteem"—were "the Saamis and the Quains" (12). Outside observers doubted whether there was a *common* North Norwegian identity: "With *contradiction* and *diversity* as key words," it is very difficult to capture "the 'soul' of North Norway" (Storå 1996, 151). "For an external observer today, it seems difficult" to imagine that the northern population has a "'common understanding of the world' because the region is quite extended in length, and not least because of the importance that ethnicity is nowadays attributed in North Norway" (Winge 1995, 64). Another outsider made a similar comment on ethnicity: "The aim is to integrate the Saamis and Finns into the account. The regional 'unity' appears in large measure to be the reactions of the different ethnic groups to centrally directed efforts at integration," including "attempts to tone down the Saami and Finnish elements" (Storå 1996, 151–52). However, a social scientist from the region meant that this work would "strengthen the perception that the North Norwegian people has of itself as a distinct type of people.... The strength of the book is in diversity and richness of details, not in overview and theoretical grip" (Brox 1995, 284).

CULTURAL HISTORIANS AS REGIONAL STRATEGISTS

The "Cultural History of North Norway" is part of what one could call a "regional identity project" within the national identity process. The public education of the people is part of the modernization project from the mid-nineteenth century by "culture pedagogues" and "national strategists" (Berggreen 1989, 23; Slagstad 1998, 93), and has involved media, arts, museums, schools, and universities (Olsen 1997, 177). Economics, technology, and sociological thinking dominated physical planning in the social-democratic welfare state until the 1970s, although social scientists also opposed the uncritical use of science as an instrument for political governing (Slagstad 1998, 367). Varying perspectives on regional identity are also found among poets and fiction writers (Engelskjøn 1999, 165). In theoretical studies of landscape, environmental planning, cultural analyses of regional stereotypes, or the cultural-heritage discourse, the author is both observer and participant. Museum curators and cultural historians are among the "culture pedagogues" and "national strategists." They confirm and strengthen premises for legal arguments for the protection of cultural heritage, including landscape. They deal with issues that in the theoretical analysis of contemporary Scottish politics have been denoted as "a kind of pervasive, second-rate, sentimental slop associated with tartan, nostalgia, Bonnie Prince Charlie, Dr. Finlay, and so on" (Nairn 1981 [1977], 114). Semiotics bridge the paradigmatic gap between sociological and humanist approaches to knowledge about human society (Olsen 1993b; 1997, 182).

The Northern Region

PRESENTING NORTH NORWAY

North Norway consists of the counties of Nordland, Troms, and Finnmark, three of the nineteen counties of Norway (including Oslo). Mainland Norway is located roughly between 58°N and 71°N, while North Norway is located between 65°N and 71°N, forming the northernmost brim of the European continent, including North Cape. Oslo is just touching 60°N. North Norway is a political-administrative region that covers about one-third of the area of mainland Norway. Yet it contains only one-tenth of Norway's total population (Table 12.1), almost the same as in the capital, Oslo.

TABLE 12.1.
Area, population, and communes (local authorities) in Norway and the three counties of North Norway, January 1, 2000.

	Area	Population	Number of Communes
Norway (excluding Svalbard)	323,758 km²	4,478,497	435
Finnmark county	48,637 km²	74,059	19
Troms county	25,984 km²	151,160	25
Nordland county	38,327 km²	239,109	45

Source: SSB 2000.

Historically a series of specific legal regulations existed for North Norway, and in particular for Finnmark, concerning not only land tenure but also fishing, customs, and trade. The issue of indigenous land rights is important in relation to the Saami population, especially in Finnmark, but also elsewhere in Norway. The Saami parliament, *Sámediggi* or *Sametinget,* was opened in 1989 by King Olav of Norway. It is headed by a president, with thirty-nine representatives from thirteen ethnopolitical electoral districts. Norwegian citizens who identify themselves as descendants of Saami speakers and formally register as Saamis can vote. The Saami parliament is advisory on all issues that it conceives as specific for the Saami population, but only in particular instances can it make legal decisions (Opsahl 1993, 90; Hætta 1994, 185; NAF 2001, 449).

The Norwegian state officially owned (until 2006) 96 percent of Finnmark county, an area about the size of Denmark. The remaining 4 percent had largely passed into private ownership since 1775. North Norwegian chieftains, probably from the ninth century, established seasonal stations for seal hunting, fishing, and other wildlife resources (Bratrein 2001). The county borders of Finnmark to the south and east correspond to the national borders with Finland from 1751 and Russia from 1826. The present southern border of North Norway is almost identical with the southern border of Hålogaland in Viking times. Hålogaland was a separate legal district at the end of the twelfth century (Hagland and Sandnes 1994, xxi, 227). In Hålogaland, all fishermen who fished in Lofoten were obliged to pay tax in fish to the king of Norway, and only the king had the right to buy fur (216–17). Nordland and Finnmark counties were established with their own county governors in the 1670s after the introduction of absolute monarchy in Denmark-Norway. In 1866, Troms was separated from Finnmark as the third county (Fladby et al. 1974, 14–18).

INTERSTICES, PERIPHERIES, AND GEOPOLITICAL PLURALITY

The identity of North Norway can be said to have been shaped at the "interstice of North and East" (Olsen 1999, 482). While the north-south axis is predominantly a nature axis, which also has a "cultural content," the east-west axis is predominantly a culture axis, but also has a "natural content." However, the cultural content of the north-south and the east-west axes is not a mirror image of the natural conditions. Elements are picked out and fit into new patterns of thought, for example, the midnight sun, the continuous summer light, the polar ice, the snowfields. The sparsely settled and even empty space represents purity, innocence, divinity, absence of human sins, absence of industrial pollution, genuineness, and so on. On the other hand, the winter darkness, the freezing cold, the snow blizzards, and the moving ice represent danger, anxiety, and death. One version of the compass rose can function as a pictorial sign for the metaphor of the interstice of North and East. It has a larger fleur-de-lis on the compass line pointing north, on the outside of the compass ring, and a smaller crosslike sign on the line pointing east, also outside the ring. This type of the compass rose is found on many well-known sea and land maps of Norden from the end of the sixteenth century well into the eighteenth (Mingroot and Ermen 1988 [1987], 13, 31, 44–45, 61, 85, 114).

An essential aspect of North Norway and northern Fennoscandia is the geopolitical competition between spheres of political interest. In 1323, Sweden and Novgorod agreed upon a boundary from the Gulf of Finland in the southeast to the Gulf of Bothnia in the northwest. In 1326, Norway and Novgorod agreed upon their common sphere of interest along the coast of the Arctic Ocean, as far west as the Tromsø region (Julku 1986, 158f., 187). Maritime resources in North Norway, rich salmon rivers, taxation of the Saamis, access from Norway to Arctic hunting, the international transport route to the White Sea, and the search for the Northeast Passage to China were economic issues of international interest. They were also national triggers for new settlement and increased population.

During World War II, when the Nazi-German army retreated in the fall of 1944 from the Arctic front in Petsamo (Petchenga), North Troms and the whole of Finnmark county were evacuated and burned. In addition, the Allied air forces had throughout the war bombed military installations in the north, while the Nazis began bombing the eastern part of Finnmark when they began their retreat. A "virgin condition was re-created" for the reconstruction plan of the Norwegian government-in-exile in London in 1944; "only ashes and remains of house foundations were left of the cultural landscape." The London government had a "*tabula rasa* and could be 'free' to plan" (Brox 1982, 94). Some families of all ethnic groups had stayed behind in the still-occupied land, hiding in caves and underground dwellings in the mountains, "and eked out a miserable existence in a devasted and impoverished country" (Vorren and Manker 1962, 153). World War II and the evacuation of Finnmark and North Troms is a current theme in cultural-heritage management. Only after the end of the Cold War has information been released about geopolitical strategies related to local cultural history and their impact on everyday life (Olsen 1997, 176f.).

MARITIME RESOURCES IN THE NORTHERN PERIPHERY

The coastal waters of Norway, like most of the coast of the Kola Peninsula in Russia, are not frozen during the winter, although the White Sea is locked by ice until June. Warm Atlantic water mixes with colder and less-salty polar water. A branch of the Gulf Stream follows the Norwegian coast northward to the western part of the Eurasian continental shelf. There it splits and passes partly west of Svalbard (Spitsbergen) and partly east into the Barents Sea. The latter is a fairly shallow sea delimited in the west by a line from North Cape across Bear Island (Bjørnøya) to the southern tip of Svalbard. Because of increased sunlight in spring and summer, the Arctic waters are biologically a highly productive area of plant plankton, the basic organic energy, which animal plankton and higher bottom organisms feed on for six or seven months, and are the ecological basis for the breeding season of sea mammals, fish and migratory birds, and the feeding of offspring (Nilssen 1989).

North of latitude 63°, two main types of cod have been important for local settlement and subsistence economy. One is the fjord cod, which is fairly stationary within limited areas and is caught year-round. The other is the migratory Northeast Atlantic cod. From January to April, the mature generations of this cod, older than seven years, leave the Barents Sea and follow the outer coast of northern Troms to Lofoten and even further south for spawning. The fry migrate back to the Barents Sea. When the cod is about four years old, it starts wandering in March and April to the coasts of the Kola Peninsula and Finnmark, following the fat capelin *(Mallotus villotus)* that spawns from the Murmansk region to northern Troms. The spawning cod is the object of the catch in the Lofoten fishery before Easter, the younger cod after that in the Finnmark fishery. From the end of the nineteenth century, the capelin has been important for the fish-oil industry but now is overexploited, like the North Atlantic cod. It is also used for baiting longlines, but not as food in Norway, and only in recent decades for the market in Japan (Nilssen 1989).

Over time, larger settlements with a more complex subsistence economy developed, using sea and land resources, combining fishing, hunting of sea mammals, trapping on land, pasturing cattle, harvesting and cultivating winter fodder, and trading with neighboring communities and migratory groups. An important source of wood into the twentieth century has been driftwood. Trees and timber from the Siberian rivers, after the ice breakup in spring, as well as timber loads from wrecked ships, have followed the ocean currents and have become stranded on the coasts of the Atlantic Ocean and the Barents Sea. Customary rules have given ownership to the finders and have been integrated into modern law.

THE ARCTIC PERIPHERY AND SVALBARD

The geographical and climatic dimension of North Norwegian identity is expressed through *vicinity* to the Arctic. In fact, almost the whole region lies north of the Arctic Circle, its coastline comprising about one-third of Norway's. Yet, viewed from inside the region, the "Arctic" has been seen as lying beyond North Norway, closer to the North Pole. The "Arctic" includes areas where sealing and whaling have played an important role: the "East Ice" (Østisen) in the Barents Sea, the "West Ice" (Vestisen) to the west and north of Jan Mayen, and the area north of Bear Island (Bjørnøya) and around Svalbard (Jakobsen 1961, 209). The "Ice Sea" (Ishavet) is an area not defined cartographically, but refers to polar waters where drift ice is common.

Svalbard, with its coal-mining communities, has a special international status under Norwegian jurisdiction by the Svalbard Treaty of 1920. The local-political administrative system of mainland Norway does not extend to Svalbard, nor does the Cultural Heritage Act of 1978, although cultural-heritage management remains Norway's responsibility. A separate Environmental Act for Svalbard came into force in 2002. "Svalbard—the northernmost Norway" was the title of an issue in 1979 of Tromsø University Museum's journal *Ottar,* in which archaeologists and cultural historians argued for more intensive registration and better protection of cultural monuments in Svalbard, arguing on behalf of the mining communities, their "identity," and "the knowledge about their own history and the concrete places and cultural memories attached to this history" (Reymert 1979, 120). On the one hand, swift and frequent aircraft transport, modern telecommunications and telemedia, global environmental issues, and mass tourism have brought mainland Norway closer to Svalbard, also in terms of the cultural-heritage discourse. On the other hand, wildlife protection has brought the "Arctic" and the "Ice Sea" closer to Norway, and these concepts have become increasingly integrated into the language of North Norwegian self-presentation.

ECONOMIC PERIPHERIES AND BORDER ZONES IN THE PAST

Several northern "borders" and "border zones" along the Norwegian coast are related to peripheries and outposts. Around 890, King Alfred of England recorded the account of Viking chieftain Ottar from Hålogaland. Ottar told that "he lived northernmost of all Northmen," and had sailed southward along "the land of the Northmen"; what he called *Norðweg* ("the north way") extended from the land of the "Finns" (the Saamis) in the north to Denmark in the south (Helle 1993, 436). Rounding the northern headland and going south again, he had sailed to the White Sea and traded with the Bjarmians, defined by some scholars as proto-Carelians (Storå 1971, 274). Ottar's home residence was most likely just west of Tromsø, on the inner fairway sheltered by large islands, with easy access to the outer fairway and the open North Atlantic. Since the political unification of the country under a central king had hardly begun, "the idea of a Norway settled by Norwegians must have been rooted in more general geographic and socio-cultural features" (Helle 1993, 436; Bratrein 2001).

Both Saami and the Finnish peasants were always strongly oriented toward Atlantic Ocean fishing, through seasonal migration work and as permanent settlers in Norway. Even for reindeer herders, the coast and the ocean have been essential for the household economy. The vital components necessary for all human survival in the northern Fennoscandian societies were fishing, hunting and gathering, and harvesting vegetation. Agriculture and cattle holding have been premises for settlement and hence for commercial fishing for the last thousand years (Bratrein 1994–95, 23).

Theoretical Approaches

Two main paradigms have structured ethnological or cultural anthropological studies in Norden since the early 1970s: the cultural-ecological model and the communication model of semiotics. Both express the holistic perspective of cultural patterns as systems of interaction between human actors and the total environment. A crucial aspect of both models is the feedback factor and the unpredictable effect of feedback. The ecological system underlines the relationship between culture and nature. In communication theory, the simple relation between the sender's message, expressed in sign systems, to the receiver and the receiver's response reflects binary opposites such as subjectivity/objectivity, insider/outsider, own/alien, self/other, we/they (Olsen 1996). Binary oppositions are also essential structures in studies of sense of place, such as home/away, affection/dislike, belongingness/fear of a place, West/Orient (Rose 1995).

Ethnicity

"Ethnicity" is an umbrella concept in mainstream Nordic research on international migration and ethnic relations (Westin 1999, 31). In international studies, the theoretical basis of the study of ethnicity before the 1970s concerned traditional, decentralized forms of ethnic relations (Barth 1996 [1994], 182). Ethnic competition over resources nowadays occurs mainly within the framework of strong, organized states, and present theory on ethnicity tends to understand all ethnic processes within the framework of the state structure. However, both types of situation are necessary for developing more general theory about ethnicity (ibid.).

A comparative study in the 1970s of linguistic minorities in postwar Europe noted that *subjective* self-identification has been important, as well as the formalization of social organizations, for *interethnic* interaction (Allardt 1979, 67). Further, *multiple identities* were typical of modern life (68). "Sweden is on its way to becoming a multicultural society. One of the large political tasks is to work for integration" (Westin 1999, 3). This is also the challenge in Norway. A theoretical approach that aims at integrating the state level into cultural analysis has to define the modern state as an actor, not only as a symbol or idea (Barth 1996, 183). The modern state delivers public goods, which it partly makes accessible for everybody, partly allocates to categories of persons, or opens up for free competition. The state also interferes directly and regulates the lives and movements of groups and categories of people (ibid.).

CULTURAL HERITAGE AND ORIGIN STORIES

In comparative sociological research on the Nordic welfare society in the 1970s, one became aware of the "cultural heritage and the feasibility of maintaining cultural identities, viewed as a condition of both well-being and a sufficient quality of life" (Allardt 1979, 5). The quality of life was "conditioned by the knowledge of one's own cultural heritage" (32). Arguments in the cultural-heritage discourse reveal contrasts and even conflicts among ethnic groups. Landscape elements and other concrete signs focused upon in the discourse become sources of information about the historical use of the natural resources as well as the users in the past, and hence the question of whose ancestors.

Norway's museum policy since the 1970s has been to support local cultural-heritage work and museums' cooperation with schools. From 1975, local and regional museums in North Norway have been granted up to 75 percent of their budgets by the state, in addition to financial support on county and commune levels. Between 1975 and 1993, 41 percent of present museums in Norway were founded, in Finnmark 71 percent, in Troms 60 percent, and in Nordland 55 percent (NOU 1996, 49). An increasing number of new museums and cultural centers have become arenas for exposing minority identities and presenting ethnic self-images, especially Saami institutions, and in the 1990s also Finnish local institutions. Nationwide cultural-heritage programs are set up to encourage and increase social solidarity within local communities through work for common aims. Public cultural-heritage funding provides a new economic resource.

The relative strength of ethnic identity has been explained as its intertwining with language, within which I include ethnic speech style, a dialect of the majority language with phonetic, semantic, and grammatical elements from the nonspoken minority language in question (Olsen 1982, 9, 38). Mythological images of the origin and lines of ancestors are conveyed by special persons through language, expressing on a symbolic level how members of a given ethnic group have a common origin (Lange and Westin 1981, 325–26). The objective of professional historians is to produce logical, coherent narratives, aiming at objectivity and reality through systematic theoretical research. According to some archaeologists, academic origin stories are "the creation myths of our time," playing "an important part in explaining the world and our place in it" (Alexandri 1997 [1995], 60; Olsen 2000; 2003, 220). The "Cultural History of North Norway" is an example of "origin stories," where ethnicity and landscape identity are only two components of what is explained as a person's *cluster* of social identities, including gender, age group, religion, nationality, class, profession, and so on (Zavalloni 1975; Lange and Westin 1981, 235; Olsen 1992b).

Human stereotypes function as schemes of expectations. They are "filters" sorting out what information is stored (Lange and Westin 1981, 346). The concept "structures of expectations" is based on people's understanding of "how things *normally, usually,* and *typically* are in the world" (Siikala 1992, 205). We interpret human behavior by ascribing it to given causes. If a planned project failed (or succeeded), it happened because the inhabitants have a disposition for behaving in a way that leads to failure (or success). Or it happened because the natural environment is unpredictable (Lange and Westin 1981, 337). The relation between ethnic identity and place identity can be reformulated as the relation between "people" and "land." Certain people behave in such and such a way because of their personal disposition (cf. ethnic, national, and regional stereotypes), or because of environmental situations, such as climate, ecological fluctuations, and macropolitical events (cf. competing regional images and metaphors such as "the diverse people" and "the rebellious land" [Drivenes et al. 1994a]).

SIGNS AS TRANSMITTERS OF CULTURAL INFORMATION

Human groups maintain their cohesiveness through culture, the totality of signifying systems, or the "totality of non-hereditary information acquired, preserved, and transmitted by the various groups of human society" (Lotman 1988a [1967], 213). The interpretation of signs is based on the way we perceive phenomena in the environment. We relate to our memory and speak with ourselves through conscious thinking and reflecting, but also unconsciously; both states of mind are active in interpretation of signs and in creating representations of outer reality. In external communication, transmitter and receiver are different bodies. In internal communication, transmitter and receiver are one and the same body; "the purpose is to retain the information one has and this includes all sorts of memoranda and reminders" (Lotman 1988b [1970], 100).

The general concept of sign as a vehicle of meaning in modern linguistics and literature studies is linked to the Swiss linguist Ferdinand de Saussure (1871–1913). In grammar and syntax of language systems, he found minor signifying elements, on phonetic and structural levels, and underlined the arbitrary aspect of the signifier, and thus the ambiguous symbols of verbal information and communication (sounds, words, grammatical endings, sentences, texts). Another pioneer was the American philosopher Charles Sanders Peirce (1839–1914). According to Peirce, a sign is "something which stands to somebody for something in some respect or capacity." The sign is not the object of reference itself, but a reference "to a sort of idea" (quoted in Nöth 1990, 42). This "sort of idea" links to "thought figure" in social psychology, that is, something in the mind that is not yet conceptualized, and underlines the individual aspect of mental processes.

Peirce's semiotic theory was introduced into Norwegian philosophy thirty years ago (Gullvåg 1972) to explain how researchers in the humanities understand research data. While the natural scientist is concerned with natural phenomena and processes and sees them as "things and events in nature," the cultural researcher looks for "traces of human activity" in a very wide meaning, or "signs," such as "letters and pictograms, texts and inscriptions, artisans' products and tools, art, literature and music, buildings and man-made changes in the landscape." The natural scientist presupposes that "events have natural causes and can be causally explained." The humanistic researcher presupposes that "actions and traces of actions have a meaning and can be understood" (Gullvåg 1982, 149; cf. Olsen 1985, 1993b). Myths, folklore, customs, literature, and figurative art have been studied as sign systems in the Tartu-Moscow school of semiotics since the 1960s, and further developed by Yuri Lotman (1922–93) in Tartu, integrating structuralist traditions after Saussure and Peirce (Lucid 1988, 1; Toorop 1998, 10).

Cognitive anthropology and comparative folkloristics have a particular interest in "common knowledge," denoted "folk models," "cultural models," and "mental models," as opposed to "knowledge codified by experts" (Siikala 1992, 203). Within a modern national society with well-educated citizens, I would include academic and theoretical knowledge in "common knowledge." As individuals, we have an "ability and way of *observing, organizing,* and *remembering* the sea of information" about ourselves (210). Images of the Northern Hemisphere as a cosmological and anthropological periphery are historically transferred through myths and observations of scholars in the past and correlated with images and facts in the present context of social knowledge and individual minds. Information may be "scattered and random," yet "still *interconnected concepts and images,*" and the basis for "a picture of the world and for actions in this world" (203–5).

One does not intuitively feel that the etymological explanations of the cardinal directions in the idiom of "interstice of North and East" have any practical meaning today. Yet it is a key explanation in the history of ideas. North and east, *nord* and *øst* in Scandinavian, can be traced back to Greek Homeric language and earlier Indo-European languages. East is "dawn," north is "on the left," when facing dawn, south is referring to the "sun" itself, and west is what is "behind" when facing dawn (Bjorvand and Lindeman 2000). Any site, standpoint, or location is described in terms of the presence or absence of the sun and the moon, and of light and darkness, of day and night, as well as in terms of temperature and weather conditions, such as cold or heat, rain, snowfall, or wind. Sites are bodily experienced, mentally remembered, and fit into known patterns of thought, as when the inhabitants of Lapland were described in the 1950s as "physically dominated by the north and mentally drawn to the south: they are apprehensive of the east and sympathetic to the west" (Mead 1974, 47). Significant, too, are the local names given to the coastal areas in Arctic Norway: South Troms and North Troms, West Finnmark and East Finnmark.

Heraldry has its own formal symbolic and pictographic language, where objects are converted into signs, or heraldic symbols, even though "symbol" is "one of the most overburdened terms in the field of the humanities" (Nöth 1990, 115), where there are several symbolic-cognitive approaches (ibid.; Sonesson 1998, 8). As a general term, "symbol" means "a concrete sign or image that represents some other, more abstract thing or idea by convention, analogy, or metaphor" (Jean 1998, 180).

The environment is sensed before we can articulate what we perceive and recognize, through a system of signs (spoken language, gestures, graphic signs, tunes, and so forth), to communicate what is on our minds; hence a "product of mind" is necessarily "processed" in individual minds. Individuals come from real places and they have autobiographies. However, what can be revealed from observation, and what is added by interpretation?

Heraldic Arms

"Circulation of Information" via Road Guides

In public planning and legislative processes that change and remodel the physical landscape, the multicultural images of a North Norwegian identity are variously expressed or ignored. Over time, people "with an allegiance to the north," such as scientists, conservationists, artists, and men of letters (Mead 1974, 45), have spread their images of a northern identity in a process similar to the "circulation of communication" in the sociology of literature. This is made up of the relation between writers of fiction, the works themselves, and the readers (or the public). Through a complicated apparatus of transmission, the "circulation of communication" unites certain defined individuals, the writers, with a collective, which is more or less anonymous (Escarpit 1972, 13).

The road guide of the Norwegian Automobile Federation, *Norges Automobil-Forbund* (NAF), is a major conveyer of local and regional information. NAF, founded in 1924, has since 1928 issued its "road book" *(veibok)*. According to the editor, the information is checked, controlled, updated, and presented by a "team of experts," consisting of "thousands of contacts in Norwegian communes, travel associations, road offices, NAF's seventy-five branches, and many of NAF's more than 420,000 individual members" (NAF 2001, 2). The book is a portable and always-available archive, a systematic storage of visual and verbal representations of the landscape, in color and black-and-white, through maps, photographs, diagrams, tables, text, traffic signs, and coats of arms.

The heraldic arms of all the counties and their communes have been depicted together with maps since 1995. The 1995 edition of the road guide was the first with heraldic arms in the town section, about a page for each of the 46 towns out the total of 435 communes, a setup that has continued in the following editions. In 1992 the legal distinction between towns and other communes in Norway was annuled (SSB 1995, 670). Since 1996, a commune can itself define its status. Table 12.2 shows the changing number of towns in Norway and North Norway presented in the road guide from 1995 with heraldic arms. The 1992 edition did not have heraldic arms in the town section, but it included more central places than the 1995 edition.

TABLE 12.2.
Towns in Norway and North Norway presented in the NAF road book, 1992–2001.

	Norway	North Norway		
		Nordland	Troms	Finnmark
1992	57	4	2	4
1995	46	2	2	3
1998	62	5	2	3
2001	79	9	3	6

Source: NAF 1992, 1995, 1998, 2001.

HERALDRY AND LANDSCAPE

"Norwegian private heraldry now leads a humble existence. . . . On the other hand, Norway has a very visible official heraldry" (Nissen 1995, 6). Armorial bearings on shields have been borne since medieval times by aristocratic and royal families and the clergy, often proclaiming the genealogical ties among families (Jean 1998, 120). In 1821 the hereditary nobility was constitutionally abolished in Norway (Mykland et al. 1989, 82). Still, family symbols often became the basis for seals and coats of arms for towns and provinces. The lion in Norway's national arms is known from the equestrian seal in the reverse of the great seal from 1243 of King Håkon Håkonsson, which was a present from the English king. From the 1280s onward, the lion is crowned and holds a battle ax. These arms were used through the unions with Denmark and Sweden. The content was officially approved in 1844, 1905, and again in 1937, with this heraldic description: "Norway's national arms are gules [red] a lion rampant or [gold] holding in its fore-paws an ax with ax head argent [silver] and ax handle or [gold]" (NKV 1987, 49). The ax has been interpreted as the martyr ax of Saint Olav, killed in battle in 1030. The crown emphasized the authority of the lion (49–50, 226).

In Norway, heraldic arms convey something about the physical landscape of the political-administrative entities of the state, the counties, and communes (Figure 12.4). Color pictures of the arms were continuously and increasingly displayed in official publications and various mass media in the 1990s. After World War II, the normal procedure has been to approve both arms and flag simultaneously. The Flag Act of 1933, which stipulated that communes could only use flags approved by royal decree, increased the interest for flags and banners (NKV 1987, 35). Heraldic arms are also the state emblems of other public administrative units, such as dioceses and military defense districts. The Church of Norway uses the arms of the Catholic archdiocese of Norway from before the Protestant Reformation in 1536, a cross with two axes, symbolizing

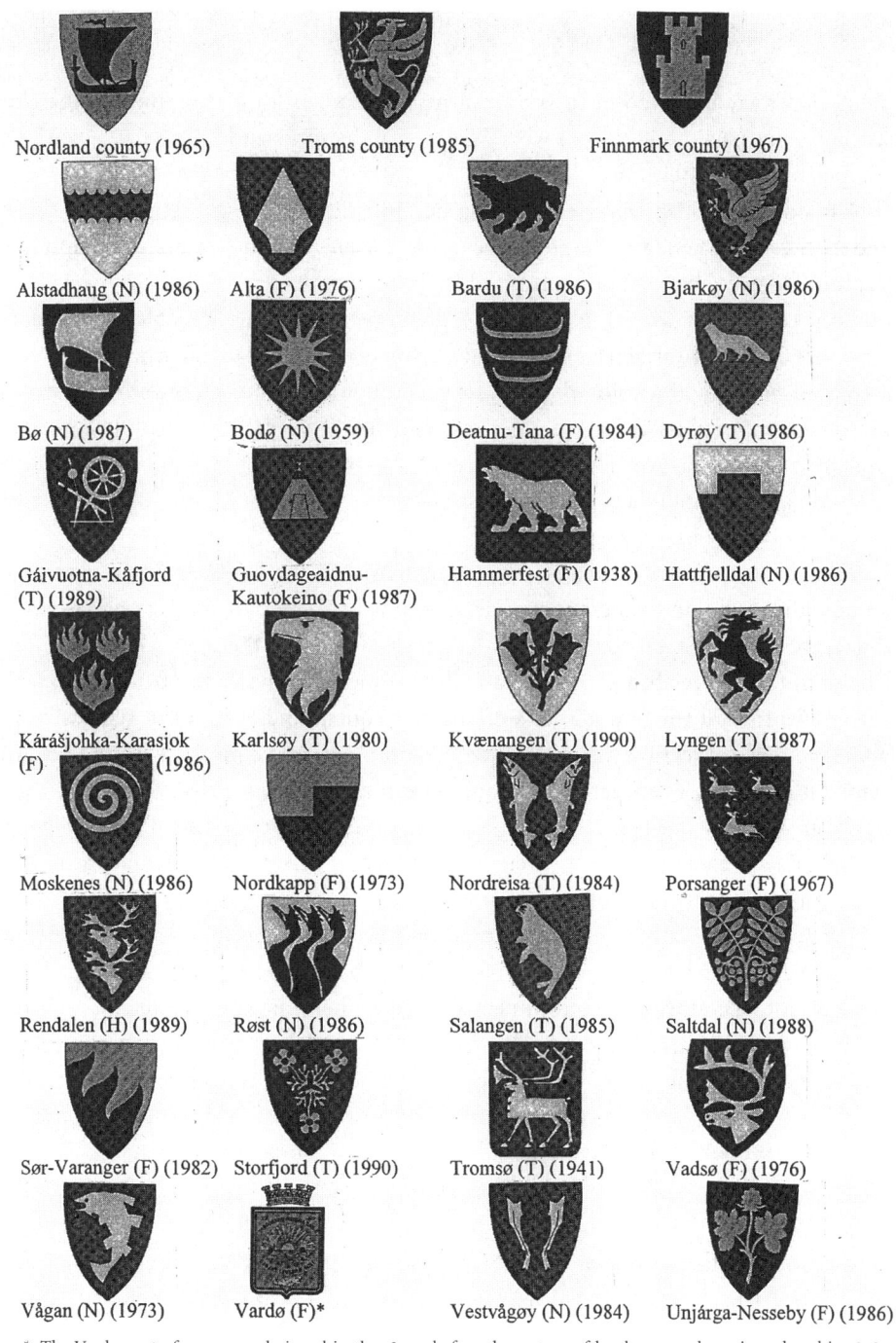

Nordland county (1965)   Troms county (1985)   Finnmark county (1967)

Alstadhaug (N) (1986)   Alta (F) (1976)   Bardu (T) (1986)   Bjarkøy (N) (1986)

Bø (N) (1987)   Bodø (N) (1959)   Deatnu-Tana (F) (1984)   Dyrøy (T) (1986)

Gáivuotna-Kåfjord (T) (1989)   Guovdageaidnu-Kautokeino (F) (1987)   Hammerfest (F) (1938)   Hattfjelldal (N) (1986)

Kárášjohka-Karasjok (F) (1986)   Karlsøy (T) (1980)   Kvænangen (T) (1990)   Lyngen (T) (1987)

Moskenes (N) (1986)   Nordkapp (F) (1973)   Nordreisa (T) (1984)   Porsanger (F) (1967)

Rendalen (H) (1989)   Røst (N) (1986)   Salangen (T) (1985)   Saltdal (N) (1988)

Sør-Varanger (F) (1982)   Storfjord (T) (1990)   Tromsø (T) (1941)   Vadsø (F) (1976)

Vågan (N) (1973)   Vardø (F)*   Vestvågøy (N) (1984)   Unjárga-Nesseby (F) (1986)

* The Vardø coat of arms was designed in the 1890s, before the system of legal approval was introduced in 1898.

FIGURE 12.4. Heraldic arms of the three counties of North Norway and thirty-two Norwegian communes. Communes indicated by county: F = Finnmark, T = Troms, N = Nordland, H = Hedmark. Dates in parentheses refer to the years the heraldic arms were approved. (Copyright: NAF, Norwegian Automobile Federation, reproduced with permission.)

Saint Olav's ax, approved in 1990. New arms and badges for all the units of the Norwegian army, navy, and air force were designed in the 1970s and 1980s (Bergersen 1992; Cappelen 1992; Nissen 1995, 7).

Heraldic rules in Norway concerning heraldic figures, the combination of colors and metals *(tinctures)*, and the structure of the *blazon*—the short verbal description of the motif[4]—were handed down from older European precepts (NKV 1987, 32). There are four colors, red (gules), blue (azure), green (vert), and black (sable), two metals, gold (or) and silver (argent), and two furs, ermine and vair; a metal ground shall have a colored figure, and a colored ground a metal figure.[5] However, tradition may be a reason for exceptions from the rules, as with the arms of Nord-Trøndelag county from 1957, which has "on silver a gold cross." The motif has direct connection to "the tradition about Saint Olav," who in the sagas had "a white shield, where the holy cross was set in with gold" (183).

Heraldic arms, seen as material signs or marks in the landscape—on road signs and buildings and on banners at special occasions—function in similar ways as graffiti did in the United States for urban street gangs, studied in the 1970s (Rose 1995, 99). The graffiti signs celebrated the gang and its exploits in the center of its area, while the graffiti toward the boundaries were more frequent and even more aggressive. An analogue sense of territory is manifested in communes and counties with the heraldic arms on buildings in centers of administration and at the borders with road signs, although without the aggressivity and insults openly expressed in the street gangs' graffiti. The street gangs are sovereign in their sense of territoriality and their claim to control people by controlling the area (100).

We are socialized into emotional reactions to such ritual symbols; for example, we often have positive feelings about the national flag, both its colors and its pattern. When we drive into a county or a commune, we meet the heraldic arms as we cross the border, and see it in various locations within the territorial unit. We may feel at home, feel very comfortable, or we may be curious or indifferent, feeling as a stranger, very different from the locals, or even uncomfortable and threatened (cf. Rose 1995).

HERALDRY MANAGEMENT IN TWENTIETH-CENTURY NORWAY

Before World War II, only towns had heraldic arms, mainly from after the 1880s, including the North Norwegian towns. The counties gained their arms and flags after 1957. The celebration of the 150th anniversary of the Local Government Act in 1987 led to an increase in the number of approved commune arms, with around 280 new ones between 1980 and 1989 (NKV 1987, 23).

Commune arms are legally approved by royal decree: the king (in practice the government) grants the arms. Commune flags can be used only if the motif is from an approved coat of arms (to restrict political-ideological messages from the local authorities, as when socialist communes before World War II wanted to use a red flag on public occasions). In 1976 general legal regulations were issued for the use of the commune arms, including the flag. Once the local authorities accept the motif, the proposal is forwarded to the county governor *(fylkesmann)*, representing the state, and then the Ministry of Local Government and Regional Development. The ministry consults the National Archives, which before 1900 had heraldic experts in connection with their collections of historical seals (NKV 1987, 21, 48).

The argument that a "product of mind" is necessarily "processed" in individual minds is valid also for the heraldic expert and the designer. The heraldic consultant at the National Archives from the early 1920s until around 1970, Hallvard Trætteberg (1898–1987), had considerable influence on official heraldic development in Norway, both as a knowledgeable designer, productive author of heraldics, and bureaucratic decision-maker. Trætteberg was born in the Lake Mjøsa district, a central historical, agricultural, and national-cultural district. In the 1930s, he had made proposals for county arms, seals, and flags in Norway (NKV 1987, 201). He designed the first eleven of the eighteen county arms granted in Norway after 1957, including the three in North Norway. He also drew around forty commune arms, revised the national arms several times, and designed King Olav V's monogram, seals for bishops and diocese councils, and badges for military units and state institutions (Riksarkivet 1998).

Trætteberg avoided compound motifs. His ambition was to achieve *plainness* in heraldry, following what he saw as medieval norms; plainness also symbolized the communes as *primary* units in public administration. He preferred heraldic motifs that alluded to the Viking Age or medieval history. An innovation in Norway in the 1890s had been a crown of bricks as a device upon the shields of towns (cf. Vardø). After 1940 such crowns disappeared (NKV 1987, 21), perhaps in Trætteberg's spirit of plainness. An additional reason for the removal of crowns on commune arms could possibly be that this underlined their low position in the institutional hierarchy of the state. Trætteberg also reformed older versions of commune arms (Bodø and Vadsø) and gave former three-dimensional images a clear two-dimensional style (Tromsø). In his choice of colors for the county arms, he did not consider a naturalistic impression but rather wanted the highest effect of contrasts, often yellow or gold combined with black (cf. Nordland and Finnmark) (NKV 1987, 215; Riksarkivet 1998).

After 1970, when still about 400 of the approximately 450 communes did not have heraldic arms, local designers became more common and were accepted as long as they followed heraldic norms. Many of them have drawn several arms, including the multiartist Arvid Sveen. He was born in 1944, like Trætteberg in the Lake Mjøsa district. After completing his education as an architect, he has worked in Vadsø, the county administrative town of Finnmark. He designed more than one-third of all commune arms in North Norway, that is, twelve of the nineteen arms granted in Finnmark (1976–88), eleven of the twenty-five arms in Troms (1984–90), and eleven of the forty-five arms in Nordland (1986–91). In addition, he has drawn several arms in southern Norway (NKV 1987; Nissen 1990, 1991, 1993; Riksarkivet 1998).

North Norwegian Coats of Arms

Nordland County: Coastal Culture beyond the Arctic Circle

The blazon of the Nordland county arms (1965) is "on gold ground a black boat with mast and sail." The explanation is that the Nordland boat is a natural collective symbol for the county because the boat has kept Nordland together. It is a reminder of people's basic living conditions and in itself is an excellent artisan product; the shape has not changed much since Viking times. Black is the color of tar, and gold (yellow) represents the sun (NKV 1987, 191, 194). The boat alludes to coastal culture, indirectly to the long history of the Lofoten fisheries and Viking ships.

Midnight sun in Bodø

The Arctic Circle crosses Nordland. The blazon for Bodø (1959), the county administrative center and cathedral town, is "on red ground a gold sun," the heraldic midnight sun marking Bodø as the first town north of the Arctic Circle. The old shield was divided horizontally, with the sun in the upper part and the hills of a local island behind a red Nordland boat in the lower part. Trætteberg kept only the midnight sun (NKV 1987, 194), which underlines the northern altitude, like the gold ground in the county arms.

Hardship in Bø

Severe climate rather than commercial fishing is symbolized by another Nordland boat in the arms of Bø commune north of Lofoten (1987). The blazon is "in black a half boat with mast and square sail in silver." The half boat denotes the boat of the sea monster *Draugen,* the terrible-looking ghost of a drowned person not buried in Christian earth, who signifies a foreboding of death if he is seen or heard because he seeks people on the sea (Nissen 1990, 50). The black ground alludes to death and mourning.

Cod Fishery in Lofoten

The Lofoten fishery is directly represented by codfish as a motif for two communes in the Lofoten area, Vågan and Vestvågøy. The spawning cod caught in the annual Lofoten fishery until Easter is commonly called the "Lofoten cod," even though spawning can occur from western Finnmark to the coast of Sør-Trøndelag. In the Finnmark fishery after Easter, the young cod is the main catch. From the nineteenth century, fishermen as far south as Trøndelag and Møre often sailed northward for a long season away from home. The Finnmark fishery became important for the Roman Catholic Church in the Middle Ages, and after the Reformation for the king. Lofoten thus became more an economic center than a northern periphery. The climate of the Lofoten region, which for its latitude has mild winters, is perfect for the preserving technique of drying fish in the air on wooden racks—still a common practice along the entire coast of North Norway. The fish racks can reach considerable size, as large as houses or small churches. Drying reduces the weight of the raw fish by three-quarters, making its transport for sale more economic.

Vågan's arms (1973) have "on blue ground a vertical forward-leaning silver cod." Trætteberg reshaped the figure, first approved in 1941, to express "vitality and strength," underlining the importance of the Lofoten cod (NKV 1987, 200). The Vågan arms expressed closeness to nature and natural resources.

Vestvågøy commune focused on economic and cultural history by depicting a dried cod (with a crown), used since 1406 in the seal and later the arms of the Hanseatic Office in Bergen. It is also known from several family arms in Bergen and North Norway in the seventeenth and eighteenth centuries. The commune arms (1984) have stockfish as their motif: "in blue two upright and counter silver dried fishes" (NKV 1987, 199). In Vestvågøy lies the new Lofoten Viking Museum at Borg, a reconstruction of "the largest, known building and chieftain estate from the Viking time" (NAF 2001, 285). The commune presents itself with symbols of national importance, in heraldic as well as in museological terms.

Landscape features in Nordland county

Of Arvid Sveen's eleven arms from Nordland, four have a nature sign as a heraldic figure. These are the profile of a mountain (Hattfjelldal: green mountain on silver [white] ground), a spiral presenting the maelstrom or whirlpool (Moskenes: silver spiral on blue), cormorants (Røst: black birds on silver) (NKV 1987, 196–98), and a twig with two rowan leaves (Saltdal: gold twig on red). Saltdal's twig "symbolizes nature and national park" and the two most important settled areas (Nissen 1991, 146). Gold and red allude to the national dimension of the protected landscape (1988).

The arms of Hattfjelldal ("hat-mountain-valley") (1986) have the profile of Mount Hatten, 1,128 meters high. It is a prominent landscape feature visible to the inhabitants of the commune center and a component in the place-name (NKV 1987, 196). What is not heraldically explained is the cultural plurality and bilingual inhabitants (cf. Lillegaard 1976, 345). In the commune is located *Sijti Jarnge,* the cultural-heritage center for South Saamis (NAF 2001, 475), among whom are reindeer-herding families in Norway and Sweden.

Idealized images of a happy, healthy, moral, and peaceful northern people, similar to ancient myths about Hyperboreans (Laureys 1992), have been repeated in travel reports and documentary literature. The arms of Moskenes (1986) stand for the opposite, the evil and monstrous. In Olaus Magnus's "History of the Nordic Peoples" in 1555 (and his map *Carta Marina* from 1539), the Moskenes current *(Mostaström)* between

Røst *(Røst)* and Lofoten *(Loffoet)* is like a "huge throat, or rather Charybdis," which "in a moment swallows the seafarers" and especially those who have "no knowledge about the character of the place" (Olaus Magnus 2001 [1555], 89). The heraldic explanation is illustrated with the original model of the current in *Carta Marina* (with the place-names *Rust* and *Lofot* and showing one big whirlpool pulling down a ship), informing that the *Moskenesstraumen* is a tidal current *(straum)*, which is about 4 km broad, where the water is pressed up to a speed of 6 nautical miles (c. 11 km) per hour and strong whirlpools emerge. Near shallows there is "boiling and grinding all the time, even in calm weather" (NKV 1987, 197). Olaus Magnus knew about the strong effects of tide and ebb (which do not occur in the Baltic Sea), and he knew that German sailors were especially knowledgeable about the effects in Lofoten (Olaus Magnus 2001, 89–90). This information implicitly assumes that local inhabitants had achieved the skill to manage the aggressive expressions of nature.

In January 1432, the Venetian nobleman and merchant Pietro Querini and his crew landed in a small boat on the outermost inhabited part of Lofoten, in the Røst archipelago. Querini's ship had got lost in a storm on the way from Genoa to the Hanseatic markets in the Netherlands. He experienced their stay on Røst with friendly and naturally behaving locals as being in paradise. Drifting toward the coast of Nordland, they had navigated by the mountains. In 1932 a monument was erected to celebrate the 500th anniversary of the landing of Querini and his crew (Eriksen 1976, 75; Wold 1991). To articulate Røst's identity, national folktale tradition rather than European scholarly narrative was chosen for the commune's arms (1986). The blazon is "in silver three rising black cormorants." The cormorant is common along the coast (and as such they are regional landscape figures), but the tale about three cormorants on *Utrøst*, "Out-Røst," is locally situated. *Utrøst* in the fairy tale appears where no land is in sight, and the cormorants, who are three brothers who can change themselves into birds, only appear for pious or visionary persons who are in danger on the ocean. The artist has given the cormorants an unreal, fairy-tale-like character (NKV 1987, 198). The periphery represented in Røst's heraldic arms is the borderland of the otherworld, the realm of supernatural beings. This confirms what is thought of as typical for North Norway: superstition and belief in ghosts of drowned people haunting the living, and positive and negative effects of Saami magic (common motifs in Scandinavian and Finnish oral tradition). However, the *Utrøst* tale also alludes to the power of an almighty God, who will save the believer when in peril. The ocean is dangerous, but God will look after his own.

The motif of the Alstahaug commune arms (1986) is the row of mountains called "the Seven Sisters" (De syv søstre). The blazon is "in silver a blue bar of engrailed lines (of partition)," alluding to the mountain formation and its mirror image in the sea (NKV 1987, 193). Alstahaug's arms share with Røst the background in oral folk tradition. "The Seven Sisters" have been a well-known sight from sailboats and coastal liners for generations of tourists. They have been seen, sketched, painted, photographed, and filmed (Aasvang and Lillegaard 1976, 375). On might have expected a heraldic motif from the writings of the clergyman and poet Petter Dass (1647–1707). His topographic description, *Nordlands Trompet* (The Trumpet of Nordland), first published 1739, has a central position in national literature (Dass 1997). Dass was born in the district and served at Alstahaug church. The local legends about his strong personality were part of oral tradition into the twentieth century. He even fooled the Devil, with whose help Petter Dass flew through the air to Copenhagen to give his Christmas sermon (Aasvang and Lillegaard 1976, 390).

In contrast to the fictitious *Utrøst*, the cormorants on Røst are real, moving in the landscape. "The Seven Sisters" in Alstahaug are real but stationary topographic features. Providing landmarks for locating fishing grounds, the topography of the coast is necessary local knowledge in fishing communities. As landmarks for sailing, such knowledge is equally important. The fairway along the Nordland coast has been the main route from further south in Europe to the polar regions and the White Sea. The mountains of the Nordland coast are familiar in national folk tradition (Aasvang and Lillegaard 1976). Mountains, islands, and characteristic traits of the landscape were transformed into humanlike beings, giants, or animals and knit together into narratives that functioned as mnemonic devices. They were easy to keep in mind, recollecting shapes, sites, and their spatial relations in the landscape. In local literature, the mountains and Alstahaug church, built in the twelfth century in the lee of a small hill, are an entity, "the hill that protects against the sea" and the church "almost humbly creeping down in the terrain with the Seven Sisters just behind as a covering wall in the east" (Weider 1976, 360).

TROMS COUNTY: A BORDERLAND

Animals as landscape figures

People in regional and ethnic costumes have been analytically denoted "landscape figures" (Berggreen 1989, 26). The term can be broadened to include birds and animals with local habitats. It might seem ironic to denote the cod a zoological landscape figure (cf. Vågan and Vestvågøy) since it lives underwater and is frequently invisible. Yet even the cod and other white fish drying on wooden racks have been prominent in the North Norwegian seasonal landscape and also a feature of the local settlement everywhere along the coast.

The walrus, although not living in Norway, was once suggested as a county symbol and refers to Ottar of Hålogaland's hunting and trading expeditions beyond the national territory before 900 as far as the White Sea, close to the habitat of the walrus. The walrus is found on Svalbard and Jan Mayen, like the polar bear, seal, reindeer, and fox—all heraldic figures in North Norway. Seals are also found along the entire coast. The seal in Salangen's heraldic arms in Troms, drawn by Sveen (1985), refers to the commune's name, meaning "seal-fjord," from *selr* and *angr* (NKV 1987, 206). However, the fox in Dyrøy's arms (1986) is not the wild fox but a cultural sign, signifying the first mutant of the silver fox, the platina fox, in 1933 on a local fox farm. The breeding of the platina fox was a commercial success (204).

The lion in Norway's national arms is just a symbolic animal in Norway, present in the city landscape of the capital, guarding the parliament, the *Storting;* this is also the origin of the metaphor "Lion's Hill," *Løvebakken,* for the *Storting,* referring to the name of the slope with the steps where the two granite lions guard the entrance. In Europe, the lion was a princely symbol, adopted by the Norwegian king just before 1200 (NKV 1987, 50; Nissen 1995). As one of five military defense regions after World War II, North Norway's heraldic arms were a rampant silver lion on black, holding a sword, with a shield on his chest, inspired by the arms of Erling Vidkunsson of the Bjarkøy dynasty, who organized the border defense of Finnmark and represented the king of Norway in the peace treaty with Novgorod in 1326 (Bergersen 1992, 243). The other four defense regions had cultural-historical symbols (arrows, axes, castle, and torches), while several military subdistricts had animal figures, including Troms and Finnmark, with a reindeer and a wolf's head, respectively. As a defense mark for Finnmark, the wolf had a special connection to this part of the country and was an applicable symbol for military divisions. By using only the head, it was possible to show such characteristic traits as "alertness, aggression, etc." (352). The largest beasts of prey in the fauna of mainland Norway are the brown bear, wolf, lynx, and wolverine. The wolverine is represented in the commune arms of Bardu, drawn by Sveen (1986), symbolizing wildlife and nature in inner Troms, which was not permanently settled before the end of the eighteenth century (NKV 1987, 203). The animal figures in military heraldic arms are either from the national fauna, such as the beasts and birds of prey, and the moose and the reindeer; or they are "imported" figures, like the lion and the fantasy animals—the griffin and dragon. All military divisions in Troms have a running silver reindeer on red or black. The badges were well known by Norway's military allies from training activity in the area (Bergersen 1992, 251) at a time when international war training was not allowed east of Troms because it was too near the Kola Peninsula and its military installations.

The reindeer: An ethnic and national animal

In commune heraldry, the reindeer is called "the most typical North Norwegian animal" (NKV 1987, 207). The seasonal migration of Saami reindeer herders from inland settlements to grazing land on the coast in summer coincides with tourist cruises along the coast, making the reindeer very visible in the coastal landscape. The reindeer is an ethnic landscape figure in Norway, representing Saami reindeer herding from the Russian border to the northern part of Hedmark county in southern Norway. It is well known from the arms of the administrative centers of Troms and Finnmark counties, in Tromsø's arms since 1870, and in Vadsø's since 1893. The present version of the Tromsø arms is "in blue a walking silver reindeer," based on a draft from 1855 and redrawn by Trætteberg in 1941 (ibid.). In 1855, Troms was still part of Finnmark county, then governed from Tromsø. It was also the cathedral town for the Hålogaland diocese (the diocese of North Norway). In 1952, the diocese was split into South Hålogaland (Nordland county) and North Hålogaland (Troms and Finnmark counties plus Svalbard).

The original arms of Vadsø from 1893 had both a seagull and a reindeer head divided by a bar. They clearly represented fishing and reindeer herding in Finnmark. Trætteberg's revised version (1976) had just "on red a silver reindeer stag's head," underlining the function of Vadsø as the administrative center and symbolizing "freedom, strength and endurance" (NKV 1987, 215).

In 1967, Trætteberg designed the commune arms of Porsanger in Finnmark "on red ground three running reindeers." His intention was to symbolize an economic activity over centuries, still important, even though agriculture combined with fishing dominated (NKV 1987, 214). During World War II, the Germans built an airport in Porsanger and it has since been a military center. The "freedom, strength and endurance" of the reindeer links up with the function of military forces. The reindeer is both a Saami landscape element in North Norway and a symbol of the national defense forces, explicitly in Troms county, while the wolf is the explicit national defense symbol of Finnmark.

A fourth commune in Norway with reindeer as a heraldic figure is Rendalen ("the reindeer valley"), not in North Norway but in Hedmark county. It was designed by Sveen (1989) and represents the place-name and common presence of reindeers over time (Nissen 1991, 187), just south of the South Saami area. The reindeer is both a regional and a national landscape-figure in southern Norway, where the only wild reindeers left in Western Europe have their habitat on the Hardangervidda mountain plateau and adjacent areas. The title character in Henrik Ibsen's play *Peer Gynt* (1867) is based on oral tradition about a reindeer-hunting farmer in Gudbrandsdalen. This was the reindeer adopted as Vågå's commune arms in 1985. Reindeer hunting is still part of common property rights of Norwegian farmers in the south, as is moose hunting in wider parts of Norway. Reindeer antlers are often found over the entrance to cabins and storehouses in farmyards near the habitat of the wild reindeer.

Viking chieftains and Bjarkøy nobility

Troms got its heraldic arms in 1960 showing "on red ground a gold griffin" (NKV 1987, 201). A colored glass mosaic of the arms was inserted as a decorative and visually prominent element on the gray concrete wall of the new county administration building from 1958 in Tromsø (Bratrein and Niemi 1994, 159). The heraldic jubilee book explains that the griffin, a combination of lion and eagle, has always been a symbol of rulers. In Norway the heraldic griffin is first known in the arms of Bjarne Erlingsson, a member of the "aristocratic family on Bjarkøy in Troms," an important location "from the time of Tore Hund" (who killed Saint Olav in 1030) (NKV 1987, 201–2). North Norwegian chieftains were culturally well integrated in the Scandinavian nobility (Bratrein and Niemi 1994, 160). Bjarkøy is at the ocean end of the sea and land route through southern Troms, connecting with the route through Sweden's sphere of interest along the Torne Lake and Torne River to the Gulf of Bothnia. The northernmost zone of annual grain cultivation is in the coastal district southwest of Tromsø, where Ottar before 900 and the later medieval Bjarkøy dynasty had their estates, although even the inner Alta fjord in Finnmark had climatic conditions for cultivating grain (Bratrein 1994–95, 11).

The griffin (half eagle, half lion) in European civilization symbolizes high-ranking guardians on the periphery of national territory against powerful intruders from the east and southeast. The tinctures of gold and red underline royal sovereignty. So does the heraldic position, the rampant griffin, like the rampant lion in the national arms. When Bjarkøy commune wanted to use the griffin, it had to choose a variety of the motif (1986), "in blue a half gold griffin" (NKV 1987, 203). The visual connection with the monarchy was reduced with the removal of the half lion, even though the gold was kept. The present half eagle is closer to the shape of a real eagle than to the original medieval griffin; hence it is closer to a symbol of local nature. However, it remains a sign and symbol of culture through the concept of griffin in the blazon and the local historical connection in the narrative text.

Karlsøy: Interstice of nature and culture

The eagle as a nature sign appears in the arms of Karlsøy (1980), a commune of many small and large islands, with high mountains. The blazon of the Karlsøy arms is "in blue a raised silver eagle's head," with the model from Nord-Fugløy ("northern bird island"), one of northern Europe's largest nesting places for the gray sea eagle. The blue ground of the eagle represents the sea, which is the link to maritime resources and fisheries, the main economy (NKV 1987, 205). In Norwegian military heraldry, the eagle is a Viking symbol, once called the "victory bird," and "the eagle is typical for coast, fjord and mountain" (Bergersen 1992, 250).

Nord-Fugløy is an elongated island of 21 km², with the highest peak 750 meters above sea level. It looks like a huge cliff in the ocean, without a sheltered harbor or landing for ships, and was mainly used as a seasonal dwelling because it had rich sources of trading goods, such as seabird eggs, eiderdown, cloudberries on the high marshy plateau, and fodder grass on the slopes. It is an old sailing landmark for voyagers from Finnmark and Svalbard, leading to Tromsø and the inner bays of the long fjords Ullsfjord and Lyngen, with their marketplaces and harbors, where traders met from Torneå on the Gulf of Bothnia. No other marketplace "in the whole area towards the North Pole" was more visited than Torneå, Olaus Magnus wrote before 1555: "Here gather namely White Russians, Lapps, Bjarmi people, Bothnians, Finns, Swedes, Tavastlanders and Hälsinglanders; in addition a great deal come from Norway across the high mountains and the wide wastelands." They traveled partly in long, narrow boats, in sledges, pulled by tame reindeers, and on skis (Olaus Magnus 2001 [1555], 938).

Nord-Fugløy was the gatepost, the landmark of safe harbors and civilized people for seafarers returning from the roaring ocean and hardships of the polar ice (and even from alien and wild peoples, cf. Rose 1995, 108–9). Nord-Fugløy is marked on several seventeenth-century Dutch maps, written as *Noor/fogele* in 1635, *Noorfogelo* and *Norfugelo* in 1662, *Noorfoeloe* in 1692, and *Noor foel Oe* in 1696, but as *Fugeloe* on an English map from 1794, without the prefix "north" to distinguish it from the southern *Fugelsoe* on the same map, a smaller nesting island, Sør-Fugløy in Karlsøy commune (Mingroot and Ermen 1988 [1987], 51, 71, 83, 91, 105, 133).

In 1975, Nord-Fugløy was protected by the Ministry of Environment as a nature reserve with special protection of birdlife. Hence the commune arms from 1980 can be seen as representing nature, topography, wildlife, and marine environment, positioned where the inhabited world meets the uninhabited world.

Trilingual communes in Troms county

Five trilingual communes in Troms county have arms designed by Sveen. Three border Finland, one borders both Finland and Sweden, and the fifth has no national border. Today the western boundary for the Finnish language is in Lyngen and Storfjord communes. From 1326 the Lyngen peninsula was the westernmost area of the legal Russian economic sphere of interest, later extending even further west to Malangen (Bratrein and Niemi 1994, 163). Finnish-speaking immigrants, especially in the eighteenth and nineteenth centuries, settled in Saami areas in the river valleys and along the bays of the large fjords from Ofoten in Nordland to Varanger in Finnmark, where fodder could be gathered for cattle breeding.

Nordreisa—A national landscape

Nordreisa commune is perhaps the most significant area of Finnish settlement in Troms. Cultural knowledge from Finland and northern Sweden about forestry, timber floating, and tar burning underline the "Finnishness." The blazon of the arms (1984) is "in green two addorsed silver salmon," since Nordreisa is known for its salmon river, the Reisa. Green denotes agriculture in a commune with a profilic nature (NKV 1987, 206). The road guide refers to the population of "Saami, Norwegian and Finnish descent" in connection with the Reisa National Park, protected in 1986. It describes the varied physical landscape of the national park, where "Saami, Quainish and Finnish culture" are "meeting," and in the Reisa Valley is a museum with "old, Quainish houses" (NAF 2001, 445).

The assimilated Lyngen horse

The Lyngen commune's arms are "in silver a standing black horse with head turned back" (1987). The "Lyngen horse" is North Norwegian stock, since 1968 officially named the "Nordland horse," yet it looks like a smaller version of the Finnish horse. It is usually red, although brown and black appear frequently. The horse represents agriculture, while silver stands for the fjord and fishing (Nissen 1990, 51). In North Norwegian tradition, the "Lyngen horse" is of Finnish stock. The ethnic plurality in Lyngen is mentioned in the road guide in connection with a northern Fennoscandian Pietist movement: "The core area of Læstadianism. The settlement characterized by Saami, Quainish and Norwegian culture" (NAF 2001, 499). This movement was founded by the Swedish vicar Lars Levi Læstadius (1800–61), who was socially opposed to the economic and ecclesiastical establishment. Although he worked within the Swedish Church, he accepted lay preachers, who wandered across the national borders preaching in Finnish and Saami as well as in Scandinavian.

Papaver læstadium of Storfjord

Storfjord commune is in the inner bay of Lyngen Fjord. The blazon of the arms (1990) is "in red three gold poppies tripartite." The three poppies, *Papaver læstadium*, symbolize that "Finland, Sweden and Norway have intersecting borders at the 'Three Countries Cairn' *(Treriksrøysa)*." The heraldic explanation of *Papaver læstadium* refers to "the famous vicar," who was also "an eminent botanist" (Nissen 1993, 334–35). The road book notes that Skibotn village had seasonal markets in November, January, and March, where among other goods sold were smithy ware and agricultural products brought from the Torne Valley. "The Skibotn valley was also the arrival gate for a large proportion of those Finns, Quains, who through the 1700s and 1800s immigrated to North Norway" (NAF 2001, 443). Combined with the traditional markets, the Læstadians have large annual gatherings in Skibotn—"mainly Norwegian, Swedish and Finnish Saamis in colorful costumes." In a valley in Storfjord, the "botanist-priest in 1831 found a mountain poppy with a special form, the Laestadius poppy, *Papaver læstadium*, today in the commune arms of Storfjord" (ibid.).

Visible women in Gáivuotna-Kåfjord

Gáivuotna-Kåfjord is a declared Saami commune, but Finnish is also spoken there. The heraldic figure in the arms (1989) is "in red a silver spinning wheel." It is a symbol for handicraft and tradition and a "typical women's tool, something rarely represented in heraldry" (Nissen 1991, 147). Weaving and knitting are considered to be a coastal Saami tradition, in addition to preparing animal skins for sewing. Large woolen carpets are sold at the seasonal markets to reindeer-herding Saamis. Some handicraft traditions originate in the Torne and Kemi valleys, where farmer-women specialized in weaving and knitting products for the annual markets in Sweden and Finland as well as Norway. Finnish-speaking immigrants to the fjords and valleys in time became assimilated into local Saami-speaking communities. Saami-speakers were also assimilated into Finnish communities, while on the outer coast Norwegian became the dominant local language. Economic and social integration has resulted in bilingual and trilingual communities in North Norway with connections to economic, lingual, and religious networks through family and kin in other communities, even across national borders.

Kvænangen: The destination of the "Lowlanders"

Kvænangen commune shares borders with Finnmark county from coast to inland and with Finland. The commune arms (1990) are "in silver a bluebell-plant," *Campanula rotundifolia,* resistant to a harsh climate and barren soil, just as "the people in Kvænangen have managed to exist in marginal nature conditions" (Nissen 1993, 337). The name comes "from *kven* and *angr,* fjord" (NAF 2001, 445): "The Norwegians used the name *kvenir* ("lowlanders") of the Finns and the name *Kvenland* ("Lowland") of the area populated by them on the east and north coast of the Gulf of Bothnia" (Vahtola 1993, 190). The three flowers represent the "meeting of three tribes," while blue and silver (white) allude to the national colors of Finland.

The present county border between Troms and Finnmark is from about 1350 and goes from the Arctic Ocean to Finland (then part of Sweden). Closest to the sea it follows a mountainous peninsula, without protecting islands north of this promontory. This is the first stretch north of the Arctic Circle with only one fairway on the way to the North Cape; from there to the White Sea, the fairway is again unprotected by islands. To avoid the exposed sea passage, people have crossed by land from Kvænangen to Alta over a narrow isthmus, *Alteidet.* The modern car road is 13 km long and the highest point is 70 meters above sea level (NAF 2001, 445).

FINNMARK COUNTY: THE EASTERN FORTRESS

The blazon of the Finnmark county arms (1967) is "on black ground a golden castle," which stands for Vardøhus fortress founded in the early 1300s as Norway's border-defending fortress to the east. Vardøhus fortress is said to have saved Finnmark for Norway and is also seen as "a mighty symbol of the rebuilding of Finnmark after the war" (NKV 1987, 209).

Merchants in Bergen and Trondheim had a trade monopoly in North Norway before free trade with Russian and Carelian merchants from the White Sea, the Pomors, was granted to Vardø and Hammerfest in 1789, Tromsø in 1794, Bodø in 1816, and Vadsø in 1833. In comparison, Archangel, on the White Sea, was founded as early as 1583 through Russian-English cooperation; Torneå, at the head of the Gulf of Bothnia, was founded in 1621. The Pomors exchanged fish for agricultural products, wooden tools, textiles, metalware, and salt after midsummer, when the White Sea was not frozen. The Pomors also fished on the Norwegian coast.

The Orthodox monasteries were trading centers. The White Sea monastery at Solovetsk had been founded around 1423, before Novgorod became subject to the rule of Moscow in 1478. Missionaries from the White Sea founded monasteries in the 1530s on the coast of the Barents Sea in Saami areas in Petchenga (Petsamo) and Kola, two important trading centers east of Varanger. An Orthodox chapel was built in 1565 on the present border between Russia and Sør-Varanger commune, and another at Neiden further west in the commune, reinaugurated in 1965 as a chapel for Orthodox East Saamis, including those in Finland (Storå 1971, 27, 142).

The Finnmark county arms are a cultural sign denoting national territorial rights in a political borderland with the eastern inhabited world as the significant other. The rebuilding of what the enemy had destroyed during World War II also expresses a will not to give up the territory. The black ground is not explained and might be interpreted as Trætteberg's aim for sharp color contrasts. Yet, in military heraldry, black stands for bravery, as seen in the rampant silver lion on black for North Norway as a unit of defense (Bergersen 1992, 254).

Local meetings of "three tribes" in Finnmark county

Since the 1820s, Alta in western Finnmark and Vadsø in the east developed into urbanized strongholds for the Finnish-speaking population, the first in connection with new copper-mining enterprises and the second in connection with modernization of fishing and the fish industry. Because of the "large immigration of Finns *(finner)*, Quains *(kvener)*, in the 1800s," Vadsø was called "the capital of the Finns in Norway. In 1875, 62 percent of the population were Finnish-speaking" (NAF 2001, 629). However, the reindeer stag's head in Vadsø's arms underlines its administrative function as the county seat (see NKV 1987, 215) although the country is known through history as a Saami region. Before 1976 the commune arms had along with the stag's head a seagull as a second figure.

The silver spearhead on a blue ground in Alta's heraldic arms (1976) symbolizes prehistoric and modern mining. Alta's great era is represented by the Kåfjord Copperworks, starting in 1826, and its importance for Finnmark, accounting "for the Finnish component of Finnmark today" (NKV 1987, 211). The road guide mentions that Alta was a Saami settlement area until 1610, when Norwegians moved in from the outer coast, and that the "Quainish (Finnish) settlement" began with "immigration from the 1690s" (NAF 2001, 514). The Saami aspect is only alluded to through mention of the "Komsa culture," the early Fennoscandian Stone Age culture named after a local hill. Blue and white (silver) hint at Finland's national colors.

The bountiful nature of Unjárga-Nesseby

The Saami and Finnish components of local culture in Finnmark are overtly expressed or just implicitly read from the visual and verbal aspects of three out of four arms, drawn by Sveen, for communes bordering Finland, all with official Saami-Norwegian double names. Three have gold and red, and one gold and blue, a significant contrast to Finnish national colors. However, the Saami flag has dark blue, red, yellow, and green, colors approved by the Nordic Saami Conference held in Sweden in 1986 (Hætta 1994, 128).

The heraldic arms of Unjárga-Nesseby (1986) in inner Varanger Fjord are the exception from a multicultural heraldic presentation and do not allude to the Saami reindeer-herding tradition. The blazon is "in red an upright gold cloudberry plant," referring to the rich occurrence of cloudberries, historically an import export product. "The marshes are characteristic for the landscape picture" (NKV 1987, 213). The marshland was not suitable for settlers in the past. Nonetheless, this did not restrict winter transport with reindeer or on skis across national borders. In the road book, the commune describes itself as a "Saami core area with Finnish and Norwegian settlement in addition. Fishing, agriculture, reindeer-herding" (NAF 2001, 451, 506).

Ethnic interaction in Deatnu-Tana

The arms of the Deatnu-Tana commune (1984) have "in red three gold boats." The long elegant riverboats are the traditional means of transport along the Tana River. The commune wanted to emphasize "the interaction between the Saami, Finnish and the Norwegian groups of people who have populated the land along the waterway." As a border commune it "wanted to use the colors of Norway's national arms" (NKV 1987, 215). The Tana River is also Norway's richest salmon river, before 1751 contested by Denmark-Norway and Sweden(-Finland).

The Saami settlement of the river valley is emphasized in the road book, although the name of the river is given in three languages (NAF 2001, 451). The information about Finnmark's school of agriculture in Tana as "the northernmost in the world" (505) underlines the peripheral climatic aspect of agriculture, which was practiced extensively by Saamis. "Basically the old form of agriculture in Finnmark and North Troms was modelled on the Finnish form, introduced by immigrant Finns" (Vorren and Manker 1962, 151).

The meeting of "three tribes" in Kárášjohka-Karasjok

The heraldic arms of the Kárášjohka-Karasjok commune (1986) have "in red three five-tongued gold flames." Fire, whether in the tent, in the house, or in the open, has been a condition for survival on the Finnmark plateau, a place to gather around and a protection against dangers. "Three fires are chosen for decorative reasons" and refer to "the meeting of three tribes" (NKV 1987, 212). Only the metaphor for multiethnicity is expressed, not the concrete groups. The road book explains the "Saami name" *Kárášjohka* from "Finnish" *Kaarejoki,* meaning "bending river." The administrative center is "the capital of the Saamis," where the Saami parliament is located (NAF 2001, 449).

Guovdageaidnu-Kautokeino: A shelter in an exposed landscape

The blazon of Guovdageaidnu-Kautokeino (1987) is "in blue a gold *lavvo.*" The *lavvo* is a Saami tent. "This tent has of old given necessary shelter on the Finnmark mountain plateau, and is still used within reindeer-herding, wildlife economy and recreation" (Nissen 1990, 51). According to the road book, Guovdageaidnu-Kautokeino was the first commune in Norway to declare itself Saami with two equal languages in the local administration. It is striving to be "a Saami center for education, research and culture," and the present University College will probably be a future Nordic Saami University. The colors, yellow (gold) and blue, the national colors of Sweden, relate to the founding of Kautokeino by "the Swedes in 1701 when Kautokeino was Swedish land" (before the border treaty of 1751) (NAF 2001, 501).

Hammerfest—"No place further north"?

Hammerfest's status as "the world's northernmost town" is emphasized in the road guide, along with its reputation for having the best ice-free harbor in the northern waters. During World War II, the town was a naval base and was burned and leveled during the German retreat. The new museum of the history of war and reconstruction in Finnmark and North Troms is located here (NAF 2001, 545).

Hammerfest's heraldic arms have "a silver polar bear on red," expressing the importance of Arctic hunting for Hammerfest, and were granted in 1938 at the celebration of the town's 150th anniversary. The commune's official seal from 1889 had contained the profile of North Cape (Nordkapp), an Arctic hunting vessel, and the midnight sun, all denoting its northern position. The Latin inscription on the seal was *VINCIT INDUSTRIA HOMINUM NATURAM* (NKV 1987, 212), "the industry of humans conquers nature," or "the human is the master of nature."[6] The Italian scholar Francesco Negri (d. 1698) wrote after his visit in 1664: "Here I am at the North Cape, in the remotest part of Finnmark, and I may just as well describe it as the remotest part of the earth, since there is no place further north that is inhabited by mankind" (quoted in Skavhaug, 1990, 37).

North Cape does not in fact lie in Hammerfest commune, but in the commune renamed *Nordkapp* in 1950, with Honningsvåg as its administrative center, and the profile of North Cape in its arms (1973), drawn by Trætteberg. The blazon is "division of gold and red by a sinister step" (NKV 1987, 214). Honningsvåg presents itself as "the town at North Cape," avoiding competition with Hammerfest as "the world's northernmost town" (NAF 2001, 545, 551). It is also possible to read "the town at North

Cape" in binary opposition to "Cape Town" in South Africa. To be the "northernmost" in Norway or on the European continent just beyond 70°N is no longer as exotic as before, when tourists can travel in cruise ships to the north of Svalbard at 80°N.

Vardø—"The End of the World"

Vardø, the easternmost brim of Norway, has also been experienced as "the End of the World." In August 1873, the Scottish lawyer John Francis Campbell "coloured from nature" a picture with this title. He sketched whale bones on the beach "cast up by the tide, sticks for drying codfish and green grass and gray slate made the background of a strange weird desolate picture of the end of the world."[7] The heraldic arms with an inscription and a compound motif (a sun, fishing boats, and a cod) were drawn in the 1890s, and although not revised according to heraldic rules they are still accepted, even with a crown of bricks, and the text: *VARDØENSIS INSIGNIA URBIS. CEDANT TENEBRÆ SOLI* ("The town of Vardø's seal. The darkness shall give way to the sun"). While "the dominating motif is a rising sun," the cod refers to the great fisheries in Finnmark in the 1840s. Vardø was for a long time the biggest fishing station in Norway (NKV 1987, 216). The road book contains information about Norway's easternmost point, 31°10'14'E, as well as the prehistoric settlement, the church, the fortress from 1307, the great witch prosecutions and burning of women in the seventeenth century, trade contacts with northern Russia, and the damaging bombing of the town by the Soviet air force during World War II (NAF 2001, 630). Local Norwegian intellectuals were significant contacts for anti-czarist revolutionaries and Vardø became an important harbor for smuggling information and persons by the sea route. The rising sun in the east becomes a subtle sign transcending the symbolism of nature in a political-ideological context.

South Varanger: Meeting place and national border

South Varanger or Sør-Varanger commune has offered varied livelihoods: fishing, cattle husbandry, reindeer herding, industrial whaling, lumbering, and iron mining. In the past it provided the inhabitants on the northern side of Varanger Fjord with winter fodder and wood for construction and fuel. It was the land of the East Saamis before Norwegian, North Saami, and Finnish-speaking Lutherans settled there from the seventeenth century onward (Storå 1971; Ingold 1976; Andresen 1989; cf. Olsen 1996, 16–17). South Varanger belonged jointly to Norway and Russia before the two countries agreed upon the border in 1826. In 1858, Sør-Varanger became a separate commune, bordering Russia, except from 1920 to 1944, when it bordered Finland (Petsamo district). A wooden church was established in 1862 in Kirkenes ("church point"), the administrative center. A second church was established in 1869, a stone building at the mouth of the eastern border river to Russia, Grense Jakobselv (*grense* means "border"). In 1873, it was named "Oscar II's Chapel" after a visit by the Swedish-Norwegian king.

In Russia, Boris Gleb chapel was completed in 1874 on a sacred Orthodox site from 1565, south of Kirkenes, on the bank of another border river, Pasvikelv. When iron mining began in 1906, the government expressed its importance as "in truth a useful border citadel" (Bratrein and Niemi 1994, 191). This was soon after the 1905 revolution in Russia. The first trade union, *Nordens Klippe*, "the Rock of the North," conveyed a different message: "We Swedish, Finnish, Lappish and Norwegian workers . . . send to you, Russian brothers, our thanks and greetings" (quoted in Andresen 1994, 107).

The commune arms (1982) were designed by a local teacher, Sissel Sildnes. The blazon is "diagonal sections of gold and red with a three-tongued flame partition" (NKV 1987, 214). The heraldic arms were also presented in 1987, after Mikhail Gorbachev's "glasnost" of 1985, in a plan for a borderland *(grenseland)* museum. The three flames were said to symbolize: "The meeting of three tribes. The meeting of three countries. The flames are nourished by the same source, the three peoples are using the economic basis in common and have become a unity. The flames symbolize, too, the destruction of war in South Varanger, but also the active growth of the commune in the past and in the future." The red symbolizes "the workers' fight for their rights" (*Grenseland* 1987, 56). The heraldic jubilee book notes that "the commune has a threefold economic base—agriculture, mining and fishing, and three border rivers—Grense Jakobselv, the Pasvik and Neiden Rivers." Here are "three countries with mutual borders," and "three population elements, Norwegian, Saami and Finnish." For a border commune, "it was natural to chose the same colours as in the national arms" (NKV 1987, 214).

While the national text marks the peripheral national border, through red and gold (yellow), the local text alludes to socialism, indirectly to the Soviet Union. While the local text uses only the metaphor "the meeting of three tribes," the national heraldic text explicitly refers to three ethnic groups, the "Norwegian, Saami and Finnish." So does the "Cultural History of North Norway" in 1994, in the caption of an illustration of Sør-Varanger's heraldic arms: "the three flames" are "symbolizing the meeting between Norwegian, Saami and Finnish" (Drivenes et al. 1994b, 11). Since the end of the Cold War and the dissolution of the Soviet Union in 1991, Russian immigration has increased the population of Kirkenes. In addition to transnational economic relations from the late 1980s and courses in Russian for local bureaucrats and private businesspeople, examples of other integrating factors are Russian books in the public library and the revitalization of the Russian Orthodox Church. Even street signs in Kirkenes appear in Russian letters. When Kirkenes adopted town status in 1998, the commune's information about the heraldic symbolism was "3 tribes' meeting; 3 countries' meeting—Norway, Finland and Russia; 3 main economic activities—agriculture, mining and fishing; 3 border rivers—Grense Jakobselv, Pasvikelva and Neidenelva."[8] None of the local texts specify who the three ethnic groups, or "tribes," are.

Vardøhus Castle in the county arms of Finnmark connotes the periphery of the national territory. The aesthetic contrast between gold and black seems secondary to the symbolic contrast between light and dark, royal sunshine and hidden powers of the dark. The local gold and red in Sør-Varanger also alludes to national territory. However, while cultural plurality is not signified in the county arms, the commune arms of Sør-Varanger denote connections, inclusiveness, plurality, and transcendence of national borders. The heraldic flames signify movement and dynamism. In the road guide, the administrative center of Kirkenes is also "in the center of the Barents Region," located "70°N and 30°E—as far east as Istanbul and Cairo, and as far north as Point Barrow in Alaska." It was in the "borderland" of Sør-Varanger that "the meeting of roads from east, south and west, and people, languages and cultures developed" (NAF 2001, 556).

THE LANDSCAPE IN THE SIGNS

Identification with the physical landscape and hard climatic conditions seems to be a common local attitude in North Norway. The sea, the mountainous topography, the barren coast, the sheltered fjords and valleys, fish, birdlife, and marginal vegetation are also concrete experiences for inland dwellers because of the economic importance of fishing. Thus frequently they are represented in heraldic figures and colors. The geography of Nordland is unique in Norway in terms of coastline and seaway communication (Alstahaug, Moskenes, and Nordland county).

Astronomical phenomena rather than seasons seem to structure annual cycles in North Norway, especially the sun as a heraldic figure, while winter darkness is expressed through verbal explanations of content of figures and colors (Bodø and Vardø; cf. Bø and Nordkapp). The physiographic periphery is presented either as the margin of the inhabited world (Karlsøy) or as the gate into polar adventures and hardships (Hammerfest). The political periphery is defined through heraldic figures and symbolism rooted in history (Finnmark and Troms counties, Bjarkøy). Cultural or ethnic duality rather than cultural plurality prevails in narrative information, in practice, in Norwegian and Saami. In the last few decades, Saami place-names as verbal signs have also become more common outside Finnmark. Nonetheless, heraldic signs depicting ethnic markers, such as the Saami tent (Guovdageaidnu-Kautokeino) and the Saami and Finnish riverboats (Deatnu-Tana), are still exceptions. The reindeer is both a Saami sign and a central-administrative sign (Tromsø and Vadsø).

TERRITORIAL ARMS OR COMMERCIAL LOGOS?

Local territorial symbols are expressions of dynamics between levels of administration and between state, county, and commune. The county authorities are the intermediate level between the local and central authorities, but direct links exist between the local and the central, for example, in primary education. The county, on the one hand, conveys national policy through the county governor, appointed by the government. On the other hand, the county is intended as an instrument for coordinating local interests through the elected county board. The Local Government Act of 1837 was based on the political ideal that local people should administer and govern themselves, because all elected representatives were in certain respects experts on local matters (Offerdal 1991, 250). The construction of the social-democratic welfare state after World War II involved the construction of the "welfare commune," building schools, health centers, and hospitals. The increased transfer of public funds from state to county and commune levels is an indicator of local public growth. The number of local employees has doubled since the 1960s on commune and county levels, while the increase has been comparatively small on the state level. Larger administrative staff and wider professional expertise were needed for economic and physical planning, which became the responsibility of the communes in the 1980s (Naustdalslid 1991, 27f.).

In the 1990s, the "ideal of lay governing" *(lekmannsidealet)* was challenged by economic "management" thinking, in which citizens are considered adaptive consumers instead of active agents (Hagen and Hovik 1991, 271). Short-term efficiency may lead to long-term negative consequences for public services, which were never thought of as commercial enterprises, but as social and political rights of civil citizens. In a welfare perspective, these rights included sufficient health care, equal opportunity for education, neutral legal procedures, democratic organization of general elections, and personal safety in public space and in traffic on land and at sea. Even military defense is among expected common goods. "Lay governing" represents political engagement, civil justice, equal opportunities, and even local identity and cultural-heritage issues. Through the twentieth century, open-air museums, elements in nation-building, have contributed to infrastructural changes in the landscape. A driving force for museum curators has been the idea of enlightenment, and "the local and popular anchoring of the many cultural-historical museums made the shaping of a local identity as equally important as the supporting of a common, national, cultural identity" (NOU 1996, 13).

The visual effect of territorial arms as landscape elements is observed only in some contexts thus far. Is the increasing visualization of local symbols a continuous articulation of local political resistance against central decision-making? Or is it just a result of trademark protection of the commune in management thinking? In the 1990s, the political-administrative aim to grant arms to all local units was attained within the state hierarchy as a territorial system. One political philosopher found heraldry uninteresting, without any social or judicial meaning, except as national and corporative symbols. Heraldry involves few if any privileges or duties, "just private ideals, social ambitions and cultural traditions," perhaps interesting for sociologists, psychologists, or historians; for the political philosopher, heraldry is just "a nice, interesting and private hobby" (Kurrild-Klitgaard 1996, 116). The intriguing question is whether this concrete articulation of visual signs combined with verbal information has any political significance. What will be the effect of official bureaucratic language when territorial symbols are denoted "logos" *(kommune-logo)* and not "arms" *(komune-våpen)*? This seemingly unimportant structural change in social or political terms reflects the tensions between defending local values and market-oriented management thinking.

Notes

All quotations from works other than in English were translated by the author.

1. Thanks are due Harald Nissen at the University Library, Trondheim, who introduced me to heraldic literature on Norwegian commune arms and kindly answered my questions; to Michael Jones, Department of Geography, Trondheim, who has commented on my manuscript while patiently correcting my English; and to Håvard Dahl Bratrein, Tromsø University Museum, for his comments on North Norwegian issues.

2. An example of ethnic cultural heritage is the coat of arms on Lerwick Town Hall, Shetland, containing various Viking motifs alongside depictions of Norse earls and Norwegian kings, with connotations to the Norse system of property rights, as opposed to the historical feudal land-tenure system of Scotland (Jones 1996, 6).

3. The International Labor Organization's Convention Concerning Indigenous and Tribal Peoples in Independent Countries (ILO Convention, no. 169), ratified by Norway in 1990 and in force in 1991.

4. The blazon is set between quotation marks when the sources are NKV 1987 and Nissen 1990, 1991, 1993, 1998.

5. For ease of reading the common English names of the tinctures are used in the remainder of this chapter instead of the precise heraldic terms. Similarly, heraldic syntax is not always followed.

6. Translation from Latin to Norwegian by Marek Thue Kretschmer, Department of History, Trondheim.

7. From Campbell's travel journals in the National Library of Scotland, Adv. MSS 50.4.8; cf. Olsen 2003.

8. Letter of September 11, 2001, from E. Pettersen, Sør-Varanger commune.

References

Aasvang, Arnt O., and Leif B. Lillegaard. 1976. "En folkeskatt." In *Bygd og by i Norge: Nordland*, ed. Leif B. Lillegaard, 372–92. Oslo: Gyldendal.

Alexandri, Alexandra. 1997 [1995]. "The Origins of Meaning." In *Interpreting Archaeology: Finding Meaning in the Past*, ed. Ian Hodder et al., 59–67. New York: Routledge.

Allardt, Erik. 1979. *Implications of the Ethnic Revival in Modern, Industrialized Society: A Comparative Study of the Linguistic Minorities in Western Europe*. Commentationes scientiarum socialium 12. Helsinki: Societas Scientiarum Fennica.

Andresen, Astri. 1989. *Sii'daen som forsvant: Østsamene i Pasvik etter den norsk-russiske grensetrekningen i 1826*. Kirkenes: Sør-Varanger Museum.

———. 1994. "Finnes det byer i Nord-Norge?" In *Nordnorsk kulturhistorie*, ed. Einar-Arne Drivenes, Marit Anne Hauan, and Helge A. Wold, 2: 100–109. Oslo: Gyldendal.

Anttonen, Marjut. 1999. *Etnopolitiikkaa Ruijassa:* Suomalaisen Kirjallisuuden Seuran Toimituksia 764. Helsinki: Suomalaisen Kirjallisuuden Seura.

Arbo, Peter. 1997. "Alternative Nord-Norge-bilder." *Nytt Norsk Tidsskrift* 14, no. 4: 310–24.

———. 1998. "Falmende Nord-Norge-bilder." *Nytt Norsk Tidsskrift* 15, no. 2: 180–84.

Balsvik, Randi Rønning, and Siri Gerrard, eds. 1999. *Global Coasts: Life Changes, Gender Challenges*. Kvinnforsk, Occasional Papers. Tromsø: University of Tromsø.

Barth, Fredrik. 1996 [1994]. *Manifestasjon og prosess*. Oslo: Universitetsforlaget.

Bergersen, Thorbjørn. 1992. "Den norske hærs avdelingsmerker." *Heraldisk Tidsskrift* 66: 241–56.

Berggrav, Eivind. 1937. *Spenningens land: Visitas-glimt fra Nord-Norge*. Oslo: Aschehoug.

Berggreen, Brit. 1989. *Da Kulturen kom til Norge*. Oslo: Aschehoug.

Bjørgo, Narve, ed. 1990. *Nybrott og gjenreisning: Ny kunnskapspolitikk for Nord-Norge*. Oslo: NAVF/ Norges allmennvitenskapelige forskningsråd.

Bjorvand, Harald, and Fredrik Otto Lindeman. 2000. *Våre arveord: Etymologisk ordbok*. Instituttet for sammenlignende kulturforskning, Serie B: Skrifter CV. Oslo: Novus Forlag.

Bratrein, Håvard Dahl. 1994–95. "Det nordnorske jordbruket—noen generelle trekk." *Jord og gjerning. Årbok for Norsk landbruksmuseum 1994–1995*: 7–24.

———. 2000. "Tre stammers møte i nordnorsk landbruk: Etniske spor i agronomi og kulturlandskap." In *Norge landbrukshistorie til år 2000*, ed. Reidar Almås and Brynjulf Gjerdåker, 73–86. Rapport 2000: 15. Trondheim: Senter for bygdeforskning.

———. 2001. "Adelsgods og krongods i Finnmark." *Håløygminne* 21, no. 3: 57–84.

Bratrein, Håvard Dahl, and Einar Niemi. 1994. "Inn i riket: Politisk og økonomisk integrasjon gjennom tusen år." In *Nordnorsk kulturhistorie*, ed. Einar-Arne Drivenes, Marit Anne Hauan, and Helge A. Wold, 1: 146–209. Oslo: Gyldendal.

Brox, Ottar. 1982. "Fem forsøk på å planlegge Nord-Norge." In *Planleggingens muligheter*, ed. Noralv Veggeland, 2: 13–44. Oslo: Universitetsforlaget.

———. 1995. "Nordnorsk kulturhistorie." *Heimen* 32, no. 4: 283–86.

———. 1998. "Peter N. Arbos Nordlandsbilder." *Nytt Norsk Tidsskrift* 15, no. 1: 91–93.

Cappelen, Hans. 1992. "Militærheraldikeren og avdelingsmerkene." *Heraldisk Tidsskrift* 66: 237–40.

Dass, Petter. 1997. *Samlede Verker: Nordlands Trompet. Leilighetsdiktning 1–3*, 2nd edition, ed. Kjell Heggelund, Inge Apenes, and Karl Erik Harr. Oslo: Gyldendal.

Davis, Dona Lee. 1999. "Changing Constructions of Gender and Community in a Newfoundland Fishing Village." In *Global Coasts: Life Changes, Gender Challenges*, ed. Balsvik and Gerrard, 145–52. Kvinnforsk, Occasional Papers. Tromsø: University of Tromsø.

Drivenes, Einar-Arne, Marit Anne Hauan, and Helge A. Wold, eds. 1994a. *Nordnorsk kulturhistorie*, 2 vols. Oslo: Gyldendal.

Drivenes, Einar-Arne, Marit Anne Hauan, Einar Niemi, and Helge A. Wold. 1994b. "Den besværlige identiteten." In *Nordnorsk kulturhistorie*, ed. Einar-Arne Drivenes, Marit Anne Hauan, and Helge A. Wold, 1: 8–17. Oslo: Gyldendal.

Edvardsen, Edmund. 1997. *Nordlendingen*. Oslo: Pax.

Engelskjøn, Ragnhild. 1999. "Local Story—Global Meaning." In *Global Coasts: Life Changes, Gender Challenges*, ed. Balsvik and Gerrard, 164–72. Kvinnforsk, Occasional Papers. Tromsø: University of Tromsø.

Eriksen, Helge. 1976. "Lofoten." In *Bygd og by i Norge: Nordland*, ed. Leif B. Lillegaard, 73–81. Oslo: Gyldendal.

Escarpit, Robert. 1972. *Bogen og læseren: Udkast til en litteratursociologi*. Copenhagen: Hans Reitzel.

Fladby, Rolf, Steinar Imsen, and Harald Winge, ed. 1974. *Norsk historisk leksikon: Næringsliv, rettsvesen, administrasjon, mynt, mål og vekt, militære forhold, byggeskikk m.m. 1500–1850*. Oslo: Cappelen.

Fulsås, Narve. 1999. "Forsøk på å gripe 'den gjenstridige nordlendingen': Bilete av Nord-Norge i dei siste 30 åras vitskaplege litteratur." In *Landskap, region og identitet: Debatter om det nordnorske*, ed. Trond Thuen, *Kulturstudier*, 3: 83–98. Bergen: Norges Forskningsråd.

Grenseland: Et internasjonalt reiselivsprodukt. 1987. Kirkenes: Sør-Varanger kommune.

Gullvåg, Ingemund O., ed. 1972. *Charles Sanders Peirce: Utvalg og innledning*. Oslo: Pax.

———. 1982. "Tegn, mening, forståelse: Grunnlagsproblemer i humanistisk forskning." *Norsk Filosofisk Tidsskrift* 17: 149–78.

Hætta, Odd Mathis. 1994. *Samene: Historie, kultur, samfunn*. Oslo: Grøndahl Dreyer.

Hagen, Terje P., and Sissel Hovik. 1991. "Mot det *andre* kommunale hamskiftet?" In *Kommunal styring: Innføring i kommunalkunnskap frå ein planleggingssynstad*, 3rd edition, ed. Jon Naustdalslid, 265–73. Oslo: Det Norske Samlaget.

Hagland, Jan Ragnar, and Jørn Sandnes. 1994. *Frostatingslova*. Oslo: Det Norske Samlaget.

Helle, Knut. 1993. "Norway." *Medieval Scandinavia: An Encyclopedia*, ed. Phillip Pulsiano, 436–40. New York: Garland.

Ibsen, Henrik. 1867. *Peer Gynt: Et dramatisk dikt*. Copenhagen: Gyldendalske boghandel.

Ingold, Tim. 1976. *The Skolt Lapps Today*. London: Cambridge University Press.

Jakobsen, Anton. 1961. "Sealing and Whaling." In *Norway North of 65*, ed. Ørnulf Vorren, 209–15. Oslo: Oslo University Press; London: Allen and Unwin.

Jean, Georges. 1998. *Signs, Symbols, and Ciphers: Decoding the Message*. London: Thames and Hudson.

Jentoft, Svein, ed. 1998. *Commons in a Cold Climate: Coastal Fisheries and Reindeer Pastoralism in North Norway: The Co-Management Approach*. Man and the Biosphere series 22. Paris: UNESCO; New York: Parthenon.

Jokipii, Mauno. 1982. "Finsk bosetning I Nord-Norge—historiske hovedlinjer." In *Suomalaiset Jäämeran rannoilla—Finnene ved Nordishavets strender*. Turku: Migrationsinstitutet–Sirtolaisuusinstituutti.

Jones, Michael. 1996. "Scots and Norse in the Landscape of Orkney and Shetland: Visible Landscape and Mental Landscape." In *Proceedings of the Permanent European Conference for the Study of the Rural Landscape. Papers from the 17th Session, Trinity College, Dublin, 1996*, ed. Frederick H. A. Aalen and Mark Hennessy, 4–10. Dublin: Department of Geography, Trinity College, University of Dublin.

———. 1999. "Perspektiver på landskap og hvordan det kan anvendes i sørsamisk sammenheng." *Seminar om sørsamisk forskning og undervisning, Tromsø, 8.–9. April 1999*. Senter for samiske studier, Universitetet i Tromsø. http://www.uit.no/ssweb/dok/seminar/sorsamisk/JONES.html.

Julku, Kyösti. 1986. *Kvenland–Kainuunmaa*. Studia Historica Septentrionalis 11. Oulu: Kustannusosakeyhtiö Pohjoinen.

Kurrild-Klitgaard, Peter. 1996. "Magt og symboler: Thomas Hobbes (1588–1679) som heraldiker." *Heraldisk Tidsskrift* 73: 111–17.

Lange, Anders, and Charles Westin. 1981. *Etnisk diskriminering och social identitet. Forskningsöversikt och teoretisk analys. En rapport från Diskrimineringsutredningen*. Stockholm: Liber Förlag.

Laureys, Godelieve. 1992. "Hyperboréernas land." In *Frihetens källa: Nordens betydelse för Europa*, ed. Svenolof Karlsson, 211–17. Stockholm: Nordiska Rådet.

Lillegaard, Leif B. 1976. "Møte med samefolket." *Bygd og by i Norge: Nordland*, ed. Leif B. Lillegaard, 337–49. Oslo: Gyldendal.

Lotman, Ju. M. 1988a [1967]. "Problems in the Typology of Culture." In *Soviet Semiotics: An Anthology*, ed. Lucid, 213–21. Baltimore: Johns Hopkins University Press.

———. 1988b [1970]. "Two Models of Communication." In *Soviet Semiotics: An Anthology*, ed. Lucid, 99–101. Baltimore: Johns Hopkins University Press.

Lucid, Daniel P., ed. 1988. *Soviet Semiotics: An Anthology*. Baltimore: Johns Hopkins University Press.

Mead, W. R. 1974. *The Scandinavian Northlands:* Problem Regions of Europe. Oxford: Oxford University Press.

Mingroot, Erik van, and Eduard van Ermen. 1988 [1987]. *Norge og Norden i gamle kart*. Oslo: Aschehoug.

Mykland, Knut, Torkel Opsahl, and Guttorm Hansen. 1989. *Norges grunnlov i 175 år*. Oslo: Gyldendal.

NAF. 1992. *NAF Veibok 1992*, ed. Erling Storrusten. Oslo: Norges Automobil-Forbund.

———. 1995. *NAF Veibok 95*, ed. Knut Evensen. Oslo: Norges Automobil-Forbund.

———. 1998. *NAF Veibok 98*, ed. Knut Evensen. Oslo: Norges Automobil-Forbund.

———. 2001. *NAF Veibok 2001*, ed. Knut Evensen. Oslo: Norges Automobil-Forbund.

Nairn, Tom. 1981 [1977]. *The Break-Up of Britain: Crisis and Neo-Nationalism*. London: Verso.

Naustdalslid, Jon. 1991. "Lokalstyret i historisk perspektiv." *Kommunal styring. Innføring i kommunalkunnskap frå ein planleggingssynstad*, 3rd edition, ed. Jon Naustdalslid, 21–41. Oslo: Det Norske Samlaget.

Niemi, E., O. Brox, O. M. Hætta, K. Jakobsen, and H. Kr. Eriksen. 1976. *Trekk fra Nord-Norges historie*. Oslo: Gyldendal.

Nilssen, Arne C., ed. 1989. "Barentshavet." *Ottar* 174.

Nissen, Harald. 1990. "Nye kommunevåbener i Norden." *Heraldisk Tidsskrift*, 62: 41–55.

———. 1991. "Nye kommunevåpen i Norge." *Heraldisk Tidsskrift* 64: 141–52.

———. 1993. "Nye kommunevåbener i Norden." *Heraldisk Tidsskrift* 68: 333–43.

———. 1995. "Norwegian Heraldry." *Heraldry in Canada* 29, no. 1: 4–11, no. 2, 4–11.

———. 1998. "Nye norske kommunevåpen." *Heraldisk Tidsskrift* 78: 77–89.

NKV. 1987. *Norske Kommunevåpen*, ed. Hans Cappelen and Knut Johannessen. Oslo: Kommunalforlaget.

Nöth, Winfried. 1990. *Handbook of Semiotics*. Bloomington: Indiana University Press.

NOU. 1984. *Om sameness rettsstilling*. Norges offentlige utredninger 1984: 18. Oslo–Bergen–Tromsø: Universitetsforlaget.

———. 1996. *Museum: Mangfald, minne, møtestad*. Norges offentlige utredninger 1996: 7. Oslo: Kulturdepartementet.

Offerdal, Audun. 1991. "Kommunalpolitikaren—rolleforventning og røyndom." In *Kommunal styring: Innføring i kommunalkunnskap frå ein planleggingssynstad*, 3rd edition, ed. Jon Naustdalslid, 249–63. Oslo: Det Norske Samlaget.

Olaus Magnus. 2001 [1555]. *Historia om de nordiska folken*. Hedermora: Gidlunds.

Olsen, Venke. 1982. *Hva er grunnlaget for bevaring og utvikling av finsk kultur I Nord-Norge? Et forsøk på situasjonsanalyse av finskættede i Finnmark og Nord-Troms som etnolingvistisk minoritet*. Arbeidsrapport fra prosjektet Finsk kulturforskning i Nord-Norge. Tromsø: Tromsø Museum/IMV, Universitetet i Tromsø/Norges almenvitenskapelige forskningsråd.

———. 1984. "Samuli Paulaharju i Nord-Norge: Bildene og mennesket bak." *Ottar* 149: 28–41.

———. 1985. *Inngruppe- og utgruppenavn i kommunikasjon mellom etniskegrupper. En teoretisk tilnærming til etnologisk analyse av kulturelle former*. Arbeidsrapport nr. 2 fra prosjektet Finsk kulturforskning i Nord-Norge. Tromsø: Tromsø Museum/ IMV, Universitetet i Tromsø/ Norges allmennvitenskapelige forskningsråd/NAVF.

———. 1986. "Northern Scandinavia: A Multi-Ethnic Society. Seen from an Ethnological Point of View." *Northern Studies* 23: 31–74.

———. 1989. "Med penn, blyant og kamera. Den finlandske kultur-skildreren Samuli Paulaharju." In *Bildet lever! Bidrag til norsk fotohistorie* 5, ed. Roger Erlandsen and Kåre Olsen, 65–81. Oslo: Norsk Fotohistorisk Forening/C. Huitfeldt Forlag.

Olsen, Venke Åsheim. 1992a. "Minority or Immigrants? A Survey of the Situation of the Finnish Population in North Norway." In *Ethnic Life and Minority Cultures,* ed. Karin Borevi and Ingvar Svanberg, 15–40. Uppsala Multiethnic Papers 28. Uppsala: Centre for Multiethnic Research, Uppsala University.

———. 1992b. "Etnisitet i tid og rom i en flerkulturell region: En kulturanalytisk tilnærming til sosial identitet." In *Kjønn, kultur og regional endring. Nye perspektiver i distriktskvinneforskning. Rapport fra seminar i Alta 21.–23. november 1990,* 64–83. Faggruppa for distriktskvinneforskning og -utvikling, ALH-forskning 1992:3 / FDH-rapport 1992:6. Alta.

———. 1993a. "Smørkven eller finsk fjøskjerring? Etniske stereotyper og bakgrunnen til kvinner som kom fra Tornedalen og Nord-Finland midt på 1800-tallet slik en reisende fra Skottland beskrev kulturmiljø og mennesker han møtte underveis." *Kvenske kvinners situasjon i et likestillingsperspektiv: Rapport fra seminar i Alta 2.–3. nov. 1991,* 17–28. Børselv: Norske Kveners Forbund/Finnmarks distriktshøgskole, Avd. for finsk.

———. 1993b. "Ethnonyms as Signs in Multicultural Societies: An Ethnological Approach." Paper presented at ISI Conference *Strangers, Eccentrics, Outsiders,* International Semiotics Institute, Imatra, Finland, July 12–16, 1993. Manuscript.

———. 1995. "Språket som kulturvern. Den finskspråklige minoriteten i Nord-Norge." In *Minoritetsspråk i Norden: En rapport från seminariet Tala eller tiga i Norden, Mariehamn den 15–16 oktober 1994,* 62–90. Meddelanden från Ålands högskola 5. Mariehamn: Ålands högskola/Ålands fredsinstitut.

———. 1996. "The Steambath on the Arctic Coast: The Finnish Sauna on the Coast of North Norway." In *Proceedings of the Permanent European Conference for the Study of the Rural Landscape. Papers from the 17th Session, Trinity College, Dublin, 1996,* ed. Frederick H. A. Aalen and Mark Hennessy, 11–18. Dublin: Department of Geography, Trinity College, University of Dublin.

———. 1997. "Where the West Ends. North Norway: The Borderland to Russia and Finland." In *The Dividing Line: Borders and National Peripheries,* ed. Lars-Folke Landgren and Maunu Häyrynen, 173–87. Renvall Institute Publications 9. Helsinki: Renvall Institute for Area and Cultural Studies, University of Helsinki.

———. 1999. "North Norway: A Multi-Ethnic Landscape." *Shaping the Land, Vol. II: The Role of Landscape in the Constitution of National and Regional Identity. Proceedings of the Permanent European Conference for the Study of the Rural Landscape, 18th Session in Røros and Trondheim, Norway, September 7th–11th 1998,* ed. Gunhild Setten, Randi Torvik, and Terje Semb, 475–88. Papers from the Department of Geography, University of Trondheim, new series A 27. Trondheim.

———. 2000. "Communication in Borderlands: An Essay on Cultural Story-Telling." In *Telling, Remembering, Interpreting, Guessing: A Festschrift for Prof. Annikki Kaivola-Bregenhøj on Her 60th Birthday 1st February 1999,* ed. Maria Vasenkari, Pasi Enges, and Anna-Leena Siikala, 469–76. Kultaneito III, Scripta Aboensia 1, Studies in Folkloristics. Joensuu: Suomen Kansantietouden Tutkijain Seura.

———. 2003. "A Highlander in the Wild: A Scotsman's Search for Origins in Northern Fennoscandia." *European Landscapes: From Mountain to Sea. Proceedings of 19th Session of the Permanent European Conference for the Study of the Rural Landscape (PECSRL) at London and Aberystwyth (UK), 10–17 September 2000,* ed. Tim Unwin and Theo Spek, 210–23. Tallinn: Huma Publishers.

Opsahl, Torkel. 1993. "Forfatningsretten." *Knophs Oversikt over Norges Rett,* 10th ed., 54–103. Oslo: Universitetsforlaget.

Reymert, Per Kyrre, ed. 1979. "Svalbard, det nordligste Norge." *Ottar* 110–12.

Riksarkivet. 1998. *Kunst med kongelig resolusjon: Kommunevåpen i Riksarkivet. Utstilling 23.10.1998–31.01. 1999.* http://www.riksarkivet.no/kommunevaapen/h_linjer.html.

Rose, Gillian. 1995. "Place and Identity: A Sense of Place." In *A Place in the World? Places, Cultures, and Globalization,* ed. Doreen Massey and Pat Jess, 87–118. The Shape of the World: Explorations in Human Geography 4. Oxford: The Open University/Oxford University Press.

Saressalo, Lassi. 1996. *Kveenit. Tutkimus erään pohjoisnorjalaisen vähemmistön identiteetistä.* Suomalaisen Kirjallisuuden Seuran Toimituksia 638. Helsinki: Suomalaisen Kirjallisuuden Seura.

Schancke, A. [Schanche, A.] 1987. "Det samiske landskap—er det kultur og landskap eller kultur i landskap?" *Fortidsvern* 13, no. 3: 17–18.

Schøyen, Carl. 1918. *Tre stammers mote: Av Skouluk-Andaras beretninger.* Kristiania: Gyldendal.

Sebeok, T. A., ed. 1994. *Encyclopedic Dictionary of Semiotics,* 3 vols. New York: Mouton de Gruyter.

Siikala, Anna-Leena. 1992. "Understanding Narratives of the 'Other.'" In *Folklore Processed. In Honour of Lauri Honko on His 60th Birthday 6th March 1992,* ed. Reimund Kvideland et al., 200–213. NIF publications 24/Studia Fennica Folkloristica I. Helsinki: Suomalaisen Kirjallisuuden Seura.

Skavhaug, Kjersti. 1990. *The North Cape: Famous Voyages from the Time of the Vikings to 1800.* Honningsvåg: Nordkapplitteratur/Nordkappmuseet.

Slagstad, Rune. 1998. *De nasjonale strateger.* Oslo: Pax.

Sonesson, Göran. 1998. *Den visuella semiotikens system och historia.* Umeå University. http://www.uk.umea.se/tegro/tegc/kkproj/semiotikk/Goran.html.

SSB. 1995. *Historisk statistikk 1994–Historical Statistics 1994.* Norges offisielle statistikk–Official Statistics of Norway C188. Oslo and Kongsvinger: Statistisk Sentralbyrå–Statistics Norway.

———. 2000. *Statitisk årbok 2000.* Norges offisielle statistikk C600. Oslo and Kongsvinger: Statistisk Sentralbyrå–Statistics Norway.

Storå, Nils. 1971. *Burial Customs of the Skolt Lapps.* FF Communications 210. Helsinki: Academia Scientiarum Fennica.

Toorop, Peeter. 1998. "Semiotics in Tartu." *Sign System Studies* 2: 9–19.

Vahtola, Jouko. 1993. "Finland." In *Medieval Scandinavia: An Encyclopedia,* ed. Phillip Pulsiano, 188–94. New York: Garland.

Vorren, Ørnulf, and Ernst Manker. 1962. *Lapp Life and Customs: A Survey.* Oxford: Oxford University Press.

Weider, Bjarne O. 1976. "Kirkeliv i Sør-Hålogaland." *Bygd og by i Norge: Nordland,* ed. Leif B. Lillegaard, 350–71. Oslo: Gyldendal.

Westin, Charles, ed. 1999. *Mångfald, integretion, rasism och andra ord. Ett lexikon över begrepp inom IMER—International Migration och Etniska Relationer.* SOS-Rapport 1999: 6. Stockholm: Socialstyrelsen.

Winge, Harald. 1995. "Bokmelding." *Dugnad* 21, no. 4: 61–69.

Wold, Helge A. 1991. *I paradisets første krets: Om drømmen om ære og rikdom, om et grusomt forlis, om et opphold på Røst i Lofoten 1432. Italieneren Pietro Querinis egen beretning i ny oversettelse, med et tillegg om en reise til Røst i vår tid for å lete etter det tapte paradiset.* Oslo: Cappelen.

Zavalloni, Marisa. 1975. "Social Identity and the Recoding of Reality: Its Relevance for Cross-Cultural Psychology." *International Journal of Psychology* 10, no. 3: 197–217.

13.
Changes in the Land and the Regional Identity of Western Norway: The Case of Sandhåland, Karmøy

ANDERS LUNDBERG

Western Norway, with its dramatic contrasts between mountain and fjord, has traditionally been seen to derive its regional identity from the physical landscape. An alternative, historical approach pays more attention to the role of human activity in shaping the area. Although the landscape superficially appears to reflect an age-old adaptation by society to nature, the region has actually undergone a series of major transformations, most recently since 1950. This study documents four major transformations of this dynamic society as manifested in an ever-changing landscape in a region that, paradoxically, derives much of its identity from its status as a symbol of Norwegian rural continuity and ecological stability. The landscapes of this area are of great importance, both in ecological terms and as an aesthetic amenity of some commercial value. The contradictions between the perception of natural stability and the dynamics of local land use must be resolved if ecological and amenity values are to be sustained.

Concepts of Region

A broad definition of region might be an area on planet Earth with one or more characteristic features. These features might be natural, the result of human activity or a combination of both. A region has a distinctive unity and differs from surrounding areas. Hence a region can be defined in cultural, economic, political, natural, morphological, or physiographic terms. Different scientific disciplines have used the concept of region in their own ways, often without paying attention to corresponding concepts within related disciplines, such as vegetation science, physical geography, and human geography. Misunderstandings and other problems can arise if the concept is used without caution, in particular if and when different disciplines take part in the same debate.

The Nordic Council of Ministers (1977) has applied the concept of region to subdivide the Nordic countries into regions for nature conservation purposes. The council suggested a system of regionalization based on biophysiographic criteria, particularly vegetation zones. The fact that vegetation is considered in relation to on-site environmental conditions was used as an argument for this decision. Superior zones of vegetation were subdivided according to differences in types of vegetation, shaped by climate, soil properties, and land use. According to the authors, the criteria used did not fully correspond to a strict scientific approach, but the suggested regional classification was promoted as a useful instrument for nature conservation. The initial criteria had to be modified and other motives considered. First, the coastal districts of western Norway needed to be separated as individual regions. Second, the altitudinal zonation of fjord districts in western Norway did not allow the use of vegetation regions because these narrow zones do not qualify as regions. Hence the complex of vegetation types from sea level to higher, alpine altitudes was considered as one unit (region). Third, in large parts of the Nordic countries, seminatural vegetation has been replaced by cultivation and other land uses, and vegetation criteria cannot be used as in natural and seminatural landscapes (70 percent of Denmark is open agricultural land). As a result, the Nordic countries were split into sixty regions, some of which were further subdivided. A revised edition was presented some years later (Nordic Council of Ministers 1984).

The Nordic Council of Ministers' reports were discussed among botanists and other scientists, and the system was further developed by Asbjørn Moen (1998) in the *National Atlas of Norway*. Here the concept of region is used for vegetational units with a regional distribution influenced by climate. The concept is also used to refer to other areas with internally similar physical environments, such as landscape regions. The major difference between Moen (1998) and the Nordic Council of Ministers (1977, 1984) is that Moen defines regions only in terms of vegetation. Perhaps this situation is functional from a purely botanical point of view, but problems may arise. Most parts of the country have been classified (regionalized) according to their potential natural vegetation. Areas less influenced by technical intervention, such as alpine areas, may correspond fairly well to the actual vegetation, but in areas heavily influenced by human activity the actual vegetation (meadows, pastures, moors, eutrophic lakes, salt marshes, and other types of seminatural vegetation) is different from the potential vegetation. In the coastal districts of western Norway, primeval forests have long since been cut and replaced by heath and moor, which has caused a radical change in species composition as well as vegetation structure. These regions have been classified according to its "actual" vegetation rather than its potential vegetation. The criteria used to define (regionalize) the coastal districts of western Norway thus differ from the criteria used for the rest of the country.

A closer look at the criteria used to identify coastal districts of western Norway reveals that the region is characterized by heathland and moor, types of vegetation that were widely distributed until the beginning of the twentieth century. Since then, extensive plantations have been introduced, the traditional use of heathland for grazing has declined significantly, and the production of peat from moors has come to a complete stop. In other words, changes in land use have considerably altered the vegetation structure of the coastal districts of western Norway. Because of the time lag between actual land-use change and vegetation response, this change has become even more apparent in recent decades. This is why I use quotation marks in "actual" vegetation above.

The pioneering work done by Nordic Council of Ministers (1977, 1984) was an important step in identifying typical and representative biophysical regions in the Nordic countries. The task now is to advance this process of detecting regional identities. It could be argued that the criteria used to characterize the coastal districts of western Norway are not in accordance with the report's own presuppositions—and if they were, the definition of region, purely based on botanical criteria, would be static and hardly capable of including present and future environmental change. Many types of vegetation in coastal districts of western Norway are now rapidly changing. Traditional types (heath and moor) are declining, while plantations and pastures dominated by grass are expanding. Thus it is impossible to understand from a purely botanical point of view how to detect and understand the human forces that trigger these processes of vegetation and environmental change. Vegetation no doubt is a characteristic feature of any region. Any effort to define regions dominated by seminatural vegetation should include vegetation and human dimensions relevant for the maintenance and development of vegetation structure. A closer look at landscape and vegetation change may indicate why and how this can be done. This chapter is meant as a contribution to such a process. The task is to introduce elements of a new approach to identify the biophysical nature of a region and the processes that contribute to its maintenance and development.

Landscape Characteristics

Landscape plays a dominant role in academic geography, and several generations of researchers have taken a special interest in the geography of landscape. In Europe, the literature has analyzed the principal processes at work in changing the landscape. Researchers have studied the processes of clearing woodlands, draining marshlands, reclaiming heathlands, and allocating land, as well as the origins and transformation of field and settlement systems, enclosures, and agrarian reforms (Baker 1988).

Many definitions of landscape focus on the dynamic relationships among climate, natural landforms, ecosystems and human cultural groups and their use, and control and impact on the landscape. Forman and Godron (1986) observed four characteristics that are similarly repeated across a landscape: (1) a cluster of ecosystem types, (2) the flow of interaction among the ecosystems of a cluster, (3) the geomorphology and climate, and (4) a set of disturbance regimes. They conclude that landscape ecology focuses on three landscape characteristics: (1) *Structure*, the *spatial relationships* among distinctive ecosystems or landscape elements present, that is, the distribution of landscape elements and their kinds, numbers, configuration, and size. (2) *Function*, the *interactions* among the spatial elements, that is, the flows of energy, materials, and species among component ecosystems. (3) *Change*, the *alteration* in the structure and function of the landscape mosaic over time. In landscape ecology, landscape elements are often defined as ecosystems, plant communities, biocoenosis, or related concepts. Judging from the literature of landscape ecology (for example, Farina 1998), the main interest is the distribution, behavior, and interaction of plant and animal species. Similar views are found in the field of biogeography (for example, Cox and Moore 2000), which is sometimes regarded as a subdiscipline of geography and sometimes as a subdiscipline of biology.

Vegetation science, landscape ecology, and biogeography have contributed much to the understanding of ecosystems and the processes of environmental change. However, landscape is more than a mix of climate, morphology, and a cluster of ecosystems. Landscape is also a cultural entity that changes over the course of history and is the result of governmental and local politics, land use, human priorities, experience, and values. A landscape ecologist may not disagree, because according to Forman (1995) landscape is a mosaic where the mix of local ecosystems or *land uses* is repeated in similar form over a kilometer-wide area. He also defines region as "a broad geographical area with common macroclimate and sphere of human activity and interest." Although they are simple, the problem is not the definitions themselves, but the way they are interpreted and put into practice. In landscape ecology, "human activity and interest" seem to be more a reference point than an object of research on its own that has to be explored in more detail.

In vegetation ecology, "the human impact" is sometimes contextualized as historical background, but it is hardly included within scientific analysis. Interesting exceptions are the works of Moen and colleagues (Moen 1995; Aune, Kubicek, Moen, and Øien 1996; Arnesen and Moen 1997; Arnesen, Moen, and Øien 1997) and Olsson et al. (Olsson 1996; Olsson, Austrheim, and Bele 1998; Austrheim, Olsson, and Grøntvedt 1999). In physical geography, increasing interest in vegetation very much fits the botanists' approach to landscape (e.g., Skånes 1996). An interesting future prospect would be to strengthen the links between the subdisciplines of vegetation ecology and human and physical geography. The landscape-geography approach could then combine the functional dimensions of ecology with the spatial and temporal dimensions of historical geography. The main focus of interest would then be processes of change as much as morphology and structure.

Historical geography studies the geographies of the past. This involves the reconstruction of a variety of phenomena, structures, patterns, and processes, such as the evaluation and use of human and natural resources (Butlin 1992). The data and information for these reconstructions are based on several different sources, such as land taxations, cadastral maps, aerial photos, population census data, written sources and personal communications, vegetation analysis (e.g., space-for-time substitution transect analysis), and more.

To explain this further, the landscape could be compared with a chessboard, a structure guiding the game (i.e., resource allocation and land use). In contrast to the chessboard with its sixty-four squares of similar size, the landscape contains far more "squares" of different sizes and values. The chessmen correspond to landholders. They differ in value and weight, and according to economic situation, heritage, and education they have different presuppositions for using the arena of the game. The rules of the game guide how the individual players can use the arena. In the real landscape there is a certain interaction among actors, and between actors and the landscape with its resources. Individuals or groups of individuals try to explore the available resources. The human community makes rules that regulate the access to and use of land and its resources (Hägerstrand 1988).

Instead of plant or animal communities, I will emphasize land-use units, such as cultivated fields, meadows, pastures, plantations, and other landscape elements used and influenced by man. Spatial analysis, for example, GIS, is considered to be relevant and useful but in itself is insufficient to give an understanding of the interactions between man and his landscape environment and why landscapes change with time. However, producing and combining digital maps showing the situation at different times can demonstrate systematically the detailed development of landscape. The static map can then be transformed to a complex and ever-changing mosaic of spatial phenomena that rise, meet, connect, separate, and disappear (Hägerstrand 1995). I will use information on land-use history and local tradition to explore possible explanations for landscape development and transformation. I will stress the historical perspective as an alternative to the worn-out dichotomization of the landscapes as either "modern" or "traditional." This is an attempt to avoid the misunderstanding that if we go back a few generations we will find patterns of vegetation and land use that have existed since the days of yore.

Sandhåland, Southwestern Norway

The aim of this analysis is to combine concepts and theories of landscape ecology with a historical-geographical analysis of structure, function, continuity, and change in the cultural landscape of western Norway. I will emphasize habitation, field systems, boundaries, use of tools, agricultural management, ecological transformation, and vegetation dynamics. I will review local and regional patterns and processes with reference to legal rights, land reforms, and national and international ideas of development. I will examine the historical development of the farm cluster of Sandhåland on the island Karmøy.

The Sandhåland cluster of farms takes its name from the sandy beach by the sea (Hålandsanden). Judging from its name and size, it may have its origin in the older Iron Age, the period before A.D. 600. The farm probably was not abandoned during the Black Death, and it has unquestionably been permanently settled since 1520 (Lillehammer 1982). During the 1700s, the clustered buildings *(tun)* of the Sandhåland farms were situated by the sea (at Nera Håland), just a few meters southeast of the sandy beach. At that time, three families lived here, each with their own dwelling houses and farm buildings. The name "Tongarden," given to the area close to the old field systems by the beach, indicates the ancient location of farm buildings. Two families had their houses side by side close to the beach, whereas the buildings of the third family were situated some forty meters up the hill. According to the land taxation records *(matrikkelen)* from 1723, the farm was seriously influenced by eolian sand drift, and it has been suggested that this may have contributed to the buildings having been moved to higher elevations at Høge Håland (Lillehammer 1982). The two farms close to the beach were moved before the third, which was situated some distance from the beach. We do not know exactly when this happened, but the event probably took place in the early 1800s. Problems caused by eolian sand drift were reported from many farms at western Karmøy, Jæren, and Lista (the major sand-dune areas of western Norway) during the 1600 and 1700s.

At least one of the three farms at Sandhåland was situated close to the south field unit (Søråkeren). According to oral tradition, several buildings (belonging to one or two farmer families) sat side by side along the main road. In 1809, a house was raised by the north field (Nordåkeren), but the sources do not tell if the house was moved from the south field or another site nearby. This house was moved again sometime during the 1800s and is not indicated on the map from 1883. The buildings close to the south field unit must have been moved sometime before 1835, and in 1857 a small patch was officially separated as a field where the ancient farm buildings had been situated. The northern part of this field used to be called *fjosåkeren* (the field by the old cow barn). The patch is clearly marked on the map of 1883. The third farm is marked as two buildings (dwelling house and barn) on the cadastral map of 1883. They were moved to Høge Håland about 1900.

Problems caused by eolian sand drift may have influenced the farmers' decisions to move their buildings from Nera to Høge Håland, but the main reason seems to be new cultivation of land at Høge Håland. When the north-facing slopes south of the sandy beach were cultivated, transport of hay must have become increasingly difficult as the distance to the barns increased. Gradually the old position of the farm buildings became inappropriate.

THE LAND CONSOLIDATION OF 1835

According to records in the Regional State Archives of Stavanger, five farmers met at Sandhåland on July 21, 1835, and by mutual agreement rearranged fields and meadows to simplify the ownership patterns of parts of the farm (south of the sandy beach): "As all the landowners were present we measured the fields to exchange all small patches in such a way that the land of each farmer would be rounded off in one single patch." Unfortunately, no map was drawn, but by means of other written, unpublished material in the Regional State Archives of Stavanger some lines can be indicated (Figure 13.1). The area included was probably equivalent to the home fields of that time. The rest of the farm, north and east of the home fields, must still have been common property (grazing grounds) and outfields (dominated by heath and moorland).

THE IMPACT OF COTTAGERS: THE GREEN LANDSCAPE

From 1840 on, several cottagers settled in the outfields of Sandhåland, and five years later seven cottagers are mentioned in the population census. Their cultivated fields were small, used for growing oats, barley, and potatoes, and fishing must have been a major occupation. Some of the cottagers had a cow and a few sheep, and they had access to peat and grazing land in the outfields for as many animals as they could feed during winter. They also had the right to collect seaweed on certain parts of the beach. Each cottager had to pay the landowner an annual rent of one *speciedaler* and he had to work on the landowner's fields for six days a year. All of the cottagers left the farm at the end of the 1800s, some emigrating to North America. Hardly any physical signs of the cottagers' settlements can be traced in the landscape today, but some place-names have survived (see Figure 13.1).

During the decades when the cottagers were present, they must have used every available patch of land for the production of grain and potatoes, for grazing, mowing, gathering heather for winter fodder and for the fireplace, peat cutting, and for collecting seaweed. These activities influenced the structure and function of the cultural landscape. The cottagers primarily had access to outfields and hence both cottagers and landowners used the more remote parts of the farm. As opposed to the brown color of today's ericaceous heath and moorland, the color of the landscape was green, reflecting the dominant types of vegetation at that time.

Intensive grazing and peat cutting affected the composition of plant species.[1] According to local tradition, juniper *(Juniperus communis)* was scarce or not present at all (Lundberg 1998) due to a combination of burning and grazing. Heather *(Calluna vulgaris)* was dominant in areas with low grazing pressure, but in areas heavily grazed by sheep, horses, and cattle, heather was displaced by gramineous plants, such as common bent-grass *(Agrostis tenuis)*, mat-grass *(Nardus stricta)*, purple moor-grass *(Molinia caerulea)*, heath rush *(Juncus squarrosus)*, and others. Some of the native plants were

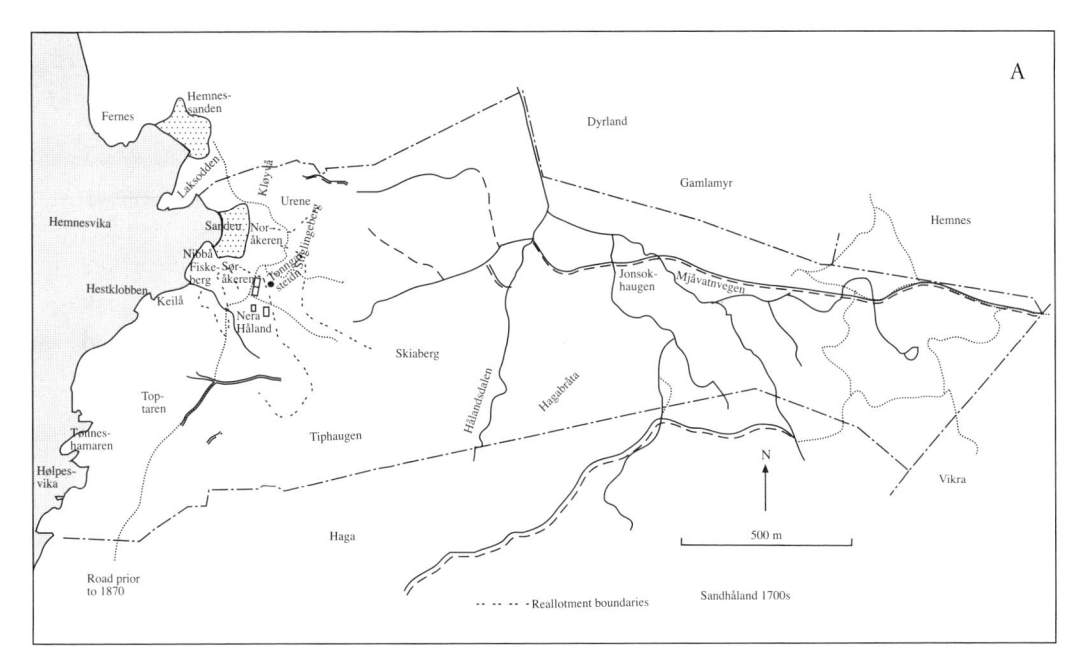

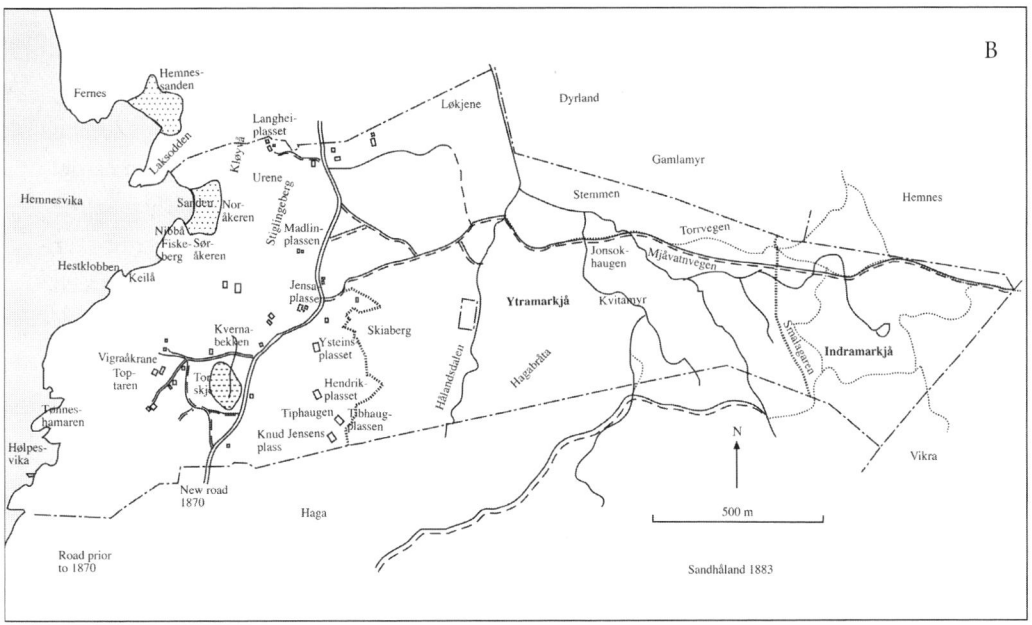

FIGURE 13.1. Sandhåland in the 1700s (A) and 1883 (B). Note the functional division of outfields in 1883: outer outfields *(Ytramarkjå)* were used for cattle grazing and inner outfields *(Indramarkjå)* for sheep grazing. Peat was cut in outfields and also at Torvskjer. The solid lines represent creeks; other lines represent roads and tracks. Open squares represent the buildings that existed at the two different time periods. The square with dashed lines in Hålandsdalen in 1883 is a cultivated field surrounded by a stone fence.

used for practical purposes. The white spongy pith of soft rush *(Juncus effusus)* was used as a wick in lamps *(koler)* containing fish-liver oil. This use of the plant was widespread in all countries bordering the North Sea until the days of kerosene lamps (Lundberg 1998).

THE FARMING SYSTEM OF THE 1800S

The land-taxation papers (deposited in the Regional State Archives of Stavanger) tell a lot about the economic situation of the region's farms. In 1668, seven barrels of grain were sown at Sandhåland. In 1723, seven and a half barrels of grain were sown and twenty-seven barrels were harvested. A yield ratio of 3:6 was among the lowest in the rural district of Skudenes. The land-taxation records of 1723 mention that the farm was heavily affected by eolian sand drift, and perhaps this was a major reason for the low yields. By 1802 the situation had improved radically. From eleven barrels of sown grain, sixty barrels were harvested, which is a yield ratio of 1:5.5. At this time, new land uphill and south of the sandy beach must have been cultivated. During the early decades of the 1800s, new fields were gradually cultivated and the potato was introduced. In 1865, the farmers sowed four and a half barrels of barley, nineteen and a half barrels of oats, and twenty-two and three-quarters barrels of potatoes. In addition, the cottagers sowed three-quarters a barrel of barley, four and a half barrels of oats, and three and three-quarters barrels of potatoes.

During the 1700s and 1800s, the number of farm animals also increased. According to the taxation records of 1668, the farms could support two horses and fifteen cattle. In 1723 these numbers had increased to three and eighteen, respectively. In 1802 the farms were considered to be able to support six horses, eighteen cattle, and twenty-four sheep. Two generations later, in 1865, these numbers had more than doubled: fourteen horses, thirty-two cattle, fifty-three sheep, and three pigs. The corresponding numbers of the cottagers were three horses, twelve cattle, and thirty-four sheep.

There is no sign of clustered buildings *(klyngetun)* on the cadastral map of 1883. The buildings of each farmer were scattered. The original farm cluster had obviously disintegrated and the original three farms were further split into smaller farm units before the first formal land consolidation of 1883. Figure 13.4 shows how and when this happened. The conclusion might be drawn that the government's land reforms of 1821 and 1857 had no triggering effect on the disintegration of the old *tun* (etymologically identical to "town") at Sandhåland. The land reform of 1821 was meant to be put into effect over an eight-year period, but the success rate in many rural districts proved to be rather low. The buildings at Nera Håland were probably moved voluntarily without government intervention. The same situation was reported from the neighboring community of Hillesland (Lundberg and Handegård 1996). At the time of reallotment, Sandhåland had twenty-one individual landowners, each with their own individual patches of land. In addition, fifteen areas of common landownership existed, remnants of a formerly widely distributed type of farming and ownership pattern.

The fourth farm at Sandhåland (the present farm 30/11) had been separated from farm 30/6 (one of the original three farms) in 1784. During the 1800s, further divisions followed. In 1883, when the original cluster of farms had been disintegrated over a long period, the ownership patterns and landscape structure were still characterized by fragmented landholdings. According to the cadastral map of 1883, Sandhåland was classified by 628 grading units *(boniteringsfigurar)*, owned by the twenty-one individual landowners (Figure 13.3). The mean size of these grading units was 0.143 hectares. The farms had 89.94 hectares of home fields *(innmark)*, most of which was uncultivated land (73.9 hectares; 82.2 percent), while the rest was fully cultivated land (14.3 hectares; 15.9 percent) and roads (7 hectares; 1.9 percent). During the land-consolidation process, the fertility of the land was evaluated. About 42.4 percent (38 hectares) of the home fields was considered to have the best-quality soils, about 22 percent (20 hectares) was considered as intermediate quality, about 18.3 percent (16 hectares) was considered

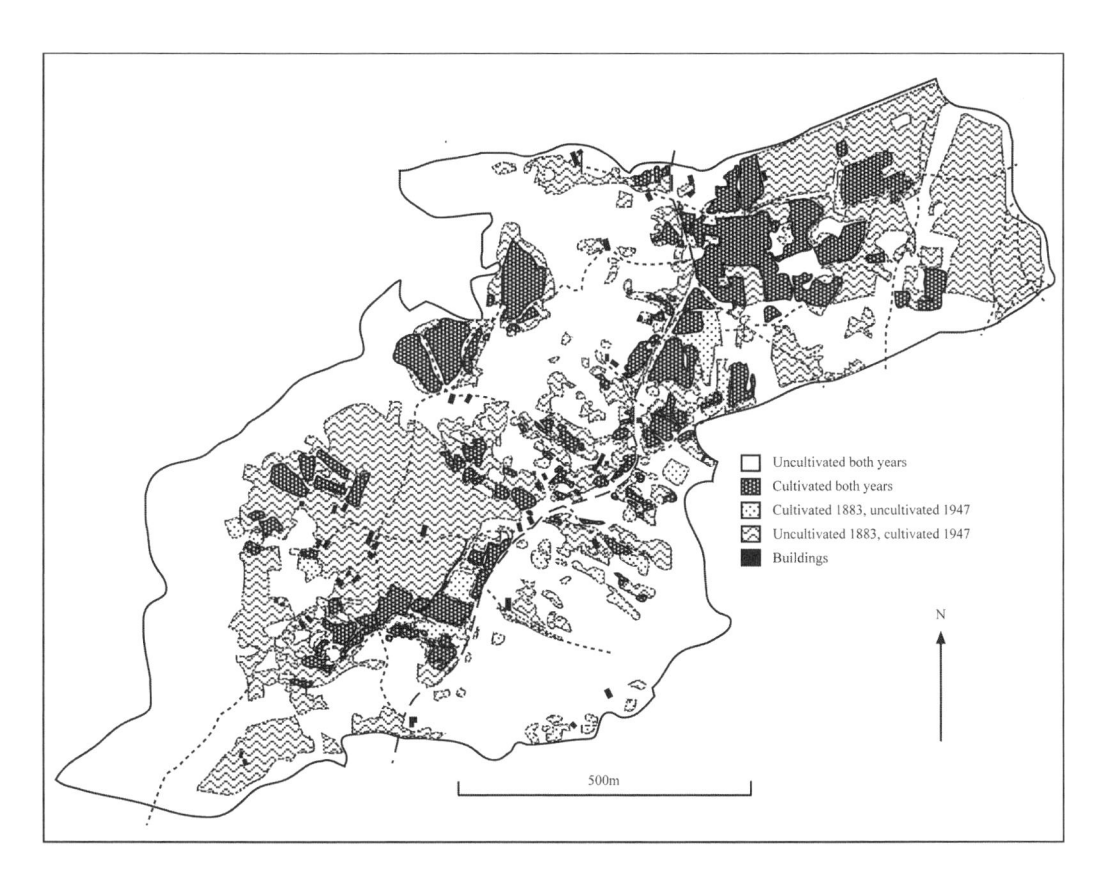

FIGURE 13.2. Land use of the home fields in 1883 and 1947 combined.

to be poor quality, and 15.4 percent (14 hectares composed of best-, intermediate-, and poor-quality soils) was not evaluated.

Although the first reallotment at Sandhåland in 1835 was by mutual agreement, a closer look at the cadastral map of 1883 reveals traces of old systems of field habitation. The original home fields were situated on the outskirts of the sandy beach with its first-quality and easily worked sandy soils. As mentioned, a small *tun* with three families was found nearby. During the 1700s, home fields were organized in two separate field units, each with their own name: Nordåkeren and Søråkeren. For a long time, these units must have been the very core of the field systems at Sandhåland. In each of the two units, the landholders had individual strips of land, an old way of organizing the field system reflected in the medieval *Gulating* Law. Although the *Gulating* Law was replaced by the Magnus Code in 1274, the traditional way of organizing the land continued until the first part of the twentieth century. This is clearly demonstrated by the cadastral map of Sandhåland from 1883. The northeastern part of the farm is called Løkjene, a moor used for peat cutting. Løkjene was a separate, functional unit, delimited from surrounding units, and each farmer had his individual strip of peatland within this unit. Peat was gradually replaced by hydroelectric power and the moor was later cultivated, initially for potatoes and oats; today it is used as a meadow. The third enclosure at Sandhåland in 1947 simplified the ownership pattern at Løkjene, but the old landownership patterns can still be recognized.

SEAWEED AS AN INSTRUMENT OF LANDSCAPE TRANSFORMATION

As in most parts of coastal western and northern Norway, the rotation of crops had never been practiced at Sandhåland, due to the use of seaweed. Every landholder had the right to collect seaweed and this was done whenever laminaria was available on the beach, most abundantly during autumn and winter. Seaweed was considered to be a valuable resource and the farmers had to be aggressive to get their share. To avoid unfairness among the farmers, the starting time was formally regulated in the reallotment papers of 1883: "The right to use seaweed drifted ashore will remain unaltered by this business because it is considered most beneficial that seaweed will be a communal resource as it has been up till now.... No one has access to collect seaweed though before 8 a.m. during wintertime, and before 6 a.m. during spring and summer."

The seaweed was transported by horses and spread on the nearby fields (Nordåkeren and Søråkeren). When seaweed was abundant, twelve to fifteen horses at a time might be needed for transport. Landowners had the right to collect seaweed anywhere, but cottagers were restricted to certain defined sections of the beach. Most of the seaweed was used on the nearby sandy fields (Nord- and Søråkeren), but the cottagers had a longer transport time to their small fields some 500 meters southeast, somewhat uphill. The seaweed was placed in large heaps to decompose over the winter. Individual heaps, one for each landholder, were placed in the western part of the field strips. Landholders without strips of land close to the sea were allowed to dry seaweed on the land of other farmers. Fresh seaweed would damage cultivated plants; hence decomposed laminaria were spread on the fields the following spring. Seaweed was also transported to Løkjene, roughly 1 km northeast, when the moor was cultivated. Seaweed was used at Sandhåland until the 1960s, when tractors replaced horses. Tractors were heavy and easily got stuck in the wet sandy beach.

Seaweed was also used for the production of kelp. Two different parts of the algae were used and both had to be dried before burning. Stipes (local expression: *tånglar,* cf. English: "tangle") were hung on wooden poles *(tånglaråter)* to dry in the wind, whereas laminaria were spread on ground to dry in the sun. The poles, connected by crosspieces, were put in lines, 10–15 m long, and placed on the high beach to avoid wave erosion.

The purpose of using seaweed on the fields was to improve the quality of the soil. The beaches of western Karmøy are made from sand of moraine origin mixed with shell fragments. Shell fragments are rich in calcium, but leaching is prominent in the porous substratum. By increasing the organic content (the humus layer) of the soil, base saturation and exchange capacity is much improved. Quantitative analysis of the chemical composition revealed extremely high pH values of the soil (sometimes above 9) and above-average values of nitrogen, phosphorus, and other plant nutrients (Lundberg 1987). In other words, by mixing eolian sand rich in calcium with decomposed laminaria, an extraordinary fertile, first-quality soil was developed. This is why farmers did not practice crop rotation.

THE LAND REALLOTMENT OF 1883

A second land reallotment was carried out in 1883, which caused a major alteration of the former landownership pattern (Figures 13.2 and 13.3). The fact that most of the farm buildings had been erected in the decades before this reallotment may have been an important reason why farmers chose a moderate land-consolidation scheme. A more radical solution would have meant extensively moving fairly new buildings. As a consequence, about 60.2 percent of the home fields was not affected by the reallotment of 1883. The process led to a simplification of ownership patterns, however, and common ownership of home fields no longer existed, in contrast to the situation before 1883.

The radical changes brought about by the enclosures in Norwegian agriculture at the end of the 1800s have been called "the Great Metamorphosis" *(det store hamskiftet)* (Krokann 1942).[2] Although the main focus of this chapter is on ecological and physical features of the landscape, the land-consolidation process also caused radical social changes. The writer Inge Krokann showed how old and close social relationships were broken by the physical disintegration of the old farm clusters. When dwelling houses were built at new locations, physically separated from the houses of neighboring farms, links to well-known places were cut. For generations, these places had been the working environment and playgrounds of the inhabitants, their ancestors and neighbors, and the physical and social setting for memories, recollections, and cultural heritage. The common social life that played an important role in the old farm clusters, whatever it was worth, was suddenly gone.

PEATLANDS IN TRANSITION

According to the land taxation of 1668, farmers at Sandhåland had peat in abundance. The reallotment that took place at Sandhåland in 1883 included home fields only. Two years later most parts of the outfields were reallocated, except for the easternmost part (remaining communal land after 1885). At that time outfields were split into two different units, outer *(Ytramarkjå)* and inner outfields *(Indramarkjå),* separated by a stone fence, and called Smalagarden (see Figure 13.1). Cows and young cattle grazed on the outer outfields (close to the home fields) and sheep grazed on the inner outfields. At that time outfields were a mixture of heathland, moor, and barren rocks; no forests existed. In 1885 outfields, corresponding to a land rent of thirty-six buckets of grain, were restructured and distributed among the seventeen landholders. Until

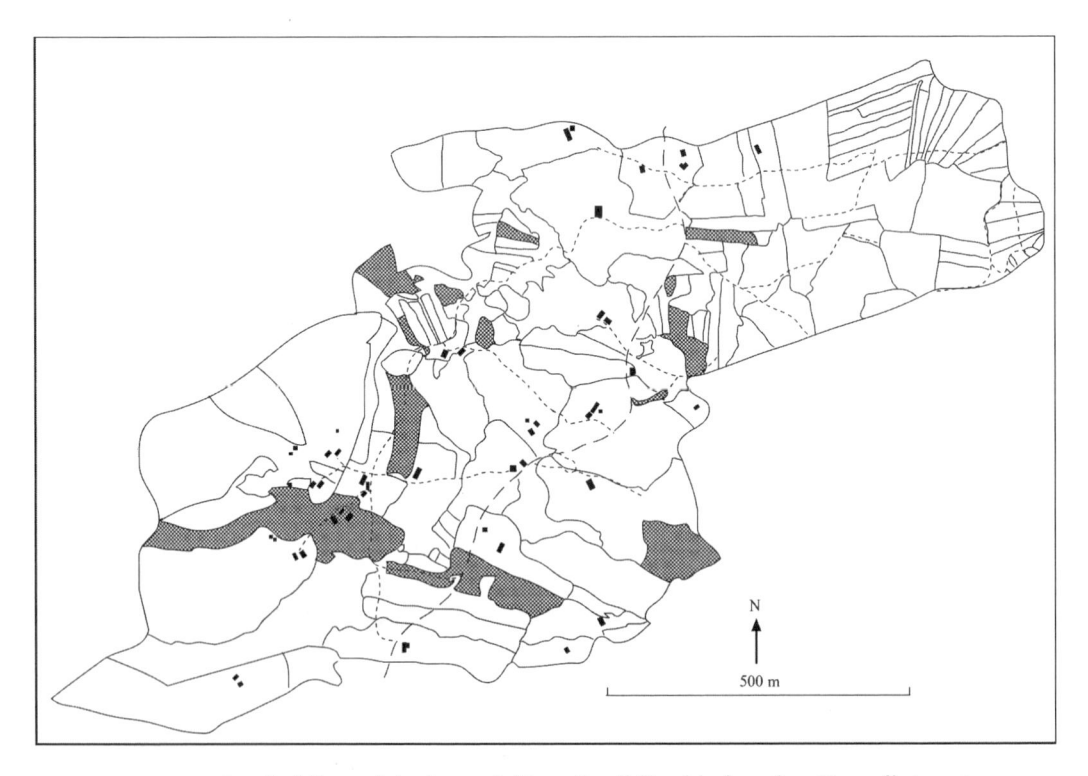

FIGURE 13.3. Landholdings of the home fields at Sandhåland before the 1883 reallotment. The thirteen plots ("teigar") belonging to landholding number 11 are screened. Of the thirty-six landholding units, twenty-one belonged to individual landholders and fifteen were forms of common ownership.

then, fragmented landholdings existed in outfields. Each landholder had his own patches of peatland mixed between the patches of other landholders. Alternatively, certain parts of the moor could be rotated between the users from one year to another. Communal peat was called *hopatorv* and was cut in common and then distributed among the participants when the peat had dried. Reallotment changed this in such a way that each landholder got his area within one piece of land. *Indramarkjå* was still common land, but according to local custom each family had their own patches where they cut peat. Neighboring families cut peat nearby one another. The functional separation of outfields for cattle and sheep, respectively, was common in many parts of western Norway.

The area of Løkjene in the western parts of the outfields (close to the home fields) was important for peat cutting during the 1800s. The peat was used for two purposes. Peat from Løkjene was considered to be poor quality and hence was called moss peat (*måsatorv*). Moss peat was not used for heating but for baking potato cakes, a regional speciality. Parts of Løkjene and other definite areas of the outfields were called mold

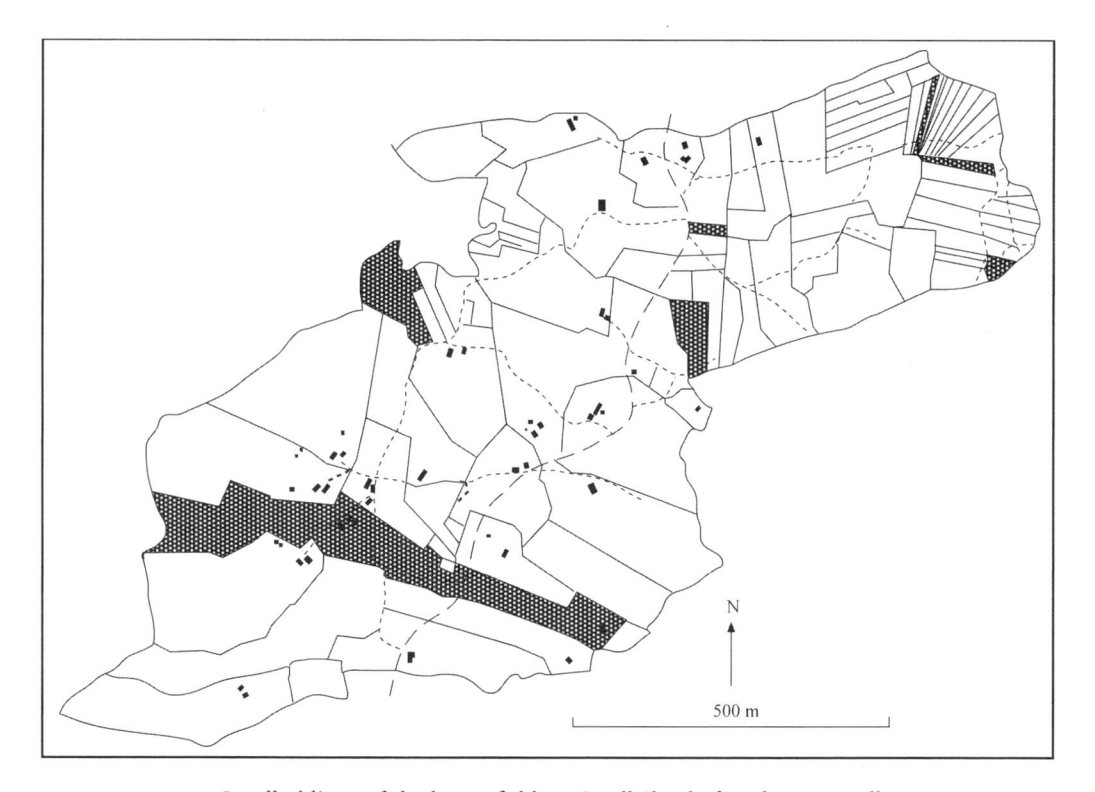

FIGURE 13.4. Landholdings of the home fields at Sandhåland after the 1883 reallotment. Plots belonging to landholding number 11 are screened, their numbers having been reduced to seven. The total number of landholding units was now twenty-four, all belonging to individual landholders with no common ownership.

land *(moldstykkjer)*. Mold of this kind was transported to special dung heaps and mixed with fish slough, heather, and other organic material. The dung was used as manure on the fields over the next spring. When Løkjene was drained, the area was used to grow potatoes and grain (oats and barley). At present it is used as a meadow for grass production. The peat found at Vondamyr and other moors was black, first-quality peat used for heating. Peat was also cut in Hålandsdalen, at Kvitamyr, at Søramyr, by Råtatjydne, and elsewhere. Some of the old place-names still exist, while some have been forgotten. Ways of extracting peat and the tools that were used have been described in detail by Anne-Berit Borchgrevink (1970, 1972). The peat could be transported to the farm by horse and cart or carried by one or two men. The carrying equipment included a willow basket *(kipe)*, a peat case *(torvkasse)*, both of which were carried on the back, or a peat stretcher *(torvbåre)*, carried by two men. A willow basket could be made of withy bands of juniper *(Juniperus communis)*, hazel *(Corylus avellana)*, goat willow *(Salix caprea)*, or creeping willow *(S. repens)*.

Until the 1950s farmers at Sandhåland, as in other parts of western Norway, used moors for the production of peat for heating (Hovde 1949; Borchgrevink 1970, 1972; Falkeid 1998). Peat cutting normally took place between spring work and the mowing season, at Karmøy from the middle of May. The reallotment in 1885 changed the ownership patterns of outfields but did not affect peat-cutting rights. Thus some farmers then had the right to cut peat on the land of other farmers within the same farm cluster, and sometimes also on neighboring farms, for example, at Noramyr ("the northern moor") on Dyrland to the north. By the same token, farmers from neighboring farms had the right to cut peat at Sandhåland. Thus people from Haga had access to peat in Kvitamyr. As can be seen on the map (Figure 13.1), a path or cart road ends in Kvitamyr from the Haga side (south). The fact that people from one farm had access to peat at neighboring farms is a well-known phenomenon from other parts of western Norway (Falkeid 1998).

There were no forests along the coast before the end of the nineteenth century, and peat was the major source of energy for heating indoors (beside driftwood found on the beach). Hence peat cutting was an important task and moors were often heavily affected. In 1949 legal restrictions were promulgated (Act on Protection against Soil Destruction) to prevent the moors from becoming completely stripped of peat. Where farmers had the right to cut peat on other farmers' land, the general rule was that at least 30 cm of soil should be left. At the time these clauses were introduced, several of the moors at Sandhåland had already been seriously damaged.

At the end of the nineteenth century, the first peat machines were introduced, at a time when the best-quality peat had been removed. They became popular in the outer districts of the counties of Rogaland and Hordaland. From that time a distinction was made between spade peat and machine peat. The latter was loose and had to be kneaded to make it stick together. In many parts of western Norway, extensive amounts of peat have been removed from the moors. The human impact radically changed the drainage, the way coastal ecosystems work, soil structure, vegetation, and the landscape appearance.

Løddesøl (1936) illustrated the extent of peat consumption in coastal districts. Each year the population in the coastal districts of Hordaland used 119,815 m^3 of peat. Peat accounted for 58 percent of their energy consumption (in calories), compared with 23.2 percent from coal, 14.4 percent from fuel wood, and 4.4 percent from oil. In houses where peat was the major source of energy, the annual consumption per person was 5.06 m^3. At that time the average number of persons per house was 5.3, and the average annual consumption per family was 25–30 m^3. This is roughly the same amount used annually by families at Karmøy (Falkeid 1998).

Although peat was the major source of energy for heating in the coastal districts of western Norway (in fjord areas and inland districts, they also had access to timber), regional differences can be traced. In some districts, the shortage of fuel was precarious and another type of peat and even turf cutting could be seen, peat or turf stripping *(torvflekking* or *bossaskjering)*, leaving nothing but bare rock. In the Hordaland communities of Herdla, Hordabø, Fjell, and Austreim, the annual loss of soil has been calculated to 7.3, 4.9, 4.7, and 4.3 hectares, respectively. In these districts, extensive areas of naked rock are still a characteristic landscape feature. The color of the rock has given rise to the name "The Gray Land" *(Det grå riket)*. At the county level, Hordaland was the worst off, with an annual loss or seriously damaged areas of 40 hectares. Over the years 2,800 hectares (1,800 hectares of moorland and 1,000 hectares on solid ground) had been lost or seriously damaged by peat stripping in Hordaland (Landbruksdepartementet 1946).

Because almost all available peat had been removed, the county council of Hordaland in 1926 proposed that further peat cutting should be declared illegal in some coastal districts. Afforestation and hydroelectric power were suggested as alternative sources of energy. The energy crisis in some coastal districts, particularly tangible in The Gray Land, triggered afforestation. Many of the plantations characteristic of the coastal districts today are first- or second-generation forests that were planted to solve the energy crisis.

At Sandhåland, peat was still available, although some moors had suffered serious devastation and forestry has been little more than a hobby. The first and only plantation, from about 1977, comprised six thousand trees of Norway spruce *(Picea abies)* and sitka spruce *(P. sitchensis)*. Elsewhere in Karmøy and other parts of western Norway, extensive plantations have been made during the twentieth century.

In 1967 ownership of the outfields at Sandhåland was restructured once again and redistributed among the twenty-two landholders. Now all parts of the outfields were included, also the easternmost parts *(Indramarkjå)*. Peat cutting had come to an end and many drainage canals had been dug for future cultivation or afforestation. Some years before, a new road, Mjåvatnvegen, had been built to the dam at Lake Mjåvatnet. The new road probably initiated the land consolidation of *Indramarkjå*. All of the farmers of course wanted land close to the road. Thus the patches were organized at right angles to the road on both sides.

1870–1950: HORSE MECHANIZATION AND NEW CULTIVATION

Comparing the arable area at Sandhåland in 1883 and 1947 shows the extent of the cultivation process following the first enclosure of 1883 (Figures 13.4 and 13.5). In 1947, 41.3 percent (37 hectares) of the home fields were arable, as opposed to 15.8 percent (14.3 hectares) in 1883. The arable area had more than doubled in two generations, an increase by a factor of 2.6. What were the main reasons for this increase? What were the ecological consequences of this transformation from uncultivated to fully cultivated land?

During the nineteenth century, the arable area on farms in western Norway should ideally be in balance with the number of cows (1 cow per decare [0.1 hectare] arable land). This relationship was gradually altered when industrial fertilizers were introduced later in the century, allowing an increase in the area under arable and meadows.

The transition from fragmented landholdings to individual farms involved changes in the use of agricultural tools. The spade had been efficient for working small, fragmented landholdings, but gradually it was replaced by horse-drawn equipment. Small fragmented fields had made the work with horse, plow, and harrow difficult, but now,

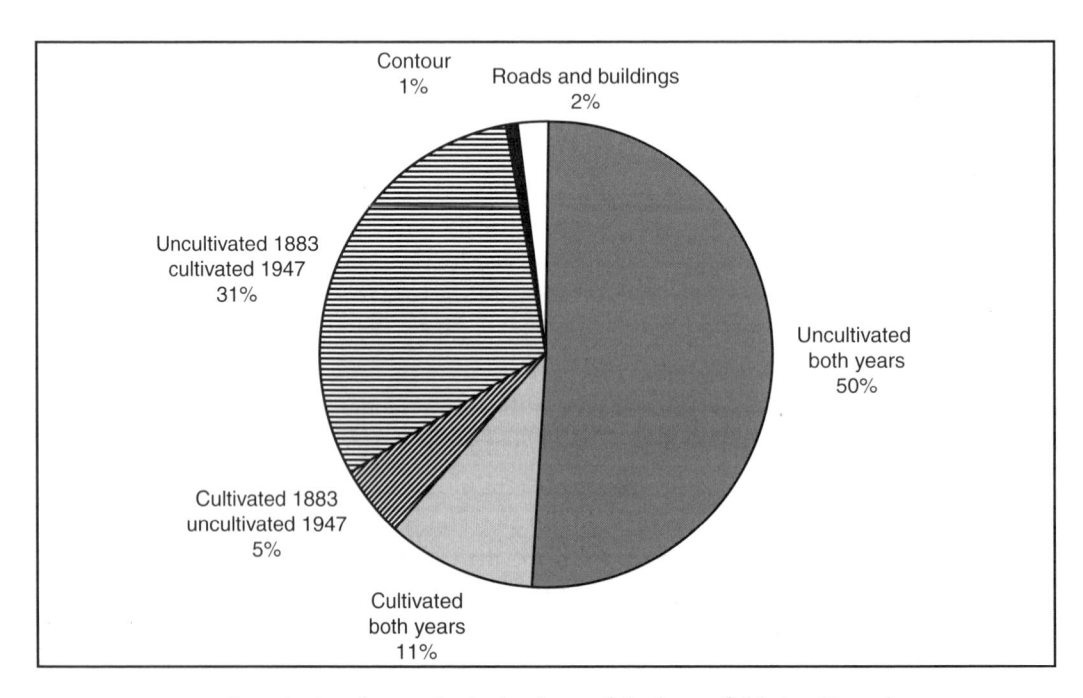

FIGURE 13.5. Quantitative changes in the land use of the home fields in 1883 and 1947.

when larger, connected fields became available after 1883, new plowing techniques were encouraged.

During the late eighteenth century, drainage ditches were introduced in western Norwegian agriculture. They greatly improved the quality and productivity of arable fields and meadows, but this transformation also altered the vegetation structure. The increase in arable land at Sandhåland from 15.8 to 41.3 percent of the home fields from 1883 to 1947 was gained at the cost of wet meadows, moist heath, and moorland.

During the twentieth century, the starting point of the hay season changed considerably. Around 1900, mowing in the study area started in mid-July, sometimes even later, and fifty years later at midsummer (June 23), depending on the weather. The general rule was that mowing could start when the grass had matured but not yet flowered. The starting point has gradually been pushed forward and today mowing often starts in the first days of June and occasionally in the very last days of May. This gradual change is one reason why the popular expressions "used to" or "traditionally" often adopted by scientists should be avoided unless explicitly related to place and time. This change has been possible because of easier access to manure and fertilizers. Before 1900 all available manure was used on the fields; today meadows and pastures get their share, too. Changes in the start of the hay season have greatly influenced the seminatural vegetation. In the early to mid-twentieth century, the fruits of many flowering plants were allowed to ripen before mowing, but today mowing starts before ripening. As a consequence, since the 1950s a group of plants that used to be numerous in seminatural habitats has declined.

THE ENCLOSURE OF 1947

The reallotment that took place at Sandhåland in 1883 considerably changed and simplified landownership patterns (Figure 13.6). As can be seen in Figure 13.3, the reallotment must have been a compromise between a radical and a moderate consolidation scheme (*arronderingsmodell*). Because of a disagreement among the farmers, ownership patterns following the enclosure of 1883 were still quite complicated. In 1943, one of the farmers (on farm 30/23) sent a formal letter to the land-consolidation judge (*jordskiftedommeren*) demanding a new enclosure. He argued that old-fashioned strip farming was still practiced and was impeding farm operations. Some farmers had ten separate patches of land, some of which they claimed were 2 kilometers from their farmsteads (the stated distance was probably an exaggeration). Due to the war, no immediate action was taken. After the war (August 1945), twenty-eight landowners sent a protest to the land-consolidation judge completely disagreeing with their neighbor who had requested another land consolidation. They argued that a new enclosure would involve the removal of many houses and the costs would be too high. Limited access to construction materials (timber) was also used as an argument.

A look at the map may give us an idea about the physical background for this

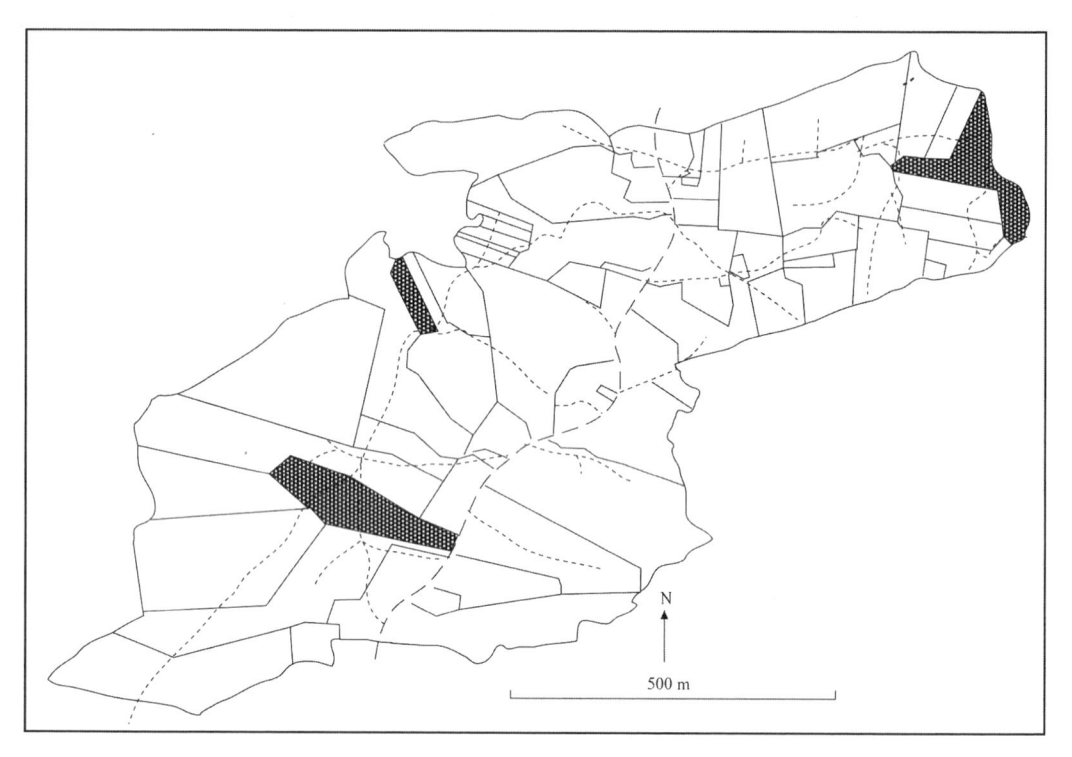

FIGURE 13.6. Landholdings of the home fields at Sandhåland after the 1947 reallotment, when the farm was split into fifty landholdings. A few farms had areas in common. Landholding number 11 now had three patches.

disagreement. Since 1883, farm 30/23 had been split into seven plots of land, widely distributed and scattered within the limits of Sandhåland. The distance between the main plot, where the dwelling house and barn were situated, and the most remote plot was about 1 km. The size of the farm was average (17 hectares); the other farms varied between 5 and 38 hectares. Two main plots and a smaller field were probably used as meadows, while a third was arable on best-quality sandy soil by the beach. The farmer had three strips of peatland at Løkjene. Until then, peat had been the major source of energy for heating and cooking. The land-consolidation judge must have agreed with the farmer of 30/23 that ownership patterns at Sandhåland were complicated and needed another consolidation. The third reallotment took place in 1947 and the ownership patterns were further simplified. As to farm 30/23, the estate was rounded off to two plots of land (as opposed to seven before 1947).

The first formal reallotment of 1883 caused a change of ownership in 37.2 percent of the home fields at Sandhåland, leaving two-thirds of the land unaffected. By the enclosure of 1947, about 88 percent of the home fields had changed ownership, leaving no more than about 12 percent of the farm's home fields unaffected by both enclosures.

The main objective for this change was to consolidate the land of each farmer as much as possible, preferably in one large plot of land. The 1947 enclosure did not fully succeed in this; the landholdings were still (and still are) somewhat fragmented, but the result was closer to the intention of the land-consolidation reform than before.

1950 TO THE PRESENT: TRACTOR MECHANIZATION, FALLOW LAND, AND CONCENTRATED INTENSIFICATION

During the 1950s, tractors were introduced. On Karmøy, the number of tractors increased from 5 in 1949 to 674 in 1989, while the number of horses decreased from 1,301 to 110 in the same period. This process also includes a change in related equipment, such as plows, harrows, and reapers. This change caused a significant alteration in land use and landscape structure (Lundberg and Handegård 1996). The land consolidation of 1947 also simplified ownership patterns, facilitating the use of tractors, and the efficiency of tractors promoted new land cultivation.

Government agricultural policies in the 1950s encouraged farmers in western Norway to concentrate on milk and beef production. The aims of this policy have been achieved, and the change in farming and land use has caused considerable changes in landscape structure and function. Using Sandhåland as an example, we can conclude that arable fields have almost disappeared. In 1997 no more than 0.7 hectares were classified as arable, being used for growing potatoes. Grain production has come to a complete stop, as in most parts of western Norway. Fieldwork in 1997 revealed that meadows make up 20.3 hectares and pasture 28.1 hectares, an increase of 30 percent compared to 1947.

In 1997, forty-five buildings at Sandhåland were used as dwelling houses. However, no more than five of the landholders practice farming, working about 4.8 man-years in total. Only one farms full-time, on a farm with a labor force of about two man-years. The remainder have full-time jobs off the farm and farm in their spare time. In recent decades many of the buildings have been given a new function. Five dwelling houses are uninhabited and some have been removed; several barns are no longer used. Three of the farmers used to keep mink or fox, but today the houses are empty due to low prices. Although some of the new houses have been built on uncultivated land, the increase in the number of buildings owned by nonfarmer residents has caused fragmentation of arable land. This is not a major reason for the decline in farming activity, however, but an alternative use of land (house building) in a situation when farming has declined for other reasons.

Most of the landowners no longer farm their land; three of the five active farmers rent land from their neighbors at Sandhåland and at other local farms. The rented land is used for pasture or meadow. This arrangement is organized informally, and the landowner is sometimes paid with half a lamb during the fall or perhaps nothing at all. To increase grass production, large quantities of manure and fertilizers are used.

A few species dominate the meadows, such as meadow fescue *(Festuca pratensis)*, ryegrass *(Lolium* spp.) and cat's tail or timothy *(Phleum pratense)*. Where grazing pressure is moderate, species-rich pastures and seminatural meadows can been found (Lundberg 1987, 1998). This is particularly true at Laksodden, the small peninsula north of the sandy beach. In this area, calcareous soils, rich in humus and shell sand, predominate, giving rise to a species-rich, calcifilous flora. Some of the plants found in abundance here are very scarce elsewhere in Norway, such as northern marsh orchid *(Dactylorhiza purpurella)*. Some of the land is now fallow, in total 7.7 hectares, some of which is best-quality soil. These areas are no longer grazed or mowed and as a result are overgrown by meadowsweet *(Filipendula ulmaria)*, reed-grass *(Phalaris arundinacea)*, and other weeds.

In recent decades, husbandry at Sandhåland has undergone distinct changes (Table 13.1). Instead of mixed husbandry with a variety of domestic animals on each farm, the farms have specialized. Only one farmer still keeps milk cows, some have oxen, and one has specialized in poultry. The poultry farmer started production of eggs about 1988 and in 1997 the farm had two thousand fowl producing fifteen hundred eggs a day. The farm has about thirty young oxen and four heifers. This farmer stopped keeping sheep some years ago. He is a part-time farmer and is employed full-time as a caretaker at a school.

The Significance of Landscape Transformations in Western Norway for Regional Identification

The historical analysis of landscape change at Sandhåland has revealed that a series of major landscape transformations have occurred during the last 250 years. Similar changes have been reported from other areas (Bjørkvik 1956; Holmsen 1956; Borgedal 1959; Jones 1982, 1988; Sevatdal 1991; Lauvås 1993; Lundberg and Handegård 1996; Lundberg 1999). The case of Sandhåland is probably typical, although local variations may be found.

During the eighteenth century, agriculture in western Norway differed from that in other parts of the country. Arable fields were sown each year, with no fallow period such as was common in eastern and central Norway. Rotation between arable and meadow was practically unknown, unlike many other areas of northern Europe. This was possible because of the use of seaweed in addition to animal dung. Oats and mixed grain were dominant cultivated plants. The fields were manured every year and the manure was spread after the seeds were sown. The most important tools were handheld, such as the spade, hoe, and rake (Skappel 1904). Farm buildings were clustered in *tun*, containing the buildings of two or more families. Fragmented landholdings *(teigblanding)* were the dominant field system. Home fields *(heimebøen)* were subdivided into many small patches *(teigar)*, some of which were cultivated and others not.

TABLE 13.1.
The number of domestic animals at Sandhåland, 1997
(data from interviews)

Winter-fed sheep	116
Milk cows	8
Other cattle	77
Fowl	2,000
Pigs	43
Horses	2

The field system was irregular, as was the organization of the *tun*. The plots belonging to one single farmer were intermixed with the plots of other farmers in the same farm cluster. Home fields were usually surrounded by a stone fence *(bøgard)*, often the only fence in use. In some parts of western Norway, different types of common property and common management were practiced. Sometimes the use of fully cultivated fields could rotate between the farmers of a farm, *årekast*. The farmers then had access to different plots or strips of the field in a three- or four-year cycle, depending on the number of farmers in the farm cluster. Land use was diverse with a mixture of arable fields, meadows, and pasture. Home fields were used as pasture in early spring and late autumn. In the summer, sheep and cattle grazed uncultivated outfields. Diversified husbandry was common and most farmers had cattle, sheep, pigs, and horses.

Farm clusters and fragmented landholdings of this type were found in the west from Lista (southwestern Norway) to outer Sunnmøre (northwestern part of western Norway), including the fjord districts of Hardanger and Sogn og Fjordane (Vreim 1936; Valen-Sendstad 1964). The system was described in the *Gulating* Law (probably promulgated before 930), which refers to (1) the division of a farm territory with a continuous area to each farm, and (2) the division into strips of equal length and breadth (Bjørkvik 1956). The latter principle led to a rather complicated system of fields and meadows, basically the same throughout the region under the *Gulating* Law, which applied to an area that roughly corresponded to present-day western Norway.

The first step in the allocation of land was the division of home fields into units of strips and plots *(teiglag)*. Each unit often had its own name and was delimited according to function or topography. Each unit was further subdivided into smaller plots, in such a way that each landholder had one plot or strip in each unit. For practical reasons, the strips ran lengthwise along the fields (Bjørkvik 1956). As noted above, there is a customary principle behind this division of land, but the actual field system found on cadastral maps from the late nineteenth century may not necessarily be very old. These field systems may have resulted from mutual reallotments that took place in the late eighteenth or early nineteenth century, as has been demonstrated by the analysis of Sandhåland.

Sometimes it is possible to detect an earlier stage of land allocation, before the subdivision of strips and plots, when entire territories were held in common. The *Gulating* Law provided for the division of jointly held land, but remnants of this old system of common ownership *(jordfellesskapet)* can sometimes be traced up to recent times, maintaining customary ownership patterns. Farmers on such multiple farms undertook the work in the fields or haymaking together, and afterward shared the grain, potatoes, and hay. Larger cultivated fields held by several landholders were not always subdivided into strips and plots. The holders could work the field as a single unit, sowing and harvesting the field in common. Examples of such common ownership have been reported up to the late eighteenth and early nineteenth centuries.

The broken topography of western Norway provided a characteristic physical setting for this type of agriculture, on the rugged strandflat or steeper slopes of the fjord districts. At Lista, Jæren, and Karmøy, characterized by sandy plains and gently undulating moraine coastal topography, more continuous farmland could be found. The size of arable fields could be somewhat larger here than in other parts of western Norway. The total size of fields belonging to one farmer could not be larger, however, than a man and his family could deal with during a short spring work period. Sometimes neighbors did voluntary communal work, but the number of available men in any case restricted the total area of cultivated fields.

This type of agriculture was labor intensive, but the strength of this form of land management was that the soil was efficiently turned by the use of spades. This work was essential to improve the productivity of the soil. In some parts of western Norway (for example, at Karmøy), farmers still say "turning the soil" *(å vende jorda)* instead of "plowing." This saying is a remnant from the past, although management practices have long since changed (plowing having replaced the use of spades).

The agriculture of western Norway was by the eighteenth and nineteenth centuries typically a *hand-dug landscape.* An earlier transition from plowing to spades had occurred, stimulated by the increasing fragmentation of land caused by the division of farms from one generation to another. According to the inheritance system known as *åsetesrett,* the oldest son had the right to take over the farm from his father, but his brothers and sisters also had inheritance rights. As fishing was formerly a major source of income, and farming produced a low surplus beyond self-sufficiency, brothers and sisters were often offered a piece of land as part of their legacy. Over the years farms became smaller and the land more fragmented. In western and northern Norway, where fishing was a major occupation, farms were continuously split up from the sixteenth century onward. In Agder, the southernmost part of Norway, seafaring and boat building later had similar results (Holmsen 1956). In such a situation, the use of horse and plow gradually became unsuitable. The plots of fully cultivated fields were too small and too dispersed.

There may be another reason for the use of the spade instead of horse and plow. As the farms and the size of cultivated fields became smaller, the farmers needed to get the most out of their fields. The plows turned the soil inefficiently compared to the spade, followed by the use of hoe and rake. This way of working the fields was labor intensive, but the fields yielded more than if they had been worked by horse and plow. This can explain why the number of plows decreased in western Norway about 1800 (Valen-Sendstad 1964; Lundberg 1999). The choice of plow versus spade is relevant for larger farms. Smaller farms could not afford a horse, and in some coastal districts horses have never had any importance in agriculture.

The subdivision of farms created small, tightly packed local communities of peasants, all roughly equal in socioeconomic status. The establishment of cottagers who cultivated the land on the outskirts of the farms caused the rise of a patriarchal community (Holmsen 1956). The use of the land by the cottagers and landowners considerably changed the structure of vegetation, creating a landscape very different from both the earlier situation and the present one. Although palynological studies have proved that the origin of heathlands of western Norway can be dated back to 1000–5000 BP, extensive areas now covered by ericaceous vegetation were transformed into gramineous pastures during the nineteenth century and the first decades of the twentieth century. Thus the present heathlands may not necessarily have an unbroken line back to the original types of heathland. These heathlands represent a mosaic of different successional stages, and their structure, function, and direction of change is closely related to past and present land use. The discussion of peatland transformation in western Norway also applies to moors and mires.

Since the end of the eighteenth century, there has been an ongoing discussion as to whether the type of agriculture described here was primitive or advanced (Strøm 1796, Hasund 1933; Borgedal 1959; Jones 1988; Lundberg 1999). Ridge plows *(ard)* and primitive plows *(treplog)* have been known in western Norway since the Bronze Age (Lillehammer 1994), and the use of the spade to work the fields was typical for the eighteenth and nineteenth centuries. Although it may appear old-fashioned and primitive compared to the more "developed" agricultural areas of Denmark and eastern Norway, the productivity could be high. Yield rates of 1:6 or 1:7 were common in districts with fertile soils, far more than could be achieved in the most "advanced" districts.

CONCLUSION

Ownership patterns, land use, and agricultural tools during the eighteenth and nineteenth centuries produced the hand-dug landscape, basically the same throughout its geographic range in western Norway from Lista to Sunnmøre. The detailed analysis of the farm of Sandhåland, southwestern Norway, indicates that landscape development in western Norway has experienced a process of continuous change during the last 250 years. The types of change and environmental consequences vary regionally and in time. Peat or turf stripping demonstrates that some of the changes were somewhat devastating, both ecologically (habitat destruction) and economically (the ruin of an energy source). Other types of change indicate that human influence can increase biodiversity (moderate grazing and mowing later than seed ripening) and improve the fertility of the soil (seaweed used to fertilize fields). The analysis suggests that we should give up the widespread Rousseauian idea about early, preindustrial societies as unquestionably skillful caretakers of their natural resources.

Western Norway has experienced radical changes in the landscape's function and structure, including soil and vegetation change, during the last 250 years. This is very glaring in the case of home fields, but it has also proved to be true for outfields and their seminatural vegetation. If regions are to be characterized by vegetation (Nordic Council of Ministers 1977, 1984; Moen 1998), structural vegetation change must be reflected in the definition of western Norway as a biophysical region. The definition should be flexible and sensitive to long-term environmental change. If not, the concept of region would be static and unsuitable to take account of the development and human impact of ecosystems. To understand the reasons for vegetation change, the "human impact" has not only to be contextualized, but it requires a detailed analysis of land use, husbandry, and the use of agricultural equipment and seasonal work as a supplement to vegetation mapping.

In western Norway, four distinct periods have been identified: (1) before c. 1870, agriculture was characterized by hamlets and fragmented landholdings; (2) formal reallotments took place from 1870 to 1910, causing the disintegration of the farm clusters and the introduction of individual farms; (3) from consolidation until c. 1950, horse mechanization replaced handheld tools, promoting new cultivation; (4) from c. 1950, tractors replaced horses as the most important tool for the farmers, allowing further cultivation of land and intensive land use. Land that could not easily be managed by tractors became fallow. This process of change dramatically altered the face of the region.

Although peat cutting came to an end in most parts of western Norway during the 1960s, traces of this activity, so important to the inhabitants of the afforested coastal districts, can clearly be seen from aerial photos and on site. These features, structures, and patterns are remnants of a regional culture producing a physical landscape very different from yet also similar to the present. The history of this landscape is essential for the understanding of its present vegetation and landscape structure and recent environmental change. It is also crucial for the definition of the region of western Norway on the basis of its cultural landscape.

The conclusion is that the definition of biophysical regions should include both natural features—climate, topography, vegetation—and human activity. All cultural landscapes and seminatural types of vegetation are embedded in a historical context. To understand fully the way regions are shaped, and why and how vegetation changes, landscapes have to be analyzed as cultural entities that change through the course of history. Cartographic mapping and GIS used in a spatial and temporal analysis have the potential to demonstrate how landscapes and regions are shaped and transformed through a complex and ever-changing mosaic of spatial phenomena that rise, meet, connect, separate, disappear, and sometimes continue. The historical context should be an object of research in its own right and has to be explored in more detail. This can be done by combining vegetation or landscape ecology and historical geography to form a new approach to landscape geography.

Notes

Research assistant Frode Skjævestad, with whom I also shared the pleasure of fieldwork during a week at Sandhåland in July 1997, prepared some of the maps. Håkon Sandhåland, Ragnvald Sandhåland, and local farmers kindly provided supporting information about the local history of Sandhåland.

1. Scientific names of plants are in accordance with Tutin et al. (1964–80); popular names of plants are those used in Clapham, Tutin, and Moore (1987).

2. Literally "the great sloughing" or "shedding of skin," using the snake as a metaphor.

References

Arnesen, T., and A. Moen. 1997. "Landscape History Coming Alive: History, Management, and Vegetation of the Outlying Haymaking Lands at Sølendet Nature Reserve in Central Norway." In *Species Dispersal and Land-Use Processes. Proceedings of the Sixth Annual Conference of IALE (UK), the UK Region of the International Association for Landscape Ecology, Held at the University of Ulster on 9th–11th September 1997*, ed. A. Cooper and J. Power, 275–82. Coleraine: International Association of Landscape Ecology.

Arnesen, T., A. Moen, and D.-I. Øien. 1997. "Changes in Species Distribution Induced by Hay-Cutting in Boreal-Rich Fens and Grasslands." In *Species Dispersal and Land-Use Processes. Proceedings of the Sixth Annual Conference of IALE (UK), the UK Region of the International Association for Landscape Ecology, Held at the University of Ulster on 9th–11th September 1997*, ed. A. Cooper and J. Power, 289–92. Coleraine: International Association of Landscape Ecology.

Aune, E. I., F. Kubicek, A. Moen, and D.-I. Øien. 1996. "Above- and Below-Ground Biomass of Boreal Outlaying Hay-Lands at the Sølendet Nature Reserve." *Norwegian Journal of Agricultural Sciences* 10: 125–52.

Austrheim, G., G. A. Olsson, and E. Grøntvedt. 1999. "Land-Use Impact on Plant Communities in Semi-Natural Sub-Alpine Grasslands of Budalen, Central Norway." *Biological Conservation* 87: 369–79.

Baker, A. 1988. "Historical Geography and the Study of the European Rural Landscape." *Geografiska Annaler* 70B: 5–16.

Bjørkvik, H. 1956. "The Old Norwegian Peasant Community, II. The Farm Territories. Habitation and Field Systems, Boundaries, and Common Ownership." *Scandinavian Economic History Review* 4: 33–61.

Borchgrevink, A.-B. Ø. 1970. "Brenntorv i Norge. En oversikt over redskaper, arbeidsliv og bruk av torv som brensel i norske landhusholdninger," 2 vols. Master's thesis, University of Oslo.

———. 1972. "Peat as Fuel in Norway." *Ethnologica Scandinavica* 1972: 58–94.

Borgedal, P. 1959. "Jordskiftets betydning for landbruk og samfunn." In *Jordskifteverket gjennom 100 år: 1859–1958*, ed. T. Grendahl, 305–77. Oslo: Det kgl. Landbruksdepartement.

Butlin, R. A. 1993. *Historical Geography through the Gates of Space and Time*. London: Edward Arnold.

Clapham, A. R., T. G. Tutin, and D. M. Moore. 1987. *Flora of the British Isles*, 3rd ed. Cambridge: Cambridge University Press.

Cox, C. B., and P. D. Moore. 2000. *Biogeography: An Ecological and Evolutionary Approach*, 6th ed. Oxford: Blackwell.

Falkeid, K. 1998. *Torv. Slit og trivsel på Haugalandet*. Aksdal: Lokalhistorisk Stiftelse.

Farina, A. 1998. *Principles and Methods in Landscape Ecology*. London: Chapman and Hall.

Forman, R. T. T. 1995. *Landscape Mosaics: The Ecology of Landscapes and Regions*. Cambridge: Cambridge University Press.

———, and M. Godron. 1986. *Landscape Ecology*. New York: John Wiley and Sons.

Hägerstrand, T. 1988. "Landet som trädgård." In *Naturresurser och landskapsomvandling. Rapport från et seminarium om framtiden*. Stockholm: Bostadsdepartementet och Forskningsrådsnämnden.

———. 1995. "Nature and Society: The Challenge of Contemporary Knowledge." In *Expanding Environmental Perspectives: Lessons of the Past, Prospects for the Future*, ed. L. J. Lundgren, L. J. Nilsson, and P. Schlyter, 165–72. Lund: Lund University Press.

Hasund, S. 1933. "Korndyrkinga i Noreg i eldre tid." In *Bidrag til bondesamfundets historie* I, ed. A. W. Brøgger and J. Frødin, 167–231. Institutt for sammenlignende kulturforskning, Serie A, Forelesninger 14. Oslo.

Holmsen, A. 1956. "The Old Norwegian Peasant Community, I. General Survey and Historical Introduction." *Scandinavian Economic History Review* 4: 17–32.

Hovde, O. 1949. "Myrene i kystherredene i Nord-Rogaland." *Det norske Myrselskap, Meddelelser* 47: 153–68.

Jones, M. 1982. "Innovasjonsstudier i historisk-geografisk perspektiv: Eksemplifisert ved spredning av jordskifte i Norden." In *Geografi som samfunnsvitenskap: Filosofi, metode, anvendbarhet*, ed. S. Strand. Ad Novas 19, 134–42, 200–201, 218. Bergen: Universitetsforlaget.

———. 1988. "Jordskiftets rolle i utformingen av kulturlandskapet." In *Nordisk Häfte i samarbete med Maankäyttö, Svensk Lantmäteri-tidskrift, Kart & Plan, Landinspektøren. XVI. Nordiska landmäterikongressen, Tammerfors, Finland*, ed. P. Lehtonen and H. Tiainen, 47–53.

Krokann, I. 1942. "Det store hamskiftet i bondesamfunnet." In *Norsk kulturhistorie* 5, ed. A. Bugge and S. Steen, 100–194. Oslo: Cappelen.

Landbruksdepartementet. 1946. "Utgreiing om jordødeleggelsen ved urasjonell torvdrift i kystbygdene på Vestlandet, i Trøndelagen og Nord-Norge." *Jordvernkomitéens innstilling nr. 10*.

Lauvås, L. 1993. "Utskiftninger på Jæren i tidsperioden 1821–1858." Norges Landbrukshøgskole, Institutt for planfag og rettslære (unpublished).

Lillehammer, A. 1982. *Bygdebok for Karmøy: Skudenes og Skudeneshavn*. Stavanger: Bygdebokutvalget i Karmøy/Dreyer Bok.

———. 1994. "Fra jeger til bonde—inntil 800 e.Kr." *Aschehougs Norges Historie*, 1. Oslo: Aschehoug.

Løddesøl, Aa. 1936. "Jordødeleggelsen ved torvstikking i våre kystbygder." *Det norske Myrselskap, Meddelelser* 2, 1936: 55–73.

Lundberg, A. 1987. "Sand Dune Vegetation on Karmøy, SW Norway." *Nordic Journal of Botany* 7: 453–77.

———. 1998. *Karmøys flora: Biologisk mangfald i eit kystlandskap*. Bergen: Fagbokforlaget.

———. 1999. "Die historische Entwicklung der westnorwegischen Kulturlandschaft im Licht der neuen Paradigmen der Ökologie: Eine Diskussion über den Einfluss des Geräteeinsatzes auf die Gestaltung der Kulturlandschaft." *Deutschen Gesellschaft für Geographie, NORDEN* 13: 77–99.

———, and T. Handegård. 1996. "Changes in the Spatial Structure and Function of Coastal Cultural Landscapes." *GeoJournal* 39: 167–78.

Moen, A. 1995. "Vegetational Changes in Boreal-Rich Fens Induced by Haymaking: Management Plan for the Sølendet Nature Reserve." In *Restoration of Temperate Wetlands*, ed. S. C. Shaw, W. J. Fojt, and R. A. Robertson, 167–81. Chichester: John Wiley and Sons.

———, ed. 1998. *Nasjonalatlas for Norge: Vegetasjon*. Hønefoss: Statens Kartverk.

Nordic Council of Ministers. 1977. "Naturgeografisk regioninndeling av Norden." *NU B 1977*, 34.

———. 1984. *Naturgeografisk regioninndeling av Norden*. Stockholm: Nordiska Ministerrådet.

Olsson, G. A. 1996. "Uröd vildmark eller fjället kulturlandskap—biodiversitetsproblematik i norska fjällområden." In *Innfallsvinkler på biodiversitet. Rapport fra flerfaglig workshop 7. desember 1995*, ed. A. E. Langaas, 12–28. SMU-rapport 6/96. Trondheim: Norwegian University of Science and Technology, Centre for Environment and Development.

———, G. Austrheim, and B. Bele. 1998. "Ressursutnytting og økologiske endringer i seterlandskapet." In *Jordbrukets kulturlandskap: Forvaltning av miljøverdier*, ed. E. Framstad and I. B. Lid, 68–98. Oslo: Universitetsforlaget.

Sevatdal, H. 1991. "Jordfellesskap og klyngetun." *Jord og gjerning. Årbok for Norsk Landbruksmuseum* 1991: 54–65.

Skånes, H. 1996. "Landscape Change and Grassland Dynamics: Retrospective Studies Based on Aerial Photographs and Old Cadastral Maps during 200 Years in South Sweden." Stockholm University Department of Physical Geography, Dissertation 8.

Skappel, S. 1904. *Træk af det norske agerbrugs historie i tidsrummet, 1660–1814*. Kristiania: Det kgl. Selskap for Norges Vel.

Strøm, H. 1796. "Forslag til Fælledsskabs Ophævelse Nordenfjelds i Norge." *Topografisk Journal* 15: 75–83.

Tutin, T. G., V. H. Heywood, N. A. Burges, D. H. Valentine, S. M. Walters, and D. A. Webb, eds. 1964–80. *Flora Europaea*, 5 vols. Cambridge: Cambridge University Press.

Valen-Sendstad, F. 1964. *Norske landbruksredskaper, 1800–1850-årene*. Lillehammer: De Sandvigske Samlinger.

Vreim, H. 1938. "Trekk fra byggeskikkens geografi i Norge." *Foreningen til norske fortidsminnesmerkers bevarings årsberetning, 1936–1937*: 33–64.

14.
The "Fjordscape" of Inner Sogn, Western Norway

INGVILD AUSTAD *and* LEIF HAUGE

A vertical topography characterizes the Norwegian fjord landscape. A typical fjord is a long, narrow inlet of the sea surrounded by mountains, often with extensive relief. Great climatic and vegetation gradients are typical both from the coast to the inner valleys and from sea level to the top of the highest mountains. This chapter focuses on the inner and middle parts of the Sognefjord area in western Norway. The distance is short from the valleys to the greatest mainland glacier in Europe, the Jostedalsbreen, in the inner fjord districts. Along the seashore, in the highlands and valleys and near the glaciers, people over time have settled and exploited the natural resources, thereby forming cultural landscapes. The landscape, by all means dramatic, has throughout history attracted hunters, fishermen, farmers, travelers, artists, and scientists. The role of human activity as an ecological factor has tended to be underestimated in the cultural landscape of Sogn, with its integrated production system, small-scale agriculture, distinctive land-use patterns, and seminatural vegetation types.

Here people have adapted to and exploited nature and have given the landscape a new and different content by the means of their everyday activities. The natural conditions have both set limitations and provided possibilities for the people living there. Steep slopes and the shape and size of arable land have, on one hand, hindered mechanization and large-scale industrial farming, and, on the other hand, kept and preserved a small-scale farming system. Furthermore, knowledge of specific farming methods, techniques, and traditions handed down from generation to generation still survives in the area. Some of these traditional farming methods can be traced far back in time.

The specific landscape conditions in this region favored an all-round animal husbandry. The comprehensive fodder collecting and extensive grazing by domestic animals through thousands of years have formed a wide range of seminatural vegetation types with biological and cultural historical values (Norderhaug et al. 1999; Austad and Øye 2001). Small-scale farming and traditional harvesting techniques still characterize the cultural landscape while contributing to its aesthetic and recreational values. Many people are attracted to these open, friendly, and varied cultural landscapes and seminatural vegetation types (Strumse 1996). However, the small-scale cultural landscape is seriously threatened by the restructuring of agriculture (Austad and Hauge 2001) and especially by overgrowing. As a result, many seminatural vegetation types in the region are critically endangered and an important part of our cultural heritage is disappearing (Fremstad and Moen 2001).

This chapter examines the strong connection between nature and man in this region and the importance of the small-scale farming system, traditional agricultural techniques and ecological processes for forming and preserving the varied plant and animal life still to be found in the landscape, and to reveal the traditions on which the cultural landscape of today is based. The chapter also focuses on the landscape as experienced by artists: painters, writers, poets, and musicians, inspired by the fjord landscape, thereby contributing to its strong regional and national identity (Sveen 1981).

Since agricultural development is affected by socioeconomic, political, and technological development in society, the distinct agricultural landscape we have come to value in the region today is the visible result of the diversity of agro-ecological and socioeconomic forces operating over hundreds and thousands of years. It is thus a landscape that has been constantly changing through time. Today's agricultural landscape, its processes and elements, are the result of a long and complicated cultural-historical development.

This presentation of the Norwegian fjord landscape, the core and heart of wooded pastures and hay meadows, pollarded and coppiced woodlands (Pushmann et al. 1999), should be seen together with the Swedish *hagmark* landscape (Ihse and Skånes, this volume, chapter 10) and the mountain agricultural pasture landscape in Hjartdal in Telemark (Norderhaug, this volume, chapter 15). All these landscapes have, despite their different scenery and physical conditions, similar types of natural and cultural values in terms of ecology and biodiversity.

Natural Conditions in Sogn:
The Outer Framework for the Pattern of Settlement and Farming

This Sogn region comprises the local authority districts (communes) situated around the innermost part of the Sognefjord (Figure 14.1). Of the total land area, only 15 percent is forest and 1.4 percent cultivated land. The rest consists of mountains, glaciers, and lakes. The Sognefjord is the deepest fjord in the world and the second longest. A steep climate gradient from coast to inland ranges from pronounced oceanic conditions in the west to a continental climate in the easternmost valleys, though the topography causes large local variations in both precipitation and temperature (Utaaker and

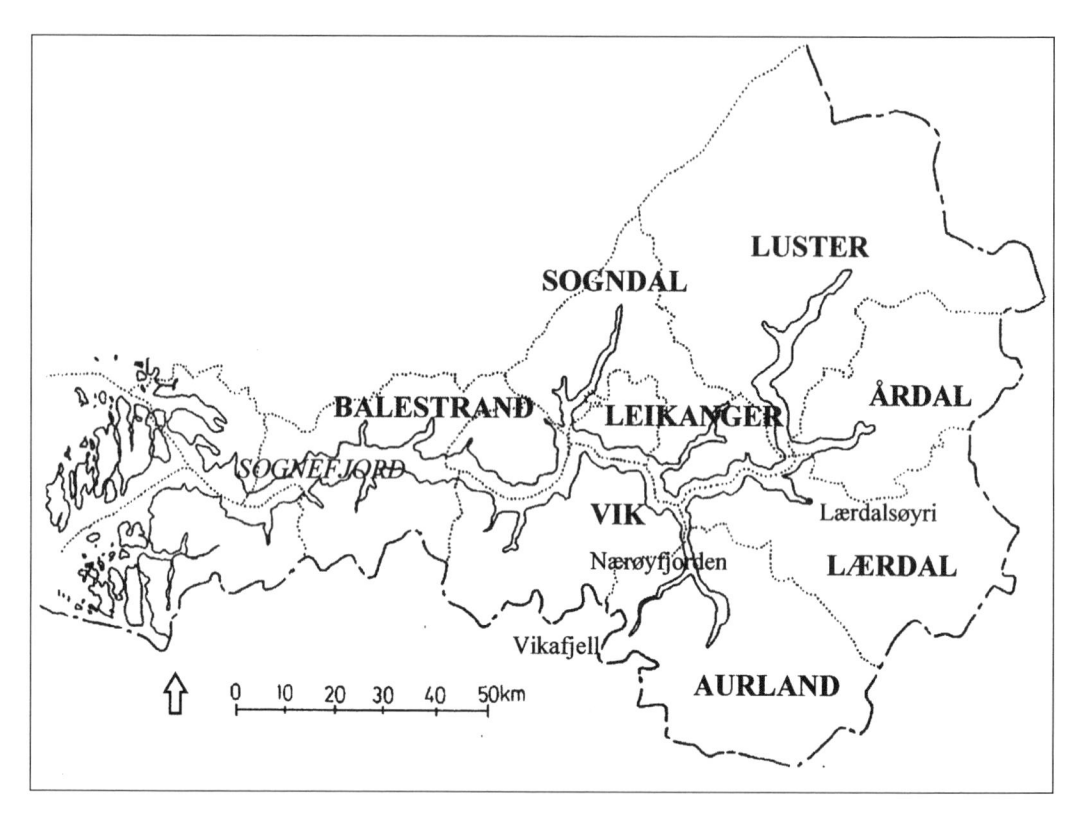

FIGURE 14.1. Map of the Sognefjord area, showing its location in Norway and the names of local authority districts (communes) in inner Sogn.

Skar 1970). In the lowlands the annual precipitation varies from 800 mm around Sogndal to 400 mm in the driest parts of Lærdal. There is also a marked climate gradient from sea level to the mountains (1,500 meters). Summer temperatures in the lowlands are fairly uniform throughout the area, around + 16°C in July, though this varies with exposure and decreases with altitude. The fjord zone has a relatively mild winter with temperatures between −2.5 and −5°C in the lowlands in January and −8 to −9°C at the forest limit. The snow cover also varies greatly, but with the exception of the narrow zone along the main fjord and the driest valley floors, there is a stable snow cover during the coldest period. The length of the growing season varies from about 150 days in the innermost valleys to about 200 days in the central areas along the fjord. The precipitation increases generally with altitude and proximity to the Jostedalsbreen glacier.

The great variations in geology, topography, and local climate result in a highly varied and complex natural vegetation. There is a mixture of natural conifers and deciduous woodlands, with Scots pine *(Pinus sylvestris)*, silver birch *(Betula pendula)*, downy birch *(B. pubescens)*, hazel *(Corylus avellana)*, elm *(Ulmus glabra)*, ash

(Fraxinus excelsior), and lime *(Tilia cordata)* as the most common species, and also wetlands and mires, river deltas, and alpine vegetation. The entire lowland zone is included in the southern boreal zone (Moen 1998). Steep mountain slopes result in a sharp zonation through middle and northern boreal and alpine zones over a short distance. The fjord zone and the lowest parts of the valleys are characterized by stands of rich deciduous forests with nutrient-rich soil. In the middle boreal zone, including the upper valley bottoms, mixed woods of birch *(Betula pubescens)* and gray alder *(Alnus incana)* dominate. Gray alder woods are common on unstable soils with ample seepage water and in alluvial zones along riverbanks and deltas. Birch *(Betula pubescens)* woodland dominates the natural vegetation at higher altitudes toward the tree line and penetrates toward the lowlands in cool, shaded parts of the valleys. The tree limit varies but can reach 1,200 meters in inner Sogn. Above the tree limit, mire complexes and poor oligotrophic plant communities are important elements.

Cultural History

Natural conditions such as topography, climate, and soil provide the basis for human exploitation of natural resources and settlement. Activities connected with the development of agriculture included burning and clearing land for farming, improving and fencing off arable land and pasture, utilization of mountain areas for grazing and haymaking, and the development of irrigation and manuring systems (Brøgger 1925; Ohnstad 1980a; Kvamme 1988; Slotte and Göransson 1996). Fuel (wood and peat) had to be collected and forests logged. In this way, people and domestic animals used and modified the natural vegetation and transformed their surroundings.

There are traces in Sogn of grazing in the mountains dating back to 2500 B.C. There appears to be evidence of foliage gathering at the start of the Bronze Age (Kvamme 1998). Traces have also been found of the cultivation of cereals dating back to 1600–1400 B.C. (Valvik 1998) and organized farms with permanent settlements, including infields and outfields, dating back to A.D. 400–500 (Valvik 1998; Åstveit 1998; Austad and Øye 2001).

The special topography of inner Sogn with its steep mountainsides meant that only small areas could be tilled. Thus the farmers had to use the resources of the outfields and mountains to a maximum. The available natural resources gave a wide variety of fodder types, the proportions of which could be varied according to need, and made it possible to survive hard times and poor harvests. Livestock husbandry was especially important in a topography with steep mountainsides as in the fjord region, and access to extensive outfields, forests, and mountain areas compensated for the lack of tilled fields (Figure 14.2). Fodder collection and animal husbandry resulted in an energy flow from the outfields, forests, and mountain areas to the infields and the main farm. Dairy cattle especially provided a large amount of manure, which was important and guaranteed a high yield on the infields.

FIGURE 14.2. The "fjordscape" of Sogn. Natural conditions make up the outer framework of human exploitation of natural resources. Photo by Leif Hauge.

VERTICAL DISTRIBUTION OF FARMS AND LIVESTOCK HUSBANDRY

Grazing resources were utilized in a dynamic production system throughout the summer season. As the grazing at different altitudes turned green in spring and early summer, the animals followed. To make the most of the natural resources, seasonal mountain farms or summer farms were located at different altitudes: along the fjord, in the valleys and on the mountainsides. These were used in a cyclical pattern. Formerly the practice of moving animals from the lowland to mountain pastures was regulated by law, for instance, the Magnus Lagabøter Code from 1274 (Brøgger 1925). A vertical distribution of farmland areas remains typical for the western Norwegian farm (Figure 14.3).

Sheep and especially goats were taken early in spring to the outfield areas with early growth. At the beginning of April, the grazing animals could be rowed along or across the fjord to sheltered and fertile grazing, where they spent some weeks until the hillsides turned green. In mid-May, dairy cows and young cattle were led from the infields up to the spring farms. These were normally situated just outside the infields. Here the cows were milked and the milk carried home to the farm both morning and evening. The cattle moved upward through the valley following the greening of the grass. The spring farm combined grazing and milking in spring and autumn and mowing in summer. In the second half of June, the animals were moved to high-lying mountain

The "Fjordscape" of Inner Sogn

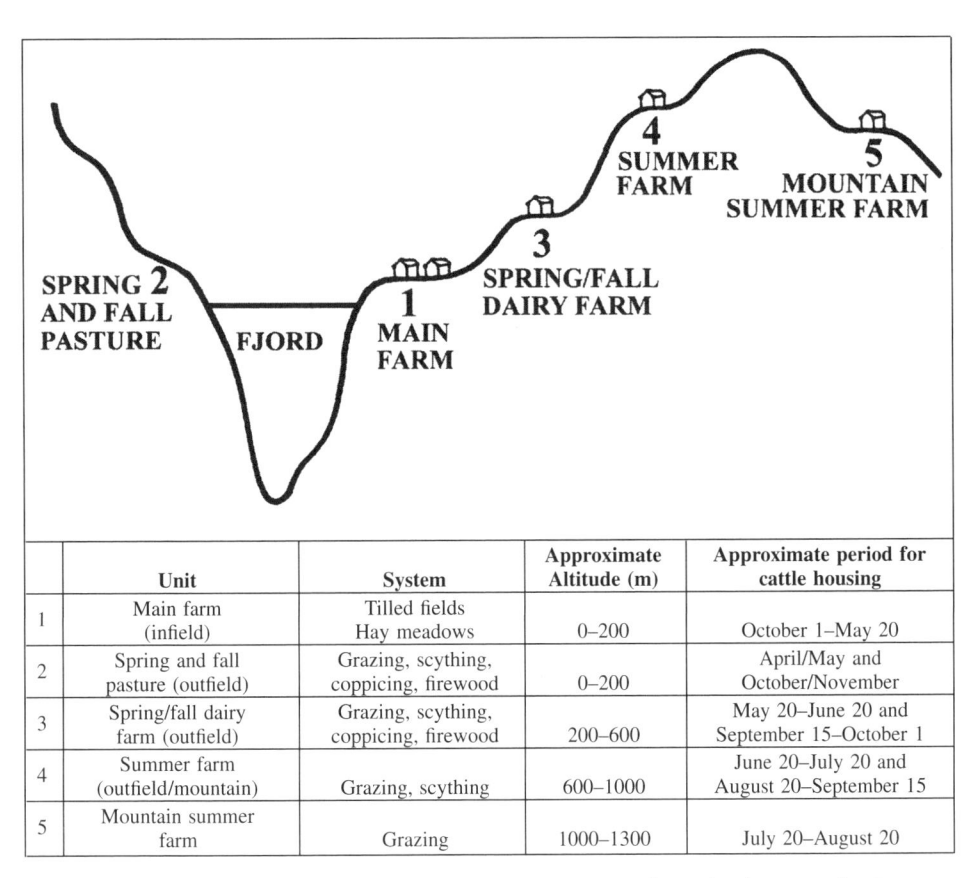

	Unit	System	Approximate Altitude (m)	Approximate period for cattle housing
1	Main farm (infield)	Tilled fields Hay meadows	0–200	October 1–May 20
2	Spring and fall pasture (outfield)	Grazing, scything, coppicing, firewood	0–200	April/May and October/November
3	Spring/fall dairy farm (outfield)	Grazing, scything, coppicing, firewood	200–600	May 20–June 20 and September 15–October 1
4	Summer farm (outfield/mountain)	Grazing, scything	600–1000	June 20–July 20 and August 20–September 15
5	Mountain summer farm	Grazing	1000–1300	July 20–August 20

FIGURE 14.3. The vertical distribution of farms and summer farms in the Sognefjord area.

farms, often up to 1,200 meters above sea level (Figure 14.4). The farms were occupied for two to three months, usually by young women, who were responsible for the daily operation. Herding was necessary to reduce attacks by beasts of prey such as bears, wolves, and wolverines. Young boys usually did the herding.

In the mountains, the landscape is characterized by wide, fertile mountain valleys with a high biological production per m². The growing season, however, is short. The milk from grazing livestock was made into cheese and butter. The long, steep path between the main farm and the mountain farm hindered daily transportation of fresh dairy products. Large amounts of fuelwood were needed to turn milk into cheese, which could be stored for a long period. Thus mountain farms were established close to the forest boundary with a short way for collecting wood for fuel. When deciding upon the location of a mountain farm, one had to take into consideration fertile grazing areas, the danger of avalanches or rockfalls, wind and snow conditions, a stable supply of water, and the danger of beasts of prey.

FIGURE 14.4. To make the most of the natural resources for agricultural production, the mountain summer farms were important elements in farm husbandry. Photo by Leif Hauge.

The cool climate of the grazing areas in the mountains made the animals healthier because they were less vulnerable to parasites and infection. The use of alpine grazing areas meant that the animals used more time consuming fodder and less time resting and as a result gained more weight.

In early autumn, the utilization profile was reversed as the animals descended, making short stays at the various levels of altitude. The animals could spend from two weeks to a month on the autumn farms in the lowland. The date at which the livestock had to be back down at the main farm varied, but custom normally set firmly defined dates for moving the animals (Figure 14.5). The dates varied according to the location of the mountain farm, the farm road itself, and conditions of weather and snow.

Thus most of the working time of the farm household was taken up with looking after the domestic animals and collecting fodder. Livestock husbandry has not only affected the daily life of the household but also influenced the organization of settlements, localization and structure of buildings, and infrastructure in the farming communities. Sale of livestock products provided the economic basis of the farm.

In the mid-nineteenth century, livestock on an average farm consisted of one or two horses, eight to twelve dairy cows, heifers, and calves, and twenty-five to thirty sheep or goats. The traditional cattle were the *vestlandsk fjordfe* (West Norwegian fjord cattle), with a weight of about 250 kg. This is about half the weight of a modern dairy cow today.

The fjord horse *fjording* is a Norwegian horse breed raised by the farmers on small farms in the northern part of western Norway. The fjord horse has a good disposition and is able to survive on marginal grazing. It is surefooted and good at making its way in rugged terrain. The horse may stand as a symbol of the western Norwegian farm, well adapted to the environment and using local resources to the utmost.

BUILDINGS, SOCIETY, AND HUMAN-MADE STRUCTURES

The location of buildings and settlements took account of natural conditions, for instance, security from rockslides and avalanches, availability of sufficient water, and good climate and soil conditions. Usually the buildings were carefully located so as not to occupy potential agricultural land, and they were often organized in clustered farm hamlets on marginal land. Old cadastral maps show small communities, often with a farm church. The buildings were well organized, with the cowsheds at the fringe of the cluster to facilitate the spreading of manure for fertilizing the infields and to avoid trampling manure in the yard. This was typical both on the main farms and on summer farms.

Stone was an important building material, particularly in areas where other materials such as timber were scarce. Stone constructions from earlier periods reflect a high

FIGURE 14.5. Early in the spring, sheep and goats were led to the outfields and later up to the summer farms. In late June, the animals were moved to higher summer farms, often to 1,200 meters above sea level. Photo Leif Hauge. Photo c. 1940, courtesy of Landbruksfilm.

level of craftsmanship. In addition to dry-stone walls, which often formed the boundary walls between infields and outfields, terraces, walls for protection against rockfalls and strong winds, piles of stones cleared from the fields, carefully constructed roads, bridges, irrigation channels, ditches, and buildings were common stone constructions. A typical feature, combining the use of stones for fences and building walls, was the cowshed placed along the border between the infields and outfields. Infrastructures such as neat constructions of cattle tracks, often paved with flagstones and fenced with dry-stone walls, led through the infields from the main farm in the lowland to the outfields and summer farms in the mountains. As cultural heritage, they are evidence of a practice that has existed since the earliest permanent settlements and up to the present.

In addition to the main farms and hamlets, there grew up a countless number of smallholdings or cottars' farms in the eighteenth and nineteenth centuries (Figure 14.6). The smallholdings could be located on the infields of the main farm generally on marginal land, at the border between the infields and outfields, or in the outfields, where the cottar had to clear agricultural land. The main farms owned the smallholdings and the cottar and his family had to pay rent for the land and work for the landowner. The agricultural land utilized by the cottar's family was small and the income was low, and these small tenant farms in particular were abandoned as a result of emigration to the United States in the nineteenth century. Another type of cottar had no land (*strandsitter*) (Engesæter 1976).

FIGURE 14.6. Traditional cottar's farm in Sogn. Photo by Leif Hauge.

TRADE

In earlier times, the farmers harvested various products from nature. Much was consumed in their own households, but specialized goods were taken to markets and sold to procure cash for the farm. What goods were sold varied considerably according to the raw materials available on the farm, demand and price, and cost of transportation.

The individual farms became the core of resource use and trade. The farms brought resources from the surrounding landscape, from fishing in the fjord, lakes, and rivers to the normal production of the farm with animal husbandry, the use of meadows and cultivated land, the use of outfields by pollarding, wood cutting and collecting of material for various purposes in outfields and pastures, including haymaking and grazing in the mountain areas. Little by little, agriculture formed a basis for constantly more specialized secondary industries and prospered through trade and the flow of business in a society where manpower was a plentiful resource. The main products from the mountains were dairy goods, but the production of cattle for slaughter or for sale as livestock was also widespread and taken to the large markets. The fjord was important for transportation.

Boat transport mainly by sailing boats *(jektefart)* had Bergen as its main destination, both as a market and as a transit port to further destinations (Thue 1980). Vast shiploads of timber were exported to Europe from the rural districts, with Bergen as the center of this international trade. Products that could withstand storage and a long transport could be sent by boats, such as dried, smoked, or salted meat, butter and potatoes, in addition to timber and firewood. The journey could last from one week to several weeks according to wind conditions. Transport by sailing boats was common until the beginning of the twentieth century, when steamers gradually replaced the traditional means of transportation. Traces from this culture can still be seen in the many rest houses and trading houses along the coast.

As a consequence, small trading communities grew up along the fjord, seashore settlements *(strandsittersteder)* that prospered in the eighteenth and nineteenth centuries. Here working people were allowed to put up their small cottages. The land was owned mainly by farmers, who also had their boathouses, wharfs, and storehouses here (Engesæter 1976). The seashore settlements grew with the expansion of trade. One of the most famous and best-kept seashore settlements in Sogn is Lærdalsøyri.

Large amounts of goods were also transported by land to the central areas of eastern Norway, crossing mountain passes at altitudes of 1,000 meters or more. This trade resulted in the growth of specific marketplaces and travel routes suited for transport by horse and cart. The authorities regulated this trade until the mid-nineteenth century. Despite this, a trade run by cattle and horse traders specializing in the breeding and sale of livestock had gradually developed in the seventeenth and eighteenth centuries. Cattle usually were bought as young animals in the spring from farms further west, driven up to fertile mountain valleys to graze there throughout the summer, and driven on to markets in eastern Norway, where the largest concentrations of people were found. Late in summer, huge droves of livestock were driven from the mountain grazing in Sogn to the southeastern part of Norway. The traffic followed specific tracks and roads and used specific grazing areas *(felæger)*, and the traders attained good prices when selling them in the marketplaces in the towns. Well-defined grazing rights in the lowlands and mountains had to be agreed upon with the landowner. This traffic came to provide important work for many people from the inner parts of the region, especially Lærdal and Aurland. An important drove route crossed the Jostedalsbreen glacier, followed the narrow valleys to the Sognefjord, crossed the fjord on rafts and small boats, and continued upward through the valleys of Lærdal and Aurland (Ohnstad 1980b). The transportation cost by road was low because the animals walked, and thus transported themselves to the markets. Cattle and horses were also transported to Bergen by boat. The building of a railroad between Bergen and Christiania (now Oslo) in the early twentieth century radically altered the old trade routes along the fjord, coast, and mountain passes.

The Agricultural Landscape of Sogn at the Beginning of the Twentieth Century

Land reallocations of the nineteenth century, the cessation of the cottar system, and the large-scale emigration to America all affected the rural population, which declined markedly. Many villages and communities further changed character as a result of mechanization, technological advances, and intensification of agriculture.

At the beginning of the twentieth century, Sogn was still characterized by small villages and hamlets. The few existing villages had developed as communication centers along the fjord for transport and trade. The majority of people in Sogn were still dependent on income from agriculture, forestry, and fishing and the fjord was the most important means of transportation. At that time, the county experienced rapid progress with new markets and the development of international trade. Individual methods of farming became more prominent as the old strip-farming system came to an end through organized land reallotment. This resulted in the dismantling of the old farm structures, which involved removing and constructing new buildings. At the same time, the buildings on each farm became gradually larger and fewer. Considerable technological advances were made. New farming equipment, methods of drainage, meadow-seed mixtures, and artificial fertilizer came into use. Agricultural colleges spread knowledge about the modern technique of crop rotation. Farming equipment such as threshing machines, seed drills, reapers, plows, and harrows were steadily improved and came into use in Sogn. The development of the cream separator in 1879 revolutionized dairies, and gradually electric farming tools and machines made farming more efficient. The development of the fodder-harvester replaced the time-consuming hay drying, which was dependent on good weather. The traditional Norwegian domestic breeds were improved in order to increase milk and meat production.

These changes were due to various state subsidies to stimulate agriculture, such as those given to bring new land into cultivation. The Sogn og Fjordane Agricultural Association allotted state subsidies for new cultivation from 1908, and from then until World War II, 25,000 acres of land were brought into cultivation; the total area of cultivated land in 1949 was 122,850 acres. In that year, only 60 tractors were registered in the whole of Sogn og Fjordane county, but by 1996 this number had increased to 9,892. The use of new machinery was widespread, both for farming and haymaking. Mechanical fertilizer spreaders meant that much larger areas of land could be fertilized, leading to a higher production on the small infields. The new system required less fodder and manure derived from the outfields. Intensive farming using modern machinery required large areas of the best soil. The land had to be cleared of stones, stone walls had to be dismantled, and ditches had to be dug to ensure adequate draining. Drainage projects, especially common in the 1960s and 1970s, were supported financially by the government. Some of the agricultural land has also been put to other uses, such as building new houses in and near villages and for road building. Altogether there has been a considerable reduction in the use of agricultural land, especially marginal land with poor soil. Areas unsuited for modern farming methods have been abandoned.

Also in the mountainous areas there have been considerable changes. The first inventory of mountain farms (summer farms) in the county in 1907 showed that there were as many as 10,558 farms in use. This census also showed that two-thirds of all the cows in the county were grazed on mountain pastures. After World War II, large areas in the mountains were used mainly for sheep grazing as well as for fishing and hunting. The long distances between the main farms and the mountain farms meant that the grazing of animals on the mountain farms involved a lot of work, and the most inaccessible of these farms gradually fell into disuse. Altogether there is no longer a need for all the buildings on the mountain farms, and these have quickly deteriorated. The pastures and meadows that had been open areas dominated by grass species have gradually been taken over by species such as nettle and ivorine, heather, juniper, and shrub. However, the more accessible summer farms are still used in the summer, especially those that are accessible by road or tractor track.

The building styles in the rural districts changed character throughout the twentieth century. Old farm buildings were modified or replaced with new buildings and adapted as specialized farms, such as for milk production or sheep and pig farming, which have replaced mixed farming. Traditional, specialized buildings, constructed in local building materials, are now seldom seen. Materials used for tools, fencing, and buildings have been standardized and are purchased. A typical feature is the replacement of the original small windows by larger standardized windows.

At the beginning of the twentieth century, the development and utilization of hydroelectric power meant competition for the rights of use of rivers and waterfalls in Sogn. Over a ten-year period, from 1910 to 1920, the rights of use of many rivers and waterways were sold. The development of the large hydroelectric resources in the Årdal Mountains started before World War II. In the postwar period, especially in the 1970s, rivers were harnessed for hydroelectricity in the mountains of the inner Sogn region, especially in Aurland, Vik, Luster, and Lærdal. In this connection, construction roads were built in the mountain areas, and large tunnel and dam projects changed the landscape considerably. Material deposited after the construction was finished created new and untypical landscape forms. Electricity produced by the hydroelectric plants was transported out of the region by an extensive network of power transmission lines. Large areas of mountain farm pasture were swallowed up by the new hydroelectric reservoirs. At the beginning of the twenty-first century, farmers started to use small rivers and waterfalls for hydroelectric energy production.

The network of construction roads eased access to and established new conditions for the use of the mountain areas. Old mountain farms were renovated and the mountain areas were exploited more for their leisure opportunities than for farming. New buildings were built on the old stone foundations, and new leisure cottages were built, often without taking into account previous knowledge concerning local climatic conditions. Many of the earlier mountain farm clusters have today become clusters of leisure cottages. The traditional agricultural mountain-farm landscape has been considerably fragmented, changed, overgrown, and partly lost.

Seminatural Vegetation Types in Sogn

Although the changes in Sogn were slower than in other parts of the country, the old social community has nevertheless been gradually dissolved and changes in farming have transformed the agricultural landscape. Many marginal farms, such as those situated on isolated mountain shelves, are no longer in use and communities without road access have been abandoned. This has resulted in a deterioration of the farm buildings, old roads, and cultural monuments. Land areas previously characterized by seminatural vegetation types are now encroached on by trees or planted with spruce. The number of domestic animals has also been reduced. Sheep have replaced cattle, and farmers now graze their goats together on common land in the mountains. The once-common fjord horse is no longer widespread. Yet animal husbandry in the region is still vigorously pursued, and the animals contribute to the upholding of the traditional landscape, both by maintaining and renewing the areas of seminatural vegetation that are dependent on regular grazing. Specialization, however, with the cultivation of a limited number of crops, and the abandonment and overgrowing of marginal areas have affected the cultural and natural diversity of the landscape. Differentiation characterizes rural districts within the county, with, for example, the cultivation of vegetables in Lærdal, fruit growing in Sogndal and Leikanger, and strawberry and raspberry growing in Luster.

Although the extent of land under cultivation in the county remains relatively unchanged, the number of small farms has been nearly halved in the last fifty years. Today, more than 90 percent of agricultural land in the county of Sogn og Fjordane is used for grass production. Agriculture is capital-intensive, and rationalization has resulted in a 75 percent reduction of farmers in the last fifty years. Fodder harvesting is now done with the help of hay bales wrapped in plastic, rows of which are clearly visible in the landscape, while the formerly characteristic hay-drying racks now are rarely seen.

However, we still find historical elements and a wide range of seminatural vegetation types and old agricultural features as relicts in the landscape, especially in areas where there are severe physical constraints on intensification, such as on hillsides and in mountainous areas. It is in these areas especially that we find the traditional harvesting techniques still in use on some farms as part of normal farming practices; these are areas where the agricultural landscape is still characterized by a small-scale pattern of variation, a mosaic of small tilled fields and hay meadows, pastures, wooded hay meadows, pollarded and coppiced woodlands, and summer farms, resulting in a great diversity of plant communities and ecological processes. The traditional agricultural techniques, handed down through generations, and still practiced in the region influence competition and favor biological diversity due to the fact that many of the wild, light-demanding grasses and herbs find niches in the agricultural landscape. As long as agriculture is mainly conducted on the basis of local conditions and without major inputs of outside resources, its impact on the environment is moderate, the effect is stable, and the seminatural vegetation types are maintained.

LEAF-FODDER COLLECTION

In a strongly seasonal climate with a nonproductive winter season as in Sogn, people had to store provisions for the lean periods. Large quantities of fodder for the animals were earlier required for the winter (grass, twigs, bark, and leaves) (Ropeid 1960, Austad 1988). The practice of collecting twigs and leaves for domestic animals is a very old form of fodder harvesting (Figure 14.7). Leaf fodder can be collected by hand or, more efficiently, with small iron knives or axes. Large quantities of fodder were required for the winter and almost all species of deciduous trees were used, as well as some conifers. Tax records from 1863 suggest that this management type was widespread. The choice of species and method of utilization varied from area to area, as did the special names given to tree management. This practice affected both the ecology and scenery of the agricultural landscape.

Pollarding refers to the process of topping trees, that is, cutting back branches above the reach of grazing animals. Lopping is the method of fodder collection. Leafy twigs were cut into smaller pieces (approximately 1 meter), then bunched and tied. The bunches were dried and later stored in barns or placed in stacks. Trees were pollarded every four to six years. Pollarding mainly took place in July and August. Young shoots were sometimes cut directly from the tree bases or as suckers. Some farmers also set aside areas that were coppiced frequently. In some areas leaves were collected for fodder by plucking them. Raking up autumn leaf-fall was practiced mostly to provide animal bedding in stalls. Branches, especially from elm *(Ulmus glabra)* and birch *(Betula* sp.), were sometimes collected during the winter for twigs and bark and later fed to the animals. Bark from elm was peeled, cut into small pieces, mixed with water and given especially to milking cows during the winter and early spring (Austad 1988). The bark of elm was also used in times of need for human nutrition, and known as bark bread *(barkebrød)* (Ropeid 1960; Aarskog 1973; Høeg 1974). Leafy twigs of elm *(Ulmus glabra)*, ash *(Fraxinus excelsior)*, goat willow *(Salix caprea)*, and rowan *(Sorbus aucuparia)* were used preferably for milking cows and calves. Birch *(Betula* sp.) and alder *(Alnus* sp.) were considered to have a lower nutritional value. Birch also had an acrid effect on milk products and was used mainly for sheep. Horses were fed on aspen *(Populus tremula)*. Leafy twigs provided valuable fodder both in terms of nutrition and digestibility. Lunde (1917) compared the nutritional value of elm, ash, birch, and gray alder with hay and red clover *(Trifolium pratense)*. The results indicated that leafy twigs were more or less as good as hay. New analyses by Nedkvitne and Garmo (1986)

FIGURE 14.7. Collecting twigs and leaves for fodder for domestic animals is an old form of fodder harvesting. Photo by Leif Hauge.

confirm Lunde's results. However, nutritional value and digestibility vary with different tree species and with the season. Leaves were found to be an important source of protein, with early summer as optimal. Leafy twigs of hazel *(Corylus avellana)*, lime *(Tilia cordata)*, bird cherry *(Prunus padus)*, and oak *(Quercus* sp.) were also used as fodder, especially for sheep, but to a smaller extent. This was partly because these species are less suitable and partly because these trees were required for other important purposes that were incompatible with fodder production (Høeg 1974). Hazel *(Corylus avellana)* sticks were used for barrel hoops and oak *(Quercus* sp.) for tanning bark and lime twigs *(Tilia cordata)* for ropes.

The normal amount of leafy twigs recorded for twenty-five to thirty sheep for the whole winter was 2,000–3,000 bundles. Each sheep had two meals a day, and the farmer calculated 120 bundles for each animal in addition to five or six burdens of hay. Lunde (1917) estimated the fodder value per bundle (dried) at 800 grams. Some farmers in the area still use a small amount of foliage in the traditional way as fodder for their sheep. Farmers are also testing new techniques for pollarding, chopping, and drying leaves and twigs for fodder (Austad et al. 2000).

The agricultural landscapes in the fjord district are still characterized by the once-widespread use of the deciduous trees for fodder, fuel, and timber. Multiple uses, with grazing and haymaking of the field layer and pollarding and shredding of the tree layer, have formed distinct seminatural vegetation types that are still to be found. The most important are birch, elm, and ash groves (wooded pastures), coppiced woodlands of gray alder and hazel, pollarded woodlands of lime, oak, and elm, and wooded hay meadows with elm and ash.

POLLARDS

On steep scree with large rocks, where grazing and haymaking yielded little fodder, the tree layer was harvested, and today pollards are to be found in the most inaccessible and rugged landscapes in the region. When the process of forming a pollard began, it was important to cut back the branches above the reach of grazing animals. Hard pruning stimulated the growth of dormant buds near the top of the stumps, and such trees therefore rapidly developed side shoots and a characteristic "candelabra" shape. The pollarding process resulted in trees with large main trunks and highly branched crowns. Some pollards (shredded elm) can have several layers of branches to ensure maximum production and can reach 12–15 meters in height (Austad and Skogen 1990). Old pollards show continuity over perhaps several hundred years and are important habitats for other organisms such as fungi, epiphytic vegetation, and insects. Thus pollarded trees contribute greatly to the biodiversity of agricultural landscapes (Moe and Botnen 1997, 2000). Cessation of regular pollarding makes aging trees with extensive foliage growing on coarse scree and loose soils vulnerable to uprooting in strong winds and heavy rain.

POLLARDED WOODLANDS

Thermophilous deciduous woodlands still found in western Norway have been strongly modified by humans (Nordhagen 1954; Skogen and Aarrestad 1986; Skogen and Vetaas 1987; Austad and Skogen 1990). Huge pollarded deciduous trees create distinctive woodland scenery forming large forests along the shores of the Sognefjord. Pollarded woodlands consist of many different types of forest (Fremstad 1997). Use of the forests, including thinning the tree layer, pollarding, and shredding and grazing in spring and autumn, results in open woodland scenery with a variety of habitat niches, related to insolation, temperature, and suitable conditions for light-demanding grassland and forest-edge species. Cessation of traditional management leads to a forest regenerative succession, the development of a luxuriant field layer, and a decrease in light-demanding species. These pollarded woodlands are endangered in Norway today (Moen et al. 2001).

COPPICED WOODLANDS

Coppiced woodlands were formerly found throughout the region. The young shoots were used both as fodder (gray alder, *Alnus incana*) and for other purposes, for instance, barrel hoops (hazel, *Corylus avellana*).

WOODED PASTURES

Several phytoecological classifications of wooded pastures are possible, both according to soil conditions and according to usage (Fremstad 1997; Austad and Hauge 1990; Austad et al. 1991; Austad 1993). The wooded pastures are most commonly found on soil that is unsuitable for arable farming and haymaking. Wooded pastures have a scattered tree layer, often with many large and old pollards (Figure 14.8). The birch grove is a characteristic example, formed by sustained grazing and pollarding. A scattered tree layer and regular pollarding (a four-to-six-year rotation for thinning the crown) provide good light conditions for the field layer, a combination that also gives a maximum harvest of fodder. Birch groves are common in marginal outfields where productivity is low, and they are most commonly found on shallow soils, gravel terraces, scree, and river plains (Ve 1941; Austad 1993). Birch groves mainly regenerate from suckers (Schübeler 1886).

Wooded pastures with stands of juniper *(Juniperus communis)* are a common seminatural vegetation type in Sogn. Juniper shows a great diversity of forms; however, to thrive and develop its characteristic shapes, it demands light and open conditions. Three main forms, ranked as varieties, can be recognized (Austad and Hauge 1990). Throughout history, juniper has had countless uses, documented in ethnological studies (Bugge 1925; Høeg 1974, 1981) and by archaeological excavations and finds. Juniper used for fences, poles, posts, and hayracks was also a commercial commodity. In western

Norway, stockpiles of juniper could be seen ready for shipping at many quays as late as 1990 (Flåten 1992), and farmers still use juniper for poles.

PASTURES

Pastures lack a tree layer in areas where the grazing pressure is heavy, where rockfalls and avalanches are frequent, and/or where the climate is severe. Typical localities are alluvial fans and talus cones in the fjord district. The field layer is dominated mainly by grasses, which have a high capacity for regeneration. Pastures are also common on summer farms in active use in the mountains. A proportion of alpine species is a typical element in the flora of such pastures.

HAY MEADOWS AND WOODED HAY MEADOWS

Herb-rich hay meadows were earlier a common type of seminatural vegetation and were widely distributed in the area. The meadows were cleared of stones, scythed, and traditionally grazed in spring and autumn. Because the meadows were grazed in spring, they were mown late, often in the second half of July. Many herbs and grasses were able to set seed before haymaking took place. Treading by grazing animals punctured the dense turf and enabled seeds to germinate and young plants to become established.

FIGURE 14.8. Wooded pastures with a scattered tree layer and regular pollarding provided good light conditions for the field layer. The pastures were important grazing areas, especially for sheep and goats. Photo by Leif Hauge.

Grazing animals also supported the dispersal of seeds through their grazing behavior and manure.

One of the most characteristic elements of the agricultural landscape in the fjord districts is the presence of colorful, relatively poor, dry hay meadows with a flora including red German catchfly *(Lychnis viscaria)*, ribwort plantain *(Plantago lanceolata)*, northern bedstraw *(Galium boreale)*, lady's bedstraw *(G. verum)*, moon daisy *(Leucanthemum vulgare)*, yarrow *(Achillea millefolium)*, yellow rattle *(Rhinanthus minor)*, and harebell *(Campanula rotundifolia)*. Field scabious *(Knautia arvensis)*, burnet saxifrage *(Pimpinella saxifraga)*, downy oat grass *(Avenula pubescens)*, and hoary plantain *(Plantago media)* are typical in the inner districts of Sogn. On generally lime-rich ground, fragrant orchid *(Gymnadenia conopsea)*, lesser butterfly orchid *(Platanthera bifolia)*, common milkwort *(Polygala vulgaris)*, and quaking grass *(Briza media)* can be found.

The wooded hay meadows in active use are characterized by a small-scale pattern of open meadows and more wooded areas, often with thermophilous tree species such as ash *(Fraxinus excelsior)* and elm *(Ulmus glabra)*, especially on good soils on south- or southwest-facing slopes. A mosaic of light-demanding and shade-tolerant species characterizes the vegetation (Austad and Losvik 1998). A combination of pollarding, grazing in spring and autumn, and haymaking in summer exploits the resources in both the tree and the field layer to the full, and the wooded hay meadow is probably a production type of high sustainability. The wooded hay meadows in Norway still remaining are critically endangered (Moen et al. 2001).

The maintenance of these seminatural vegetation types representing a high biodiversity is dependent on management. This cannot be done without substantial support from the farming communities and other local people who care for their environment and have a meaningful part to play in food production. New roles for multifunctional agriculture and a pronounced demand for high-quality food production will support the upholding of such habitats.

The Artist's Landscape of Sogn

The dramatic nature, the always-shifting weather conditions, the strong and distinct human–nature interaction, the small-scale farms, the hard work, the aesthetic elements of pollarded trees and stone constructions, and the biologically diverse and colorful hay meadows have always attracted artists. Painters, writers, poets, and musicians have contributed to a strong regional and national identity.

Pictorial Art

Until the end of the eighteenth century, wild and raw nature was looked upon as threatening, with few traits of beauty, in contrast to the peaceful and cultivated landscape that was seen as ideal nature. As late as 1820, Bishop Claus Pavels (1769–1822) wrote about Nærøyfjorden (the first fjord system in Norway suggested for protection): "Compared to the fjords I have seen in Søndmøre, Nærøyfjorden . . . is certainly second to none in narrowness, danger and ugliness" (Pavels 1904, our translation).

In the early 1800s, Norwegian painters created pictures of the magnificent fjord landscape filled with romantic and patriotic symbols (Melkild 1993). Johannes Flintoe (1787–1870), the first to see the romantic side of the western Norwegian landscape, was especially moved by the mountains, waterfalls, glaciers, historic sites, and farmers in their various local costumes. Other painters, such as Johan Christian Dahl (1788–1857), portrayed gale-blown trees and fjordscapes in deep perspective, with strong effects of light and shadow (Nasjonalgalleriet 1988) (Figure 14.9). Motives related to Norse mythology were popular, one example being Carl Peter Lehmann's (1794–1876) painting *Fridhiof Kills the Two Trolls on the Ocean.*

FIGURE 14.9. The fjord landscape was rich in ritual symbols from the past. J. C. Dahl painted *Winter by the Sognefjord* with the stone monument in 1827. Photo: National Gallery, Norway, reproduced with permission.

These early romantic paintings expressed the strong patriotic movement dominating Europe at that time. Norwegian artists eager to create a national identity for the reborn state of Norway after five hundred years in union with Denmark sought traces and symbols from before this union. Urban culture had none of this heritage, having been dominated and affected by Continental influences over a long period of time.

The fjord landscape was "original and true." Among the inhabitants here, one could find traditions that had survived through hundreds of years of foreign dominance. Besides, the fjord landscape was exceedingly rich in what was perhaps the most special symbol of former times—the stave churches. Here also was a profusion of historic evidence that came to be classic in the popular range of motives for the romanticists. The tall stone monuments from the Viking Age, monumental burial mounds, mythological heroes *(Fridtjof den frøkne)*, and peasant culture were central elements in their art. History gave life and character to nature; the fjord landscape became gradually the landscape for patriotic romanticism in Norway.

After this first generation of romantic fjord landscapes focusing on dramatic nature and supplemented with historic motives, Sogn became popular among painters from the great art centers of Europe, especially from Düsseldorf in Germany. They traveled in Sogn, making sketches, which they used as the basis for vast canvases. The Düsseldorf painters defused the drama in the landscape, painting in warm colors, in mellow light and even, shiny surfaces of water, turning the landscape into idylls (Melkild 1993). Cooperation between painters to maximize the effect of the work of art was typical for this period. The painter of everyday peasant life, Adolph Tidemand (1814–76), and the landscape painter Hans Gude (1825–1903) worked together on several great works, the most famous being *The Bridal Journey in Hardanger* (although probably it was painted after sketches from Sogn).

As the number of illustrated magazines and reproductions of well-known paintings increased, the nature of western Norway lost some of its mystique for painters. However, the romantic tradition of fjord painting lasted far into the twentieth century. Best known were the painters Adelsteen Normann (1848–1918) and Hans Dahl (1849–1937) (Sveen 1981). Both built in Balestrand huge wooden villas in "national" style with the heads of dragons and rich ornaments. They lived there in summer, forming part of the Continental elitist set that had searched out Sogn and Balestrand.

Nikolai Astrup (1880–1928) is perhaps the artist who best captures the changing moods and mysticism of the western Norwegian landscape. His neo-Romantic paintings focus on details of everyday life, buildings, and small-scale farming, but he also attempted to represent the cycles of nature and life, weaving symbolic elements into his paintings (Figure 14.10). The elements of nature—earth, fire, water, and air—often gain a humanlike character in his paintings (Loge 1986).

The painter Knut Rumohr (1917–2002) is the most well-known contemporary artist in the county. His abstract landscapes capture the moods and colors of nature

(Figure 14.11). His paintings demand much of the viewer, who needs to be familiar with the variations in nature to recognize its moods. A trained eye will identify the countless nuances from spring green shoots, varicolored lilac, dewy moss, lichen-covered stones, glittering autumn colors, and playful fjord surfaces. Rumohr reinvented Norwegian landscape painting (Askeland and Ljøsne 2000).

Some artists use the landscape as a gallery or scene of performance. In the 1990s, different landscape art projects were carried out in the region. Perhaps the most comprehensive took place in the dramatic setting of Utladalen in 1999 (Eriksen et al. 1999). The valley, which cuts into the heart of the mountainous Jotunheimen, drains verdigris green glacier water through a narrow valley with several majestic waterfalls. In this landscape, a group of artists played on the variations of nature, placing mirrors strictly arranged geometrically in scree to reflect the sun and shadow between the mountainsides, or long flowing blue swathes of silk to mark old thoroughfares. They placed small signs with words to give pause for thought for the walkers along paths and bridges. A central contribution was also the reconstruction of cone-formed, traditional stacks created for the storing of hay, foliage, wood, and tar production (Figure 14.12).

FIGURE 14.10. In the painting *Martzmorgen*, Nikolai Astrup captured the landscape with its mystic elements related to superstition. Photo: National Gallery, Norway, reproduced with permission.

FIGURE 14.11. The abstract works of Knut Rumohr, an important modern Norwegian painter, clearly evoke the changing colors and seasons in the fjord landscape. Collection of Ingvild Austad. Photo by Leif Hauge.

FOLK MUSIC

A close connection exists between the nature of the fjord landscape and traditional Norwegian music. Dramatic nature, with vigorous rivers, cascading waterfalls, snow-covered mountaintops, and the glassy fjord surface has been an inspiration for both hypnotizing and lyrical dance tunes. The fiddlers' talents derive inspiration from the natural elements (Kvestad 1981). The fiddler learned from the waterfall but had to throw a cured ham into the water as an offering. The bigger the offering, the better the fiddler. Several of the best fiddlers were said to be in close contact with the darker elements of nature.

Many traditional dance tunes are melodies that have been formed through generations of village life and taught or handed down. Many tunes have no particular composer and have been passed down through long lines of tradition. Thus, they may be called real folk music, created by common people. The music expresses what common people are engaged in and provides associations to rituals and sentiments. Some tunes were related to weddings and burials; others were developed for social activities and dancing.

Folk music can draw lines far back in history, and much of it can be played on fairly simple instruments. The human voice has been the most important instrument. Folk music was an element in daily life, in work, and in festive occasions. Goat- and cow-calls were developed to bring the animals together for milking. The calls were usually short and high-pitched so that the sound would carry far into the distance and be easily recognizable to the cattle. The *lur*, a trumpet with a long tube, or a horn were effective in frightening off bears and wolves and also could serve as a means of communication over long distances. Gradually, more complicated instruments were introduced, such as the Jew's harp and the Norwegian zither and fiddle.

Melodies were often titled after daily activities or special areas. One of the best-known tunes from Sogn is "The Dairy Maids on Vik Mountain," which describes life on a summer farm. Here the composer wove in various natural sounds with both a gentle atmosphere and dynamic or sorrowful parts meant to symbolize the feelings of the three dairymaids. In its original version it is said to have contained the mooing of cows to enrich the music.

Fiddlers' tunes are handed down from one generation to another, and the Hardanger fiddle is the most common instrument. This fiddle is closely related to the violin and was likely developed through the close trading activity that western Norway has had with the Continent. The Hardanger fiddle developed into an instrument for Norwegian

FIGURE 14.12. A cooperative land-art project in Utladalen, Årdal, among painters, textile artists, photographers, musicians, and writers. Leif Hauge constructed elements from the traditional landscape and follows the changing processes of winter. Photo by Leif Hauge.

music tradition, the characteristic resonance from the double set of strings perhaps imitating natures's own sentiments.

Modern musicians are also influenced by the nature and cultural landscape of Sogn. Young musicians have sought their roots in Norwegian traditional folk music, converting it into a modern sound by playing both traditional and modern instruments. This music reaches a new and international audience. The most well known is the saxophonist Karl Seglem (1960), who has made a number of CDs.

LITERATURE AND POETRY

In contrast to sea and river landscapes, the fjord is seldom a subject in literature or poetry. Like other dramatic and changing elements of nature, the fjord played a central role in popular belief and tradition. Faith in destiny was widespread; people and animals could do little to counteract the elements of nature. Thus, people had a somewhat passive attitude toward the natural elements because they believed that destiny was preordained. Along the fjord, the spirits called *vetter* had their special haunts, especially in dangerous and shifting waters. Here the seafarers had to sneak past the spirits undetected.

Popular local folklore related to myths and legends included figures of both fantasy and religion. Often much wisdom and morality were distilled into simple sayings that were easy to remember and hand down. Many stories circulated about supernatural beings who could cause sickness, death, and tragic destinies. People firmly believed in trolls *(hulder* and *underjordiske)* and considered their influence when they planned work and activities.

The fjord landscape did not have the same status among patriotic writers as among painters. There is, however, *Frithiof's Saga,* a cycle of poems based on the Norse saga tradition. Frithiof lived in Sogn, down by the fjord, but the fjord is spoken of as a shore or bay. This suggests that the author, the Swedish poet Esaias Tegnér (1782–1846), never visited the fjord (Beyer 1981).

Henrik Ibsen (1828–1906) used elements from the fjord landscape in his literary works. The play *Brand* from 1866 is built on notes from a journey made to the fjords of Sogn and Sunnmøre. The play *Ghosts (Gengangere)* takes place "by a long fjord in Western Norway" where the dramatic final scene has "the glaciers and the mountain tops" as its background (Ibsen 1881, 1890). Patriotic writers such as Bjørnstjerne Bjørnson (1832–1910) and Jonas Lie (1833–1908) also wrote about life by the fjord, with a rather melancholy atmosphere that ran parallel to the emotions of the characters. The fjord has often been the symbol of something narrow and locked up in human beings. Sogn does not have a regional poet, but Jacob Sande (1906–1967) and Olav H. Hauge (1908–1994) expressed sentiments in their poems about daily life along the fjord that all western Norwegians will recognize (Beyer 1981).

Henrik Wergeland (1808–1845), who instituted the celebration of May 17, the Norwegian constitution day, hailed from Sogn. He wrote several poems inspired by his travels in western Norway. His poem "Sognefjord" illustrates the harsh elements of the fjord:

> He has been the guest of death,
> He has sailed on thunder,
> He has been baptized in horrors,
> Who has sailed the Sognefjord
> From Forthun to Sygnefest.

(Wergeland 1842, our translation)

Conclusion

The landscape of Sogn still has the same outward physical and topographical framework as before. The mountains are still as majestic and covered in snow and glaciers; the valleys are still as expansive and the fjords still as narrow. This spectacular landscape attracts thousands of tourists every year, both foreign and Norwegian, who visit the region. The landscape, however, has changed: there are fewer waterfalls, the mountains are pierced by numerous tunnels, new built-up areas and new industries occupy gradually more and more land area, and increased air and boat traffic bears witness to a more vibrant region than was the case before. The fjord is still important economically as a means of transport, fishing, and new, industrial fish farming (aquaculture). In the future, the consequences of climatic changes will affect the natural conditions in Sogn, as in other parts of the world, and the diminishing of the glaciers might be a result.

Today the flow of information and the numerous impressions from the outside world affect us more and more; laws, regulations, and subsidies govern our activities to an increasing extent. People's values become more uniform, while the cultural landscape (buildings, villages, plantations, and agricultural activity) becomes gradually more homogenous. "Development" often implies a discontinuance of the elements, techniques, and processes that no longer fit in and thus many traces of the past are erased. "Preservation" is often understood as having a museumlike character or being synonymous with stagnation. The cultural landscape today contains important elements and agricultural production that may be useful in the future. It also contains valuable features of biological and cultural-historical value that is underestimated by many.

For some, this leads to a search for what is unique in the landscape. This search for the "true" and "original," the distinctiveness of a place, focuses on both natural and cultural conditions. We are not necessarily searching for "our own roots," but rather for the "roots and soul of the landscape," created by the people who lived and worked there for centuries. To discover, preserve, and comprehend fragments from another time and existence provides a richer understanding of the cultural landscape of today. We can do this by collecting and integrating material and information from various sources and disciplines such as archaeology, vegetation history, ethnology, history, and social anthropology, as well as geology, climatology, biology, geography, and landscape ecology. In contrast to the scientific approach, artists capture the moods and impressions of the very same landscape and interpret and present it within their individual frame of reference.

While the managers of cultural monuments and vegetation ecologists document and present knowledge about the landscape's historical development, highlighting vulnerable and threatened elements, species, and landscape types, artists are inspired by the drama and vulnerability of the landscape. Undoubtedly the amazing fjord landscape will continue to inspire everyone—artists as well as scientists.

References

Aarskog, H. 1973. *Om lauving slik den vart bruka i desse bygdene ved hundreårsskiftet. Balestrand herad.* Sogndal: Historielaget for Sogn.

Askeland, J., and H. Ljøsne. 2000. *Knut Rumohr.* Bergen: Nord 4.

Åstveit, J. 1998. *Ormelid—marginal eller sentral? En arkeologisk punktundersøkelse av Ormelid i Sogn og Fjordane.* Hovedfagsoppgave i arkeologi. Universitetet i Bergen.

Austad, I. 1988. "Tree Pollarding in Western Norway." In *The Cultural Landscape: Past, Present, and Future,* ed. H. H. Birks, H. J. Birks, P. E. Kaland, and D. Moe, 13–29. Cambridge: Cambridge University Press.

———. 1993. "Wooded Pastures in Western Norway: History, Ecology, Dynamics, and Management." In *Soil Biota, Nutrient Cycling, and Farming Systems,* ed. M. G. Paoletti, W. Foissner, and D. Coleman, 193–205. Boca Raton, Fla.: Lewis Publishers.

Austad, I., and L. Hauge, L. 1990. "Juniper Fields in Sogn, Western Norway, a Man-Made Vegetation Type." *Nordic Journal of Botany* 9: 665–83.

Austad, I., and Skogen, A. 1990. "Restoration of a Deciduous Woodland in Western Norway Formerly Used for Fodder Production: Effects on Tree Canopy and Floristic Composition." *Vegetatio* 88: 1–20.

Austad, I., A. Skogen, L. Hauge, T. Helle, and A. Timberlid. 1991. "Human-Influenced Vegetation Types and Landscape Elements in the Cultural Landscapes in Inner Sogn, Western Norway." *Norsk Geografisk Tidsskrift* 45: 35–58.

Austad, I., and M. H. Losvik. 1998. "Changes in Species Composition Following Field and Tree Layer Restoration and Management in a Wooded Hay-Meadow." *Nordic Journal of Botany* 18: 641–62.

Austad, I., A. Braanaas, and M. Røysum. 2000. "Lauv som ressurs—ny bruk av gammel kunnskap." In *Det vestnorske kulturlandskapet. Rapport fra seminar i Sogndal 11.–12. oktober 1999,* ed. E. Ådland, I. Austad, and S. Indrelid. S. Bergen Museums Skrifter, no. 6, 25–30. Bergen and Sogndal: Bergen Museum, University of Bergen and Sogn og Fjordane University College.

Austad, I., and L. Hauge. 2001. "Sognefjord, Norway." In *Threatened Landscapes: Conserving Cultural Environments,* ed. B. Green and W. Vos, 57–64. London: Spon Press.

Austad, I., and I. Øye. 2001. "Den tradisjonelle vestlandsgården som kulturbiologisk system." In *Kulturminner og miljø. Forskning i grenseland mellom natur og kultur*, ed. B. Skar, 135–205. Oslo: Norsk institutt for kulturminneforsking.

Beyer, E. 1981. "Fjordene i litteraturen." In *Fjordheimen: Om Vestlandets fjorder, folk og samfunn før og nå*, ed. K. Fægri, G. Hagen Hartvedt, and F. P. Nyquist, 161–75. Oslo: Grøndahl & Søn Forlag A.S.

Bugge, A. 1925. *Den norske trelasthandels historie*. Skien.

Brøgger, A. W. 1925. *Det norske folk i oldtiden*. Oslo: Institutt for sammenlignende kulturforskning.

Eriksen, H., E. Haarr, L. Hauge, K. Seglem, and M. Vangsnes. 1999. *UTL: Landskapskunstprosjekt i Årdal*. Årdal: Årdal kommune.

Engesæter, Aa. 1976. *Sogndalsfjøra 1801–75: Trekk av den sosiale og økonomiske historia i ein strandstad*. Hovedfagsoppgave i historie, Universitetet i Bergen.

Flåten, I. 1992. *Kulturlandskap og kulturmarkstypar i Gloppen kommune*. Rapport no. 26. Sogndal: Sogn og Fjordane Distriktshøgskule, Avdeling for landskapsøkologi.

Fremstad, E. 1997. *Vegetasjonstyper i Norge*. NINA Temahefte 12. Trondheim: Norsk institutt fornaturforskning.

———, and A. Moen. 2001. *Truete vegetasjonstyper i Norge*. Rapport botanisk serie 2001–4. Trondheim: Norges teknisk-naturvitenskapelige universitet, Vitenskapsmuseet.

Høeg, O. A. 1974. *Planter og tradisjon. Floraen i levende tale og tradisjon i Norge 1925–1973*. Oslo: Universitetsforlaget.

———. 1981. "Eineren i norsk natur og tradisjon." *Norsk skogbruksmuseums særpublikasjon*, no. 5. Elverum.

Ibsen, H. 1866. *Brand: Et dramatisk dikt*. Copenhagen: Gyldendalske Boghandel.

———. 1881. *Gengangere: et familjedrama i tre akter*. Copenhagen: Gyldendal.

———. 1890. *Ghosts: A Drama of Family Life in Three Acts*, translated from the Norwegian by Henrietta Frances Lord. London: Griffith.

Kvamme, M. 1988. "Pollen Analytical Studies of Mountain Summer-Farming in Western Norway." In *The Cultural Landscape: Past, Present, and Future*, ed. H. H. Birks, H. J. Birks, P. E. Kaland, and D. Moe, 349–67. Cambridge: Cambridge University Press.

———. 1998. "Sluttrapport, vegetasjonshistoriske undersøkelser." In *Den tradisjonelle vestlandsgården som kulturbiologisk system. Modellområder Havrå, Grinde, Lee og Ormelid*, ed. I. Austad. NFR-MU-prosjekt107807/730, Utvidet sluttrapport. Sogndal: Høgskulen i Sogn og Fjordane, Avdeling for naturfag.

Kvestad, J. 1981. "Natur, tradisjon og folkemusikk." In *Fjordheimen: Om Vestlandets fjorder, folk og samfunn før og nå*, ed. K. Fægri, G. Hagen Hartvedt, and F. P. Nyquist, 189–95. Oslo: Grøndahl & Søn Forlag.

Loge, Ø. 1986. *Nikolai Astrup. Gartneren under regnbuen*. Oslo: Dreyer Forlag.

Lunde, J. 1917. *Lauv som hjelpefôr*. Christiania: Grøndahl & Søn Forlag.

Melkild, A. 1993. *Kunstnarliv*. Leikanger: Skald.

Moe, B., and A. Botnen. 1997. "A Quantitative Study of the Epiphytic Vegetation on Pollarded Trunks of *Fraxinus excelsior* at Havrå, Osterøy, Western Norway." *Plant Ecology* 129: 157–77.

———. 2000. "Epiphytic Vegetation on Pollarded Trunks of *Fraxinus excelsior* in Four Different Habitats at Grinde, Leikanger, Western Norway." *Plant Ecology* 151: 143–59.

Moen, A. 1998. *Nasjonalatlas for Norge: Vegetasjon*. Hønefoss: Statens kartverk.

———, T. Alm, I. Austad, J. Kielland-Lund, M. Losvik, and A. Norderhaug. 2001. "Kulturbetinget engvegetasjon." In *Truete vegetasjonstyper i Norge*, ed. E. Fremstad and A. Moen,

68–98. Rapport botanisk serie 2001–4. Trondheim: Norges teknisk-naturvitenskapelige universitet, Vitenskapsmuseet.

Nasjonalgalleriet 1988. *Johan Christian Dahl 1788–1857: Jubileumsutstilling 1988*. Oslo: Nasjonalgalleriet.

Nedkvitne, J., and T. Garmo, T. 1986. "Conifer Woodland as Summer Grazing for Sheep." In *Grazing Research at Northern Latitudes*, ed. O. Gudmundsson. New York: Plenum Publishing.

Norderhaug, A., I. Austad, L. Hauge, and M. Kvamme. 1999. *Skjøtselsboka for kulturlandskap og gamle norske kulturmarker*. Oslo: Landbruksforlaget.

Nordhagen, R. 1954. "Om barkebrød og treslget alm i kulturhistorisk belysning." *Danmarks geologiske undersøgelse* 2-R., no. 80: 262–308.

Ohnstad, A. 1980a. "Glimt frå fylket si soge." In *Bygd og By i Norge: Sogn og Fjordane*, ed. N. Schei, 166–98. Oslo: Gyldendal Norsk forlag.

———. 1980b. "Driftehandel." In *Bygd og By i Norge: Sogn og Fjordane*, ed. N. Schei, 226–41. Oslo: Gyldendal Norsk forlag.

Pavels, Claus. 1904. *Claus Pavels's Dagbøger for Aarene 1817–1822*, vol. 2: *1820–1822*. Udgivne for den norske historiske Forening af Ludvig Daae. Christiania: Grøndahl & Søn.

Pushmann, O., J. Hofsten, and A. Elgersma. 1999. *Norsk jordbrukslandskap—en inndeling i 10 jordbruksregioner*. NIJOS-rapport 13. Ås: Norsk institutt for jord- og skogkartlegging.

Ropeid, A. 1960. *Skav. En studie i eldre tids fôrproblem*. Oslo: Universitetsforlaget.

Schübeler, C. F. 1886. *Viridarium Norwegicum 1. Norges væxtrige. Et Bidrag til Nord-Europas Natur og Culturhistorie*. Christiania: Aschehoug forlag.

Slotte, H., and H. Göransson. 1996. *Lövtäkt och stubbskottsbruk. Människans förändring av landskapet—boskapsskötsel och åkerbruk ved hjälp av skog*, Del I–II. Stockholm: Kungliga Skogs- och Lantbruksakademien.

Skogen, A., and P. A. Aarrestad. 1986. *Botaniske undersøkelser og vurderinger av Flekke-Guddals-, Os-, Naustdals-, Gjengedals-, Gaular-, Jølstra-, Breims- og Sværefjordsvassdragene i Sogn og Fjordane*. Rapport 43. Bergen: Botanisk Institutt, Universitetet i Bergen.

Skogen, A., and O. R. Vetaas. 1987. *Flora og vegetasjon ved Olden- og Hornindalsassdragene i Nordfjord med vurdering av deres verneverdier i distriktet*. Rapport 45. Bergen: Botanisk Institutt, Universitetet i Bergen.

Strumse, E. 1996. *The Psychology of Aesthetics: Explaining Visual Preferences for Agrarian Landscapes in Western Norway*. Thesis. Bergen: Research Center for Health Promotion, University of Bergen.

Sveen, D. 1981. "Fjorden og billedkunsten." In *Fjordheimen: Om Vestlandets fjorder, folk og samfunn før og nå*, ed. K. Fægri, G. Hagen Hartvedt, and F. P. Nyquist, 177–87. Oslo: Grøndahl & Søn Forlag.

Thue, J. 1980. "Jektefart, handelsferder og marknader." In *Bygd og By i Norge: Sogn og Fjordane*, ed. N. Schei, 266–79. Oslo: Gyldendal Norsk forlag.

Utaaker, K., and E. Skaar. 1970. "Local Climates and Growth Climates of the Sognefjord Region." *Acta Agriculturae Scandinavica* 1: 1–218.

Valvik, K. A. 1998. *Lee—en tradisjonell vestlandsgård? En arkeologisk punktundersøkelse av gården Lee, Vik, Sogn og Fjordane*. Hovedfagsoppgave i arkeologi. Universitetet i Bergen.

Ve, S. 1941. "Bonden, buskapen og skogen i gamle Vestlandsbygder." *Tidsskrift for Skogbruk* 49.

Wergeland, H. 1842. "Sognefjorden." *Christiansandsposten*, December 12, 1842.

15.
The Agropastoral Mountain Landscape in Southern Norway: Museum or Living Landscape?

ANN NORDERHAUG

During the twentieth century, land-use changes caused major landscape transformations all over Europe (Baldock 1990). In Norway, rapid and extensive changes in the rural landscape took place particularly in the second half of the century. These changes have partly been caused by urban expansion and partly by the general process of intensification and specialization in agriculture, leading to the abandonment of marginal agricultural land and to the expansion of forests. This is not a new phenomenon. Landscapes are dynamic and continuously changing. There is, however, a considerable difference between earlier processes of change and the alteration of the rural landscape that has taken place in the twentieth century. Until 1950, agriculture in Norway was still partly based on ancient techniques and seminatural vegetation in both infields *(innmark)* and outfields *(utmark)*. These vegetation types are the products of utilization over a long period of time, and were therefore seen as both stable and natural. The area of infields and outfields, as well as of the different vegetation types, has changed through history due to variations in population pressure, political factors, and ecological conditions. The seminatural vegetation used for grazing and fodder production has, however, been fundamental for human land use since agriculture began in Norway. Today, these landscape elements are rapidly disappearing. Together with this we are losing a rich biological diversity and unique evidence of living cultural history. In this process we also lose other parts of our cultural heritage, such as ancient buildings, stone walls and other boundaries, clearance cairns, and old paths and roads. Our landscape is getting poorer.

In Norway, these changes have led to an increasing interest in the cultural landscape and the conservation of its natural and cultural assets (Norderhaug 1997). Between 1992 and 1994, valuable rural landscapes were systematically registered throughout Norway in a national inventory (Norderhaug 1992). As Ihse and Skånes point out in chapter 10 (this volume), the complexity and many different values associated with the cultural landscape underline the need for a holistic approach to landscape management as well as to landscape analysis. In Norway, efforts were made to integrate both natural and cultural landscape values in the registration process. The aim was to identify areas where these qualities still make up an integrated whole and give the landscape a historical perspective. It was also hoped that this registration of valuable cultural landscapes would initiate a process for a better understanding of this important part of our natural and cultural heritage and the need for conservation and a more holistic management of valuable areas.

The Hjartdal–Svartdal Area

One of the areas given priority in the national inventory comprised the valleys of Hjartdal and Svartdal in the inland part of Telemark county in southern Norway (Det sentrale utvalget for nasjonal registrering av verdifulle kulturlandskap 1994). The interactions between the natural environment and sustained human use over a long period have created a landscape with high biodiversity at landscape, biotope, and species levels (Norderhaug, Ihse, and Pedersen 2000), which makes it of high conservation value. It also contains a variety of cultural monuments, including several medieval wooden buildings (Wagn 1992). The relationship between cultural monuments and their surroundings can still be distinguished, and the landscape structure makes it possible to trace the history of the use of the countryside and how people have lived (Figure 15.1). The area is therefore of great value for educational and research purposes. In addition, the area has aesthetic values and considerable potential in terms of tourism and outdoor recreation. These values received recognition when the Ministry of Environment nominated Hjartdal–Svartdal as Norway's candidate for UNESCO's cultural landscape prize for 2003 (Melina Mercouri International Prize for the Safeguarding and Management of Cultural Landscapes).

Natural Conditions

The Hjartdal–Svartdal area stretches westward from lake Hjartsjåvannet along the narrow Ambjørndalen valley (in Hjartdal local authority district) to Svartdal (in Seljord local authority district). Two rivers, the Svorteåi and the Hjartdøla, run from the mountains in the north, where they meet before flowing into Hjartsjåvannet.

The area is representative of one of the physical-geographical regions defined by the Nordic Council of Ministers (Nordiska Ministerrådet 1984), the forest area of Øvre Setesdal and Telemark. This classification of the Nordic regions, mentioned also by Ihse and Skånes and by Lundberg in their chapters in this book, aimed at providing a sound basis for conservation planning. The classification is primarily based on differences in

FIGURE 15.1. (A) The cultural landscape of the Hjartdal–Svartdal area in 1912 was open and extensively used with small arable fields for crops and large seminatural hay-producing and grazing areas. Courtesy Tuddal museumslag, reproduced with permission. Photo by K. Asland. (B) In 1996, most of the seminatural grasslands were abandoned and overgrown. The fertilized arable land was used as hay meadows. Although the forest is gaining ground, the agricultural history can still be read in this landscape. Photo by Ann Norderhaug.

vegetation, climate, and geomorphology. The forest area of Øvre Setesdal and Telemark is in this classification system described as northern boreal vegetation consisting of mountain birch and coniferous forests. The regional classification is, however, on a small scale and of a general nature and does not give a detailed description of variations within each region. The Hjartdal–Svartdal area stretches from more heat-demanding, deciduous-dominated, boreal-nemoral forest (Moen 1998) in the valley bottom (200 meters above sea level), with deciduous trees such as oak *(Quercus robur)*, wych elm *(Ulmus glabra)*, ash *(Fraxinus excelsior)*, lime *(Tilia cordata)*, and hazel *(Corylus avellana)*, through different vegetation zones up to the alpine area. Above the boreo-nemoral zone on the slopes, there is a zone containing a mosaic of deciduous and coniferous forest, including elm and hazel (the southern boreal zone), then the middle boreal zone, dominated by coniferous forest, the northern boreal zone, dominated by mountain birch *(Betula pubescens)*, and stretching up to the alpine zones.

The geology of the area is varied and dominated by bedrock of the Seljord group, including quartz, schist, and conglomerate (Dons and Jorde 1978). The bedrock is mostly nutrient-poor, but calcareous schist and sandstones are also found. A sill-like intrusive body of gabbroic rocks stretching through part of the area provides better conditions for the vegetation. The Hjartdal–Svartdal area is situated above the highest postglacial shoreline. There are, however, till and fluvial deposits in the valley bottom and moraine of varying thickness on the mountainsides. At higher altitudes in the north of the area there are large peat-mire complexes. The winter is relatively cold and the summers relatively warm. The normal monthly mean temperature (1961–90) is –7.0°C for January and 11.4°C for July. The mean annual precipitation for the nearest weather station (Øyfjell–Trovatn at 715 meters a.s.l.) is 1050 mm, which results in slightly humid conditions. Precipitation is highest from August to November (Norwegian Meteorological Institute, unpublished data). However, the local climate is influenced by the topography.

The farms are located in the valley bottom and on the slopes from about 200 meters to 600 meters above sea level, that is, from the boreal-nemoral to the middle boreal zone. The summer farms *(setrer)* are found at higher altitudes, 600–900 meters a.s.l., in the middle-boreal to north boreal zones. The older farms are largely concentrated where the climate is most favorable on the south-facing slopes at about 300–600 meters a.s.l. on veins of more nutrient-rich bedrock, while the cottars' farms were located in less favorable areas. In aerial photographs, some places may be seen to have a sharp boundary between poor and rich bedrock, corresponding with the old boundary between infields and outfields.

Agricultural History

Human influence on and use of this landscape has varied. The first people who used the Hjartdal–Svartdal area were Stone Age hunters. Hunting and fishing continued to play an important role in this part of the country after the development of agriculture. The first permanent agricultural settlements in Telemark were established in the coastal and lowland areas, but from about A.D. 400 there are traces of permanent settlement in inland parts of the county. Many archaeological finds from the period A.D. 500–800 have been made in the area, but the richest discoveries are from the Viking Age, A.D. 800–1050. From then until the Black Death in 1349, there was a period of expansion; the population continued to grow both here and in the rest of the country, and much new farmland was cleared. After the ravages of the Black Death, many farms were abandoned, and it was not until the eighteenth century that there was renewed expansion (Tables 15.1 and 15.2) (Flatin 1942; Martens 1975; Wagn 1992; Braathen 1995).

TABLE 15.1.
Number of farms and cottars' farms in Seljord, 1664–1989.

Year	Farms	Cottars' farms	Total
1664	131	25	156
1723	190	33	223
1801	230	194	424
1865	393	315	708
1939	603	—	603
1989	232	—	232

Source: Flatin (1942) and Statistics Norway (unpublished data).

TABLE 15.2.
Number of domestic animals in Seljord 1657–1989.

Year	Horses	Cattle	Sheep	Goats	Pigs
1657	132	759	1114	350	45
1723	132	927	854	25	—
1784	222	1929	3169	—	—
1835	222	2015	3688	347	141
1855	249	2901	5484	1221	161
1890	255	2296	3370	1702	249
1907	252	2054	2612	1614	670
1939	341	2370	2450	1609	653
1959	245	1447	3347	650	744
1989	51	736	3212	70	239

Note: The number of goats and pigs in the eighteenth century is uncertain.
Source: Flatin (1942) and Statistics Norway (unpublished data).

The history of the area and the way its natural resources have been used during the past three centuries have been well documented. The priest Hans Jacob Wille (1989 [1786]) wrote about his parish of Sillejord (Seljord). The merchant Engelbret Michaelsen Resen Mandt (1989 [1777]) wrote a general description of upper Telemark. The topographer Jens Kraft (1826) wrote in detail about Bratsberg county (roughly equivalent to the modern Telemark). Magnus Brostrup Landstad (1985 [1880]), another priest in Seljord, described the district and collected songs and stories. In addition, there is a great deal of literature on local history, including documentation of summer farming, marginal farms, cottars' farms, mills, and sawmills (Aakre 1991, 1996; Nes 1991) that no longer exist. The Hjartdal–Svartdal area is also mentioned in Bishop Jens Nilssøn's descriptions of his travels in 1574–97 (Nilssøn 1885 [1574–97]).

The geology, soils, climate, and topography of the area have been of decisive importance for the way its natural resources have been used. Land use has been based on animal husbandry. Wille wrote in 1786 that the parish of Seljord contained 188 farms, 184 cottars' farms, and 515 summer farms. The infields, consisting of arable land and hay meadows, were small, but the outfield areas were very large. The arable land was mainly used to grow barley, oats, potatoes, hemp, and flax. In addition, winter rye and barley were cultivated using slash-and-burn techniques in the outfields. Wille himself owned 4 hectares of arable land, 22 hectares of hay meadows, and large areas of outfield in the mountains. He also established gardens with fruit trees. In favorable years he produced eighty barrels of grain. He kept three horses, twenty-four head of cattle, and thirty sheep. Most farms had a smaller area of infield than the priest's, but the total area of hay meadows and pasture, forest and mountain, was usually a hundred times larger than the area of arable land. In preindustrial times, this was important because farmers could keep many animals and thus get enough manure for the infields and food production. The landscape around the farms was open and forested areas were mainly in the outfields higher up toward the mountains. In the mountains, there were large areas of pasture and outlying hay meadow. Many farms had both spring and autumn farms and one or two mountain farms (Figure 15.2), which made it possible to use land at varying altitudes at different times of year. Spring and autumn farms could be close to the settlements or up to 20 km away (Nes 1991). The grazing areas in the mountains were so extensive that several hundred cattle from other parts of the county were also allowed to graze there in summer. It was more difficult to provide enough winter fodder. To supplement the hay, farmers used straw and leaves, bark and twigs of elm, ash, birch, rowan *(Sorbus aucuparia)*, and sallow *(Salix caprea)* as winter fodder (Wille 1989 [1786]; cf. Resen Mandt 1989 [1777]).

The traditional date for starting the spring work was May 1, according to Wille (1989 [1786]). Earth or ashes were often scattered on the snow to make it melt more quickly and allow plowing to start. The plows used were small, sometimes of a primitive type *(ard)*. Small stones and pebbles were left in the fields because they retained heat and

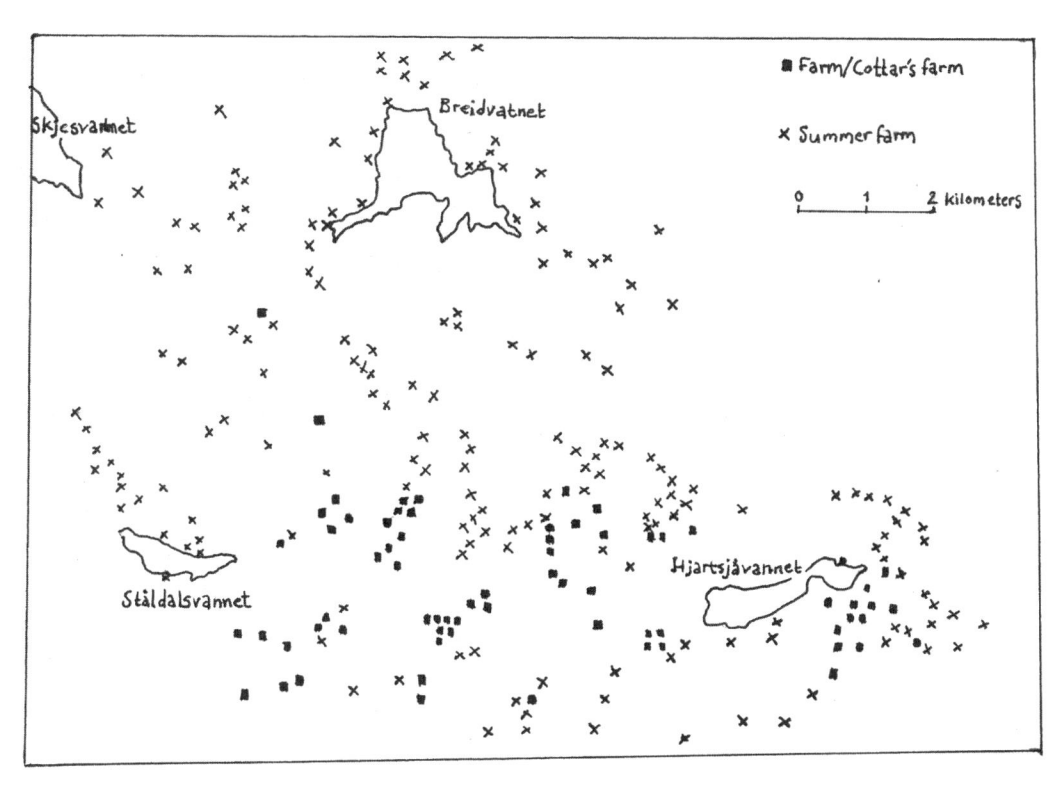

FIGURE 15.2. Map of Hjartdal showing the location of farms, cottars' farms, and all the (earlier) spring/autumn farms, that is, summer farms used before and after the cattle were moved to summer farms higher in the mountains (Nes 1991). Courtesy Trygve Nes.

made the corn ripen more quickly. Once the seed was sown, the fields were harrowed. When the corn germinated, calves were often allowed to graze in the fields for a short period of time because this was supposed to improve the growth. The corn was harvested early, bound in sheaves and dried on poles. Manure from the winter was spread on the fields after they were sown. Alternatively the cattle were penned in small enclosures while the fields lie fallow. The enclosures were moved from one part of the fields to another in order to manure the whole area. A general opinion among the farmers was that the cornfields should rest every fourth year, but such a rotation was rarely practiced.

In spring, the animals grazed in the infields until it was possible to move them up to the spring farm for a fortnight before moving on to the summer farm higher up in the mountains. Here the animals were kept for the rest of the season. In some cases, all the people from the farm moved up to the mountain farm for some time for haymaking in the mires and outlying hayfields. At other times, there were generally two girls at the summer farm, one to herd the animals and the other to milk the cows and goats in the morning and evening and then make butter and cheese during the day.

The small breeds of cattle and sheep did well in rugged terrain. During the night they were kept in small enclosures or in a cow house to protect them against bears and wolves. At the beginning of September, both people and animals moved down to the spring/autumn farm and stayed there for a fortnight before continuing down to the main farm. Food production on the summer farms was very important for the farmers' self-sufficiency. Products from the summer farms were brought down to the farm on horseback.

Both infield and outfield meadows were scythed at the end of July or the beginning of August. The hay was dried on the ground for a few days before being carted back to the farm. Hay from outlying meadows was stored in barns or haystacks until it could be transported to the farm by sledge in winter. Hay meadows in the infields were grazed in spring and autumn before and after the animals were moved to the spring and summer farms. When the animals were moved to the spring farm, the meadows were raked and cleared of stones. However, they were not plowed, nor were they manured, because the manure usually was needed for the crops. Between haymaking and the corn harvest, and well into the autumn, as much fodder as possible was collected by pollarding trees (lopping) or by leaf-plucking as described by Austad and Hauge in chapter 14. Bunches of twigs were dried slowly in the barn. Trees were also pollarded at the end of winter if there was a fodder shortage and the branches were used as fodder.

The Nineteenth Century

A description dating from about fifty years later (Kraft 1826) gives a similar picture of the area and its agriculture. During the nineteenth century, however, the population grew quickly. By 1865, the number of farms in Seljord had more than doubled to 393 and the number of cottars' farms had risen to 315. The landscape and its resources were more or less fully utilized. The importance of summer farms increased with population pressure. Many of the farms were still of considerable size and had rich grazing areas at the beginning of the eighteenth century. During the eighteenth and nineteenth centuries, however, these farms were divided once or several times. In this connection the importance of grazing and fodder harvesting in the outfields increased considerably. Each farm had a particular grazing area in the mountains of about 300–400 hectares. On the subdivision of a farm, outlying grazing areas could also be split or kept as a common grazing area. In Hjartdal, the spring and autumn farms were utilized through a particular system of manuring *(hevdeturprinsippet):* four or five spring farms were located together within a common grazing area, but only one of them was used at a time, that is, all the animals belonging to the main farms were gathered for milking and for the night at one of the spring farms. This made it possible to manure the infields *(setervollen)* of this spring farm. On the other spring farms, the hay on the infields was cut. The next year one of the other spring farms was manured (Nes 1991).

During the last part of the nineteenth century, emigration and industrialization gradually reduced the population. A number of small, marginal farms were then abandoned. Along the old road, Skarsveien, connecting Svartdal with another valley, Åmotsdal, most of the cottars' farms were abandoned during this period. Many went to the United States and Australia, while others moved to other places in Norway. In general, all these cottars' farms were poor, small in size, and recently established. Another reason for the abandonment of these sites was that Skarsveien, which had earlier been an important route for both people and cattle, lost its importance as a result of the construction of a new road through Dyrlandsdalen during the 1860s.

At the end of the nineteenth century, new agricultural equipment and techniques were developed. Fertilizing, better plows, iron harrows, reapers, and threshing machines were introduced. Some infield hay meadows that had not previously been cultivated were plowed and sown with grass seed, and in the 1880s crop rotation was also introduced (Table 15.3). Animal husbandry was improved by better feeding and systematic breeding. The agricultural show in Seljord, which has been an annual event since 1866, played an important role. In addition, land reallocation made it easier to introduce new methods. This was, however, a long process in the Hjartdal–Svartdal area, stretching over several decades (Karlsrud 1998).

The Twentieth Century

Abandonment of farms, as well as agricultural improvements, led to changes in the rural landscape. This process continued in the twentieth century and the number of farms and animals decreased. This trend was temporarily reversed in the 1930s as a

TABLE 15.3.
Crops sown in Seljord, 1835–1907.

	1835 Barrels	1855 Barrels	1875 Barrels	1890 Hectoliters	1907 Hectoliters
Wheat	4	8	10	1	8
Rye	1	8	1	1	33
Barley	521	726	772	891	693
Mixed grain	2	0	1	12	8
Oats	54	134	110	167	176
Peas	0.5	1	0	1	3
Potatoes	1,395	2,213	2,387	3,362	2,911
Grain fodder	—	—	2	40	54
Legumes	—	—	1	4	—
Grass	—	—	62	35,000	100,800

Note: One barrel (korntønne) = 145 liters.
Source: Flatin (1942).

result of the difficult economic situation. New land was then cleared and some abandoned farms and cottars' farms were taken into use again. However, since World War II, the pace at which farms have been abandoned has accelerated. The topography of the area limits the extent to which agriculture can be rationalized, and the farms are small and inconvenient by today's standards. Many farmers therefore have another paid job, and full-time farmers often must have access to land belonging to several farms to obtain large enough area for haymaking and grazing. Because of the climate and topography, agriculture in the area is still based on animal husbandry. The arable land is now used mainly as cultivated meadow, and most of the old, formerly unfertilized hay meadows have been fertilized and are used as pasture. Grass is cut from June onward. The mires and outlying hay meadows are no longer used and trees are no longer pollarded to provide winter fodder. The abandoned farms, old meadows, and pastures are becoming overgrown by forest. From about 1910, the use of summer farms gradually decreased, even if the summer-farm activities increased again during World War II. Some of the summer farms even had a larger number of animals during this period than in former times. In the 1960s most of the traditional use of summer farms came to an end. However, young cattle and sheep still graze the outlying pastures and other grazing areas in the forests. Some of the farms have introduced a new kind of summer farming: people are hired on a summer farm with a new big cow house to take care of cows from several farms. A new road allows the milk to be transported by tankers to the dairy. This kind of "modernized summer farming" has also led to a limited expansion of cultivated hay meadows around the summer farms.

Forestry, Mining, and Hydropower

In addition to animal husbandry, forestry has been very important in the Hjartdal–Svartdal area. The timber trade started in the sixteenth century (Flatin 1942). Large areas of forest were also used for charcoal production. In the mountainous parts of inland Telemark, large quantities of charcoal were used to produce bog iron from the Iron Age (500 B.C.–A.D. 600) and until mining activities started. The first ironworks began operating in Telemark in the sixteenth century and mining for copper started in 1538 in the vicinity of Hjartdal–Svartdal. Both logging and charcoal production severely depleted the forests. The 1723 land register as well as district governors in the nineteenth century mentioned this fact. Timber prices rose steeply in 1870, thus putting even more pressure on the forests. However, during the twentieth century, the forest in the area again improved in quality and expanded, partly because conservation regulations were implemented and partly because farms and agricultural land were abandoned and pressure on natural resources was reduced.

There have also been mining operations in Svartdal itself (Telnes 1994). A French company started the Blika goldmine in 1882 and many people from Svartdal and

Hjartdal found work there. A smelting works, machine room, and a road up to the mines were built in 1897, but in 1916 the mine was sold and closed down. In the 1930s, when unemployment was high, a new attempt was made to open the mines, and in 1937–38, eighteen men were employed there. The mines were closed again at the beginning of the war. The total landscape impact of this activity was not significant, but possible new trial operations in the area may be more visible.

Norway's first hydropower plant providing electricity for the general public was built in Skien in Telemark in 1885. However, hydropower developments began to gather momentum only at the turn of the century, and it was not until after World War II that large-scale exploitation of river systems occurred and the electricity grid was established (Straume 1975). Electricity reached the Hjartdal–Svartdal area at the end of the 1940s. The Hjartdal watercourse was developed for hydroelectric power in the 1950s, when the lakes Skjesvatnet, Breidvatnet (Figure 15.2), Kovatnet, Bonsvatnet, and Vindsjåen were regulated and 500 hectares of land and a number of summer farms were submerged (Nes 1991).

Cultural History

According to Norwegian myths, the people of Telemark were hard and rebellious, and only too happy to kill the clergy, tax collectors, and officers (Friis 1632). In 1540, they rebelled against King Christian III and the mining company, in which Germans were employed by the king to run the two copper mines in inland Telemark. The mining company had been taking the law into its own hands and plundering the local people until the farmers finally lost patience and drove the company's employees away. The king answered by sending an army to Telemark. The farmers gathered in Ambjørndalen to meet the army but were tricked into laying down their weapons. The king's men broke their promise of peace and the rebels were given harsh sentences. Six men were executed at Hjartdals Prestegård (the parsonage). After that, according to hearsay, the people of Telemark never again trusted the state or the authorities (Landstad 1985 [1880]; Aanderaa 1990).

Perhaps the people of inland Telemark were not particularly pugnacious or selfwilled. However, they repeatedly resisted the pressure from central authorities. They opposed the taxation introduced by King Magnus Lagabøter in 1277 and continued to pay their tax by an older regime. As mentioned by Sporrong (in chapter 8 in this volume), farmers who owned their land perhaps had a stronger feeling of independence. Many of the farmers of inner Telemark owned their farms even at times when most farmers in the rest of the country were tenants. They seem always to have had a strong sense of identity, which has left rich local traditions of poetry, folktales and songs, and local costumes. It was mainly in Telemark and Hallingdal that the special Norwegian acanthus or rose painting *(rosemaling)* was developed during the rococo period. Many of the Norwegian folk songs from the Middle Ages originate from Telemark, and Telemark provides a rich source of folk music (Østvedt 1975). Wood ornamentation *(treskjærerkunsten)* reached an advanced level in Telemark. In the Hjartdal–Svartdal area, examples of this impressive artwork can still be observed in a number of old storehouses.

Historically, the Hjartdal–Svartdal area and other inland districts of Telemark were relatively isolated. This is presumably one reason why agricultural and cultural traditions have been maintained here until recently. The priest M. B. Landstad wrote in 1848:

> The inner part of Telemark is due to its isolated location and the character of its people, one of the districts in our country where ancient language and traditions have been maintained. Arriving over Midtheien to Hittedal or Gransherred, foreigners will be surprised by meeting the ancient buildings, dresses, language and life style. Foreigners may get the impression that they are transported many centuries back, to the Viking age. People love the old and traditional and express mistrust with the new, resulting in a slow process of improvements and change. (Landstad 1925 [1853], IX; my translation)

Norway has a highly varied nature (Moen 1998), and there are considerable regional differences. Human use of nature resources has contributed to the rich landscape diversity. Accordingly, a number of landscapes could be described as "typically Norwegian." The striking fjordscape described by Austad and Hauge in chapter 14 of this volume is one example; the inner parts of Telemark are another. Telemark was "discovered" by tourists at the beginning of the nineteenth century and landscape painters followed (Østvedt 1975). The first was the Danish Johannes Flintoe, followed by the founder of Norwegian landscape painting, J. C. Dahl. Later came several well-known Norwegian painters such as Adolph Tidemand, August Cappelen, Erik Werenskiold, and Fritz Thaulow. In the 1890s, Halfdan Egedius developed the Telemark identity in his painting by combining a new psychological insight into the individual's mind with the mood expressed in the surrounding landscape (Storm Bjerke 1998).

It is primarily in the twentieth century, however, that the inland districts of Telemark came to be regarded as a typically Norwegian landscape. The clusters, rows, or courtyards of buildings in the farms were a particular source of inspiration. This can be seen clearly in the work of the large group of painters, including Henrik Sørensen, Harald Kihle, Kai Fjell, Kaare Espolin Johnson, and Per Rom, who formed the "Telemark school" of Norwegian art from the mid-1920s to about 1945 (Storm Bjerke 1998). Much of their art is consciously Norwegian, with a national romantic touch (Willoch 1937). The painting of the typically Norwegian found in the periphery was to some extent also a polemic against the center, a kind of antiurbanism. The inner parts of Telemark formed a preindustrial landscape shaped by a lifestyle now in flux due the "agricultural revolution," hydropower development, and industrialization. The Telemark school of painting can thus be regarded as a wish to strengthen and preserve a lifestyle concentrated on the farm (Storm Bjerke 1998). The Hjartdal–Svartdal area played a central role in this period and is depicted, for instance, in Henrik Sørensen's painting *From Hjartdal* from 1925 (Figure 15.3).

The Current Landscape

Even though the landscape in the Hjartdal–Svartdal area has changed, especially in recent decades, elements of the old structure can still be seen, for example, dry-stone walls and other boundaries, and many paths, tracks, and roads of varying age and width. Although the forest is gaining ground, the numerous cairns of stones cleared from fields and hay meadows are evidence that the landscape once was more open. Remains

FIGURE 15.3. *From Hjartdal,* painting by Henrik Sørensen, 1925. Courtesy Holmsbu Billedgalleri, reproduced with permission. Photo by O. Væring.

of small mills and sawmills, house sites, and the ruins of outlying barns also bear witness to the earlier utilization of natural resources and the fact that the landscape has at times been intensively used.

Both functional and economic adaptations are very clear in this landscape, and it is possible to see how changing requirements have resulted in different kinds of adaptations. For instance, many of the older farms were located where the local climate was most favorable and within easy reach of outfield resources, but now they have been abandoned because they are too far from the main road in the valley. Farms that had the richest and extensive outfield resources have also been abandoned because by today's standards there is too little arable land.

Since the climate is relatively dry in the Hjartdal–Svartdal area, wooden buildings have a long life. Many of the houses have been well maintained, especially the storehouses, which were often decorated with beautiful carvings. Formerly the farms had several small buildings with separate functions, usually arranged around a courtyard. On some farms this arrangement remains well preserved. There is a wealth of historical buildings and many are protected, most of which are medieval or from the eighteenth century (Wagn 1992). However, some have been destroyed in fires or sold and moved and many of the outlying buildings are dilapidated.

Interviews with farmers in the Hjartdal–Svartdal area in 1990 showed that, despite the modernization process, traditional techniques such as pollarding, summer farming, and the use of seminatural pastures and old unfertilized, unplowed hay meadows were commonly practiced as recently as the 1940s. Even in the 1990s, there were a few farmers who ran traditional summer farms, had Telemark cattle, mowed the seminatural hay meadows, and pollarded elm trees.

The landscape thus still consists of a mosaic of human-influenced vegetation types. Most of these are intensively farmed, but there are still many old pollarded trees scattered in the woodland as well as remnants of various types of grazed forests and pastures that were created by preindustrial agriculture. A unique feature of the Hjartdal–Svartdal district is the existence of many small hay meadows with a species composition created by ancient techniques and a long period of sustained management. They are highly species-rich and of vital importance for biological diversity (Figure 15.4). Their moisture content varies, and they can be identified in plant-sociological terms as the *Nardus stricta* (mat grass), *Geranium sylvaticum* (wood cranesbill), or *Avenula pubescens* (meadow oat grass) meadow types (Kielland-Lund 1992). Most of them are identified as the *Avenula pubescens: Hypochoeris maculata* (spotted catsear) type (Sjørs 1954; Kielland-Lund 1992), characterized by species like *Rhinanthus minor* (yellow rattle), *Polygala vulgaris* (common milkwort), and *Botrychium lunaria* (moonwort), and orchids such as *Gymnadenia conopsea* (fragrant orchid) and *Dactylorhiza sambucina* (elder-flowered orchid). The latter is the county flower of Telemark. Others are of the *Avenula pubescens: Hieracium pilosella* (hawkweed) type (Kielland-Lund 1992), with species like *Lychnis viscaria* (sticky catchfly) and *Antennaria dioica* (mountain everlasting), or as the *Geranium sylvaticum: Trollius europaeus* (globe flower) type. The latter is a rare species in this part of Norway.

Many species depend on seminatural vegetation types. They disappear when management is intensified or changed. For instance, many plant species disappear from the old hay meadows if they are fertilized, cut earlier than before, or grazed instead of cut. As the seminatural vegetation types are steadily decreasing due to land-use changes, some of these species are endangered, and remnant habitats become increasingly important. To secure the survival of these species, it is necessary to protect and maintain some of the seminatural habitats.

The Future

As was the case fifty years ago, the Hjartdal–Svartdal area may be seen as a land where the links to premodern Norway are still alive. It is a landscape that has changed through

FIGURE 15.4. Seminatural hay meadows are rare and the conservation value of remaining habitats is increasing. The colorful meadows often have 30–40 species of grass and flowers per m^2 (Norderhaug, Ihse, and Pedersen 2000). Photo by G. Horn, reproduced with permission.

the centuries but that still contains remnants of ancient agrarian ecosystems important for Norwegian history and biodiversity.

Conservation and management of valuable cultural landscapes of this kind is a difficult task. The old buildings, dry-stone walls, roads, and other cultural monuments can be maintained if sufficient money is available. It is more complicated to maintain the landscape structure and the interplay between the cultural monuments and the environment. This requires a dynamic countryside where communities have both the knowledge and the interest necessary for maintaining traditional landscape elements. The old hay meadows and other seminatural vegetation types play an important role in this connection. To maintain the high biodiversity, they must be managed more or less in the traditional way. Furthermore, just selecting a small meadow and managing it appropriately cannot maintain biological diversity. It is probably necessary to preserve several meadows relatively close to each other, preferably as part of a landscape containing other types of human-influenced vegetation and elements of importance for ecological processes.

In other words, conservation of the valuable cultural landscape in the Hjartdal–Svartdal area will require a strategy in which the entire landscape is considered as an ecosystem. To develop such a strategy, we must learn as much as possible about the history of the landscape and traditional farming methods and develop a scientifically based conservation and management plan. However, such a plan will not be successful unless local people are interested and willing to maintain the unique qualities of the landscape, and society as a whole is willing to pay for and promote the necessary conservation measures. Thus, maintenance of the valuable cultural landscape of the Hjartdal–Svartdal area will require the sustainable development of agriculture, based not only on modern knowledge and ideas but also on knowledge of traditional farming methods.

The Hjartdal–Svartdal area is a vital, rural society. We hope that it will remain so. To transform this area into a museum would not make it easier to maintain the values of the cultural landscape; on the contrary, it would be an almost impossible task. Instead, cooperation between the authorities and the landowners has been initiated. Furthermore, local people have established a limited company to maintain this unique cultural landscape in a way that could also strengthen agriculture and contribute to the development of the local society in the twenty-first century.

REFERENCES

Aakre, H. 1991. *Skarsvegen*. Bø i Telemark: Bø Trykk AS.

———. 1996. *Husmannsplasser i Svartdal*. Mimeographed.

Aanderaa, D. 1990. *Telebondens ære og Kongens makt*. Notodden: Hjartdal kommune, Teledølens boktrykkeri.

Baldock, D. 1990. *Agriculture and Habitat Loss in Europe*. WWF International Cap Discussion Paper no. 3. Gland.

Braathen, H. 1995. "Krisetider og busettingsspor: Mellomaldergarden i Telemark." *Årbok for Telemark* 41: 47–53.

Det sentrale utvalget for national registrering av verdifulle kulturlandskap. 1994. *Verdifulle kulturlandskap i Norge. Mer enn bare landskap! Sluttrapport.* Trondheim: Direktoratet for naturforvaltning.

Dons, J. A., and K. Jorde. 1978. *Geologisk kart over Norge, berggrunnskart Skien 1:250000.* Trondheim: Norges geologiske undersøkelse.

Flatin, T. 1942. *Seljord*, vol. 1. Oslo: Johansen and Nilsen.

Friis, P. Claussøn. 1632. *Norriges oc omligende Øers sandfærdige Bescrifuelse*. Kiøbenhaffn: Melchior Martzan, Jochim Moltken Bogfører.

Karlsrud, G. K. 1998. *Hjartdalsoga: Gard og ætt* III B. Notodden: Hjartdal kommune, Telemark Trykk AS.

Kielland-Lund, J. 1992. *Nasjonal registrering av verdifulle kulturlandskap. Håndbok for feltregistrering—viktige vegetasjonstyper i kulturlandskapet, Øst-Norge.* Oslo: Norsk institutt for naturforskning.

Kraft, J. 1826. *Topografisk-statistisk beskrivelse over kongeriket Norge* 3. Christiania: Grøndahl.

Landstad, M. B. 1925 (1853). *Norske folkeviser fra Tellemarken*. Nyutgåve ved Knut Liestøl. Christiania: Aschehoug.

———. 1985 (1880). *Gamle sagn om Hjartdølene*. Nyutgåve. Notodden: Hjartdal museumslag, Teledølens boktrykkeri.

Martens, I. 1975. "Fra veidemark til bondebygd." In *Bygd og By i Norge. Telemark,* ed. J. E. Holand, 94–119. Oslo: Gyldendal Norsk Forlag.

Moen, A. 1998. *National Atlas of Norway: Vegetation.* Hønefoss: Statens kartverk.

Nes, T. 1991. *Med lokk og lur. Stulshistorikk fra Hjartdalsfjella.* Bø i Telemark: Bø trykk.

Nilssøn, J. 1885 (1574–97). *Biskop Jens Nilssøns visitatsbøger og reiseoptegnelser 1574–1597*, udgivne efter offentlig foranstaltning ved Yngvar Nielsen. Kristiania: A. W. Brøggers Bogtrykkeri.

Norderhaug, A. 1992. *Nasjonal registrering av verdifulle kulturlandskap. Håndbok for feltregistrering—omfang og skjema.* Oslo: Norsk institutt for naturforskning.

———. 1997. "Registrering og forvaltning av verdifulle kulturlandskap i Norge." In *Rapport og forelesninger fra Nordisk seminar om kulturlandskab. Nordens Hus i Reykjavik, Iceland 19–21 September 1997,* ed. B. Spur, 50–55. Reykjavik: Reykjavik Fjølritunærstofa.

———, M. Ihse, and O. Pedersen. 2000. "Biotope Patterns and the Abundance of Meadow Plant Species in a Norwegian Rural Landscape." *Landscape Ecology* 15: 201–18.

Nordiska Ministerrådet. 1984. *Naturgeografisk regionindelning av Norden.* Stockholm: Nordiska Ministerrådet.

Østvedt, E. 1975. "Kulturliv i fortid og nåtid." In *Bygd og By i Norge. Telemark,* ed. J. E. Holand, 120–37. Oslo: Gyldendal Norsk Forlag.

Resen Mandt, E. M. 1989 (1777). *Historisk beskrivelse over Øvre Tellemarken,* ed. O. Solberg. Espa: Lokalhistorisk forlag.

Sjørs, H. 1954. "Slåtterängar i Grangärde Finnmark." *Acta Phytogeographica Suecica* 34: 1–135.

Storm Bjerke, Ø. 1998. *Tun og vassdrag: Telemarks typologier. Kunstnernes hus, Oslo, 18. april–31. mai 1998, Telemark fylkesgalleri, Notodden, 12. juni–26. juli 1998/Farm and Waterway. Telemark Typologies.* Oslo: Kunstnernes Hus.

Straume, A. 1975. "Næringslivet før og nå." In *Bygd og By i Norge. Telemark,* ed. J. E. Holland, 138–68. Oslo: Gyldendal Norsk Forlag.

Telnes, B. 1994. "Gruvedrift i Seljord." *Seljord Sogelag: Årsskrift for kultur og historie* 1994: 24–42.

Wagn, A. H. 1992. *Bygningsarv i Hjartdal.* Notodden: Hjartdal kommune, Teledølens boktrykkeri.

Wille, H. J. 1989 (1786). *Beskrivelse over Sillejords Præstegield i Øvre-Tellemarken i Norge tilligemed et geographisk Chart over samme.* Espa: Lokalhistorisk Forlag.

Willoch, S. 1937. "Telemark i Nasjonalgalleriet." *Den norske turistforenings Årbok* 1937: 32–39.

Finland

16.
Reflections on the Historical Landscapes of Finland

W. R. MEAD

MAISEMA

In 1932, J. G. Granö published a series of maps of settlement in Finland that collectively served to create the impression that the country was divided into two halves. To the south and west lay the territory of the *Kulturlandschaft*, while to the north and east was *Naturlandschaft*. German concepts influenced Granö. To reflect upon the historical landscapes of Finland is to consider the changing shapes of these two territories. To the outsider, Finland still looks as though it is the natural landscape that prevails over the cultural—or, as it might be more satisfactorily defined, the humanized landscape. In fact, there are few areas of the Finnish countryside that have not been subject to modification, indirectly, if not directly, by the hand of man, though in some parts of the country the visible impact has been relatively recent. Changes in the cultural landscape of Finland have been analyzed by Tarja Keisteri (1990), who provided an invaluable statement on landscape terminology, including references to the origin and acceptance of the Finnish word for landscape itself—*maisema*.

The following reflections upon the historical landscapes of Finland by an outsider are intended to complement the chapters of the Finnish contributors. They begin with a selection of laconic comments by the earliest British travelers who passed through Finland and a brief section on the source materials that are available for helping to construct the essential features of the country's past landscapes. Observations on the transformation of the natural landscape into a cultural landscape are balanced by illustrations of the reversion of the cultural landscape to the wilderness. Such changes are complemented by the persistence in the present scene of "landscaped" legacies from the past. J. G. Granö's unique model for the representation of "landscape as perceived by the senses" is recalled (Granö and Paasi 1997), and there is a final comment on the response in Finland to the contemporary concept of landscape as a commodity for sale to consumers.

A Land of Nuances

Wood, water, rock, and swamp—these are the basic components of the old peneplain that constitutes the basis of the landscape of Finland. They enter all of the early observations by British travelers about the Finnish scene. Most travelers passed through in summer and "monotonous" and "repetitious" are the two adjectives most commonly employed to describe the landscape. "The sameness of the appearance" throughout this "wild and watery" land were John Barrow's words in 1834. Traveling in 1799, Edward Clarke (1824) had been apprised by his Swedish friends that across the Baltic Sea he would find "one continuous dark forest." In fact, over extensive tracts, he found the countryside to be "most beautiful and picturesque" (it was probably the first time that the fashionable English word "picturesque" was applied to the Finnish scene). Sir John Bowring, writing in 1823 to Johan Jacob Julin, lamented that he could not return to Finland's "charming wilds." In 1856, Selina Bunbury reported "fir, granite and water and wide, barren swampy places." Both Clarke and Bunbury put their fingers on more significant facts. Clarke observed that in the countryside at large "the gradations of civilized life are marked" and were reflected in the increased or diminished number of painted houses. In Ostrobothnia, coppery red—*rödfärg*—prevailed. In towns, "the ochre uniform of Swedish towns was repeated." Selina Bunbury had a keen eye for what she called "the generic features" of the local landscapes. "Only a plain John Bull" she declared "could dismiss the countryside as monotonous" (Bunbury 1856, 55). There was always "some new shape, some curious appearance, some changing effect" in the water or on the land to attract attention. It was an astute comment because it pinpointed the essential character of the local scene. A French aphorism may be quoted—"Ce sont les nuances qui comptent." The Finnish landscape is one of nuances. Nowhere is this characteristic more evident than in the landscape of the southwestern archipelagoes—"the Cyclades of the North" as one traveler described them.

As the nineteenth century advanced, the new reading public of the British Isles and much of Western Europe were treated to a variety of popular encyclopedias (Mead 1968a). Entries on Finland seem to have been gathered from a handful of travelers' accounts. Wood, water, rock, and swamp were inevitably inhabited by a race of hardy peasants. But the landscape of Finland and the Finns themselves tended to be a caricature of reality. And another element was added to the Finnish scene—climate, a feature that was invariably subject to exaggeration. Of the early travelers, Joseph Acerbi (1802) and Edward Clarke (1824) first published accounts of the winter landscape.

Sources for the Reconstruction of Past Landscapes

While these accounts of Finland were being circulated in the outside world, a variety of records of the country were being made in Finland itself. The materials, diverse in character and sometimes surprisingly comprehensive, were collected essentially for economic reasons. Sequentially, they consist of the cadastral surveys, the topographical surveys, parish descriptions, and the preliminary statistical surveys. Collectively, they provide a notable source for the reconstruction and appraisal of past landscapes, urban as well as rural (Mead 1968b; Jutikkala 1977).

In Finland, which for such purposes was an integral part of Sweden until 1809, the large-scale cadastral maps of the settled areas are the earliest materials. From the 1630s onward, the land surveyors gradually recorded the occupied land, often providing remarkable vignettes of particular places and properties. Moreover, as a result of successive changes in land redistribution *(isojako* and *uusijako)*, snapshots of the local scene through time can be constructed. The study of Maxmo commune in coastal Ostrobothnia by Michael Jones (1987) illustrates the potential of these materials for studies of the changing Finnish landscape.

Large-scale topographical mapping was undertaken between 1776 and 1805 with a view to providing coverage of extensive areas of the country, on a scale of 1:20,000. The territory surveyed was divided into a southern and a northern reconnaissance. The resulting maps—*Kuninkaan kartasto Suomesta/Konungens kartverk från Finland* (Alanen and Kepsu 1989)—offer a highly detailed picture of the form and the basic use of the land. The working materials, retained in Krigsarkivet, Stockholm, supplement the printed maps (Mead 1968c). They consist of many detailed sketches, made in the field incorporating many local place-names for the first time and offering precise evidence of the character of particular parts of the country. They include descriptions of defensive sites, strategic isthmuses in the lake areas, and the character (including the fordability) of rivers and streams.

Complementary to the cartographic surveys are the population records. In addition to the demographic details from individual parishes, many of the quinquennial returns contain descriptive materials about the character of particular areas. In some respects, this supplementary information anticipates the product of the experimental statistical surveys that were conducted by the Finnish Economic Society in the early nineteenth century. The surveys were initiated by Carl Christian Böcker, its first active secretary. The printed forms that he eventually circulated provide information for some eighty items and are bound in three major volumes—*Böckers Statistiska Uppgifter* (Mead 1953). Already before the surveys were set in motion, a series of parish descriptions had been provided by authors, most of whom were members of the Society. The best of these descriptions approach in quality some of the county descriptions printed in England at the time of the creation of the British Board of Agriculture.

The earliest depictions of the Finnish landscape are the stylized etchings produced for Erik Dahlberg's *Suecia antiqua et hodierna* (1716) (Figure 16.1). In 1747, Augustin Ehrensvärd included a number of india ink sketches in his *Anteckningar under en resa i Finland* (1882) (Figure 16.2), but the dawn of Finnish landscape painting awaited the

FIGURE 16.1. Kastelholm, an etching signed "J.v.d. Areelen sc.Holmiae 1705," in Erik Dahlbergh's *Suecia antiqua et hodierna* (1716), one of the first landscapes of Finland. Courtesy Museovirasto, Helsinki.

nineteenth century. Meanwhile, in a lecture to the Swedish Academy, the Finnish cartographer C. P. Hällström (1815) urged that landscape painting should be encouraged in order that the appearance of the countryside from times past might be recorded for future generations. Rather more for aesthetic than for utilitarian reasons, the first landscape artists to record something of the Finnish scene came from outside the country, for example, Elias Martin, Louis Belanger, Carl von Kügelgen, and Valerian Galyamin (Figure 16.3). They were soon followed by Magnus von Wright, Werner Holmberg, and Hjalmar Munsterhjelm. Examples of their work were included in the first publications to provide a "picture gallery" of the landscapes of Finland for popular consumption. Lithographs of their paintings were used to illustrate *Finland framstäldt i teckningar* (1845–52) and *En resa i Finland* (1873), the accompanying texts of which were written by Zachris Topelius (Hirn 1988). In the lectures he gave as the first professor of Finnish history at the Imperial Alexander University in Helsinki (Helsingfors), he included landscape appraisals for the historic provinces of the Grand Duchy (Tiitta 1994). Legend has it that the lectures were given against the background of A. W. Eklund's wall map of the physical geography of Finland (Hjelt 1903). Doubtless Topelius was also familiar with the first qualitative map of the distribution of

Finland's forest resources. The map (Figure 16.4) was the work of C. W. Gyldén, who eventually would become the first director of the Statistical Bureau of Finland. The map clearly reflected the inroads made upon Finland's woodlands by the production of pitch and tar (especially in western Finland) and by slash-and-burn agriculture in eastern Finland. In the process, the map indicated the extent to which the Finnish economy was becoming a *Raubwirtschaft*. In the early years of the century, a series of colored reproductions of typical Finnish landscapes by a dozen or more Finnish artists were hung on the walls of many classrooms. They were produced by the Tilgmann printing house and must have left a lasting impression upon a full generation of children about the geography of their homeland and its economic activities (Figure 16.5). That the country's natural resources were being overexploited was as yet a concept without meaning (Hakkulinen and Yli-Jokipii 1983).

From *Naturlandschaft* to *Kulturlandschaft*

All of these sources that posterity may use to build up its pictures of the past scene tell us little of the way in which most Finns of the time looked upon the landscape

FIGURE 16.2. India ink drawing of Magnusilla village, dated August 15, 1747, by Augustin Ehrensvärd from his *Anteckningar under en resa i Finland år 1747*. Courtesy Kungliga Biblioteket, Stockholm.

FIGURE 16.3 Watercolor of Hålvik by V. Galyamin, 1827. Courtesy Helsinki City Museum.

that was the setting for their daily lives. Doubtless, it was simply accepted for what it was. The couplet of Alexander Pope (1733) sums it up: "What happier natures shrink at with affright / The hard inhabitant contends is right" (*An Essay on Man*, Epistle II, lines 229–30). The townsman I. K. Inha (1909), with his camera at the ready for photographs that were to become historical documents, looked at the interior forests in what might be called *Kalevala* romantic terms. "I had seen the forest as through the eyes of a folk poem," he wrote, imbued with "the spirit of Tapiola." On the spot, he continued, "I viewed it with the eyes of a toiler.... I saw it as a hostile force, the foe of man and his cultivation." Pioneers going into the wilderness *(erämaa)* were at first exhilarated by the generous supplies of game, of berries and mushrooms, by the lakes that were full of fish, and the undisturbed timber that was available for building, heating, and other purposes. By one or another methods of slash-and-burn agriculture *(kaskipoltaa)*, grain crops were produced annually, though weather conditions could still ruin the harvest. Then, bark bread became the symbol of pioneering penury. And with failing human energy, another clearing in the woods *(aho)* might return to nature. As in Larin-Kyösti's short story "Man and Moss" (1905, 1937), the humble plant would gradually return to take back its own.

In contrast to the primeval woodlands of the interior, long occupation and the increasing diffusion of improved techniques had created a humanized landscape around the coastal areas. The Danish novelist Martin A. Hansen (1956, 143) has written of "Christianized fields." The early church encouraged the creation of open cultivated areas—light and sacred places in contrast to the dark and profane places of the primeval forest where wood demons dwelt. The cultivated land became the "blessed plot," its broader expression—the *pays, hembygd*—a distinctive constituent in the human landscape. It took time to create the patrimonial farmsteads. And for Martin Hansen, the entire activity had another dimension. His thesis was that art is more than the transformation of the natural landscape into a humanized landscape. To quote from

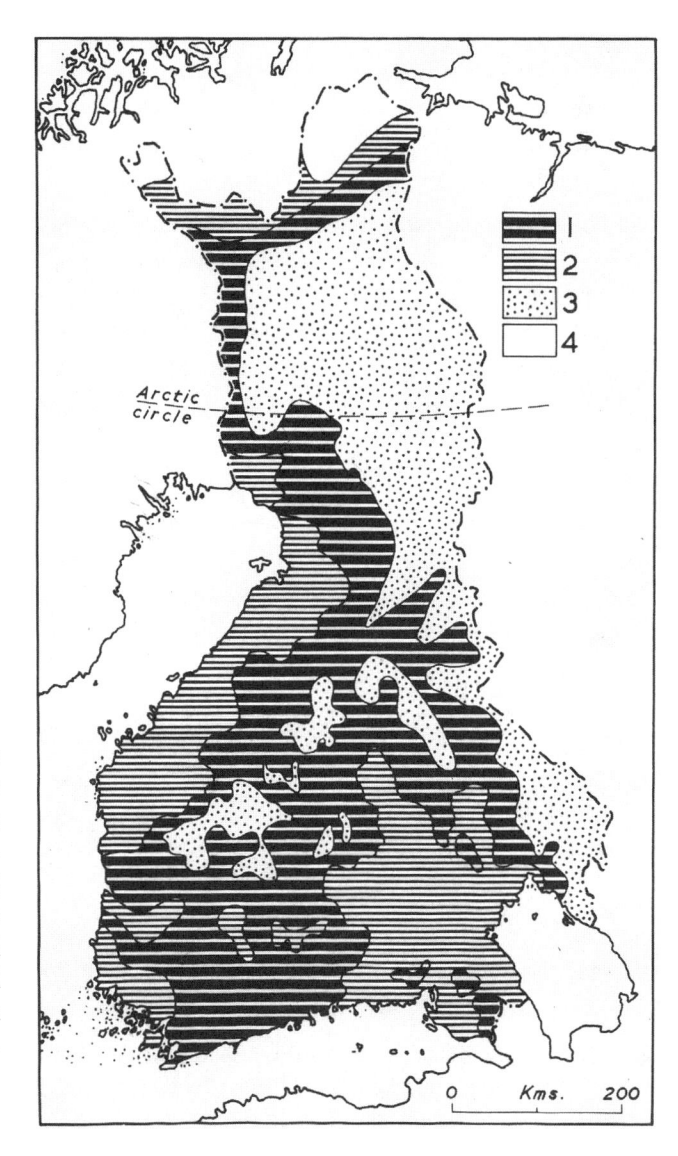

FIGURE 16.4. Qualitative assessment of Finland's forests by C. W. Gyldén in 1850 indicating (1) high-quality timber to (4) absence of timber. Map redrawn from the original in the Finnish National Archives (see also Åström 1978, 124–25).

his "Legends in September" *(Sagn i September)* (1947), "What man has experienced, believed and thought—including legend and localized story—have not merely animated the landscape but have created it" (translated in Ingwersen 1976, 26). For the landscape of unspoiled nature, beautiful though it may be to those who behold it, "is foreign, even threatening, to the spirit" (Ingwersen 1976, 27). It is a landscape of fear—unfriendly and alien until it has been tamed and given a new shape by the hand of man. The monastic foundations of the Catholic West set an example in contrast to those of the Orthodox East. With the coming of the Protestant Reformation, the Lutheran priests of the north acquired their church farms and many fulfilled in a latter-day role that of the monastic orders, with the church farm becoming the center of diffusion of ideas for the improvement of agriculture. The "garden" was extended into the wilderness (land breaking was even regarded as a redemptive pursuit). Then machines entered the garden and the primeval wilderness retreated at an increasing pace.

The ultimate consequence has tended to be that the historic islands of cultivation in the *Naturlandschaft* have been succeeded in the broad picture by islands of residual wilderness and wildwood in a predominantly *Kulturlandschaft*. Whereas in the former state, humanity had to be protected from a primeval world, in the latter state the landscape of the residual *erämaa,* with all of its significance ecologically and "spiritually," had to be protected from human destruction. Moreover, as one of "the last wildernesses" in Europe, it is of more than Finnish concern. Thus in broader terms the

FIGURE 16.5. *Punkaharju* by V. Ylinen, 1914, one of the Finnish landscapes produced by Tilgmann printing house in 1914 for display in schools (see Hakulinen and Yli-Jokipii 1983). The original painting is in the Museovirasto, Helsinki.

roles seem to have been reversed. Between the two extremes, of course, there are everywhere transitional zones. They are mostly forested and to the eyes of most people they look as though they belong to the *Naturlandschaft*. But because they support managed timber crops, they are essentially a part of the cultural landscape. All in all, much of the Finnish landscape is managed wilderness.

In the case of Finland, there is another consideration. Because of the delayed transformation of its landscape in a European context, primitive myths and ideas associated with it were retained by its inhabitants longer than elsewhere. J. R. R. Tolkien, an admirer from his youth of Kalevalan mythology, wrote of such myths metaphorically as a part of that "primitive undergrowth that the literature of Europe has on the whole been steadily cutting and reducing for many centuries" (Carpenter 1977). He believed that England once had the same archetypal mythology, destroyed as the natural landscape was destroyed through the centuries. Not without reflecting upon the Finnish model, Tolkien sought to replace its loss with his own set of core myths in *The Hobbit* (1937) and *The Lord of the Rings* (1954–55).

Landscape and Memory: East and West

Finland has present-day landscapes in different parts of the country that for various reasons are reverting from a cultivated to a wild condition. In westernmost Finland, in particular in the Åland Islands, there is a withdrawal of settlement from the outer islands. In this climatically most favored part of the country, with the harvest of the sea to add to the potential of the land, settlers pushed to the outermost fringes of the archipelago as late as the nineteenth century. What at first had been a successful subsistence economy was slowly eroded. In more recent years, the desertion of many isolated fisher-farming settlements has proceeded steadily for social as well as economic reasons. Abandoned houses, byres, and fishing huts are to be seen amid gardens, orchards, and cultivated fields, which have been invaded by the rich native flora. The result is a place taken over by nature from man, a sort of Sleeping Beauty landscape, which is awakened seasonally where the dwellings have become the summer homes of townsfolk with their speedy motorboats. The process is illustrated in Stig Jaatinen's memoir (1994) on Vårdö, the island background also to the homely novels of Anni Blomquist.

At the easternmost extremities of what was formerly Finland, extensive tracts of land have also reverted to the wilderness. Relics of the sophisticated and unsophisticated alike are fast disappearing. To the former belong the homes and gardens of what constituted the summer villa culture around Viipuri (Viborg) bay and in the Karelian isthmus (Figure 16.6). The *haute bourgeoisie* of St. Petersburg left its legacy to the citizenry of Viipuri, who cherished and extended it. Its character and atmosphere are recalled in the autobiographies of many Finns who were brought up in the area,

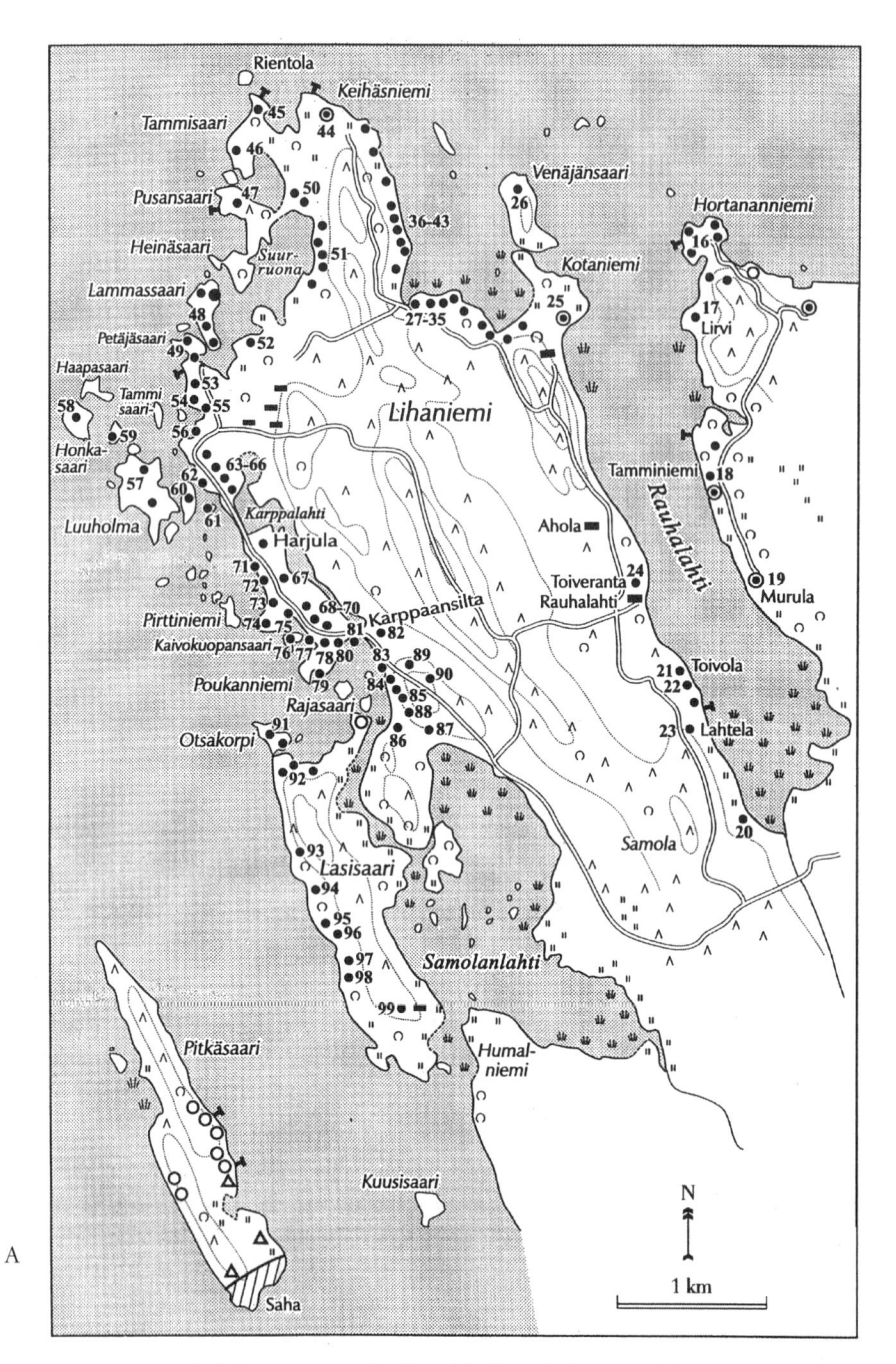

FIGURE 16.6. (A) Estates around Viipuri (Viborg) that would have displayed evidence of landscaping in its parks and gardens, and (B) summer villas on the southeastern side of Viipuri Bay, which was a distinctive landscape in the prewar years. Reproduced from Jaatinen (1997, 46 and 132) with the permission of *Finska Vetenskaps-Societeten*.

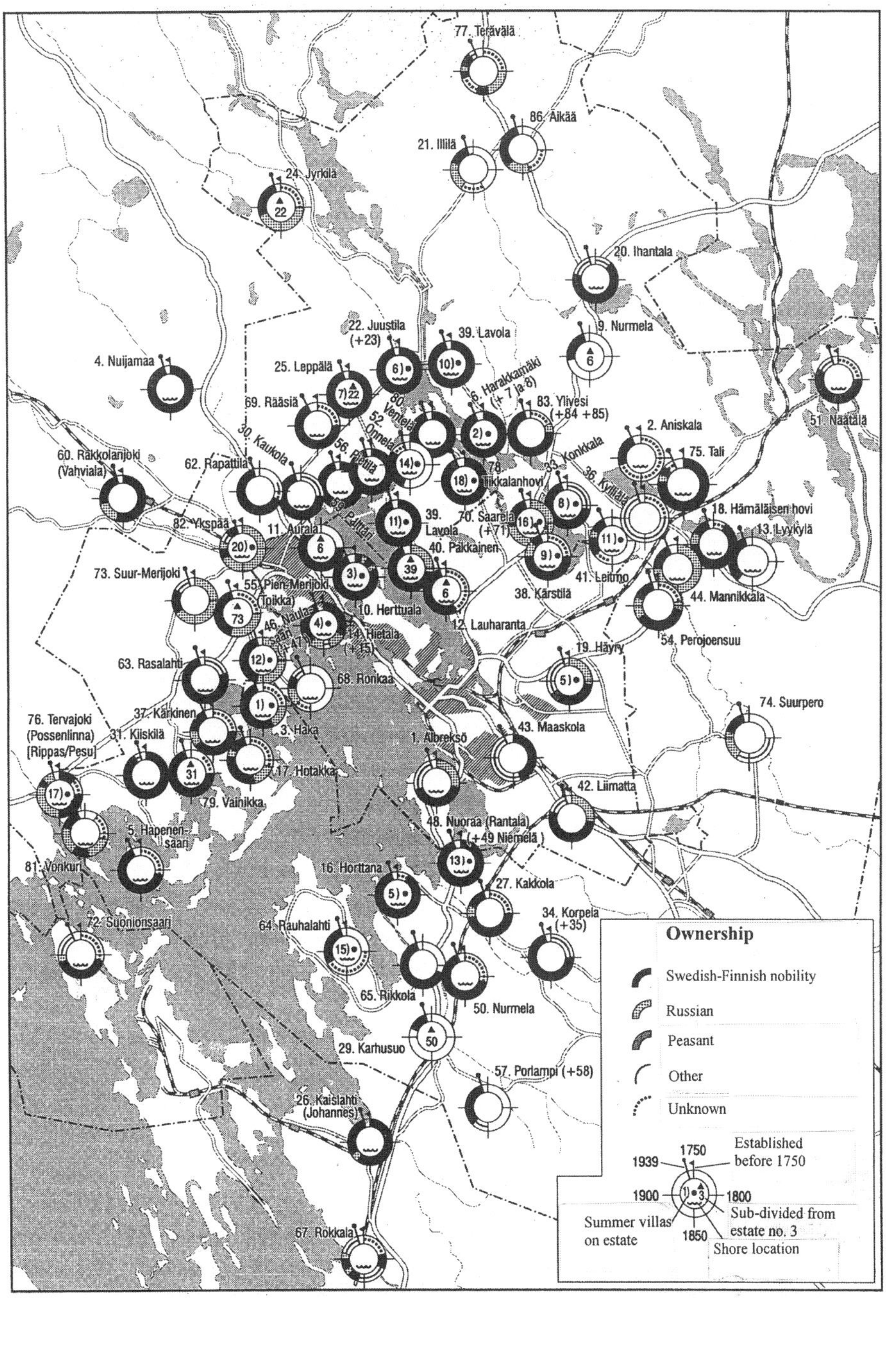
Ownership
Swedish-Finnish nobility
Russian
Peasant
Other
Unknown
Established before 1750
Summer villas on estate
Sub-divided from estate no. 3
Shore location
1750
1800
1850
1900
1939
77. Terävälä
86. Aikää
21. Illilä
24. Jyrkilä
20. Ihantala
9. Nurmela
4. Nuijamaa
22. Juustila (+23)
39. Lavola
25. Leppälä
69. Rääsiä
83. Ylivesi (+84 +85)
2. Aniskala
75. Tali
51. Näätälä
60. Rakkolanjoki (Vahviala)
62. Rapattila
82. Ykspää
11. Auralá
73. Suur-Merijoki
18. Hämäläisen hovi
13. Lyykylä
40. Pakkainen
38. Kärstilä
41. Leimo
44. Mannikkala
54. Perojoensuu
10. Herttuala
12. Lauharanta
19. Häyry
63. Rasalahti
68. Ronkaa
74. Suurpero
76. Tervajoki (Possenlinna) [Rippas/Pesu]
31. Kiiskilä
37. Kärkinen
3. Haka
17. Hotakka
79. Vainikka
1. Albreksö
43. Maaskola
42. Liimatta
5. Hapenensaari
81. Vonkuri
48. Nuoraa (Rantala) (+49 Niemelä)
16. Horttana
27. Kakkola
34. Korpela (+35)
72. Suonionsaari
64. Rauhalahti
65. Rikkola
50. Nurmela
29. Karhusuo
57. Porlampi (+58)
26. Kaislahti (Johannes)
67. Rokkala

B

especially in the nostalgic writings of Oscar Parland (1991) and Benedikt Zilliacus (1990). Stig Jaatinen's *Elysium Wiburgense* (1997) documents the history of this distinctive landscape feature, the relics of which he had the doubtful pleasure of visiting as they were sinking into decay. The mood is that of Simon Schama's major work, *Landscape and Memory* (1995).

Beyond Lake Ladoga, reversion to the landscape of the wildwood has proceeded faster among the former unsophisticated farming settlements. *Siirtolaiset* (evacuees) from the former parishes of Salmi, Suistamo, and Impilahti—evocative place-names—were evacuated to northern Savo after their home territory was lost to the USSR in 1944. Here, in virtual pioneer holdings, they retained memories of a land where summers were warmer, crops were heavier, and fish were fatter than beside the waters of the latter-day Babylon of Savo. Landscapes held in the mind's eye by the next generation, born of received memories, encouraged visits to the old homesteads once the frontier was reopened. As often as not, the home farms were no longer identifiable on the ground—indeed, often they were no longer even accessible through half a century's growth of forest. As with people, so with places—and, as Shakespeare's (1709) couplet reminds us: "Golden lads and girls all must / As chimney sweepers, come to dust" (*Cymbeline* IV.2, line 258).

Elsewhere, in the northeastern interior of Finland, landscapes are reverting to the wilderness around the edges of the oecumene. Grandparents who had pioneered holdings have sometimes survived long enough to desert them because they no longer provide a livelihood (Ingold 1988; Abrahams 1991). In Finland at large, lands of deficit were becoming lands of surplus. The irony has been that overproduction has led to land being taken out of cultivation for food and fodder crops. Sometimes it has been set aside, sometimes planted with timber crops. The open landscapes of plowland may not yet have become closed by timber stands, but with varying speeds in different parts of the country the process has begun. It reflects a changing system of values. Landscapes of financial profit must prevail over landscapes of loss.

Landscape Legacies: East and West

These macro developments in the Finnish scene are complemented by the tenacity with which two distinctive minor features retain their place in the cultural landscape. They are the old landscaped estates of the *kartanot (herrgårdar)* (Nikander 1917; Nikander et al. 1928–30), most of which continue to be effectively maintained, and the development of a more widely distributed Orthodox cultural landscape (again springing out of the consequences of the Peace Settlement of 1944).

Easternmost and westernmost Finland are linked by the most traveled route in the country—that between the old capitals of Stockholm and St. Petersburg, with Turku (Åbo) and Helsinki in between (Gardberg and Dahl 1991). Ideas from the practical to the aesthetic, as well as traffic, traveled along it between the manor houses, the administrative offices, the military centers, and the vicarages.

Thus, the English fashion for landscaping country estates seems to have reached its easternmost eighteenth-century expression in southwest Finland. The concept was probably transmitted from Sweden, though J. C. Linnerhjelm's influential text on the subject (1816, 1932) was not published until after the first attempts at landscaping were under way in Finland. Fagervik and Åminne estates offer two pioneering examples. It is mildly ironic that, while in England coniferous species were introduced to parks and gardens as exotic elements (with intent to foster particular moods and sensations), in Finland the native conifers were removed and replaced where possible by deciduous species. Thus, from the terrace at Åminne, it is the carefully nurtured oak and other deciduous trees, so-called "noble trees," that close the vista. At Fagervik, the "English park" incorporated features that were at the height of fashion in England. Jacob Bonsdorff published what was to all intents and purposes a sentimental journey in *Sommarresan* (1799). From Fagervik's Chinese pleasure garden, one followed "skilfully arranged footpaths [to] a hut which counterfeits a hermit's simple abode." Bonsdorff even affected the language of the day to describe the atmosphere that he sensed in this "mournful place, where silence and gloom prevail."

Farther east, outside Viipuri (Figure 16.6), the landscaped park reached its fullest expression somewhat later on the estate of Baron Ludwig von Nicolay, who may well have derived certain of his ideas from the font of English landscape ideas, Humphrey Repton (whose *Enquiry into the Changes of Taste in Landscape Gardening* [1806] was in his library). The baron wrote his own sentimental poem about the property that he had created, *Das Landgut Monrepos in Finland 1804* (1806).

In other instances, landscaping was associated with the development of early industrial enterprise. As in Sweden, southwest Finland has a scattering of *bruk*, mostly concerned with mineral or softwood processing. They had differing antecedents, but as their owners accumulated wealth, fine patronal residences grew up beside the industrial plant. The running water that determined the industrial site also provided possibilities for water features in the park. Svartå, now a hotel and conference center, provides an example. When he set about "improving" the Fiskars estate, Johan Jacob Julin already was familiar with the parks and gardens of British industrialists. His library, even more than that of Monrepos, contained a number of books relevant to the laying out of a park. J. C. Loudon's celebrated *Encyclopaedia of Gardening* (1822) was foremost among them.

By then the first urban park was established in Finland, at Kaisaniemi in Helsinki. Landscaping as a component of town planning in Finland has not ceased to be refined and developed. Thus, from being a rural refinement for the favored few, landscaping and the provision of public parks have become common urban features for the many.

The other distinctive minor feature to make a mark on the Finnish landscape has been the impact of the Orthodox Church. It may be modest, but it is a unique phenomenon. Historically, the Orthodox faith was an essential characteristic of Karelia—Gamla Finland as it was after the annexation by imperial Russia in 1721 until its restoration to the Grand Duchy in 1812. When international tourists began to visit Finland between the wars, the island monastery of Valamo in Lake Ladoga became a high point. Elsewhere, it was most evident in the distinctive churches that were built mostly in the administrative and garrison centers. They were dramatically alien elements in a sober Lutheran environment. But after the cession of southeastern Finland to the USSR in 1944 and the dispersal of displaced Karelians among the reception areas, a new Orthodox constituent was added to local landscapes. Provision had to be made for Orthodox burial grounds in existing cemeteries. New parishes had to be created, new churches built, a new administrative center established. Eventually a new Valamo emerged on a Heinävesi estate. Scattered though these constituent features were, they were a source of spiritual solace as well as satisfaction for the Orthodox Karelians. It was natural that these iconographic features should disturb the inhabitants of old-established Lutheran communities—indeed, to evoke a response disproportionate to their numbers. Petri Raivo (1996) has produced the definitive statement on the Orthodox element in the human landscape of Finland.

The topic raises the question as to whether there are other definable "ethnographic" landscapes. The very mobility of the Lappish population has limited their legacy of permanent features in the landscape. The infinity of minor features patiently gleaned and painstakingly mapped by M. Salmela in *Suomen Perinne-Atlas* (1994) has become so well established in the mind's eye that they stimulate the search for evidence on the ground. In general, however, it is elusive.

Island of Light

In the 1920s, J. G. Granö developed a unique methodological study for the representation of landscape features. It has remained a neglected study in the history of the topic from the time when *Puhdas maantiede* was published in 1929–30 until its English translation finally appeared as *Pure Geography* (Granö and Paasi 1997). It was based on the idea that a fundamental object of geographical research should be "the environment as perceived by the senses." In this, it anticipated the later concept of environmental perception and the behavioral studies in geography that began to appear forty years later.

Granö was concerned with the representation of data perceived and recorded on the spot, a distinction being made between observation of the immediate locality and the broader field of vision. The procedure was to record the topological features quantitatively and qualitatively and then to consider them chronologically (as features in process of change). Landscapes generate their particular *genius loci* or spirit of place, so that even the "spiritual" qualities of landscape are relevant. The German geographer E. Banse went so far as to write of "the soul of the landscape": for the French author H. F. Amiel, landscape produced "un état de l'âme." Bo Carpelan in his novel *Urwind* has intimated that a landscape can generate its own language: "Landskap har ju sina egna språk" (Carpelan 1993, 54). Thus, for any particular area or region, the entire assemblage of observable features—and indeed of sensations—will combine to constitute the character of the landscape. Accordingly, geography becomes at least in part "a theory of human perceived environments" (Granö and Paasi 1997, 31)—and the perception naturally changes with time.

It was on the island of Valosaari (happy name, Island of Light) that Granö demonstrated his method. His patient and detailed observations in the "field of proximate vision" are given cartographic representation in a series of maps covering the visual, olfactory, and auditory. The visual called for a record of the variable and invariable elements—atmospheric changes, the changing quality of light (sunshine and shadow), color changes through the seasons (chromatology), mobile elements in the field of vision. William Wordsworth's "recollections in tranquillity," "after image" as he called them, were not for him.

The two-dimensional records that he made (and, in some instances, gave more extended form to in distribution maps of Estonia) have become historic documents in their own right. They constitute a layered image—visual, olfactory, auditory—and a personal one. But the methodology offers a model for use in other places and for other purposes. Thus, it lends itself to the task of landscape assessment that can have practical value for conservational ends. It is exemplified in the communities of Finnby and Saxby near the Gulf of Finland in the south. In the latter, against the background of a physically diverse terrain and favorable landownership conditions, a particularly attractive human landscape has evolved through the generations—a "heritage" landscape that has attracted artists past and present. By employing certain of the representational methods developed by Granö, a detailed and imaginative study has been undertaken that makes a strong case for the protection from development in an area sensitive to the pressures of the nearby expanding historic town of Borgå (Porvoo) (Tuhkanen et al. 1990).

Landscape as a Commodity

Landscapes, with their associations from the past, have acquired a new value with the rise of the tourist industry (Prentice and Guerin 1998). All countries have their tourist promoters and for a full century they have steadily fostered the kind of image that they hope will attract visitors. The founders of the Finnish Tourist Board in 1888 had a strong geographical connection. It was the unsophisticated character of the countryside that they sought to sell for the first half century of their existence—the woods and the lakes that were "off the map." For the handful of interwar visitors, these qualities were appealing. "A hundred yards from the roadside . . . the horizon closes in . . . and the world outside is abolished . . . one universe gives place to another . . . rich in rewards for the senses near to the spirit . . . good things are restored . . . silence, coolness, peace." These reactions, which Claude Lévi-Strauss relished elsewhere (*Tristes tropiques*, 1955), still attract the minority, but for the majority of landscape "consumers" the exotic and the spectacular have greater appeal.

The search for and the promotion of the unusual in Finland have not been in vain. Thus, it was discovered that, invisible though it might be, the Arctic Circle had drawing power. To its north was Lapland, though no one knew exactly how to present its landscape. It did not seem to occur to tourist agencies that to promote the Lappish element in their literature was to strengthen the old-established confusion of Finn and Lapp. Even now, tourist literature is only slowly replacing the name Lapp with Saami.

Reindeer were a part of the Lapland scene and, their being traditionally linked with Santa Claus, it was but one step forward to a Father Christmas theme park, giving a new twist to the parkland concept. The sale of the winter landscape followed—from the experience of the *aurora borealis* and *kaamosaika* (the dark time, a nice contrast to the white nights of summer) to icebreakers and ice castles.

Finland's chromatology was increasingly and studiously presented. Color must be sought in the landscape. It was Edward Clarke who first observed the rainbow coloring in the white winter landscape—in the curtains of icicles at the entrances to Åland's ice caverns. Topelius directed attention to the prismatic effects produced by the sun as it shone through ice: winter had color. The contemporary tourist is presumed to want color so that tourist literature must provide it, often with the strongest colors that the palette can offer. The camera obviously detects a vividness that the eye often fails to register. The natural landscape of Finland is commonly pastel colored. It is doubtful if Granö would recognize what color film reveals in the landscape of Valosaari.

For it is the camera that has taken over the landscape today, with the tourist often more interested in photographing it than in looking at it. In another context, the Finnish photographer Jorma Puranen has written: "Taking a picture is like a ceremony or ritual where the photographer struggles to reach beyond the landscape as it appears in its everyday guise" (Puranen 1999, 50). He has conceived his own unusually imaginative way of joining together the past and the present in landscape photography. Having discovered a collection of late nineteenth-century photographs from Lapland, he has grafted the faces of Lapps from the past on to his own photographs of contemporary Lapland's subarctic fell country. Thus "the eyes of the past, so to speak, scrutinize the immense changes that have befallen the northern landscape during this dying century" (54).

All of these considerations, in their various ways, represent the gradual fulfillment of Karl Ritter's belief that the earth might one day become "the playground of mankind," with leisure taking precedence over labor in the daily round. In Finland, landscapes of pleasure have certainly been created out of what for many years were so long landscapes of hardship, even of pain.

REFERENCES

Abrahams, Ray. 1991. *A Place of Their Own: Family Farming in Eastern Finland.* Cambridge Studies in Social and Cultural Anthropology 81. Cambridge: Cambridge University Press.

Acerbi, Joseph. 1802. *Travels through Sweden, Finland, and Lapland during the Years 1798 and 1799,* 2 vols. London: Joseph Mawman.

Alanen, Timo, and Saulo Kepsu. 1989. *Kuninkaan kartasto Suomesta, 1776–1805.* Suomalaisen Kirjallisuuden Seuran toimituksia 505. Tampere and Helsinki: Suomalaisen Kirjallisuuden Seura.

Åström, Sven-Erik. 1978. *Natur och byte: Ekologiska synpunkter på Finlands ekonomiska historia.* Ekenäs: Söderström & Co.

Barrow, John. 1834. *Excursions in the North of Europe: Through Parts of Russia, Finland, Sweden, Denmark, and Norway in the Years 1830 and 1833.* London: Murray.

Bonsdorff, Jacob. 1799. *Sommarresan.* Åbo: Johan C. Frenckell.

Bowring, Sir John. 1823. A letter, October 26. Fiskars archive, E a 2.

Bunbury, Selina. 1856. *A Summer in Northern Europe, Including Sketches in Sweden, Norway, Finland, the Aaland Islands, Gothland etc.* London.

Carpelan, Bo. 1993. *Urwind.* Stockholm: Bonnier.

Carpenter, Humphrey. 1977. *J. R. R. Tolkien: A Biography.* London: George Allen and Unwin.

Clarke, E. D. 1824. *Travels in Various Countries of Europe, Asia, and Africa, Part 3: Scandinavia.* London.

Dahlberg, Erik. 1716. *Suecia antiqua et hodierna.* Stockholm.

Ehrensvärd, Augustin. 1882. *Anteckningar under en resa i Finland år 1747.* Helsingfors.

Gardberg, C. J., and K. Dahl. 1991. *Kungsvägen från Åbo till Viborg.* Helsingfors: Schildts.

Granö, J. G. 1929. "Reine Geographie: Eine methodologische Studie beleuchtet mit Beispielen aus Finnland und Estland." *Acta Geographica* 2.2: 1–202.

———. 1930. *Puhdas maantiede: Tutkimusesimerkeillä Suomesta ja Virosta valaistu metodologinen selvitys.* Porvoo: WSOY.

———. 1932. *Suomen maantieteelliset alueet.* Porvoo: WSOY.

———, and Anssi Paasi, eds. 1997. *Pure Geography* by J. G. Granö, translated by Malcolm Hicks. Baltimore: Johns Hopkins University Press.

Hakulinen, Kerkko, and Pentti Yli-Jokipii. 1983. *Maammekuvat.* Espoo: Oy Wellin & Goos.

Hällström, C. P. 1815. "Tal om den tillvext Faderneslandets Geographie." *K. V. Academien* 19, February 1812. Stockholm.

Hansen, Martin A. 1947. "Sagn i September." *Nationaltidende,* October 11.

———. 1956. *Tanker i en korsten.* Copenhagen: Gyldendal.

Hirn, Marta. 1988. *Finland framställt i teckningar.* Skrifter utgivna av Svenska Litteratursällskapet i Finland 551. Helsingfors: Svenska Litteratursällskapet i Finland.

Hjelt, R. 1903. "A. W. Eklund." *Finsk Biografisk Handbok,* ed. Tor Carpelan. Helsingfors: G. W. Edlund.

Ingold, Tim, ed. 1988. *The Social Implications of Agrarian Change in Northern and Eastern Finland.* Suomen Antropologisen Seuran toimituksia 22. Helsinki: Suomen Antropologisen Seura.

Inha, I. K. 1909. *Suomen maisemia. Näkemänsä mukaan kuvaillut*. Porvoo: WSOY.
Ingwersen, Faith, and Niels Ingwersen. 1976. *Martin A. Hansen*. Boston: Twayne.
Jaatinen, Stig. 1994. *Vårdö i skärans, liens och mulens landskap genom seklerna*. Bidrag till kännedom av Finlands natur och folk 146. Helsingfors: Finska Vetenskaps-Soceteten.
———. 1997. *Elysium Wiburgense: Villabebyggelsen och villakulturen kring Viborg*. Bidrag till kännedom av Finlands natur och folk 151. Helsingfors: Finska Vetenskaps-Soceteten.
Jones, Michael. 1987. *Landhöjning, jordägoforhållanden och kulturlandskap i Maxmo*. Bidrag till kännedom av Finlands natur och folk 135. Helsingfors: Finska Vetenskaps-Soceteten.
Jutikkala, Eino. 1977. *Scandinavian Atlas of Historic Towns*, Vol. 1: *Finland: Turku–Åbo*. Vol. 2: *Finland: Borgå–Porvoo–Odense*. Odense: Odense University Press.
Keisteri, Tarja. 1990. "The Study of Changes in Cultural Landscapes." *Fennia* 168–6: 31–11.
Larin-Kyösti. 1905. "Inehmo ja sammal." In *Leipä ja laulu. Kokoelma kertomuksia ja taruja*, ed. Larin-Kyösti, 85–90. Helsinki: Otava.
———. 1937. "Man and Moss." In *Northern Lights: A Collection of Short Stories*, by Larin-Kyösti, 193–97. Hämeenlinna: Suomalainen Kirjapaino Oy.
Lévi-Strauss, Claude. 1955. *Tristes tropiques*. Paris: Librairie Plon.
Linnerhjelm, Jonas Carl. 1816. *Bref under senare resor i Sverige*. Stockholm: A. Gadelius.
———. 1932. *Brev under resor i Sverige*, ed. Harald Schiller. Stockholm: Wahlström & Widstrand.
Loudon, J. C. 1822. *Encyclopaedia of Gardening*. London.
Mead, W. R. 1953. "Land Use in Early Nineteenth-Century Finland." *Annales Universitatis Turkuensis*, A XIII–2: 1–23. Turku.
———. 1968a. "The Delineation of Finland." *Fennia* 97: 1–18.
———. 1968b. "Géographie de la Finlande au dix-huitième siecle: Examin de quelques matériaux de base." *Norois* 59: 247–85.
———. 1968c. "The Eighteenth-Century Military Reconnaissance of Finland." *Acta Geographica* 20: 255–71.
———. 1993. "The Changing Rural Landscape of Sweden." *Bebyggelsehistorisk Tidskrift* 26: 122–28.
Nicolay, Ludwig Heinrich von. 1806. *Das Landgut Monrepos in Finland 1804*. St. Petersburg: F. Drechsler.
Nikander, Gabriel. 1917. "Finlandska herrgårdar under den Gustavianska tiden." *Historisk Tidskrift för Finland* 1–3: 67–87.
———, et al., eds. 1928–30. *Herrgårder i Finland*, 4 vols. Helsingfors: Söderström & Co.
Parland, Oscar. 1991. *Kunskap och innlevelse: Essayer och minnen*. Helsingfors: Schildts.
Pope, Alexander. 1950 (1733). *An Essay on Man*, ed. Maynard Mack. London: Methuen; New Haven: Yale University Press.
Prentice, R., and Sinead Guerin. 1998. "The Romantic Walker: A Case Study of Users of Iconic Scottish Landscape." *Scottish Geographical Magazine* 114–3: 180–91.
Puranen, Jorma. 1999. "Kuvitteellinen kotiinpaluu." *Books from Finland* 1999–1: 48–54.
Raivo, Petri J. 1996. *Maiseman kultuurinen transformaatio ortodoksinen kirkkosuomalaisessa kultuuurimaisemassa*. Oulu: Nordia Geographical Publications.
Repton, Humphrey. 1806. *Enquiry into the Changes of Taste in Landscape Gardening*. London.
Salmela, Matti. 1944. *Suomen perinne atlas*. Suomalaisen Kirjallisuuden Seuran toimituksia 587. Helsinki: Suomalaisen Kirjallisuuden Seura.
Schama, Simon. 1995. *Landscape and Memory*. London: HarperCollins.
Shakespeare, William. 1709. "Cymbeline." *The Works of Mr. William Shakespear*, vol. 6. London.

Tiitta, Allen. 1994. *Harmaakiven maa: Zacharias Topelius ja Suomen maantiede.* Bidrag till kännedom av Finlands natur och folk 147. Helsinki: Suomen tiedekunta.

Tolkien, J. R. R. 1937. *The Hobbit, Or, There and Back Again.* London.

———. 1954–55. *The Lord of the Rings,* 3 vols. London: Allen and Unwin.

Topelius, Z. 1845–52. *Finland framstäldt i teckningar.* Helsingfors: A. W. Gröndahl & A. C. Öhman.

———. 1873. *En resa i Finland.* Helsingfors: F. Tilgmann.

Tuhkanen, S., S. Gronlund, S. and T. Keisteri. 1990. *Natural and Cultural Landscapes in Physical Planning.* Helsinki: Center for Urban and Regional Studies.

Zilliacus, Benedikt. 1990. *Båten i vasser: En berättelse om en förlorad ö.* Stockholm: Gedin.

17.
Landscape Territory, Autonomy, and Regional Identity: The Åland Islands in a Cultural Perspective

Nils Storå

This contribution to the study of the landscapes of historical regions focuses on the Åland Islands, situated roughly halfway between Sweden and Finland (Figure 17.1). "*Landskapet*" Åland is the only political "landscape" in Norden where the term *landskap* is still used to denote an administrative region, one that since the beginning of the 1920s has had the status of an autonomous region under Finland's sovereignty. Åland's population in 2007 was 26,900, 94 percent of which are Swedish-speaking and scattered over some sixty of the islands. The largest island is that of mainland Åland *(Fasta Åland)*, where Mariehamn, the "capital" and only town of the islands, is situated. It was founded in 1861, and its background as a harbor and seaside resort is still seen in the architecture. At the time of the introduction of autonomy, the town had fewer than 1,500 inhabitants or less than 6 percent of the total population of the islands. Today nearly half the population lives in the town.

The historical agrarian economy, based on a multiple use of local natural resources and small-scale carrying trade, has been replaced by international cargo shipping. Maritime transport and the related service industry, including tourism, provide the base of Åland's successful economic development (Figure 17.2). Transport constraints and the limited supply of labor prevented a traditional industrialization process. Thus local manufacturing industries do not employ more than about 10 percent of the active population, whereas 70 percent are engaged in the service industry. Åland's geographical position, close to the sea border between Finland and Sweden, contributed to the success of the local shipowners, private entrepreneurs who laid the foundations of modern passenger traffic in the late 1950s (Lindström 2000). Today an intense traffic along the sea route between Finland, Åland, and Sweden is carried out with the support of tax-free sales, permitted thanks to Åland's constitutional status.

A CULTURAL VIEW OF LANDSCAPE-REGIONAL IDENTITY

In size, a *region*—and thus also a *landscape-region*—is often defined as falling between the nation and the local community (e.g., Salomonsson 1996, 13). The same hierarchy gives *regional identity* an intermediate position in relation to *national identity* and *local identity*. In a cultural perspective, this hierarchy is primarily concerned with interacting levels of cultural communication rather than separate geographical areas. Accordingly, the same levels are found in definitions of *culture as national, regional, and local* (e.g., Brück 1988, 85). A supposition is that people have much more in common on the local level than on the regional and national levels. In cultural studies, identification is often conceived as a process of selection in relation to the elements or traditions of the culture of a group. Certain cultural elements are singled out and made representative of a group. This may involve traditions referring to language, history, costume, food, music, myth, or geographical location, each of them also having a symbolical meaning. The symbols selected (consciously or unconsciously) serve the sense of unity and togetherness of the group (Honko 1988, 11). From this point of view, landscape-regional identity refers to the part played by territory in the identity

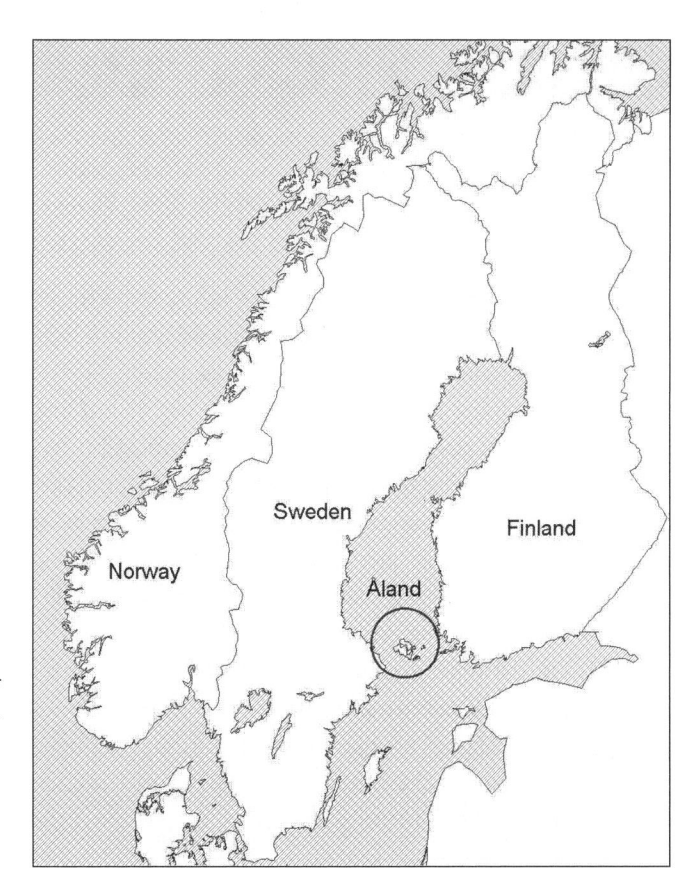

FIGURE 17.1. Map showing the location of Åland. Courtesy Provincial Government of Åland.

FIGURE 17.2. Large car ferries carry tourists to Åland from Sweden and Finland. The old sailing ship *Pommern* (left) is now part of the Maritime Museum of Åland. Courtesy Maritime Museum of Åbo Akademi University, Turku. Photo by P.-O. Sjöström, 1995.

experienced by a group of people living in the same region. Cultural-ecological studies, focusing on the human-environment interaction, relate to people who not only live in a region but also use its natural resources.

Identity-shaping factors are not always visible but are mentally experienced, and therefore difficult or—when unconscious—impossible to verbalize by representatives of the group. In trying to reveal identification criteria, the researcher frequently takes the position of an outsider constructing the identity of the group. One of the main problems the researcher faces is the interaction between individual and group identities (Honko 1988, 14f.). In a social context, an individual may identify himself or herself with more than one group, implying that the individual may have more than one identity. On the other hand, identity is not only a matter of self-identification. The sense of group identity is also affected by "characteristic" qualities attributed to the group by "the others," people outside the group or outside the region. Identity may therefore be *natural*, or *constructed* by neighbors or other outsiders.

A study of Åland identity clearly shows that natural identity and constructed identity cannot always be separated. The local government, after the introduction of autonomy, was particularly interested to shape and communicate the cultural identity of the region. Regions are not, however, closed entities. Mobility is one of the outstanding features of Åland society. There were ample opportunities for the Ålanders themselves to communicate their identity and for people in "the world outside"—above all the people of Sweden and mainland Finland—to receive and interpret the signals and shape their own picture of the Ålanders. The fact that the inhabitants of the neighboring mainland archipelago belong to the Swedish-speaking part of Finland's population means that the closest eastern social contacts have not in general been faced with language problems.

The Åland Islands as a Historical Region

The first attempt at a general description of the Åland Islands was made in 1795 by the provincial medical officer F. W. Radloff. In his introduction, the doctor states that Åland, like other "landscapes" in Sweden, deserved a description of its own, in spite of the fact that it was the smallest of all. Åland made a small island territory, intersected by deep inlets, with a minute population of (then) fewer than 11,500 scattered on some eighty inhabited islands. There were numerous small lakes, but no rivers, except for a few brooks. The Ålanders lived mainly on fishing, which was a major commodity in the carrying trade with Stockholm. Every spot of arable land was used for farming. Only a few of the eight parishes and seven small congregations had forest enough for exporting firewood to Stockholm. According to Radloff (247), the Ålanders were "sensible, active, obliging and merry, known as fearless and highly skilful at sea." The strategic importance of the islands, as a buffer territory between Western and Eastern Europe, is enforced by their position as a "threshold" between the Baltic in the south and the Gulf of Bothnia in the north. As "a crossroads of the Baltic" (Mead and Jaatinen 1975, 92), the islands have been involved in repeated war operations, starting in the sixteenth century. However, this crossroads position has also made Åland open to cultural influences from several directions. As an island bridge between Sweden and Finland, the Åland Islands have in many ways served as a melting pot of cultural influences from both west and east. The Ålanders frequently refer to their position on "a cultural bridge" between the two countries, which were united until 1809 (Storå 1988, 13f.).

The Åland Islands were populated in prehistoric times. Christianity reached the islands from the West in the eleventh century. The framework of a regional administration can be traced as far back as the early Middle Ages. In the fourteenth century, two "Ålanders," Peter Älänning and Gunder Älänning, are known to have acted as merchants in Stockholm. No doubt these two were not the only Ålanders who had settled in Stockholm. At that time the Swedish stronghold on the islands, the castle of Kastelholm, was erected. An "Åland" seal was used as a *sigillum terrae* on an official

Act of 1322 concerning a legal assembly, *ting*. This assembly apparently represented the territorial administration of the islands, which since 1309 ecclesiastically were part of the diocese of Åbo (Turku). The oldest of the eleven medieval stone churches still existing were built in the twelfth century.

As a province, Åland did not differ very much from the other provinces or landscape territories *(landskap)* of Sweden. For tax collecting and war service, the medieval province of Åland was divided into three districts *(tridungar)*, each ruled by a *lagman* or legal officer appointed to conduct the *ting* or local assembly. Some of the assemblies were common to all three districts (Voionmaa 1913, 8f.; Dreijer 1968, 20f.). According to a tax book of 1413, there were 526 "peasants" or farm owners *(bönder)* in Åland (Sundwall 1958, 18), no doubt combining farming with fishing. Matters of landownership were settled in accordance with Swedish laws and legal practices. The rights to land and water were defined in the common principle expressed as: "He owns the water, who owns the land." This expression implied that the landless had no immediate access to the fishing waters, that is, the inshore waters. The lists of taxable products included in sixteenth-century tax books (Mikander 1964) reflect the material bases of Åland culture: stock raising, agriculture, fishing, and hunting. The lists include cattle, sheep and lambs, swine, geese, fowl, hares, fish, butter, eggs, rye, malt, and flour, as well as timber, wood, and charcoal.

A main interest of the Crown concerned Åland's central position on the route from Sweden (Stockholm) to Finland (Åbo). The duty of transporting people and goods along this route was a heavy burden not only for the Ålanders but also for their western and eastern island neighbors. Traveling in the late autumn and early spring—when the ice was too thick for boat traffic but not strong enough to support the horse-drawn sledges—could make the crossings both tedious and hazardous. The duty to forward the mail was not abolished until the end of the nineteenth century, when steamships took over.

In 1593, Åland as a county *(grevskap)* received a coat of arms, still used, depicting a yellow stag on a blue field. After 1634, when Åland was united with the province *(län)* of Åbo and Björneborg, in the southwestern part of mainland Finland, the castle of Kastelholm, the seat of the local government, lost its former significance. During the Great Northern War, 1709–21, when the Åland Islands were occupied by Russian troops, the Ålanders, in 1714, had to flee to Sweden, where they stayed for about eight years. In 1742, Åland was again invaded by Russian troops. In accordance with the peace treaties of 1721 and 1743, the Åland Islands, like the rest of Finland, remained under Swedish sovereignty. During the Napoleonic Wars in 1808–9, a Russian army of seventeen thousand men, at the beginning of March 1809, marched across the ice once again to conquer Åland, the island outpost in the west. The Swedish troops were forced to withdraw across the frozen Åland Sea to the Swedish mainland, and again a large number of Ålanders had to flee to Sweden. After a peace treaty was signed in 1809,

Åland and the rest of Finland were ceded to Russia as part of the Grand Duchy of Finland. The efforts of the king of Sweden, in 1812, to retain the Åland Islands were in vain.

In 1830, the new rulers started building the fortress of Bomarsund. The fortress was only half erected when, in 1854, during naval operations related to the Crimean War, it was destroyed by a British and French fleet. Although not engaged in the war against Russia, Sweden put forward a claim to the Åland Islands at the end of the war, but had no success. As a consequence of the peace treaty of Paris in 1856, they remained under Russian sovereignty, but Russia was forced to agree to the demilitarization of the strategically important islands. In spite of this international agreement, however, the outbreak of World War I called forth new Russian fortifications in the islands. The demilitarization of the islands was confirmed in 1921, together with a declaration of Åland's neutrality (Björkholm and Rosas 1989, 17f.).

After the collapse of Russia toward the end of World War I, Sweden once again made efforts to regain the Åland Islands by approaching the German government in 1917, but without success. By this time, a local separatist movement had arisen in the islands (Modeen 1973, 178f.), fearing Finnish dominance in an independent Finland. The position of the Swedish language in the new state, it was thought, would not be strong enough to prevent Finnish suppression of the islands. A majority of the Ålanders wanted reunion with Sweden, the old motherland. One of the most influential leaders of the Åland movement, Julius Sundblom, was the founder and publisher of the local newspaper, *Åland,* which was first issued in 1891. The publisher, known as the "Åland King," was an important spokesperson of autonomous Åland until 1945.

The introduction of autonomy in the islands was influenced by the intervention of the Council of the League of Nations, which after World War I was engaged in questions concerning the protection of national minorities. The autonomy was the result of a compromise. Sweden and Finland both claimed the islands, and a majority of the Ålanders wanted reunion with Sweden. In 1917, representatives of the rural districts and the township of Mariehamn petitioned the king and the government of Sweden to take action for reunification, which would secure the national identity of the Ålanders. The parliament of Finland adopted an Autonomy Act for Åland in 1920, which provided wide autonomy. In 1921, the Council of the League of Nations recognized Finland's sovereignty over the islands, but Finland had to guarantee their national identity, that is, "the Swedish language, culture and local traditions" (Modeen 1973, 178f.).

According to the autonomy legislation, the Åland parliament *(landstinget,* today the *lagtinget),* led by a "speaker" *(talman),* was the highest provincial authority. The first assembly of the parliament was held on June 9, 1922, a date now celebrated as the Day of Åland Autonomy. The executive council (today the *landskapsregeringen)* is frequently referred to as the government of Åland. A governor represents the government of Finland. In 1951, a new Autonomy Act was adopted, which was replaced in 1991 (Act 1997).

This outline of the history of the Åland Islands illustrates an administrative landscape territory at the center of western and eastern political interests. It also shows that by the time of the introduction of autonomy, there was already a regional "Åland" identity supported by the provincial administration. Even if there is reason to believe that the concept of "Åland" denoted mainland Åland, the inhabitants of the smaller islands at least to some extent identified themselves as "Ålanders." That local people in the islands called themselves "Ålanders" was often witnessed by eighteenth- and nineteenth-century visitors traveling along the route between Stockholm and Åbo or by ship along one of the sea routes touching the islands.

In his diary, a peasant in one of the eastern, more isolated island communities in the 1850s made notes of local boats going east to "Finland" and west to "Åland"—to mainland Åland. This "worldview" was not unique among islanders living in the parishes outside the Åland mainland. There were also instances, however, indicating a double identity: Ålanders in Stockholm accepted being identified as "Finns" (i.e., representatives of Finland), and Ålanders visiting Finnish-speaking neighbors in southwestern Finland brought greetings from "Sweden." The identification with Sweden occurred exclusively in contacts with the Finnish-speaking, and not with the Swedish-speaking, population of Finland.

The Ålanders had permanent contacts in the west and east as the result of their commercial seagoing. Except for the period of the Russian occupation, when there was a great demand for local products, the home market was insignificant. Moreover, the resource base was limited, particularly arable land. By the end of the Middle Ages, the Ålanders carried out an active trade westward to Stockholm and other towns in the district of Lake Mälaren, and eastward to Åbo and further along the Finnish south coast, and even across the Gulf of Finland to Reval (Tallinn). A large proportion of the Ålanders was engaged in these seagoing activities not only as crew members but also as shareholders in the sailing vessels. The development of a small-scale carrying trade *(bondeseglation)*, based on cooperatively owned sailing vessels, led to increased pressure on the local resources. This could easily lead to overexploitation. During periods when the local forest resources were exhausted, the timber and wood for trade had to be acquired from the Finnish mainland. Even in the days of self-sufficiency, the Åland Islands were dependent on products from outside.

The Many Landscapes of Åland

The name of the islands, *Åland* (in Finnish, *Ahvenanmaa*), has been interpreted as "Waterland," the first part relating to the Latin *aqua*. The name first occurred in an itinerary, written about 1250. Since more than 75 percent of the "territory" is covered by water, Åland is to a great extent also a *seascape*, particularly experienced by the local sailors and fishermen. In essence, land and sea both occur in the concept of *skärgård* (skerry guard or archipelago), used (not only in Åland) to denote a cluster of islands or "skerries" separated from the mainland by open water. Conceived in relation to the sea, such a "fence" of islands rendered a shelter against the open sea.

The physical landscape, the unique *archipelago* landscape, includes more than 6,500 small islands and rocks marked on the map (Hustich 1964, 18), extending 110 km from west to east and approximately 70 km from north to south. To the west the Åland islands are separated from the archipelago on the Swedish coast by the Åland Sea, 40–50 km of open sea. In the east, the outposts almost reach the outer islands off the Finnish mainland, separated only by the narrow channel of Skiftet, formerly called Vattuskiftet, the "Water Shift." The varying width of Skiftet, from 5 to 30 km, makes the eastern border rather vague. Therefore the archipelago landscape of the Åland Islands makes a fairly unbroken part of the larger archipelago along the Finnish coast.

There is a great difference between the *summer landscape* and the *winter landscape*. During the time of ice cover, from about the end of December until the middle of April, although with great annual variations, the landscape appears as far less broken up than in the summer. During "good ice winters," the islands were connected to the busy winter road between Sweden and the Finnish mainland. The summer landscape is the landscape of one million tourists who visit the islands. The recreational use of the island landscape has also grown as a consequence of the increasing town population.

The cultural landscape of Åland reflects the dispersed inhabited area. The internal mobility of the population is illustrated by the varying number of inhabited islands: in the 1790s about 80, in 1855 nearly 100, and in the 1890s some 150. The growing number was above all a result of the growing population, particularly among the landless. The emigration of Ålanders, in particular to Sweden, and the migration of others to Mariehamn, gradually expanding as the economic center of the islands, have been reasons for the diminishing number of permanently inhabited islands in the twentieth century. From 1905 to 1960, about 60 islands were deserted, especially small islands isolated from the Åland mainland. In addition to the 60 islands inhabited today, many islands are inhabited only in the summer.

Because of the lack of heavy industry in the islands, there has been no industrial landscape replacing the old landscape of self-sufficiency, as could normally be expected. The cultural landscape is still marked by the traditional use of natural resources. The period of self-sufficiency in the islands prevailed until the end of the nineteenth century and even longer in the outlying parishes.

In a dispersed landscape territory like that of Åland, one would a priori expect little of the unity implied in regional identity. Naturally, the autonomy had a strengthening influence on the landscape-regional identity of the islands. They had constituted an administrative region even before the 1920s, but after the introduction of autonomy "*Landskapet*" Åland was no longer a landscape territory administrated by provincial officials, but an autonomous "landscape," governed by a democratically elected parliament with a growing intention of building "a state within the state," in accordance with the Act on Autonomy. In the cultural perspective of this chapter, therefore, a main interest concerns the effects of autonomy on the cultural identity of the Ålanders. Autonomy brought forth a "nation-building" process, which strongly affected the cultural identity of the Ålanders. The roots of this "Ålandishness" have to be traced to the period before the autonomy of the 1920s.

The Archipelago Landscape and the "Island Culture"

In geographical research on the island coast of Finland, the archipelago is seen as a transitional area with a gradual change from a terrestrial to a maritime environment (Jaatinen 1982, 43f.). The islanders themselves have often distinguished the *inner*, sheltered, and fairly fertile archipelago from the barren *outer* archipelago, which is much more affected by maritime conditions. Geographically the differences between the inner and the outer archipelago have been described in terms of *zones* running parallel to the coast, and starting with the *innermost* zone, with similar conditions as the mainland, and ending with the *outermost* zone of barren rocks, closest to the sea. The geographical zones are determined by features such as vegetation, soil, and climate, as well as economic features strongly influenced by environmental factors (Granö 1981). Some of the large islands have conditions so similar to those of mainland Finland that they may have their own surrounding outer zones. This is particularly the case with mainland Åland.

Although conditions may vary in different parts of the same zone, this division offers a schematic picture of the variations in the distribution of natural resources. In essence it seems to correspond to the inhabitants' own view of their island environment. For example, the diminishing area of arable land and forest toward the sea has played an important part in the adaptive processes of the island settlers. Schematically, it is possible to distinguish between mainland Åland as the center and the two surrounding zones of the inner and outer archipelago. In view of the distribution of resources, the Ålanders of mainland Åland have been able to base their economy on agriculture and forestry, as *farmers*. Their forests were formerly also used for pasture, and by providing timber for local shipbuilding the farmer could at the same time obtain a share in the ships. To a great extent, the inhabitants of the inner archipelago zone had to combine farming and animal husbandry with fishing. This group of islanders therefore lived as *farmer-fishermen*. The "real" islanders, the *fishermen*, were above all found in the outer archipelago (Figure 17.3), where they mainly lived by fishing—bartering fish for grain. Various groups of nonlandowners made their living as crofters, handicraft workers, or hands in the service of the landowners, among which a few were owners of large estates (Storå 1993, 191f., 199f.). The three types of adaptation, corresponding to the three resource zones of the islands, provide a schematic picture of the traditional self-sufficient economy of the Ålanders. The islanders could not rely on local resources alone. The shortage of arable land as a whole had to be compensated with barter trade, in particular by the fishermen and farmer-fishermen. The fishermen on barren islands were forced to obtain timber for their buildings from mainland Åland and sometimes also the wood needed for the fish barrels. At least to some extent, the shortage or total lack of a resource could be compensated through exchange between the zones.

The islanders use a rich terminology in identifying and classifying islands and rocks according to varying qualities (Mead and Jaatinen 1975, 34) as part of what is called *folk classification*. The island typology, which reflects the islanders' adaptation to and ability to cope with the environment, involves a number of concepts distinguishing islands and rocks by characteristic features such as size, shape, height above the water surface, vegetation, and so on. The terminology partly reflects the use of resources, both in the proper names of each island and in classifying islands. Islands and islets offering different resources have often been referred to as *lands*, constituting the

FIGURE 17.3. The old village harbor of a small settlement in the outer Åland archipelago, formerly a base for small-scale fishing. Courtesy Åbo Akademi University, Department of Ethnology, Turku. Photo by Nils Storå, 1967.

resource landscape. Some islands were viewed as *pasture lands, egg lands,* or *down lands* with regard to their main economic value. The egg lands and down lands referred to the collecting of seabirds' eggs and eiderdown. In the outer archipelago, every little island and skerry was used as pasture, provided it had some spot of grassland. In 1775, one of the settlers in this zone made use of fourteen skerries as pasture and for harvesting winter fodder, in addition to eleven spots on the home island yielding fodder, and four tiny enclosed pastures for calves. There were six small pieces of arable land mainly for growing potatoes (Storå 1993, 193, 206). The island profiles were of particular importance in fixing the landmarks needed to determine positions in the sea, for example, the location of underwater fishing grounds. A researcher trying to reconstruct the traditional island typology is faced with the problems caused by the constant changes that occur in the natural landscape, even without human influence. An important factor in this connection is shore displacement as a result of land uplift, in this area about 55 centimeters per century.

In view of the seagoing activities of the Ålanders, the bordering waters not only separated but also united the islands with the world outside. This was particularly the case during cold winters when landscape and seascape merged and land transport was possible along the ice roads. The islanders have used the ice not only for transport of wood and stored hay from more distant islands but also for visiting relatives and neighbors. Their knowledge of ice that was strong enough for traveling included a terminology of different types of ice (Storå 1990, 213f.), corresponding to the classification of the islands. The knowledge of ice of different qualities was of crucial importance to winter communication between the island communities—before the time of ice-going ferries.

The native conceptions refer to the perceived island environment, the mental archipelago landscape (cf. Jones 1992). The terminology enabled the islanders to exchange experiences with others in matters of vital importance to island life. In this respect the traditional culture of the Ålanders is an "island culture," based on the economic pluralism of groups here identified as farmers, farmer-fishermen, and fishermen. In general, this was the culture to some extent still characteristic of the Åland islanders at the time of the introduction of autonomy and the culture implied in the protection of culture and local traditions involved in the autonomy acts.

The Autonomous "Landscape" and the Protection of Åland Identity

Officially, the landscape-regional identity of Åland has the character of a national identity, enforced by Åland citizenship. Since the preservation of the Swedish language, culture, and local traditions closely concerns the cultural identity of the Ålanders, their cultural identity thus has legal protection. However, apart from the Swedish language, there are no hints in the Autonomy Acts of what is contained in the "culture and local traditions" of the Ålanders. The "island culture," based on multiple uses of natural resources, was not only typical of the Ålanders but also of their island neighbors east and west. The Ålanders shared the Swedish language with the Swedish-speaking population in the archipelago east of Åland and on the Finnish mainland. The guarantees given the Ålanders for the preservation of their language and culture did not concern a national minority in the usual sense because they form only a small part of the Swedish-speaking minority in Finland (Modeen 1973, 184f.). The Swedish language of the Ålanders is supported by the local mass media of Åland: two newspapers, *Åland* and *Nya Åland* ("New Åland"), as well as radio and television.

As a result of autonomy, "Ålandishness" (De Geer Hancock 1985) has a stronghold in this regional citizenship. The "region of domicile" *(hembygden)* takes the place of a native country in the identity of the Ålanders. The protection of the national character of the area implied restrictions on the ability of non-Ålanders to purchase real estate (Modeen 1973, 181f.). The idea behind the restrictions was to keep land on Åland in the hands of the Ålanders, or those having the "right of domicile in Åland." This right can be obtained by a person who has been a resident of the islands for an unbroken period of at least five years. One is granted the right of regional citizenship, "Åland citizenship," on application by the provincial administrative council, the local government. Regional citizenship implies exemption from military service, as the result of the international agreements on Åland's demilitarization and neutrality. In this respect, Åland today is often referred to as "the Islands of Peace."

The autonomous "landscape" differs from the island landscape of the traditional small-scale archipelago economy discussed above. One difference is due to depopulation of peripheral areas and abandonment of formerly inhabited islands—changes that have taken place in the archipelago areas of Finland in general (Jaatinen 1982, 48f.). The same applies to the changes in the island landscape when islands or island clusters, sometimes a whole island parish, are connected to the mainland through bridges and road banks, or through the establishment of car-ferry routes. Many old ice roads today occur only in the mental landscapes of the older generation. In the administrative landscape, the *skärgård* has been reduced to denote only the eastern, offshore parishes because of the lack of a bridge connection with the Åland mainland.

Gradually Åland ships also sailed outside the Baltic. The development of shipping made the town of Mariehamn the main trade and service center of the islands, attracting a growing part of the population from the countryside. The building of bridges and road banks (Figure 17.4) and the creation of ferry connections gradually made it easier for islanders to work in town. In spite of urbanization, Mariehamn is still a small town, and like the rest of the islands it lacks heavy industry. No industrialization has taken place because of the difficult transport conditions and the limited local labor supply. Today only 10 percent of the active, working population is engaged in manufacturing industries. The small workshop industry, often based on traditional handicrafts, has brought little change to the urban landscape.

FIGURE 17.4. An archipelago landscape with bridges and road banks connecting distant islands to mainland Åland. Courtesy Åland Tourist Association, Mariehamn. Photo by Hannu Vallas.

The growing concentration and capitalization marking the economic development of the twentieth century gradually brought great changes, above all after World War II. The importance of agriculture increased as a result of the expansion of capital-intensive wheat growing and the cultivation of sugar beets, onions, and gherkins. The intensification of farming meant greenhouse cultivation, which was possible even in areas where arable soil was scarce. Thanks to a number of local dairies founded at the beginning of the twentieth century, the importance of animal husbandry increased. Yet by 1970 almost all local dairies had been closed and dairy production was concentrated at the central dairy in Mariehamn (Jaatinen 1981, 95f.). The introduction of trawlers meant capital-intensive fishing based on a few harbors. The traditional small-scale carrying trade has been supplemented by international cargo shipping.

To a great extent these changes were supported by the local parliament and by a growing awareness among the Ålanders of the possibilities that autonomy offered. The outward sign of autonomy today is the "House of Autonomy" *(Självstyrelsegården)*, the Åland parliament building, inaugurated in 1978. Earlier plans to build a parliament house, in 1928, did not meet the approval of the Finnish government.

As a consequence of the efforts of the parliament to build an economy based on regional balance, the peripheral areas became more and more integrated in the autonomous landscape region. The economic regulations, however, are under the control of the Finnish government.

The "National" Landscape and the Symbols of Åland Identity

The Ålanders were not unaffected by the nationalist movement among the Swedish-speaking population on the Finnish mainland, which arose in the 1850s and 1860s. A main reason for the rise of the "Finland-Swedish" movement in Finland was the perceived threat the Finnish nationalist movement and the growing strength of the Finnish language—the same threat the Åland separatist movement experienced later on. The aim of this movement was to unite the Swedish-speaking people in Finland, including the Ålanders, by making them aware of their cultural identity. The roots had to be traced among the ordinary people, the *folk,* with its cultural heritage of folklore and folk songs. In this context, the first national costumes, representing Åland parishes, were worn at the folk-song festival held at Mariehamn in 1908. Some of the early cultural, ideological, and economic organizations in Åland were inspired by the organizations found among the Finland-Swedes of the mainland. This also applied to the school for educating the youth of the countryside, the "folk high school" *(folkhögskola),* founded in 1895 in one of the mainland parishes as a supplement to the primary schools.

Autonomy summoned a new national movement on the islands in the 1920s. The strong emphasis on language and culture underlined the construction of Åland identity that was implicit in the nation-building process related to Åland. In 1922, an Ålander living in Stockholm wrote the lyrics to "The Song of the Ålander," which, with music composed by a Finland-Swede, the Ålanders soon adopted as their national anthem. No doubt the Ålanders still identify themselves with the picture of their homeland and its beauty that this song evokes: *"The land with thousands of islands and skerries, shaped by a womb of waves of the sea."* Åland is frequently referred to as the "island realm." The island landscape makes the *national* landscape of the Ålanders. This is a summer landscape, like that of the tourists. The aesthetic values of the Åland landscape, the beautiful (summer) scenery with its wealth of flowers, were also depicted in the paintings of the international colony of artists that existed from 1886 to 1914 in Önningeby, a village outside Mariehamn.

One element above all that makes the national landscape a summer landscape is the "midsummer pole" *(midsommarstången)* (Figure 17.5), one of the Åland traditions borrowed from Sweden. The Åland midsummer pole is a maypole that the youth of most villages raise at midsummer, when the leaves of the trees have grown large enough to adorn the pole. The pole, around which the people gather to dance, sing, and make merry, is one of the most distinctive features of the summer landscape and is mentioned also in the anthem. In addition to the aspen leaves, which are fixed on threads strung between the transverse arms of the garlanded pole, a number of symbolic wooden figures are attached to the pole, such as a sun and a cock. Small sailing ships are often fixed to the ends of a rotating cross arm so that they sway in the wind. On top of each pole there is the figure of a man with waving arms dressed in a sailor's or folk costume. Although the poles are raised to celebrate the arrival of summer, the

FIGURE 17.5. A "midsummer pole," the main symbol of Ålanders. Courtesy Department of Ethnology, Åbo Akademi University, Turku. Photo by H. Tegengren, 1962.

height of the agrarian year, they also can symbolize traditional seafaring. The poles vary in height from 10 to 25 meters and resemble the masts of ships, with their arms giving the impression of the yards of a mast. The midsummer poles are left to stand until the next midsummer, when they are taken down and redecorated. The high poles are often visible far out at sea. They therefore also could serve as beacons (Midsommar 1991).

Since autonomy, the poles have often flown the national flags of Sweden and Finland—and, since 1954, also the flag of Åland. The first Åland flag, yellow and blue, appeared at a folk-song festival at Mariehamn in 1922. There were other flags, all of them unofficial until 1954. The Åland flag resembles the flag of Sweden, a yellow cross on a blue bottom, but it has a narrow red cross inserted in the yellow cross.

Decorated with the Åland colors, the midsummer poles constitute the main national symbol. In 1955, a local author in his book on Åland gave Åland the epithet "Land of the Midsummer Pole" (Nyman 1955). In 1984, Åland began to issue stamps of its own, and one of the first stamps featured the midsummer pole.

The development of the local school system was one of the early tasks of the local parliament. Naturally, the local schools offered an important way of educating young Ålanders in the spirit of the autonomy. There was a growing need to acquaint the young Ålanders with Åland history. A major step in this direction was the introduction of the book *Åland and the Ålanders (Åland och ålänningarna)* as a common school textbook. The book was published in 1943 by Matts Dreijer, who held an influential position within the local administration as head of the department of antiquities and superintendent of the Åland Museum, the "national museum" founded in 1934. His book was to be read by all young Ålanders, which made it an important factor in shaping the Åland identity. A new edition had to be published only two years later. In the foreword to the first edition, Dreijer notes: It is "written for the Ålanders, who are attached to their landscape with strong ties." The book aims at "arousing an even stronger love of the homeland" by spreading knowledge about Åland among children, youth, and adults, and also informing strangers about the "unique conditions" of the landscape region.

Dreijer's book begins with a survey of the physical landscape of the islands, followed by an outline of the history and a survey of the Ålanders and their culture. The book ends with the biographical data of a number of "remarkable" Ålanders. It gives a vivid, often highly romantic, picture of Åland and the Ålanders, sometimes referred to as the "Åland people." A comprehensive, geographical chapter entitled "The Map of Åland" stresses the beauty of the thousands of islands, skerries, and rocks of "the island realm of Åland, girdled by the blue ribbon of the sea." The outline of the autonomy period ends with the presumption that the most important result of the autonomy has been that Ålanders more than ever feel to be "a people of its own." The national romanticism contained in the book no doubt contributed to the awareness among Ålanders not only of their history but also of their national identity.

In a questionnaire on Åland identity, sponsored by the Department of Statistics and Economic Research in Åland at the beginning of 1999, 820 Ålanders over the age of eighteen were asked to choose among fifteen given "factors of importance to identity" and arrange them in order of importance. Completed forms were received from 427 persons, only 11 of whom stated factors not included on the form (Radar 1999). The most important identity-shaping factors in rank were, according to an average evaluation, the Swedish language, the right of domicile, the natural environment, and autonomy. Lower-ranking factors were Åland history, Åland culture, demilitarization, and the cultural environment. Although the questionnaire reflects the designer's own view on Åland identity, it is interesting to note the top position of the Swedish language and, in this connection, the third place of the natural environment. The importance of the archipelago landscape as a resource landscape concerns a far smaller part of the population than before. The resource landscape has been reduced as former resources have lost their economic value. Part of the resource landscape is perceived only as a mental landscape of older generations. To nearly half the population, representing an urbanized culture, the island landscape is above all a recreational landscape, but with great symbolic value, apparently common to all Ålanders.

REFERENCES

Act 1997. *Act on the Autonomy of Åland.* Mariehamn.
Björkholm, Mikaela, and Allan Rosas. 1989. *Ålandsöarnas demilitarisering och neutralisering.* Meddelanden från Ålands kulturstiftelse 3. Mariehamn.
Brück, Ulla. 1988. "Identity, Local Community, and Identity." In *Tradition and Cultural Identity,* ed. Lauri Honko, 77–92. NIF-Publications, no. 20. Turku: Nordic Institute of Folklore.
De Geer Hancock, Yvonne. 1985. *"Åländskhet": En studie i nationalism och nationsbygge.* Stockholm: Institutet för folklivsforskning vid Stockholms universitet.
Dreijer, Matts. 1943. *Åland och åländingarna.* Helsingfors: Söderströms förlag.
———. 1947. *Ålands självstyrelse 25 år.* Mariehamn: Landstingets presidium.
———. 1968. *Glimpses of Åland History.* Mariehamn: Ålands Museum.
Granö, Olavi. 1981. "The Zone Concept Applied to the Finnish Coast in the Light of Scientific Traditions." *Fennia* 159, no. 1: 63–68.
Honko, Lauri. 1988. "Studies on Tradition and Cultural Identity." In *Tradition and Cultural Identity,* ed. Lauri Honko, 7–26. NIF-Publications, no. 20. Turku: Nordic Institute of Folklore.
Hustich, Ilmari. 1964. *Finlands skärgård: En ekonomisk-geografisk översikt.* Skrifter utgivna av Svenska Handelshögskolan 10. Helsingfors.
Jaatinen, Stig. 1981. "Development of the Surface Communications in Åland, SW Finland." *Fennia* 159, no. 1: 93–101.
———. 1982. "Maamme saaristomaisemien luonne ja muuttuminen." *Terra* 94, no. 1: 43–51.
Jones, Michael. 1992. "Persepsjon og landskap—landskap som sosial og kulturell konstruksjon." *Papers from the Department of Geography, University of Trondheim,* no. 116. Trondheim.
Lindström, Bjarne. 2000. "The Culture and Economic Development in Åland." In *Lessons from the Political Economy of Small Islands: Resourcefulness of Jurisdiction,* ed. Godfrey Baldacchino and David Milne, 107–20. New York: St. Martin's Press.

Mead, W. R., and S. H. Jaatinen. 1975. *The Åland Islands*. Newton Abbot: David and Charles.
Midsommar 1991. *Midsommar*. Glimtar ur Ålands folkkultur 6. Mariehamn.
Mikander, Kaj. 1964. "Åländska handlingar 1530–1634," 1: 2. *Ålands urkundssamling* II. Mariehamn: Ålands kulturstiftelse.
Modeen, Tore. 1973. "The International Protection of the National Identity of the Åland Islands." *Scandinavian Studies in Law* 17: 177–210.
Nyman, Valdemar. 1955. *Åland: Midsommarstångens land*. Helsingfors: Schildts förlag.
Radar 1999. *Radar*, no. 1, 1999. Mariehamn: Ålands högskola.
Radloff, F. W. 1795. *Beskrifning Öfver Åland*. Åbo: J. C. Frenckells boktryckeri.
Salomonsson, Anders. 1996. "Regionaliteten som problem." In *Att skapa en region—om identitet och territorium*, ed. Markus Idvall and Anders Salomonsson, 13–20. NordREFO 1996: 1. Stockholm: NordREFO.
Storå, Nils. 1988. "Åland—kulturområde mellan väst och öst." In *Åländskt språk mellan öst och väst*, ed. Åsa Stenwall-Albjerg, 13–25. Ålands högskola 1988: 1. Mariehamn.
———. 1990. "Kognitiva system och maritim anpassning: Om skärgårdsbornas iskunskap." In *Folklore and Folkkultur*, ed. Jón H. Adalsteinsson, 205–22. Reykjavik: Háskóli Íslands.
———. 1993. "Adaptive Dynamics and Island Life: Resource Use in the Åland Islands." In *Resurser, strategier, miljöer: Etnologiska uppsatser av Nils Storå*, 181–228. Åbo.
Sundwall, Johannes. 1958. "Ålands medeltidsurkunder II. 1400–1450." *Ålands urkundssamling* II. Helsingfors: Ålands kulturstiftelse.
Voionmaa, Väinö. 1913. "Studier i Ålands medeltidshistoria." *Finska fornminnesföreningens tidskrift* 27.

18.
Landscapes of Domination: Living in and off the Forests in Eastern Finland

ARI AUKUSTI LEHTINEN

Late Wednesday evening, on June 17, 1998, a highly dramatic event took place in Ruokolahti, a rural municipality in southeastern Finland. A wild bear killed a man, only two kilometers from his home. The bear, with her one-year-old offspring, had accidentally come across the unlucky man, an experienced outdoor hiker who was out for his usual evening jog. The marks on the trail told the searchers later that he, after having seen the bear, had hurriedly turned around, but in a few seconds she had caught him. His body was found later, in the early morning, close to the trail. Shortly thereafter, the police and frontier guards shot the bear and her cub.[1]

Consequently, during the midsummer weeks and throughout the rainy summer of 1998, the whole of Finland was talking about the danger imposed by the gradually growing number of bears learning to live close to settled areas. Today, anyone who wants to go out in the forests to pick berries or mushrooms, or only to wander around in nature, has to keep in mind the increased risk of facing a bear. The number of wild bears in Finland now exceeds one thousand, and, because of the successful control of hunting, the number has doubled in a decade. In the early 1970s, no more than about one hundred bears were known in Finland, and the average had been even below that number throughout the century. However, only a few of the present bears have turned into unnaturally fearless individuals who let humans observe them (Pulliainen 1994; Helle 1998).

In the summer of 1997, a similar kind of debate on bears emerged in Finland. Then someone noticed a lonely female bear in southwestern Finland, wandering slowly over the course of two months, the distance of 200 kilometers from the capital city region to the town of Rauma on the western coast. Before being shot by the police, the "city bear," as it was soon named, had terrified the guests of a garden party in Vantaa, killed a lamb in Kirkkonummi, and feasted on honey in an apiary in Masku.

The debates surrounding these accidents and, in general, the growing number of observations of bears have strengthened the criticism against nature conservation in Finland. Bears are symbols of the wild nature of the boreal north (Pentikäinen 1994) and their increased number is easily connected to the recent extension of conservation of the old-growth forests. In addition, the bear reminds the Finns, as perhaps most northern people, of those days when their ancestors lived under the constant pressure of the hostility of the backwoods, threatening their very existence as individuals and as a civilization (Pulliainen 1984).

Because of these recent accidents, recreational hunters now have good reason to apply for more licenses to kill bears during the hunting seasons and whenever they find them close to human settlement. In this respect, enthusiasm is high and the moral support of laypeople is strong, regardless of the statistics expressing the actual dangers to humans in Finland. In comparison, several hundred Finns are killed annually in traffic accidents, for example, 400 to 650 road deaths every year during the first half of the 1990s, but no strong voice has emerged against the growing number of private cars. Moreover, even though traffic accidents with the semi-wild elks result in approximately ten human lives lost in Finland each year, no agreement on reducing the elk population has been reached (*Raportti liikennevahinkojen . . .* 1996).

In this chapter I shall examine the background and reasons for this concern with bears, first by discussing the historical-cultural mechanisms involved in the present debate on wild nature, and, second, by specifying the areal variations of wilderness appreciation and use—in other words, landscaping—in contemporary Finnish forests.[2] I will make an etymological and comparative conclusion by evaluating the specific cultural features of Finnish forest landscape in a northern European context. However, this excursion into Finnish wilderness forests begins with an excerpt from a classic novel on Finnish backwoods mentality, the final scene from Ilmari Kianto's (1946) *Punainen viiva* ("The Red Line").

The Lore of the Backwoods

> It was Topi himself, whilst tending his cow and cutting wood for gate posts at the edge of the forest, who had first been startled by the fence mysteriously crashing and Jysky, barking, fleeing back to the farmyard. In the flick of an eye, Topi had taken in the scene and charged at the bear with his raised axe.
>
> "What the hell!"
>
> Terrible anger washed over him, and the emergency brought the hero forth. Just as the beast was climbing over the fence with its prey, the axe whacked it in the chest. But the beast defended itself. In an instant it released its claws from the cow and hurled itself over the man.
>
> No time for Topi to raise his axe for the second time. Spurting blood, he fell behind the fence with a thump, and the axe flew from his hand. On the

other side of the fence, Ämmikki, in the throes of death, lay with her hooves in the air.

The commotion and cry of anguish had been heard up at the cottage. Riika ran down along the side of the field, agonising, fearing the worst, clutching at her heart . . .

The bleeding bear was by then vanishing into the forest. Jysky dared not follow but howled and trembled, tail between legs, behind the sauna.

"Topi, hoy, Topii . . . !"

But Topi did not answer. Unmoving, he lay where the snow had melted, and across his neck ran blood—in a red line.

Behind the shingle-mended windowpane could be seen the pale, terrified face of a girl.

(Ilmari Kianto 1946, 210–11, translated by Sisko Porter)

Ilmari Kianto, the celebrated author of the deep backwoods of eastern Finland, draws in "The Red Line" a thrilling picture of the early twentieth-century frontier life in the Nordic east. Kianto paints for us a landscape of the "northern hunger land," striving for survival on the margins of civilization. For the generations after, the book has functioned as a memory source, or a collective logbook of the experiences based on the permanent threats from the external world, from both the wilderness enveloping the small islands of human settlement and also from the vast eastern continent neighboring the Finnish frontier culture. Kianto's bear came from the eastern wilderness, that is, from the other side, away from the familiar Finnish world, and symbolized the hostile nature characteristic of forest life. Without doubt, it contained a hidden reference to the threatening expansive acts of the contemporary Russian Pan-Slavists, as symbolized by the Russian bear (see Jutikkala and Pirinen 1984, 211) (Figure 18.1).

The novel, finished as a manuscript in 1909, recalls the first experiences of democratic parliamentary elections in Finland in March 1907. A radical political reform had only recently been carried out, which gave the disenfranchised classes the right to vote. Women were included in the extended suffrage, too. A rather radical unicameral system of representation was adopted, where the whole parliament, not only the lower chamber, was to be elected by democratic elections. The reforms were remarkable, keeping in mind the geopolitical setting of Finland during those days when Finland was part of the Russian Empire, with a fragile status as an autonomous grand duchy.

"The Red Line" illustrates well the elements of the Finnish mentality rooted to traditions of forest life, but it also introduces the geopolitical and economic motives that were gradually made concrete in the form of forest-based modernization and welfare. The book shapes the frames of the Finnish model: the way of coping with the north, and the way of building a welfare society on the historical demarcation zone between the East and the West.

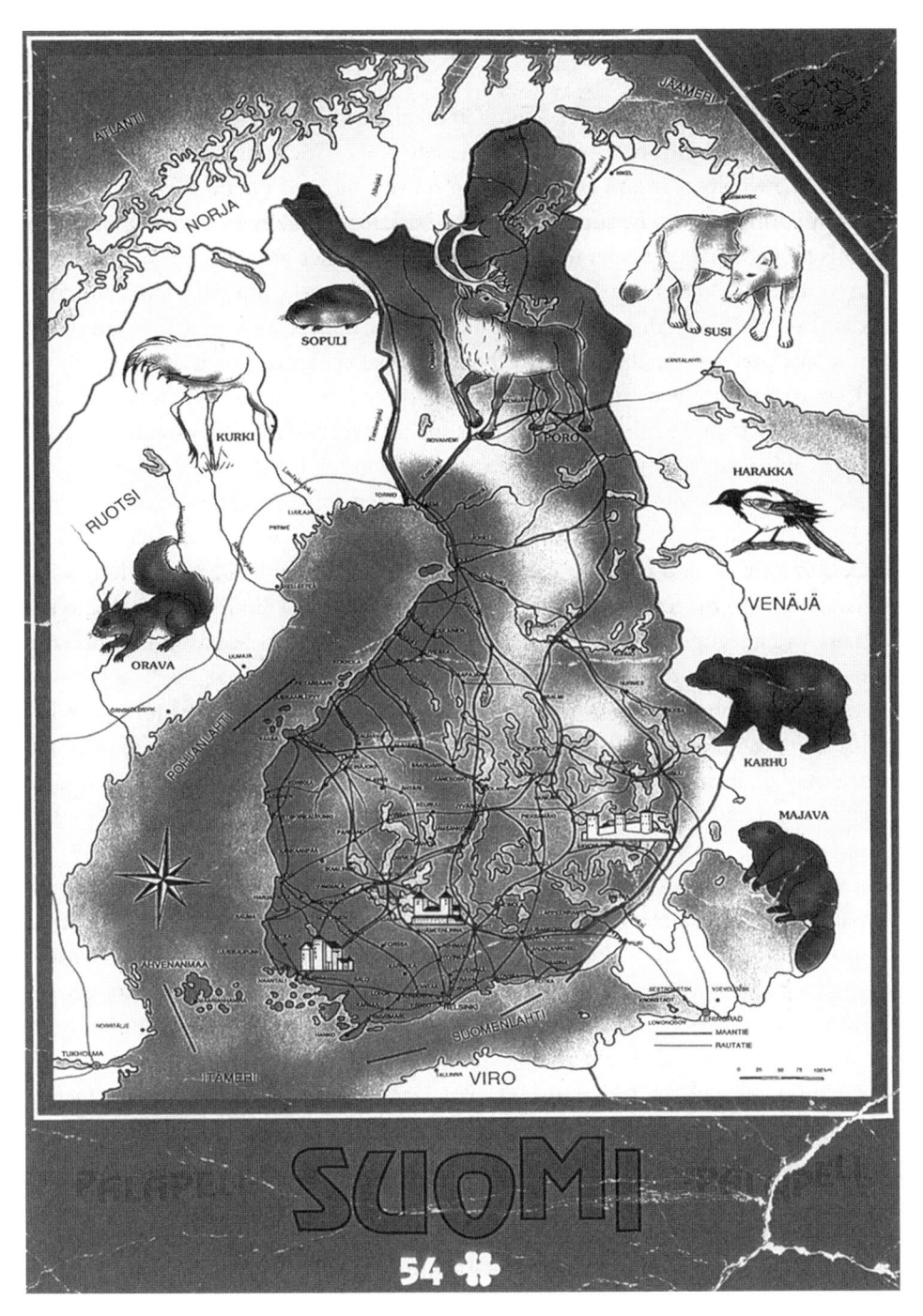

FIGURE 18.1. This traditional fifty-four-piece jigsaw puzzle of the map of Finland features its native animals. The bear emerges from the east, its right paw extended. The reindeer is the only animal inside Finland (actually Finnish Lapland). The medieval castles of Olavinlinna (Nyslott), Hämeenlinna (Tavastehus), and Turku (Åbo), representing southern Finland, symbolize the territorial defenses of the historical interface between East and West. Reprinted by permission of Peliko-pelit Ltd.

The north completely surrounds human life in Kianto's hunger land. In particular, the severe winter is strongly present in the landscape. The snow extends her fingers through the windowpanes. The kitchen shelves run short of rye flour already during the early spring, only to be supplemented by bark from pine trees. Topi Romppainen and his family struggle desperately to prepare for another growing season. However, the severe climate of the north, together with the wild beast from the east, finally concludes the play. The fatalism of the marginal north stands in stark contrast to the early democratic practices boldly played at the municipal center during those days.

Kianto's message is clear: life in the backwoods means hard and tireless work and immense efforts against the enemy with its superior force. Only by defending against the environing threats and, consequently, only by taming the surrounding wilderness would cultural continuity be secured. The experience of constant insecurity beyond the known world was countered by expansive modification of the lands within reach. Success in farm work during the short growing season was crucial, and the long winter could only be overcome by hunting. The old practice of killing bears in their winter lairs was an important means of survival in the northern backwoods, especially toward spring, when daily nourishment was often critical.

Understandably, the hunter Martti Kitunen is the hero of the forest lore, with his record of 198 adult bears killed and countless numbers of their cubs (Figure 18.2). An entire chapter, including this painting, is devoted to Kikunen in *Maamme kirja* (*The Book on Our Land,* Topelius (1981 [1875]: 162–63). According to Klaus Ekman (Honkanen 1998) of the Central Association for Hunters, hunting is a hobby for about 290,000 men in Finland. "Almost one and all of the men in eastern and northern Finland participate in seasonal hunting." The real enemy for the wilderness hunters today comes from the European Union (EU) in the form of nature directives demanding specific grounds for killing wild or endangered animals, such as wolves, lynx, and bears. Therefore, the national annual quota for killing bears, which in recent years was about 100 licenses, is no longer acceptable.

Living in and off the forests meant organized hunting by the villages, extractive slash-and-burn cultivation, tar production, and, later, modern industrial forestry. The Finnish success story, the tale of those who were lucky enough to succeed, is based on the initial challenge of taming the forest, hence turning wildernesses into a source of livelihood. The authorities throughout the centuries have supported the challenge. Already in 1550, King Gustav I Vasa of Sweden wrote about the necessity of colonizing the eastern backwoods: "The wilderness in Finland must be cleared and settled in order to augment the revenue due the Crown of Sweden" (Jutikkala and Pirinen 1984, 60–61). The geopolitical interests of consolidating the frontier conditions of the East, of course, motivated this aim. Only later, during the nineteenth century, did the support of colonization turn into official fear of forest devastation by the peasants. Then, however, the concern was focused on exploitation of the woodlands close to the settled

FIGURE 18.2. Martti Kitunen (1747–1833), the great bear killer. Kitunen's courage became a heroic myth. The original painting by Alfr. Wahlberg and J. W. Wallander is in the Nationalmuseet, Stockholm. Reprinted with permission.

areas, not in the backwoods (Leikola 1987). Hence, the virgin backwoods, those forests mostly untouched by human beings, have broadly served as a security problem both for families and the authorities, and, in economic terms, as not fully utilized property (see Hustich 1965). In other words, the gradual loss of the old-growth forests has been the outcome, and price, of Finnish cultural continuity and independence.

Forest Ethos and the Layers of Landscape

Cultural continuity in Finland has always been focused on the question of the success of living in and off the forests. Every generation has been forced to face this challenge: a realm of necessities caused by conditions in the northern backwoods. The Finnish forest landscape is a mirror of historical experiences, and it reflects and contains the Finnish attitudes toward forest, land, and nature in general. This historical forest ethos, sedimented in the landscape, is a kind of largely agreed and adopted mental orientation, within which human creativity is mainly aimed toward expanding control over the natural environment (see Lehtinen 2001).

The forest ethos was initially developed as the driving force for the farmers settling down and clearing forests in different parts of Finland. Forest devastation was the only means of earning one's living in the backwoods. There was no alternative: taming the land was a question of life and death for the early invaders. Deep in the minds of those pioneers, nature was a cause of both awe and threat. This had of course grown from the bitter experiences of the unforeseen and fierce nature: severe climate was a constant threat, wild beasts were regular visitors to the grazing cattle, and frosty nights during the growing seasons resulted in dramatic hunger years. In the northern hunger land, the uncleared forest, from whence all the setbacks beyond human control originated, was a perpetual challenge for the peasant (see Virtanen 1994).

The land reform, or the Great Land Enclosure *(Isojako)*, from 1750 to 1950, gradually restructured the traditional economy based on the clearing of arable land. The common lands of the villages were divided and shared among the farms. Those who received large forest areas soon became the fortunate suppliers of industrial timber. Consequently, the land ethos originating from traditional agricultural practices was expanded to cover the use of the privatized forests. The limits of intensive land use were cast far beyond the hunting grounds—close to the extreme wilderness forests belonging to the Crown (see Linkola 1987).

Toward the end of the nineteenth century, a need for state-led forest management became apparent to the Grand Duchy of Finland. This was a direct response to the growing official concern about the devastation of the private forests by the peasants. However, the actual outcome was more visible in the state forests: an intensive forest-management culture—the "German model"—was introduced to state foresters (von Berg 1988 [1859]). The German model was easily assimilated into the Finnish forest ethos; in fact, it essentially supported the domestic idea of expanding control and use of nature. As a result, efficient forest management soon covered the whole country, occasionally reaching the belt of old-growth forests on the eastern border as well (see Leikola 1987).

During the twentieth century, Finland was deeply integrated into global pulp and paper markets, which put pressure on domestic forest management and utilization. The conceptual tools for official forest management, both in private and state forests, have been grounded on official national forest inventories and intensive forest cultivation methods *(metsänviljelytalous)*. Both methods are in fact derivations from the old peasant ideology, and they explain much of the mechanism by which forests are treated as agricultural land. Consequently, national forest policy has been based on an idea of full mobilization, which means an ideal situation where the annual level of cuttings "balances" the wood increment. Not an inch of the annual gross increment of wood is deliberately allowed to be lost to nature in the form of rotting timber (Holopainen 1968; *Metsä 2000* 1985; Lehtinen 1991).

In recent years, controversies connected to the conservation of the old-growth forests are again good examples of the viability of the old peasant ethos. Still, the economic effects of the conservation plans are estimated from the point of view of full mobilization. The annual growth rates of wood in the planned conservation areas form the basis for economic calculations—as sources of direct and indirect losses to the national economy. Therefore it is easy to dramatize the negative results of conservation decisions and exclude the positive potential connected, for example, to the credibility of Finnish forestry for those asking for certificates of ecological sustainability (Lehtinen 1996, 2003).

Even today, the Finnish forest landscape continues to reflect historical peasant attitudes against unmanaged and untamed nature,[3] threatening our basic existence. The deep and "overaged" forests symbolize backwardness and economic idling, which necessarily need to be eliminated by more efficient management measures—even at the risk of losing markets among the environmentally aware consumers of Central Europe.

The Marginal Lands at the Center: Shadow Forests

It is no exaggeration to say that the expansion of control over forests is an important element in Finnish cultural continuity. Hence, the scenic diversity of Finnish forests is a result of intense domination of the nature within reach, reproduced through generations and gradually becoming more and more mechanized. Today we have trouble finding a place in Finland that does not have a strong visual message of apparent human domination (Figures 18.3–18.5). The current debate over the last remaining pockets of old-growth forests is therefore a competition over the control, and fate, of the last indigenous wildernesses forming a scenic connection or, say, memory from the past to the present.

It seems that the pockets of indigenous forests serve as memories of the historical landscapes of fear,[4] where naturalness symbolizes wildness and external unpredictability and life always at risk. Consequently, these pockets remind the Finns of a past from which they wish to distance themselves—something to be forgotten, or at least to be reduced to a historical curiosity. However, the authenticity of the pockets is related to their qualitative uniqueness: nowhere else has ecological and cultural continuity been preserved to the same degree as within these marginal areas, which have never been subjected to intensive utilization. The perception of these areas as untouched wilderness has caused much confusion in Finnish forest policy. Virginlike in nature is indeed a matter inciting conflicting interpretations.

The debate on the areal extension of the old-growth forests in Finland has clearly illustrated the difficulties of defining the indigenous in nature. The research director

FIGURES 18.3–18.5. The image of the endless forests in Finland is based on fact: 87 percent of Finland is forest land (Kuusela 1976) and about 95 percent of that land is controlled by commercial logging. Thus the forests at ground level often appear as plantations of even-aged pine trees. It is difficult to find a forest area in Finland without a strong visual message of apparent human domination (Elo 1983). Reprinted with permission of artist Markku Tanttu (Figure 18.3) and photographer Erkki Oksanen, Forest Research Institute in Finland (Figures 18.4–18.5).

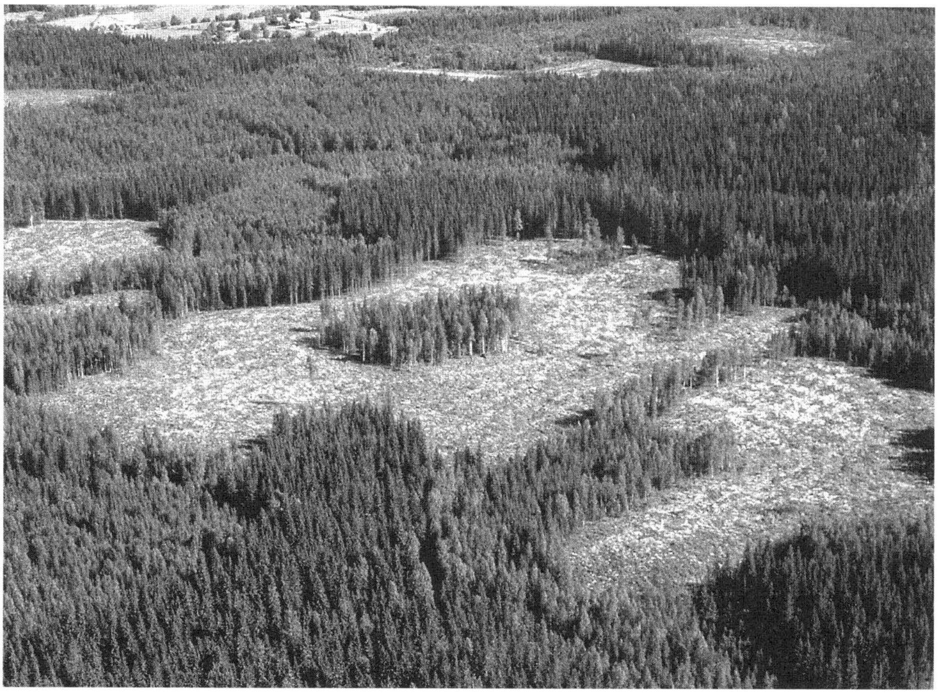

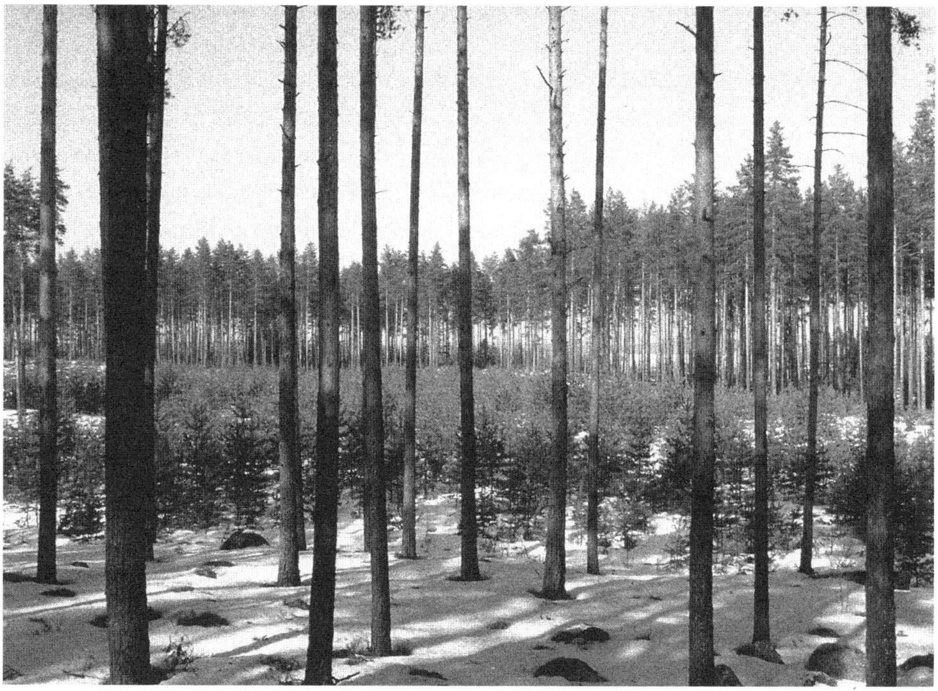

of the Finnish Forest Research Institute, Professor Matti Kärkkäinen, has classified all forests with dominant trees reaching the average age of 120 years as old-growth forests. This definition thus includes all (over)aged forests with falling rates of annual wood increment and with delayed final cuttings. Therefore, it is correct to show that the coverage of old-growth forests, that is, the share of aged forests, has more than doubled during the last seventy years in the southern part of Finland. This means a rise from 0.7 percent up to 1.6 percent of the total area of forest land (Kärkkäinen 1995). However, the qualitative difference between an aged economic forest and an authentic old-growth forest is significant. The share of the latter covers only a few percent of the total area of aged forests. From a forest conservational perspective, it is as correct to say that these aged forests are ecologically poor because of the breaks in species continuity caused by the cuttings in the past and subsequent forest management. In addition, from a biodiversity perspective, the economic forests of the 1920s were much more valuable than the aged economic forests where the key biotopes have disappeared (Simola 1995).

The indigenous virgin forests in Finland can be found only on the margins of Forest Lapland, Kainuu, and North Karelia. These shadow forests *(katvemetsät)* have been saved from the historical effects of slash-and-burn cultivation, tar production, and industrial forestry as a result of their peripheral location. The areal extension of these shadow forests is of course an ecological question and, moreover, a biodiversity problem, but it is also a cultural challenge, a question about the admittance of the Finnish past—both as individual experiences and collective identity constructions.

The embeddedness of ecological and cultural elements within forest landscapes varies greatly in different parts of Forest Finland. The regional histories are different and consequently the regional landscapes and identities are different. The dividing line runs along the Suomenselkä, the back of Finland, which is the main dividing zone for the natural waterways of the southeastern lake area and the northwestern "land of eleven rivers." A remarkable belt of old-growth forests exists within the dividing zone (Figure 18.6).

The present landscape of the Finnish lake area in eastern Finland is, in broad terms, a product of forest industrial division of labor using the main drainage system and watercourses. The macroscale regional structure was originally based on the floating of logs, on the mobilization of the energy potentials of the rapids, and on the market options at the seashore, and this history of land use still conditions landscape formation and also the internal configuration of regional economic and political problems. The southern part of the lake area is coping with restructuring in the forest industry mostly caused by organizational merging, redevelopment, and the automation of the industrial units. Intensified public pollution control is also shaping the modernization of the production processes. The northern part, as a regional resource periphery, has by contrast suffered from low industrial refining, as well as from socioecological limits

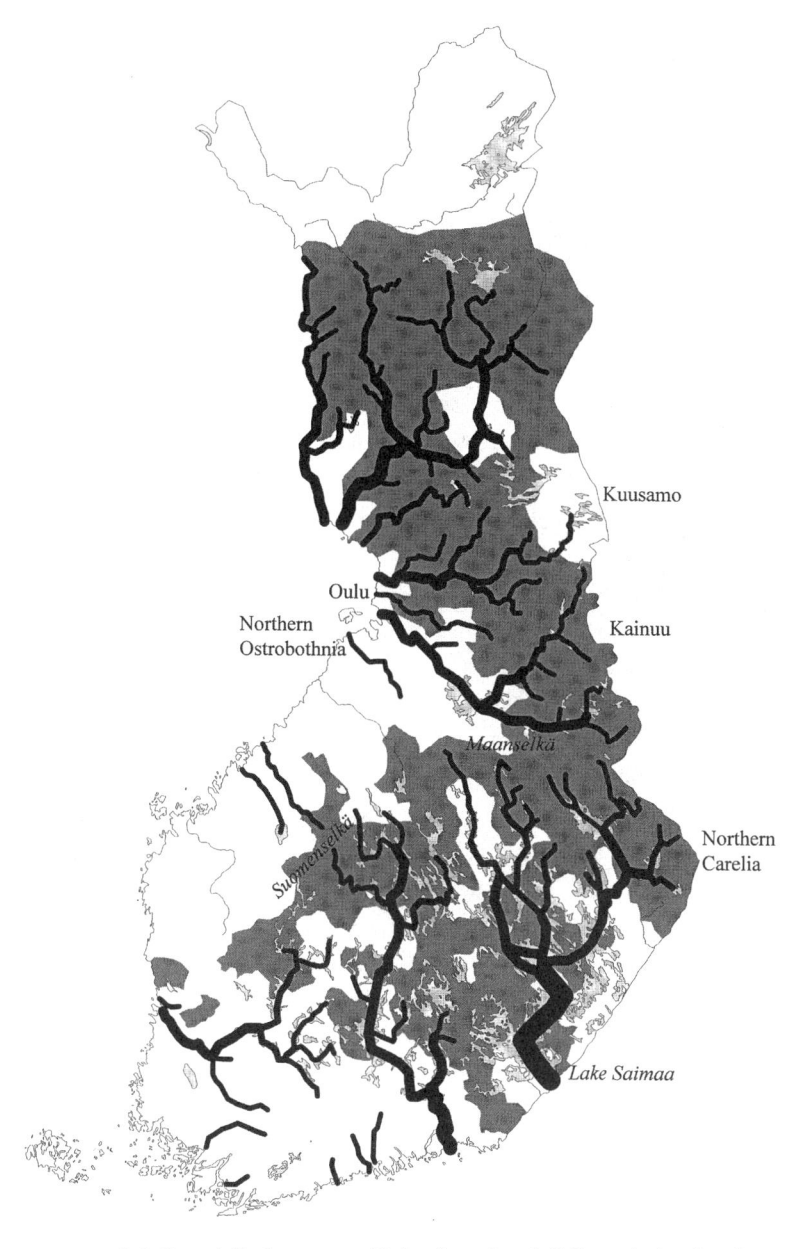

FIGURE 18.6. Especially in eastern Finland, regional differentiation has been modified by forest exploitation, log transport, and wood processing. The map describes the setting between 1951 and 1955: the bold lines represent the volume of logs floated via the lake and river systems (the maximum is about 3 million cubic meters in Lake Saimaa). The shaded areas represent backwoods, where forest work accounted for more than 20 percent of labor in one year (1950) (Kortelainen 1999). Reprinted by permission of the Geographical Society of Finland.

to wood mobilization, such as forest conservation, green image problems, and wood certification. Export orientation resulted in a regional division of labor where the backwoods served as resource peripheries for the forest industry located on the lower courses of the rivers. Floating was the initial mode of log transport.

The north of Lakeland Finland carries the traces of dominant land uses of the past and present, especially marks of slash-and-burn cultivation and today's industrial forestry. During recent decades, the forestry-based regional landscape has turned into a political battleground for logging corporations and forest conservationists. The deep (national) economic recession of the 1990s was brought to a head in the resource peripheries, and this made it easy for the logging interests to blame expanding forest conservation for the growing unemployment. The falling regional economic options have in a way favored a more aggressive criticism of official conservation policy, which is regarded as insane, turning forests into outdoor museums and of course places for bears and wolves to increase in number. Collective insecurity caused by the perception of the external threat to the provincial economy has, by old custom, provided the basis for full-scale criticism against any conservation initiative.

The forest landscape north of the Maanselkä ridge, that is, Northern Ostrobothnia, Kainuu, and Kuusamo, is largely shaped by the historical effects of tar production and, currently, expanding industrial logging. The town of Oulu on the northern shore of the Gulf of Bothnia has functioned as a trade center for the developing regional economy. Today the details of the landscape connected to tar production are disappearing due to ecological succession and the expansion of industrial forestry. Here the scenic remnants of tar production are important, not so much as landscapes of authentic indigenous nature, or as outdoor museums, where natural evolution is limited, but as gradually vanishing memories of layers of land-use history.

During the 1990s, however, the land-use confrontations focused on the last remaining shadow forests (existing behind the "tar woods") became increasingly heated. The arguments carried two contrasting backwoods landscapes: the "great logging frontier" was severely challenged by a broader Europe-wide vision of an ecological "Green Belt" along the Finnish–Russian border (Lehtinen and Rytteri 1998).

The backwoods provinces are chronically troubled by serious economic problems. When causes of this or scapegoats are sought, the environmental groups are often the first to be blamed. Moreover, this is fed by the broad and continuing publicity about the visits of wild beasts of the wilderness, that is, a constant state of alertness under the pressure of externalized nature. It is thus easy to understand criticism from the backwoods against proposed enlargements of nature conservation, now more and more in the shape of directives from the European Union. The current environmental strategies of the EU aim both at the unification of national conservation criteria (the *Natura* Program) and at setting limits to hunting practices. Within the EU, the bear is considered a species that needs specific protection, and this in turn is increasingly limiting the licenses for killing bears. The backwoods experience is, quite understandably, opposed in principal to the EU regulations. This became clear in the referendum of 1994 (Figure 18.7). The current environmental regulation of the EU is hardly calming the conflict, even though the EU has opened several new alternatives for funding economic activities in the backwoods villages.

INTERNAL AND EXTERNAL LANDSCAPES

The currently dominant ethos of industrial forest management in Finland can be seen as a logical outcome of the historical culture of forest clearing. Today, the intensity and areal coverage of this practice has achieved its extreme development stage. Both the marginal shadow forests and old economic forests are under intensive regulative control. The landscape of fear has finally turned into a safe and known forest and, thus,

FIGURE 18.7. The national referendum on EU membership in 1994 split Finland into urban areas, which approved the EU, and rural areas, which did not approve. The backwoods provinces of Finland belong mostly to the *Nejden*, the anti-EU block of the new Nordic countries *(det nya Norden)*. Reprinted by permission of *Ordfront Magasin*, no. 1 (1995).

became a cultivated land, where the relationship between the annual wood increment and cutting, calculated in cubic meters, is considered sustainable when the former exceeds the latter.

The conserved forests cover between 2 percent and 9 percent of the total area of forest land in Finland.[5] Within the backwoods provinces, each extension of conservation is perceived as a threat to social welfare expectations. It seems the men of the provinces will calm down only after the last pockets of indigenous nature have become transformed into economic land; only after the last backwoods have become useless for the wild bear, the earlier guardian and ruler of the deep forests. The fact that the economic forests, as they become poorer and poorer habitats for the wilderness fauna, push the bears closer to settled lands for feeding has only a marginal influence on the debate.

The northern forest landscape in Finland is primarily a resource base subjected to intensive exploitation, a means to economic growth. However, it is also a frame for Finnish self-reflection and identity construction: a kind of memory of the past, in the same way that, for example, a photograph in a family album shows the ancestors of the past, with hollowed cheeks and deep-grooved faces. The past is kept safely in the album (as a curiosity), for the purpose of remembering, but otherwise it is considered something primitive and poor—something from which completely to free ourselves.

The uncontrollable and the unpredictable in the Finnish past/nature are considered a threat and a risk, whereas the expansive manipulation of nature is thought of as an option for a better future. However, because of geopolitical changes within the environing world, several new challenges have emerged. The question concerns the future of the Finnish model: How useful and competitive is it in today's globalizing world? How to cope with the legitimation and credibility problems connected to the continuous struggle for expansive domination of nature? How to evaluate the costs of the current highly uneven distribution of risks and benefits? And finally, how to react against the growing foreign conservation interests focused on Finland's backwoods—which will gradually turn them, perhaps, into the EU's Green Belt of the northwoods? In any case, both Kianto's hunger land and the bear have become "hot spots" for international environmentalism and therefore their future largely depends on the paper consumers of Europe.

The bear has repeatedly emerged from the deep backwoods. It has also come from the east and from the south (in the form of nature conservation), but no doubt the bear has grown from the inside, too, appearing as a remnant from the Finnish past, an internal landscape of fear producing cultural confusion and defensive positions—in an era of radical global time-space compressions. Old positions have gradually lost much of their validity and evidently a reorientation or at least a reevaluation based on historical experiences seems necessary.

Kianto's hungry bear, awakening from her long winter sleep to take those bloody actions, represents the whole genesis of threats from the outside, burdening the small farmers of the northern hunger land. In other words, the surrounding threat, together with the constant risk from the east, is the question of life and death. Indeed, the geopolitical location is adopted as a popular explanation for all the setbacks of the past—and present. Therefore, the self-image of the Finns is reproduced in a dual process: by underlining the apparent external dependencies resulting in bad years and, simultaneously, by practically turning the immediate factors at hand into objects for replaced domination (see Siltala 1996; Lehtinen 2000).

The marginal northern setting and the border location of Finland are continually reproduced by actions that emphasize and even deepen the dominant macroscale geopolitical and geo-economic division of labor. Location on the margins is paradoxically confirmed by acts aiming at the opposite. The Finnish versions of modernization and westernization have kept Finland in its place and strengthened the elements that have always constituted it. The country's fate seems to be anchored in the success of "forest fundamentalism": "The forest industrial cluster is the stem of the Finnish economy" (*Pro Silvis* 1995). "In the latter half of the 20th century, [the forest cluster] has formed the key to Finland's economic competitiveness" (Reunala and Tikkanen 1999, 11). Within the framework of the international division of labor and in the borderland between the East and the West, Finland lives in and off her forests—now and in the future, as if immune to the emergence of challenging alternatives.

During the 1990s, however, the pressures for economic diversification in Finland grew large enough to burst out in the form of information technology. Paper production with its capital-intensive machinery was challenged by the expansion of new technology based on the paperless modes of communication. In Finland, the export share of forest goods dropped from 34 percent to 25 percent in a five-year period (1995–2000). The change was a dramatic one in production but also at the level of everyday life. A major shift toward new industrial agglomerations in Finnish regional economic infrastructure was witnessed and, simultaneously, a new networking of social life took place via mobile phones and in connection with the World Wide Web and the Internet (Vartiainen and Antikainen 1998; Kopomaa 2000; *Forest Industries Finland* 2001).

"The Red Line" reflects experiences of parallel social efforts aimed at breaking the deterministic and authoritarian construction of Finnishness. In the dramatic epilogue, the red line grows on the throat of the poor Topi Romppainen, but, earlier the same spring, the red line was drawn on a ballot paper, voting for the social reforms mentioned at the outset. "The Red Line" is also a symbol of political awakening, a sign of willingness to free the country from the fate of determinism. While the historical self-image was based on naturalization and externalization of the reasons for bad years, the emerging political activism was focused on internal causalities and, consequently, on self-made reforms. In fact, the modern independent Finland was constructed by those social movements that believed in their own efforts to change the poor life conditions (see Alapuro 1988).

The break and the promise were revolutionary. The idea of social (self)production of societal problems was to become an alternative to the "religion" of fate. The external enemy and neo-Romantic national totalities were gradually being replaced by articulations of social inequality and injustice as an outcome of our own actions. However, both of those explanations—the deterministic and reflexive ones—are still decisively conditioning the political life in Finland by producing confronting and competing arguments for today's environmental and (geo)political orientations.

The official and popular westward orientation of Finland (e.g., Jakobson 1996) is a telling example of unintended consequences. Urban Finland seems to become easily integrated into the international Anglophonic community, while the rest of the country is absorbing one more layer of suspicion in its landscape. In eastern Finland, parallels are repeatedly drawn between the evacuation of the frontier populations during World War II and the current emigration from the rural areas because of unemployment and nature conservation (Lehtinen 1996). The aggression against any kind of environmental concern is, once again, a reaction to the bad years of high unemployment in the 1990s. However, only by an extremely selective interpretation of the most recent past can the expansion of nature conservation be considered an essential reason for the chronically high level of unemployment (varying between 20 percent and 30 percent in the 1990s) in the eastern provinces (see Rannikko 1999). The EU's *Natura* nevertheless became a swearword for many in the forest provinces of the northern hunger land.

The current disagreement concerning the key criteria for sustainability within forestry and wood-based production can be seen against the background of the need to make good the lost economic well-being of the population. The introduction of ecocertificates, as well as the evaluation of the life cycles of the forest-based products, could be taken as an opportunity for enlarging the export markets, for example, in a form of flexible ecological modernization of the Finnish forest industry. However, this opportunity is instead repeatedly regarded as an external threat impelled by forest conservationists and environmentalist consumers in Central Europe. Image campaigns designed to promote Finland as an ecologically aware paper producer have been easily shot down by critics, with the consequent eroding of the credibility of the green products and images of the northern backwoods.

The Finnish forest landscape of domination, shaped by the economic policy integrating Finland as a resource periphery into international markets, and characterized by the deep regional economic decline of the 1990s, stands in clear opposition to the landscape image of the green campaigns, where social sustainability and biodiversity are represented as key elements of the green welfare of the European North.

The Nordic East as a Landscape

The concept of landscape used in this chapter is applied in a specific way. Landscape is used as a bridging tool for two deep-rooted dualisms of modern world viewing—

that of nature/society, on the one hand, and that of materiality/experience, on the other. Landscape refers to the basic embeddedness of natural and cultural elements in our terrestrial environment. Landscape is, indeed, an artifact based on the materiality of nature. In other words, landscape refers to those terrestrial processes conducted by human efforts resulting in continuing transformation of land, that is, landscaping (see Olwig 1996). Therefore, for humanity, there is nothing in nature that would not be marked by culture. The same is conversely true, too: there is nothing in culture that is not built upon nature (see Haila 1995, 168).

As an artifact, landscape is a human creation—a result of social production and human imagination. Landscape is both terrestrial and experiential, made of physical materiality but, simultaneously, becoming part of our life world as a mental mindscape, a way of seeing and perceiving the environing world. There is no way out of this embeddedness. In essence, landscape is beyond dualisms (see Jones 1991). Jakob Donner-Amnell, a Finnish forest sociologist, expressed this clearly in his concluding speech at a forest conference in Joensuu in eastern Finland: "While talking about forests we, Finns, talk about ourselves" (Donner-Amnell 1996).

In this chapter, the connections (bridging the dualisms) have been introduced as a question of mentality. In modern Finland, as elsewhere in the modern world, the conceptual definition of culture as being external to nature has been a historical necessity. The initial national emergence and self-reliance was based on this distanciation (Lehtinen 1991). Nature as the original reason for the bad years has functioned as an excuse for setbacks, the *force majeure* beyond human control. The causal links of the hunger years and recessions had to be externalized and here the natural realm has functioned as a useful explanation. Thus, the external(ized) threats were, partly purposefully, projected in extreme natural conditions. But, as I argued earlier, nature has also become a symbol of the past, that is, life at natural risk, to be used as a negation to those much-welcomed processes of modernization and civilization. Nature, as the primitive past, was to be left behind. The dualism was constructed as a compensation, a message of relief, for the deep awareness of the indelible dependence on nature. Therefore, in Forest Finland, to talk about nature is to talk about the Finns themselves.

Landscape is here used as a framework for constructing environmental phenomena as meaningful wholes, a way of seeing and producing space or place-bound orientations. Landscapes are continually evolving as cultural code systems, helping us to notice, select, interpret, and renew our world(view)s as meaningful spatial categories (see Burgess 1990). Landscaping is accordingly a process emerging where an environmentally loaded impression or expression takes place; it is a process where our mind produces ties for regional identity formation or "gathers" place-related views from the messages we face in our daily lives. Landscaping is a process where spatial connections and contexts are constructed. Hence, landscape is the environmental fix in the (re)production of cultural continuity. It is the spatial basis for social commemoration and, therefore, a tool for drawing territorial lines and borders between us and them. Landscape is, finally, an ever-developing product of competing projections of domination (Mitchell 1998, 2001).

Here landscape refers both to (1) the aesthetic appreciation of terrestrial land and life interrelations, and (2) the practices of territorial integration and control. In the northern European context, landscape emphasizes the importance of both visual (picturesque) and scenic attributes in human place orientation (see, e.g., Häyrynen 1994 and chapter 19 in this volume) and, perhaps even more fundamentally, the terrestrial roots of regional identity formation and territorial division of lands. Kenneth Olwig (1996) has argued that the historical concepts of *Landschaft* and *landskap* have referred to shared terrestrial and territorial values belonging to the local traditional communities and regional administrative complexes. In this respect, *Landschaft* is a territorial unit with shared values, customs, and, moreover, collectively accepted practices of justice. In his own words: "The link between customary law, the institutions embodying that law, and the people enfranchised to participate in the making and administration of law is of fundamental importance to the root meaning of *Land* in *Landschaft*" (Olwig 1996, 633).

In the Finnish context, the dual character of Scandinavian *landskap* and German *Landschaft* has emerged in a specific way. The aesthetic *landskap* has found its Finnish synonym in *maisema*, whereas the administrative *landskap* is *maakunta* (Granö 1996). Both words are derivations of a Finnish root concept *maa*, "land," applicable in several spatial connections extending from soil stratification to land areas and world pictures, and from countries and countryside to global dynamics and earthly cosmologies. *Landskap*, as a translation of *maakunta*, is fully developed as an administrative practice on the island of Åland (Ahvenanmaa) in the Baltic Sea. Åland has an autonomous status as an administrative *Maakunta* and linguistically belongs to Scandinavia (see Olwig 1993). In present-day mainland Finland, *maakunta* refers to provinces or counties, derived both from the early settlement and administrative history of the historical provinces and from the more recent functional areal system of economic provinces (see Paasi 1984, 9–23, and chapter 20 this volume).

The *maisema*, as a synonym of landscape, is generally connected to visual aesthetics, but it also refers to regional morphological categories and spatial scales in geographic research (see, e.g., Granö 1997 [1929]; Keisteri 1990), as well as in descriptive and encyclopedic literature (Alalammi 1993). Sociohistorically, however, the Finnish *maisema* is especially tied to regional identification of areal particularities and provincial interests. The provincial landscapes of Finland were originally constructed by Zacharias Topelius in his famous *The Book on Our land* (*Maamme kirja*, Topelius 1981 [1875]), which became the geography book in schools and homes for generations.[6] Topelius emphasized the scenic values of cultivated landscape where harmony between nature and culture existed. In the spirit of Enlightenment, local events and histories were published to teach the Finnish folk to learn from the past.

For Topelius, the landscapes were tools for education: ideal events and sceneries with which to construct the territorial extension of Finnishness. The message was widely read and adopted in Finland. The message was connected not only to the territorial construction of Finland but also to the way of making that construction. Landscape was a framework to be used for identity formation. From then on, the Finnish landscapes have carried the double emphasis of visual aesthetics and territorial identity formation. Landscape is for the Finns a central medium with which to evaluate, orientate, and even compete in the environing world. In this respect, the link between the Topelian landscapes and present-day image campaigns, designed by the exporters of Finnishness and Finnish products, is clear.

This interpretation of the role of Finnish landscapes directs our attention to three basic theoretical and, simultaneously, contextual elements of landscape formation. First, Finnish landscapes seem to serve as a spatial framework by which the familiar and the well known within the environing world are identified as a home region or arena and, hence, as a territory under domestic control. Second, landscape is also a collective means to exclude the elements of the environment that deviate from learned codes. It is a tool for (re)producing the difference between the known and the unknown. Landscape divides "us and them" and helps us to make and maintain the distinction between the familiar and the rest. Accordingly, every scene becomes a landscape through a continuous process of exclusion and inclusion. The central content of landscape is therefore permanently under intense competition and commodification. The image campaigns with impressive commercial landscapes are mostly export-oriented, but they simultaneously have an internal impact on Finnish self-realization. The familiar is reproduced in connection to the external(ized) world via export strategies.

Third, Finnish landscapes can be identified on various levels of abstraction, from short-lived "export landscapes" down to archetypal basic landscapes of self-reflection and differentiation. From every image, or scenery, several levels can be distinguished. Throughout the history of settlement and state formation, the complementary motifs of dependence and self-reliance have lived as continuity factors in Finnish culture (see Alapuro 1997, 192–200). This has regularly surfaced in the debates on Finland's geopolitical position within the circumboreal forest zone and between the East and the West. The basic spatial orientation of Finnishness has indeed been shaped by the growing awareness of the eastern-ness and backwoodsy intimacy in the Finnish landscape. The East and the Forest are necessities, fixed dimensions repeatedly faced and, therefore, needing spatial antipodes: Westernization and modernization. The Finnish bear, however, has not yet become extinct. It has just moved closer to the cultural cores of the West.

NOTES

1. In Finland, no confirmed killing of a human being by a bear (before the drama of June 17, 1998) occurred during the twentieth century, even though some rumors and tales of such incidents are common in the forest provinces. These are mostly connected to wartime experiences or memories from the early years of this century (Paavolainen 1946; Pulliainen 1984; Helle 1998).

2. On the concept of landscape and landscaping, see the concluding section of this chapter, "The Nordic East as a Landscape."

3. The chapter focuses on the historical changes and continuities in land-use practices, extending from traditional to late modern ones. The scope is from the early years of settlement to contemporary (high) industrialization and territorial governance. I pay only minor attention therefore to deep-rooted archaic attitudes connected to land beliefs and uses. However, I argue here that even the current late modern landscaping of Finland is conditioned by both archaic and traditional layers of memories. The past is profoundly intermingled with the contemporary life modes and it continuously surfaces in our daily routines and customs. According to Paul Connerton (1989, 2–3), (a) "concerning memory as such, we may note that our experience of the present very largely depends upon our knowledge of the past," and, moreover, (b) "concerning social memory in particular, the images of the past commonly legitimate a present social order." According to Connerton, remembering takes place mostly through habitual practices, that is, by habit-memory, conveyed and sustained by (more or less ritual) commemorative actions and within our bodily automatism (see also Schama 1995, as well as Birkeland 1999 and Lehtinen 2000 in a northern European context).

4. The landscape of fear is one of the background concepts of this chapter. It refers to the shared collective experience of threat reflecting the continuous demarcation or dialogue between the external(ized) dangers from the outside and the self-production of fear elements from inside (see the medieval roots for this demarcation in Europe: Fumagalli 1994). The landscape of fear mirrors both the learned and unreflected threats as well as the collective inscriptions of threats conditioning our basic experience of fear against the unpredictable both within ourselves and in our environment (see Tuan 1979; Short 1991, 6–9; Emel 1995; Pile 1996, 88–95; Siltala 1996). Derivatively, the landscape of domination covers the elements of landscape that emerge both in the form of symbolic defense systems and as a compensative expansion of sociospatial control (replaced superiority) over the surrounding lands (see Leiss 1994, 178–90; Cronon 1996; Fumagalli 1984, 67–69; Denham 1997). In Finland, the landscape of domination refers to those features of landscaping that are found in the variations of the Finnish "success story"; paving its way from the backwoods miseries toward late modern welfare security and also managing to build a mixed economy benefiting from the differing market dynamics of the East and West. The landscape of domination thus includes those geopolitical and commercial images by which Finland and its products and know-how are exported. Consequently, the landscape of domination emphasizes those elements of landscape construction that support the Finnish success model—including the ecological and social consequences of this success.

5. Variations in the proportions of the land area under conserved forests are due to different methods of calculation and of course depend on the ideological standpoint. The official estimate of conserved forests (*Criteria* . . . 1997) is as much as 9 percent of the total forest area, but it includes much of the so-called stunted lands of marshes and high-latitude forests beyond the "zero limit" of economically feasible cuttings. This official percentage also includes forests

under nature-imitating loggings with options of exploitation. According to the Nature League of Finland, only 420,000 hectares (2.1 percent of forest-land area) are protected on the basis of the Conservation Acts and over 95 percent of forest land is in commercial use (Pennanen 1993). Both the official and Nature League's readings were estimates before the acceptance of the *Natura* Program by the European Union, aiming to cover 12 percent of Finland's surface area (Heikkinen 1998).

6. The 1981 edition of *Maamme kirja* was the sixty-first edition in Finnish; "in all, several million copies of the book has been printed" (Mäkinen 1981). Therefore, it is no exaggeration to say that *Maamme kirja* is one of the most popular and "adopted" books of homes and schools in Finland.

References

Alalammi, Pentti. 1993. *Maisemat, asuinympäristöt: Suomen Kartasto 350.* Helsinki: National Board of Survey and Geographical Society of Finland.

Alapuro, Risto. 1988. *State and Revolution in Finland.* Berkeley and Los Angeles: University of California Press.

———. 1997. *Suomen älymystö Venäjän varjossa.* Helsinki: Tammi.

Birkeland, Inger. 1999. "The Mytho-Poetic in Northern Travel." In *Leisure/Tourism Geographies,* ed. David Crouch, 17–33. London: Routledge.

Burgess, Jacquelin. 1990. "The Production and Consumption of Environmental Meanings in the Mass Media: A Research Agenda for the 1990s." *Transactions, Institute of British Geographers* 15: 139–61.

Connerton, Paul. 1989. *How Societies Remember.* Cambridge: Cambridge University Press.

Criteria and Indicators for Sustainable Forest Management in Finland. 1997. Helsinki: Ministry of Agriculture and Forestry.

Cronon, William. 1996. "Introduction: In Search of Nature." In *Uncommon Ground: Rethinking the Human Place in Nature,* ed. William Cronon, 23–66. New York: W. W. Norton.

Denham, Helen. 1997. "The Cunning of Unreason and Nature's Revolt: Max Horkheimer and William Leiss on the Domination of Nature." *Environment and History* 3: 149–75.

Donner-Amnell, Jakob. 1996. "Metsän eri merkitykset—haaste vuoropuheluun." [Concluding speech at the conference "Forest Conflicts" at the University of Joensuu, June 19, 1996.] Mimeo.

Elo, Kirsi. 1983. *Tämä vihreän kullan maa.* Helsinki: Suomen Luonnonsuojelun Tuki.

Emel, Jody. 1995. "Are You Man Enough, Big and Bad Enough? Ecofeminism and Wolf Eradication in the USA." *Environment and Planning D: Society and Space* 13: 707–34.

Forest Industries Finland. 2001. http://www.forestindustries.fi/statistics, January 13, 2001.

Fumagalli, Vito. 1994. *Landscapes of Fear: Perceptions of Nature and the City in the Middle Ages.* Cambridge: Polity Press.

Granö, Johan G. 1997 (1929). *Pure Geography,* ed. Olavi Granö and Anssi Paasi. Baltimore: Johns Hopkins University Press.

Granö, Olavi. 1996. *Tieteellisen maisemakäsityksen muodostuminen ja tulo Suomeen.* Department of Geography, Publications 154. Turku: University of Turku.

Haila, Yrjö. 1995. "In Search of the Wilderness." In *Ikijää—Permafrost,* ed. Yrjö Haila and Marketta Seppälä, 161–71. Pori: Pori Art Museum.

Häyrynen, Maunu. 1994. "The National Sets of Landscape in Finland." In *Landscape and Northern*

National Identity, ed. Maunu Häyrynen, 16–22. Helsinki: Renvall Institute of Historic Research, University of Helsinki.

Heikkinen, Ilkka. 1998. "Luonnonsuojelu lähemmäs kansalaisia." *Helsingin Sanomat,* August 26, 1998, 2.

Helle, Timo. 1998. "Karhu pysyköön korpiemme kontiona." *Suomen Luonto* 57, no. 8: 23.

Holopainen, Viljo. 1968. "The Forest Policy of Independent Finland." In *The Forest Industry in Independent Finland,* ed. Juha Laurila, Eino Mäkinen, and Heikki Vuorimaa, 76–103. Helsinki: The Central Association of Finnish Woodworking Industries.

Honkanen, Veijo. 1998. "Suomalainen metsästäjä saa kulkea pää pystyssä." *Karjalainen,* August 13, 1998, 13.

Hustich, Ilmari. 1965. *Suomi tänään.* Helsinki: Tammi.

Jakobson, Max. 1996. "On pohdittava, mikä on turvallisempaa—sitoutuminen vaiko yksinjääminen?" *Helsingin Sanomat,* May 7, 1996, A8.

Jones, Michael. 1991. "The Elusive Reality of Landscape: Concepts and Approaches in Landscape Research." *Norsk Geografisk Tidsskrift* 45: 229–44.

Jutikkala, Eino, and Kauko Pirinen. 1984. *A History of Finland.* Helsinki: Weilin & Göös.

Kärkkäinen, Matti. 1995. "Missä vanhojen metsien uhanalaiset lajit piileskelivät 1930–luvulla?" *Karjalainen,* January 23, 1995, 13.

Keisteri, Tarja. 1990. "The Study of Changes in Cultural Landscapes." *Fennia* 168: 31–115.

Kianto, Ilmari. 1946. *Punainen viiva.* Helsinki: Otava.

Kopomaa, Timo. 2000. *Kännykkäyhteiskunnan synty.* Helsinki: Gaudeamus.

Kortelainen, Jarmo. 1999. "Metsäteollisuus karttakuvan muokkaajana." In *Suomen Kartasto,* ed. John Westerholm and Pauliina Raento, 64–67. Helsinki: Geographical Society of Finland and WSOY.

Kuusela, Kullervo. 1976. "Metsätalous." In *Atlas of Finland, Folio 234 on Forestry,* ed. Pentti Alalammi, 2–12. Helsinki: National Board of Survey and Geographical Society of Finland.

Lehtinen, Ari Aukusti. 1991. "Northern Natures." *Fennia* 169, no. 1: 57–169.

———. 1996. "Tales from the Northern Backwoods: Fairy Tales and Conspirators in Finnish Forest Politics." In *Approaching Nature from Local Communities: Security Perceived and Achieved,* ed. Anders Hjort-af-Ornäs, 37–57. Linköping: Environmental Policy and Society, University of Linköping.

———. 2000. "Mires as Mirrors. Peatlands: Hybrid Landscapes of the North." *Fennia* 178, no. 1: 125–37.

———. 2001. "Modernization and the Concept of Nature." In *Encountering the Past in Nature,* ed. Timo Myllyntaus and Mikko Saikku, 29–48. Athens: Ohio University Press.

———. 2003. "Mnemonic North: Multilayered Geographies of the Barents Region." In *Encountering the North: Cultural Geography, International Relations, and the Northern Landscape,* ed. Frank Möller and Samu Pehkonen, 31–56. Aldershot: Ashgate.

———. 2006. "Postcolonialism, Multitude, and the Politics of Nature: On the Changing Geographies of the European North." Lanham, Md.: University Press of America.

———, and Teijo Rytteri. 1998. "Backwoods' Provincialism: The Case of Kuusamo Forest Common." *Nordia* 27, no. 1: 27–37.

Leikola, Matti. 1987. "Metsien hoidon aatehistoriaa." Summary: "Leading Ideas in Finnish Silviculture." *Silva Fennica* 21, no. 4: 332–41.

Leiss, William. 1994. *The Domination of Nature.* Montreal: McGill-Queen's University Press.

Linkola, Martti. 1987. "Metsä kulttuurimaisemana." Summary: "The Forest as a Cultural Landscape." *Silva Fennica* 21, no. 4: 362–73.

Mäkinen, Vesa. 1981. "Zachris Topelius ja Maamme kirja." In *Maamme kirja*, ed. Vesa Mäkinen: III–XIV. Helsinki: Werner Söderström.

Metsä 2000. 1985. Helsinki: Talousneuvosto.

Mitchell, Don. 1998. "Writing the Western: New Western History's Encounter with Landscape." *Ecumene* 5, no. 1: 7–29.

———. 2001. "The Lure of the Local: Landscape Studies at the End of a Troubled Century." *Progress in Human Geography* 25, no. 2: 269–82.

Olwig, Kenneth Robert. 1993. "Sexual Cosmology: Nation and Landscape at the Conceptual Interstices of Nature and Culture, or: What Does Landscape Really Mean?" In *Landscape: Politics and Perspectives,* ed. Barbara Bender, 307–43. Oxford: Berg.

———. 1996. "Recovering the Substantive Nature of Landscape." *Annals of the Association of American Geographers* 86, no. 4: 630–53.

Paasi, Anssi. 1984. *Suomen väliportaan aluejärjestelmän kehitys ja hahmottuminen suomalaisten aluetietoisuudessa.* Publications of the Association of Planning Geographers 14. Helsinki: University of Helsinki.

Paavolainen, Olavi. 1982. *Synkkä yksinpuhelu.* Helsinki: Otava.

Pennanen, Juho. 1993. "0.27% Protected!" In *Finland and Forest: A Success Story?* ed. Harri Karjalainen et al., 13–15. Helsinki: World Wide Fund for Nature Finland, Finnish Association for Nature Conservation, Greenpeace, Finnish Wilderness Movement, Nature League, and Finnish Forest Action Group.

Pentikäinen, Juha. 1994. "Metsä suomalaisten maailmankuvassa." In *Metsä ja metsänviljaa,* ed. Pekka Laaksonen and Sirkka-Liisa Mettomäki, 7–23. Helsinki: Suomalaisen Kirjallisuuden Seura.

Pile, Steve. 1996. *The Body and the City.* London: Routledge.

Pro Silvis. 1995. *Metsien puolesta: Metsäpoliittisen keskustelun tueksi.* Helsinki: Metsämiesten Säätiö.

Pulliainen, Erkki. 1984. *Petoja ja ihmisiä.* Helsinki: Tammi.

Rannikko, Pertti, 1999. "Forest Work as the Cause of Settlement and Depopulation in the Remote Parts of Finland." In *The Green Kingdom,* ed. Reunala and Tikkanen, 222–26.

Raportti liikennevahinkojen tutkijalautakuntien tutkimista moottoriajoneuvossa kuolleiden onnettomuuksista vuonna 1995. 1996. The Report on Death Cases in Traffic Accidents in 1995. Helsinki: Liikennevakuutuskeskus.

Reunala, Aarne, and Ilpo Tikkanen. 1999. "To the Reader." In *The Green Kingdom,* ed. Aarne Reunala and Ilpo Tikkanen, 9–11. Helsinki: Otava and the Metsämiesten Säätiö Foundation.

Schama, Simon. 1995. *Landscape and Memory.* London: Fontana Press.

Short, John Rennie. 1991. *Imagined Country.* London: Routledge.

Siltala, Juha. 1996. "Yksilöllisyyden historialliset ja psykologiset ehdot." In *Yksilö modernin murroksessa,* ed. Antti Hautamäki et al., 117–204. Helsinki: Gaudeamus.

Simola, Heikki. 1995. "Land-Use History Explains the Value of the North Karelian Biosphere Reserve in Preserving Old-Growth Forest." In *Karelian Biosphere Reserve Studies,* ed. Timo J. Hokkanen and E. Ieshko, 15–20. Joensuu: Biosphere Studies, University of Joensuu.

Topelius, Zacharias. 1981 (1875). *Maamme kirja.* Helsinki: Werner Söderström.

Tuan, Yi-Fu. 1979. *Landscapes of Fear.* Minneapolis: University of Minnesota Press.

Vartiainen, Perttu, and Janne Antikainen. 1999. *Kaupunkiverkkotutkimus 1998*. Publications of the Committee for Urban Policy 2. Helsinki: Ministry of Interior.

Virtanen, Leea. 1994. "Suomen kansa on aina vihannut metsiään." In *Metsä ja metsänviljaa*, ed. Pekka Laaksonen and Sirkka-Liisa Mettomäki, 134–40. Helsinki: Suomalaisen Kirjallisuuden Seura.

Von Berg, Edmund. 1988 (1859). *Kertomus Suomenmaan metsistä 1859*. Helsingin Yliopistin metsänhoitotieteen laitoksen tiedonantoja 63. Helsinki: University of Helsinki.

19.
A Kaleidoscopic Nation: The Finnish National Landscape Imagery

Maunu Häyrynen

Abstracted stereotypical landscape or isolated "landscape monuments" may be seen as collective landmarks, tokens of nationhood. I argue here that such emphases ignore the way landscape becomes constructed and used as a national symbol, or indeed as any kind of symbol. I define "landscape" here as a societal phenomenon, produced by an interpretive interaction between culture and the environment. The modern Western concept of landscape entails the historical transposition of an intricate representational system into the external world and its encoding into value-laden sets of symbols. These in turn may become incorporated into nationalist discourses. Finland offers a case in point to prove how exactly such development has taken place during a relatively short interval and in a relatively small country.

I became intrigued by the relationship between landscape and the Finnish cultural identity in the late 1980s, when a new expression—or rather a new use for an old one—spread rapidly in Finland. All of a sudden, one started to refer to a number of well-known Finnish scenic sites as "national landscapes." Earlier this term would have covered the stereotypical, idealized landscape of Finnish art, literature, and popular media, basically the panoramic lake-and-forest landscape. The use of the term expanded quickly and at present may occur in connection with any landscape considered to be of value, usually with a nostalgic emphasis (see, e.g., Lowenthal 1989). The unique importance of a landscape is explained by its connotation to the past, understood as a teleological formative process with the Finnish nation as its outcome.

At first, one site was primarily presented as "the" national landscape of Finland: the heights of Koli in North Karelia. The site had allegedly been "discovered" by a group of National Romantic artists in the 1890s. Subsequently it became a well-known visual symbol, associated with the cultural resistance against the attempted Russification at the turn of the century. During the boom years of the 1980s, the landscape of

Koli became threatened by large-scale tourist development, triggering unexpectedly a nationwide intellectual "hands off" movement for its rescue. As a result, the government called off development plans and, instead, eventually established a new national park there in 1991.

Besides Koli, other sites associated with the history of nation-building began to be referred to in a highly evaluative manner as national landscapes, especially when perceived to be under threat—as, for instance, the prehistoric hill fort of Rapola or the national writer Aleksis Kivi's birthplace in Nurmijärvi, the former jeopardized by highway construction and the latter by a quarry project. As an official response to these debates, the Ministry of the Environment entrusted a committee with the task of identifying the national landscapes of Finland. The committee set about with the task and ended by drawing up a catalogue of sites based on a mix of scientific, cultural, and historical criteria, which was motivated by a wish to "promote discussion" (*National Landscape* 1994). In the discussion that followed, the top-down definition of national landscape was more or less taken for granted. Questions about by and for whom the listing had been made, or whether and how the national value of the landscape could be determined in the first place, were not asked.

To find sites of national value, the committee had ventured to establish "objective" criteria, based on specifically devised scenic regions (see also Paasi's chapter in this volume) as well as on the broad lines of national history. Instead of just a few valued sites, a representational system was refashioned, its structural logic and much of its content deriving from the nineteenth-century nation-building period. The committee explicitly expressed a wish to continue this tradition, no one seriously objecting to this goal. This revealed a wide cultural consensus on the deeply rooted national importance of landscape and its general understanding as a representative set of sites.

How to Study National Landscape Imagery

Initially my own interest directed me to the study of the "national landscapes" to find out when, how, and why they had acquired their exceptional cultural value in the Finnish context. I gradually began to see landscape as a part and product of a signifying system, constructed for the conceptualization and propagation of the nation-state. Thus my focus shifted from isolated sites and their representation into the entire discourse (in a loosely Foucauldian sense) of landscape images: territorial divisions, travel accounts, maps, and tourist guides that had generated their meaning, and their intertwining with national ideology.

Instead of national landscapes, I then proceeded to study the totality of landscape images concerning Finland in order to establish their origin and the structural and narrative logic, still so vividly present in the 1993 committee report. It seemed possible to describe the development in terms of national landscape imagery, which could be outlined as a systematic imagined topography of Finland consisting of sets of landscape and other images that had gradually emerged from the late eighteenth century onward and become fully established by the end of the nineteenth century, then disintegrated during the twentieth century and bureaucratically resuscitated in the 1990s.

How should one approach such a phenomenon? Landscape has been studied as a national symbol and a collective "realm of memory." Attention has also been paid to different kinds of symbolic inventories as part and parcel of "banal nationalism" (Billig 1995; cf. Daniels 1993; Cubitt 1998). Most of these studies have followed the lines of history, art history, cultural studies, or cultural geography. They unmask a visual vocabulary of everyday nationalism, continuously applied for ideological inculcation. Against this background, the cohesive force of landscape and other national symbols becomes understandable: they provide a sense of identification across society. In the case of landscape, this force is enhanced by a sense of nature-based verisimilitude: for a member of the national community, symbolic landscape includes familiar traits easy to pinpoint to one's own surroundings. This has a double effect: the nation-state is presented as an existing place and, at the same time, is naturalized (see Meinig 1979; Mitchell 1994).

Among the most notable attempts to describe the historical evolution of landscape imagery was that by Andrew Hemingway, who used as his material the paintings included in the nineteenth-century Royal Academy exhibitions and their criticism. Among Hemingway's aims was to produce social history of landscape art untied to the history of painting, privileged in traditional art history, by concentrating instead on pictorial imagery as communication of social meanings. He attempted to reconcile two different approaches, the art-historical, dealing with the interpretation of unique artworks, authors, and oeuvres, and the art-sociological, interested in popular imageries within a social context (Hemingway 1992).

Until the last two decades, the study of Finnish landscape imagery had mostly been the business of art or cultural historians, who have concentrated their efforts on painting or heraldic symbols, ending up with a picture of a young nation striving to build its identity on its particular natural characteristics. Some researchers have paid attention to the popular printed images as a means of conveying an idea of the nation-state. These have revealed the existence of a larger imagery with a more complicated representational logic, comprising changing artistic ideals and conventions as well as varying articulations of areal, social, and ethnic differences (Klinge 1993, 1981; Reitala 1983; Hirn 1988; Knapas 1993).

My own study focuses on published landscape images representing the Finnish landscape. The reasons for this choice of subject area are the following. Landscape images are a part of the social representation of Finnishness. Following Serge Moscovici (1984), they evoke collectively shared feelings and convey meanings and metaphors from the past, directing the way the present-day Finns observe and categorize national space and objectivate it into a part of their social environment. As a theory of seeing, the national landscape imagery permeates the everyday life of the Finnish society, shaping the very environment it claims to represent.

The sociologist Jorma Anttila argues that the concept of Finland as an imagined community may be due more to its rapid nation-building process and its strong cultural unity and is more "imagined" than the older nation-states, where alternative bases of identification are available. The few basic elements applicable for the nineteenth-century nation-builders—the folklore, the nature, and the peasant population—still dominate the mental landscape in spite of modernization. At least they have continued to do so until now. This has provided Finland with a strong mutual cross-society code of understanding, a common matrix of cultural interpretation, maintaining the national identity (Anttila 1993).

Finnish landscape imagery, gradually established during the nineteenth century and becoming a standard during the twentieth, has been a key element in conditioning the Finnish community to perceive the national territory, along with its boundaries, internal divisions, and social differences, as an organic and rational whole. It has further played a crucial part in the inculcation of national values and emotions: love for the country, feeling of belonging, reverence of the past, and expectations about the future. It has provided the citizens with an ideological package and an emotional map of the homeland.

The published sets of landscape images have constituted a part of what Benedict Anderson (1991 [1982]) calls "print capitalism," and thus have directly accessed the entire national community, creating unity in personal perceptions of and responses to certain types of landscapes or their representations. In the case of Finland, its short national history and small population are for once an advantage for the researcher striving for an overall view, if such an expression may be allowed.

I should point out that I am aware of the risk involved with a nationally defined study, in that it can become celebratory of its subject. Yet I consider it possible, if not necessary, to study the evolution of national representational systems, provided that it is done in a critical and historically context-sensitive way. Ideally this should happen in the form of a cross-national comparative study, which, however, is not possible here. Let me state at any rate that my intention has been to track down the process that has led to the emergence of national landscape imagery in order to historicize its meaning and thus to de-center its underlying assumption of the naturalness of the nation-state. A different reading, though, might interpret this as value-added to the notion of national landscape.

A closer observation of landscape imagery reveals that its overall structure is repeated in a very uniform way. The pattern is well recognizable, enabling the variation and reproduction of imagery. Imagery becomes formed conforming to fixed rules of representation. Thus landscape imagery cannot consist of random representations of landscapes but is constructed out of preselected landscape images. Each image is allotted a certain narrative role, supporting the thematic content, while at the same time imagery is claimed to give a reliable and exhaustive visual account of its subject area. To paraphrase W. J. T. Mitchell (1994), imagery attempts to conceal its own representational character in the same manner that a singular perspective image does.

Is imagery an authentic historical phenomenon or a research concept? Is it intentionally produced? Is there an overlying generalized imagery, to which particular subsets relate? It may be best to attempt a provisional definition for the purposes of the study. At its most concrete level, imagery is a group of pictures thematically joined together. At a more abstract level, it is a pool of interrelated images organized along certain representational practices. On the whole, imagery may be seen as a discursive technique of visualization and naturalization.

Landscape imagery may have different ideological contexts, of which nationalism constitutes only one. As far as the national discourse is concerned, the same game may also be played with more than one set of images; there are national historical, ethnographic, and architectural imageries. In practice these may overlap, and one may speak about an umbrella of national imagery, which in turn intermingles with national or transnational representations.

The Emergence of Finnish Landscape Imagery

The development of the imagery shows a constant redefinition of a center-periphery relationship, where the periphery has typically appeared as an extracultural liminal zone, a "place on the margin." It is hardly a coincidence that in each phase the outstanding landscapes were located in peripheries (Shields 1991). One may discern at least three successive stages (not counting the prenational imageries, where the entire present area of Finland appeared as a periphery): first, the early national landscape imagery, concentrating on South Finland and promoting by now-canonized lake-and-forest landscape; second, the imagery of National Romanticism, projecting the source of national identity into the Karelian borderlands; and third, the popularized imagery of Independence, when Lapland gradually rose into the position of the last heroic wilderness of Finland (and now also of Europe). These stages bear a certain resemblance to the North American Frontier myth in that they reflect simultaneous phases of economic expansion and modernization (Conzen 1990).

The first, "prenational" imageries (in the seventeenth and eighteenth centuries) presented a characteristically external approach to the area (Figure 19.1). Finland was then considered an ethnic and administrative concept rather than a national unit by itself. The motivation for its representation lies mostly in its perceived character as an exotic, wild, and little-known periphery. For the same reason, Finland was usually lumped together, and confused, with Lapland. In the *voyages pittoresques* of the time,

Finland was mostly seen as an interlude on the way to Russia or Lapland. Some sites became standard issues, but a certain randomness would reflect in most of the landscape images. The sites were located along the most frequented routes. The focus was on the natural and the sublime, whereas urban settlements were rarely shown, and the local elite was at best portrayed as quaintly peripheral. At this stage, the Finnish (and Lapp) landscapes were for the first time categorized and evaluated against the established characters of landscape art. Their foremost quality was seen to be their sublimity, highlighting the rugged, desolate, or otherwise overwhelming natural sights such as fells or rapids. These were combined with the recording of ethnographic curiosities: the saunas, reindeers, and Saami huts still so loved by publishers of postcards.

FIGURE 19.1. An 1802 engraving portrays the "sublime" Finland as a faraway country approached by a perilous journey over the frozen Baltic (Acerbi 1802).

During the eighteenth century, the strategic importance of South Finland increased as a result of the foundation of St. Petersburg; it also attracted foreign visitors (e.g., Clarke 1829). The area was thoroughly mapped, providing a conceptual framework for determining the geographically typical qualities of the landscape. Characteristically enough, the first published artistic rendition of an inland lake-and-forest landscape was on the title page of an atlas of Finland in 1799 (Hirn 1988).

When Finland became an autonomous part of the Russian Empire in 1809, its mental map was still mostly void, chiefly consisting of two contradictory elements: the cultivated south coast and the eastern part of Lapland, by chance left to the Finnish side of the new border. Neither from the outside nor by its inhabitants was the new autonomy regarded as a ready national entity, but it certainly became more distinct as a geopolitical concept (Manninen 1997, 129–35).

The new masters were interested in establishing an areal Finnish identity that would replace the former Swedish allegiance with a politically loyal and neutral one. Thus it is not surprising that the production of the first representative imagery of Finland was an Imperial project. This was the first serious attempt at making Finland (or at least its southern parts) visible as a territory in its own right. What was left from the preceding state of the *voyages pittoresques* was the anonymity of the ordinary landscape: only monuments and towns were recognized as individual sites, whereas outside them the picturesque but nameless nature prevailed.

The explicit intention of Carl von Kügelgen's *Vues pittoresques de Finlande* (1823) was to reject existing doubts about the value of the new conquest, but its subscribers appear to have been mostly Finns. The motifs were systematically searched from around South Finland, and the picturesque potential of the lake-and-forest landscape was for the first time fully appreciated by the author, a professional artist of German origin. An embryonic structure of landscape imagery thus predated the birth of the Finnish national movement, which took part during the following decades.

A new mode of landscape art was applied in the representation of the Finnish landscape: the Picturesque. Although the idea of the Picturesque had been explicitly mentioned only once or twice in the context of Finland, as a manifestation of new landscape aesthetics it certainly contributed to the sudden taste for the (until then) ignored lake-and-forest landscape.[1] The Picturesque differed from both the topographical landscape tradition, preoccupied with the exact rendition of facts, as well as from the landscapes of the Sublime, which concentrated on the extraordinary and striking. Contrary to them, the Picturesque relied on the average scenery and the traits of everyday life, skillfully converting them into aesthetic objects. This involved a systematic exploration and assessment of the subject area in order to find out its scenic values.

Like the Sublime, the Picturesque offered a formal approach to the landscape without requiring previous knowledge about its history or cultural meanings. By the mid-nineteenth century, however, the anonymous landscape was transformed into a meaningful national territory by the budding Finnish national movement. Among its first protagonists was Zacharias Topelius, who created an elaborate representational system based on the historic-ethnographic regions of Finland (following Swedish models). Another major innovation was the association of landscape imageries with narratives. By using landscape imagery (Figure 19.2), among other media, Topelius carefully constructed a nationally defined way of looking at the land and investing it with collective memories and shared feelings (see Paasi, chapter 20 in this volume).

The publication of Topelius's book, *Finland framstäldt i teckningar* (*Finland in Pictures*, 1845–52), may be the formative moment of the national landscape imagery, in which the preceding discontinuous texts were joined together into a new canonized discursive system, building on some parts and rejecting others (Baehr and O'Brien

FIGURE 19.2. Driven by a peasant in a horse-drawn carriage, a gentleman passes through a gate on a public highway as local children obediently open the gate. The lake landscape in the background represents the ideal of Finland's scenic beauty. From Topelius, *En resa i Finland* (*A Journey in Finland*), 1872.

1994). Not only were monuments and cultural, economic, and technological achievements as well as the national borders and other extremities paraded in front of the reader, but each image was attributed with an individualizing story, defining its character and emotional tone (Paasi 1996).

The book had a number of traits that turned out to be less conspicuous in later imageries. Its geographical emphasis gravitated to the south, and, among its motifs, estates and manufactures were overrepresented. The people were depicted in a paternalistic way as anonymous extras, sometimes clad in folk costumes. Otherwise folk traditions were practically absent, and history was presented in terms of nobility and warfare. Natural landscape, when seen at all, was strongly pastoralized; the preferred solution was nevertheless to show nature in a utilitarian way, as exploited and reclaimed. The Saami, as opposed to their visibility in the earlier imagery, were not shown at all, obviously to discourage their prevalent association with Finnishness. Neither were there any references to the Russian power centers or their presence in Finland.

The early national landscape imagery was all-embracing. It incorporated both the Sublime landscapes of the early travelers and the later Picturesque imagery. What was deliberately left out was the ethnographic depiction of the Finns and their saunas, whereas Lapland imagery was presented as a visible proof of the alleged cultural inferiority and otherness of the indigenous Saami. The pastoral overtones present in the Picturesque imagery were continued. The more rugged scenes were as if muffled by an idyllic filter, and signs of human activity or Finnish cultural achievement were added into practically every image, underscoring the national project of reclamation and exploitation that Topelius envisaged.

Topelius also initiated the promotion of particular landscapes as the holy sites of the nation. Their origins were equally motley as those of the entire imagery: some of them were already established tourist attractions, others historic sites, others again had been introduced to the popular imagery by landscape painters. Not all of them became generally recognized, but, more important, the national landscape imagery assumed its present-day hierarchical structure, centered on certain outstanding sites considered to embody the very essence of the nation-state and its nature. These sites became more or less regular items, their meanings becoming historically stratified and thus further emphasized. Over time, some of these landscapes acquired the status of a national benchmark for scenic beauty, metonymically standing for the entire country (Baehr and O'Brien 1994).

Topelius was addressing an as-yet-thin layer of the Finnish administrative and clerical elite, which at the time was entirely Swedish speaking as opposed to the majority of the population. This elite, settled in the coastal towns, saw the Finnish-speaking inland as an inaccessible and slightly exotic area.

One may argue that by the end of the nineteenth century, national landscape imagery had grown from an intentional educational project to an allegedly natural way of representing the Finnish national territory as well as a crucial instrument in defining Finnish nationhood. Topelius was followed by a host of new publications presenting increasing numbers of sites. From this point on, it is possible to quantify the imagery in order to follow its further development and changes.

The Aims and Scope of the Study

In previous studies of Finnish landscape, the focus has been either on its artistic interpretation or on the histories of particular scenic sites or areas. When attempting to explain the position of landscape in the national culture or its role in constituting national identity, it is essential to examine popular landscape imagery, which has in many respects differed from the artistic representation of landscape. As for studying certain sites invested with national symbolic value, such an approach easily leads to essentialism or "place fetishism" regarding the meanings attached to a landscape as somehow intrinsic or autonomous.

To operationalize the concept of imagery, my study has concentrated on published landscape images (disregarding, for instance, literary imagery or the moving image). Until the 1960s, the principal medium by which landscape imagery would reach its national audience was books and printed images (posters, brochures, postcards). Books focusing on Finnish landscape had existed from the late eighteenth century onward (though they reached only a limited public at first). Studying these publications offered a way to understand the inner workings of the national landscape imagery. Since the range of publications throughout the period was wide, and the sets of landscapes varied from one publication to another, my approach was to include a representative array of different types.

Finally, the aim of the study was not to create a new ranking of Finnish landscapes or to establish exact criteria for their appreciation. It was rather to produce an extensive preliminary analysis of the popular landscape imagery, ending with a periodization and the highlighting of certain categories of landscapes, comparable with earlier studies on representations of Finnish landscape. This would enable a more traditionally humanistic examination of landscapes and landscape types emerging from the material.

Data and Methods

I examined landscape images from publications that presented Finnish landscape(s) as their principal or substantial topic.[2] These were published both in and outside Finland, mostly in Finnish or Swedish but also in other European languages, and their publication dates ranged from 1744 to 1958. Before 1744, only isolated images representing

Finnish landscapes had been published, whereas after 1958 both the character of Finnish landscapes and their mediation transformed rapidly—the former because of the deep structural change of the Finnish society and the latter above all due to the diffusion of television.

To qualify for inclusion, the publications had to aim at the presentation of Finland as a geographical entity, thus excluding publications that would be devoted only to some particular Finnish landscape or region. Another criterion was a known publication date and a booklike character, hence excluding anonymously published unbound collections of views or other more sparsely circulated publications as well as periodicals. Finally, I did not include reprints, new editions, or other books clearly based on previously published materials in order to prevent repetition. It is, however, argued that the material amounts to a Greimasian *corpus,* covering enough exemplars from the chosen field to provide an understanding of the phenomenon.

The forty-nine publications I included comprised luxurious collections of prints as well as school readers and sourcebooks, travel guides and accounts, even cheap souvenir books, covering a considerable portion of publications in the field. The different publication types tell us something about how the national landscape imagery penetrated the national community: first, by reaching the middle class by means of tourism and travel literature, and second, by being mediated to the rank and file via the primary school system and the press. Despite the different social biases in the various publications, their influence was parallel: a feeling of belonging based on a shared imaginary landscape, or a set of landscapes, promoting horizontal cohesion among the different social or ethnic groups.

Detailed information about every image from each publication was recorded. Among the information obtained from the images were title, source, year, subject matter, author(s), technique, and pictorial type. From landscape images, subject landscape and location as well as landscape type were recorded. Location was primarily recorded according to the present or 1939 local authority district or region, when known, but the areal references given in the publications were separately noted. Altogether 4,447 images were recorded, of which 73 percent could be termed "landscape images," while the remaining 27 percent consisted of portraits, single objects such as artworks, handicrafts, animals or plants, architectural details, or vignette-type illustrations.

The material thus collected represented different stages of history. The character as well as status of publications about Finnish landscape clearly varied during the entire period of study, not to mention the change of the landscape. Thus one cannot draw conclusions directly from the total material, but it had to be divided according to historic time limits. In particular, the increased publication activity of the post-Independence era distorted the results by causing the landscapes representative of this period to dominate the entire material, eclipsing those important during earlier phases.

Periodization

The material was divided roughly into three historical periods: pre-Independence (1744–1917); prewar (1917–39); and postwar (1939–58). The material fell into the following period groups: from the pre-Independence period, 1,272 images, of which 874 were landscapes; from the prewar period, 1,469 images, of which 1,049 were landscapes; and from the wartime and postwar period, 1,706 images, of which 1,105 were landscapes (Table 19.1). Thus divided, the period groups would be more or less equal in size, with historically justifiable time limits, whereas equal-length periods would have been strongly biased toward the twentieth century.

Of the period groups, the first one, pre-Independence, is problematic in that most of its images derive from the more voluminous publications from the turn of the twentieth century. As this period group is already the smallest, and as the earliest publications as well as their landscape images would on their own have formed a sparse and fragmented material, further division seemed unnecessary. I will deal with the emergence of Finnish landscape imagery in the early publications by looking at each one individually.

Three-quarters of all landscape images could be located into a present-day or, in the case of the areas ceded in 1940–44, into a prewar local authority district and region. In addition, 253 landscape images, or 8 percent of all landscapes, could at least be associated with some region, if not with a certain locality. A considerable number, 558 (17 percent), consisted of totally unidentifiable or generic landscapes.

The structure of the national landscape imagery relied, save its earliest occurrences, on the so-called historic regions of Finland. The regional division used here was based on a Finnish government decision in 1998, but the regions themselves may be described as historic-ethnographic (see Paasi, chapter 20 in this volume). I should note that the regional division, as well as its official status, has changed over time, and especially during the earlier phases the regional association of certain areas was unclear or ambiguous. Many of the present-day regions would not have been recognized, or their boundaries would have been elsewhere. Since no indisputable regional division existed throughout the period under study, I have used a somewhat modified present-day division for the sake of unity. Some recent administrative subdivisions seemed to have little relevance to the material and I have ignored them here.[3]

I have also recorded information about the regional association of a landscape, if given in its context, enabling a comparison between the historical and the present-day concept of Finnish regions. During the first period, the difference was clear: first, regions such as Kymenlaakso or Central Finland were not generally recognized; second, now-established subdivisions of historic regions (South Ostrobothnia, North Karelia) were then not commonly and uniformly made; and last, some areas had a different regional association from the present. Without taking this into account, one would have failed

TABLE 19.1.
National landscape images by region and periods.

Helsinki and present-day provinces	Pre-Independence	Percent	Given provinces	Percent	Prewar	Percent	Postwar	Percent
Helsinki	117	13%	117	13%	127	12%	267	24%
Uusimaa*	115	13%	118	14%	115	11%	57	5%
Uusimaa as a whole	232	27%	235	27%	242	23%	324	29%
Varsinais-Suomi	105	12%	106	12%	51	5%	105	10%
Satakunta	23	3%	51	6%	6	1%	39	4%
Häme**	106	12%	89	10%	100	10%	146	13%
Kymenlaakso	37	4%	28	3%	13	1%	31	3%
Central Finland	8	1%	1	0%	20	2%	42	4%
South Karelia	83	9%			104	10%	39	4%
Ladogan Karelia (pre-1939)	29	3%			68	6%		
North Karelia	10	1%			37	4%	22	2%
Karelia as a whole	122	14%	113	13%	209	20%	61	6%
South Savo	32	4%			67	6%	44	4%
North Savo	32	4%			32	3%	28	3%
Savo as a whole	64	7%	72	8%	99	9%	72	7%
South Ostrobothnia	42	5%			13	1%	33	3%
North Ostrobothnia***	30	3%			77	7%	54	5%
Kainuu	17	2%			36	3%	9	1%
Ostrobothnia as a whole	89	10%	97	11%	126	12%	96	9%
Lapland	49	6%			116	11%	187	17%
Petsamo (pre-1939)	2	0%			51	5%		
Lapland as a whole	51	6%	45	5%	167	16%	187	17%
Åland	37	4%	37	4%	16	2%	2	0%
N =	874	100%	874	100%	1049	100%	1105	100%

*Uusimaa, containing Uusimaa and Itä-Uusimaa
**Häme, containing Häme, Pirkkala, and Päijät-Häme
***North Ostrobothnia, containing North and Middle Ostrobothnia

to notice obvious differences between the nineteenth-century mental map of the Finnish regions and the present-day one.

STAGE-BY-STAGE OVERVIEW

In the general overview, the three periods did not differ from one another in any dramatic manner.[4] Each period showed a similar areal distribution, more or less based on the historic regions and thus apparently supporting the claim of representativity inherent in the national landscape imagery. On the other hand, the dominance of the capital and the other major centers was clearly revealed. During the first period, Helsinki accounted for 13 percent of all landscape images—more than any other single site and more than most regions—and together with Turku and Viipuri (now under Russia, known as Vyborg), the percentage was 23 percent. During the prewar period, Helsinki provided 12 percent of all landscape images and the four major cities—Helsinki, Turku, Viipuri, and Tampere—altogether 20 percent.

In the postwar period, Helsinki alone generated about a quarter of all landscape images, thus becoming more depicted than any other site or region in the country, except of course the region in which it is situated, Uusimaa (which otherwise was insignificant). Even without the ceded Viipuri, the three major cities held altogether one-third of all landscape images. Thus the urban centers could be said to have been overrepresented throughout the period, although this could naturally be explained by their demographic, political, and economic weight. All in all, the strong urban bias of the imagery contradicts the traditional view of the stereotypical Finnish landscape as predominantly rural or natural. This of course says nothing about the historical change occurring in the landscapes themselves: in each phase the townscapes had a wholly distinct character and the town-country relationship was articulated in a different way.[5]

Another major difference between the periods was the change in artistic and printing techniques, as well as that of the overall character and quality of the publications. Although photography had already made its breakthrough by the end of the first period, enabled by the breakthrough of halftone printing in 1890s, the traditional engraving and lithographic methods still prevailed during its course. From the second period onward, cheaply produced and reproduced photography eventually became the exclusive medium for landscape representation. At the same time, the general outlook of the publications ranged from high-quality exhibit pieces to standardized cheap editions—often collections of pictures first published in magazines.

THE PRE-INDEPENDENCE PERIOD

The first period, comprising fifteen publications, showed the most equal regional distribution of landscape images: the only salient region was Uusimaa and that is because of Helsinki. Perhaps more notable was the relative absence of the more peripheral regions, such as Ladogan Karelia and North Karelia, North Ostrobothnia and Kainuu, as well as Lapland. Overrepresented instead were the allegedly civilized southern and midland regions. South Ostrobothnia was seen mostly in terms of its Swedish-speaking coastal area, as three-quarters of the landscape images depicting it originated from there. This would indicate a later origin for the present-day inland image of the region (the Plains, *lakeus*), considered regionally representative but often also quintessentially Finnish. Lapland, by contrast, was to an exceptional level represented by generalized regional landscapes that could not be accurately located.

During the first period, there were fewer prominent individual sites than during the later ones. The images were not only more evenly distributed among the regions but also among the different localities. This shows that "cults" of certain landscapes or sites had not yet established themselves. Neither would important sites solely become introduced through certain publications, meaning that the set of landscapes changed relatively little from one book to another.

Apart from the major cities, only the historic towns and administrative centers of Porvoo and Hämeenlinna, as well as the still-evolving industrial center of Tampere, emerged as important landscapes during the first period. Other notable towns were the southern ports of Kotka and Hanko as well as the provincial centers of Kuopio and Oulu. Sund from the Åland Islands featured as a historic site because of the medieval ruins of Kastelholm and the fortress of Bomarsund, whereas the rapids of Imatra were the only salient natural sight. In spite of its importance for the nation-builders, the lake-and-forest landscape surprisingly was not represented by any particular site in any significant extent. The Saimaa Canal (opened in 1855), however, attracted some attention as a major technological achievement.

Before photography became a primary medium, landscape imagery was produced by traditional artists, first foreign "picturesque travelers" and printmakers, then more or less established Finnish artists (either as graphic artists or the original painters of reproduced paintings). Among the best-represented artists were several household names of nineteenth-century Finnish art: Gunnar Berndtson, Albert Edelfelt, Eero Järnefelt, Akseli Gallen-Kallela, Johan Knutson, P. A. Kruskopf, and Berndt Lindholm. The biggest input, however, was by the prolific photographer Into Konrad Inha, who reproduced 181 pictures in his book *Finland i bilder (Finland in Pictures)*, published in leaflets in 1895–96 (Figure 19.3). These alone formed 14 percent of the entire pre-independence imagery. Inha's book was the first to take full advantage of the new photographic reproduction methods, but it had no immediate followers.

Prewar Period

The number of publications from the prewar period was the highest: twenty-four. The given regional associations reflected more closely the present-day ones (not counting the most recent administrative subdivisions), the percentile differences being insignificant. The most marked change linking with the period is the emergence of Karelia with one-fifth of all landscape images, mostly from South and Ladogan Karelia. Other conspicuous regions were—apart from Uusimaa, swollen by Helsinki—Ostrobothnia and Lapland together with the newly acquired land corridor of Petsamo in the northeast. Lapland was again the region most often represented by anonymous landscapes. The prominence of Karelia and North Ostrobothnia depended partly on Inha's continuing activity.

Inha's influence is confirmed when looking at the individual sites. Now several natural sights or scenic areas were prominent in the imagery, among them certain localities in southern and mid-Finland (Lohja, Virrat, and Padasjoki), in Ladogan Karelia (Sortavala and Korpiselkä), and in Kuusamo by the eastern border. The majority of landscape images representing each were photographs Inha took and published in his one book, *Suomen maisemia (Finland's Landscapes)* (second, illustrated edition 1925). At least three of the areas can be explained by Inha's personal interest: Virrat as his birthplace and Lohja and Padasjoki as his one-time places of residence.

Inha also paid attention to the inland water connections, which were already losing their practical significance in favor of tourism. Waterway traveling was reflected in the number of his landscape images linked with well-known routes such as Lake Päijänne or Lake Ladoga near Sortavala. Another important route he documented was the northern River Oulujoki, which had a particular association with the historic tar trade, and increasingly also with the shooting of the rapids by tourists.

Without Inha, the prewar imagery would have looked quite different. Otherwise conspicuous landscapes were—again apart from the major cities—the already-familiar rapids of Imatra (though harnessed in 1922), the Punkaharju nature park as well as the towns of Kuopio and Savonlinna in Savo (the last one because of the medieval Olavinlinna castle), and some areas in eastern Lapland. The most important area of all, not counting Helsinki, would have been Petsamo, sometimes considered a part of Lapland and sometimes a region of its own. Its status as a new land acquisition, as well as the access to the Arctic Ocean it provided, undoubtedly added to its attractiveness.

Other notable sites were the old town of Porvoo, the Hogland island in the Gulf of Finland, the "Karelianist" Koli heights, and the venerable old Orthodox monastery of Valamo. Whereas Inha deliberately sought remote and untapped wilderness scenery, other salient sites were partly inherited from the preceding period, and partly a new layer was added by the increased interest in the eastern and northern border areas. This had been introduced by the turn-of-the-century Karelianism and the establishment of the 1920 national border to the east, creating a string of "outpost" landscapes. Open references to the 1918 civil war and to the ensuing conflict with Soviet Russia were scarce, though (usually they were limited to particular politically motivated publications about the border or Russian Karelia).

FIGURE 19.3. The Karelian *Urwald* frontier. Peasants from the eastern region of Karelia are portrayed as mythological, primeval Finns emerging from the forest (Inha 1896).

On the regional level, Inha's activity mostly explained the importance of North Ostrobothnia and Ladogan Karelia, significantly contributing to the exceptionally high one-fifth share of all Karelian landscapes in the national landscape imagery. By contrast, Lapland proper—inspiring less interest in Inha—showed the highest proportional increase of all regions from the preceding period (about one-tenth growth to 16 percent, the absolute number of landscape images becoming more than threefold).

Overall, Inha's book withheld 112 images. They had been taken mostly before Independence, from the 1890s to the 1910s, and thus partly overlapped with those in Inha's earlier book. As for the other photographers, many of them had originally contributed to the magazine *Suomen Kuvalehti,* which in the 1920s began to do high-quality gravure printing. The magazine republished in book form images resulting from landscape photography contests, often taken by amateurs. Toward the end of the period, the traditional landscape photography, inspired by painting, was complemented by pictorial reports. This, together with the general influence of Modernism, was reflected in the growing share of experimental landscape images, details, and action shots.

The Postwar Period

Only ten of the publications dated from the postwar period, but each one contained a much larger number of images than during the preceding periods, mostly because of the inflation of the traditional solitary "portrait landscapes" in favor of series and pictorial reports. The scarcity of publications combined with the high amount of landscape images is somewhat problematic, as a single book may affect the outcome more than those linking with the other two periods.

In addition to the dominance of Helsinki and other major cities, the last period was characterized by the abrupt disappearance of all the ceded areas—having accounted for 16 percent of the prewar imagery—and by various attempts to fill the gap they left. Neither could many references to the two wars be found among the landscapes (Figure 19.4), although there were numerous publications separately devoted to war and, later, to the ceded Karelia. On the other hand, many of the sites already introduced during the first period were still being depicted, having by now acquired a more or less permanent status as landscape classics. Also some of the borderland landscapes and areas survived from the preceding period, even though some of them were truncated by the new border. The most important region apart from Uusimaa—actually Helsinki—was Lapland, which in spite of the loss of the exotic Petsamo comprised 17 percent of all national landscape imagery.

The highest increase of all regions was that of Uusimaa. This was caused entirely by the more than doubling of the volume of Helsinki townscapes, while the rest of the region decreased in proportion. When looking at the sites, one predictably meets first the major cities and established tourist attractions such as the town of Hämeenlinna, now boosted by the popularity of the 1930s Functionalist Aulanko hotel and its surrounding nineteenth-century park. The wilderness landscapes of Kuusamo, which Inha introduced as popular imagery, have now become an established sight, despite the loss of the key site, Lake Paanajärvi, behind the new border.

Other sights were Heinola with its Functionalist sports institute, the harnessed Imatra rapids, now gravitating toward an industrial landscape, and Punkaharju as well as Kuopio and Savonlinna with its castle. The River Oulujoki disappeared from the imagery after its regulation as a part of the postwar reconstruction effort. Postwar Lapland boasted a large number of scenically important sites and areas as well as with unattached "regional" landscapes, but here again the work of a single photographer published in a single book lay behind their visibility: Matti Poutvaara in his *Suomi–Finland* (1952).

Like Inha, Poutvaara had his favorites. His influence was best seen in those areas other photographers had not yet exploited, most clearly in the extremities of Lapland (Figure 19.5). Poutvaara took more than half of his pictures from Inari, Enontekiö with Finland's newly discovered extremity of Kilpisjärvi, Utsjoki, Rovaniemi, and Kittilä with the Pallas-Ounas national park and ski resort. Further, he was the only photographer to capture the landscapes along the northern border River Teno and in Pyhätunturi, the other of the first national parks remaining after the war. Without his efforts, the only salient area in Lapland would have been the already-familiar scenery of Inari.

Poutvaara undoubtedly was the most prolific author of his period as well as of the entire study period, contributing 470 pictures. He was seconded by Heikki Aho and Björn Soldan, stepbrothers working jointly, with their 422 pictures. Together, the three photographers took care of more than half of all the landscape images from the period. Whereas Poutvaara's emphasis on Lapland was obvious, Aho and Soldan concentrated more on Helsinki and on generic landscapes, as well as on "cinematic" action pictures. (The stepbrothers also produced a number of tourist films.)

Discussion

The quantification of the national landscape imagery presented here indicates its overall structure and dynamics, but the picture remains superficial without deeper study.

FIGURE 19.4. Two border guards pose against the menacing Soviet territory shortly before World War II (Peltoniemi 1938).

Some features appear to be obvious: first, the structure of the imagery—in broad terms based on historic regions—largely remains the same, even when the geopolitical, social, and artistic emphases change. Each period appears to have a distinct profile, which to a degree is explained by the contributions of particular authors. As for the national landscapes, each period clearly has its celebrated landscapes and types of landscape, some of which may become classics that reappear as an indisputable canon during the further stages (provided that the sites are preserved and remain within the country's borders). The center-periphery dualism is present throughout the study period, parallel with the tension between modernization and tradition (see Lehtinen, chapter 18 in this volume).

The first period can be described as Topelian, because of the seminal influence of the composite athlete of Finnish culture, the author and historian Zacharias Topelius. Even though his two publications provided only one-tenth the volume of landscape images, he established the structural patterns and social biases of the national landscape imagery, consciously building a spatial model of the Finnish nation-state on the narrative roles he skillfully allocated to different historic regions and sites. From his influential works was also inherited the predominantly urban and southern view

FIGURE 19.5. An unnamed spot on a ski trip showing Lapland as a country of winter recreation (Poutvaara 1952).

of the country during the period, promoting Finland as a civilized nation and a land of opportunities, while signs of poverty and backwardness were rationalized, idealized, or censored from the imagery. Of natural sights, only the Imatra rapids were routinely presented. As an internationally renowned tourist attraction, they had a cosmopolitan character, which somewhat compromised their nationalist readings.

The second period clearly shows a shift of emphasis toward the peripheries, embodied by Inha, whose principal attempt to translate into photography the artistic imagery of National Romanticism was printed only after Independence. In spite of his influence, his activity alone would not determine the position of a particular site in the imagery. Although Inha sought to avoid the beaten path, a commercial tourist imagery developed before World War II, even if it was still imbued by nationalist ideology. The Karelianist landscapes of Koli and Ladogan Karelia turned into tourist attractions, whereas Lapland was promoted by the budding winter tourism. Both areas were made accessible by improving traffic connections, and the establishment of nature conservation areas encouraged tourism.

Typical for the prewar period was the repeated presentation of certain towns, areas, or routes in the publications, often in nearly identical images. These were also echoed in numerous publications solely devoted to them. Their images gradually seemed to lose whatever connection they had had with the original associations of the site—local, regional, historical, natural-historical, or other—and turn into an obligatory part of the national imagery, becoming interpreted only in its context. This was made possible by the effective and persistent proliferation of the images, making them a common and collectively felt heritage, whereas during the preceding stage the landscapes were more evenly distributed so as to allow as much local identification with them as possible.

The third period could be described in terms of polarization between Helsinki and the urbanizing south, on the one hand, and the "last wilderness" of Lapland, on the other one. The same general tendencies were still present from the preceding period, but now the search for and incorporation of new landscapes assumed a more businesslike character. The prewar patriotism had yielded to commercial tourism, leaving less room for the earlier didactic inventory of every nook and corner of the nation-state. In the material one may also note a widening rift that separates tourist and scientific-conservationist literature, the latter concentrating mostly on the natural characteristics of landscape and on nature conservation areas.

The postwar emphasis on Lapland was shared by both strains. After the improvement of traffic connections before and during the war, Lapland had become more accessible for tourism, and its natural resources were ruthlessly used to fuel the postwar recovery. At the same time, the network of national parks gravitated into Lapland, where it had been initiated by Pallas-Ounas, Pyhätunturi, and Heinäsaaret (Petsamo) in 1938, followed by Oulanka and Lemmenjoki (Inari) in 1956. The landscape of Lapland became sharply divided into spheres of production—large-scale forestry, open mining, hydroelectric plants—and of consumption, indicated by holiday resorts and national parks.

The logic of representation in the landscape imagery is above all rhetorical. By the choice of a set of landscapes, the complexity, redundancy, and arbitrariness of the national territory and its borders are swept aside. The selection process of individual landscapes has been guided by the various international and national standards of beautiful or sublime, as well as by the accessibility of the sites, the level of knowledge about them, and their political status. The sites then have become intertextually linked with preferred areas or nationalities (Greece, Italy, the Alps), while unwanted characteristics such as poverty or Easternness may have been ignored or censored (see Osborne 1996, 23–40).

The deciphering of the national space appears to have occurred on an emotional basis. Especially in early imagery, landscape images conformed to classical landscape genres and characters, whereby particular physical or historical features generated standard responses (Hunt 1992). By means of landscape imagery, geographical, social, and cultural data were presented within a uniform and generally recognized psychological matrix. On the other hand, the sites represented were an actual part of the everyday environment, which lent verisimilitude to the ideological notions linked with them. The abstract concepts of homeland and nationhood could thus reach the sphere of personal experience and become an integral part of it (see Cubitt 1998, 6–7, and Lehtinen, chapter 18 in this volume).

The emotional content of a landscape culminated in the *in situ* experience in which the physical environment and its representation coincided (Frow 1991, 123–51). The imagery has corresponded to a network of existing sites, an archipelago of model landscapes, forming halting stations for real or imagined routes. As a printed set of images, an imagery enabled the reader to sense the different aspects of the nation-state from a distance, but on site the reader's emotions were amplified by the feeling of authenticity, of "being there." Landscapes were expected to correspond to the preconceived image, and alterations of their outlook were often strongly objected to, ultimately leading to their protection by state purchases and so separating them from their surroundings.

Landscapes were intended as socially neutral symbols, representing the nature that allegedly welded the different social strata or ethnic groups together into one nation. The landscape imagery was supposed to reflect every level of society and to address every social group, each one in its own way.

As a result of the educational process that occurred from the mid-nineteenth century onward in the school system, in popular patriotic literature, prints, and postcards, and in tourist guides and brochures, the landscape imagery formed a mnemonic frame-work of feelings about the country. A picture would evoke a geographical area, an area, or a narrative, triggering national pride, melancholy, or aesthetic appreciation, depending on the particular context. These emotions would then constitute an unquestioned shared feeling, the textual origin or constructed nature of which might necessarily no longer be recognized.

In published sets of landscape images, much attention was paid to meticulously listing regions and localities, which allowed the predominantly peasant population to identify their everyday surroundings with the country as a whole. This was vital for the Finnish nation-builders, who lacked a solid and unambiguous historical footing and had to work their way from discrete areal, ethnic, and class identities. In a pointed way, one could characterize national landscape imagery as a means of turning a nation-state into a coffee-table book (or a mail-order catalogue), implying a network of physical places and spaces in between that incarnate the very essence of the homeland.

There was no direct correlation between the imagery and the sites involved, as only a part—even if major—of the landscape images were linked with identifiable sites, and even fewer ever attracted tourism. The rest were a diverse mixture containing recordings of fleeting moments and anonymous roadscapes as well as more rigorous scientific illustrations of regional features. A clear division was made into unique (outstanding, rare, exceptional) and typical (ordinary, average, representative) landscapes. The latter included "mediating" landscapes that created images of distance and areal extent by representing movement between the core areas and the outmost reaches of the country.

The first category of unique landscapes became the symbolic core of the imagery. These were natural formations and geographical extremities as well as historic sites and monuments. They often became permanent members of the imagery, their longevity reflecting in, as it were, biographical place narratives. These landscapes constituted the national landscapes, the commonly recognized and undisputed classics and national benchmarks for scenic beauty, which metonymically represented the entire country (see Baehr and O'Brien 1994). The role of the "lesser" landscapes was crucial as well: they embedded the core images in nested areal scales, creating associative links between the local, the regional, and the hallowed national sites. The elaborate hierarchy of imagery enabled the national subjects to feel their everyday surroundings as an integral part of a national spatial continuum.

To reduce the dynamics of the multifarious landscape imagery into one generalist model would be an oversimplification, disregarding its intricacy and sensitive balances. A closer analysis of the imagery will bring into light a delicate rhetorical play with images and symbols as well as effectively programmed emotional qualities and narratives, which still have not lost their evocative force.

The Present Situation?

The official landscape inventories of the 1980s and early 1990s are put into new perspective by the findings above. These were based on representativity, employing a variety of scientific, historical, and administrative criteria, and could thus give an idea about the present spatial image of the nation-state, at the same time pointing out how the national landscape imagery now has become actively sustained by the state authorities. The foremost example of this are the twenty-seven sites declared to be national landscapes by the Ministry of the Environment (*National Landscape* 1994). Even without the status of official designation, the listing has often been perceived as such, giving heed for protection and management planning as well as for marketing efforts.

Another important report was that of the Ministry of the Environment's Working Group on Landscape Areas, *Important Landscape Areas* (*Maisema-aluetyöryhmän mietintö* 1993). A third report was published jointly by the Ministry of the Environment and the National Bureau of Antiquities on the built cultural environments of national interest (*Rakennettu kulttuuriympäristö* 1993). The former comprised 150 large landscape areas of national interest—covering altogether 2 percent of the country's total areal and about 11 percent of all cultivated land—as well as a preliminary listing of what were termed "traditional landscapes" (landscapes produced by largely abandoned land-use practices), while the latter listed altogether 1,772 properties ranging from individual buildings to cultural landscapes. Because each inventory targeted different types of landscapes (the first, the nationally representative symbolic landscapes; the second, the regionally representative landscape areas; and the third, historic buildings and sites of national interest), they are not comparable, but each one may be set against the background given by the results of my study.

Most of the sites listed in *National Landscape* (1994) have been salient in national landscape imagery during one or several of the periods under study here. Those that were not had not broken into the public awareness (or existed, as in the case of the Modernist Tapiola garden city) before the 1960s. The sense of continuity with the past seems strongest in the context of sites such as Helsinki and the other larger cities, the historic Porvoo, Hämeenlinna, and Olavinlinna, the natural sights of Imatra, Punkaharju, and Koli, or the tourist areas of Kuusamo, Pallas, and Utsjoki. Some of the choices of the committee actually caused some controversy: for example, the absence of the Swedish-speaking coast or that of Central Finland could be seen as flaws in the intended national representativity.

The inclusion of industrial heritage and Modernist environments into canonized landscapes clearly represented newer strata of landscape imagery in *National Landscape*. The debate around the national landscapes had also highlighted literary and artistic landscapes. Missing from the listing were the rapids of the River Oulujoki, harnessed after World War II. Imatra, on the contrary, made it onto the list despite the fate of its rapids, now counted as industrial heritage. The landscape of Sund with its historic ruins reemerged from long oblivion, asserting the sometimes-disputed Finnishness of the Åland Islands. Lake Köyliö was for the first time presented as landscape rather than merely a commemorative site.

In the report on landscape areas, the choice was rather different because of the emphasis on extensiveness, scientific criteria, and preservedness. A division of the country into scenic regions was specifically drawn up for the inventory, based on both natural and cultural landscape features, and the landscape areas were chosen to exemplify these regions. Although several of the landscape areas of national interest also had stood out in the national landscape imagery—as well as in the national landscape listing—this was not the case with most of them. The inventory of built cultural environments, by contrast, mostly relied on architectural, antiquarian, and overall historical value. Of these two inventories, above all the geographic distribution of the latter is easily comparable with the study results, showing less weight on the capital but more on the rest of southern and western Finland. This apparently corresponds with their rich building heritage and its relatively good preservation.

One may venture to suggest two lines of development exemplified by the reports. First, the previously ubiquitous anonymous agricultural landscape has now turned into a scenic and ecological asset and an object of some kind of protective measures, practically wherever still found unspoiled by progress. Second, the present concept of heritage has increased the value of what is left of vernacular architecture and cultural environments, and has also integrated environments produced by modernization. Heritage landscapes are selected, assessed, and managed by experts. Bearing in mind that the various landscape designations tend to increase the publicity and tourist value of the sites, it would seem that the latest additions to the national landscape imagery are the heritage landscapes linking with the agrarian past—earlier taken for granted—as well as with the history of the nation-building process, now seen in a nostalgic light.

An interesting trait in itself is, in the light of Anssi Paasi's chapter in this volume, the intertwining of the cultural and administrative concepts of landscape in the committee reports, as well as its public acceptance. Since Topelius, the national landscape imagery has had a "semi-official" status in school curricula, even if the actual administrative policies directly relating to valued landscapes have been until recently relatively few.

The most intriguing fact seems to be the survival of national landscape imagery to this day, partly resulting from conscious efforts by the environmental and heritage administration. According to Paasi (1986 and this volume), the stereotypical regional landscape images are still generally recognized by the Finns. The preservation movements for Koli and other national landscapes would also suggest that their cohesive power still exists, perhaps even amplified by a reaction to the recent globalization and European integration processes. One could even venture to say that the official interest in national landscapes may have been triggered by previous spontaneous landscape conservancy movements.

National landscape imagery may no longer condition the perception of Finnish national space, or at least it is becoming increasingly interpolated with other imagery and discourses. On the other hand, this is hardly anything new, Finland's history has rather been one of constant tension and struggle between national and international, traditional and modern, central and peripheral. Landscape is not going away as a privileged way of seeing, no more than nationalism as a spatial discourse. There is, however, a marked increase of alternative readings for national landscapes—global, consumerist, artistic. The cultural hegemony of national landscape imagery has not been dissolved, at least not in Finland, but it is becoming challenged more than ever.

NOTES

This study has been made possible by several research grants from the Academy of Finland. I have published the results elsewhere in: "The Kaleidoscopic View: The Finnish National Landscape Imagery," *National Identities* 2, no. 1 (2000); "A Periphery Lost: The Representation of Karelia in Finnish National Landscape Imagery," *Fennia* 182, no. 1 (2004); "Countryside Imagery in Finnish National Discourse," in *European Rural Landscapes: Persistence and Change in a Globalising Environment,* ed. Hannes Palang et al. (Dordrecht: Kluwer Academic Publishers, 2004); and *Kuvitettu maa: Suomen kansallisen maisemakuvaston rakentuminen (The Illustrated Land: The Construction of Finnish National Landscape Imagery)* (Helsinki: Finnish Literature Society, 2005). For recent publications on related topics, see Ville Lukkarinen and Annika Waenerberg, *Suomi-kuvasta mielenmaisemaan (From the Image of Finland to Mindscape)* (Helsinki: Finnish Literature Society, 2004); Helen Soovāli, *Saaremaa Waltz: Landscape Imagery of Saaremaa Island in the Twentieth Century* (Tartu: Tartu University Press, 2004); Paul Wilson, "Banality as Critique: Contemporary Photography and Finnish National Landscapes," *Journal of Finnish Studies* 9, no. 2 (2005); and Derek Fewster, *Visions of Past Glory: Nationalism and the Construction of Early Finnish History* (Helsinki: Finnish Literature Society, 2006).

1. De Saint-Morys (1802) quoted William Gilpin; see Ramsay 1999 [1807]).
2. The figures discussed in this chapter are preliminary and are suggestive results from the study.
3. For instance, Päijät-Häme and Pirkanmaa are not historic regions but are recent administrative constructions.
4. Toponyms will be given in their Finnish form, unless an established international form exists (such as Karelia, Ostrobothnia, Hogland).
5. Interestingly enough, Andrew Hemingway (1992) ended up with a similar notion of urban predominance concerning nineteenth-century British landscape imagery, and Nicholas Green (1991) about the contemporary French one.

REFERENCES

Acerbi, Giuseppe. 1802. *Travels through Sweden, Finland, and Lapland to the North Cape in the years 1798 and 1799.* London.

Anderson, Benedict. 1991 (1982). *Imagined Communities: Reflections on the Origin and Spread of Nationalism.* New York: Verso.

Anttila, Jorma. 1993. "Käsitykset suomalaisuudesta—traditionaalisuus ja modernisuus." *Mitä on suomalaisuus,* ed. T. Korhonen, 108–34. Helsinki: Suomen Antropologinen Seura.

Baehr, Peter, and O'Brien, Mike. 1994. "Founders, Classics, and the Concept of a Canon," *Current Sociology* 42, no. 1: 1–151.
Billig, Michael. 1995. *Banal Nationalism*. Manchester: Manchester University Press.
Clarke, Edward Daniel. 1829. *Travels in Various Countries of Europe, Asia and Africa*, Part III: Scandinavia. London.
Conzen, Michael P., ed. 1990. *The Making of the American Landscape*. Boston: Unwin Hyman.
Cubitt, Geoffrey, ed. 1998. *Imagining Nations*. Manchester: Manchester University Press.
Daniels, Stephen. 1993. *Fields of Vision: Landscape Imagery and National Identity in England and the United States*. Cambridge: Polity Press.
De Saint-Morys, Etienne Bougelin Violart. 1802. *Voyage pittoresque de Scandinavie*. Paris.
Frow, John. 1991. "Tourism and the Semiotics of Nostalgia," *October* 57 (Summer): 123–51.
Green, Nicholas. 1991. *The Spectacle of Nature*. Manchester: Manchester University Press.
Hemingway, Andrew. 1992. *Landscape Imagery and Urban Culture in Early Nineteenth-Century Britain*. New York: Cambridge University Press.
Hirn, Marta. 1988. *Finland framställt i teckningar*. Helsingfors: Svenska litterratursällskapet i Finland.
Hunt, John Dixon. 1992. *The Garden and the Picturesque: Studies in the History of Landscape Architecture*. Cambridge, Mass.: MIT Press.
Inha, Into Konrad. 1896. *Finland i bilder*. Helsinki.
———. 1909. *Suomen maisemia: Näkemänsä mukaan kuvaillut*. Porvoo: WSOY.
Klinge, Matti. 1981. *Suomen sinivalkoiset värit: Kansallisten ja muidenkin symbolien vaiheista ja merkityksistä*. Helsinki: Otava.
———. 1993. *Finnish Tradition: Essays on Structures and Identities in the North of Europe*. Helsinki: Finnish Historical Society.
Knapas, Rainer. 1993. *Historiallisia kuvia: Suomi vanhassa grafiikassa*. Helsinki: Suomalaisen Kirjallisuuden Seura.
Lowenthal, David. 1989. "Nostalgia Tells It Like It Wasn't." In *The Imagined Past: History and Nostalgia*, ed. Christopher Shaw and Malcolm Chase, 18–33. Manchester: Manchester University Press.
Maisema-aluetyöryhmän mietintö. 1993. I *Maisemanhoito*: II *Arvokkaat maisema-alueet*. Työryhmän mietintö 66/1992. English abstract: "Report of the Working Group on Landscape Areas. I Landscape Management. II Important Landscape Areas." Helsinki: Ympäristöministeriö, ympäristönsuojeluosasto.
Manninen, Ohto. 1997. "Finland Takes Shape." *Historiallinen Aikakauskirja* 95, no. 2: 129–35.
Meinig, Donald W., ed. 1979. *The Interpretation of Ordinary Landscapes: Geographical Essays*. New York: Oxford University Press.
Mitchell, W. J. T. 1994. "Introduction." In *Landscapes and Power*, ed. W. J. T. Mitchell, 1–4. Chicago: University of Chicago Press.
Moscovici, Serge. 1984. "The Phenomenon of Social Representations." In *Social Representations*, ed. R. M. Farr and Serge Moscovici, 3–70. New York: Cambridge University Press.
National Landscape. 1994. Helsinki: Ministry of the Environment.
Osborne, Brian S. 1996. "Figuring Space, Marking Time: Contested Identities in Canada." *International Journal for Heritage Studies* 2, nos. 1–2: 23–40.
Paasi, Anssi 1996: *Territories, Boundaries, and Consciousness: The Changing Geographies of the Finnish-Russian Border*. Chichester: John Wiley and Sons.
Peltoniemi, U. 1938. *Itärajan kuvia*. Helsinki.
Poutvaara, Matti. 1952. *Suomi—Finland*. Helsinki.

Rakennettu kulttuuriympäristö: Valtakunnallisesti merkittävät kulttuurihistorialliset ympäristöt. 1993. Museoviraston rakennushistorian osaston julkaisuja 16. Helsinki: Museovirasto & Ympäristöministeriö.

Ramsay, Carl Gustaf. 1999 (1807). *Resejournal 1807: En resa i Södra Finland,* ed. Sampo Honkala. Helsinki: WSOY.

Realms of Memory: Rethinking the French Past. 1996–98 (1984–94). Volumes 1–3, under the direction of Pierre Nora, ed. Lawrence D. Kritzman. New York: Columbia University Press.

Reitala, Aimo. 1993. *Suomi-Neito: Suomen kuvallisen henkilöitymän vaiheet.* Helsinki: Otava.

Shields, Rob. 1991. *Places on the Margin.* New York: Routledge.

Topelius, Z. 1845–52. *Finland framställdt i teckningar.* Helsingfors: A. W. Gröndahl & A. C. Öhman.

———. 1873. *En resa i Finland.* Helsingfors: F. Tilgmann.

Von Kügelgen, Carl. 1823–24. *Vues pittoresques de Finlande.* Saint Petersburg.

20.
Finnish Landscape as Social Practice: Mapping Identity and Scale

Anssi Paasi

Landscape and Territory:
From Tradition to Reinvention of the Links

"Few words in the language have had so tyrannical an effect on our ways of thinking and acting as landscape," wrote the well-known landscape specialist J. B. Jackson (1964). This provocative comment points to the fact that landscape is a complicated word but also to how landscapes structure social practices and the ways people organize their relations with their environment. This complexity becomes even more obvious when one compares the referents of this category in different languages. The use of the word "landscape" fuses several different aspects contextually: its vernacular origin (landscape as a collection of small areal domains or administrative units); later multiple meanings, such as a picture of the surroundings of farm life, as discussed mainly by critics and artists (Jackson 1964); or the understanding of landscape as a text or center of subjective identification. At times landscapes are comprehended as objectively existing elements and backgrounds in human life, at times as subjective, experienced and emotional entities. In current academic discourse the metaphoric dimensions of landscapes are also increasingly being evaluated and mapped.

Different dimensions of landscape are underlined in different languages. In Swedish, for instance, the word *landskap* means both "landscape," "landscape painting," and a concrete territorial unit, "province." This is true also of the German *Landschaft*. Of course geographers, particularly those inspired by the German *Landschaft* tradition—such as Carl Sauer (1963 [1925]) in the United States or J. G. Granö (1997 [1929]) in Finland—emphasized areal connections of landscapes. The Finnish word for landscape, *maisema,* came into use in the 1830s to correspond to a variety of dialectal words for soil, land, terrain, district, place, or locality. The words *landskap* and *Landschaft* were employed in Finnish scientific discourse in a broad sense at the turn of the twentieth century (Keisteri 1990, 34). *Landskap* is a particularly important category for Finnish geography, since the first academic geographers, who created the visions of what landscape actually means, published their original works in Swedish (Hult 1895).

Landscape thus seems to refer at times to visual impressions, stressing the "power of the eye," at times to concrete areas or territorial units, or to literary or artistic images of these, that is, aesthetic representations. Landscape seems to connote larger spaces than the geographical idea of place, usually understood as a center of human experience, particularly in humanistic and interpretive geographies. Cosgrove (1989) reminds us that for landscape we are always outsiders because landscape is always seen and viewed beyond us. Landscapes are, he writes, pictorial images whose history is completely bound up with the inscription of environmental images by various media. Landscapes are typically painted on canvas, photographed on film, or gardened on the earth's surface (104). Cosgrove's comments take us back to the English use of "landscape," which refers mainly to the visual outlook on the world.

Jackson (1964) pointed out that in traditional usage the words and phrases that have been employed to define landscape almost always have been those used by artists and art critics and not by "farmers or geographers." Raymond Williams (1973) also argued that it is mainly outsiders, such as estate owners, industrialists, artists, or improvers, who have had recourse to the notion of landscape, and not the people who in fact live in the area in question. Hirsch (1995) criticized this view as "romanticism" and pointed out that, like place and space, the ideas of inside and outside are not mutually exclusive but may shift in national and wider contexts. Even if Hirsch does not discuss this topic, his comment points to the importance of *spatial scales* and *distinctions* (inclusion/exclusion) in the construction and reconstruction of the images and representations of landscapes. This reveals the multiple links between landscape and *identity*.

Ideas of landscape are not developed *in vacuo*. Hirsch (1995) in particular reminds us of the close link between the idea of landscape and other related concepts, such as space and place, inside and outside, image and representation. The first of these pairs of concepts refers to the context and form of everyday life and nonreflexive forms of experience. The second is crucial in the creation of distinctions, boundaries, and identity (see Newman and Paasi 1998). The third, for its part, has more to do with the context and form of collective representations communicated, for example, by the media or by literature, although these may also be linked with personal identity.

The eminent Swedish geographer Torsten Hägerstrand (1995) made a rather skeptical judgment regarding the lack of interest in landscape studies in the mid-1980s. Nevertheless, the idea of landscape seems to have revived, to become a significant object of academic discourse not merely within geography but also in fields such as anthropology, architecture, planning, and sociology. Instead of pure "objective mapping" of landscapes, scholars have increasingly become interested in the symbolic roles of landscapes and in their roles as signifiers or symbols of power, gender, or race relations (Zukin 1991; Rose 1993). Landscape images typically create, transform, or reconstitute spaces so that they fit in with human ideas of order, truth, beauty, harmony, or fear—and consequently include a clear moral, normative dimension (Cosgrove 1989, 104). In practical social life, this means simply that ideas of landscapes are contested, with some voices dominating the interpretations within the "limits of tolerance," while others are suppressed (see Raivo 1997).

Some authors are ready to take one more step toward understanding the idea of landscape. Zukin (1991, 22), for instance, points out that "the spatial consequences of combined social and economic power suggest that landscape is the major cultural product of our time." Mitchell (1994), for his part, reminds us that landscape is an instrument of cultural power, maybe even an agent of power. Thus, instead of asking what landscape "is" or "means," he wants to know what landscape *does,* or "how it works as a cultural practice." He wants to change landscape from a noun to a verb. In this sense, a landscape is a body of cultural, economic, political, or administrative practices that are crucial in the making of history both in the represented and the real environment. The most powerful and long-standing images of landscapes have typically been produced for the powerful, and they imply social and technological power (Harley 1988; Cosgrove 1989). Accordingly, landscape may be understood as a kind of synthesis and expression of the meanings of physical surroundings and various material and social practices (such as law, justice, and culture) and their representations (Olwig 1996).

The Purpose of the Chapter

I shall formulate a framework here to facilitate an understanding of the relations between landscape as a visual and territorial category in the Finnish context, and, further, to map the roles of landscape, understood as a set of social practices and discourses. I will begin from the two meanings of the Finnish idea of landscape. In most languages the word "landscape" includes two dimensions: the idea of a field of vision or perceived environment, and the idea of an areal unit with a territorial or bounded extension. These meanings are more complicated in Finnish, so that the word *maisema* points typically to the visual dimension of landscape, and the word *maakunta* points to the areal, vernacular, and administrative dimension. A specific Finnish combination of these dimensions is the word *maisemamaakunta* (literally "landscape province"), which typically refers to the products of scientists by which they aim at spatial classification of the visual elements of nature and culture. The idea of a specific regional identity has typically been connected both with these products of scientists and with the vernacular *maakunta* areas. This implies that the ideas of distinction and identity are closely linked with both dimensions. Hence landscape implies a dialectic between closure and opening or between inclusion and exclusion, that is, certain natural, cultural, and symbolic entities are understood as coming together in a "landscape," making it different from other landscapes. The practice of distinction and classification is thus implied in this category, both in ordinary daily experience and academic studies.

The idea of landscape will therefore be contextualized here in relation to certain major social and cultural practices, particularly administration and government, the action of geographers in the production of landscape provinces and national landscape ideals, and the role of the media in the production and reproduction of these ideas. These social practices and discourses produce and reproduce specific "readings" of landscapes and turn them into sets of (spatial) representations and meanings, whether visual (cartography, paintings, photography, film), verbal (written texts), or audible (music). All these practices and discourses are characterized by power relations, just as these representations are manifestations of power. These are produced in the context of the social and spatial division of labor, and some actors are always more powerful than others in the production of representations. Furthermore, one object of crucial interest in the following discussion is vernacular, daily life *in* the landscape and the collective representations of the media that literally wrap people inside media discourses. In all of these practices and discourses, either visual, regional, or symbolic dimensions of *maisema*, landscape, or all of these, are crucial.

In this chapter, I will first discuss the relations between landscape, territory, and identity. Then I will develop a conceptual framework to trace the meanings of landscape and territory in various social and cultural practices, and the power relations that have been involved in the production of the ideas of landscape and the production of their identities on various spatial scales. Finally, I will look at the Finnish landscape from six perspectives:

- The nationalistic ideological Finnish landscape
- Landscape as a medium of administrative power and governance
- Landscape as a product of academic discourse
- Landscape as a constituent of regional images and identities
- Landscape as a form of regionalism and provincial identity
- Landscape as an instrument of media discourses

The links between previous conceptions of landscape are illustrated in Figure 20.1. This framework is in a sense heuristic, in that it aims at mapping some links that are important for understanding the complicated meanings of landscape in Finnish social practice and discourse. This means that the links between the concepts in the framework are *not* to be understood as expressions of causal relations but rather as showing the directions of the connections between the different practices and discourses that the concepts depict. The figure identifies two major dimensions in this process—

the regional and the visual. It shows how landscapes and provinces are made into meaningful entities by means of administration, politics, and governance, as well as by the actions of geographers (and other scholars) and journalists. Some other key dimensions are not included in the figure because of their complexity. One of them is the link between individuals and social groups. A major question, then, is how individuals are linked with landscapes/provinces and how social communities and the whole of society are involved in the social and cultural production of these elements. A significant concept is identity, which mediates the links between memory, territory, individuals/collectives, and landscapes.

LANDSCAPE, TERRITORY, AND IDENTITY

The idea of identity is linked closely with the concept of landscape on various spatial scales. On the local and regional scales, it may be argued that both natural and cultural landscapes characterize places and regions and give them a "personality," which may also be significant in the process whereby people individually and collectively identify themselves with these spatial contexts. Identity, boundaries, and distinction are linked together. The idea of a specific local or regional identity is often used to

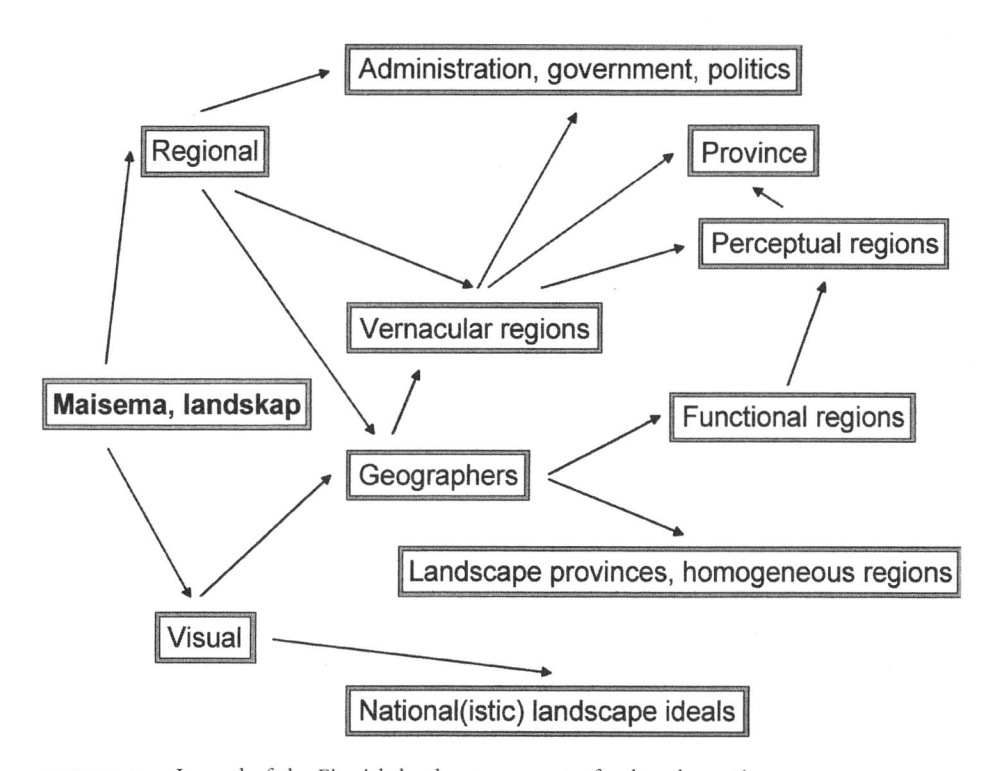

FIGURE 20.1. Legend of the Finnish landscape as a set of cultural practices.

depict the link between people, memory, and territory. Identity is one of the contemporary catchwords, but scholars rarely provide any explicit definition of what this complicated phrase means. A typical point of departure is to understand identities in an essentialist fashion, as being stable and permanent. I should note, however, that identities are nested and networked: people typically identify themselves with numerous social groupings (ethnic, religious, gender, generation), and territory and landscape may be only one part of such an identity. This means that boundaries and identities are not inevitably exclusive. It is also important to remember that regions themselves are not eternal: they emerge, exist for some time, and may vanish in the transformation of the regional structure (see Paasi 1996, 1999).

While keeping these comments in mind, it is nevertheless obvious that at least some of the following elements are crucial for understanding what identity means or how it is constituted in the case of specific places and regions:

- The material-morphologic basis: physical nature and landscapes
- The history of human and nature relations, that is, the history and organization of work
- Communal institutions, particularly in the spheres of politics, administration, economics, culture, and tradition, where local, regional, or national "flagging" may occur and collective memories will be maintained
- Systems of symbolism (language, dialects)
- Values and norms
- Identities and symbolic/physical boundaries between "us" and the "other," that is, the nature of imagined communities
- Narratives of "us" and our identity and tradition: literature, newspapers, education

These elements may create a link between territory, people, and history/memory. Landscapes may also be major material elements in local and regional identities, but they may also be used in more symbolic ways as signs or "names," which can convey a sense of place, a sense of belonging. As Karjalainen (1986, 143) points out, "Landscape functions as a means of placing oneself somewhere, of getting an image of one's lived environment." For a resident, therefore, a landscape is typically known by its familiarity, and for a tourist by its novelty. The latter of course may also result from diverging media representations and advertising, and not merely from personal experience.

Palmer (1998), who has traced the banal forms of nationalism, argues that landscape is, together with body and food, a concrete material medium for a national identity, not merely a symbolic one, even though the body and landscape have by tradition been employed in allegorical and metaphorical ways in national discourses. This becomes obvious in such expressions as the "Maid of Finland" or "Svea mamma," for instance, which point to specific female allegories that have come to characterize national images (Paasi 1996). Body and landscape also have material dimensions. Palmer reminds us that "how we use our bodies, in the foods we consume and in our relationships with the landscape there is a continual reminder of who we are and what we believe in" (see Paasi 1999).

In many cultures, *naming* gives a person social existence, a specific identity (and position) in a social community (Fitzgerald 1993). The creation and maintenance of landscape discourses occupies a crucial position in the creation of regional identities. Territories are normally named in these discourses, and the names are usually major symbols of identity, together with more material elements such as particular features of nature and culture. Names are crucial, whether we are discussing administrative areas or regional divisions produced by geographers. In his book *Mapping the Invisible Landscape,* K. C. Ryden (1993) discusses how the sense of place achieves its clearest articulation through narrative, and he is ready to argue that the oral historical narrative may well be the most common genre found in any place. Places accumulate folk names and also have a close relation to personal life histories and the oral histories of communities. Place-names and other symbols provide the most significant fixed points in this sense. Ryden (79) quotes Mary Hufford (1987, 21–22), who points out that "place names, linked with landscape features, encode the shared past, distinguishing members of one group from another," and, further, that "places and their names are sources of identity and security."

Naming is significant in the emergence or *institutionalization* of regions or territories (Paasi 1986a, 1991, 1996). The symbolic shaping of regions is a crucial element in the institutionalization process, together with territorial and institutional shaping. Administrative, "geographical," and cultural regions may overlap or not, but all these types of region become institutionalized in various social and cultural practices that take place on different spatial scales. Geographers and other scientists produce boundaries and symbols for various purposes, as does administration and government. It may be argued that landscapes are a form of *professional knowledge.* Cultural regions may emerge spontaneously from the action of social collectives through time.

My argument in developing this analytical framework is that when an institutionalized region achieves an established position in the territorial system, it gains a specific regional identity, which may, again analytically, be divided into two dimensions: the identity of the region and the regional identity of the inhabitants, that is, regional consciousness. This holds only in the case of regions that have emerged out of the social life of their inhabitants. Regions, regional divisions, boundaries, and namings that geographers produce as a result of their research processes may, on the one hand, be completely external to the daily life experience of ordinary people. On the other hand, as Figure 20.1 indicates, "perceptual regions" that geographers have produced and shaped on the basis of interviews with people living in various areas are typically a combination of vernacular and functional regions (e.g., Shortridge 1984).

Collective, particularly national, identities have a close relation to what has been labeled as *ideology.* I have argued elsewhere that ideology is not merely a set of ideas or beliefs but rather a complicated set of social practices and rituals that manifest themselves in the ideological state apparatus (Paasi 1999). National ideologies are structured in numerous institutions that mediate individual and social life, among which religious, educational, jurisdictional, political, professional, and cultural institutions are crucial. This does not simply mean that ideologies are given to us from above. It may be emphasized, using the categories of Bourdieu (1998 [1994]), that these institutions operate in the field of *symbolic economy*. Symbolic work is, Bourdieu notes, concomitantly an act of both giving form and following rules. Symbolic change is possible only if the participants have identical categories for their observations and appreciation. In this change, people acquire ontological security and continuity for their lives and become committed to broader social collectives and their rules. The ideas of regional or national landscapes may also be crucial sources and outcomes of symbolic exchange, and therefore of security and identity.

The major dimensions of regional identity are conceptualized in Figure 20.2, which makes the distinction between the identity of regions and the regional identity (or regional consciousness) of individuals living in a region. Regions have a specific image not only in the minds of the inhabitants but also for outsiders. The community is part of identity in two senses. On the one hand, people may be active members of associations that strive to promote the region in question. On the other hand, the community may be merely a set of collective representations and images put forth by the media, idealized images that resemble what Anderson (1991) labeled *imagined communities* in his discussion of the development of nationalism. Regional identity thus consists of several dimensions and practices, which partly reflect the material basis of the region, and partly the symbolic processes and representations of the region, which are mediated by the media and literature, for example.

Geographers have been active in producing "identity regions" when constructing their regional divisions based on nature and/or culture, but these "regions" or identities have perhaps not been very significant in shaping the vernacular identities of provinces. Rather, they have been typical illustrations of objectifying tendencies of scientists and their efforts to classify the complexity of the world.

The above discussion leads us to the following points. First, regions, landscapes, and identities are social and cultural constructions. Second, they are constituted on various spatial scales, which are typically connected with each other, that is, local and regional images are linked with national images and vice versa. Third, landscape images, practices, and identities are created in many different discourses, the most important of which include academic, administrative, and popular discourses, which all produce different meanings for these categories. Fourth, the ideas of landscape and identity are contextual and historically contingent, that is, while being crucial for the maintenance of history and memory, their meanings may vary a lot between times and places. This means that identities are dynamic and subject to change.

I will now discuss various dimensions of landscape discourse and their links with social practices in the Finnish case. It is obvious that there exist very different practices and discourses in which ideas and images of landscapes have been produced and have been transformed as part of identity narratives. As Cosgrove (1989) pointed out, the moral dimension is often very obvious in connection with landscape images, and this means that images are constantly open to challenge, alteration, and reinterpretation.

When analyzing the construction of Finnish landscape images, I will make an analytic distinction between several manifestations of landscape and identity practices and discourses. These discourses are partly independent, partly overlapping, and in most cases are connected with various spatial scales, from local and regional features to national landscapes. My intention is to show how territorial symbolism, boundaries, and divergent institutional practices have been constructed in each discourse.

Landscape and Ideology:
The Shaping of the Finland of Runeberg and Topelius

Meinig (1979) points out that all "mature nations" have their national landscapes, which are significant in their national iconography. This notion should not be understood in an evolutionist sense, but rather as a note of the importance of landscapes

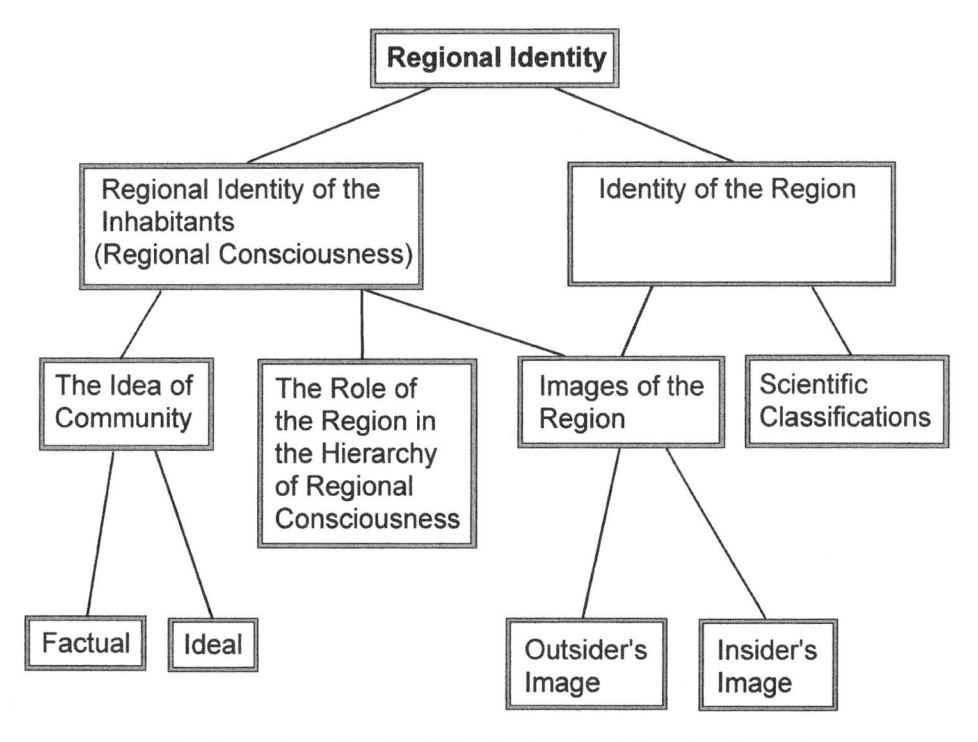

FIGURE 20.2. Key dimensions of regional identity (modified from Paasi 1986a).

as an instrument in nation-building. While the internal organization of landscapes is typically characterized by a specific discontinuity and in most nation-states landscapes are characterized by huge variation, the representations of national landscapes usually serve the purposes of national integration and depict these landscapes as being homogeneous and *typical*. After becoming accepted, they are usually used effectively in art, education, and the media.

In the nineteenth century, soon after Finland became an autonomous Grand Duchy of Russia, a specific literary and artistic landscape discourse was developed to represent a material and aesthetic basis for the emerging national feeling. Since myriad diverging landscapes exist in Finland, as in other countries, a specific landscape type emerged as a "generalization" of this diversity on the national scale. The poet Johan Ludvig Runeberg and the historian Zacharias Topelius occupied a crucial position in the social and cultural construction of these ideal landscapes (Paasi 1996). The ideal national landscape of the emerging Finnish consciousness came to be a view of *Järvi-Suomi*, Finland's Lakeland, located in the central part of the country. Since the nineteenth century, lakes and forests have been depicted as the key elements in the national landscape. Lampén (1923, 134, my translation) describes this landscape as follows, drawing a clear distinction between it and others:

> When we are looking at our land from a hill, we will see in this vast green forest white, glittering stripes, like a silver texture in green and blue velvet. Our countless lakes and waterways draw these stripes in the landscapes. There is hardly any country in the world that would beat Finland in the richness of lakes. Our greatest poet, J. L. Runeberg, has given a well-known favorite name to our country, "the land of a thousand lakes." Neither he nor later poets can cease praising the excellent grace of the lakes of our country.

Lampén then painted a very metaphorical view of the national landscape, dressing it, literally, in human clothes:

> It is true that our lakes have been created almost exclusively with an eye to their decorative qualities. . . . Green forests are the uniform of Finland; our blue lakes are the decorations on its costume, its bright collars, its strings of pearls, its hems, its belts, and the small heath ponds are its shining buttons. This costume distinguishes Finland from all other countries on our globe.

Topelius did not restrict his description merely to the national scale, however. Instead he provided, alongside a detailed description of the landscapes and the character of the inhabitants of each province, a very persuasive framework that combined various spatial scales, territorial distinctions, and features of religious rhetoric and "national character." In his influential *The Book of Our Land (Maamme kirja)* (1981 [1875]), he effectively wove together a discussion of the roles of various spatial scales and religious elements in identity formation. First he reflected the links between the nation and the people living in various Finnish provinces:

> This nation has grown up as one, like many trees form a large forest. Pines, spruces and birches are different timber species, but together they make up a forest. The people in one province differ from those in another as regards their appearance, clothing, character, habits and standard of living. We can easily distinguish a person from Häme from a Karelian, a person from Uusimaa from an Ostrobothnian; we can even distinguish the inhabitants of one parish from those of neighboring parishes.

Then he expanded the spatial scale to look how the members of the nation belong together also in broader territorial contexts:

> But when we travel abroad and meet our fellow countrymen from other parts of our country, we find out that in many respects they are similar. And when a foreigner travels to Finland, he finds that the inhabitants of this country have much in common as regards their character. For those who have lived in the same country since their childhood, subordinated to the same laws and living conditions, must have, with all their differences, much in common.

Finally, Topelius used deep religious rhetoric to depict the "natural" basis of a nation and its "character":

> Such a sense of belonging or quality *(national character)* is much easier to feel than to explain. It is the stamp that God has impressed on all nations by letting their inhabitants live together for a long time. (Topelius 1981, 124, my translation)

The images projected by Runeberg and Topelius were literary, poetic, and pictorial, that is, written landscapes (of national and provincial identity). They created an idea of a Finnish landscape dominated by lakes. The second influential territorial context for national identity was *Karelia,* the area that has been a contested territory between Finland and Russia for hundreds of years. The landscapes of this region have also been represented for more than a hundred years as a crucial part of the national imagery of Finland. At the end of the nineteenth century, the Karelian culture in general and the Karelian landscape in particular were the main themes in the thinking of the National Romantic movement of Finnish artists and intellectuals. This situation changed dramatically after World War II, when Finland had to cede a huge part of its Karelian territory to the Soviet Union. After the war the images of the ceded Karelian areas were removed from the national landscape gallery, and soon the hilly lakeside landscape of

Northern Karelia (particularly the *Koli* area), the part of Karelia still existing on the Finnish side of the border, came to be treated as a new icon of the Karelian landscape. This is a fitting illustration of the moving heritage of national landscapes (Raivo 1996; Paasi and Raivo 1999).

Images of nature and a specific national cultural heritage have hence been important in determining how Finland and its landscapes have been represented. Both elements have been present both as factors explaining the "national character" of the Finns and as major constituents in the representations of national identity (Paasi 1996, 1997). The *content* of these images has been much more dominating than their spatial *form*, and it may be argued that the visual dimensions of the landscape idea, and their inclusive or integrative content, have clearly dominated over the territorial or exclusive dimension that they also contain. This does not detract from the fact that national landscapes have always been political landscapes as well, constructed to create distinctions among territories. Landscapes are often crucial elements for social integration, since their representations may be a significant part of the nation's shared memories, ideas, and feelings. Concomitantly, landscapes may provide a concrete, material basis for the abstract national symbolism represented in the formal national iconography (such as flags) as well as in paintings, poems, and novels (Paasi 1997). By contrast, chosen collective memories and cultural signs from the past may be transformed as part of landscapes by constructing monuments and establishing rituals in specific places. Landscapes may also become a significant part of the national socialization of the citizens of a state. In Finland, for instance, children's geography textbooks have effectively been used to mediate the image of "our" land and its landscapes. These have served as key instruments in the *institutionalization* of images.

Administrative Landscapes of Power as Instruments of Governance

The Finnish provinces *(maakunta)* and counties *(lääni)* have for a long time been significant instruments of governance and an important spatial scale for regional identification along with the local-level administrative units *(kunta)*. Both have had different connotations over time. A county has most often been understood as an "outpost" of the state administration, whereas a province has been interpreted as an area created through the actions of ordinary people. Although the meanings of provinces and counties have been perceived differently, the areas in question have often been largely the same.

We have come a long way from the castle counties of the Middle Ages to the present ways of defining county divisions, and many changes have taken place in the number of regions and in their borders. These developments continued into the twentieth century, with several new counties being formed, and the formation of further counties was still seriously being considered as late as the 1980s. The county reforms of 1997 nevertheless reduced the number from eleven to five (Figure 20.3).

Provinces have originally referred to the historical provinces, units corresponding to the medieval castle counties. The borders of these took shape gradually, as people began to utilize areas and to divide the uninhabited regions into hunting lands. The borders were based on customary law and were not particularly exact, but they did provide a basis for determining the limits of each province. The division of labor, the construction of centers and spheres of influence and also the strengthening of regional civil activity, the establishment of regional associations and the development of the mass media that took place in the nineteenth century created new areas and, concomitantly, a new regional division, the present provinces. Provincial associations were formed as early as the 1920s to advance the interests of these units, and they began actively to create images for the provinces. For decades they acted to develop the educational and financial aspects of life in each region. Nature and landscapes provided

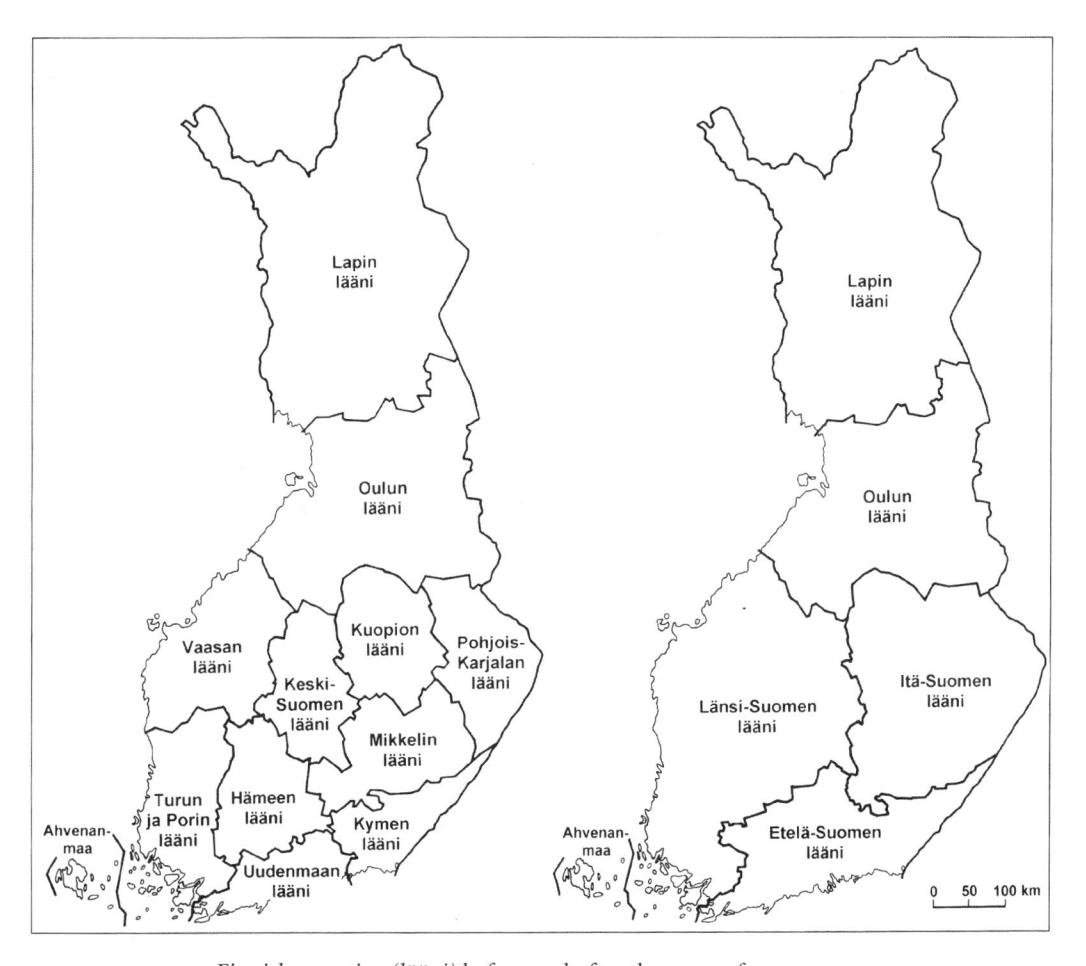

FIGURE 20.3. Finnish counties *(lääni)* before and after the 1997 reform.

a material and symbolic basis for this work, together with cultural features. These were "drawn together" in readers compiled by the provincial associations (Paasi 1986b).

In the case of the administrative landscapes of power—the counties—the spatial form implicit in the idea of landscape has been much more prominent than the visual landscape dimension that is their content. The discourse has been very much characterized by interpretations of landscape as sections of space and continuity, while the discourse on provinces in particular has implied that landscapes are actually "places," centers of human and collective identity and meaning. The provinces *(maakunta)* as cultural units have never become administrative units, but they have often been considered important from the point of view of regional identity. The strength gained from identifying with them varies greatly from one area of Finland to another, however. The counties *(lääni)* were and still are instruments of the state for regional government.

Since the provinces have not had any specific administrative role in the Finnish system of government, people tend to identify themselves with smaller districts than this. There are probably several reasons for the insignificance of the provinces. With only a few exceptions, the current Finnish provinces are not clear cultural units. Their areas have not had any political, financial, or administrative status that represents and enhances their importance in society and in people's everyday lives (Paasi 1986b).

In 1997 the number of administrative counties was reduced from eleven to five, and this may give the provinces a stronger position in the future, at least as far as people's regional identification is concerned. If we begin from the number of provincial associations, there are twenty provinces in Finland today, and their areas differ greatly from the historical division of the country into provinces (Figure 20.4). History books are now being written more and more on the basis of the present provinces, typically the territories covered by the provincial associations. This shows that not only the areas but also the meanings attached to them have changed in the course of time.

Regional identification is also being weakened by the current high regional mobility. Less than half of the Finns live in the local authority district in which they were born, although there are great regional differences in this respect. Before the reform of the counties in 1997, proportions of inhabitants living in the county of their birth varied from 86 percent in Uusimaa to less than 60 percent in Mikkeli and Kuopio.

In spite of these tendencies, the Finnish provinces have their images. Apart from the stereotypic descriptions of the "character" of their inhabitants, many Finnish provinces are identified by specific landscape images, often following Topelius's descriptions of their character. Landscape, in a way, has become one of the constituents of the representations of the provinces. It should also be noted that names have been significant elements as far as the identities of Finnish provinces are concerned. A fitting illustration of this is the heated debate regarding the names of the forthcoming counties in 1997, when the number was reduced. This shows the power and importance of naming in the construction of territorial identities.

Provincial divisions based on landscapes have also been crucial for territorial governance, particularly in the case of conservation. The most typical examples are found in the committee reports on landscapes where the territorialization of the national space is based on heritage and conservation (KM 1980). The continued importance of landscapes in the "official" national imagery also becomes obvious in the efforts to map the national landscapes (see Häyrynen, chapter 19 in this volume).

Landscapes and Provinces as Academic Constructs: The Case of Finnish Geography

The landscape provinces *(maisemamaakunta)* are typical results of the scientific endeavor of academic scholars. Their studies have transformed the landscapes as a result of *professional knowledge*, which has tended to naturalize the cultural and social in a specific territorial framework, that is, they have tended to emphasize both the content

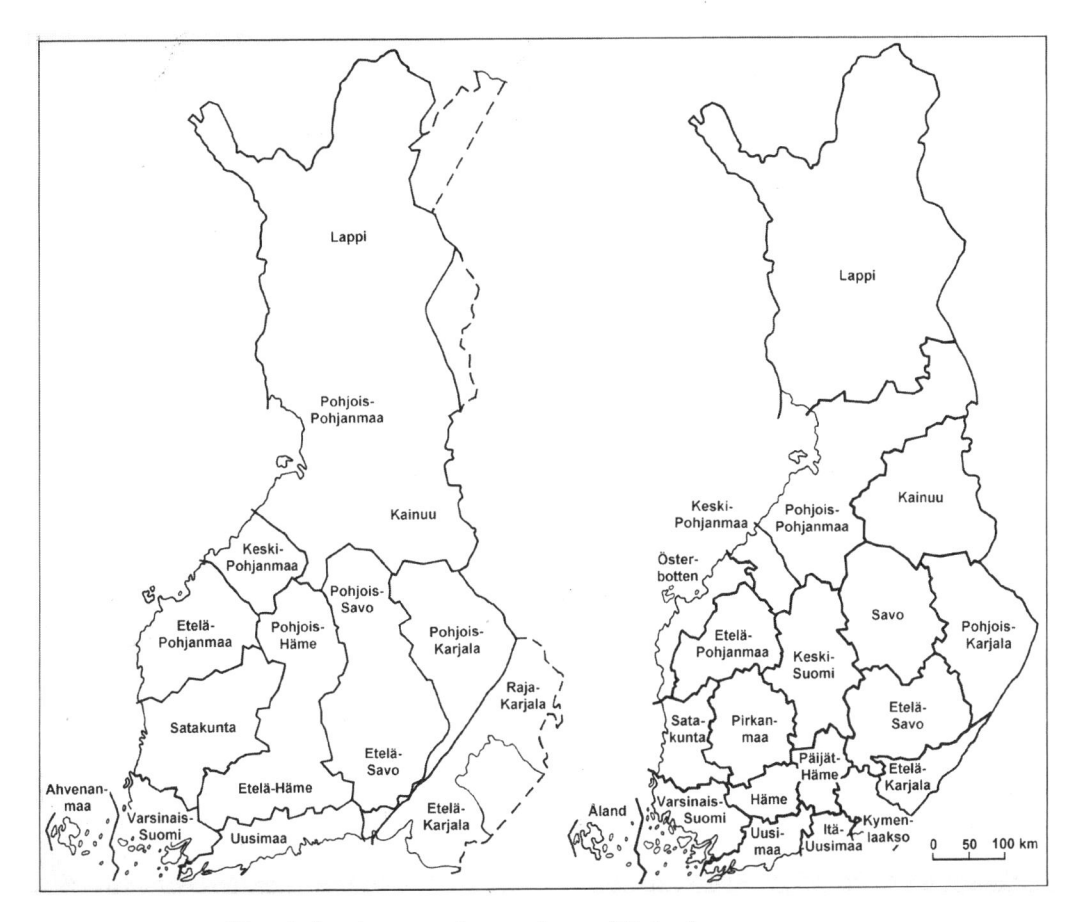

FIGURE 20.4. Historical and present-day provinces of Finland.

and the form of landscape. Like administrative regions, these "academic" landscape units tend to objectify, rationalize, and delimit the continuity of space and transform it into distinguishable, exclusive territorial units. This is certainly not a new feature in geography. In his inaugural lecture, August Tammekann (1953), professor of geography at Helsinki University from 1953 to 1959, discussed how Strabo used historical provinces, tribal areas, or state and other administrative regions when depicting the features of regions. Dickinson (1970, 10) argued that no matter how sophisticated today's concepts and procedures have become, the essential and unique content of the regional concept goes back to the classical ideas of links between "the natural attributes of place" within a framework of relations to other places on the surface of the earth.

Tammekann speculated on how historical provinces and various "tribal areas" have often been relatively homogeneous in terms of their landscapes and separated by frontiers. As we saw, this has been the point of departure in the construction of the Finnish landscapes of administrative power in the form of counties. He came to the conclusion that the expansion of state areas and the various boundary changes have gradually made these regions—which Bernhard Varenius, for instance, regarded as the major objects of special or particular (that is, regional) geography in the mid-seventeenth century (see Dickinson 1970, 11)—less suitable as geographical regional units.

The need to develop new interpretations of the spatiality of the world in new historical situations provides one background for understanding the efforts of many Finnish geographers (and not only Finnish geographers!), who have tried to construct and map regional divisions in the course of the last hundred years or so. It is not a full explanation, however. As Wood (1992) has pointed out, maps work by serving interests—whether politico-administrative, academic, or both. A sociological explanation would perhaps see the delimitation of regions as deriving from the efforts of academic geographers to construct their own approach or method as an instrument of *social distinction* in the field of academic disciplines. Thus regional divisions are also the results of academic competition. This explanation is evident in the debates of two Finnish geographers, J. G. Granö and Iivari Leiviskä, for instance, who were involved in a hot polemic on the justifications for different regional divisions in the Finnish geographical journal *Terra* in the mid-1930s—even though the divisions they suggested did not differ very much from one another! (see Granö 1935).

Thus it is perhaps not a surprise that several Finnish geographers (and other scholars) have constructed their own "objectifications" and classifications of the national space in the form of a specific territorial division. Major instruments in the construction and visualization of these divisions have been maps, which have been crucial in producing the idea of the existence of exclusive territories. Since maps and their explanations were in many cases published in foreign languages as well, images of Finland's divisions into regions have also become familiar to foreigners.

Wood (1992) noted that maps construct the world and not only reproduce it; they naturalize the cultural and culturalize the natural. This becomes obvious in the case of the representations of landscapes and provinces. As Ryden (1993, 21) pointed out, the map in a way "freezes" the landscape in stasis and compresses its ambiguities into an arbitrary and simplified flatness, a surface lacking in depth. He points out that maps have nothing to do with the quality and character of human existence as it is lived and felt on the surfaces that the maps describe. The names that geographers have given to the regions that exist in their divisions have often been "above" the identity matrices of ordinary people. Tammekann (1953) in fact reminded geographers that they should keep themselves at a distance and not carry out surveys by which they might penetrate the daily lives of ordinary people!

The complexities of language have been of great help in mapping. The Finnish word *maisema* refers explicitly to visual landscapes, but Finnish geographers have often linked a spatial dimension with landscape—an idea of the existence of a specific regional framework. The regional concept that has been used is *maakunta*, which means literally a territory occupied by a human community and is conventionally translated into English as "province." Several Finnish geographers have discussed "landscape provinces" *(maisemamaakunta)*, that is, specific regions that are mainly constructed and generalized on the basis of their physical and visual appearance. The first geographer to depict the "natural provinces" of Finland *(Finland's naturliga landskap)* was Ragnar Hult (1895). He constructed a regional division—to be used in elementary education—that aimed at identifying natural regions that would be independent of historical and administrative divisions. Based on various sources and his "personal experience," he identified altogether twenty-four areas (Figure 20.5A).

While several Finnish geographers have produced regional divisions, perhaps the most influential of them was J. G. Granö, who also developed a specific approach to landscape geography *(Pure Geography)*, which was based on the perceived environment of the scholar (Granö 1997 [1929]). He published several studies of landscape provinces and also evaluated the specific features of the Finnish landscape (Granö 1910, 1927, 1930, 1932). The landscape provinces he outlined were prominent in the third edition of the *Atlas of Finland* (1925). His approach was based on physical features and landforms, generalizing from which he distinguished nineteen landscape areas (Figure 20.5B) and named them on the basis of their physical appearance. This harks back to the fact that maps can be inadequate to express the depths of human life, but they can nevertheless inspire imagination and emotion (Ryden 1993), both in the creation of the maps themselves and in the naming procedure that are often a part of mapping.

The regional division put forward by Leo Aario (1949) is well known in Finland and consists of sixteen regions. It closely resembles that of Granö, but Aario did not discuss geographical *provinces* explicitly in the manner of Granö (Figure 20.5C). Also, his aim was to construct a map of Finland's division into "natural regions," based on both natural and human features. Aario's regional division remained a powerful instrument

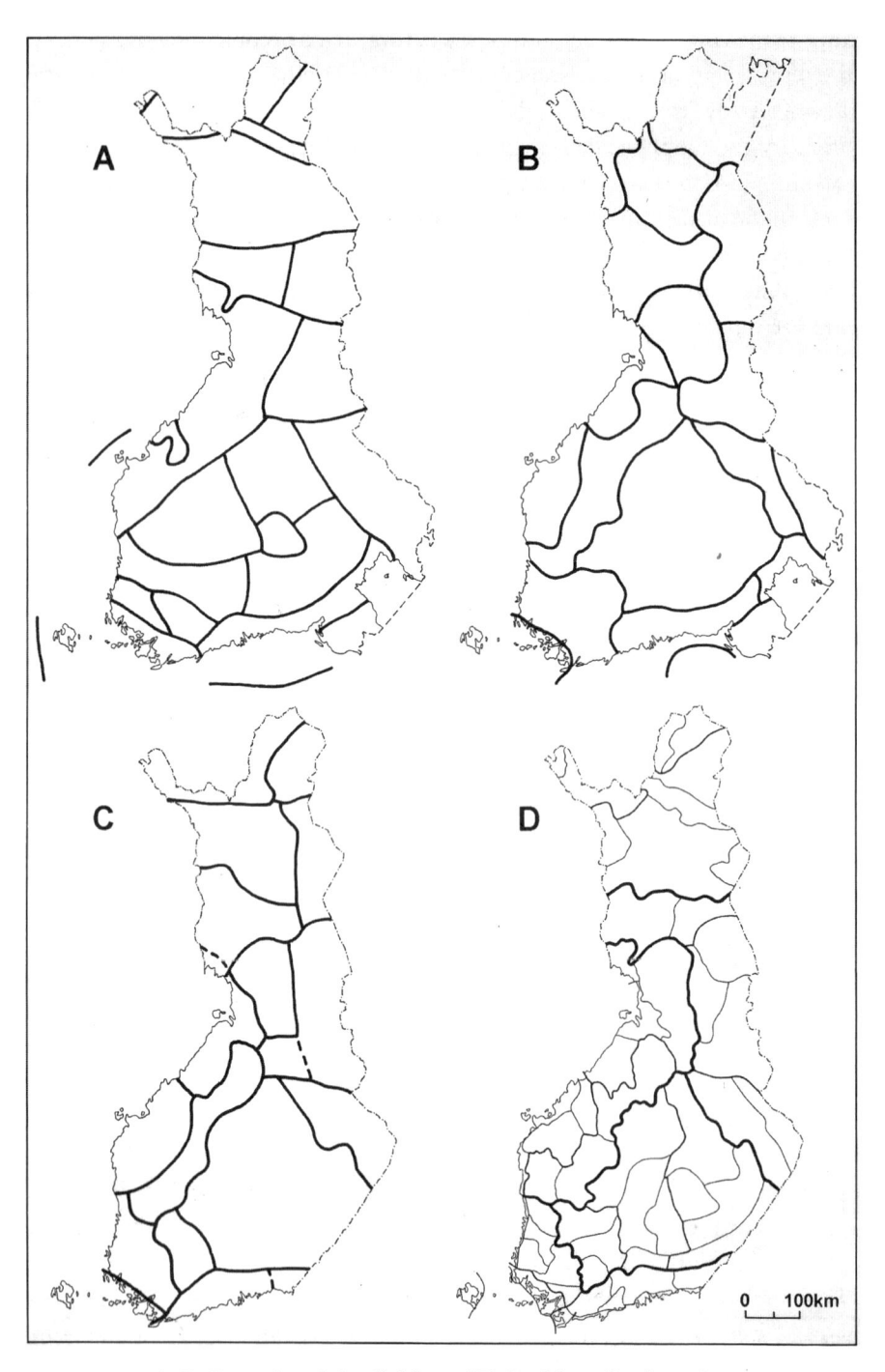

FIGURE 20.5 A–D. Examples of the division of Finland into (landscape) areas proposed by Finnish scientists (A. Ragnar Hult 1895; J. G. Granö 1930; C. Leo Aario 1949; and *Atlas of Finland* 1994).

for a long time in postwar Finland for naming sections of the national space. Following mainly the area names invented by Granö, Aario made effective use of names that also included a specific image of the visual landscapes of the regions in question. In some cases the connotations of the names were negative, as in the case of *Kainuun nevalakeus* (the marsh plains of Kainuu) or *Suomenselän suomaa* (the bog area of Suomenselkä). These connotations emerged from the fact that these areas were economically peripheral regions of out-migration. Some names were connected with culture, particularly agriculture.

The most recent illustration of the marriage of the ideas of landscape and province is the division into fourteen landscape provinces used, among other regional divisions, in the newest edition of the *Atlas of Finland* (1994). The names follow to a great extent the names of the administrative-historical provinces or provinces proper, while names reflecting more explicitly the various forms of physical landscape are employed in the case of subareas, altogether fifty-one regions (Figure 20.5D).

In most cases these "academically constructed" regional divisions—which differ from one another in almost every case—do not correspond to the provinces proper, defined on historical and cultural grounds, in terms of their area. These provinces are typically the areas with which discussions about "regional identity" in Finland have been connected. This idea goes back in history to the time when the "Finnish tribes" occupied certain areas, from which historical provinces began to shape. In old maps, like the *Carta Marina,* Olaus Magnus's map of Norden from 1539, these provinces were identified by their names, such as Finlandia, Tavastia, Botnia, Lapponia, and Carelia but were not represented as clearly bounded units. The first real provincial maps were those of Samuel Gustaf Hermelin published at the end of eighteenth century. These maps remained in use for the next fifty years (*Atlas of Finland* 1984). The first map of Finland and of the Finnish provinces to contain Finnish place-names and texts was published in 1846 by the Student Union *Savo-Karjalainen osakunta* at Helsinki University (Paasi 1986b, 1996). Shortly after that, several maps and school atlases were published in Finnish.

Finnish geographers have also studied functional or economic regions and the identification of the Finns with these from the 1930s onward. The research of this kind carried out since the 1960s points to considerable differences in the intensity of regional identification between the various provinces, so that some (such as Southern Ostrobothnia, Northern Karelia, or Kainuu) are relatively clear entities, whereas others (Lapland in particular) prove to be very diffuse (Paasi 1986b).

Overall, the regional divisions in Figure 20.5A–D indicate that various scholars have come to quite different conclusions when mapping the landscape areas of Finland. These divisions resemble one another in some respects but not in others. Ryden (1993) observes that maps are "segments of the professional autobiographies of their makers." Evaluation of the regional divisions made by geographers suggests that the production of maps and regional divisions may be understood partly on the basis of Bourdieu's (1985 [1979]) theories of distinction and symbolic capital in the scientific field. Hence regional divisions are not merely the results of pure objectifying academic effort at creating order on the earth or constructing identity, but the ensuing maps are also instruments of academic distinction and prestige.

Landscape as Part of a Regional Image: The Case of Four Finnish Provinces

Landscape is one important dimension in both the constitution and representation of regional and provincial identity, whatever elements this "identity" may be said to contain in different contexts. The Finnish provinces are not static, homogeneous entities but are territorial units that exist sometimes in the form of well-known cultural areas with vague boundaries (like southern Ostrobothnia), and at other times merely in the form of relatively vague ideas of an integrated territory that may have clear administrative boundaries (like central Finland).

We have seen that the number of provinces has varied considerably in the course of history, as has their function in Finnish society and later in the nation-state. This shows that regions and their identities are social and cultural products made by human beings and organizations for various purposes. Ideas on the roles of provinces have been highly contested as far as state administration and government are concerned, and the provinces have not been self-evident territorial units in the spatial experiences of their inhabitants or of outsiders. A "province" means different things to different people and can vary from one part of Finland to another. Some people understand the Finnish communes, or local administrative units, as provinces, while some have other regional units in mind. When asked to name their province, people are apt to answer in very different terms. Also, the number of provinces has increased from eight or nine to twenty since the nineteenth century as a part of the perpetual regional transformation (Paasi 1986b; see Figure 20.4).

Vernacular and popular images of landscape emerge from the fusion of daily life, personal experience, education in schools, and media representations. These images tend to emphasize the *content* of images and lean on stereotyping. The spatial form and explicit territorial framework is inevitably secondary. In these images, landscapes and place are fused as an experienced context of day-to-day life rather than abstract territorial space. The images may be characterized by local noncontinuity and exclusion, but simultaneously these may also constitute broader spatial representations and therefore be symbols of national continuity, as in the case of the Finnish Lakeland. The processes and practices of national socialization (media, education) in particular produce collective images and identities of landscapes.

In order to evaluate the meanings of landscapes as constituents of regional images,

I will return to the results of a study carried out in the mid-1980s. A questionnaire was sent to random samples of people living in four Finnish provinces—Uusimaa, Southern Ostrobothnia, Lapland, and Northern Karelia—to map differences in spatial images and identities (Paasi 1986b). I will discuss the results here from the viewpoint of landscapes and nature to show how these elements were represented in the external and internal images of these provinces. People had to consider these images independently since no ready-made alternatives were given to them. Some respondents depicted the regions in terms of one attribute each, while some mentioned several attributes. The respondents living in each province were asked what elements came to mind first when they thought about all the above areas, that is, the others as well as their own.

The answers were arranged first into 18–19 classes and then were further classified and generalized into 6–9 classes. This approach, although technical, gives a good impression of the meanings of landscapes (and nature) in the case of each region, particularly compared with one another. The classes used follow:

Northern Karelia

1. Landscape and nature (aesthetic features and natural elements)
2. Culture (transformation processes in culture, dialects, Karelian culture)
3. Economy of the province
4. Tourism
5. Stereotypes depicting the inhabitants (liveliness, happy disposition, melancholy)
6. Underdeveloped character of the province (unemployment, out-migration)
7. Poor health of the inhabitants
8. Personal contacts with the province
9. Other

Southern Ostrobothnia

1. Nature and landscapes (plains, floods)
2. Culture (music, buildings, sports, bilingualism, local dialect)
3. Economy (small enterprises, economic wealth)
4. Stereotypes depicting the inhabitants (fiery temper, patriotism, enterprising spirit, frigidness)
5. Personal contacts with the province
6. Other

Lapland

1. Nature and landscape (aesthetic role, single elements, especially mountains)
2. A negative vision of nature (snow, cold climate, polar night, barrenness of nature)
3. Lappish symbols (Saami culture, reindeer, Father Christmas)
4. Tourism

5. Regional structure (disintegrated regional pattern, distant location)
6. Economy
7. Stereotypes depicting the inhabitants (perseverance, easily satisfied character)
8. Personal relations with the province
9. Other

Uusimaa

1. Nature and landscapes (sea landscapes, beauty, verdure)
2. A negative vision of nature (pollution)
3. Culture (central position, bilingualism, lack of identity)
4. Urbanization (good services, crime, alienation, shortage of housing, traffic jams)
5. Stereotypes depicting the inhabitants (pride, arrogance)
6. Economy
7. Personal relations with the province
8. Other

Figure 20.6 presents a summary of the internal and external images of each region and indicates how important the roles are that landscape and nature play in the images of each region. Nature and landscapes dominate both the internal and external images of three provinces. As many as 40 percent to 50 percent of all descriptions belonged to this category in the case of Northern Karelia, and even more in Southern Ostrobothnia.

The province of Uusimaa, in which the capital, Helsinki, is located, is an exception, since here the positive and negative elements of urbanization are the most important factors in the images, whereas nature and landscapes seem to be relatively unimportant. What is interesting, however, is the fact that nature is not merely a positive factor in the images of the provinces: it may also carry negative meanings as well, as is the case in Lapland and Uusimaa. In the image of Lapland, as many features connected with nature are negative as positive. People living in the provinces of Northern Karelia and Uusimaa characterize Lapland more often in negative terms than in positive ones. These negative views are, for instance, based on pollution and the extreme climate.

Another significant feature is the fact that the internal and external images seem to be quite similar and the variations between the image profiles of the people living in different areas are relatively small. One explanation for this must be the fact that the role of personal experiences seems to be quite limited in the images, so that the stereotypical expressions and images communicated by the media, education, novels, newspapers, and so forth are very important. This becomes obvious in the case of regional stereotypes. In the same study, it was found that regional newspapers seem to continually recycle the stereotypes that were popularized by Topelius in his "Book of Our Land" (Paasi 1986b), effectively maintaining the criteria for *ideal* regional identities by telling the readers "who they are" and "how they behave" as a regional collectivity.

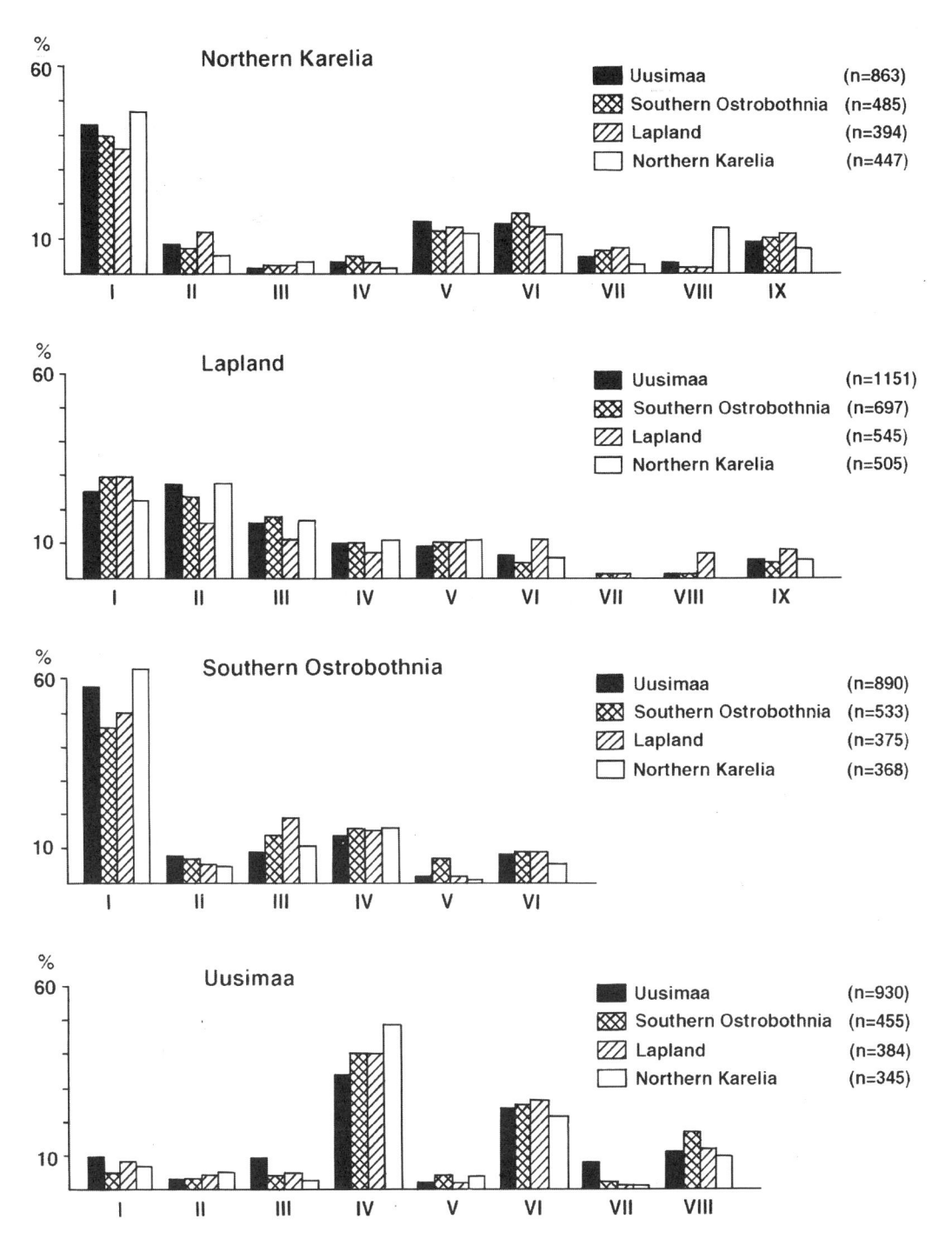

FIGURE 20.6. Internal and external images of the Finnish provinces of Northern Karelia, Lapland, Southern Ostrobothnia, and Uusimaa (modified from Paasi 1986b).

Landscapes of Regionalism

Regionalism is a complicated phenomenon that manifests itself in the fields of art as well as in the activities of political, cultural, and social movements. Regionalism commonly stresses the personality of regions, and the roles of dialects, and typically emphasizes the distinction between one region and all the others. In the field of architecture, the regionalistic approach is typically seen in opposition to international rationalistic trends in that it celebrates and conserves local and regional traditional values. In politics and history, regionalistic thinking usually stresses the need for internal development (or autonomy) in a region, which may imply ideas of separatism. This is often linked with peripheral areas—in spite of the fact that peripheries are typically created by centers! National differences in the forms of regionalism may also be considerable. Whereas in some areas it is mainly a cultural movement that aims at promoting the culture or traditions in a region, in other contexts it may turn into a violent political strategy.

Regionalism is expressed in many fields. Within the arts, it has been particularly strong in literature, where the aims of authors to build up a bond between the text and a particular territorial unit are often very apparent. The link between the political and literary dimensions can also be very close. It has been argued that in many European countries novelists have effectively created regional images and identities and have therefore played a significant political role (Gilbert and Litt 1960).

In Finland, regionalism has been mainly linked with literature, but it has been a weak phenomenon politically and no real separatist movements have existed. Regionalism emerged in Finnish literature during the nineteenth century in the form of realistic stories of folk life. In the 1970s, a renaissance in regionalism took place. The social context for this was the rapid change that occurred in socioeconomic structures. A more specific background was perhaps the declining role of the traditional agrarian culture—one of the key constituents of Finnish national identity—as well as rapid urbanization and increased spatial and social mobility. Perhaps the best known of the novelists who began to write in a regionalist tone is Heikki Turunen, with several books in the 1970s that often emphasized the differences between the largely rural province of Northern Karelia and the more urbanized areas of southern Finland, particularly around the capital, Helsinki. Some of his books also became famous in film versions (Karjalainen and Paasi 1994). Turunen describes particularly the landscapes, villages, and people of Northern Karelia, and as a regionalist he contrasts the modest local life in this province with that of Helsinki and its region, often painting gloomy views of urban alienation. As a regionalist, he builds strongly on rather "black and white" dichotomies such as rural/urban, genuine/artificial, traditional/modern.

Landscapes are one significant element in regionalist writings. This is based on the fact that most provinces have their unique landscape features, which novelists can transform into meaningful narratives and contexts where the imagined and real worlds of their texts come alive. Most Finnish provinces have in the course of the years had their own novelists, who are typically linked with regional life in all forms of cultural publicity. Thus Antti Tuuri is closely linked with Pohjanmaa (Ostrobothnia), Heikki Turunen with Northern Karelia, Eino Säisä with Savo, Kalle Päätalo with Koillismaa, and Veikko Huovinen with Kainuu.

It seems that Finnish regionalism and the provinces are perhaps going through a renaissance. This is because Finland joined the European Union (EU) at the beginning of 1995 and questions concerning the images of regions and provinces are currently topical in Europe. The position of the provinces in the intermediate stratum of government also became stronger in the 1990s, partly through the uniting of the provincial associations with the regional planning associations to form a network of regional councils representing the provinces. One of the most important tasks of these councils was preparatory and program work connected with EU regional policies and control.

Dialects also seem to be experiencing a renaissance in Finland. A number of associations supporting dialects have been established, and various provincial dialect readers have been published. The provinces are developing new images and presenting them more actively to the outside world. Together with various institutions and organizations, the provinces are also becoming linked with numerous symbols through which they are able to display their distinctive features. In addition to traditional symbols such as coats of arms, each province has been assigned a typical bird, flower, fish, and rock.

Landscapes of Publicity

Even if Finnish regionalism has been strongest in literature, it has not been merely an artistic movement. As far as the major constituents of territorial identity are concerned, the role of the media and—in the case of the provinces—that of newspapers in particular is crucial for exploiting regionalist ideas in Finland. The provincial newspapers have in many cases been active promoters of regional identity and "self-consciousness" in the peripheral areas. Particularly significant forms of regionalism seem to have emerged on the occasions of structural changes in society such as have occurred in the 1960s and 1970s and are currently taking place again.

Content analyses of Finnish newspapers suggest that it is possible to discuss specific territorialized *landscapes of publicity*. A large proportion of the news and articles in newspapers are based on a regional framework at different spatial scales. It is hard to assign any explicit causal role to newspapers in the construction of identities, but rather, following the *agenda-setting* approach, we can think of newspapers as disseminating a collection of regionalized information that provides the readers with numerous choices (Paasi 1986b). On the other hand, the newspaper discourses themselves can easily promote regional feelings and ways of thought. This is particularly obvious in articles that stress the collective characteristics of the region, its landscapes or its people in comparison with other regions and their inhabitants. Articles may also personalize regions and present them as collective *actors* engaged in competition with other regions. It is even possible to discuss the *language of anthropomorphism*. These discourses provide a strong basis for a feeling of "us" (Paasi 1996). The importance of provincial newspapers as creators of time-space-specific regionalistic feelings is especially pronounced since they have particularly wide circulations within their own provinces, to the extent that an 80 percent coverage of homes in the region is not a rarity. Newspapers often provide a specific written identity that seems to be relatively stable and may be linked with both the inhabitants and the region and its landscapes.

A comparative study of the emergence and identity of four Finnish provinces has demonstrated that nature and landscapes are significant elements in the regionalistic writing style that the regional press follows. These elements are part of a broader regional way of thinking and are often used as indicators of the regional or local uniqueness represented in the promotion and marketing of the area for the purposes of tourism, for instance (Paasi 1986b).

Landscapes may also have a more specific, instrumental role in the regional and national representations maintained by the media. The media may exploit diverging landscapes in advertising at the local, regional, and national levels—and use them for the promotion of a national identity, for example. The landscapes concerned may vary from national to provincial ones and the motives behind the advertisements may vary from promoting tourism to selling whatever products can conveniently be linked with the ideas of landscapes or nature. For Finland, it is often the case that food and agricultural products, for instance, are advertised in a framework provided by stereotypical landscapes, such as open rural spaces with green meadows, blue lakes and green forests, all of which aim at creating visions of a good-quality environment in which the products have been grown and processed (see Kuusinen 1998). This means that concrete landscapes and mindscapes with certain national "structures of expectations" effectively coincide in the sphere of advertising.

Conclusions

This chapter has had two main aims. The first was to conceptualize the idea of landscape as a social practice, that is, to map the major dimensions and meanings of landscapes in terms of social life. This means that landscapes are understood as being results of the action of social processes such as vernacular life (and meanings emerging from this life), the media, administration, and government and academic scholarship. The second aim was to evaluate the social construction of Finnish landscapes. Landscapes include two major dimensions that are particularly significant in the Finnish case: the territorial dimension and the visual or experienced dimension. These dimensions are present in different landscape discourses, but they vary in importance, the functional territorial dimension being more important in administration and government, whereas the idea of producing formal, homogeneous regions has been the major source of inspiration in geographers' attempts to construct landscape provinces. The aesthetic and experiential dimension has been strongest in the vernacular interpretations of landscapes and in regionalistic literature, and projects aimed at constructing specific national landscape images have also been effective in exploiting these dimensions.

This study makes clear that the ideas of landscape are present in numerous ways in social practices, which partly produce these ideas and partly are reproduced by them. The most typical examples are the ideas of national landscapes, which are common in discourses on national identity. In a way, social life and landscapes are constitutive of each other.

The major conclusion of this chapter is that landscape and identity are contextual and contested categories—they can be used in different ways in different contexts. This has been illustrated by analyzing different conceptualizations of landscape in Finland and how these are linked with different social practices and discourses. In the light of this chapter, it is at least reasonable to ask, "Whose landscapes and identities" are in question and "for what purpose" are they of concern?

Acknowledgment

The author is grateful to the Academy of Finland for the grant (No. 121992) that has been significant in finishing this study.

References

Aario, Leo. 1949. *Suomen maantiede*. Helsinki: Otava.
Anderson, Benedict. 1991. *Imagined Community*. London: Verso.
Atlas of Finland. 1925. Helsinki.
Atlas of Finland. 1984. "Mapping of Finland." Helsinki.
Atlas of Finland. 1994. "Finland's Landscapes and Urban and Rural Milieus." Helsinki: National Land Survey of Finland and Geographical Society of Finland.
Bourdieu, Pierre. 1985 (1979). *Sosiologian kysymyksiä* (originally published as *Questions de Sosiologie*). Tampere: Vastapaino.
———. 1998 (1994). *Järjen käytännöllisyys* (originally published as *Raisons practiques: Sur la théorie de l'action*). Tampere: Vastapaino.
Cosgrove, Denis. 1989. "Power and Place in the Venetian Territories." In *The Power of Place: Bringing Together Geographical and Sociological Imaginations,* ed. John Agnew and James S. Duncan, 105–23. Boston: Unwin Hyman.
Dickinson, Robert E. 1970. *Regional Ecology: The Study of Man's Environment*. New York: John Wiley and Sons.

Fitzgerald, T. K. 1993. *Metaphors of Identity: A Culture-Communication Dialogue.* Albany: State University of New York Press.

Gilbert, E. W., and B. Litt. 1960. "Geography and Regionalism." In *Geography in the Twentieth Century,* ed. G. Taylor, 345–71. London: Methuen.

Granö, J. G. 1910. "Maantieteellisen maakuntajaon perusteista." *Terra* 22: 309–14.

———. 1927. "Suomalainen maisema." (Referat: "Die Finnische Landscaft.") *Publicationes instituti geographici universitatis Aboensis,* no. 2. Turku.

———. 1930. "Maisematieteellinen aluejako." *Publicationes instituti geographici Universitatis Aboensis,* no. 4. Turku

———. 1932. *Suomen maantieteelliset alueet.* Porvoo: WSOY.

———. 1935. Maantieteellinen aluejako vielä kerran. *Terra* 47: 66–74.

———. 1997 (1929). *Pure Geography,* ed. Olavi Granö and Anssi Paasi. Baltimore: Johns Hopkins University Press.

Hägerstrand, Torsten. 1995. "Landscape as Overlapping Neighbourhoods." In *Geography, History, and Social Sciences,* ed. G. B. Benko and U. Strohmayer, 83–96. Dordrecht: Kluwer Publishers.

Harley, J. B. 1988. "Maps, Knowledge, and Power." In *The Iconography of Landscape,* ed. Denis Cosgrove and Stephen Daniels, 277–312. Cambridge: Cambridge University Press.

Hirsch, E. 1995. "Landscape: Between Place and Space." In *The Anthropology of Landscape: Perspectives on Place and Space,* ed. E. Hirsch and M. O'Hanlon, 1–30. London: Clarendon Press.

Hufford, Mary. 1987. "Telling the Landscape: Folklife Expressions and Sense of Place." In *Pinelands Folklife,* ed. Rita Zorn Moonsammy, David Steven Cohen, and Lorraine E. Williams. New Brunswick, N.J.: Rutgers University Press.

Hult, Ragnar. 1895. "Finlands naturliga landskap." *Geografiska Föreningens Tidskrift* 7, nos. 1 and 2: 64–94, 126–54.

Jackson, J. B. 1964. "The Meanings of 'Landscape.'" *Kulturgeografi* 88: 47–52.

Karjalainen, P. T. 1986. "Geodiversity as a Lived World: On the Geography of Existence." *University of Joensuu Publications in Social Science,* no. 9. Joensuu.

———, and A. Paasi. 1994. "Contrasting the Nature of the Written City: Helsinki in Regionalistic Thought and as a Dwelling Place." In *Writing the City,* ed. P. Preston and P. Simpson-Housley, 58–79. London: Routledge.

Keisteri, Tarja. 1990. "The Study of Changes in Cultural Landscapes." *Fennia* 168: 31–115.

KM. 1980. "Maisematoimikunnan mietintö." *Komiteanmietintö* 44/1980. Helsinki.

Kuusinen, Kaisu. 1998. "Hyvää Suomesta: Kulttuurimaantieteellinen tutkimus mainosten kansallisista mielikuvista." M.S. thesis, Department of Geography, Oulu University.

Lampén, Ernst. 1923. "Suomen maisemien kauneus." In *Suomi: Maa, kansa, valtakunta,* ed. A. Donner et al., 131–39. Helsinki: Otava.

Meinig, Donald W. 1979. "Symbolic Landscapes: Some Idealizations of American Communities." In *The Interpretation of Ordinary Landscapes,* ed. Donald W. Meinig, 164–92. Oxford: Oxford University Press.

Mitchell, W. J. T. 1994. *Landscape and Power.* Chicago: University of Chicago Press.

Newman, David, and Anssi Paasi. 1998. "Fences and Neighbours in the Post-Modern World: Boundary Narratives in Political Geography." *Progress in Human Geography* 22: 186–207.

Olwig, Kenneth. 1996. "Recovering the Substantive Nature of Landscape." *Annals of the Association of American Geographers* 86: 630–53.

Paasi, Anssi. 1986a. "Institutionalization of Regions: A Theoretical Framework for Understanding the Emergence of Regions and the Constitution of Regional Identity." *Fennia* 164: 105–46.

———. 1986b. "Neljä maakuntaa: Maantieteellinen tutkimus aluetietoisuuden kehittymisestä." *Publications in Social Sciences,* University of Joensuu, no. 9. Joensuu.

———. 1991. "Deconstructing Regions: Notes on the Scales of Spatial Life." *Environment and Planning A* 23: 239–56.

———. 1996. *Territories, Boundaries, and Consciousness: The Changing Geographies of the Finnish-Russian Border.* Chichester: John Wiley and Sons.

———. 1997. "Geographical Perspectives on Finnish National Identity." *GeoJournal* 43: 41–50.

———. 1999. "Rationalizing Everyday Life: Individual and Collective Identities as Practice and Discourse." *Geography Research Forum* 19: 4–21.

———, and Petri J. Raivo. 1999. "Boundaries as Barriers and Promoters: Constructing the Tourist Landscapes of Finnish Karelia." *Visions in Leisure and Business* 17: 30–45.

Palmer, C. 1998. "From Theory to Practice: Experiencing the Nation in Everyday Life." *Journal of Material Culture* 3: 175–200.

Raivo, Petri. 1996. "Maiseman kulttuurinen transformaatio: Ortodoksinen kirkko suomalaisessa kulttuurimaisemassa." *Nordia Geographical Publications* 25, no. 1. Oulu.

———. 1997. "The Limits of Tolerance: The Orthodox Milieu as an Element in the Finnish Landscape, 1917–1939." *Journal of Historical Geography* 23: 327–39.

Rose, G. 1993. *Feminism and Geography: The Limits of Geographical Knowledge.* Cambridge: Polity Press.

Ryden, K. C. 1993. *Mapping the Invisible Landscape.* Iowa City: University of Iowa Press.

Sauer, C. O. 1963 (1925). "The Morphology of Landscape." In *Land and Life: A Selection from the Writings of Carl Ortwin Sauer,* ed. J. Leighly, 315–50. Berkeley and Los Angeles: University of California Press.

Shortridge, J. 1984. "The Emergence of 'Middle West' as an American Regional Label." *Annals of the Association of American Geographers* 74: 209–20.

Tammekann, A. 1953. "Maantieteellinen aluejako." *Terra* 65: 117–25.

Topelius, Z. 1981 (1875). *Maamme kirja.* Porvoo: WSOY.

Williams, R. 1973. *The Country and the City.* London: Chatto and Windus.

Wood, D. 1992. *The Power of Maps.* London: Routledge.

Zukin, S. 1991. *Landscapes of Power: From Detroit to Disneyland.* Berkeley and Los Angeles: University of California Press.

Norden

21.
The Nordic Countries: A Geographical Overview

Michael Jones *and* Jens Christian Hansen

The five northern European countries of Denmark, Finland, Iceland, Norway, and Sweden denote themselves collectively as Norden—meaning the North. Norden includes the autonomous island territories of Greenland (Kalaallit Nunaat) and the Faeroe Islands, belonging to Denmark, and Åland (Ahvenanmaa) in Finland (Figure 21.1). Internationally the Nordic countries are often referred to as Scandinavia, although the latter term properly encompasses only Sweden, Norway, and Denmark. The five Nordic countries share a cultural heritage and have many common interests in international and economic affairs. The Scandinavian languages, Danish, Norwegian, and Swedish, are mutually comprehensible. Icelandic and Faeroese are more distantly related. Finnish belongs to the distinct Finno-Ugrian language group, as does Saami (Lappish), spoken by an indigenous minority in the central and northern parts of Norway and Sweden and in northern Finland. The Swedish language was historically important in Finland's public life and Finland still has a sizable Swedish-speaking minority. The indigenous Inuit language is spoken alongside Danish in Greenland.

Much of Norden rests on the ancient crystalline rocks of the Fennoscandian Shield. Greenland lies on a similar ancient shield. In some areas, in particular Denmark, there are younger sediments. Iceland and the Faeroes consist largely of basalts of volcanic origin. Active volcanoes remain a feature of Iceland. Most of Norden has been glaciated and acidic, morainic soils dominate. Glaciers are found in Norway and Iceland and ice covers the greater part of Greenland.

Climatically, there are marked contrasts within Norden. The climate is maritime in the west and continental in the east. The snow cover lasts on average only a few weeks in the south but up to seven months in the north. The mean midwinter temperature differs by almost 15°C between northern Finland and Denmark. Annual precipitation is 2,000–3,000 mm in the coastal mountains of West Norway and 500–700 mm east

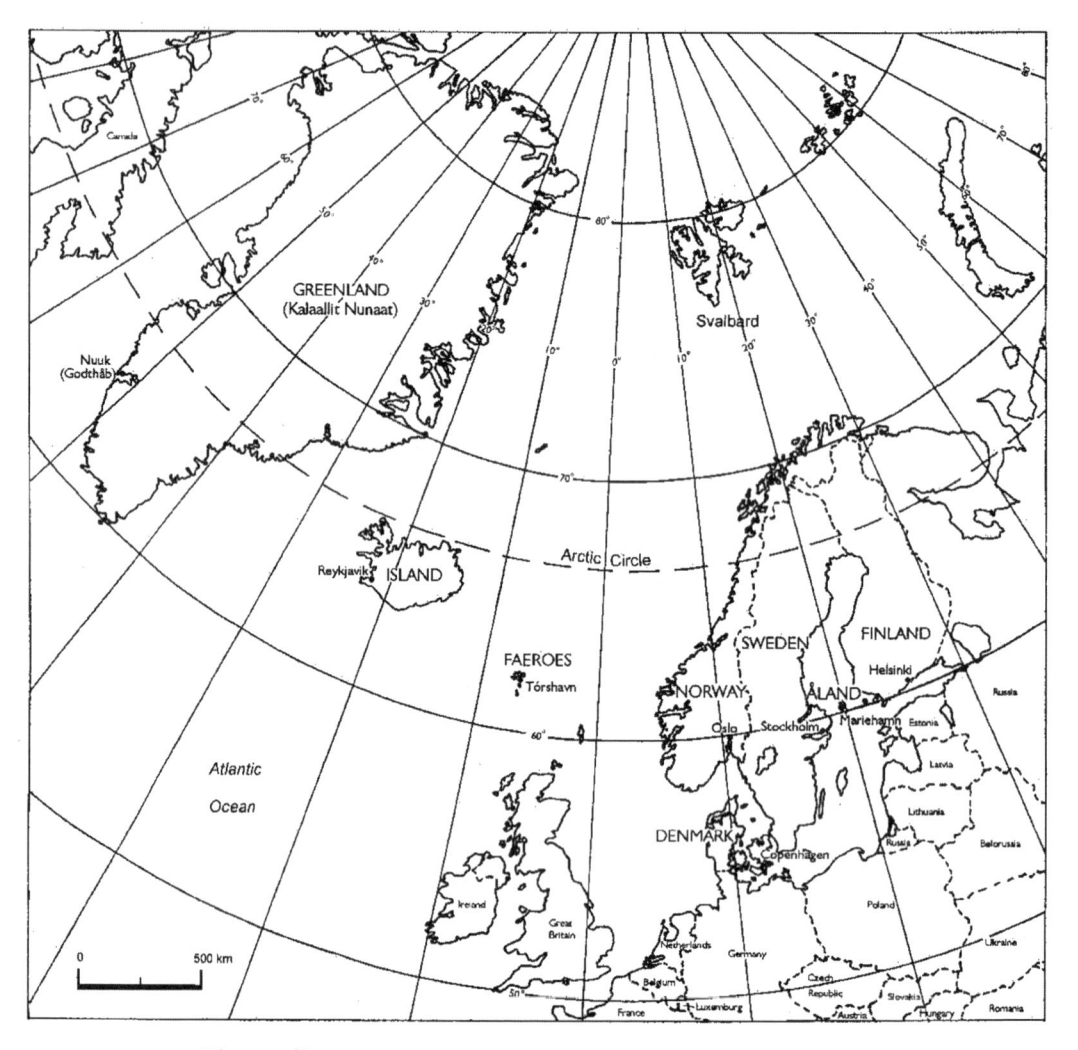

FIGURE 21.1. The Nordic countries.

of the mountains. Despite these variations, Norden has a harsh winter climate—although the influence of the Gulf Stream makes it milder than other comparable latitudes—and a rugged topography.

Rural settlements were historically isolated from one another by lakes, bogs, forests, mountains, and fjords; towns were small and far apart and, except in Denmark, national peripheries were distant from the capital cities. Time distances have been reduced through the development of comprehensive transport systems, but distance still counts in the spatial organization of everyday life as well as in regional development. Center-periphery tensions still exist.

Although plans for military cooperation and a customs union were abandoned after

World War II, the Nordic Council was set up in 1952 as a consultative forum to formalize cooperation in a wide range of other fields. A Nordic passport union dates from 1952 and a Nordic labor market was set up in 1954.

Denmark was the first Nordic country to join the European Union (EU), in 1973, followed by Sweden and Finland in 1995. After membership was rejected in referendums in 1972 and again in 1994, Norway has remained outside the EU, as has Iceland, although both countries cooperate with the EU through membership in the European Economic Area (EEA). Greenland and the Faeroes, internally autonomous within the Danish kingdom, have also opted to remain outside the EU.

Denmark

Denmark proper (54° 33'–57° 45' N, 8° 04'–15° 12' E) consists of the Jutland peninsula and more than four hundred islands, eighty of them inhabited. The coastline is 7,300 kilometers long, while the land frontier with Germany is only 68 kilometers. Early in its history, Denmark became a center of trade between the Baltic and the North Sea, as well as between the Nordic countries and central Europe.

Denmark has been a kingdom for a thousand years, although with changing frontiers and fortunes. Between 1150 and 1250, Denmark extended its political control over much of the Baltic. Through dynastic intermarriage, Denmark–Norway was united in 1380; by 1397 it had joined Sweden to form the Kalmar Union. As Finland belonged to Sweden and Iceland to Norway, Norden thus became a single political unity, with Copenhagen as its political and economic center. A toll collected from all foreign ships passing through the Sound (Øresund), linking the Baltic and the North Sea, provided an important source of income. With the end of the Kalmar Union in 1523, Sweden again became an independent kingdom. Competition with Denmark and Norway often led to wars. Denmark's economic and political power declined, partly as the result of unfortunate European political alliances. In 1645 and 1658, Denmark–Norway was forced to cede provinces in southern and central Scandinavia to Sweden.

A policy of mercantilism lasted until the end of the eighteenth century, when Denmark–Norway became a colonial power with trading stations in the West Indies, West Africa, and India. Export industries were encouraged and fish, forest products, metals, and minerals were beginning to be exported. At the same time, an active colonization began in Greenland.

Denmark's sympathies with Napoleon led in 1807 to the bombardment of Copenhagen and the capture of the Danish fleet by the British fleet. After the defeat of Napoleon, the Danish king had to cede Norway to the king of Sweden by the Treaty of Kiel in 1814. Denmark was reduced to a small state, and it became even smaller when, in 1864, it lost Schleswig-Holstein (Slesvig-Holsten) as a result of war with Prussia and Austria.

Denmark is physically an extension of the northern German lowlands. Quaternary deposits cover sedimentary bedrock. The climate is milder than elsewhere in Norden and the growing season is longer. In contrast to the other Nordic countries, the greater part of the land can be cultivated. In 2007, 66 percent of an area of 43,500 km² was under cultivation, 23 percent was covered by forest, meadow, pasture, heath, marshland, bogs, sand dunes, and lakes, while built-up and traffic areas made up the remaining 13 percent. With 5.4 million inhabitants, Denmark's population density is 126 per km², compared with between 3 and 22 per km² in the other Nordic countries.

Agriculture and fishing have been the main sources of income through most of Denmark's history. Around 1700, 75 percent of a total population of 700,000 lived on farms. Copenhagen had 30,000 inhabitants. Grain and cattle were exported, with the Netherlands as the most important destination. By 1814, the population had increased to one million. Denmark was still very much a farming country, but it was also engaged in international shipping and trade. Danish ships were in great demand, and overseas goods were channeled to European markets through Copenhagen, by then a city of 100,000 inhabitants.

New manufacturing industries appeared in the 1840s. The construction of railroads and a telegraph system began in the 1850s and 1860s. While provincial towns expanded as skilled trades, services, and home-market industries grew, Copenhagen's dominance increased, and by 1900 the city housed one-sixth of Denmark's population.

Farming underwent great changes during the second half of the nineteenth century. More land was cultivated and production intensified. Livestock farming increased in importance, accelerated by falling grain prices. Denmark became an importer of grain and exported livestock products, in particular to Britain. Agricultural products accounted for 85 percent to 90 percent of the country's exports. The estates of large landowners were replaced by family farms and agricultural cooperatives became important.

Despite economic growth, Denmark could not provide work for all of its rapidly increasing population. Emigration amounted to almost three hundred thousand persons between 1870 and 1914, mainly to the United States.

Danish society became transformed during the second half of the nineteenth century. Agricultural industries developed, manufacturing food and beverages as well as farm machinery. Shipbuilding expanded and diesel engines became an important innovation. Limestone was used for cement production. Economic growth was accompanied by social reforms, such as a more equitable tax system, a better secondary-school system, and labor-market regulation.

After Germany's defeat in 1918, the Treaty of Versailles stipulated that the future of Schleswig should be decided through plebiscites. The outcome was that North Schleswig was returned to Denmark in 1920 as Sønderjylland. In 1918, Iceland became an independent nation, but until 1944 it accepted the Danish king as king of Iceland.

World War I was followed by a period of economic difficulties, with the collapse of banks, serious labor disputes, and high unemployment. However, agriculture and manufacturing were modernized. Ford set up Europe's first car-assembly plant in Denmark. The Great Depression hit Denmark in 1930. The export-dependent agriculture suffered from shrinking markets and falling prices and many farmers went bankrupt. The agricultural crisis had negative effects on the whole economy, but state intervention and the immediate prewar boom in Europe led to a gradual recovery.

Germany occupied Denmark in 1940. Britain seized two-thirds of Denmark's merchant navy and occupied the Faeroe Islands. In 1941 the United States set up military bases in Greenland. These outlying parts of Denmark became important Allied strongholds in the battle of the Atlantic. A Danish resistance movement gained force after 1942. German troops in Denmark surrendered to the British in 1945.

Denmark is one of the world's most prosperous countries. In 2006, 57 percent of its foreign trade was with other EU countries. Around 1950, agricultural products accounted for 70 percent of all exports; in 2006, the figure was 9 percent, while manufacturing accounted for 70 percent.

Danish agriculture employs 3 percent of the workforce, but it can feed 15 million people, three times the national population. Farm numbers have declined from 200,000 in 1950 to 47,000 in 2006. One-third of the farms cultivate more than fifty hectares of land. These farms control 80 percent of the cultivated land. More than 17,000 offer full-time family employment and farming has become highly specialized. Over 90 percent of plant production is used to feed animals. EU milk quotas have contributed to the number of dairy cows declining by a half, from one million in 1980 to 550,000 in 2000. The production of pork increased by 40 percent in the same period. The number of pigs increased from 10 million in 1980 to more than 13 million in 2000. Two-thirds of agricultural production is exported. In addition to 90,000 people working on farms and in fishing, 130,000 people are employed in industries, supply, transport, and other services related to agriculture. Between 1991 and 2001, the proportion of farms practicing organic farming increased from 1 percent to 6 percent; they occupy 5 percent of the agricultural area.

Denmark is among the world's top fifteen fishing nations, and about 75 percent of the catch is taken in the North Sea. The fish-processing industry takes some of its fish from the Faeroes, Greenland, and even Norway.

Apart from food-processing, Danish manufacturing cannot rely on the country's natural resources but is dependent on processing imported raw materials and semi-manufactured products. While most firms are small, there are examples of successful large firms. The Carlsberg Group, dating from 1847, now brews beer in 48 countries and employs more than 30,000 people. LEGO toys are sold in more than 130 countries, and LEGOLAND is one of Denmark's main tourist attractions. Novo, a pharmaceutical firm, exports 98 percent of its output and employs 19,000 people in 69 countries.

Denmark's largest enterprise is the A. P. Møller–Mærsk Group, a conglomerate with activities in offshore oil and gas and shipping, employing more than 110,000 people in 130 countries. To survive, both small and large firms have to compete abroad because the home market is so small.

Initially, manufacturing industries were located mainly in the capital region, but from the 1960s there has been strong industrial growth in small provincial towns and the countryside. Jutland has profited from this decentralization and new furniture, textiles, and clothing industries have emerged. As in most European countries, these industries have shed labor, but firms survive by designing products for niche export markets. Secondary industries—manufacturing, building, and construction—employ one-fifth of the workforce; more than three-quarters are employed in services. The public sector increased from 25 percent of the workforce in 1970 to 34 percent in 2005.

The main objective of postwar economic policy has been to provide work for all. Economic growth has increased buying power and generated tax income to pay for the modern welfare state. Until the first oil crisis in 1973, Denmark, like the rest of Europe, experienced a period of growth that generated a demand for labor. Whereas much of the growth in the 1960s took place in manufacturing industries, most of the employment increase during the 1970s and 1980s was in services. Between 1940 and 2007, the population increased by 1.6 million, while the workforce increased by 1.1 million; 700,000 were women, most working in services. Today, almost three-quarters of women between the ages of sixteen and sixty-six are working, one of the highest participation rates in the world.

Due to North Sea oil and gas, Denmark has been self-sufficient in energy since 1997. Rising oil prices that resulted from the Gulf crisis in 1991 led to an improved balance of trade and reduced foreign debt. The export of energy and energy products is twice that of pork, traditionally a major Danish export product.

Compared to the other Nordic countries, rural settlement in Denmark is relatively dense. Land reforms at the end of the eighteenth century moved farms out of villages. Some villages developed into market towns. Links between different parts of the country were by boat, and even today more than seventy ferry lines are in operation. Sea transport was gradually replaced by a railway system that serves almost all towns with more than ten thousand inhabitants. The nonagrarian population clustered around ports and railway stations, where manufacturing industries were located.

Since the 1960s, much of the urban growth has taken place in suburbs and in small housing estates near former villages and market towns. With 1.65 kilometers of public roadways per kilometer, Denmark's road network is among the densest in the world. Denmark has a well-developed collective transport system and an extensive system of bicycle paths.

While some provincial towns have declined because of job losses in manufacturing and farming, others have grown as new industries have become established. In the

1950s, over half of jobs in manufacturing were in the Copenhagen region; in the 1990s, the figure was only one-quarter. Regional policy measures introduced in the 1960s encouraged the establishment of enterprises in areas that needed new jobs, such as Jutland. These tend to employ many unskilled workers. High-tech, research-based industries employing highly paid employees have remained in the capital region. There is thus a certain regional division of labor. Employment in public services doubled between 1970 and 1985. Welfare services were decentralized, which led to growth in the local administrative centers.

In the 1990s, net immigration contributed more to Danish population growth than natural increase. Increasingly, immigrants came from distant countries with different cultural backgrounds. In 1984, foreign citizens comprised 2 percent of the population; in 2007, it was 5 percent. More than a half of the foreign nationals live in the Copenhagen region, which explains some of the recent growth of this area. Greater Copenhagen now has 1.75 million inhabitants, or one-third of the population of Denmark. A challenge is to give immigrants access to the labor market. Since the total number of jobs in the Danish economy remains almost constant, some people feel that foreigners are taking jobs from Danes, and political trends have emerged calling for a more restrictive immigration policy.

Large bridge and tunnel projects have consolidated Copenhagen's position as a transport hub. In the 1930s, the islands of Falster and Lolland were linked to Zealand (Sjælland) by bridge, while Funen (Fyn), Denmark's second largest island, was linked to Jutland. A rail link between Funen and Zealand was opened in 1997 and a road link in 1998. Transport between the capital, the provinces, and the Continent can now take place without ferries. In 2000, a 16-kilometer road and rail bridge across the Sound (Øresund) was opened, linking more closely the Malmö-Lund urban region of southern Sweden with the Copenhagen region. Almost 70 percent of air transport in Denmark goes through the Copenhagen airport, which is also the most important Nordic airport, a transit hub for Swedish and Norwegian passengers going overseas and to major European destinations.

Energy consumption increased tenfold between the 1930s and the 1980s. Gas and oil have been substituted for coal, but Denmark does not generate nuclear power. A growing awareness of environmental problems led to the passage of the Environmental Protection Act in 1974. Important measures of the Act included the development of new forms of energy, more efficient energy-production systems, less energy-consuming production, housing insulation, and pollution control in manufacturing, transport, and agriculture. Denmark is a leading nation in wind-power technology, which has become a significant export. Wind power produces about 5 percent of Denmark's electricity. A system of district heating uses natural gas, recycled wood waste, and other waste products. Denmark's energy consumption has remained stable for a considerable number of years.

Greenland and the Faeroes

The former Norwegian territories of Greenland and the Faeroes, like Iceland, remained with Denmark after 1814. The Faeroe Islands (61° 20'–62° 24' N, 6° 15'–7° 41' W), with an area of 1,500 km², are located in the Atlantic Ocean, midway between Norway, Iceland, and Scotland. They formed a Danish county until 1948, when they attained home rule. The island of Greenland (59° 45'–83° 20' N, 12° 30'–73° 00' W), east of northern Canada, covers 2.2 million km², although the ice-free area is only 410,500 km². The island extends 2,650 km from south to north. Greenland attained home rule in 1979. The population of Greenland is 56,900 and that of the Faeroes is 48,400. Nuuk (formerly Godthåb), the capital of Greenland, has 15,000 inhabitants, while Tórshavn, the capital of the Faeroes, has 19,000.

More than 94 percent of the Faeroes' total exports are fish and fish products, while the corresponding figure for Greenland is 87 percent. Fishing rights are hence a dominating economic and political issue, and the main reason why the Faeroes and Greenland are not part of the EU. Their economies would suffer if Faeroese and Greenland waters were opened for EU fishing vessels. Problems in the fishing industry, including overcapacity of the fishing fleet and overfishing, have led to the depletion of several species. The Faeroes suffered an economic crisis in 1989 and many enterprises went bankrupt. The population declined from 48,000 in 1989 to 43,000 in 1995. Since then, the economy has recovered and the population is back to its former level.

The Inuit population of Greenland underwent a radical transformation after World War II. They used to live in small hunting and fishing settlements, but now four-fifths of the population live in small towns, mainly in central West Greenland. Greenland is still very much a natural resource economy. A government-owned company controls almost the entire fishing industry. In many settlements, underemployment and unemployment rates are high and welfare and pension payments make up a large share of the total income.

The northernmost settlement in Greenland is Thule (77° 30' N). As a result of the deterioration of hunting and fishing after the establishment of the Thule Air Base by the United States in the early 1950s, the settlement was moved 130 km northward to its present location.

The Faeroes and Greenland are vulnerable because they are dependent on international resource management and market forces. The local economies have been restructured in a way that has marginalized parts of the population, who have responded either by migration or by early retirement.

Iceland

Iceland is the westernmost country in Europe, lying just south of the Arctic Circle (63° 18'–67° 10' N, 13° 30'–24° 32' W), 1,000 km from Norway, 800 km from Scotland,

and 300 km off the Greenland coast. The land area is 103,000 km² and its maritime economic zone covers 780,000 km². Only 312,000 people live on the island. It lies astride the Mid-Atlantic Ridge and the southwest and central parts form an intensely active volcanic zone. The climate is arctic and maritime, with average January temperatures between 0 and −4°C and July temperatures between 8 and 10°C. Two-thirds of the area is barren highlands and mountains, with recent lava flows, sandy outwash plains, and glaciers. Only one-third of the land is covered by continuous vegetation. Birch woods and scrub cover a mere 1 percent of the lowlands.

Although Irish monks may have reached Iceland earlier, the first settlers known with any certainty were Vikings in the ninth century, mainly from western Norway. Iceland established its own parliament in 930. In 1262, the Icelanders accepted Norwegian sovereignty and with Norway joined Denmark in 1380. When the union between Denmark and Norway was dissolved in 1814, Iceland remained with Denmark. In 1918, Iceland became a sovereign state, with the Danish king as monarch. In 1944, after a referendum, it became a republic.

Historically Iceland had to import all its timber and grain. Stockfish became the most important export commodity. Most Icelanders lived on farms and were bound to the land by law. Fishing and farming were combined. Fishing alone was generally not a viable way of living because of fluctuating catches. The Icelanders fished in open boats near the coast. Danish merchants controlled trade. Deepwater fishing was in the hands of foreigners, mainly the English, Dutch, and French. Iceland became increasingly isolated and barred from legal contact with other nations. Climatic changes—the "Little Ice Age" (sixteenth to nineteenth centuries)—marginalized Iceland even more. One hundred years ago, Iceland was one of the poorest and technologically most backward nations of Europe, and as late as the 1880s it received famine relief from abroad.

Trade regulations were gradually abolished in the nineteenth century and the steamship transport revolution helped Iceland out of its isolation. At the first national census, in 1703, there were 50,000 Icelanders. In 1901, the population was barely 80,000, of which 7,000 lived in the capital, Reykjavik.

The motorization of the fishing fleet and the introduction of trawlers around 1900 transformed the economy. Home rule from 1904 and the development of a banking system opened for industrial investment and the development of roads, fishing ports, and telecommunications. Iceland was hit hard by the 1929 world depression. As a result of the Spanish Civil War, it lost its most important market for fish, accounting for one-third of the export trade.

One month after the German occupation of Denmark in 1940, Iceland was occupied by British troops and in 1941 these were replaced with U.S. troops. Roads and airfields were constructed and services had to be provided for 25,000 soldiers in a country with a workforce of 50,000. Unemployment vanished almost overnight. An increasing demand for fish products, in particular from Britain, led to high prices. The war affected the social fabric and the way of life of the Icelanders, with increasing in-migration to the capital, which in 1950 had 40 percent of a total population of 144,000.

As a result of the Korean War, the airfield of Keflavik became an important NATO base, run by the Americans. Aid from the Marshall Plan boosted industrial investment. The fishing industry boomed during the 1950s and most of the 1960s. Cod was abundant, and herring, which almost disappeared in the 1950s, came back in the 1960s. Toward the end of the 1960s, a new crisis in the herring fisheries and plummeting fish prices weakened the Icelandic economy. Fish and fish products made up over 90 percent of exports. In order to diversify the economy, the country began to use hydroelectric power for aluminum smelting in 1969, and later for the production of ferroalloys. Iceland was hit by the oil crisis in 1973. The same year, a volcanic eruption in the Westman Islands (Vestmannaeyjar) covered the fishing town of Heimaey with ashes and lava and 5,300 people were evacuated. Iceland, still very dependent on its fish resources, extended its fishing limits to fifty nautical miles in 1972 and two hundred miles in 1975 to secure resources for its expanding fishing fleet. Icelandic fishermen contributed to overfishing in the North Atlantic. A serious resource crisis in 1982 hit the national economy and inflation rose sharply. New economic measures were introduced, such as the deregulation of financial markets and the liberalization of trade.

The economy of Iceland is vulnerable, partly because it is affected by natural and human-induced changes in marine resource stocks and partly because of fluctuating prices for its export products. However, Iceland in 2004 was thirteenth among the world's fishing nations in catch. Marine products account for about 51 percent of the total export value, while aluminum and ferro-alloys account for 26 percent. Fishing and fish-processing provided work to 15 percent of the workforce in 1950; in 2006 the figure was 3.8 percent. Agriculture employed 25 percent of the workforce in 1950, and in 2006 only 2.7 percent. The female participation rate in the workforce is very high, 78 percent.

Iceland is now a service economy. The service sector employed 33 percent of the workforce in 1950 but 72 percent in 2006. Services are concentrated in Reykjavik and other towns in southwest Iceland, where almost 80 percent of the country's population lives. Tourism has become an important source of income. Iceland's four national parks cover 5,000 km^2, while nature reserves cover another 4,000 km^2.

Iceland produces three-quarters of its energy requirements from domestic sources, 80 percent geothermal and the remainder from hydroelectric power. Imported oil, gas, and coal account for 23 percent of energy consumption.

Small, peripheral communities exist on the west, north, and east coasts. Some local authority districts have only a dozen inhabitants. With some exceptions, these regions are declining and it is difficult to see how they can be revitalized. In order to develop more of Iceland's hydroelectric potential, one of Europe's largest aluminum smelters has been constructed on the east coast. The government and the local population favor it as a means of stemming out-migration from the area. Despite the mobilization of public opinion against the prospect of pollution from emissions and the damming of a large wilderness area, the Icelandic government signed an agreement in 2003 with Alcoa, Inc., for the construction of a smelter. It opened in June 2007 and has an annual capacity of 322,000 metric tons. A second smelter is under construction.

Iceland provides its inhabitants with a good education and health system. Demographically, it is one of the few European nations with a fertility rate high enough to ensure a natural increase in population. One-quarter of the population is below the age of fifteen and only one-tenth is above the age of sixty-four. The population more than doubled during the second half of the twentieth century.

Norway

Norway proper (57° 57'–71° 11' N, 4° 30'–31°10' E) extends 1,750 km from south to north as the crow flies. Approximately one-third of the country lies north of the Arctic Circle. The mainland coastline is 21,000 km long; 57,000 km when its islands are included. A population of 4.7 million inhabits 324,000 km²; 75 percent live on the coast. The average population density is 14 per km². Urban settlements cover 0.7 percent of the area, arable land 2.7 percent, productive forest 22 percent, nonproductive forest 10 percent, and lakes and rivers 5 percent. The remaining 60 percent are mountains, glaciers, bogs, and heaths. Norway is a country of great natural contrasts and communications have to overcome a difficult and rugged topography. Investment costs are high and a collective traffic system in sparsely populated areas is expensive.

Despite the country's northerly location, the climate is milder than elsewhere on the same latitude due to the influence of the Gulf Stream. Along much of the west and north coast, mean February temperatures remain close to 0°C. Hence, the Arctic coasts of North Norway have ice-free harbors.

Much of the Norwegian coast is shielded by the skerry guard, an archipelago of large and small islands that protects the main coastal navigation route. The coastal liner, established in the nineteenth century, sails daily between Bergen and Kirkenes, close to the Russian border in the far north, and carries freight as well as being popular with tourists. Along much of the west and north coast, settlement is concentrated on the strandflat, a low-lying flat strip, including islands and islets, between the sea and the mountains.

Offshore, Norway controls rich marine and petroleum resources in its maritime economic zone, which covers more than 800,000 km² of the North Atlantic. Norway controls fisheries in a further one million km² around the Arctic archipelago of Svalbard and the island of Jan Mayen in the Greenland Sea.

Norway became a unified kingdom in the Viking period, when Christianity was also introduced. The oldest-known description of Norway, made by Ottar from North

Norway, described the livelihoods of the north and trade route along the Norwegian coast to Denmark. The Norwegian realm reached its greatest extent in the 1260s, when, in addition to mainland Norway, it encompassed Orkney and Shetland, the Hebrides and the Isle of Man, the Faeroes, Iceland, and Greenland. The Hebrides and the Isle of Man were lost to Scotland in 1266, and Orkney and Shetland came under Scottish rule in 1468–69. Norway was hit severely by the Black Death in 1349, which reduced the population by half. After periods of personal union with Sweden in the fourteenth century, Norway came into a dynastic union with Denmark in 1380 and after this gradually lost its independence.

Norway has been a seafaring nation since Viking times. The seasonal cod fisheries in the Lofoten Islands have produced stockfish for export for eight hundred years. Later, herring supported many coastal communities.

Forest resources in Norway entered European trade in the sixteenth century, earlier than in Sweden and Finland because they were closer to the main markets around the North Sea. From the seventeenth century, copper, silver, and iron mines in Norway contributed to the national economy of Denmark–Norway. The two economies were in many ways complementary.

When the union with Denmark was dissolved in 1814, it was replaced by a personal union with Sweden, which in turn was dissolved in 1905. Although sharing the Swedish king, Norway gained its own constitution, parliament and government, central bank, and university. The capital, Oslo (then called Christiania), although far from the main maritime axis of the west, was situated near the major agricultural and forest regions of southeastern Norway. Timber was floated down the major rivers to sawmills and other wood-processing plants in coastal settlements. New roads and, from 1854, railways linked Oslo with its hinterland and it became the center of finance and higher education. Commerce, international trade, and manufacturing led to further growth. Oslo grew from a small city of 9,000 inhabitants in 1801 to 77,000 inhabitants in 1875. However, in 1875 only one-quarter of Norway's population lived in urban settlements. The total population of Norway doubled, from 0.9 to 1.8 million, between 1815 and 1875. Improved health care and a better and more regular provision of food were the main causes of the rapid population increase. Increasing pressure on agricultural land and few alternative jobs in the towns before the end of the century resulted in the emigration between 1865 and 1915 of 700,000 Norwegians, mainly to the United States.

Industrialization started slowly after 1850. The home market for industrial products was small. Fish, wood, and mineral resources were exported unprocessed or semi-processed, often on Norwegian ships. A shipbuilding industry developed, but Norway was late in converting from sail to steam. Pulp and paper mills, often foreign-owned, were established after 1870. The development of hydroelectricity allowed for energy-consuming, export-oriented industries after 1900. To some extent they processed local minerals and metals, but the raw materials were mostly imported. The factories often created small one-company towns in isolated locations. As Norway built up its modern merchant fleet, it established shipyards in major ports. Links between shipbuilders and shipowners were close. Although shipbuilding has declined, in 2001 the Norwegian merchant shipping fleet remained the world's seventh largest in tonnage.

After World War I, Norwegian export industries and shipping ran into difficult times. Slow recovery in the interwar period was temporarily stopped by the 1929 world recession and its aftermath. Small manufacturing industries oriented themselves toward the national market. Market towns extended their hinterlands as communication systems were developed. Public services expanded and the education system was enlarged and improved. The embryo of a welfare state emerged.

Despite its declared neutrality in 1939, Norway's strategic maritime position and natural resources attracted the attention of the warring nations. In 1940, Germany invaded Norway. A government-in-exile was set up in London. The war left numerous coastal towns in ruins after bomb attacks, and in 1944 the far north was devastated by Germans retreating before Russian troops crossing the border. Costly reconstruction dominated Norway far into the 1950s. However, economic growth permitted the development of a modern welfare state. There was work for all, with many new jobs in manufacturing and service industries.

One-quarter of the workforce was in primary occupations in 1950, when more than 200,000 small farms shared one million hectares of farmland. The average farm was a part-time business. Women did much of the work, whereas men found cash income in fishing and forestry or in jobs elsewhere. By 2007, agricultural restructuring had reduced the number of farms to fewer than 50,000. The cultivated area has not been reduced, but farming is becoming more machine-intensive and less labor-intensive. The number of dairy cows declined from 750,000 in 1950 to 260,000 in 2007. By contrast, the number of pigs doubled to 700,000, while there are almost one million sheep and nearly 50,000 goats. Agriculture is heavily subsidized; in Europe, only Switzerland has the same level of subsidies. In addition, agriculture is protected through import duties and quotas. Surplus production of animal products is a problem because the national market is small and exports are insignificant, given the high price level of Norwegian agricultural produce. The most important goal of regional policy has been to prevent the drain of people from marginal communities. Today, agriculture can no longer guarantee the preservation of marginal Norway. The young are educating themselves out of primary occupations and manufacturing into knowledge-based service jobs.

Reindeer herding is carried on by the indigenous Saami (Lapp) people in northern and central Norway. The Saami population is estimated at between 25,000 and 40,000, including many who live in Oslo and other towns. There is also a smaller but long-established Finnish-speaking minority on the Arctic coast.

Overfishing by trawlers from European nations, including Norway, triggered an ecological crisis in the North Atlantic in the late 1980s, hitting coastal communities in northern Norway. Fish farming, however, after a slow start in the 1970s, accelerated after 1980. In 2004, Norway had the world's tenth largest fish catch and was the ninth largest aquaculture producer. The production of salmon and trout, fewer than 10,000 tons in 1980, reached 150,000 tons in 1990 and passed 600,000 tons in 2004. The export value of fish farming is now higher than that of traditional fisheries. Norway is the leading world producer of farmed salmon and exports almost all of it, mainly to the EU. Fish farmers in the EU have agitated for restrictions on imports of salmon from Norway, and the inclusion of Eastern European markets in the EU since May 2004 may worsen market conditions for Norwegian salmon.

Norway became a rich oil- and gas-producing nation in the 1980s. The first North Sea oilfield came into production in 1971. Pipelines facilitate exports to the main European buyers. The EU receives 30 percent of its natural gas from Norway. After Saudi Arabia and Russia, Norway is the third largest exporter of oil in the world, and after Russia and Canada the third largest exporter of natural gas. In 2005, half of the value of Norwegian exports came from oil and gas, accounting for one-fourth of the BNP. Norway produces the equivalent of 1.7 million barrels a day. Employment effects are more modest: 3 percent to 4 percent of the national workforce is employed in oil-related activities. The Stavanger area has profited most, followed by Bergen. Oil has made Norway one of the richest countries in the world. The Gross National Product (GNP) is 50 percent above the OECD average. Yet oil is a mixed blessing. Most Western European countries went through fundamental economic restructuring, triggered by rapidly increasing oil prices, in 1972. Norway for several years used oil income to protect existing industries until it became clear that restructuring was inevitable. Employment in manufacturing industries was reduced by 25 percent in the 1980s. The oil industry could offer high wages in a labor market with little unemployment. High wage levels made many exporting firms less competitive on international markets. Some closed down, while others expanded abroad, closer to their customers or in countries with low labor costs.

Less than 3 percent of the working population are employed in farming, forestry, and fishing; 13 percent in mining, oil-related activities, and industry; 7 percent in construction, and 77 percent in services. Female participation in the workforce is 72 percent for the age group 16–74. Since 1960, immigration to Norway has exceeded emigration. Well-paid jobs in the oil industry have attracted Europeans and North Americans, while lower-paid work has attracted immigrants from Asia and Africa, bringing in their wake problems of discrimination. In 2006, 222,000 foreign citizens lived in Norway.

Several Norwegian resource-based companies have developed into worldwide concerns. Norske Skog, founded in 1962, now has eighteen factories in fourteen countries and is one of the world's leading producers of newsprint, supplying 10 percent of the world market. Norsk Hydro was founded in 1905 for industrial production of fertilizers using hydroelectric power to extract nitrogen from air, a method developed in

Norway. In the 1960s, it branched out into aluminum production and North Sea oil and gas. The state-owned oil-and-gas company, Statoil, established in 1972, and the oil-and-gas activities of Norsk Hydro were merged into StatoilHydro in October 2007, employing 31,000 persons in forty countries.

Norway's rich endowment of natural energy resources—waterpower, oil, and natural gas—and its long tradition of energy-demanding industries mean that energy consumption per capita is among the highest in the world and ten times the world's average. Almost half of Norway's energy consumption is in the form of electricity, 99 percent of which is hydroelectric power. Another 37 percent of energy consumption is based on oil and gas.

Between 1962 and 1991, nineteen national parks were established on the Norwegian mainland. A new plan for national parks is being implemented and several parks have been established since 2002. In 2007, there were twenty-nine national parks on the Norwegian mainland covering 14.3 percent of the land area. A similar area is included in other protected areas.

NORWEGIAN TERRITORIES

Norwegian sovereignty over Svalbard (74° 20'–80° 45' N, 10° 30'–33° E) was established in 1925 in accordance with the Spitsbergen Treaty of 1920, which guarantees all signatory nations equal rights in the exercise of economic activity on the island group. Sovereignty over Jan Mayen (71° N, 8° 30' E) was declared in 1939. Both Svalbard and Jan Mayen are regarded as integral parts of Norway while retaining distinct administrative arrangements. Svalbard and Jan Mayen together comprise 61,400 km². The resident population of Svalbard is 3,000, of which 1,700 are Norwegian and the remainder mostly Russian or Ukrainian. Norwegian coal mines are operated at Longyearbyen and Svea, producing 2.9 million tons annually, and there is a small Russian coal mine at Barentsburg, producing 100,000 tons. Otherwise, scientific research and, increasingly, tourism are the mainstays of Svalbard; 65 percent of Svalbard's area (40,000 km²) is protected in seven national parks, six nature reserves, and six bird sanctuaries. The administrative center is Longyearbyen. The research station at Ny-Ålesund is the world's northernmost settlement (78° 55' N). Norway maintains a weather station on Jan Mayen.

In the Southern Hemisphere, Norway declared sovereignty over Bouvet Island (54° 26' S, 3° 22' E) in 1928, Peter I Island (68° 50' S, 90° 35' W) in 1931, and Queen Maud Land (Dronning Maud Land), a sector of the Antarctic continent lying between 20° W and 45° E, in 1939. These three territories have the status of Norwegian dependencies. None is permanently inhabited, although several nations maintain research stations in Queen Maud Land. Norway is signatory to the Antarctic Treaty of 1959, by which all territorial claims south of 60° S are held in abeyance in the interests of international scientific cooperation.

Sweden

Sweden (55° 20'–69° 4' N, 10° 58'–24° 10' E), occupying the eastern part of the Scandinavian peninsula, is with 450,000 km² Europe's fifth largest country and the third largest in the EU. The distance from south to north is 1,600 km. The greater part is covered by coniferous forest, the main arable areas lying in the southern third of the country. Although the climate shows continental influences, the prevailing westerly winds and the effect of the Gulf Stream make Sweden warmer than its northerly position would suggest. The Gulf of Bothnia and the Baltic Sea are, however, partly frozen in winter.

With nine million inhabitants, Sweden is the largest Nordic country, although one of Europe's smaller countries in terms of population. The average population density is 22 persons per km². More than 85 percent of the population live in the southern part of the country and about 20 percent in the region of the capital, Stockholm. Sweden's second largest city, Gothenburg (Göteborg), has 485,000 inhabitants. The Saami form a minority of between 10,000 and 15,000 people in the north. Besides old-established Finnish settlements in the far north, extensive Finnish settlement occurred in the forested areas of central and western Sweden in the seventeenth century. Since the 1950s, the high standard of living in Sweden has attracted migrant workers from Finland as well as other European countries. In 2005, there were nearly half a million non-Swedish citizens living in Sweden, 5 percent of the total population.

In contrast to Denmark and Norway, which are members of NATO, Sweden in the twentieth century maintained a policy of neutrality, supported with a powerful military force. Most of its weapons, including Saab fighter jets and Bofors antiaircraft guns, are supplied by Sweden's own armaments industry.

Sweden is known for its well-developed welfare system and policies that promote equal rights for women. More than 70 percent of all men and women between the ages of sixteen and sixty-four are gainfully employed.

The Swedish kingdom emerged as a unified entity in Viking times, centered on Lake Mälaren. Stockholm dates from the thirteenth century. The Swedish crown established its rule over Finland during the twelfth and thirteenth centuries. In the sixteenth century, Sweden became a major power, controlling much of the Baltic region, and reached its greatest extent in the mid-seventeenth century. In 1645, Sweden gained Jämtland from Norway and in 1658 Bohuslän, Halland, Blekinge, and the fertile Scania (Skåne) from Norway and Denmark. After this, Sweden gradually declined as a great power. It lost territory on the eastern side of the Baltic to Russia as a result of the wars in the eighteenth century, and Russia annexed Finland in 1809. By contrast, Norway was in personal union with Sweden under the Swedish crown from 1814 to 1905.

The nineteenth century was a period of economic, social, and political reform. The liberalization of trade, the modernization of agriculture, the rise of the middle classes, and the improvement in education transformed Sweden from a poor rural country into a modern industrial one. The Göta Canal, linking Stockholm with Gothenburg across central Sweden via Lakes Vättern and Vänern, was completed in 1832. In 1854, the government initiated a program of railway-building. In 1866, the parliament was reformed. Favorable markets led to an upswing in the timber trade in mid-century, followed by international demand for pulp after about 1870. In Bergslagen, Swedish iron manufacturing, an important export industry from the Middle Ages, was modernized. New techniques enabled exploitation of iron-ore reserves in the Kiruna area of northern Sweden. Swedish steel had become a major export by 1900.

The population of Sweden grew from 2.5 million in 1815 to more than 5 million in 1900. People moved from the countryside to the towns. Whereas 90 percent of the population earned its living from primary occupations in 1850, the figure was below 50 percent fifty years later. Considerable emigration to the United States occurred, particularly during periods of recession.

Social Democratic governments since 1932 established a welfare state and stimulated the economy by public investment, especially in housing. Collective bargaining, introduced in 1938, led to a long period of industrial harmony, contributing to economic growth after World War II. Social reforms after 1945 consolidated Sweden's position as one of the world's most advanced welfare states. Low unemployment and generous welfare benefits gave way in the 1990s to rising unemployment and cuts in the welfare system, after which the economy began to pick up again.

The main agricultural areas are Scania in the south, with its Cretaceous rocks, and the plains of central Sweden. In the 1920s, agricultural land covered 11 percent of Sweden, but today accounts for less than 7 percent. The main decline has been in the less fertile areas of the country. In 1944, 25 percent of the employed population worked in agriculture, forestry, and fishing; now, that figure is only 2 percent. The number of farms decreased from 296,000 in 1946 to 75,000 in 2005, and the number of dairy cattle dropped from 1.4 million to 393,000. Yet increased efficiency means that Swedish agriculture produces as much as it did fifty years ago. Milk and meat account for 70 percent of agricultural production. Grain and other arable crops are cultivated mainly in southern Sweden. Agricultural policies protecting Swedish farming against competition from imports were introduced in the 1930s. Since 1995, EU agricultural policies have applied in Sweden.

Half of Sweden is covered by forest. Active reforestation increased the forest area during the twentieth century. After World War II, forestry became highly mechanized and road transport replaced river transport. The main demand for timber today is from the pulp and paper industries. This has resulted in forestry practices giving priority to rapid-growth rather than high-quality timber. About 70 percent of the annual growth is felled each year. While forestry has long been important in northern Sweden, its importance has increased in the south, where timber growth is more rapid. In 2004, exports of wood, pulp, paper, and other wood products accounted for 12 percent of Sweden's earnings. In 1996, the Swedish forest industry had 140 factories abroad, employing more people than in Sweden.

Important for Sweden's industrial revolution toward the end of the nineteenth century was cheap hydroelectric power from Swedish rivers. This compensated for Sweden's lack of coal resources. By the end of the 1950s, hydroelectric power could no longer meet the country's requirements. In addition to thermal-power stations based on imported coal, coke, and oil, Sweden turned to nuclear power. Between 1963 and 1985, twelve nuclear reactors were built. Because of strong public opposition, these are to be phased out by 2010. Today approximately half of Sweden's electricity is nuclear power and about 45 percent is hydroelectric power. Sweden intends to develop wind- and bioenergy, combined with energy-saving measures, to replace nuclear power and to reduce emissions from coal- and oil-fired power stations.

Sweden has always been rich in mineral resources, particularly copper, iron ore, and silver. Copper was mined in the Falun area of Bergslagen from at least the eleventh century. At times during the seventeenth and eighteenth centuries, the Bergslagen region was the world's largest producer of iron and copper. In 1989, the last mines in Bergslagen were no longer able to compete on the international market and were closed. Sweden's iron-ore mining is now concentrated in Kiruna and Malmberget in Lappland. Copper is mined in the Skellefteå area. The Swedish steel industry is concentrated in the southern-central part of the country. Ores and metals accounted for 10 percent of Sweden's exports in value in 1998, nearly half of this from iron and steel.

Manufacturing industries processing natural resources are still important for the Swedish economy, but mechanical and electrical engineering have become more important. Swedish innovations have generated new industries. Alfred Nobel, chemist and industrialist, invented dynamite in 1865 and gelignite in 1875. These inventions were important for the development of Sweden's roads and railways over difficult terrain, but they were also used for destructive purposes. Nobel left his fortune to finance the Nobel prizes in physics, chemistry, medicine, and literature, as well as the Nobel Peace Prize, which, unlike the others, is awarded by the Norwegian parliament. In the 1920s, Electrolux, a manufacturer of household machines, SKF, the world's biggest manufacturer of ball bearings, and Swedish Matches entered the world market. Nearly half of Sweden's export income is derived from machinery. EU countries take about 60 percent of Sweden's exports by value. Ericsson is Sweden's largest enterprise. It began manufacturing telephone handsets in 1879 and exported them to European countries before World War I. Ericsson has been a world pioneer in the development and marketing of mobile telephones since 1979. Only 4 percent of Ericsson's production is sold in Sweden. Ericsson's toughest competitor is Nokia in Finland. In 1999, Ericsson employed about 100,000 people in 130 countries, including 45,000 in Sweden, but cutbacks reduced the number in 2006 to 64,000 internationally, of which 19,000 are in Sweden.

Manufacturing and mining employ about 16 percent of Sweden's workforce, down from 34 percent in 1965. Services employ three-quarters of the workforce.

Government policies since the 1960s have attempted to counteract the rapid growth in population and employment in the main metropolitan regions. However, employment has continued to grow more slowly in the assisted areas than in Sweden as a whole. Since 1995, Swedish regions have been entitled to grants from the EU's regional development fund.

Since the Environmental Protection Act was passed in 1969, Sweden has invested heavily in environmental measures. Swedish forests and lakes have suffered damage from acidic rain due to airborne sulfur dioxide from industry in Western and Central Europe. As a result of international agreements, most countries in Europe have agreed to reduce sulfur emissions. Sweden was the first Nordic country to establish national parks, when eight parks were designated in 1909. Sweden now has twenty-eight national parks, covering 7,000 km^2.

Finland

Finland (59° 30'–70° 5' N, 19° 8'–31° 35' E) is known for its vast forests and for its sixty thousand lakes that form an inland archipelago of numerous islands. An extensive archipelago, the skerry guard, is also found along much of the Finnish coast. Finland extends 1,150 kilometers from south to north. The terrain is generally low-lying; the mean altitude is 132 meters above sea level. Nearly one-third of the country lies north of the Arctic Circle.

The climate of Finland shows a mixture of maritime and continental influences. The south is cold in winter and mild in summer. Mean temperatures for Helsinki are minus 5° in January and plus 17°C in July. North of the Arctic Circle, the climate shows extreme seasonal variations. Mean temperatures for Sodankylä range from minus 15° in January to plus 14°C in July. The maritime influence of the Gulf of Bothnia and Gulf of Finland is offset in winter, when the sea is frozen. Frozen seas and lakes change patterns of communication in winter. Compared with countries farther south, snow and ice in Finland produce substantial additional costs for fuel, housing, snow-plowing roads and railways, and ice-breaking on waterways.

Shore displacement, which resulted from the uplift of land after the weight of the ice sheet was removed at the end of the last Ice Age, is still marked along the low-lying shores of Finland. In the north of the Gulf of Bothnia, the land is rising by nearly 90 cm per hundred years. As a result, Finland's area is increasing by between 1,000 and 2,000 km^2 a century. Land emergence has historically provided an important agricultural resource, but it causes problems for navigation and fishing because ports and harbors must be continuously moved to new locations

Despite difficult terrain, Finland has an extensive road network and railway links to Sweden in the north and St. Petersburg in Russia to the southeast. The lakes have historically been important for freight transport, although today they are more significant for tourism and recreation. The Saimaa Canal, completed in 1856, links the eastern lakes to the Gulf of Finland near Viipuri (Viborg), enabling seagoing vessels to reach the lake town of Kuopio. Most of the canal became Russian when the Viipuri area was ceded to the Soviet Union in 1944. However, under Finnish-Soviet cooperation after World War II, Finland leased back the Saimaa Canal in 1963. After being modernized, it was reopened in 1968.

Finland has 5.3 million inhabitants. The country's total area of land and lakes is almost 340,000 km², giving a population density of 15 persons per km². Almost one-fifth of the population lives in the Helsinki conurbation and almost four-fifths in the southwestern third of the country.

Both Finnish and Swedish are official languages, except in Åland, where Swedish is the only official language. The Saami once lived over most of the area that is now Finland but now form only a small minority (0.03 percent), living on the northern fringes of the country. Less than 6 percent of the population is Swedish-speaking, living mainly on the islands and adjoining coast of southern and western Finland. The proportion speaking Swedish as their first language has halved since 1900. Finland has two state churches, Lutheran and Orthodox, reflecting Finland's historical position on the cultural and political border between Eastern and Western Europe.

Bordering Sweden on the west and Russia on the east, Finland throughout history has been affected by the competing interests of these two countries. A treaty in 1323 recognized Finland as a part of the Swedish kingdom. A new treaty in 1595 extended Swedish sovereignty to what is now northern Finland. As an integral part of Sweden, Finland had representatives in the Swedish parliament and Swedish laws applied. Swedish was the language of government, although in most of the country Finnish remained the language of everyday life. Finland was occupied and plundered by Russia in the wars of the early eighteenth century and southeastern Finland was lost to Russia. In 1809, Finland became an internally self-governing grand duchy under the Russian czar. The new boundary with Sweden in the north followed the Tornio River, leaving a Finnish-speaking minority on the Swedish side. In 1811, the land earlier lost to Russia in the southeast also became part of the grand duchy. Helsinki became the capital in 1812.

The population of Finland grew from one million in 1810 to 1.8 million in 1870, then doubled to 3.5 million in 1930. Increasing demand from the Western European market stimulated the growth of the timber industry during the second half of the nineteenth century. Pulp and paper manufacturing developed toward the end of the century and became Finland's major export industry. Tampere became a center for the textile industry. Industrial development led to migration from the countryside to the towns. The urban population grew from 7.5 percent in 1870 to 20.6 percent in 1930. Emigration, primarily to North America, was also significant.

The liberalization of trade during the second half of the nineteenth century benefited the small commercial and industrial class. Population growth resulted in the emergence of an industrial and rural proletariat, which was unrepresented in parliament. A process of Russification in 1899 and attempts to impose the Russian language met strong opposition. Parliamentary reform, providing for a unicameral parliament and universal suffrage, came in 1906. Finnish women were the first in Europe to gain the vote.

Finland declared independence in 1917. A bitter civil war broke out between the Reds and the Whites, ending with victory for the Whites. In 1919 a republican constitution was adopted (in contrast to the monarchies of Sweden, Denmark, and Norway). The Soviet Union recognized the borders of Finland in 1920, when Finland received Petsamo as an ice-free harbor on the Arctic Ocean.

The Winter War of 1939–40 occurred after the Soviet Union attacked Finland, which had refused demands to cede territory. By the Treaty of Moscow, Finland was forced to cede more than one-tenth of its territory. The Winter War cost Finland 25,000 dead. When Germany attacked the Soviet Union in 1941, Finland entered the war and reconquered the lost territory. Known as the Continuation War, this war resulted in a further 86,000 dead.

In 1944, Finland signed an armistice and was forced once more to cede one-tenth of its territory to the Soviet Union, including Petsamo, the Karelian isthmus, and Finland's second city, Viipuri. (The old frontier lay only 30 km from the edge of Leningrad.) Some 400,000 people left these areas and were resettled elsewhere in Finland. The Soviet Union established a military base at Porkkala, west of Helsinki, which it held until 1955. Finland had to pay $300 million in reparations to the Soviet Union, which it succeeded in paying off by 1952. Finland paid in goods, half of which consisted of machinery and shipping. Finland's subsequent trade with the Soviet Union arose out of the adjustments forced by the war reparations.

After World War II, Finland accepted a close relationship with the Soviet Union as the price of retaining its independence. The policy of accommodation included self-censorship and regular political consultation. Militarily, Finland has followed a policy of neutrality, although in 1948 the country signed a treaty with the Soviet Union whereby it agreed to resist any attack on the USSR made through Finland by Germany or its allies. This treaty was terminated in 1991.

An economic boom in the 1980s led to a rise in the standard of living, but the boom came to an abrupt end in 1991 with the collapse of the Soviet Union, which had taken a substantial part of Finland's exports. Finland realigned and four years later joined the EU. Finland's loss of stable markets resulting from bilateral trade with the USSR and the Eastern European bloc caused a recession that the country emerged from only in the late 1990s. Unemployment reached record levels (20 percent of the workforce) and the welfare system came under pressure. Austerity measures included welfare cuts, higher taxation, and wage restraints. The strength of Finnish industry helped the country recover from the worst effects of the recession. Alone of the Nordic countries, Finland has joined the EU's single currency.

Much of Finland's bedrock is granite, covered by morainic soils and extensive peat lands. Slash-and-burn, a form of shifting cultivation practiced in the deep forest, continued in some areas until the early twentieth century. Until the mid-twentieth century, forestry and agriculture provided the mainstays of the Finnish economy. The climate, however, places limitations on crop production. Cereals are grown in southwest and central Finland, while in the east and north crops are limited to grass for fodder. In 1917, two-thirds of the population was dependent on agriculture, and in 1970 still 18 percent were. In 2006, less than 5 percent of the workforce was employed in agriculture, forestry, and fishing. Since the end of the 1960s, the arable area has declined due to agricultural rationalization and farm abandonment and now covers less than 7 percent of Finland. The number of farms has decreased substantially since the late 1960s and now there are less than 70,000 farms. The number of dairy cattle decreased from 890,000 in 1970 to fewer than 310,000 in 2006, but, as the mean yield per cow has increased, milk production has remained stable.

Fur farming became a specialty in western Finland after World War II, and Finland is among the world's leading exporters of fox fur and mink, although this industry has met criticism from animal rights groups in recent years. Reindeer husbandry is significant in northern Finland, where, unlike in Sweden and Norway, it is not restricted to the Saami population. High self-sufficiency has been an aim of Finnish agricultural policy. However, since 1995, Finland has adjusted to the EU's Common Agricultural Policy.

Most of Finland lies in the northern coniferous forest belt and commercial forests cover 60 percent of the area. The state owns about 20 percent of the forest. Until the mid-1960s, Finland's forest area was declining as land was cleared for new farms. Since then, the forest area and growth volume have increased through state-supported programs of forest cultivation, including afforestation of abandoned farmland. Timber, sawmill products, pulp, and paper accounted for 80 percent of exports in the 1920s, whereas forest products now account for one-quarter. Finland is one of the world's largest exporters of pulp and paper and pulp and paper producers have consolidated their position through a series of mergers. Finnish and Swedish companies have merged to form important transnational companies, which in turn are establishing new partnerships with North American companies.

Historically, wood, peat, and waterpower supplied Finland's energy needs. Bog ore, charcoal, and waterpower gave rise to an iron industry in the seventeenth century. It declined in the nineteenth century, when coke replaced charcoal and mined ore replaced bog ore. The modern copper-mining industry dates from 1910, based on copper ore from Outukumpu in eastern Finland. Other ores include iron, nickel, chromium, and cobalt. Finland has significant hydroelectric power resources but is nonetheless dependent on imported fossil fuels. Between 1948 and 1990, Finland imported most of its oil on favorable terms from the Soviet Union in return for manufactured goods. This arrangement ended when the Soviet Union collapsed, leading to a marked rise in Finland's oil import costs. The energy requirements of industry are met to an important extent by thermal power, including nuclear energy. Finland's first nuclear power plant was built in 1977, and there are now four nuclear reactors. Despite growing public concern about nuclear safety, the Finnish parliament voted in 2002 in favor of constructing a fifth reactor. With priority given to energy efficiency, more than 40 percent of homes are connected to district heating schemes.

Since World War II, mechanical engineering has developed into a significant branch of industry, triggered by the payment of war reparations to the Soviet Union. At its peak, Finland's bilateral trade with the USSR accounted for one-fifth of its foreign trade. This led to the diversification of Finland's industry not only through engineering but also in stimulating the manufacture of textiles and footwear. Finland manufactures machinery for the wood-processing industries, particularly paper-making machines. Shipbuilding, concentrated in the south, has been important. At the height of trade with the Soviet Union, Finland regularly exported purpose-built vessels such as icebreakers. With the loss of the Soviet market and Finland's entry into the EU, with its own competing shipbuilding industry, Finnish shipbuilding has suffered a decline, but it has found new markets through the construction of cruise ships and marine engineering. Finland has its own vehicle-building industry, especially trucks and buses, partly in cooperation with Swedish manufacturers. The chemical industry is also significant, especially in the production of fertilizers. One of Finland's largest industrial enterprises is the state-owned Neste oil company, which after 1957 built several oil refineries in response to the imports of cheap oil and gas from the Soviet Union. Finland's oil-refining capacity is greater than its domestic requirements and refined oil has become a significant export. Finland is known for its innovative design, and Finnish products such as Marimekko textiles, Iittala glass, and Arabia ceramics have become hallmarks of quality on the world market. In 2006, 19 percent of the Finnish workforce worked in manufacturing and mining.

Finland's small domestic market, lack of energy sources, and specialized production mean that the country is highly dependent on foreign trade. Its chief trading partners are Germany, Sweden, and Britain. Although pulp, paper, and wood products remain important exports, heavy machinery and metal products now account for an equally large share. The main export harbor is Kotka, while Helsinki handles most imports.

Nokia, a major producer of mobile phones, is an example of Finnish industrial enterprise. The company began in 1865 as a wood-pulp mill but over the years expanded into making paper, rubber footwear, tires, and cables. Diversifying into telephones in the 1990s, Nokia provides a strong challenge to Sweden's Ericsson on the world market. Electrical and high-tech products, which accounted for only 2 percent of Finnish exports in 1970, now account for more than 45 percent.

Northern and eastern Finland have long suffered strong out-migration to southern Finland and Sweden. In an attempt to counteract this, Finland has practiced regional-development policies similar to those of Sweden's. The northern and eastern two-thirds of Finland are similarly now eligible for EU funding for sparsely populated regions. Finland has thirty-five national parks, covering 8,800 km^2; other environmentally protected areas cover double this area.

Åland

The Åland Islands (59° 45'–60° 40' N, 19° 8'–21° 20' E), with a land area of 1,500 km^2, lie in southwest Finland, midway between the mainland of Finland and Sweden. Of the sixty-five thousand islands and islets, sixty-five are inhabited. Åland is Swedish-speaking and constitutes an internally autonomous province within Finland. The population is 27,000 and more than 40 percent live in the capital, Mariehamn. In 1921 an international convention recognized Åland as part of Finland with internal autonomy, even though a majority voted in favor of becoming part of Sweden. Åland's status as a demilitarized area, dating back to 1856, was upheld. Åland took advantage of its geographical location and developed an economy in which merchant shipping became the mainstay. Its shipping has become global and 4 percent of the world's tanker tonnage is registered there. Shipping has given rise to a prosperous economy with one of the highest BNP per capita in Europe. When Åland joined the EU, along with Finland, a special protocol recognized the special political and economic position of the islands.

Conclusion

The Nordic countries lie at the northern margins of Europe. Despite environmental restraints, their inhabitants enjoy one of the world's highest standards of living. Although individual Nordic countries have chosen some divergent policies, such as membership in the EU and NATO, they continue to share a high degree of political, social, cultural, and economic cooperation through the Nordic Council and related institutions. While there are marked geographical variations among the countries and considerable regional differences within them, Norden shows a higher degree of integration than most other comparable groups of countries.

References

Alvstam, Claes Göran, ed. 1995. *Manufacturing and Services*. In *National Atlas of Sweden*. Stockholm: SNA Publishing.

Bailly, Antoine, and Armand Fremont, eds. 2001. *Europe and Its States: A Geography*. Paris: Datar, La Documentation Française.

Bernes, Claes. 1993. *The Nordic Environment: Present State Trends and Threats*. Nord 1993: 12. Copenhagen: Nordic Council of Ministers.

———, and Claes Grundstein, eds. 1992. *The Environment*. In *National Atlas of Sweden*. Stockholm: SNA Publishing.

Denmark. 2003. Copenhagen: Royal Danish Ministry of Foreign Affairs, Department of Information.

Hansen, J. C. 2001. "Iceland." In *Europe and Its States: A Geography*, ed. Bailly and Fremont, 138–40.

Hansen, J. C., and M. Jones. 2001. "Denmark, Finland, Sweden: The Nordic Countries." In *Europe and Its States: A Geography*, ed. Bailly and Fremont, 47–67.

———. 2001. "Norway." In *Europe and Its States: A Geography*, ed. Bailly and Fremont, 141–44.

Hálfdanarson, Guðmunður. 1997. *Historical Dictionary of Iceland*. In *European Historical Dictionaries*, no. 24. Lanham, Md.

Helmfrid, Staffan, ed. 1996. *The Geography of Sweden*. In *National Atlas of Sweden*, Stockholm.

Historisk statistikk 1994. Historical Statistics 1994. 1995. Norges offisielle statistikk—Official Statistics of Norway C 188. Oslo and Kongsvinger: Statistisk sentralbyrå–Statistics Norway.

Landshagir 2007. Statistical Yearbook of Iceland 2007. 2007. Hagskýrslur Íslands—Statistics of Iceland. Reykjavik: Hagstofa Íslands–Statistics Iceland. http://www.statice.is/Publications.

Maude, George. 1995. *Historical Dictionary of Finland*. In *European Historical Dictionaries*, no. 8. Lanham, Md.

Nordic Statistical Yearbook 2007. Nordisk statistisk årbok. Nord 2007: 1. Copenhagen: Nordic Council of Ministers.

Scobbie, Irene. 1995. *Historical Dictionary of Sweden*. In *European Historical Dictionaries*, no. 7. Lanham, Md.

Statistical Yearbook—Annuaire Statistique. 2006. New York: United Nations/Nations Unies.

Statistisk årbog 2007. Statistical Yearbook. 2007. Copenhagen: Danmarks Statistik. www.dst.dk/aarbog.

Statistisk årbok 2007. 2007. Norges offisielle statistikk—Official Statistics of Norway. Oslo and Kongsvinger: Statistisk sentralbyrå–Statistics Norway. www.ssb.no/aarbok/emne10.html.

Statistisk årsbok för Sverige. Statistical Yearbook of Sweden 2007. 2006. Örebro: Statistisk centralbyrån–Statistics Sweden. www.scb.se.statistik/publikationer/.

Suomen tilastollinen vuosikirja 2007. Statistisk årsbok för Finland 2007. Statistical Yearbook of Finland 2007. 2007. Helsinki: Tilastokeskus—Statistiskcentral—Statistics Finland.

Thomas, Alastair H., and Stewart P. Oakley. 1998. *Historical Dictionary of Denmark*. In *European Historical Dictionaries*, no. 33. Lanham, Md.

22.

Features of Nordic Physical Landscapes: Regional Characteristics

ULF SPORRONG

The distinctive image of a landscape depends on a number of mutually independent historical contexts; features that are the products of cultural development have seldom arisen as the result of planned and uniform strategies. Frequently, the only common factor among phenomena characteristic of a particular landscape is that they are to be found in the same place. Depending on economic strategies and technical advances, human influence has varied during the time the landscape has been cultivated or exploited in other ways. A landscape can therefore be regarded as the result of a series of historical processes that have often brought about changes in the relationship between people and their physical environment. Agriculture, fishing, forestry, mining, ironworks, communications (roads, railways), and not least settlements have made their mark on the appearance of the landscape in historic times. Even though the population today is mainly urban in character, it is precisely these earlier livelihoods that have been instrumental in forming our image of the Nordic landscape in a physical sense, as well as leaving a heritage of place-names and administrative districts (Figure 22.1). In the future, however, the image of the landscape more and more will be associated with the "modern" environments of the expanding built-up areas.

When considering both historical and modern aspects of landscape, one needs to understand the decisions made as to how the landscape is exploited. This is dependent on various strategies that determine the right to own and use land. The way these rights have been passed on from generation to generation has been important in determining the details of the landscape. These factors are integral to the social organization and values of society, and this is why these immaterial links, relationships, and boundaries have played so large a part in the formation of the landscape. Finally, external factors are significant for how the landscape has been exploited in any given region. Important here are the organization of the use of landscapes resources and

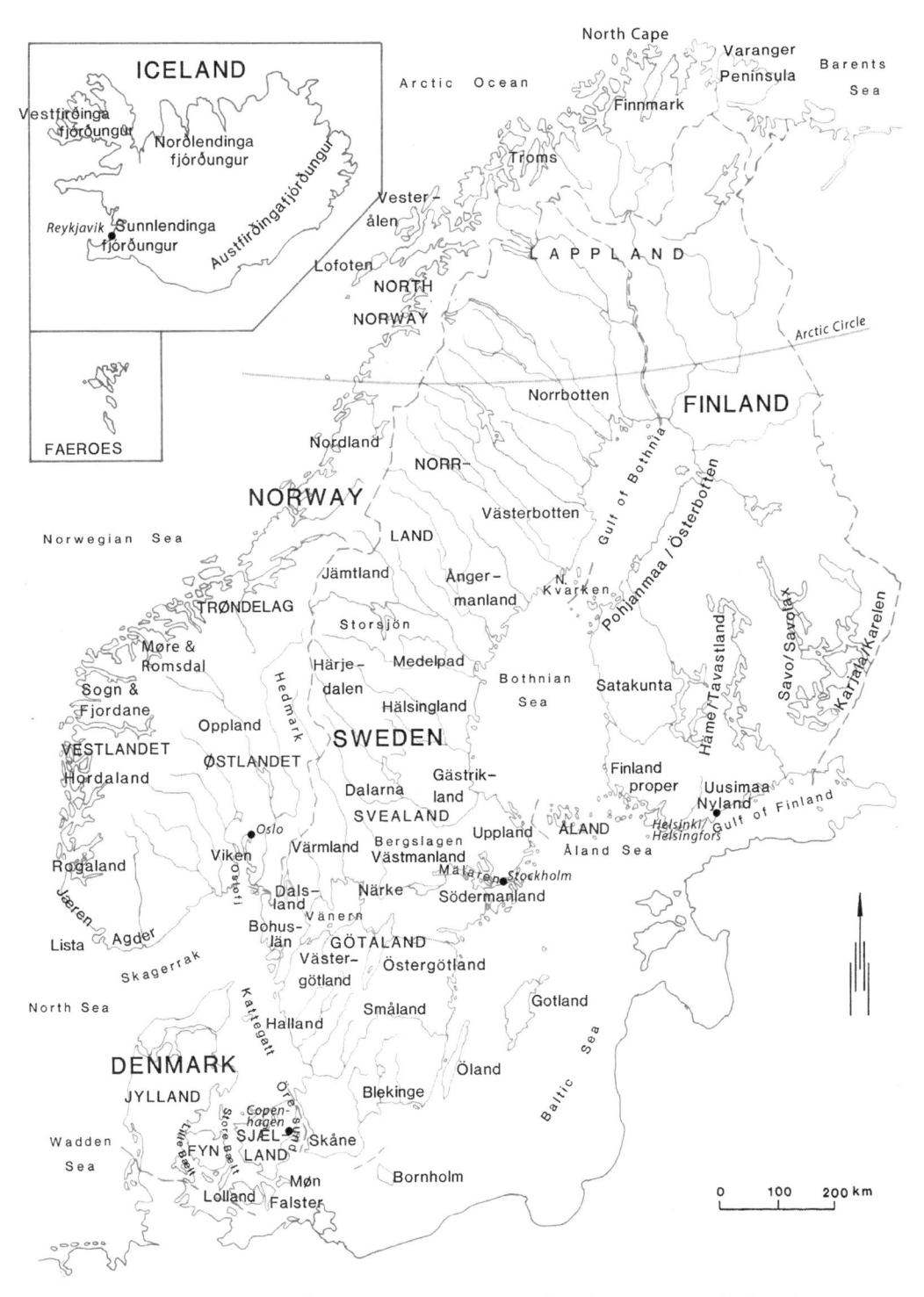

FIGURE 22.1. Regions of Norden. Place-names mentioned in the text are marked on the map.

how technical knowledge has allowed them to be exploited. As the pace of globalization increases, technology becomes more important. A landscape is not static but is in a continuous state of flux. In brief, a physical landscape is an indivisible environmental whole, which in most cases has been strongly influenced by human decisions and activities. This chapter is intended to give the reader an idea of the character and physical resources of the Nordic landscape, and how these resources have formed the basis for human survival in historic times. How have people exploited them, and what have been the obstacles to this exploitation?

THE PROCESSES

A number of abiotic processes were originally the principal agents in creating the Nordic landscape. A continuous interaction between constructive and destructive forces has reshaped the landscape, the latter forces being mainly running water, wind, waves, and ice. It was during the most recent geological eras that the details of the landscape were chiseled out, above all during the Quaternary ice ages. Pieces of bedrock were torn loose, crevices and weak zones were hollowed out, the surface of the bedrock was ground down, and ice carried away the loose material and deposited it in other places in the form of till, glaciofluvial and oceanic sediment. The numerous eskers in Sweden and Finland are the results of such processes, as are the large terminal moraines and vast areas of outwash plains in the southernmost parts of Scandinavia, especially in Denmark (Figure 22.2), but in more recent times also in Greenland and Iceland.

The shoreline was affected at different times by the melting of the ice sheets (resulting in isostatic uplift). The water released by the melting of the ice led to a rise in sea level (eustatic change). In some parts of northeastern Sweden and northwestern Finland, around the Gulf of Bothnia, the earth's crust has risen almost 800 meters since the land ice was at its maximum thickness, more than ten thousand years ago. At the same time, the release of the weight of the ice resulted in a rebounding of the earth's crust depressed by the ice sheet. The result of this combined process has been displacement of the shoreline. The highest shoreline (the highest marine limit reached by the sea at the end of the Ice Age) is now about 300 meters above the present level (Figure 22.3). Around the Baltic, the coastline is still changing as a result of land uplift. Because of this phenomenon, the highest shoreline has been important, for below this line the main areas of cultivable soils in the form of sea sediment are found. Depending on the character of the till, cultivable land can also be found in some places above the highest shoreline.

Abrasion and sedimentation are responsible for the more detailed features of the landscape. Bedrock formations are often covered with superficial deposits, mostly deposited during and after the last Ice Age. It is mainly soils on these deposits that have

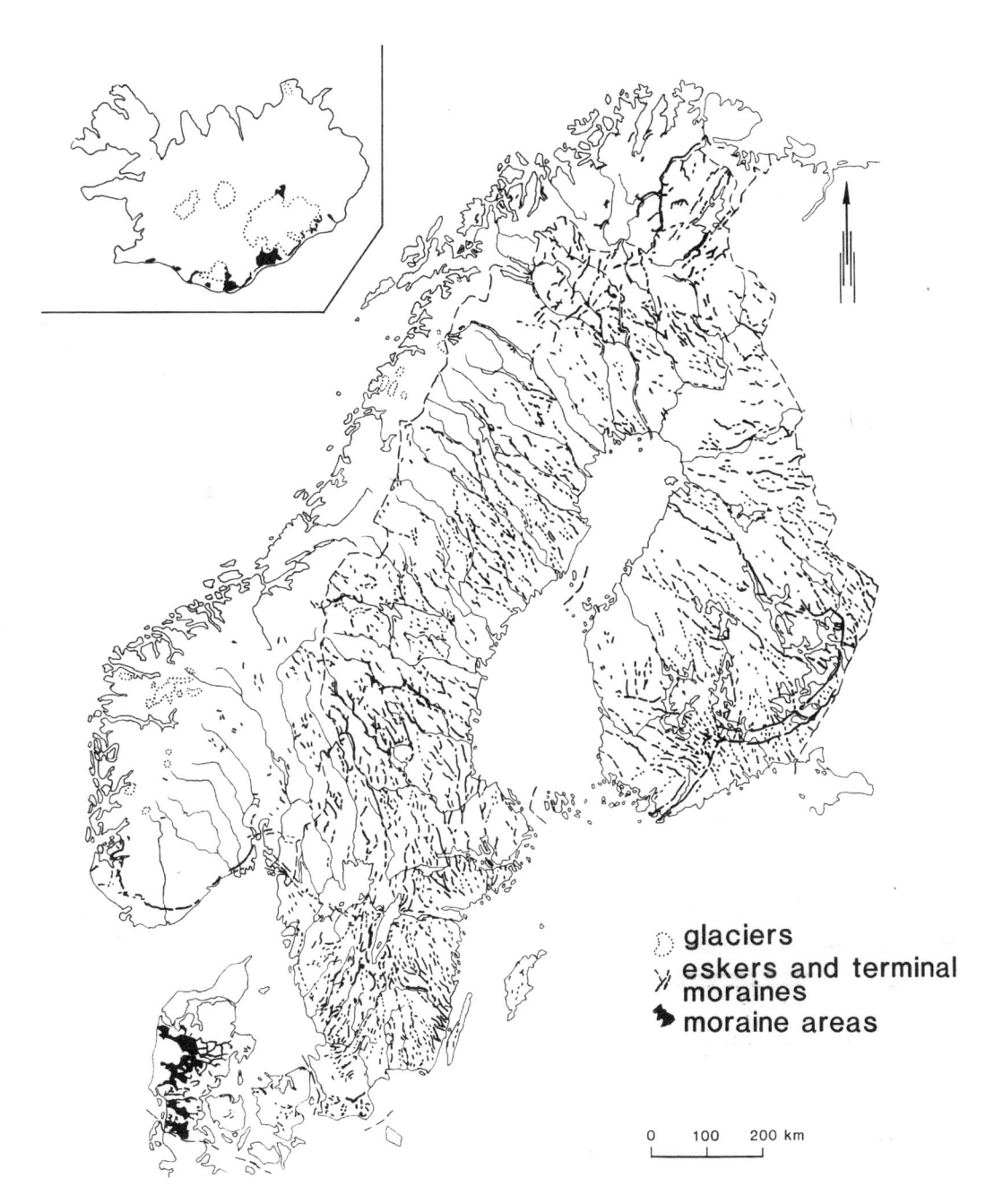

FIGURE 22.2. Quaternary deposits throughout Norden.

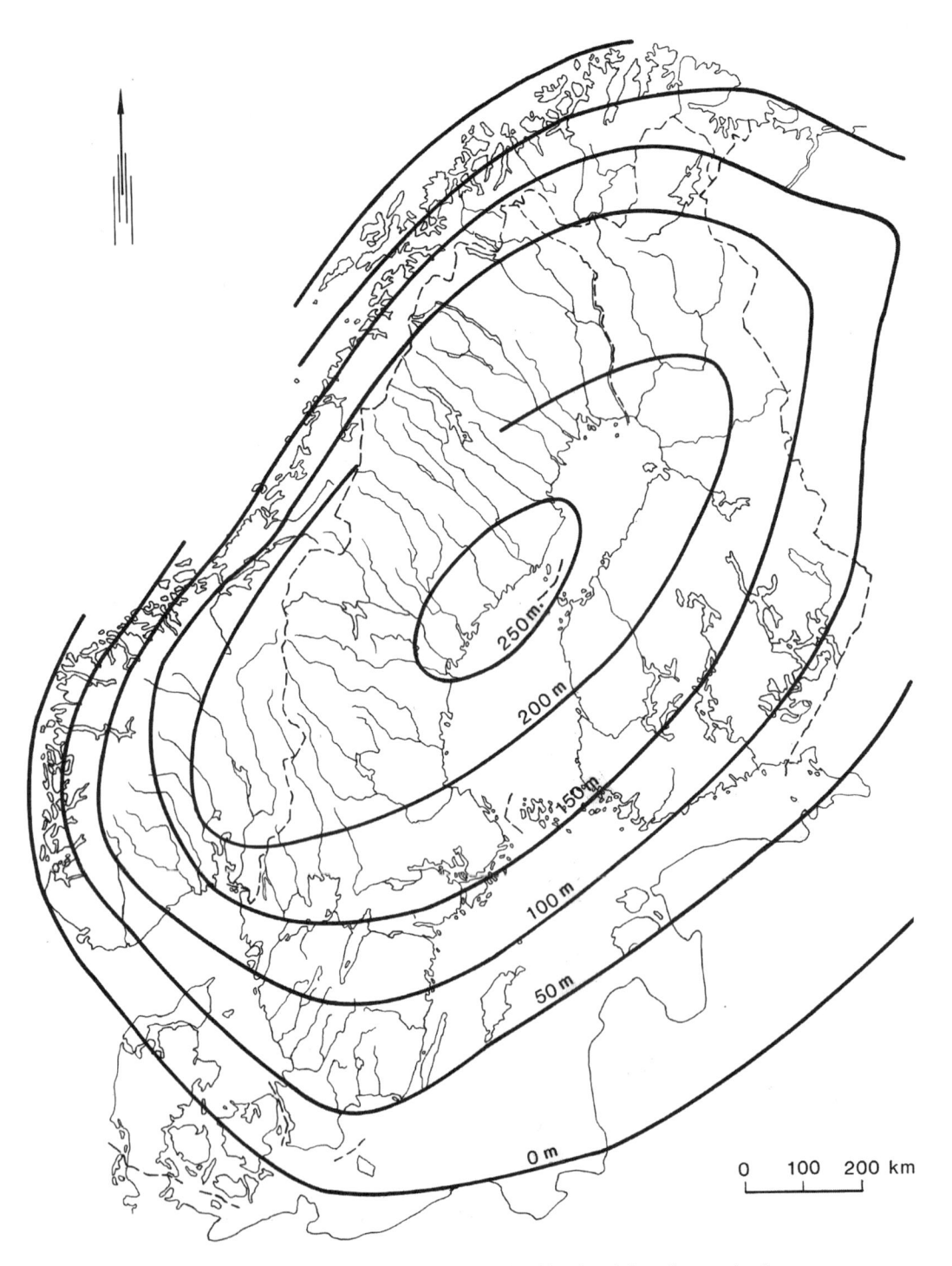

FIGURE 22.3. Shore regression in Norden as a result of land uplift and eustatic changes (the postglacial upper marine limit or highest shoreline, in meters).

been the object of cultivation. Depending on the type of the bedrock, but also on climate and altitude, the soil offers different prerequisites for cultivation. Thus agriculture varies greatly according to the condition of the soil as well as to local climatic conditions. In Denmark, more than 60 percent of the land is under the plow, while in the other Nordic countries the cultivated area covers just a few percent of the land. The natural vegetation is affected by the same factors.

The Bedrock

A geographical border separates the southern and southwestern parts of the region from the rest of Norden. The two large islands in the Baltic, Öland and Gotland, as well as the southeastern corner of Sweden around the region of Kalmar, northwestern Scania (Skåne), and Denmark, differ from the rest of the Nordic countries. In this area, we find stratified rock, around the Baltic mainly consisting of limestone. In Scania and Denmark, deposits from the Cretaceous period (chalk) support an open, cultivated landscape. The remainder of Norden consists principally of the Fennoscandian Precambrian bedrock (the Baltic Shield) underlying Finland, southern Norway, and most of Sweden. Here the till is coarser and the soils are more acid. This is why large areas of forest predominate on the eastern side of the Caledonides, the mountain range that forms the Swedish-Norwegian frontier. It is a relatively young mountain ridge, structured by large transported sheets of rocks pushed up and onto the Baltic Shield. Another extensive shield of Precambrian origin is found around Greenland.

These Precambrian shields were once covered with sedimentary rock (Cambrian, Ordovician, Silurian), and here and there these rocks have escaped erosion. Such areas are rich in clayey till deposits, which have been used for cultivation and permanent settlement in Scandinavia for thousands of years, even above the highest shoreline. Such areas look like open windows in the forest landscape. The greater part of the soil in northern and central Scandinavia and Finland, however, consists of coarse till that is difficult to cultivate, though it is possible to do so in places, especially above the highest shoreline, where finer fractions were created by the ice. Sometimes large moraine ridges dammed up valleys and low-lying ground and created lakes and wetlands, where cultivable sediment soil developed.

The southernmost parts of Scandinavia—Scania and Denmark—consist of younger Mesozoic and Tertiary rock, especially sandstone and limestone strata from the Cretaceous period. These areas consist of clay, marl, and sand, but they were also formed to a great extent during the Quaternary ice age, when large quantities of sand were deposited close to the edge of the ice, creating a hummocky landscape.

From a geological point of view, the landscapes of the Faeroe Islands and parts of Greenland and Iceland differ from the rest of Norden in that they are characterized by Tertiary basalt and lava plateaus belonging to the North Atlantic basalt area. Iceland belongs to a boundary zone along the Central Atlantic Ridge with volcanic activities. The eruptive bedrock produces poor soils and the landscape gets its color from dark basalt and lava stones.

Coast and Seas

In many places the Nordic coastland is markedly archipelagic in nature. On the west coast, the archipelagoes often consist of naked rock. Partly they were once covered with vegetation but later were denuded through human use. The majority of the population of these archipelagoes earned its living by fishing and shipping rather than farming. However, fishing is still frequent in many places along the coast, for instance, in the Lofoten area in northern Norway. But there are also exceptions from the general description of the coastal landscape, such as Jæren, south of Stavanger. Where the Scandinavian mountain range meets the Norwegian Sea, a fjord landscape has evolved. Next to the coast, inside the archipelago, we find low-lying land just above the continental shelf, the *strandflat,* which is suitable for settlement. This is also true along the coast of southwestern Greenland in an impressive fjord landscape.

In Denmark, particularly in northern Jutland, and along the southern part of the Swedish west coast there is a *klittkust* (a coast whose morphology is marked by sand dunes). South of the *klittkust* in Middle Jutland, the coast is affected by tidal movement with accumulation of material that reshapes the coastline. In the far south of western Denmark, the Wadden Sea (Vadehav) begins, shallow and 10–50 kilometers broad. In most places, however, the coasts of Denmark and Scania lack any kind of archipelago or sedimentary accumulation.

Along the coast of the Baltic and the Gulf of Bothnia, the archipelago consists mainly of islands covered with vegetation. They once had relatively large permanent settlements with a diversified means of livelihood. These areas are special because the continuous land uplift has prevented the till from being abraded and totally washed away. The Baltic is the world's largest sea of brackish water. Together with the distinctive archipelagos, this aquatic environment is one of the most singular of the Nordic landscapes.

In addition to the countries along the Baltic coast, the drainage area of the Baltic includes Ukraine and Belarus. This brackish sea is thus fed by a large flow of fresh water, which constantly dilutes the salt content of the sea. The salinity is about one percent in Öresund and less than 0.3 percent in the northern part of the Gulf of Bothnia and the inner part of the Gulf of Finland. The Baltic is thus dependent on an inflow of saltwater from the Norwegian Sea in order to maintain its biological balance. During the twentieth century, the Baltic suffered from heavy pollution.

Finally, limestone cliffs are found, mainly in Denmark but also in Öland and Gotland. Here the coast is straight and normally without an archipelago. As a result of the basalt bedrock, the Faeroes and Iceland are surrounded by a rough, rocky coast with vertical cliffs, which often rise to dizzying heights.

Climate

The length of the Nordic countries from north to south, and their proximity to the Atlantic in the west and to vast inland regions in the east, give the region a variable climate. Added to this, there are great variations in altitude. The average temperature drops by 0.6 degrees C for each rise of 100 meters. The climate is influenced by the Atlantic, especially in Iceland and western Norway. The prevailing winds are from the southwest and contribute to a humid climate, particularly in coastal regions, where winters are comparatively mild and summers are cool, the characteristics of a maritime climate.

Diametrically opposed conditions reign in regions closest to the Eurasian continent in the east. In these areas, winters can be very cold and summers hot and dry, especially in central and eastern Finland. The climate here can be termed "continental," and the same applies to Lapland in northern Finland and Sweden. The low mountains inland from Trondheim in central Norway, however, allow westerly winds to sweep in over central Jämtland, creating oceanic conditions there, thus underpinning the Storsjön district's unique position in the world.

Denmark, Norway, and Sweden, and especially the Atlantic realms of the Faeroes, southern Greenland, and Iceland lie within the sphere of influence of the Gulf Stream and the western cyclone belt. In low-lying areas the precipitation is relatively moderate. However, the altitudes found on the western side of the southern Swedish highlands and particularly in the Norwegian mountain range and in parts of southwestern Iceland, the southern part of Greenland, and the Faeroes are sufficient to give a precipitation of between 1,000 and 3,000 mm per year, sometimes more (Figure 22.4). The rest of Greenland has an arctic climate, which is the reason 80 percent of Greenland is covered by an ice cap—the ice-free land consists only of a coastal strip nowhere more than 200 kilometers wide.

Northern Sweden and Finland form a markedly dry region, as are the inner parts of the Norwegian fjord landscape and northeastern Iceland. Here, however, the mean temperatures are low, resulting in extensive mires. The east coast of Sweden and the island of Gotland in the Baltic have a dry climate. The southern part of the island of Öland, also in the Baltic, has almost a steppe climate with a precipitation of less than 400 mm per year.

The length of the growing season is a crucial factor in cultivation. This is most clearly seen farthest north, where the number of sunshine hours in the summer greatly exceeds the norm in the rest of Norden. Arable farming is possible in certain river valleys, and grassland can be cultivated at modest altitudes far above the Arctic Circle.

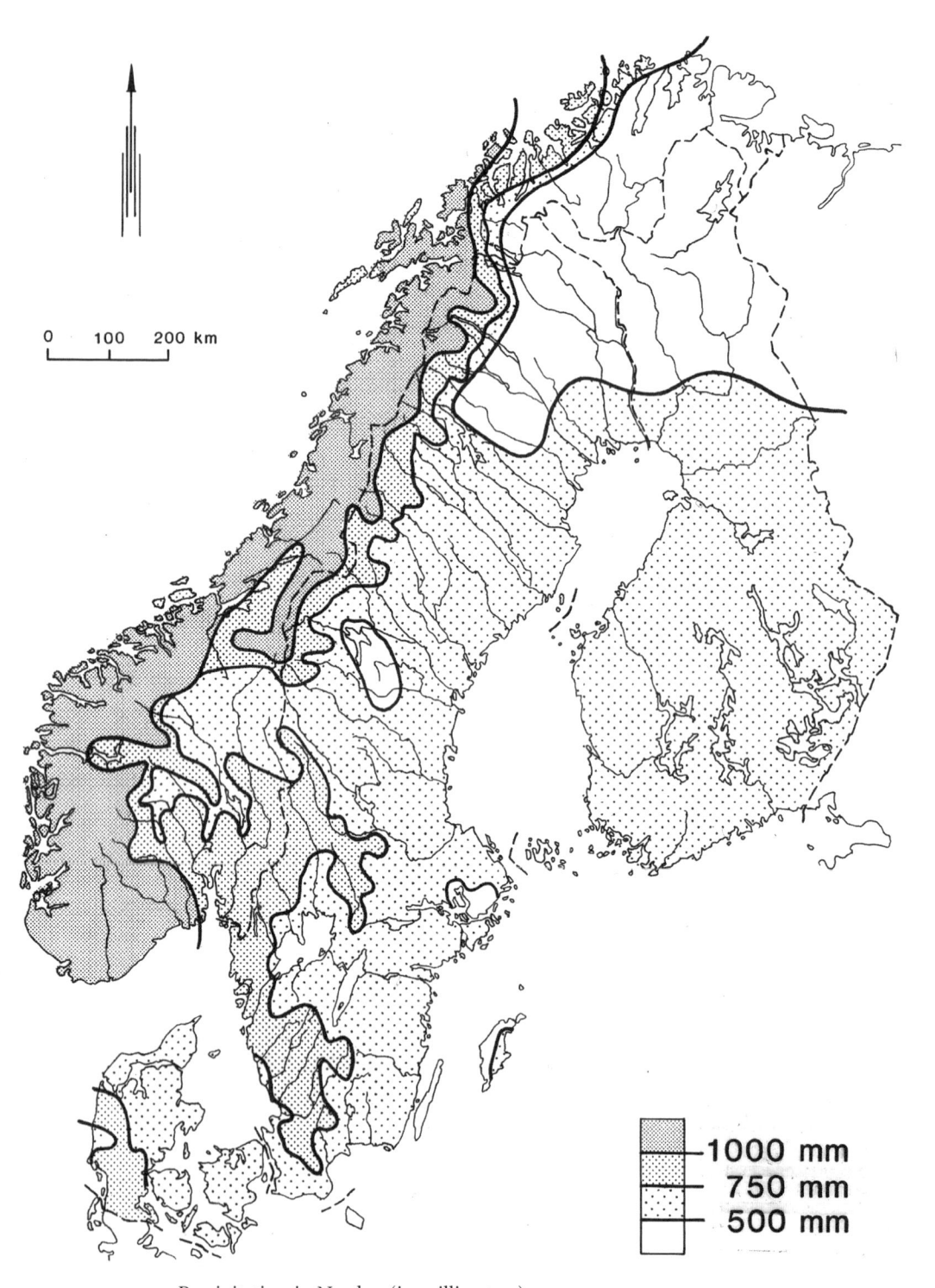

FIGURE 22.4. Precipitation in Norden (in millimeters).

In winter, large areas of the Nordic countries are covered with snow, particularly the southern Swedish highlands and the northern part of Sweden, Finland, and the interior of Norway. The Baltic is often frozen from its northern end and as far south as the Åland Sea and the Gulf of Finland, sometimes even further south.

Natural Types and Zones of Vegetation

Denmark and the southernmost part of Sweden, including the west coast and the far south of Norway, belong to the central European deciduous region (Figure 22.5). The northern limit of this region is the natural line to which spruce extends southward. Beech is the most characteristic tree in the region, but oak, elm, linden, and ash are also plentiful. The flora is rich and extensive grasslands still exist in parts of the southern Swedish highlands. The whole landscape is strongly marked by cultivation, but vast treeless areas are also to be found, for instance, the heaths in Jutland and in western Sweden, although they are rapidly decreasing. In southeast Scania and in Öland, there is steppelike vegetation. Due to extensive drainage projects in the nineteenth century, wetlands are mainly small.

A large region of mixed forest covers the rest of southern Sweden and the southern tip of Norway. The predominant trees in these forests are pine and spruce interspersed with deciduous trees. In the proximity of cultivated areas, hazel was characteristic of meadows and groves. Birch, elm, linden, and ash were used for fodder and many other purposes in the subsistence economies of earlier times. Fairly extensive wetlands can be found in this region. The mixed forest region terminates at the very distinct southern limit of the taiga, where differences in relief often exceed 100 meters.

North of the belt of mixed forest, the enormous Eurasian coniferous forest (the taiga) covers large parts of Norway, Sweden, and Finland. The boundary of the taiga can be very sharp. In Sweden it is known as *limes norrlandicus*. Within this boundary, spruce and pine predominate, spruce growing on the moister land. Birch and some alder are almost the only deciduous trees here. Birch grows in the whole of the coniferous forest region. It also marks the lower limit of the bare mountain and alpine zone, which covers the north of Finland and Sweden and most of inner Norway along the mountain ridge. The timberline varies from an altitude of about 1,000 meters in the south to about 500 meters in the north. In Finland, the proportion of birch increases. The taiga is comparatively poor in species and contains extensive mires, particularly in the far north. Once in this forest region there was a great deal of cultivated land in the form of hay fields and grazing, improved by the water-meadow technique. Today cultivation is mainly in river valleys and coastal districts, but some agrarian settlements are still to be found inland, often on high ground. In addition, this region has long been the home of the formerly nomadic Saami, who now practice a more stationary form of reindeer husbandry.

In Iceland, there are two natural vegetation zones. On lower ground there are some trees, mainly birch and juniper, but human activity and the farming of sheep have reduced the arboreal vegetation. Grasslands dominate the physical landscape, which is also the case in the Faeroes. In both countries, woods are to be found only occasionally in small areas. In Iceland, on higher levels—above the timberline, 600 meters above sea level—the vegetation is alpine. Mires are common above as well as below the timberline. In southern Greenland, the vegetation is of the same character, but on the ice-free areas of mid-Greenland the vegetation has adapted to the extreme

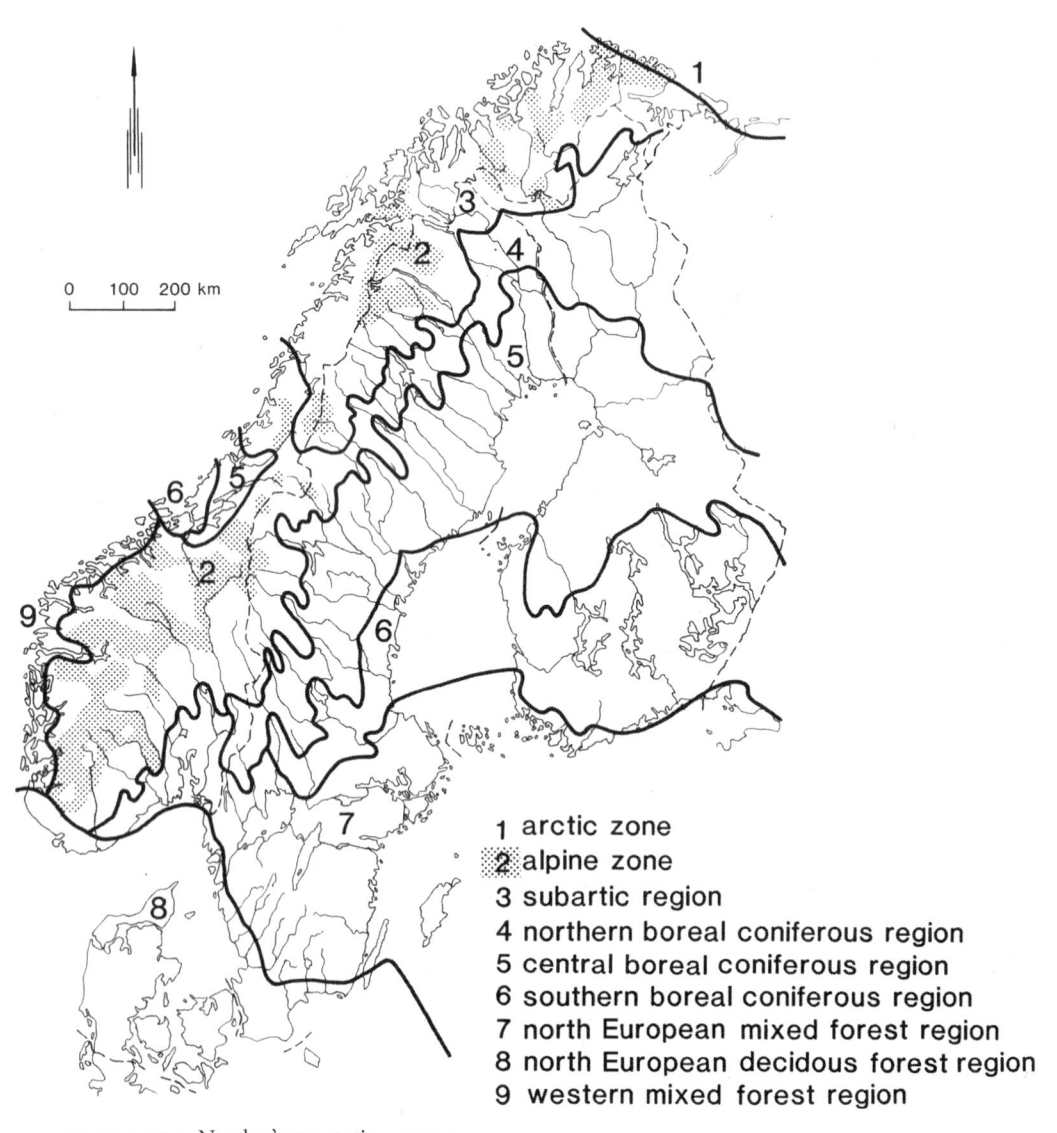

1 arctic zone
2 alpine zone
3 subartic region
4 northern boreal coniferous region
5 central boreal coniferous region
6 southern boreal coniferous region
7 north European mixed forest region
8 north European deciduous forest region
9 western mixed forest region

FIGURE 22.5. Norden's vegetation zones.

arctic conditions with permafrost and long, dark winter periods alternating with a bright summer with long days. Thus, southward-facing slopes often have a short but profuse flowering of plants.

Topography

In southernmost Sweden and in Denmark, the superficial deposits shaped by running water have given the landscape its hilly profile (Figure 22.6). The geomorphologic similarity to parts of Western Europe and the British Isles is sometimes striking. By Nordic standards, the degree of cultivation is high. Other parts of southern and central Sweden and most of Finland consist of what is geologically a plain. Differences in height are seldom more than 50 meters. In the interior of Finland there is a plain with residual hills, with a relief of up to 100 meters. The landscape here is of highland character and was colonized late. These inland regions were not settled until the Scandinavian Middle Ages, from A.D. 1050.

The central Swedish depression and southwestern Finland consist of fissure valley terrain. Tectonic movement over millions of years has broken down the even surface of the bedrock and produced fissures. These fissures were then eroded by the ice and today form a pattern of straight, often cultivated valleys running through a landscape covered with coarse till—not seldom in the form of meandering eskers—and forest. The whole of this landscape lies below the highest shoreline and fine material has thus formed sediment in the lower parts of the terrain, especially in the central Swedish depression and Östrobothnia in Finland.

Hilly terrain recurs in the inner parts of northern Finland and Sweden, but here the differences in height are greater—more than 100 meters. Here we find an immense undulating forest region, which was long characterized as wilderness and colonized fairly late as farmland. Hunters and gatherers had long used this region, but it was only in the seventeenth and eighteenth centuries that it became settled, among others by Finns, who supported themselves by slash-and-burn farming. The nomadic Saami people have continuously used large parts of this region.

An exception in these forests is the Storsjön district in Jämtland, for here the calcareous sedimentary bedrock did not undergo any changes when the mountain ranges were formed—the bedrock of this particular section was not metamorphosized. Thus, around Lake Storsjön, we have a rich agricultural district whose roots go back to prehistoric times. The district lies on the 63rd parallel and more than 300 meters above sea level, which makes the region unique in a sense, although with parallels in the neighboring province of Trøndelag, west of the Norwegian border. The rest of the extensive Norwegian-Swedish border consists of forestland or the Caledonides mountain range. Especially in Norway, we meet an alpine region characterized by relatively young mountains, often dramatic in shape.

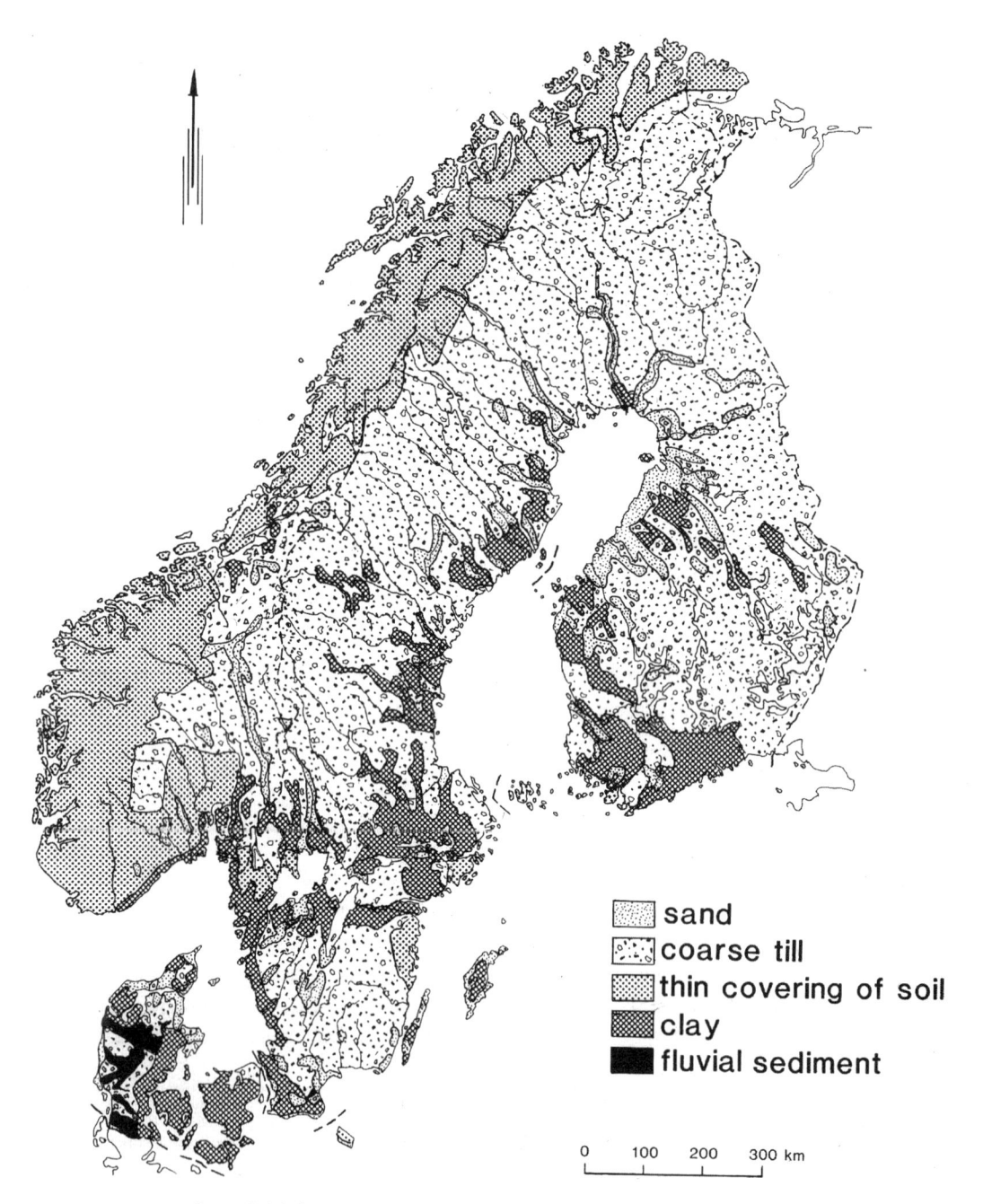

FIGURE 22.6. Superficial deposits in Norden shaped by running water.

The greatest part of Iceland as well as the Faeroes consists of Tertiary basalt. Morphologically these countries form a plateau-and-fjord landscape. The mean level of this lava plateau is generally 300–500 meters above sea level, often with a dramatic morphology, especially on the western side of the Faeroes.

Iceland is divided in two parts through a rift in which vulcanism is concentrated. Iceland has thirty more or less active volcanoes, of which Askja, Eldfjell, Hekla, and Katla are among the most famous, having had several eruptions in very recent times. More than 10 percent of the land surface is covered with glaciers, of which Vatnajökul is by far the largest—8,400 square kilometers. In the valleys, the glaciers often create deposits in the form of *sandurs,* extensive areas of sand and running water.

Settlement and Cultivation

Because of the unfavorable conditions for cultivation in much of Norden, the population is still concentrated in those areas that were colonized and cultivated early on. Today, between 70 percent and 85 percent—in Iceland more than 90 percent—of the population lives in built-up areas. These areas are often clearly linked to the earliest cultivated rural regions: Zealand (Sjælland), where Copenhagen (København) is situated, Funen (Fyn), and South Jutland (Søndrejylland) in Denmark, southern Finland, the Reykjavik region in Iceland, the areas around Oslo, Stavanger, and Trondheim in Norway, and, in Sweden, eastern central Sweden, the region around Göteborg, and the Öresund region in the south. This process of urbanization means that newly colonized country districts, especially in the north, are extremely sparsely populated. On average, about 35 percent of the total area of Norden consists of mountains and lakes and about 55 percent forest. The exceptions are Denmark, where the cultivated area exceeds 60 percent, and Iceland and the Faeroes, which have practically no forest at all. Built-up areas form just a few percent of the total area. Except for Denmark, the cultivated area varies from just 1 percent (Iceland) to not more than 8 percent (Sweden) (Figure 22.7). To sum up, these natural conditions once made cultivation in the broadest sense possible:

1. Ice has ground down the calcareous bedrock to fertile till clay. This soil is found in Denmark and Scania but also in the islands of Öland and Gotland, where we find remains of Cambro-Silurian rock and chalk. A large part of the rural population still lives in these regions today. Minor districts in which these conditions prevail are to be found, for instance, in central and southern Norway, the Storsjön district in Jämtland, and in Östergötland and Västergötland in southern Sweden. Apart from the clay till, Denmark is very much characterized by terminal moraines and, in Jutland, outwash plains.
2. Below the highest shoreline there are postglacial clays, mainly along the coasts,

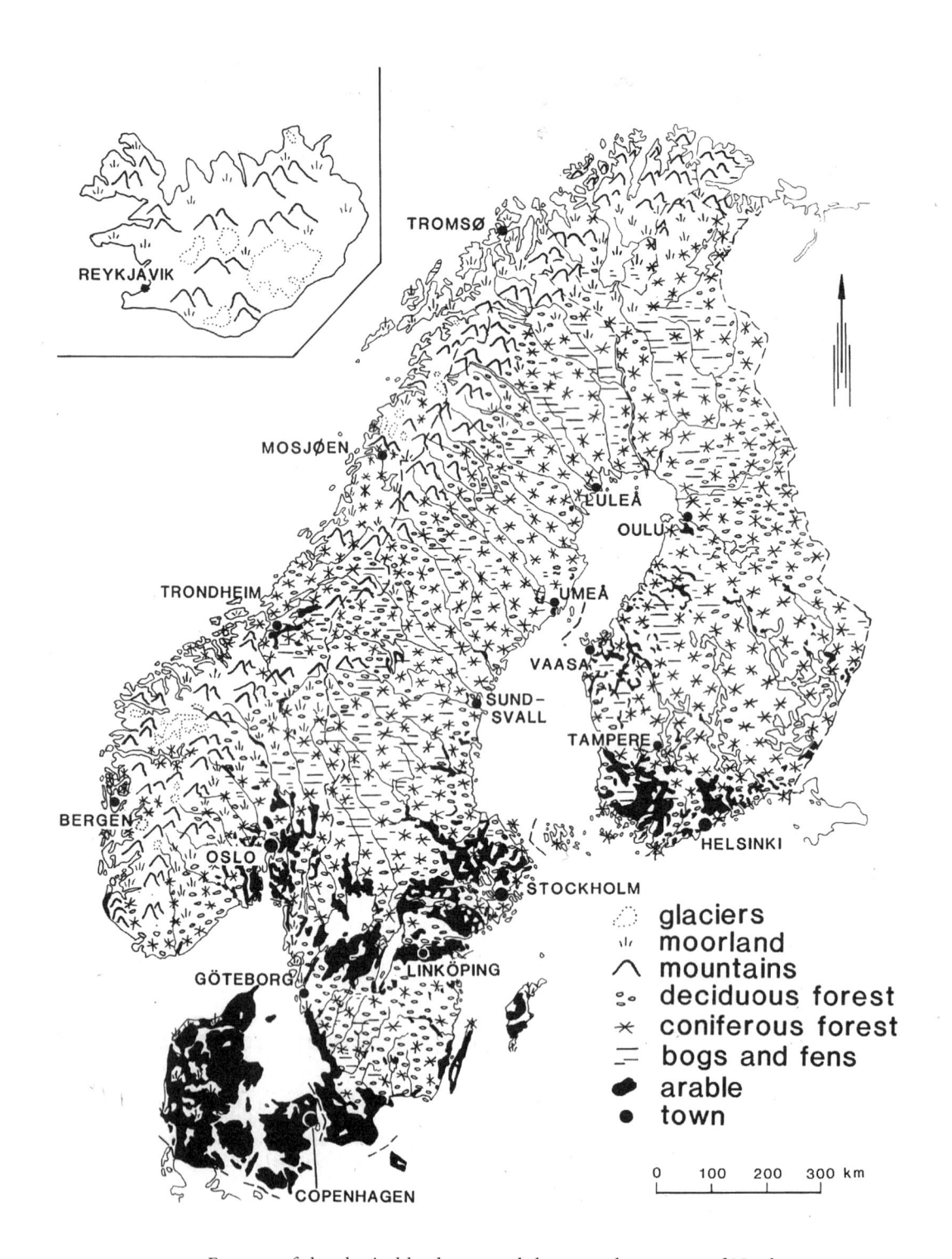

FIGURE 22.7. Features of the physical landscape and the natural resources of Norden.

and in a region extending from southern Norway to the central Swedish depression and southwestern Finland. The majority of the population of Denmark, Norway, and Sweden lives in or near these low-lying areas and not far from the coast.

3. Most of the fluvial material has been deposited along the banks of the river valleys and created conditions suitable for cultivation, mainly below the highest shoreline. This is particularly typical of northern Finland and Sweden and to some extent of Norway.
4. Cultivable soil is to be found in certain places above the highest shoreline, often on high ground, which also gives certain climatic advantages. Settlement of this kind is to be found frequently in Norway and Sweden.
5. In the nineteenth century, the water level of practically every lake and watercourse was lowered in those parts of southern and central Scandinavia that were then under cultivation, with the object of increasing the cultivable area. Now most of this cultivated ground lacks importance because the organogenic soils have oxidized. Sometimes, however, the lowering of the water level of a lake was successful. One of the best-known cases is Lake Hjälmaren in Sweden, where the water was lowered by 1.9 meters, thus reclaiming about 190 square kilometers, which are still under cultivation.

These five prerequisites have shaped the historical rural Nordic landscape, consisting of scattered hamlets, small villages, and sometimes single farms (Iceland and Norway). The transformation of the rural landscape brought about by land reforms in the nineteenth century obliterated many features of the historical physical landscape, especially grasslands and meadowlands. Inspired by movements in Britain and the Netherlands, a new field layout was introduced in central districts such as Denmark and Scania. Farms thus became dispersed over large areas. This coincided with the beginning of the industrial revolution in northern Europe. Most of the early nineteenth-century preindustrial landscapes are gone. The effects of modern agriculture with intensified production are a shrinking acreage, while the size of the farms is increasing. Enlarged fields have led to the disappearance of impediments and old transition zones in the rural landscape. The use of artificial fertilizers affects the metabolism of nature. With the exception of Denmark, just a few percent of the population in the Nordic countries are still involved in rural production.

On the other hand, shortly after World War II, urbanization accelerated in the former central rural districts and a large number of new towns and suburbs were planned, often in the form of satellite towns around old urban centers, parallel with growing industrial districts. Suburbs created by a policy of welfare and social non-segregation have in recent decades nonetheless become highly segregated, especially in Denmark and Sweden.

At the same time there has been a dramatic change in the main city centers. Competition for the best locations has led to an internal differentiation in the use of land and buildings. City centers are occupied by offices, central services, and specialized shops. Housing areas near the city center have become attractive and, in a sense, a new era of segregation has been initiated, among other things because of a belief in a free housing market. Thus, during the last hundred years, most parts of the Nordic countries have changed from a rural society into a strongly urbanized part of the industrial world.

REFERENCES

Bernes, Claes. 1993. *The Nordic Environment: Present State, Trends, and Threats.* Nord 1993: 12. Copenhagen: Nordic Council of Ministers.

Fredén, Curt, ed. 1994. *Geology.* In *National Atlas of Sweden.* Stockholm: SNA Publishing.

Helle, Knut, ed. 2003. *Cambridge History of Scandinavia,* vol. 1. Cambridge: Cambridge University Press.

Nationalencyklopedin: Ett uppslagsverk på vetenskaplig grund utarbetat på initiativ av Statens kulturråd. 1989–96. 20 vols. Höganäs: Bokförlaget Bra Böcker.

Nationalencyklopedin. 1990. "Danmark," 4: 379–404.

Nationalencyklopedin. 1991. "Finland," 6: 265–91.

Nationalencyklopedin. 1992. "Island," 9: 599–609.

Nationalencyklopedin. 1994. "Norge," 14: 241–62.

Nationalencyklopedin. 1995. "Sverige," 17: 499–538.

Naturgeografisk regionindelning av Norden. 1977. NU B 1977:34. Stockholm: Nordiska Ministerrådet.

Natur- og kulturlandskapet i arealplanleggingen. 1987.1. Regioninndeling av landskap. Miljørapport 1987:3/NORD 1987:29. Nordisk ministerråd.

Påhlsson, Lars, ed. 1994. *Vegetationstyper i Norden.* TemaNord 1994: 665. Köpenhamn: Nordiska Ministerrådet.

Sjöberg, Björn, ed. 1992. *Sea and Coast.* In *National Atlas of Sweden.* Stockholm: SNA Publishing.

Sömme, A., ed. 1961. *A Geography of Norden: Denmark, Finland, Iceland, Norway, Sweden.* Oslo: Cappelen.

Contributors

Ingvild Austad is professor in landscape planning at Sogn og Fjordane University College in Sogndal, Norway. She is educated as a landscape architect, with further qualifications in planning and vegetation ecology. She has written on the management of human-influenced vegetation types and other features of historical landscapes in western Norway.

Gabriel Bladh is associate professor in human geography at Karlstad University, Sweden. His areas of specialization include historical geography, environmental and human influences on forests, theoretical perspectives connected to nature–society relations, and the history of geographical ideas. He has studied the migration of Finns to Sweden in the sixteenth century, especially Finnish settlement and the changing landscape of the region of Värmland in western Sweden from the sixteenth century to the present.

Tomas Germundsson is associate professor in human geography at Lund University, Sweden. His research concentrates on landscape and settlement changes in southern Sweden, especially Scania (Skåne), since the eighteenth century. He coedited an atlas of Scania as part of *The National Atlas of Sweden*, and his current research focuses on landscape changes, culture, and power relations on the landed estates of Scania.

Jens Christian Hansen is professor emeritus in geography at the University of Bergen, Norway. He has written extensively about one-company industrial towns, regional problems and regional development policies, and settlement patterns and population change in Norway.

Kirsten Hastrup is professor of anthropology at the University of Copenhagen. She has written three monographs on Icelandic history and culture and has published widely on general theoretical and methodological issues, drawing not only from her regional study but also from research in theater and human rights.

Leif Hauge is associate professor of landscape ecology at Sogn og Fjordane University College in Sogndal, Norway. He has written on the historical exploitation of natural resources and its influence on the landscape, including how different fodder-collecting techniques and grazing have produced distinct vegetation types, buildings, and structures. He works with photographic and video documentation of disappearing landscapes.

Maunu Häyrynen is professor of landscape studies at the University of Turku, Finland. He has published on the history and conservation of urban parks, on garden history, on lived maritime landscape, and on Finnish national landscape imagery. He has also worked as a consultant in garden and landscape conservation.

Margareta Ihse is professor of ecological geography at Stockholm University, Sweden. Her research interests are the study of landscape and vegetation with respect to biodiversity and nature conservation, environmental monitoring, and physical planning. She has published on present and historic ecosystems in the Nordic countries, including boreal forest, agricultural landscape, grasslands, mires, and mountain ecosystems.

Michael Jones is professor of geography at the Norwegian University of Science and Technology, Trondheim. He is author of *Finland, Daughter of the Sea* and has published on landscape change in Norway, landscape and planning, agricultural policies and environmental management, land tenure, cartographical history, and the concept of "cultural landscape."

Ari Aukusti Lehtinen is professor of geography at the University of Joensuu, Finland. His special interests are environmental conflicts in the European north, northern boreal forestry, and forest-based development in the Russian taiga. His books include *Northern Natures: A Study of the Forest Question Emerging within the Timber-Line Conflict in Finland*, *The Fall of the Forest Villages: Ecological and Cultural Conflicts in the Russian Taiga*, and *Postcolonialism, Multitude, and the Politics of Nature: On the Changing Geographies of the European North*. He is the coeditor of *Politics of Forests: Northern Forest–Industrial Regimes in the Age of Globalization*.

Anders Lundberg is professor of geography at the University of Bergen, Norway. His scientific interests are in landscape geography, historical geography, ecology, cartography, GIS, and environmental science. He has published on coastal biodiversity and ethnobotany, vegetation and soils, cultural landscapes, and nature conservation in Norway, as well as on environmental problems in Peru.

Contributors

W. R. Mead is emeritus professor of geography at University College London. His books include *Farming in Finland, An Economic Geography of the Scandinavian States and Finland, Finland, An Historical Geography of Scandinavia, An Experience of Finland,* and *A Celebration of Norway.*

Ann Norderhaug is head of cultural landscape research at the Norwegian Institute for Agricultural and Environmental Research and works at the Kvithamar Research Center, near Trondheim. She has written on rural landscape dynamics in the Nordic countries, including ecology, environmental history, resource use, landscape change, and landscape management.

Venke Åsheim Olsen is an ethnologist and has worked as museum curator and cultural researcher in northern and central Norway. She has written about the Finnish minority and ethnic relations in North Norway, women and cultural history, and multicultural relations and ethnicity in the Nordic countries.

Kenneth R. Olwig is a geographer and professor in landscape planning, specializing in landscape theory and history of landscape, at the Swedish University of Agricultural Sciences, Alnarp. Among his books are *Nature's Ideological Landscape* and *Landscape. Nature, and the Body Politic.* He has written extensively on conceptions of nature and landscape, the ideology of national parks, children and the environment, literature and landscape, and heritage interpretation.

Anssi Paasi is professor of geography at the University of Oulu, Finland. He has published on the history of geographical thought, theories of region and place, and the links between territories, boundaries, and individual and social consciousness. His books include *Territories, Boundaries, and Consciousness: The Changing Geographies of the Finnish–Russian Border.*

Helle Skånes is assistant professor in ecological geography at Stockholm University, Sweden. Her research experience covers the dynamical interface between physical geography, human geography, and ecology using remote-sensing techniques, GIS, and cartography. She has published in the field of ecological geography, focusing on spatial changes in the rural landscape over time and their implications for biodiversity on a landscape level.

Bo Wagner Sørensen is an anthropologist and associate professor at the Danish Research Center on Gender Equality at Roskilde University, Denmark. His research interests include cultural identity and ethnicity, cultural politics, gender, alcohol use, material culture, and migration. He has published on contemporary social phenomena in Greenland.

Ulf Sporrong is emeritus professor in geography at Stockholm University, Sweden. He has published in historical geography and historical cartography, focusing on the long-term evolution of rural landscapes and settlement patterns in Sweden.

Nils Storå is emeritus professor of ethnology at Åbo Academy University in Turku (Åbo), Finland. His books include *Burial Customs of the Skolt Lapps*. He has written on a variety of topics in cultural anthropology and cultural history, including preindustrial hunting, fishing, and gathering cultures in northern Europe and the cultural history of ice.

Arne Thorsteinsson is an archaeologist and former director of the National Museum of the Faeroe Islands. He has published on Viking Age archaeology and history, and the history of settlement, farming, land tenure, and land valuation in the Faeroes.

Index

The Danish and Norwegian letters æ, ø, and å are alphabetized in the international manner as ae, o, and a.
The Swedish letters å, ä, and ö are alphabetized as a, a, and o. The Icelandic Þ is alphabetized after t.

Figures and Tables are indicated by italics.

Aabel, Wilhelm 204
Aakjær, Jeppe 45n
Aarhus 16, 25
Aario, Leo 527–29
abandonment of farmland: Åland 429; Faeroes 82, 96; Finland 564; Norway xxii, xxiii, 380, 383, 384, 401, *403*, 405, 409–10, 414; Sweden 153, 187, 209, 256, 257, 262, 272, 275
Abelsen, Emil 125
Åbo *see* Turku
Absalon (bishop) 165
absolutism 29, 35, 39, 43n, 175, 184, 222, 307
Acerbi, Joseph 422, *488*
acid rain 561
administrative units: Åland 444; Denmark 43n; Faeroes 86–87; Finland 522–25, *523*, *525*, 530; Greenland 132n; Iceland 62; Norway 307, *307*, 316, 335–36; Sweden 146, 157–58, 263
ærgi (Faeroes) *see* summer farms

aerial photography 254, 255, 258, 263, 348, 368, 404; aerial photographs *257*, *261*, *265*; CIR (color infrared) photography 257, 258, 261, 265
Agder 366
Agenda 21: 252
agricultural landscape 283; Finland 507; hand-dug landscape (Western Norway) 366–67; Norway 283–84, 286–87, 294, 295; Scania 160, 166, 167, 179, 186; Sweden 253, *256*, 259, 266, 267, 268, 270, 272, *273*; Western Norway 373, 382–85, 387, 390
agricultural policies: Denmark 547; Finland 564; Norway 284–86, *285*, 294, 363, 367, 555; Sweden 559; *see also* agricultural subsidies
agricultural societies: Norway 383; Royal Agricultural Society (Denmark) 39; Sweden 234; *see also* Danish Heath Society

agricultural subsidies: Area and Cultural Landscape Payments (Norway) 286, map 285; Norway 284–86, 383, 397, 555; Switzerland 555; *see also* agricultural policies

Agricultural University of Norway 287

agriculture 166, 345, 568, 573, 577, 581, 583; Åland xxv, 429, 443, 444, 446, 448–50, 452; Dalecarlia xix, 196, 198, 203–204, 206, 207–209, 211–17; Denmark xv, 4, 5, 345, 367, 546, 547; Faeroes xvii, 77, 81–84, 86, 88–91, 97, 99, 100, 101, 103; Finland xxiii, 426–27, 428, 432, 489, 506, 529, 536, 564; Iceland 59, 63, 552; North Norway xx–xxi, 286, 291, 292, 294, 295, 311, 325, 326, 327, 328, 331, 334; Norway xx–xxi, 283, 286, 287, 295, 355, 367, 401, 555, 556; Scania 149, 158–59, 160, 162, 163–64, 165, 167, 171, 174, 176–86; southern Norway xxiii, *403, 404–10, 409*, 414; Sweden xv, xviii, xx, 143, 145, 147, 149, 150, 152, 223, 224, 252, *256*, 258–59, 263, 266, *273*, 558, 559; Swedish *hagmark* 251, 257, 259, 260, *260, 261*, 263, 266, 268, 269, 270–77, maps *262, 264*; Värmland 225, 226–27, 232, 233, 234–35; Western Norway xxii–xxiii, 348, 349, 350–55, 358, 359, 360–61, *360*, 362, 363–65, 366–68, 372–73, 375–80, 381, 382–83, 384–90, 392, 397; *see also* abandonment of farmland; agricultural landscape; agricultural policies; agricultural societies; agricultural subsidies; animal husbandry; agropastoral landscape; autumn farms; crofters; cottars; fodder; field systems; grazing; infield(s); meadows; open-field system; outfield(s); pastures; seasonal movement of livestock; spring farms; summer farms

agriculture, crop rotation in 89, 176, 185, 233, 274, 354, 355, 364, 382, 407, 409

agriculture, fallow 149, 176–77, 226, 364, 368, 407

agriculture, fertilization in: artificial fertilizer 195, 270, 275, 360, 361, 363, 382, 383, 404, 409, 410, 415, 583; flooding 163; guano 91; industrial production of fertilizers 556, 565; manure 39, 89, 176–77, 182, 358, 361, 363, 364, 375, 379, 383, 390, 406, 407, 408; seaweed 86, 89, 354–55, 364, 368

agriculture, growing season 575; Denmark 546; Finland 462, 464; Western Norway 374

agriculture, intensification of 583; Denmark 39, 40, 546; Norway xxii, xxiii, 286, 287, 363–64, 368, 382, 383, 385, 401, 414, 415, 555; Scania 158, 163, 176; Sweden 275

agriculture, marginal: Faeroes 82; North Norway 286, 291, 294, 295; Norway 287, 401, 555; southern Norway 406, 409; 275, 276, 295; Sweden 275, 276; Western Norway 379, 380, 383, 384, 388

agiculture, mechanization of: Denmark 546; Norway 295, 409, 555; Sweden 152, 185, 258; Western Norway xxii, 355, 358, 360–61, 363, 368, 372, 382–83

agriculture, modern 583; Denmark 547; southern Norway 409–10, 414; Sweden 178, 272, 558; Western Norway 360, 378, 382–83

agriculture, multifunctional 251, 390

agriculture, organic 547

agriculture, overproduction in: Norway 286, 555

agriculture, restructuring of: Norway 295, 356, 359, 373, 385, 555; Finland 564

agriculture, slash-and-burn: Finland 425, 426, 462, 468, 470, 564; Finns in Sweden 227, 235, 579; southern Norway 406; swidden (Denmark) 15

agriculture, specialization of: Denmark 547; Norway xxiii, 364, 381, 383, 384, 401; Sweden 258, 273, 275

agriculture, subsistence xx, 58, 295, 429, 447, 577

agropastoral landscape: southern Norway xxiii, 401–18; Swedish hagmark xx, 251–80; *see also* pastures

Aho, Heikki 501

airports: Copenhagen 549; Keflavik (NATO base, Iceland) 551–52; Porsanger (North Norway) 325; Thule Air Base (Greenland) 550

Åland (Ahvenanmaa) ix, xiv, xxiv–xxv, xxiii,

429, 436, 440–57, 476, *495,* 497, 507, 543, 562, 566, map *441*
Åland Museum 455
Åland Sea 444, 447, 577
Ålänning, Gunder 443
Ålänning, Peter 443
Alaska 113, 335
alcohol 15, 117, 547
Alfred (King of England) 310
Alfvén, Hugo 201
Alps 504
Alstahaug 322–23, 335
Alta 329, 330
Altafjorden 326
Althing *see* parliaments
aluminum production 552–53, 557
Amager 187
Ambjørndalen 402, 411
America (North) xii, xiii, xiv, xxviii, 153, 562, 564; Norse discovery of xvi, 55, 59; *see also* United States of America
Amiel, H. F. 435
Åminne 433
Åmotsdal 409
Andersen, Hans Christian 12–13, 14, 15–16, 17, 21–23, 24–25, 27, 35, 38, 40, 42n, 201
Anderson, Benedict 134n, 486, 518
animal husbandry: Åland 444, 448, 452; Denmark 546, 547; Faeroes 77, 81, 82–84, 88, 89–91, 94, 97, 101, 103, 104; Finland 564; Jutland 15, 39, 45n; livestock trade 381–82; livestock droves 382; North Norway 291, 333; Norway 555; Scania 149, 164, 165, 167, 176–77, 185, 226, 227, 232, 235; southern Norway *405,* 406, *407,* 407–10; Sweden 150–51, 154; Swedish *hagmark* 251, 274, 275, 277; Western Norway xxii–xxiii, 350, 352, 364, *365,* 375–79, *376, 377, 378, 379,* 381–82, 383, 384, 385–87, *389,* 389–90, 395; *see also* agropastoral landscape; autumn farms; cattle; fodder; goats; grazing; horses; meadows; pastures; pigs; reindeer; seasonal movement of livestock; sheep; spring farms, summer farms
Ankarcrona, Gustaf 200
Antarctic Treaty 557

anthems xxv, 22–23, 112, 133n, 228, 453–54
anthropology, landscape in xiii, 53, 108-10, 115, 133n, 231–32, 288–91, 512
Anttila, Jorma 486
Appadurai, Arjun 110
aquaculture *see* fish farming
Arabia ceramics 565
Arason, Jón 71–72
Arcadia 167–68, 195
archaeological sites: Jutland 16; Scania 162; southern Norway 405; Sweden 268, *268,* 270, 272; Western Norway 388
Archangel 329
archipelagoes 574; Åland xxiv–xxv, 429, 446–47, 448–51, *449, 452,* 456, 566; Baltic 143, 574; Danish islands 3–4, 7, 8–9, 24, 159, 545, 549; Faeroes xvii, 77, *80;* Finland, coast 422, 443, 451, 561; Finland, inland archipelago 561; Gulf of Bothnia, 574; Norway 292, 310, 322, 326, 553, 574; Skagerrak coast 143; skerry guard *(skärgård)* 446–47, 451, 553, 561; Svalbard 553
architecture 53, 64, 153, 440, 487, 493, 500, 507, 513, 534
Arctic 106, 109–10, 112, 113, 115, 126, 300, 308–10, 314, 332, 553, 554
Arctic Circle 291, 310, 320, 329, 436, 550, 553, 561, 575
Arctic Ocean 308, 310, 329, 498, 563
Årdal 384, *395*
Ari (Ari inn fróði) 56, 74n
art xi, xii, 109, 172, 173, 221, 222, 314, 485, 511, 512, 514, 534; Åland xxv–xxvi, 4, 453; Dalecarlia xix, 199–200, 201, 208; Denmark 4, 23, 24, 25, 27; Dresden school of painters 43n; Düsseldorf painters 392; Finland xxv–xxvi, 423–24, *424,* 425, *425, 426,* 428, 435, 462, 463, 466, *466,* 483–510, *488, 490,* 520, 521, 522, 534, 535, 536; Greenland 111, 128, 134n; land art 393, *395;* North Norway 315, 319, 322, 323; rose painting (Telemark) xxiii, 411–12; Scania 170, 172–73; Sweden 141, 147, 159, 162, 172–74, 222, 244, 268; Telemark school of painters 412–13, *413;* Värmland 222; Western Norway xxiii, 372, 373, 390–94, *391, 393, 394, 395,* 396, 398;

wood carving (Telemark) xxiii, 412, 414; *see also* landscape painting; national romanticism
Askja 581
Astrup, Nikolai 392, *393*
Atlantic Ocean xv, 4, 79, 309, 311, 547, 550, 575; *see also* North Atlantic
atlases 3, 434, 489, 529; *Atlas of Finland* 527–529; *National Atlas of Norway* xxii, 345; *National Atlas of Sweden* 155n, 247n, 266; *Pontoppidan's Danish Atlas (Pontoppidans Danske Atlas)* 20, 43n
Aulanko 500
Aurland 382, 384
Austad, Ingvild 412
Australia 58, 59, 409
Austreim 359
Austria 545
authenticity xiii, 487; contested (Greenland) 106; Finland 465, 468, 470, 504; Iceland 58
autonomy ix, xix, 22, 534; Åland xiv, xxiv–xxv, 440, 442–43, 445–46, 447–48, 450–51, 452–53, 455, 456, 566; Faeroes xv, 543, 545, 550; Grand Duchy of Finland 489, 520; Greenland (Home Rule) xv, 112, 114, 123, 125, 131, 132, 133n, 543, 545, 550; House of Autonomy *(Självstyrelsesgården)* (Åland) 453; Iceland 551
autumn (fall): colors 313; fodder collection 386, 408; grazing 82, 365, 376, *377*, 378, 388, 389, 390; herding 82; *see also* seasons
autumn farms: southern Norway 406–408, *407*; Western Norway *377*, 378; *see also* spring farms; summer farms
avalanches 377, 379, 389

backwoods 459–60, 462–64, *469*, 470–72, 474, 477, 478n
Balestrand 392
Baltic: Baltic Shield 573; Baltic Sea (Østersøen, "Eastern Sea") xv, xxiv, 4, 13, 16, 23, 42n, 143, 157, 263, 322, 422, 443, 451, 476, 545, 558, 574, 577; climate 575, 577; countries ix, 5, 254, 574; frozen *488*, 558, 577; islands 573, 574, 575; region xix, 162, 165, 169, 174,
545, 558, 573, 574, 575; shoreline changes 570, *see also* shore displacement
Banse, E. 435
Bardu 324
Barentsburg 557
Barents region 301, 335
Barents Sea 5, 291, 292, 309, 310, 330
Barrow, John 422
barrows *see* grave mounds
Basso, Keith H. 109
Bateson, Gregory 72
beach *see* seashore
Bear Island (Bjørnøya) 309, 310
bears 324, 377, 395, 408, 458–63, *461, 463*, 470–71, 472, 477, 478n; polar bears 69, 323, 332
bedrock 254, 258, 570, 572–74, 575, 581; Denmark 546; Faeroes 573, 575; Finland 564, 579; Greenland 573; Iceland 573–74, 575; Norway 404; Sweden xviii, 141, 143, 160–61, 224, 260, 266, 270, 271, 272
Belanger, Louis 424
Belarus 574
Belt, Great and Little 24, 168, 174
Benediktsson, Hreinn 74n
Benediktsson, Jakob 74n
Bergen xxiii, 283, 321, 329, 381, 382, 553, 556
Bergsäng 203
Bergslagen 146, 148–49, 195, 226, 231, 232, 238, 559, 560
Berndtson, Gunnar 497
Berntsen, Arent 99
Bertelsen, Reidar 293–94
Berthelsen, Jess 14, 134n
binary oppositions 311, 333
biodiversity 251, 252–55, 267–68, 276–77; Convention on Biological Diversity 252; Denmark 26; Finland 468, 474; Norway xx, 286, 287, 294, 401; southern Norway 402, 414–15, 416; Sweden 258, 266; Swedish *hagmark* 257, 261, 262, 275–76; Western Norway xxii, xxiii, 368, 385, 387, 390
biographies 59, 65, 455; autobiographies 315, 429, 529; biographical place narratives 505
biotopes 253–54, 255, 266–67, 269, 275, 286, 402, 468; key 267, 270; small 255, 259–60, 262, 267

Bjarkøy 324, 325–26, 335
Bjerregaard, Hans (son) 22
Bjerregaard, Hans Jensen (father) 22
Björnberg, Niklas 233
Björn riki (Björn "the rich") 65
Bjørnson, Bjørnstjerne 396
Black Death: Faeroes 82; Iceland 66; Norway 82, 349, 405, 554; *see also* plagues
Black Forest 30
Black Mads (Sorte Mads) 37–38, 41
Blekinge 6, 157, 161, 174, 263, 558
Blicher, Niels 43n
Blicher, Steen Steensen 14, 27, 29, 30, 31–32, 41, 42n, 43n, 45n, 46n; and Dalgas, 39–40; on landscape 34–38
Bloch, Marc 32
Blomquist, Anni 429
Böcker, Carl Christian 423
Bø (Nordland) 320, 335
Bodø 319, 320, 329, 332
Bofors 558
Bohuslän 144, 145, 558
Boisen, F. E. 28
Bomarsund 445, 497
Bonsdorff, Jacob 433
Bonsvatnet, lake 411
Borchgrevink, Anne-Berit 358
borderlands x; Åland xxiv, 440, 443, 447, 450; Denmark–Germany 9, 17, 45, 545; Denmark–Sweden 159, 168; East–West ix–x, xxi, xxv, xxvi, 443, 460, 461, 462, 473, 562; Finland–Norway 307, 327, 328, 331, 333–34; Finland–Russia 432, 458, 470, 474, 489, 494, 498, *501*, 521, 562, 563; Finland–Sweden 7, 155, 440, 562; Finnish frontierlands xxv, xxvi, 460–63, 470, 474, 476, 487, 498, *499*, 500, *501*, 526; geographical 224, 225, 573; Norden ix–x, 141, 545; North Norway 307, 308, 310–11, 323–24, 327, 329, 330, 333–35; Norway–Russia 5, 7, 307, 324, 330, 333–34, 553, 555; Norway–Sweden xix, 7, 225, 227, *236*, 238–40, 296, 327, 328, 332, 579; Novgorod 308, 324; Swedish frontiers 141, 144–45, 155; Wales 40–41; *see also* boundaries; frontiers

bördsjord (inherited ancestral land; udal land) (Sweden) 195, 214, 216
Borg 321
Borgå *see* Porvoo
Boris Gleb 334
Bornholm 4, 162, 175
Botanical Museum, Bergen 283
Bothnia, Gulf of 144, 145, 308, 326, 329, 330, 443, 470, 558, 561, 570, 574
boundaries 37, 63, 166, 176, 270, 274, 348, 401, 413, 512, 515–16, 517, 519, 568, 574; administrative 266, 270, 274, 318, 329, 530; forest 377, 577; graffiti signs 318; infield–outfield 82, 88, 90–91, 380, 404; language 327; national ix, 3, 7, 17, 23, 112, 114, 159, 308, 486, 494, 562; natural xvi, 7, 8, 63, 90; physical 207, 516, 574, 577; property 91, 207–208, 211, 212, 259, 274, 568; regional 161, 163, 494, 517, 526, 530; settlement 87, 90, 126; social 213; *see also* borderlands; seashore
Bourdieu, Pierre 44n, 518, 530
Bowring, John 422
Braun, George 42n, 169–70, 188n
Breiðamerkurjökull (glacier) 69–70, *70*
Breidvatnet, lake 411
Bremen 20
Bremer, Fredrika 194
bridges: Åland 451, *452;* Denmark 43n; Faeroes 94; Norway 380, 393; Öresund xix, 8, 186–87, 549
Britain ix, xiii, xxiii, xxiv, xxviiin, 4, 5, 16, 20, 173, 179, 181, 423, 433, 445, 545, 546, 547, 551, 565, 583; *see also* England; Scotland; Wales
British Isles xxviiin, 55, 422, 579
Bronze Age: Western Norway 367, 375; Sweden 147
Bruntland Commission (World Commission on Environment and Development) 252
building styles: Åland 449; Dalecarlia 199, 201, 202; Denmark 4, 8; Faeroes 78, 86, 87, 90; Finland 422; Greenland 116–17, *117*, 130, 135n; Iceland 62, 64; North Norway 321; Scania 158–59, 176; southern Norway 402, 412, 414, 416; Sweden 143, 147, 223,

269; Western Norway 349, 352, 355, 363, 364, 378, 379–80, 382, 383, 384, 392; *see also* smoke cottages
Bunbury, Selina 422
burial mounds *see* grave mounds
bygd (countryside; cultural landscape) 221, 223; Faeroes (settlement; township) xvii, 87, 95; Värmland 220, 226, 230, 231, 236, 238, 242, 243; *glesbygd* (sparsely populated areas) 223; *landsbygd* (countryside) 223; *mellanbygd* (semi-open mixed landscape) 259, 262–63, map *262*; see also *hagmark*
býlingur (-ar) (group of houses; hamlet) (Faeroes) xvii, 87, 88, 96

cadastral maps *see* cartography
Cairo 335
Caledonides 573, 579
Campbell, Åke 160
Campbell, John Francis 333, 337n
Canada ix, 59, 113, 550, 556
Canute (Knut the Great) (king) 5, 20, 164
Cape Town 333
capitalism 67, 172, 184, 188n, 221, 486
Cappelen, August 412
Carlsberg Group, beer manufacturers 547
Carpelan, Bo 435
cartography xv, xxiv, 7, 10, 72, 112, 178, 204, 308, 348, 513, 526–27; agricultural areas map (Virestad) *264*; Åland map *441*; cadastral mapping xix, 31, 151–52, 213, 348, 349, 352–54, 365, 379, 423, maps *152*, *257*; *Carta Marina* 321–22, 529; Dalecarlia map *193*; Denmark 17–18, 20–21, 42, map *18*; Faeroes map *80*; Finland 423, 424–25, 525–26, 527–30, maps *427, 461, 523, 525, 528*; Finnskogen hundred map *239*; fishing areas map (Røst) *290*; Greenland 133n, map *107*; Hjartdal map *407*; Iceland 55, 63, map *60*; land reorganization maps *180, 183*, 356, 357, 362; North Norway 289–90, 305, 327; Norden maps *303, 441, 544, 569, 571, 572, 576, 578, 580, 582*; North Norway maps *304, 305*; Norway 315, map *284*; Odense map *21*; Sandhåland maps (Western Norway) *353, 356, 357, 362*; Scania 159–62, 182–83, map *160*; Sognefjord area map *374*; Sweden 7, 141, 151–52, 213, 215, 216, 239, map *142*; Swedish *hagemark* map *262*; Western Norway 348, 349, 350–57, 361–62, 365, 368, 369, 379; Värmland map *225*; Viipuri area maps *430, 431*; see also atlases; mental maps
Carus, Carl Gustav 43n
Caspian Sea 147
cattle: Denmark 15, 45n, 546, 547; Faeroes 81, 82–83, 87, 88, 89, 90, 91, 96, 97, 100, 103; Finland 459, 464, 564; North Norway 309, 311, 327, 333; Norway 555; Scania 149, 167, 176, 177; southern Norway 405, 406, 407–408, 409, 410, 414; Sweden 150–51, *268*, 274; Värmland 226, 227, 232, 235; Western Norway 350, 351, 352, 356, 357, 360, 364, 365, *365*, 375, 376–77, 378, 380, 381–82, 383, 384, 386, 395
Central Finland 494, *495*, 506, 530
centralization 10; Denmark 5, 10, 20, 29, 159; Greenland 124–25, 131, 134n; Sweden xviii, 159, 222
Certeau, Michel de 57
children: books for, 149, 150, 151, 455; children and landownership (Dalecarlia) 201, 206, 211; children in Greenland 116, 119, 128, 129, 130, 134n; children in farming 295; Finnish children 425, *490*, 521, 522; schoolchildren (Finland) 425, (Sweden) 150, 162, 174, 194, 201, 242, 243; *see also* schools
China 308
chorography 17, 19–20, 33, 35, 42n, 170
Christian III 18–19, 411
Christian IV 5, 93
Christian V 93, 94, 95
Christiania (Kristiania; Oslo) 238, 382, 554; *see also* Oslo
Christianity 103, 427; Åland 443; Denmark 19–20, 163; Greenland 110–11; Norway 553; Scania 166; Värmland 237
Christiansen, Thue 113
Church 111, 181; Church of Norway 316; Lutheran 333, 428, 434, 562; Orthodox xxiv; 330, 334, 335, 428, 432, 434, 498, 562;

Protestant 19, 316, 428; Roman Catholic 6, 316, 320, 428; Swedish Church 175, 328
church(es): Åland 444; cathedrals 45n, 164, 320, 325; chapels 330, 333–34; Denmark 19, 37, 43n; Finland 434; Greenland 128; Iceland 61, 62, 71; Norway 323, 333, 379, 392; Scania 163, 164, 166, 167, 170, 171, 174, 176; stave churches 392; Sweden 147–48, 196, 199, *199*, 210, 240, 272
church land: Faeroes xvii, 96; Finland 428; Norway 406; Scania 171
cipher xvi, 12, 13–14
CIR (color infrared) photography *see* aerial photography
cities *see* towns and cities
Clarke, Edward 422, 436
climate xxii, 313, 345, 346–47, 369, 404, 543–44, 573, 575–77, 583, precipitation map (Norden) *576;* Åland 448; Baltic 575, 577; climate change 397, 551; Denmark 4, 546; Faeroes 79; Finland 422, 429, 462, 464, 531, 532, 561, 564; Greenland 110, 127, 129; Iceland 551; North Norway 291, 310, 320, 321, 326, 329, 331, 335; Norway 553; Western Norway xxii, 372, 373–74, 375, 378, 379, 384, 385, 389; southern Norway 404, 406, 410, 414; Sweden 224–25, 235, 254, 260, 266, 270, 271, 558; *see also* Gulf Stream
coal-mining: Svalbard 310, 557
coasts 217, 329, 570, 574–75, 577, 581; coastal liner (Norway) 323, 553; Denmark 4, 159, 545, 583; Faeroes xvii, 83, 86; Finland 423, 427, 446, 447, 448, 458, 489, 491, 497, 506, 561, 562; Frisia 9; Greenland 55, 116, 124, 132, 551, 575; Iceland 58, 62, 69, 71, 551, 552; Ireland 78; *klittkust* (coast with sand dune morphology) 574; North Atlantic 77–78; North Norway xx, 283, 288, *289,* 291, 295, 301, 308, 309, 310, 311, 314, 320, 321, 322, 323, 324, 326, 329, 330, 335, 553, 555; Norway 5, 7, 283, 288, 291, 295, 296, 309, 310, 320, 326, 366, 405, 553, 554, 555, 556, 583; Scania 157, 162, 164, 165, 168; Svalbard 55; Sweden xviii, 7, 141, 143, 144, 145, 147, 149, 150, 154, 192, 202, 263, 447, 575, 577, 583; Western Norway xxii, 345–46, 358, 359, 366, 367, 368, 372, 373, 381, 382, 543; *see also* archipelagoes; Coast Saami; seashore; shore displacement
coats of arms *see* heraldry
Coke, Edward 31
Cold War 309, 334
Collingwood, R. G. 63
Cologne 42n, 188n
colonialism: Greenland xviii, 110, 111, 123, 125, 127, 130, 133n, 134n, 545; *see also* India; West Africa; West Indies
color infrared photography (CIR) *see* aerial photography
commons xvii, 31, 45n, 66, 104, 270; common goods 336; common fishing areas *290;* common property regimes xx, 294; common resources (Faeroes) 95, (North Norway) xx, 292–93, 294; *see also* common land
common land: Dalecarlia 207, 209; Denmark 187; Faeroes xvii, 77, 87, 89, 90, 104; Finland 464; Greenland 126; Jutland 31, 45n; southern Norway 408; Sweden 270, 272; Western Norway 350, 353, 355, 356, 357, 362, 365, 366, 384
communalism 188n, 221–22, 247n, 516
communication theory 187, 311, 313, 315, 441, 485
community (-ies), human x, 31, 44n, 166, 127, 173–74, 113, 114, 247n, 348, 416, 441, 476, 515, 517, 518; Åland 446, 450; Dalecarlia xix, 204, 209; Faeroes xvii, 78, 79, 81, 85, 89, 97, 99, 101, 102; Finland 434, 485, 486, 493, 527; Greenland 122, 126, 128, 131, 134n; Iceland 552; Jutland xvi; North Atlantic 77–78; North Norway 295, 301, 302, 309, 310, 312, 323, 329, 554, 556; Norway 102, 555; Scania 165; Western Norway 352, 359, 367, 378, 379, 381, 382, 384, 390; Värmland 233; *see also* imagined communities; plant communities
Connerton, Paul 478n
conservation *see* landscape conservation; nature conservation
constitutions: Finnish 441, 563; French 181; Norwegian 240, 303, 316, 397, 554; Swedish 208, 240

construction, cultural and social: Danish state 20; "other" 237; past (Greenland) 124, 134n; past (Iceland) 72–73; past (Scania) 167; past (Värmland) 238; plant communities 267; space (Iceland) 69; Sweden, "national construction of the provinces" (*landskapens nationsbyggande*) 222–23; *see also* identity, construction of; landscape, cultural and social construction of; regions, cultural and social construction of
Continuation War 563
Convention on Biological Diversity (1992) 252
Conzen, Michael xii, xiv
cooperation: agricultural (Denmark) 39, 546; agricultural (Sweden) 209, 217, 226; art (Western Norway) 392, *395;* Denmark–Sweden 8; European Economic Area (EEA) 545; Finland–Soviet 562; Finland–Sweden 565; fishermen (North Norway) 289–90, 293; Greenland 120; international scientific 557; landscape (southern Norway) 416; museums–schools (Norway) 312; reindeer herders (North Norway) 293; Russian–English 329; sailing vessels (Åland) 446; *see also* Nordic cooperation
Copenhagen (København) xv, xvi, 4, 5, 6, 7, 8, 9, 12, 16, 22, 29, 38, 39, 43n, 115, 132n, 134n, 174, 187, 545, 546, 549, 581; University of 42n, 43n
copper: copperplate engravings 43n, 169–70, *169;* mines 146, 195, 200, 330, 410, 411, 554, 560, 564
coppicing: Sweden 167, 268; Western Norway xxiii, 373, *377,* 385, 386, 387, 388; *see also* fodder; pollarding
Cosgrove, Denis xii, xiii, xiv, 172, 221, 223, 512, 519
Coster-Waldau, Nukâka 134n
cottars (cottagers): southern Norway 404, *405,* 406, 408, 409, 410, map *407;* Western Norway 350, 352, 354, 367, 380, *380,* 382; *see also* crofters
Coyet, Vilhelmina Eleonora 181
Crimean War xxiv, 445
crofters: Åland 448; Scania 174, 175, 184; Sweden 215; Värmland 229, 235; *see also* cottars (cottagers)
crown land: Faeroes xvii, 85, 96, 100; Finland 464
cultural ecology 311, 422
cultural heritage xi, 306, 312, 543; Åland 453; cultural-heritage center for South Saamis 321; Finland 522, 525; Greenland xviii, 125; management 252, 253, 254, 255, 268, 309, 310; North Norway 302, 309, 312–13; Norway xx, xxi, 286, 287, 294, 309, 310, 312, 336, 356, 392, 401, 402; Saami 287–88; Scania 186; Shetland 337n; Sweden 147, 251, 252–53, 259, 268–69, 276–77; Värmland 241, 247; Western Norway 373, 380; *see also* heritage
cultural landscape 185, 221, *256,* 269, 276, 515, 568; Åland 447; concept 223, 247n, 283–84, 287–88, 291–92, 296; "cultural landscape of agriculture" 283, 284–87, *285,* 294, 295; Dalecarlia 196, 202, 203–204, 213, 216, 217; Finland xxiii, 421, 428–29, 432, 506–507; Greenland 118, 120, 123, 125, 128, 131–32; Iceland 65; government payments (Norway), 286, map *285; Kulturlandschaft* 421, 428–29; Norway xx–xxiii, 283–99, 302, 308, 401–402, map *285;* policy 286; Saami 287–88; Scania xix, 160–61, 168, 178, map *160;* sea as cultural landscape 288–94, *289,* map *290, see also* seascape; southern Norway xxiii, 402, *403,* 416; Sweden 144–49, 151, 266, 268, 269, 271; Western Norway xxii–xxiii, 348, 350, 368–69, 372–73, 396, 397–98; Värmland 226, 241; UNESCO's (Melina Mercouri) cultural landscape prize 402
cultural landscapes, inventories of: Finland (national inventory) 506–507; Norway (National Registration of Valuable Cultural Landscapes) xx–xxi, xxiii, *285,* 286–87, 294, 401–402
culture, definition 166
custom(s) xi, xii, xiv, xv, 6, 29–32, 41, 44n, 46n, 125, 173, 174, 188, 296, 314, 476; customary law 30–31, 32, 85, 93, 94, 216, 226, 293, 309, 365–66, 476, 523; Dalecarlia

xix, 194–95, 198, 201, 204, 209–10, 215, 217; Faeroes 78–79, 84, 85, 93–95, 125; Finland 470, 478n, 523; Frisia 9; Jutland xvi, 30–32, 34, 38, 39; Norway 216, 293, 307, 309, 356, 357, 365–66, 378; Värmland 226

Dahl, Hans 392
Dahl, Johan Christian 391, *391*, 412
Dahlberg, Erik 423, *424*
Dahlgren, F. A. 228
Dala Commission 208–209
Dalälven, river 192, 194, 196, 202; Österdalälven 192, 202; Västerdalälven 192, 202
Dalarna *see* Dalecarlia
Dalby Hage 167–68, 187
Dalecarlia (Dalarna) xix, 146, 150, 192–219, map *193*
Dalgas, Enrico 25–27, *26*, 40, 41; and Blicher 39–40
Dalsland 153, 154
Daniels, Stephen xii, xiii, 16
Danish Heath Society (Det danske Hedeselskab) 25, 40
Dass, Petter 323
Davidsen, Agnethe 134n
Deatnu-Tana 331, 335
Debes, Lucas 93, 99
defense 5, 7, 102, 169, 324, 326, 336, 461, 478n; heraldry 316, 319, 324, 325, 330; *see also* military landscape
Degn, Anthon 101
demilitarization (Åland) xxiv–xxv, 445, 451, 456, 566
democracy x, xxv, 4, 306, 336, 447–48, 460, 462
Denmark ix, xiv, xv–xvi, xvii, xviii, xix, xxvii, 1–49, 93, 94, 102, 110, 112, 114–15, 116, 119, 121, 123, 124, 125, 127, 129, 130, 133n, 134n, 144, 145, 146, 157, 159, 164–66, 167, 168–72, 174–76, 179, 184, 187, 194, 207, 216, 217, 307, 310, 345, 367, 543, 544, 545–50, 551, 554, 558, 563, 570, 573, 574, 575, 577, 579, 581, 583, map *18*; *see also* Faeroes; Greenland; Jutland
Denmark–Norway xv, 4, 9, 22, 83, 307, 316, 331, 392, 545, 551, 554, 558

depopulation: Åland xxiv, xxv, 429, 447, 451; Faeroes 83; Finland 432; Greenland 124; North Norway 301; Sweden 153
dialect(s) xi; Dalecarlian 146, 198, 201; ethnic speech style 312; Finnish 511, 531, 535; identity and dialects 158, 312, 516; Jutland 34–35, 43n; Norwegian 9; region and dialect xi, 534; Saami dialect regions, map *303*; Scanian 6, 8, 146, 158; Swedish 146–47; Telemark 412; Värmland 146, 240; Zealand 8
Dickinson, Robert E. 526
dictionary definitions 6, 10, 11, 16, 29, 42n, 44n, 188n, 247n
Dicuil (Irish monk) 55
discourse x, xxvii, 32, 33–35, 38, 223, 247n; Finland 248n, 511, 513–14, 520, 524, 535–37; Greenland 106, 126, 123, 132n, 164; Iceland 53, 58, 71; landscape xiii–xiv, xx, xxvi, 106, 224, 511, 512, 513–14, 516–19, 520, 524, 535–37; modernity 188; national(istic) 106, 126, 164, 516; nature 168; Scania 186; urban 109, 131; Värmland 247
Dithmarschen (Ditmarsken) 9, 42n
Djurmo 192, 202
Donner-Amnell, Jakob 475
Dreijer, Matts 455
Drekkingahylur ("drowning pool") 64
Dresden 43n
drift ice *see* sea ice
driftwood 69, 91–93, 99, 309, 358
Dronning Maud Land (Queen Maud Land) 557
Düsseldorf 392
dwelling x, xiii, 6, 30, 32, 46n
Dyrlandsdalen 409
Dyrøy 324

eastern Finland xxiv, xxv, 421, 425, 429–33, 434, 458–71, 474, 475, 477, 489, 498, *499*, 562, 564, 566, 575; *see also* Karelia; Northern Karelia
East Greenland 132n
ecology xii, xxvi–xxvii, 252–55, 276, 313, 347; Finland 428, 465, 468, 470, 474, 478n, 507; Greenland 110, 124; Iceland 53; North

Atlantic 556; North Norway 289, 293, 300, 301, 309; Norway xxii, 284, 286, 347, 401, 416; Sweden 176, 178, 211, 217, 258–59, conceptual model 256; Swedish *hagmark* xx, 251, 259, 261, 266–68, 270, 275, 277; Western Norway 344, 348, 355, 360, 368, 369, 372, 373, 385, 398; see also biotopes; ecosystems; landscape ecology
economic policy 305, 474, 548
economic zones, maritime 551, 557
ecosystems xxii, 253–54, 256, 267, 269, 346–47, 359, 368, 416
Edelfelt, Albert 497
egalitarian society 4, 209, 215, 217, 221, 226, 336, 367, 558
Egedius, Halfdan 412
Egilsstaðir 62
Ehrensvärd, Augustin 423, 425
Einarsson, Magnús 64
Eklund, A. W. 424
Ekman, Klaus 462
Eldfjell 581
Electrolux household machines 560
Elkan, Sophie 245, 248n
emigration: from Åland 447; Denmark 546; from Faeroes 550; from Finland 562, 566; from Greenland 550; from Iceland 59; from Norway 55, 77–79, 350, 380, 382, 409, 554, 556; from Sweden xix, 153, 208, 209, 232, 240, 559; to North America 59, 153, 350, 382; to USA 232, 240, 409, 380, 546, 554, 559, 562; see also immigration; migration
employment: Åland 44, 448, 451; Denmark 547–48, 548–49; Finland 469, 564, 565; Greenland 111, 116, 118, 124; Iceland 551, 552; Norway xxi, 291, 295, 311, 336, 411, 555, 556–57; seasonal 198, 202, 240, 311; Sweden 198, 558, 559, 560–61; see also unemployment
enclosure see land reorganization
encyclopedias 422, 443, 476
energy: Denmark 548, 549; Finland 564, 565; Iceland 68, 552; Norway 554, 557; Sweden 560; see also hydroelectric power; nuclear power; oil and gas; peat cutting; water power; wind power

Engelbrekt Engelbrektsson 194–95
Engelsberg's Ironworks 149
England xiv, xxviiin, 20, 41, 43n, 147, 148, 172, 184, 185, 206, 216, 310, 423, 429, 433; see also Britain
Enlightenment 4, 178–79, 181, 184–85
Ennis 70
Ennisfjall 70
Enontekiö 501
enskifte see land reorganization
environmental change xxii, 32, 346, 347, 368
environmental determinism x
environmental goods 286
environmental policy 284–86, 285, 294, 470
environmental planning and management 246, 253, 306; Denmark 549; Finland 465, 507, 566; Norway xx, 283, 284, 286–87, 292, 294, 327, 390, 402, 414, 415, 557; Sweden 159, 178, 246, 252, 253, 259, 261, 561; see also landscape planning and management
eolian sand drift 349, 352, 355
erämaa (wilderness) (Finland) 426, 428 see also wilderness
Ericsson telephones 560, 566
Erik (of Pomerania) (king) 3
Eriksen, Thomas Hylland 112
Erlingsson, Bjarne 325–26
Eskimology 110
estates, landed 512; Åland 448; Dalecarlia 206, 210, 214, 217; Denmark 22, 38, 546; Faeroes xvii, 96–97; Finland xxiv, 430, 432–33, 434, 491, 512; Norway 240, 321, 326; Scania 159, 161, 171–74, 179–84; Sweden 270; Värmland 225, 228, 230, 233; see also manors
Estonia ix, 5, 144, 435
ethnicity 311–13, 516, 523; Denmark 20; ethnopolitics 106, 114, 133n, 307; Finland 485, 487, 493, 504, 505; Greenland xviii, 112–14, 120, 133; North Norway xxi, 301–303, 306, 307, 324, 328, 331–32, 334–35; multiethnicity xxi, 20, 302–306, 328, 331–32, 334–35; Shetland 337n; Sweden 222; United States xxviiin; Värmland 207, 230, 235, 238, 241
European Economic Area (EAA) 545

European Union (EU) ix, xix, 10, 545, 550, 566; Åland 566; Common Agricultural Policy 547, 559, 564; environmental directives 462, 470–71, 472, 474, 479; Denmark 545, 547; Finland 462, 470–71, 472, 474, 479, 535, 545, 563–64, 565, 566; Finnish referendum 471, *471;* hunting policy 462, 470–71; Norwegian referendum 545; regional policies 146, 154, 535, 561, 566; Sweden 146, 154, 545, 558, 559, 560, 561; trade 547, 556, 560

eustatic (sea-level) change *see* shore displacement

exclusion xi, 220, 231, 232, 247, 477, 512, 514, 531

Eysturoy *88*

Eythórsson, Einar 293

fäbod (Sweden) *see* summer farms

Faeroes ix, xiv, xv, xvi, xvii, xviii, xxvii, 4, 9–10, 59, 77–105, 216, 543, 545, 547, 550, 554, 573, 575, 578, 581, map *80;* "Sheep Islands" 77, *78*

Fagersta 149

Fagervik 433

Falster 549

Falu Copper Mine 195, 200

Falun 560

farming *see* agriculture

feitilendi (fattening land) (Faeroes) 83, 89, 90–91, *91,* 103

Fell 70, 71, 72

Fennoscandia 308, 311, 328, 330; Fennnoscandian Shield 543, 573

festivals 35, 38, 200, 453–55

feudalism 171, 184, 216, 217, 238, 337n

field systems 346; Faeroes 82, 88–91; Western Norway 348, 349, 354, 364–65, map *351; see also* infield(s); land fragmentation; land reorganization; open-field system; outfield(s)

Fienup-Riordan, Ann 110

Filipstad 227

film(s) 129–31, 134n, 242, 323, 501, 514, 534

Finland ix, xii, xiv–xv, xxiii–xxvi, xxvii, xxviiin, 10, 141, 144, 216, 227, 238, 247n, 248n, 254, 302, 307, 327, 328, 329, 330, 331, 333, 334, 419–539, 543, 545, 554, 558, 560, 561–66, 570, 573, 575, 577, 579, 581, 583, map *525; see also* Åland; eastern Finland

Finland, Gulf of 144, 308, 435, 446, 498, 561, 562, 574, 577

Finnby 435

Finnish civil war 563

Finnish Economic Society 423

Finnmark 293, 306, 307, *307,* 308, 309, 312, *316, 317,* 319, 320, 324, 325, 326, 327, 329–35, maps *305;* East Finnmark 308, 314; West Finnmark 314, 320, 330

Finns in Norway: Ålanders 446; Forest Finns 227, 234, 236, 238–41, *239,* 303; North Norway xxi, 295, 302, 303, 305–306, 311, 327–29, 330–32, 333–34, 555, map *305;* Saami ("Finns") 310

Finns in Sweden: Forest Finns xix–xx, 227, 234–41, 303, 558; northern Sweden 155, 558, 562, 579

Finnskogen 220–21, 227, 228, 229, *230,* 232, 234–41, *236, 241,* 242, 243, 247; Finnskogen hundred 228, 238–41, map *239*

First Grammarian 56, 74n

fisher-farmers: Åland xxv, 429, 448–49, 450; Iceland 551; Norway 291, 295

fish farming: Norway 295, 296, 301, 397, 556

fishing xx, 568, 574; Åland xxv, 443, 444, 448, *449,* 450, 452; Denmark 546, 547; Faeroes 81, 84–86, 92, 92–93, 103, 550; Finland 561, 564; Finnmark fishery 309, 320; fishing limits 552; fishing rights 550; Greenland xviii, 111, 118, 124, 128, 550; Iceland 58, 59, 67, 68, *68,* 71, 551–52; Lofoten fisheries 289, 291, 307, 309, 320–21, 554, map *290;* North Norway xx–xxi, 283, 284, 287, 288–296, *289, 290,* 301, 307, 309, 311, 320–21, 323, 325, 326, 330, 333, 334, 335; Norway 404, 553, 555, 556; overfishing 67, 292, 295, 301, 309, 550, 552, 555–56; Sweden 216, 559; Western Norway 350, 366, 381, 382, 383, 397

Fiskars 433

Fitjaannál 66

Fjall 69–70

Fjell 359

Fjell, Kai 412
fjords (fiords) 544; Faeroes 79, 81, 85, 92, 581; Greenland 111, 128, 547; Iceland xvi, 55, 69, 581; boat transport 381–82; fjord cod 309; fjord horse *(fjording)* 379; fjordlike landscape (Sweden) 144; North Norway 292, 296, 326–31, 333, 335; Norway xiv, 296, 574, 575; plateau-and-fjord landscape 581; Western Norwegian fjord landscape ("fjordscape") xxii–xxiii, 277, 344, 345, 359, 365, 366, 372–400, *376*, 412
flags: Åland xxv, 455; Denmark ("Dannebrog") 5; Faeroes 84; Finland 522; Greenland 112; Iceland 45; Norway 316, 318–19; Saami 331; Scania 187
Fleischer, Ono 113
Flintoe, Johannes 391, 412
Floda 213
Flóki Vilgerðarson 55, 69
fodder 577; Åland 450; Denmark 39; Faeroes 81; Finland 432, 564; North Norway 309, 326, 327, 333; Norway 401; southern Norway xxiii, 406, 408, *409*, 410; Sweden 163, 232, *256*, *268*; Western Norway xxii, 350, 372, 375, 378, 383, 385–88, *386*; *see also* coppicing; meadows; pollarding
folk costumes 441; Åland 453, 454; Dalecarlia 194, 198, 199, *199*, 201–202; Denmark 20, 41, 42n; Finland 491; Norway 323, 328, 391, 411, 412
folk culture: Åland 449; Dalecarlia 146, 199–200; Finland 534; Iceland 67; North Norway 293, 322–23; southern Norway 411–12; Sweden 147; Western Norway 396; Värmland 243; *see also* folk costumes; folklore; folk music; poetry; stories
folklore 314; Åland 453; Dalecarlia 192, 198–200; Denmark 37; Finland 248n, 486, 562; Iceland 63, 71; Norway 322, 392, 396; Sweden 146, 222, 241, 268; *see also* stories; trolls
folk music: Åland 453, 455; Dalecarlia 199–201; southern Norway xxiii, 411, 412; Western Norway xxiii, 394–96
food and identity 441, 516–17; Finland 536; Greenlandic xviii, 116, 118–19, 121, 122, 128

Forchhammer, Søren 124–25, 126
Forest Finns xix–xx, 227, 234–41, 303, 558
forestry 568; afforestation 22, 39, 40, 262, 275; Åland 448; Finland 429, 432, 462, 464–65, 466, 468, *469*, 470, 474, 503, 564; German model 464; Jutland 15, 22, 25, 26, 39, 40, 545; Norway xxi, 295, 296, 327, 346, 348, 359, 368, 382, 410–11, 553, 554, 556; plantations 271, 273, 276, 346, 348, 359, 397, 466; Sweden xviii, 154, 155, 177, 178, 252, 270, 271, 273, 275, 276, 559, 560
forest(ry) landscape 575; Finland 463, 465, 468, 470, 472, 474, 483, 489, 497; Jutland 25; Norway xxi, 283, 295, 296; Sweden 143, 155, 236, *236*, 240, *260*, 263, 266, 275; *see also* forestry; forests
forests 212, 251, 573, 577, 579, 581; Åland 443, 446, 448; Dalecarlia 196, 198, 203, 204, 207, 209; Finland xxv, 422, 425, 426, 429, 432, *427*, 458–82, 466–67, *469*, 483, 487, 489, 497, 499, 520–21, 536, 561, 564; forest policy (Finland) 464–65; Jutland 12–13, 15, 17, 21, 25, 26, 27, 36; Norway 283, 553; Scania 159–61, 165, 167–68, 175–76, 177, 186, 187; shadow forests 468, 470, 471; southern Norway 401, 402, *403*, 404, 406, 410, 413, 414; Sweden 143, 146, 148, 150, 151, 153, 154, 266, 558; Swedish *hagmark* 251, *257*, 259–63, *260*, *261*, 270, 271, 272, 275, 276; treeline 375, 577, 578; Värmland 220, 225, 226, 227, 229–31, 232, 235, 236, *236*; Western Norway 345, 358, 373, 374–75, 377, 388
förloppslandskap ("process-landscape") 246
Forman, R. T. T. 346, 347
Forslund, K. E. 201
fowling: Åland 450; Faeroes xvii, 77, 81, 84, 85, 90–91, *91*, 94, 99, 103; North Norway 326
fragmentation, landscape ecological 255, 256, 268, 271, 274, 276, 384; *see also* land fragmentation
framing 108–10, 115
France xxiv, 58, 157, 181, 184
Frederik III 175
Frederik VI 23

French Revolution 181, 184
Freya 24
Fridegård, Jan 155
Friedrich, Caspar David 43n
Fries, Carl 164, 251
Friis, J. A. 305
Frisia 9; Frisians 20
Fröding Gustaf 227–28, 230, 231
frontiers *see* borderlands
Fryken 225
Fryksdalen 220–21, 224, 226, 228, 230, 231–34, *233*, 237, 242, 243, 245, 247
Fryxell, A. 228
functionalism 500
Funen (Fünen; Fyn) xv, 4, 9, 12, 16, 36, 40, 93, 549, 581
Funningur 88
fur farming: Finland 564; Norway 324, 363
Fyn *see* Funen

Gáivuotna-Kåfjord 328–29
Gallen-Kallela, Akseli 497
Galyamin, Valerian 424, *426*
Gamla Finland 434
Ganander, Christfrid 237
Gästrikland 148
Geijer, Erik Gustaf 227
gender xi, 167, 301, 312, 513, 516; gender-blindness 284, 294
genealogy 17, 42n, 74n, 113, 205, 210, 312, 316
Genoa 322
geology 254, 267, 573–74, 579, 581; Baltic Shield 573; Denmark 4, 543, 546, 570, 573–75, 579; Faeroes 79, 543, 573, 575. 581; Fennoscandian Shield 543; Finland 564; Greenland Shield 543, 573; Iceland 543, 573–74, 575, 581; Norway 374, 404; Quaternary deposits map (Norden) *571;* Sweden xviii, 141, 143, 160–61, 203, 224–25, 258, 260, 266, 270–71, 272, 559; *see also* volcanoes
geometry 166, 170, 179, 182
geomorphology *see* landforms
geopolitics: Finland xxv, 460, 462, 472, 473, 474, 477, 478n, 489, 502; Hansa 168; North Norway xxi, 308, 309; Scania 159, 165, 169

Germans 17, 20, 22, 187, 204, 411, 489, 547; geographers 221, 435
Germany x, xv, xxviiin, 3, 9, 17, 22, 30, 42n, 45n, 145, 157, 159, 168, 392, 445, 545, 546, 547, 551, 563, 565; German–Danish frontier 9, 17, 545; occupation of Denmark 4, 547, 551; occupation of Norway 308, 325, 332, 555; *see also* Prussia; Schleswig–Holstein (Slesvig–Holsten)
geysers 53 *see also* hot springs
GIS (geographical information systems) 246, 348, 369
Gjógv *92*, 94
glaciation 225, 543, 570, 573, 581, 591; Faeroes 79; Finland 561; Sweden xviii, 141, 143, 145, 160, 270–71; *see also* glaciers
glaciers 4, 15, 46n; Greenland 110, 113, 543, 575; Iceland 53, 69–72, *70*, 543, 551, 581; Norway xxii, 372, 373, 374, 382, 391, 393, 396, 397, 404, 543, 553; Sweden 141
Glafsfjorden 224, 225
globalization ix, xix, 10, 472, 476, 507, 570; economy 67, 301, 464, 566; environmental issues 252, 253, 310; landscapes 508; networks 127, 223
goats: Faeroes 84; Norway 376, 378, *379*, 384, *389*, 395, 405, 407, 555
Godron, M. 346
Goethe 40
gold: Blika mine (Norway) 410
Göngu-Hrólfr 58
Gorbachev, Mikhail 334
Gorm (the Old) (king) 19–20, 45n
Gösta Berling's Saga (Lagerlöf) 220–21, 223, 228–35, 237, 238, 242–43, 245, 247n, 248n
Göta Canal 559
Göta, River 233
Götaland *261*, 266, 275
Gothenburg (Göteborg) 153, 227, 233, 558, 559, 581
Gotland 143, 145, 573, 575, 577, 581
Gottlund, Carl Axel xx, 220, 228, 234–43n, 248n
graffiti 318
Grafström, Anders Abraham 206

grannastevna (neighborhood council) (Faeroes) xvii, 93–95
Granö, J. G. xxiv, 421, 434, 435, 436, 511, 526, 527–29
grass(lands) 575, 577, 578, 583; Åland 450; conceptual model *260;* Faeroes 79, 81, 84, 88, 89, 90, *91;* Finland 564; North Norway 286, 326, 333; southern Norway xxiii, 404, 409, 414, *415;* Western Norway 345, 350, 358, 361, 363–64, 377, 383, 385, 388, 389–90; Sweden xx, 149, 150, 251, *257,* 258, 259–60, 262, 263–64, 266–68, 269–70, 272, 273, 274, 275, 276, 277; *see also* grazing; *hagmark;* meadows; pastures; seminatural vegetation
grave mounds (barrows; burial mounds; tumuli): Denmark 4, 15, 19, 24, 25, 27–28, 29, 31, 33, 37, 38, 43n, 46n; Norway 392; Sweden 147, 268
grazing 15, 577; Dalecarlia 198, 203, 212, 226; Denmark 15, 21, 35; Faeroes 77, 78, 82–83, 86, 87, 88, 90–91, 93, 94, 95, 99, 103, 104, *see also feitilendi, hagi;* Finland 577; Iceland 58, 72; North Norway 283, 292, 301, 324; overgrazing 292, 295; Scania 161, 164, 167, 177, 182; southern Norway xxiii, 401, 403, 406–408, 410, 414, 415; Sweden *144,* 148, 149, 150, 267; Swedish *hagmark* xx, 251, 259, 268, 270, 272, 275, 277; Western Norway xxii, 346, 350, 351, 356, 364, 365, 368, 373, 375–79, *376, 377, 378,* 381, 382, 383, 384, 385, 387, 388, 389–90, *389*
Great Depression 547, 551, 555
Great Northern War 444
Greece 73, 504
Greenland (Kalaallit Nunaat) ix, xiv, xv, xvi, xviii, xxvii, 4, 9–10, 55, 59, 106–38, 543, 545, 547, 550, 551, 554, 570, 573, 574, 575, 578, map *107*
Greenland Labor Union 114
Greenland National Museum and Archives 106, 107, 115
Greenland Sea 553
Greenland shield 543, 573
green movement 187, 474
Green, Nicholas 508n
Grense Jakobselv 333, 334

Grønlykke, Jacob 129
Grundberg, O. 263
Grundtvig, N. F. S. 23
Gudbrandsdalen 325
Gude, Hans 392
Guðrun Sjúrðardóttir 97
Gulf Stream 79, 291, 309, 544, 553, 558, 575
Guovdageaidnu-Kautokeino 332, 335
Gustav I Vasa 193–94, 200, 204, 462
Gustav II Adolf 222
Gustav III 181
Gýlden, C. W. 425, *427*

habitat destruction: Norway xxii, 361, 368, 415; Finland 472
habitus 6, 31, 35, 40, 44n
Hafnarfjörður 66
Hägerstrand, Torsten 246, 247n, 248n, 255, 512
hagi (-ar) (hill grazing land) (Faeroes) 89–91, 94, 95, 99, 101, 104; *hagaleys* (hill-less) land 89, 96, 99; *húshagi* (grazing area) 89–90, 103; *lambhagar* (lambing enclosures) 83, 91
hagmark (wooded grassland) (Sweden) xx, 259–80, *257, 261, 262, 265, 268,* 373; concept 259–63, conceptual model *260*
Hakkarainen, Anna (Pasu-Anni) 236–37
Håkonsson, Håkon (king) 316
Hákun (duke) 95
Hald, Kristian 45n
Halland 6, 144, 145, 157, 161, 174, 558
Hallingdal 411
Hällström, C. P. 424
Hålogaland xxi, 307, 310, 323, 325
Hälsingland 150, 215
Hålvik 426
Häme *495,* 521
Hämeenlinna (Tavestehus) 461, 497, 500, 506, 508n
Hamlet's grave 13, 16, 27, 42n
Hammerfest 329, 332, 335
Hanko 497
Hansa 168, 322; Hanseatic Office in Bergen 321
Hansen, Martin A. 27, 427–28
Hanssen, Bjørg Lien 286–87
Harald Bluetooth (king) 19, 20, 45n, 163

harbors: Åland 440, *449*, 452; Copenhagen 5; Faeroes *92, 93*; harbors and shore displacement (Finland) 561; ice-free harbors 332, 553, 563; Hammerfest 332; Kotka 565; Mariehamn 440, *442*; North Norway 292, 326, 327, 332, 333, 553; Nuuk 118, *119*; Petsamo 563; Vardø 333
Hardanger 365, 392, 395
Hardangervidda 325
Härjedalen 215
Hartshorne, Richard 247n
Hastrup, Kirsten 11
Hatten, Mount 321
Hattfjelldal 321
Hauge, Leif *395*
Hauge, Olav H. 396
Hazelius, Artur 198, 246
heath(s) 15; Alhede (Jutland) *36*; Denmark 546; Jutland 12–13, 15–16, 17, 21, 22, 25–27, 36–37, 38, 39, 577; Norway 553; Öland 143–44, *145*; Sweden 143–44, 259, 577; Western Norway 345–46, 350, 356, 361, 367; *see also* Danish Heath Society
Hebrides 554
Hedmark *317*, 324, 325
Heidegger, Martin 30, 33, 34, 35, 46n
Heimaey 552
heimrust(ir) (area for farm buildings) (Faeroes) 87, 88, 89, 90, 95, 103, 104; *see also býlingur*
Heinävesi 434
Heinäsaaret national park 503
Heinola 500
Hekla 581
Helsinki (Helsingfors) 424, 432, 433, *495*, 496, 498, 500, 501, 503, 506, 532, 534, 561, 562, 565; Helsinki University 526, 529; Imperial Alexander University 425
Hemingway, Andrew 485, 508n
heraldry 314–15; Åland 444; Finland 485, 535; Greenland 112; North Norway xxi, 300, 302, 315–37, *317*, 337n; Shetland 337n
Herdla 359
heritage xxviii, 170, 348, 568; Dalecarlia 206; Danish 157; Finland 435, 503, 506–507, 522, 525; Greenland 115; industrial 506–507; Jutland 16, 27, 31, 41, 43n; natural heritage 269, 276, 402; Norway 402; North Norway 300, 312–13; Sweden 269; Värmland 228, 236, 238, 241, 243; *see also* cultural heritage; World Heritage
Hermelin, Samuel Gustaf 529
High Coast 144
highest shoreline (highest marine limit) *see* shore displacement
high-tech 154, 549, 566
Hillesland 352
Hirsch, Eric 108–109, 231–32, 246, 512
Hjälmaren, Lake 583
Hjartdal-Svartdal xxiii, 277, 402–16, map (Hjartdal) *407*
Hjartsjåvannet, lake 402
Hobsbawm, Eric 46n
Hogenberg, Franz 169–70, 188n
Hogland 498, 508n
holism xxvii, 252–53, 269, 276, 311, 402; *see also* integrated landscape analysis
Holmberg, Werner 424
Holstein (Holsten) *see* Schleswig–Holstein
Holy Roman Empire 17, 22
Home Rule *see* autonomy
Honningsvåg 332–33
Hordabo 359
Hordaland 358–59
Hornafjörður 62
horses: Åland 444; Faeroes 92, 100, 103; Finland *490*; fjord horse 379; Lyngen horse (North Norway) 328; southern Norway *405*, 406, 408; Sweden 149, 150; Western Norway xxii, 350, 352, 354, 355, 358, 360, 363, 365, *365*, 366, 367, 368, 378, 379, 381, 382, 384, 386
Hoskins, Walter G. xxviiin
hot springs 68
housing 584; Denmark 548, 549; Finland 532, 561; Greenland 117, *117*, 124; Sweden 153, 559
Hrollaugr 58
Hufford, Mary 517
Hult, Ragnar 527–28
human rights 303
Hund, Tore 326

hunter-gatherers 162, 147, 579
hunting 579; Åland 444; Denmark 38; Faeroes 77, 81, 84–86, 91, 93, 94; Finland 458, 459, 462, *463*, 464, 470, 523; Greenland xviii, 111, 118, 119, 120, 121, 122, 124, 128, 129, 130, 550; Iceland 67, 69; Norway 284, 287, 293, 294, 307, 308, 309, 311, 323, 325, 332, 372, 383, 404–405; Sweden 147, 162, 171, 216, 234; *see also* sealing; whaling
Huovinen, Veikko 535
Húsavík 97
Hvalfjörður 66
hydroelectric power: Finland 504, 565; Iceland 552; Norway 384, 354, 359, 384, 411, 413, 554, 556–57; Sweden 143, 560

Ibsen, Henrik 325, 396
ice 308, 327, 436; costs 561; icebreakers 436, 561, 565; ice-going ferries 450; ice roads 447, 450, 451; icescape xxv; *see also* glaciation; glaciers; sea ice
Ice Age(s) *see* glaciation; Little Ice Age
Iceland ix, xiv, xv, xvi–xvii, xviii, xxvii, 4, 9, 43n, 45n, 53–76, *54, 61, 68,* 69, 79, 86, 100, 102, 104, 119, 543, 545, 546, 550–53, 554, 570, 573–74, 575, 578, 581, 583, map *60*
iconography xii–xiii, 244, 519, 522
identity ix–xii, xxvi, 10, 54, 119, 312, 441–42, 512–19, 530, 537; class 505; ethnic 114, 120, 312–13, 505; local xxiv, xxv, 43n, 111, 122, 322, 336, 441, 515; multiple xi, 311, 442, 446; place x–xii, xvi, xx, 10, 41, 313, 517; Saami 287–88, 292; *see also* food and identity; identity, construction of; national identity; regional identity
identity, construction of x–xii, 10, 300–301, 442, 518; Åland 543; Finland xii, xiv–xv, xxv–xxvi, 468, 472–73, 522, 530, 535; Norway xii, xxi, 9; Scania xix, 158; Värmland 227
ideology ix, 485, 517–18; Åland 453; Denmark 29; Finland xxvi, 464, 478n, 484, 486, 503, 504, 514; Germany x, 247n; Greenland 120, 124, 134n; Iceland 72; Norway 318–19, 333
Igelösa 171
Ihse, Margareta 253, 401

Iittala glass 565
Ilulissat 124, 125, 129, 131, 132
imagined communities 228, 242, 486, 516, 518
Imatra 497, 498, 500, 503, 506–507
Immigration: Finns to Norway 302–303, 327, 328, 329, 330, 331, 334, 556; Russians to Norway 334; to Denmark 549; to Sweden 145, 147, 148, 153, 188n, 447, 566, 558; *see also* emigration; migration
Impilahti 432
Inari 501, 503
India: Danish colonies xv, 4, 545
Indigenous and Tribal Peoples Convention (1989) 303, 337n
industrialization xx, 583; Norway 409, 413, 554; Sweden 147, 153, 185, 195, 222, 227, 232; industrial revolution 152, 583
industry, manufacturing: Åland 440, 451; Denmark 546, 547, 548–49; Ford cars 547; Finland 433, 562–63, 564–66; Iceland 551, 552; Norway 291, 295, 296, 301, 309, 330, 381, 397, 554–55, 556–57; Sweden 146, 147, 148–49, 152, 153, 154, 185–86, 187, 226–27, 240, 243, 558, 559, 560–61; *see also* iron; mills
infield(s): Faeroes (homefield; *bøur*) 78, 88–89, *88;* Faeroes (infield; *innangarðs*) 78, 82, 83, 87, 88–89, 90, 95, 97, 98, 99, 103; Norway *(innmark)* 401; southern Norway 404, 406–407, 408, 409; Western Norway (home field; *heimebøen*) 362, *362,* 364–65, 368, 375, *376,* 379, 380, 383; Sweden *(inäga)* 176, 204, 256, 257, 272; *see also* outfield(s)
Ingold, Tim xiii, 288
Inha, Into Konrad 426, 497–500, 503
inheritance *see* land inheritance
innovations 35, 46n, 225, 233, 243, 546, 560, 565
insider perspective xiii, 40–41, 245, 301, 311, 512, 518; Finland 530, 532, *533;* Greenland 109; Jutland 14, 27, 40, 41; North Norway 301, 310; *see also* outsider perspective
institutions xi, xviii, xxvii, 238, 240, 287, 296, 348, 516, 518; administrative 319; cooperative management 289, 295; cultural 8, 110, 115, 127, 128, 240, 312, 518; ecclesiastical 95, 240; legal xviii, 6, 29, 32, 33, 44n, 77, 95,

226, 476; Nordic 566; Saami 288, 312; *see also* landscape, institutional; regions, institutionalization of
integrated landscape analysis xx, 251–59, 266, 269, 277, 402
International Labor Organization 337
Inuit (Eskimos) xiv, 20, 109, 110, 113, 126, 550
Ireland 78, 79, 147, 551
Iron Age: Denmark 4, 15; Norway 102, 294, 349, 410; Sweden 147, 272
iron: bog-ore 159, 410; iron mining 226, 333–34, 554, 560; iron ore xviii, 146, 149, 159, 178, 232, 559, 560, 565; iron production 146, 159, 178, 559, 560, 564; iron tools 385, 409; ironworks 148–49, *148*, 152, 226–28, 230, 231, 232–34, 237–38, 239, 240, 410, 568
irrigation 375, 380
island culture (Åland) xxv, 448–50
Íslendingabók (Book of Icelanders) 55, 56, 74n
Istanbul 335
Italy 504

Jaatinen, Stig 429–32
Jackson, J. B. (John Brinckerhoff) xxviiin, 511, 512
Jæren 349, 366, 574
Jämtland 144, 145, 192, 558, 575, 579, 581
Jan Mayen 310, 323, 553, 557
Janssen, Carl Emil 135n
Japan 309
Järnefelt, Eero 497
Järta, Hans 212
Järvsö 215
Jelling *19*, 19–20, 27, 31, 33, 43n, 45n, 163
Jensen, Marianne 125
Jentoft, Svein 292–93
Jews 303
Joensuu 475
Johansen, Mayvi B. 288–90
Johnson, Bengt-Emil 245
Johnson, Kaare Espolin 412
Johnson, Matthew 185
Jones, Michael 423
Jönköping 275
Jordan, Marcus 18, 42n

Jorma, Puranen 436
Jósephsson, Þorsteinn 61
Jostedalsbreen 372, 374, 382
Jotunheimen 393
Julin, Johan Jacob 422, 433
justice 43n, 221, 222, 336, 476, 513; injustice 235
Jutes 28
Jutland (Jylland) xiv, xv, xvi, 3–4, 6, 7, 8, 9, 10, 12–49n, 163, 174, 545, 548, 549, 574, 577, 581
Juva 238; "Juvaniemi" 239
Jylland *see* Jutland

Kåfjord Copperworks 330
Kainuu 468, 470, *495*, 497, 529, 535
Kaisaniemi 433
Kalaallit Nunaat (Greenland) 112, 126, 133n, 543; *see also* Greenland
Kalevala see poetry
Kálfafelsstaður 71
Kalix Älv, river 143, *144*
Kalmar 263, 275, 573
Kalmar Union (Nordic Union) xv, 169, 545
Kangersuatsiaq 121
Kárášjohka-Karasjok 332–33
Karelia 434, 487, *495*, 498, 499, *499*, 500, 503, 508n, 521, 522; Karelian isthmus 429, 563; Ladogan Karelia *495*, 497, 498, 499, 503; *see also* Northern Karelia
Kärkkäinen, Martti 468
Karlfeldt, E. A. 201
Karl Johan (king) 239
Karlsøy 326–27, 335
Karlstad 226, 227, 232; University 246
Karl XII 181
Karmøy xxii, 349–58, 359–64, 366
Kastelholm *424*, 443, 444, 497
Katla 581
Kattegat 42n
Kautokeino *see* Guovdageaidnu-Kautokeino
Keflavik 552
Keisteri, Tarja 421
Kemi valley 328
Kianto, Ilmari 459–60, 462, 472–73
Kierkegaard, Søren 38
Kihle, Harald 412
Kilpisjärvi 501

King's River (Kongeåen) 17, 29
Kirkenes 333–35, 553
Kirkkonummi 458
Kiruna 559, 560
Kittilä 501
Kitunen, Martti 462, *463*
Kivi, Aleksis 484
Klarälvdalen Valley 224
Klarälven River 226
Kleggjaberg *91*
Klepp, Asbjørn 288
Klopstock, Friedrich Gotlieb 22
Knudsen, Samuel 134n
Knut *see* Canute
Knutson, Johan 497
Koillismaa 535
Kola 330; Kola Peninsula 309, 324
Koli 483–84, 503, 506, 507, 522
Komsa 330
Kongeåen *see* King's River
Korean War 552
Korpiselkä 498
Kotka 497, 565
Kovatnet, lake 411
Köyliö, Lake 507
Kraft, Jens 406, 408
Kristiania (Christiania; Oslo) 238, 382, 554; *see also* Oslo
Kristinehamn 227
Kristjánsson, Haukur 70
Kristjánsson, Jónas 57
Krokann, Inge 355
Kronoberg county 263, map *264*
Kruskopf, P. A. 497
Krylbo 195
Kügelgen, Carl von 424, 489
Kuopio 497, 498, 500, 524, 562
Kuusamo 470, 498, 500, 506
Kvænangen 329
Kvívík 92
Kymenlaakso 494, *495*
Kymmen, Lake 236, 237
Kymsberg 237

Ladoga, Lake 432, 434, 498
Lærdal 374, 382, 384, 385

Lærdalsøyri 381
Læstadius, Lars Levi 328
laga skifte see land reorganization
Lagerlöf, Daniel 237
Lagerlöf, Selma xviii, xx, 149, 150–51, 220–23, 227–34, 236–38, 242–47, 247n, 248n; Nobel Prize, literature 150, 228, 247n
Lagerroth, Erland 244, 246, 248n
lakes 545, 573, 581, 583; Åland 443; Denmark 546; Finland 432, 426, 432, 436, 468, 469, 470, 483, 487, 489, 490, 497, 498, 520, 521, 530, 536, 561, 562; lake-and-forest landscape 483, 489, 497; Norway 345, 360, 373, 381, 402, 411, 553; Sweden 143, 150, 192, 196, 198, 202, 207, 220, 224, 225, 229, 230, 231, 236, 236, 237, 259, 263, 264, 270–71, 274, 326, 558, 559, 561, 579, 583
Lampén, Ernst 520
land division: Dalecarlia 204–16; Faeroes xvii, 86, 88, 90, 95, 96–97, 98, 99, 104; *sämjodelning* (informal division) 206–209, 211; Scania 165, 176, 182; southern Norway 408; Western Norway 352, 353, 364–66, 367; Sweden 215–16, 235; *see also* land fragmentation; land inheritance; land reorganization
land dues *(landskuld)*: Faeroes 81, 82, 96, 97–98, 99, 100, 102; *see also* land evaluation
landed estates *see* estates, landed
land evaluation 172; Dalecarlia 213–14; Denmark 102; Faeroes 77, 97–99, 101–103; Iceland 102; Norway 102; Orkney 102; Shetland 102
landforms xviii, 225, 254, 270, 288, 346, 404, 527, 570, 573, 579, map of Quaternary deposits (Norden) *571*, map of superficial deposits (Norden) *580*
land fragmentation: Dalecarlia 198, 204–15; Faeroes 95–96; Western Norway xxii, 353, 357, 360, 361–63, 364–65, 366, 368, 384; *see also* fragmentation, landscape ecological
land inheritance: Dalecarlia xix, *193*, 204–206, 208–14, 215–17; Faeroes xvii, 95, 96, 97, 101, 102, 103; primogeniture 153, 216–17; Sweden 153; West Norway 366; *see also* land division; land fragmentation

Index

landless people: Åland 444, 447; Sweden 198, 235

landmarks: Åland 450, 455; Finland 483; Greenland 129, 133n; Iceland 53; North Norway 288, 292, 323, 326, 327; Scania 166

landnám (settlement): Faeroes xvii; Iceland xvi, 55–56, 58–59, 71, 74n

Landnámabók (The Book of Settlements) 55, 58, 74n

landownership 255, 348, 568; Åland 444, 448, 451; Dalecarlia 195, 198, 201, 202, 204–17; Denmark 546; Faeroes xvii, 65, 81, 84–86, 88–93, 95–97, 99, 103–104; Finland 435, 564; Iceland 58; Norway 287, 307; Scotland 337n; Shetland 337n; southern Norway 406, 411, 416; Sweden 147, 151, 259, 260, 266, 270, 271, 272, 274; Western Norway 350, 352–58, 359, 361–67, 380, 381, 382, maps *356, 357, 362; see also* church land; crown land; estates, landed

land reorganization (enclosure; consolidation; reallocation; reallotment; redistribution; reform) 346, 583; Dalecarlia 198, 204, 206–209, 211, 213; Denmark 21, 179, 548; *enskifte* (Scania) 179–87, maps *180, 183;* Faeroes 94; Finland 423, 464; *laga skifte* (Sweden) 211; Norway xxii, 240, 409; *solskifte* (Sweden) 214, 216; Sweden xix, 151–53, 214, 216, 258, 272, 274; Western Norway 349, 350, 352–54, 355–57, 361–63, 382, maps *356, 357, 362*

land rights 32, 217, 568; Åland 444; beach usufruct (Iceland) 68–69, (Western Norway) 350, 354; Dalecarlia 204, 209, 210, 212, 213–14, 216, 217; Faeroes xvii, 77, 81, 84, 85–86, 87–93, 95, 101, 103–104; Jutland 39; Norway 325, 350, 354, 358, 382, 384; right of public access 187; Saami 154, 307; Scania 163, 165, 171; Shetland 337n; Sweden 147, 154, 259, 260, 266, 270, 272, 274; *see also* land division; land inheritance; landownership

landscape: action landscape 35, 187, 224, 236, 536; contested landscapes xvi, xxvi, 41, 63, 106, 109, 122, 513, 537; institutional landscape 224, 226, 238; landscape as commodity xxiv, 421, 435–37, 477; landscape as way of seeing 41, 167, 170, 172, 173, 221, 223, 224, 232, 246, 247, 475, 486, 508, 512; landscape figures 322, 323–25; landscape of signification 224, 227, 236, 237, 240–41; landscape policy xxvi, 26, 286; landscapes of domination xxv, 168, 171, 226, 284, 291, 294, 465–67, 471–74, 476, 478n, 486, 496, 500, 508n, 513, 522; landscapes of fear xxv, 428, 465, 471–72, 478n, 513; landscapes of hardship 437; landscapes of pleasure 437; landscapes of publicity 535–36; landscaping 223, 246, 430, 431, 433, 459, 475, 478n; local landscape knowledge xx, 63, 78–79, 235, 245, 284, 287–89, 292, 293, 294–95, 323, 416; spatio-legal landscapes 56; *see also* agricultural landscape; agropastoral landscape; cultural landscape; forest(ry) landscape; maritime landscape; mental landscape; military landscape; moral landscape; national landscapes; natural landscape; physical landscape; political landscape; "process-landscape"; regional landscapes; rural landscape; Saami landscape; townscape; urban landscape

landscape aesthetics xxvii, 244, 476, 512; Åland 453; Denmark 27; Finland xxv, xxvi, 476–77, 489, 505, 520, 531, 537; Norway 344, 373, 402; Sweden 268, 269

landscape concept xii–xiv, xx–xxi, xxv, 6–10, 13, 16, 32, 44n, 46n, 108–109, 161–62, 173–74, 186, 220–24, 246–47, 254–55, 283–84, 287–88, 292, 294, 346–48, 421, 474–77, 483, 511–15, 536–37; cultural practices model *515;* ecological-geographical approach model *256;* landscape characteristics defined 346–48; landscape development model *273; Landschaft* 9, 42n, 44n, 221, 247n, 476, 511; *landskab, landskap* xiv, xviii, 6, 157–58, 161, 220–28, 230, 232, 245–47, 435, 444, 476, 511–12, 527; *Landskapet* Åland xiv, 440, 446, 447; *landskapr* 31; *maakunta* 476, 513, 522, 524, 527; *maisema* 421, 476, 498, 511, 513–14, 527; *maisemamaakunta* 513, 525, 527; Nordic xiii–xiv, 221, 223 landscape conservation: Finland xxvi, 435, 508, 525; Norway 401, 402, 416; Sweden 252; *see also* nature conservation

landscape, cultural and social construction of x, xxvi, 8, 40, 53, 173, 223, 300, 483, 487; Denmark 16; Finland 475, 484, 490, 505, 512, 518–20, 522, 526–30, 536–37; Iceland 59; Jutland xiv; North Norway xvi; Scania 186

landscape ecology xxii, xxvii, 254, 268, 270, 347, 348, 355, 369, 398; *see also* ecology; ecosystems; fragmentation

landscape gardens and parks 433; Denmark 22; Finland xxiv, *430*, 432–33; Värmland 237

landscape imagery xii, xx, 512–13, 518, 568; Åland xxv; Denmark 15–16, 39; Finland xxv–xxvi, 435, *466*, 474, 477, 483–510, *495*, 519, 521–22, 524–25, 529, 530–32, 537; Norway 301; Scania xix, 158–59, 161, 169–70, 178; Sweden 223, 248n, 254, 268; Värmland 227, 228, 231, 232, 243, 245, 248n; *see also* heraldry; landscape painting

landscape laws: Åland xxiv; Central Norway (*Frostating* Law) 210; Dalecarlia 210, 213–14, 216, 217; Denmark 93; Faeroes 95; Frisia 9; Funen 93; Jutland 10n, 29–31, 44n, 45n; Iceland xvi–xvii, 65; Scania 6, 165–66; Sweden 146, 210; Uppland 210; Värmland 226; Western Norway (*Gulating* Law) 210, 354, 365, 366

landscape painting(s) 43n, 109, 172–74, 221, 485, 511, 512, 514; Finland *424*, *425*, *426*, *428*, *488*, *490*, 491, 522, 536; Norway 390–94, *391*, *393*, *394*, 412, *413*; Sweden 222; Telemark school xxiii, 412–13, *413*

landscape parks *see* landscape gardens and parks

landscape photography 512, 514; Finland 426, 436, 496, 497, 498–99, 500–501, 503; Norway 302, 315, 323; *see also* aerial photography; landscape imagery

landscape planning and management 253, 512; Finland 506; management agreements 286; Norway xx–xxi, xxiii, 283, 286, 287, 294, 402, 416; Sweden 178, 213, 253, 260, 276, 277; *see also* environmental planning and management

landscape regions xii, xiv, xx, xxiv, 220; Finland, map *528*; functional *(hagmark)* 251, *261*, *262*, 263, 277

landscape values xii, xxvii, 347, 348, 483, 568; Åland xxiv, 450, 453, 456; Finland xxvi, 432, 435, 476, 483–84, 486, 489, 492, 507; Greenland xviii; Norway xx–xxi, 372–73, 397, 401–402, 415, 416; Sweden xx, 148, 153, 155, 268–69, 270, 274, 275, 276–77; *see also* land evaluation

Landschaft see landscape concept

landskab; landskap; landskapr see landscape concept

Landstad, Magnus Bostrup 406, 412

landsting (representative body; legal court) *see* parliaments

land taxation 348; Åland 444; Dalarna 198, 204, 210; Faeroes 97–98, 99, 100, 101–102; Norway 348, 349, 352, 356, 385, 411; Sweden xix, 232

land uplift xviii, 141, 143, 144, 145, 202, 450, 561, 570, 574, map *572*; *see also* shore displacement

Lange, Ulrich 177

language(s) xii, xiv, xxvii, 5, 6–7, 14, 34, 35, 56, 116, 172, 237, 245, 300, 301, 310, 313–15, 335, 336, 492, 511, 513, 516, 526, 527, 536, 543; bilingualism 133n, 321, 329, 332; Celtic 79; Danish in Iceland 56, 543; Danish in Greenland 116, 120, 122, 130, 543; Danish in Scania 174; English ix, 162, 223; Esperanto 241; ethnic 312; Faeroese 77, 78–79, 543; Finnish xiv–xv, xxviiin, 446, 453, 476, 491, 492, 543, 555, 562; Finnish in Sweden 154–55, 235, 238, 240–41, 562; Finnish in Norway 291, 302, 305, 307, 327, 328, 329, 330, 332, 333, 527, 543, 562; Frisian 9; French 34; German 33–34, 511; Greek 14, 33, 314; Greenlandic xviii, 111, 113, 116, 120, 129, 133n; Icelandic xvi–xvii, 56–57, 79; identity and language xi, xviii, xxv, 111, 112, 312, 441, 450, 453, 456, 516; Inuit 113, 543; landscape and language 109, 222, 247, 435, 511, 513; Latin 33–34; lingual networks 302, 329; Norwegian 9, 291, 543; pictorial 222, 314; Russian 334, 563; Saami (Lappish) 154, 291, 292, 296n, 302, 305, 307, 328, 329, 543, map *303;* Scandinavian xiv, xxviiin, 6, 162, 221, 314, 328, 476, 543; Swedish 7, 174–75, 240–41, 511, 543; Swedish in Åland ix, xiv, xxiv, 440, 445, 450–51, 453, 456, 562, 566; Swedish in Finland 443, 445, 446, 451, 453, 491, 492, 497, 506, 543, 562; trilingual region of North Norway 291, 302, 305, 327–29, 330–32, 333–34; *see also* dialects

Lapland 314; Finnish xxiv, 434, 436, 461, 468, 487–88, 489, 491, *495,* 497, 498, 499, 500, 501, *502,* 503–504, 529, 531–32, *533,* 575; Swedish 144, 154

Laponia 144

Lapps *see* Saami

Larin-Kyösti 426

Larsen, Hanne Aa 46n

Larsson, Carl 201

Latin 6, 17, 18, 20, 33–34, 42n

Latvia 144

law(s) xii, xiv, xv, 9, 10, 29, 33–34, 43n, 44n, 174, 210, 217, 306, 397, 513; Åland xxiv, 443–44, 445, 448, 450; Christian IV's Norwegian Law of 1604 93; Christian V's Danish Law of 1683 93, 94; Christian V's Norwegian Law of 1687 93, 94, 95; common law 30; customary law *see* customs; Dalecarlia 201, 210, 213–17; Denmark xvi, 10, 549; Faeroes 77, 85, 87, 93–95, 96; Finland 479n, 521, 562; Iceland 56–57, 64–65, 74n, 551; land reorganization laws 94, 182, 274; medieval xviii, 147, 210, 217; natural law 10, 29, 46n, 181; nature conservation laws 143, 253, 310, 479n, 549, 561; North Norway 303, 307, 309, 315; Norway 302, 303, 310, 315, 316, 317, 318–19, 336, 358–59; Roman law 6; Scania 6, 182; Sweden 143, 201, 274, 561; *see also* constitutions; landscape laws; Magnus Code; parliaments; *Seyðabrævið; ting*

Law Rock (Lögberg) 64–65

League of Nations xxiv–xxv, 445

legends 428; Faeroes 82; Iceland xvi, xvii, 61, 66, 71; Jutland 29, 42n; Norway 323, 396

LEGO toys 547

Lehmann, Carl Peter 391

Leikanger 385

Leiviskä, Iivari 52

Leksand 196, 198, 200, 201; church *199;* Leksand Art and Crafts Association 200; maypole *200;* parish records *205,* 210; Sjugare village 201, *203;* Käringberget outlook tower 197, *197;* Tibble mountain *208*

Lekvattnet 241

Lemmenjoki national park 503

Leningrad 563

Lerwick 337n

Lévi-Strauss, Claude 436

Lie, Jonas 396

Limfjord 15, 46n

liminality x, 69, 487

Lindgren, Astrid xviii, 149, 151

Lindholm, Berndt 497

Linnaeus *see* Linné, Carl von

Linné, Carl von (Carolus Linnaeus) xviii, 149, 179

Linnerhjelm, J. C. 433

Lista 349, 365, 366, 367

Lítla Dímun 99

Little Ice Age 551

Ljusnan Valley 215

Lockarp map *180*

Löfgren, Orvar 245

Lofoten 288–90, 291, 307, 309, 320–21, 322, 554, 574

Lofoten Viking Museum 321

Lögberg *see* Law Rock

logos 337

Lohja 498

Loki 13, 15–16, 24

Lolland 4, 9, 43n, 549

London 308, 555

Longyearbyen 557

Lotman, Yuri 314

Loudon, J. C. 433

Louis XIV 181

Louisiana Art Museum 8

"Lövsjö hundred" (Fryksdals Härad) 228–31, 237, 238, 242

Lowenthal, David 237, 553
Ludvika 201
Lund xxvii, 163, 164, 166, 171, 549; University of 175
Lunde, J. 386–87
Lynge, Hans 128–29
Lyngen 326, 327, 328

maakunta (administrative landscape) (Finland) *see* landscape concept
Maanselkä 470
Macklean, Rutger 179, 181–82, 184, 188n; map of Svaneholm *183*
Magnus Code *(Magnus Lagabøters Landslov)* 87, 93, 95, 354, 376
Magnus Lagabøter (king) 411
Magnus, Olaus 321–22, 326–27, 529
Magnusilla 425
maisema (visual landscape) (Finland) *see* landscape concept
maisemamaakunta (landscape province) (Finland) *see* landscape concept
Malangen 327
Mälar, district (region) 195, 198, 215, 216, 263; Lake Mälaren 195, 198, 446, 558; Valley 145, 148, 153, 154, 198
Malkki, Liisa 112
Malmberget 560
Malmö 149, 153, 169, 186, 549
Maltesholm 172–73, *172*
Man, Isle of 554
manors: Denmark 4, 12, 37; Finland 433; Scania 159, 171, 182, 184; Värmland 227, 228, 237, 243
mapping *see* cartography
Mårbacka 231, *234*, 236, 238, 242, 243, *244*
Margaret (Margrete) I 3, 9
marginality x, 217; Faeroes 82, 550; Finland 462, 465, 471, 473; Finnskogen (Värmland) 243; Greenland 550; Iceland 69, 73, 551; Jutland 8; North Norway 329, 335; Norway 555; Sweden 272, 275, 276; *see also* agriculture, marginal
Mariehamn xxvii, 440, 445, 447, 451, 452, 453, 455, 566
Marimekko textiles 565

marine limit *see* shore displacement, highest shoreline
maritime landscape: North Norway 284, 288–94, *289*, map *290*
Maritime Museum of Åland 442
markets: Åland 446; Denmark 44n, 45n, 546, 547, 548; Faeroes 550; Finland 464, 465, 469, 474, 478, 562, 563, 565–66; Greenland 550; Iceland 551; North Norway 295, 301, 309, 322, 326, 328, 329–30, 381, 382; Norway 554, 555, 556; Scania 164, 165, 169–70, 171, 176, 185; Sweden 559, 560; Western Norway 381–82
Marryat, Horace 196, 202, 206
Marshall Plan 552
Marstrand, Wilhelm 201
Martin, Elias 424
Masku 458
Mathisen, Stein 293
maypoles (midsummer poles) xxv, 200, 453–55, *454*
Maxmo 423
meadows 345, 348, 577, 583; Denmark 546; Faeroes 89, 99; Finland 536; Iceland 58; Jutland 37, 39, 40; Dalecarlia 198, 203, 207, 212, 213, 214, 216; north Sweden *144*; Scania 161, 163–64, 165, 167, 176, 177, 182; southern Norway *403*, 406, 408, 409, 410, 413, 414–15, *415*, 416; Sweden xx, 150, 153; Swedish *hagmark* 257, 259, 266, 267, 268, 269, 270, 272, 273, 274–75, 277; water meadows 39, 163–64, 577; Western Norway xxii, 350, 354, 358, 360, 361, 362, 363–64, 365, 373, *377*, 381, 383, 385, 387, 389–90; *see also* seminatural vegetation
Mecklenburg 168
media 242–43; Åland 451; Finland xxvi, 483, 490, 512, 514, 516, 518, 520, 523, 530, 532, 535–36; Norway 306, 310, 316; Värmland 245; *see also* newspapers
Meinig, Donald W. 519
Meløe, Jakob 283, 291–92, 293
Mels, Tom 168
memory x, xii, xvii, 54, 313, 432, 478n; Åland 429; Denmark xvi, 10, 16, 20, 32, 33, 34, 42n, 45n; Faeroes 98; Finland xxiv, 432,

460, 465, 470, 472, 475, 478n, 485, 490, 505, 507, 515, 516, 518, 522; Greenland 128, 130; Iceland xvii, 53–54, 55, 57–61, 64, 66, 67, 68, 71, 72–73; Norway 288, 310, 356; Sweden 195, 222, 226, 236, 238, 277; *see also* nostalgia

mentality: Finnish 459, 460, 475; Greenlandic 114, 122–23

mental landscape 284, 288, *291*, 450, 451, 456, 475, 486

mental maps 293, 489, 496

mercantilism 172, 178, 545

Mercator, Gerhard 42n

metaphors 308, 315, 429, 511, 516; Finland 485, 520; Greenland 110, 113, 114, 122, 123; Iceland 65; Jutland 13, 14; North Norway xxi, 302, 305, 313, 332, 334; Norway 324, 369n; Scania 184; Värmland 230, 231, 237, 245

Mid-Atlantic Ridge 551, 574

Middle Ages xiv, 216, 221, 579; Åland xxiv, 443–44, 446, 497; Dalecarlia 194, 195, 201, 210, 213, 216–17; Denmark 159; Faeroes xviii, 81, 82–87, 89, 95–104; Finland 461, 498, 523; Iceland 62, 65, 68; Jutland 17, 45n; North Norway 294, 321, 326; Norway xiv, xxi, 55, 210, 316, 319, 326; Schleswig-Holstein 17; southern Norway 402, 412, 414; Sweden xviii, 143, 146, 147–48, 159, 165, 167, 168, 187, 214, 216, 221, 272, 274, 559; Värmland 225–26; Western Norway 354

midnight sun 308, 320, 332

midsummer *see* summer

migration 127, 222, 309, 311; evacuees (Finland) 432, 434; migrant workers 113–14, 128, 198, 202; out-migration 232, 529, 531; to towns (Åland) 447, (Finland) 474, 562, 566, (Greenland) 117, 121, 122, 124–25, 126, (Iceland) 552, 553, (Sweden) xix, 153, 232n, 559; migratory birds and fish 309; Vikings 55, 77–79; *see also* emigration; immigration; seasonal migration

Mikkeli 524

military landscape 295; military bases 332, 547, 550, 552, 563; military installations 308, 324, 325

mills 44n, 185, 406, 414; paper mills 154, 554; pulpmills 227, 565; sawmills 152, 233, 406, 414, 554, 564; windmills 170; *see also* industry

mining 568; Finland 504, 564, 565; Norway 295, 330, 333, 334, 410–11, 554, 556; Svalbard 310, 557; Sweden 146, 150, 195, 200, 226, 231, 275, 560, 561

minorities: indigenous 284, 312, 303, 543, 558, 562; linguistic 154–55, 311, 303, 312, 543, 555, 562; national 303, 445, 451; *see also* ethnicity

Mitchell, W. J. T. 487

Mjøsa, Lake 319

modernism 500, 506; modern art *394;* modern music 396

modernity 8, 10, 16, 40, 41, 188n, 311–12, 314, 348, 474–75, 483, 568, 583; Åland 440; Dalecarlia 195, 207; Denmark 5, 10, 17, 24, 548; Faeroes xviii; Finland xv, xxv, 462, 473, 475, 478n, 508, 534; Greenland xviii, 106, 117, 118, 125, 131, 132; Iceland xviii, 73; Jutland 8; North Norway 289, 290, 295; Norway xiv, 295, 555; Scania 8, 164, 171, 173, 174, 178–86; Sweden 152, 243, 272, 276; Western Norway 382–83; Värmland 226–27, 237, 240; *see also* modernism; modernization

modernization 487; Denmark 547; Finland xxv, 460, 468, 473, 474, 475, 477, 486, 502, 507, 562; Greenland 123; North Norway 300, 330; Norway 306; southern Norway 410, 414, 416; Svalbard 310; Sweden 558–59; Värmland 222; *see also* modernity

Moen, Asbjørn 345, 347

Møller–Mærsk Group 547–48

Møn 4

Montesquieu x

monuments: Dalgas monument (Jutland) *26;* Finland 483, 489, 491, 505, 522; Iceland 45n, 53, 64, 72; Jutland 19–20, 26, 31, 33; Norway 310, 322, 384, *391,* 392, 398, 402, 416; Sweden 176, 200, 259, 266, 268

Mora 200, 201

moral economy 31, 35, 44n

moral landscape 172, 513, 519

Møre 320

mørk (merkur) (land or sheep evaluation unit) (Faeroes) 96–103
mosaic, landscape 255, 256, 259, 263, 271, 272, 276, 347, 367, 385, 390, 404, 414
Moscow 314, 330
Moskenes 321–22, 335
Motzfeldt, Jonathan 126, 131–32
Motzfeldt, Josef 126
multiculturalism 301–302, 305, 311, 315, 321, 327, 328, 329, 330–35; *see also* ethnicity; immigration
Munkeliv Monastery 100
Munsterhjelm, Hjalmar 424
Murmansk 309
museums xi; Åland 442, 455; Denmark 8; Finland 470; Greenland 106, 107, 115; North Norway 310, 312, 321, 327, 332, 334; Norway xiii, 283, 306, 312, 336, 416; Sweden 155, 174, 198–99, 200, 223, 246
music xi, 141, 314, 441, 514; Åland 453; Denmark 12; Finland 531; Norway xxiii, 373, 390, 394–96, 412; Sweden 149, 199–201, 234; *see also* anthems; folk music; songs
Mykines 78, 82, 100
myths x, 46n, 176n, 194, 312, 314, 321, 429, 441, 487; Dovre witch 237; Finland 429, 463, 499; Iceland 56, 57, 68, 72; mythical landscape (Saami) 288; Norse mythology 14, 16, 391, 392; Norway 396, 411; Värmland 222, 231, 236–38, 242; *see also* trolls

Nærøyfjorden 391
naming 517; Finland 524, 528–29; Iceland xvii, 54, 56, 59, 62, 63, 235; Norway 288–89; Värmland 235; *see also* place-names
Nanortalik 124
Napoleonic Wars 22, 444, 545
Närke-Värmland 227
narrative(s) 312, 516, 517, 519; Finland 484, 487, 490, 502, 505; Greenland 120, 128; Iceland 60, 61, 68; Norway 288, 322, 323, 326, 335
Narsarsuaq 116, *117*
Nås 213
Nash, Catherine 188n
Näsström, Gustaf 194–95

nation 103, 112, 113, 221, 441; Finland 483, 488–89, 491, 504, 521, 522; Greenland 111, 112, 119; Iceland 56, 544; Norway 300, 305; Sweden 246
national atlases *see* atlases
national identity x, xxiv, 112, 113, 284, 441, 516, 517; Åland 445, 450, 453, 455; Denmark 5, 20, 22–23; Finland xiv–xv, xxv, 248n, 472, 477, 483, 485, 486, 487, 489, 492, 521–22, 534, 536, 537; Greenland xviii, 106, 111, 114, 117, 118, 129–30, 132n, 133n; Iceland xvi–xviii, 55, 56–59, 60, 72, 73; Norway xiv, xxiii, 286, 306, 336, 373, 390, 392; Sweden xix; *see also* identity; regional identity
national image 268
nationalism x, xii, xxvi, 106, 112, 485, 516, 518; Denmark 22; Finland 453, 487, 503, 514; Greenland 114, 118, 125, 126; Iceland 64; Sweden 168, 222
nationality xi, 175, 504; *see also* "right of domicile" (Åland)
national landscapes 110, 514, 518, 519–20; Åland 453; Denmark 22; Finland xxvi, 483–510, 519, 520, 522, 525, 537; Greenland 106, 112–14; Norway 327; Sweden 247
national minorities *see* minorities
National Museum of Folklore (Sweden) 198
national parks: Finland 484, 501, 503–504, 566; Iceland 552; Norway 321, 327, 557; Sweden 561
national romanticism: Åland 455; Denmark 4, 21–24, 27; Finland 248n, 483, 487, 503, 521; Greenland 111; Iceland 64; Norway 391–92, 412; Sweden 194–95, 222–23
nation-building x, 30, 159, 164, 520; Åland xxv, 448, 453; Denmark 159, 164; Finland xiv, xxv, 484, 486, 497, 505, 507; Germany x; Greenland 110–12; Norway 336; Sweden 159
nation-state ix, 3, 8, 10, 485, 520; Denmark xvi, 21–22; Finland xiv, xxv–xxvi, 484, 485–86, 491, 502, 503, 505, 506, 530; Norway xiv; Sweden 192, 246, 247n
natural landscape (natural environment) 53, 254, 269; Åland 450; Denmark 13, 26, 427; England 429; Faeroes 79, 81, 103; Finland 421, 428–29, 436, 491; Greenland 109, 125;

Naturlandschaft 421, 428–29; Norway 78, 345, 402; Sweden 143, 145, 146, 167–68, 202–203, 251, 272; *see also* physical landscape

nature conservation 345; Finland 459, 465, 468, 470, 472, 474, 478n–79n, 503, 566; Iceland 552; Norway xxii, 401, 402, 410, 415, 416, 557; Sweden 143, 251, 252–53, 259, 266, 268, 276, 561; *see also* landscape conservation; national parks

nature-culture relations *see* society-environment relations

nature reserves 327, 552, 557

Negri, Francesco 332

Neiden 330; Neiden River (Neidenelva) 334

Neolithic 225

Neste oil company 565

Netherlands xiv, xxviiin, 42n, 221, 322, 546, 583

New Guinea 60

New Iceland 59

newspapers: Åland xxv, 445, 451; Denmark 129; Finland 516, 532, 535–36; Greenland 111, 113, 125–26, 126–28, 129, 133n, 134n

Nicolay, Ludwig von 433

Nils Holgersson (Lagerlöf) 150, 222, 245–46, 248n

Nilsson, Eja 120

Nilssøn, Jens 406

Nobel, Alfred 560

Nobel Prizes 150, 228, 247n, 560; Nobel Peace Prize 560

nobility 173; Denmark 12, 27, 29, 37–38, 171, 175; Finland 491; Norway 95, 96, 316, 325–26; Scania 165, 171, 175, 181, 188n; Schleswig–Holstein 17; Sweden 221–22

Nokia telephones 560, 565–66

Nólsoy 94

Norden ix–x, xii, xiv, xv, xix, xxiv, xxvi–xxviii, 162, 164, 308, 311, 440, 529, 541–84; definition ix, 543; Nordic countries map *544;* physical landscape and natural resources map *582;* place-names and regions map *569;* precipitation map *576;* Quaternary deposits map *571;* shore regression map *572;* superficial deposits map *580;* vegetation zones map *578; see also* Kalmar Union (Nordic Union); Nordic cooperation; Scandinavia

Nord-Fugløy 326–27

Nordic cooperation xxvi–xxviii, 266, 345–46, 402, 544–45, 566; Nordic Council 545, 566; Nordic Council of Ministers xxii, 345–46, 402; Nordic Saami Conference 331; Nordic Saami University 332; Nordic Seminar for Landscape Research xxvii–xxviii

Nordland 306, 307, *307,* 312, 316, *317,* 319–23, 325, 327, 335

Nordling, Johan 196–97

Nordreisa 327

"Nordscapes" xxvii–xxviii

Norðstreymoy 96

Nord-Trøndelag 318

Norðuroy 100

Normann, Adelsteen 392

Norrland 143, 147, 154, 195–96, 203, 232, 238

Norse xviii, 13, 31, 55–56, 59, 77–79, 81, 85, 95, 102, 133n, 337n, 391, 396; law xvi, 56, 85, 95; cosmology 14, 55

Norsk Hydro 556–57

Norske Skog, newsprint manufacturers 556

North America *see* America (North)

North Atlantic ix, xiv, xv, xvi–xviii, xix, 3, 4, 9–10, 51–138, 214, 216, 301, 309, 310, 550–53, 556, 573; *see also* Faeroe Islands; Greenland; Iceland; Orkney; Shetland

North Atlantic Treaty Organization (NATO) 552, 558, 566

North Calotte (Nordkalotten) 300

North Cape (Nordkapp) 307, 309, 329, 332–33, 335

Northeast Passage 308

Northern Karelia 468, 483, 494, *495,* 497, 522, 529, 531, 532, *533,* 534–35

Northern Ostrobothnia 470, *495,* 497, 498

North Friesland 9

North Greenland 132n

North Jutland 17, 18, 31

North Norway xx–xxi, 283–84, 286, 287–95, 296, 300–43, map *305, 307,* 553, 554–55, 556, 574; "Cultural History of North Norway"

(*Nordnorsk kulturhistorie*) 291, 302, 305–306, 312, 334; *see also* Finnmark; Nordland; Troms
North Sea xiv, xv, 4, 13, 16, 42n, 352, 545, 547, 548, 554, 556, 557
Norway ix, xii, xiv, xviii, xix, xx–xxiii, xxvii, xxviii, 5, 7, 9, 10, 19, 22, 45n, 56, 58, 74, 77–79, 82, 102, 144, 145, 163, 210, 214, 216, 225, 227, 234, 236, 237, 238–40, 277, 287–418, 543, 545, 547, 550, 551, 553–57, 558, 563, 564, 573, 575, 577, 579, 581, 583, map *284*; *see also* Denmark–Norway; North Norway; southern Norway; Svalbard; Sweden–Norway; Western Norway
Norwegian Automobile Federation *(Norges Automobil-Forbund)* (NAF) 315–16; *see also* road book
Norwegian Sea 574
nostalgia 44, 58, 127, 306, 432, 483, 507
novels: Åland 429; Denmark 427; Finland 435, 459–60, 473, 522, 532, 534–35; Sweden xviii–xix, 194, 196–97, 236, 242; Wales 40–41; *see also* Gösta Berling's Saga; Nils Holgersson
Novgorod 308, 324, 330
Novo pharmaceuticals 547
nuclear power 549, 560, 565
Nurmijärvi 484
Nuti, Lucia 170
Nuttall, Mark 112, 12–122, 134n
Nuuk (Godthåb) 106, *108*, 109, 110, 114–24, *115*, *119*, *121*, 126, 128–29, 131, 132, 132n, 134n, 135n, 550; *see also* Narsarsuaq; Qoornoq
Ny-Ålesund 557

óðalsjørð (allodial or freehold land) (Faeroes) 95–96, 100
Odarslöv 171
Odense, map *21*
Odin 14, 16
OECD 556
Oehlenschläger, Adam 23–25
Ofoten 327
oil and gas, offshore 296, 301, 548, 556–57
Öland 143, 144, 145, 216, 573, 574, 577, 581; Ölands Stora Alvar *145*

Olav, Saint 316, 318, 326
Olav V 307, 319
Olavinlinna (Nyslott) 461, 498, 506
Olsson, G. A. 347
Olwig, Karen Fog 127
Olwig, Kenneth R. xxviiin, 166, 174, 221, 476
Øm Monastery 102
Önningeby 453
open-access resources 292
open-field system 8, 165, 176–77, 179, 182, 213, 216
Optand 152
Öræfa 70
Örebro 227
Öresund (Øresund; the Sound) 149, 168, *169*, 170, 186–87, 545, 549, 574, 581; *see also* bridges
orientation xxv, 57, 62, 73
Orkney ix, 59, 102, 216, 554
Ornässtugan 200
Ørsted, Anders Sandøe 29
Ørsted, Christian 23, 24, 25, 29
Ortelius, Abraham 42n
Oscar (Crown Prince, Sweden) 239
Oscar II 333
Oslo 238, 287, 306, 307, 382, 554, 555, 581
Oslofjorden 296
Östergötlund 145, 154, 216, 263, 275, 581
Östmark 240
Ostrobothnia (Pohjanmaa) 422, 423, *495*, 498, 499, 508n, 521, 535; *see also* Northern Ostrobothnia; Southern Ostrobothnia
Ottar of Hålogaland 310, 323, 326, 553–54
Oulanka national park 503
Oulu 470, 497
Oulujoki, River 498, 500, 506
outfield(s): Faeroes *(uttangarðs)* 78, 82–83, 86, 87, 88, 89–91, 93–94, 96, 97, 98, 99, 101, 103, 104; Norway *(utmark)* 401; southern Norway xxiii, 404, 406, 407–408; Sweden *(utmark)* 176–77, 179, 212, 256, 257, 272, 273; Western Norway xxii, 350–52, *351*, 356–60, 368, 375, 376–78, *379*, 380, 381, 383, 388; *see also* infield(s)
outsider perspective xii, xiii, 40–44, 109, 311, 442, 512, 518; Åland xxiv; Finland 530, 532,

533; Greenland 109, 132n; Jutland 14, 16, 40; North Norway 301, 305–306; *see also* insider perspective

Paanajärvi, Lake 500
Paasi, Anssi xi, 238, 239, 508
Päätalo, Kalle 535
Padasjoki 498
Päijänne, Lake 418, 498
Päijät-Häme 295, 508n
painting *see* art
Pallas-Ounas national park 501, 503, 506
Pallin, Brita 213–14
Palmer, Catherine 119, 516–17
Pálsson, Gisli 53
paper industry *see* pulp and paper industry
parish registers (Sweden) xix, 151, 205, 210
parks: deer parks 167, 187; nature park 498; theme parks 151, 436; urban parks 433, 500; *see also* landscape gardens and parks; national parks
Parland, Oscar 432
parliaments: Åland *(landsting; lagting)* xxv, 445, 447–48, 452–53, 455; Britain 179, 181; Denmark *(Folketing)* 29, 35, 43n, 44n; Faeroes (Althing; *alting; løgting*) 85, 86–87, 95; Finland 445, 460, 563, ,565; Iceland (Althing) 45n, 64, 551; Jutland *(Landsting)* 27–35, 41, 45n; Norway *(Storting)* 240, 287, 324, 554, 560; parliament *(Sámediggi; Sameting)* 307, 332; Schleswig–Holstein (Assembly of the Estates) 29; Sweden 181, 239, 240, 559, 562; Värmland 221; *see also ting*
pastures 345; Åland 448, 450; Denmark 15, 45n, 546; Faeroes xvii, 77, 83, 90–91, 103; Iceland 58; pastoral economy xxiii, 166; North Norway 292, 309; Scania 164, 165, 167, 171, 173, 176, 182; southern Norway xxiii, 406, 410, 414; Sweden 186, 207, 259; Swedish *hagmark* xx, 251, 257, 259–60, 266–70, 272, 273, 274, 275, 277; Western Norway xxiii, 346, 348, 361, 363–64, 365, 367, 373, 375, 376, 377, 381, 383, 384, 385, 387, 388–89, 389; *see also* agropastoral landscape; grazing; seminatural vegetation
Pasu-Anni *see* Hakkarainen, Anna

Pasvik River (Pasvikelva) 334
paths and tracks: Denmark 548; cattle paths xxiii, 87, 88, 268, 274, 382, 380; Faeroes 87, 88; Finland 433; Iceland 54, 57, 60–61, 73; Norway xxiii, 352, 358, 377, 380, 382, 383, 393, 401, 413; Sweden 151, 268, 274
Pavels, Claus 390–91
peat cutting: Faeroes 86, 91, 95, 99, 104; Finland 564; Jutland 37; mechanization 358–59; Sweden 149, 272; Western Norway 346, 350, 351, 354, 356–59, 362, 367, 368, 375
Peirce, Charles Sanders 313–14
perception xi, 16, 64, 141, 162, 166, 313, 315, 475, 513; Åland 450, 456; Faeroes 104; Finland 453, 465, 470, 472, 484, 486, 487, 506, 508, 522; environmental perception xxiv, 421, 434, 435, 450, 513, 527; Greenland 106, 109, 131; Iceland 62, 73; Jutland xvi, 13, 22, 45n, 46n; Norway xxii, 287, 291, 296, 301, 306, 344; Sweden xx, 232, 246, 251, 260–61, 263
periphery (-ies) 314, 326, 534, 544, 552; Åland 451, 453; Arctic 310; Finland 468, 470, 474, 487–88, 496–97, 502, 503, 508, 529, 535; Greenlandic settlements 115; Icelandic coastal communities 552; Jutland 8, 39; North Norway xxi, 291, 300, 302, 308–11, 321, 322, 331, 334, 335; Norway 412; Scania 8, 159; Sweden 251, 260–61, 263; Värmland 220, 228, 230
permafrost 579
perspective drawing 18, 41, 170, 172, 391, 487
Persson, Jones Mats 198
Peter I Island 557
Petersen, Tove Søvndahl 127
Petsamo (Petchenga) 308, 330, 333, 495, 498, 500, 563
phenomenology xiii, 41, 46n
photography *see* aerial photography; landscape photography
physical landscape xi, xxvi, xxvii, 109, 141, 221, 224, 254–55, 256, 258, 300, 475, 568–84, maps (Norden) 571, 578, 580, 582; Åland xxv, 447, 455; Dalecarlia xix, 192, 202–204, 209; Jutland xvi; Finland xxiv, 516, 527, 529; North Norway 301, 315, 327,

335; Norway 316; Scania 159–61, 172, 176; Sweden xviii, 141, 143; Western Norway xxii, 344, 368; Värmland xx, 227, 243; *see also* natural landscape

picturesque 195, 234, 422, 476, 489, 491, 497; *voyages pittoresques* 222, 488–89

pigs: Denmark 547; Faeroes 83–84, 100, 103; Norway 352, 365, *365*, 383, *405*, 555; Sweden 149

pilgrimage 45, 129, 131, 134n

Pirkanmaa 508n

Pirkkala *495*

Pite Älv, river 143

place(s) ix, x–xiv, xxvi–xxvii, xxviiin, 10, 12, 14, 16, 30–31, 32–34, 41, 59, 63, 170, 312, 315, 397, 427, 432, 475, 476, 485, 511–12, 515–17, 518, 526; attachment to 127, 129, 355–56; Certeau, Michel de 57; Finland xxv–xxvi, 487, 492, 505, 511, 522, 524, 530; fishing (North Norway) 288–89, 293; *genius loci* (spirit of place) 434; Greenlandic 106, 120, 122–23, 125, 127, 129, 131, 132; Iceland xvi, 58, 59, 61, 63, 64–65, 69–71, 72–73; Jutland xvi, 12, 16, 17, 28, 31, 33, 34, 35, 38, 39, 41, 43n, 44n, 45n, 46n; meeting places 224, 246, 333–35; "out of place" 133n; Norway 333, 355; place and time 361; place fetishism 492; Saami 287–88; Scania 6, 8, 173, 187; sense of place 311, 516, 517; Svalbard 310; Sweden 263, 269; Värmland 231, 235, 247; *see also* identity, place

place-names (toponyms) xi, 361, 517, 568, map (Norden) *569;* Åland 446, 449; Dalecarlia 202; Denmark 5; Faeroes 77, 82, 83–84, 89, 99, map *80;* field names (Faeroes) 88, 98, 99, (Sweden) 212, 259, (Western Norway) 354, 365; Finland 423, 432, 435, 509n, 524, 527, 529; Greenland 106, 129, 132, map *107;* Iceland xvi–xvii, 54, 55–56, 58, 59, 61, 62–63, 64–66, 69–70, map *60;* Jutland xvi, 17, 29, 31, 42n, 43n–44n, 45n; North Norway 287, 288–89, *289,* 314, 321, 322, 324, 325, 329, 331, 332, 333, 335; Norway 324, 325; Saami 287, 332, 335; Scania 162, 187n; Värmland 235; West Norway 349, 350, 358, 359

plagues 66–67, *see also* Black Death

planning xxviii, 174, 187, 263; land reorganization plans *180,* 182–83, *183;* Marshall Plan 552; national park plan (Norway) 557; nature conservation plans 253, 259; North Norwegian reconstruction plan 308; physical planning 306, 315, 336; planned settlements 216; regional planning 187, 535; town planning 63, 131–32, 134n, 433, 583; *see also* environmental planning and management; landscape planning and management

plant communities 266, 267–68, 347, 348, 375, 385

plebiscites: Iceland 5, 551; Schleswig 546

plurality: North Norway xxi, 302, 308–309, 321, 328, 335

poetry 166, 173; A. Oehlenschäger's 22–23; Dalecarlia xix; Denmark 4, 167, 172–73; F. G. Klopstock's 22; Finland 433, 520, 521, 522; folk poetry (Finland) 426, (Värmland) 236–37, 238, 241, 248n; Greenland 111; H. C. Andersen's 12–13, 14, 15, 16, 17, 22, 23–24, 27, 38; H. Wergeland's 397; *Kalevala* 248n, 429; Maltesholm (Scania) *172,* 172–73; Norway 306, 323, 373, 390, 396–97, 411; Poland 174; Scania 167, 172–73, 187, S. S. Blicher's 35–36; Sweden 268; Western Norway xxiii, 396–97

Pohjanmaa *see* Ostrobothnia

Point Barrow 335

political landscape xiii, xxviii, 6, 10, 29–32, 522; Åland xxiv; Denmark xv; Iceland 65; Jutland 6, 27, 30, 33, 38, 39; Sweden xviii; *see also* polity

polity xiii-xiv, xxviiin, 33, 40, 46n; Åland xiv; Denmark xv, 6, 10; Jutland xvi, 8, 27–33, 38, 40–41; Iceland xvi; Scania 6, 7, 9; Schleswig–Holstein 9; *see also* landscape concept, *Landschaft*; political landscape

pollarding: southern Norway 408, 410, 414; Sweden 268, 269, 277; Western Norway xxiii, 373, 381, 385–88, *389,* 390; *see also* coppicing; fodder

pollution 301, 308, 468, 532, 549, 553, 574

Poltava 181

Pomors (Russians trading with North Norway) 329–30
Pontoppidan, Erich 20–21, 41, 43n; *see also* atlases
population 568, 581, 583; Åland 440, 443, 447, 452, 566; Dalecarlia 202, 212; Denmark 6, 546, 548, 549; Faeroes 81, 88, 550; Finland 423, 562–63, 564; Greenland xviii, 115–16, 122, 124, 132, 134n, 550; Iceland 58, 66, 551, 552, 553; North Norway 291, 307, *307*, 308, 330, 334; Norway *307*, 382, 553, 554, 555, 556; population censuses 259, 348, 350; population records 210, 423; Saami 555; Scania 176, 186; southern Norway 405, 408, 409; Svalbard 557; Sweden 153–54, 176, 271, 274, 558, 559, 561; Värmland 226, 227, 232, 235, 240; *see also* depopulation; urban population
Porkkala 563
Porsanger 325
Porvoo (Borgå) 435, 497, 498, 506
postmodernism 237
Poutvaara, Matti 500–501
poverty 44n, 173; Dalecarlia , 201, 208; Finland 473, 503, 504; Iceland 66, 69, 551; Jutland 37; Sweden 559; Värmland 235
power: academic xxviiin, 527–28; administrative 514, 524, 526; balance of, 174, 212; central 159, 168, 491; Denmark 164, 545; female 66; foreign 127; landed estates 171–72, 217; landscape xvi, 57, 63, 73, 164, 167, 173, 185, 507, 512, 513; legal 166; literary 247n; nature 71; place 34; political xi, 14; relations xi, xxvi, 224, 228, 232, 240, 514; struggle for 164, 165; Scania 186; Sweden 144, 157, 178; towns 131, 132, 167, 168, 170, 174
practice, social x, xi, xxvi, 31, 40, 44n, 165–68, 109, 114, 127, 221, 222, 223–24, 245, 476, 478n, 487, 511–37; Certeau, Michel de 57; Iceland xvi–xvii, 53, 54, 56, naming practices 59, 62, 72, 73; Scania 165–68, 178; Sweden xx; Värmland 226, 227, 228, 231–32, 235, 237, 238, 240, 243, 245, 246, 247
Pred, Allan 185, 248n
primogeniture *see* land inheritance

"process-landscape" *(förloppslandskap)* 246
prospects, landscape 10, 18, 20, 29, 173; Odense *21*; town prospects 43n, 170
Prussia 17, 545
Ptolemy (Claudius Ptolemaeus) 42n, 170
pulp and paper industry: Finland 562, 564, 564; Norway 554; Sweden 559
Punkaharju *428*, 498, 500, 506
Puranen, Jorma 436
Pyhätunturi national park 501, 503

Qaqortoq 131
Qoornoq 121, 128
Quains *(kvener)* 303, 305, 327, 328, 330; *see also* Finns in North Norway
Quaternary deposits 254, 546, 570, map *571*, 573
Queen Maud Land (Dronning Maud Land) 557
Querini, Pietro 323
questionnaires: Åland identity 456; images of Finnish provinces 530–32, *533*

Radloff, F. W. 443
railroads 568; Denmark 25, 546, 548, 549; Finland 561–62; Norway 554; Sweden 185, 201, 559, 560
Raivo, Petri 434
Randers 22
Rantzau, Henrik 7, 17–19, 20, 21, 41, 42n–43n
Rantzau, Johan 18–19, 42n
Rapola 484
Rasmussen, Knud 113
Rauma 458
records, historical *see* sources, historical
recreation 168; Åland xxv, 447, 456; Finland 459, *502*, 562; Iceland 64; Norway 294, 296, 332, 373, 382, 384, 402; Sweden 154, 269; *see also* second homes
referendums *see* European Union; plebiscites
Reformation 71, 87, 96, 316, 320, 428
reforms, social x; Denmark 23, 546; Finland 460, 473; Sweden 184, 558–59; land reform *see* land reorganization
region, concept 221, 344–46, 347, 368, 517–18, 526, 527

regional associations 523, 535
regional descriptions: Sweden 155n, 247n
regional development xxi, 544, 561, 566; *see also* regional policies
regional divisions 517, 518; Nordic regions xxii, 266, 345–46, 368, 402, 404; Norwegian vegetation regions xxii, 345–46, 368; Finland 494, 522–30, *528*
regional geography 20, 42, 244, 245, 526; *see also* chorography
regional identity ix–xii, xiv, xxvii, 441, 475–76, 515, 517–18, *519;* Åland xxiv–xxv, 442–43, 446, 447–48, 450–56; Dalecarlia xix, 212, 217; Denmark 8–9; Faeroes 77; Finland xiv, 441–43, 453, 468, 513, 514, 521–22, 524, 527, 529, 53–32, 534, 535–36; Frisia 9; Jutland xvi, 8, 17, 41, 43n; North Norway xx–xxi, 283, 284, 291, 293, 294, 296, 300–302, 306, 308, 310, 312–13, 315; Norway xiv, 9, 238; Scania xix, 6, 157–58, 165, 174, 186–87; Sweden xviii–xix, 146–47, 222–23, 246, 247n, 276; Telemark 411, 412; Värmland xx, 220, 224–28, 232, 238, 241, 242, 245, 247; Western Norway xxii–xxiii, 344, 346, 364–67; *see also* identity; national identity
regional images xxvi, 306, 313, 514, 518, 529, 530–33, 534, 535
regional integration 164–65, 187, 306
regionalism xii, xxvi, 222, 247, 247n, 514, 534–37
regional landscapes ix, x, xix, 158–59, 322, 500, 507, 514–15, 518, 521–22; *see also* landscape regions
regional newspapers 533, 535–36
regional personality xi, 161, 515, 534, 536
regional policies: Denmark 549; Finland 535, 566; Norway 555; Sweden 146, 154, 561
regional strategy: North Norway 306
regions: administrative xxiv, 146, 157, 186, 307, 517, 522–25, 526, *see also* administrative units; agricultural 266; biophysical xxii, 345, 346, 368–69; border 159, 224, 236; coastal 296, 345, 346; construction of xi, xxvi, 238–43, 518, 525–30; cultural 159–61, 202, 517; dialect regions, map *303;* economic 529; functional xix, 517, 529; historical ix, 440, 443–46, 490, 494, 496, 502; homogeneous (uniform) 141, 526, 537; imagined 238; institutionalization of 238, 517; perceptual 518; natural 527; scenic 484, 507; vegetation xxii, 345, 346, 368

regions, cultural and social construction of xi, xiv, xxvi, 518; Finland 476–77, 478n, 526–30; Sweden 266; Värmland 228, 231, 238–39, 242–43

Rehnberg, Mats 196, 209, 222

reindeer: Finland 436, *461,* 488, 531, 564; heraldry 323, 324–25, 330, 331, 335; Saami reindeer herding xx, 154, 283, 287, 291, 292–96, 301, 311, 321, 324–25, 328, 331, 332, 333, 555, 564, 577; transport 327, 331

Reisa, national park 327; river 327; valley 327

remote sensing 254; *see also* aerial photography

Renaissance xv, 4, 7, 10, 17, 30, 41, 167, 170, 173

Rendalen 325

reparations, Finland to Soviet Union 563, 565

representation, political 22, 29, 30, 41, 42n, 181, 307, 336, 444, 460, 562, 563

representation(s) xii, 158, 167, 173–74, 187, 224, 246, 284, 301, 313, 315, 441–42, 483, 512–13, 514, 516, 518; buildings 164; cartographic 261, 165, 178, 435, *461,* 529; film 130, 133n; Finland 472, 474, 484, 485–505, 506, 520, 521–22, 524, 527, 530–31, 536; Granö's method 421, 434–35; heraldic 320–37; literary 16, 20, 40, 109–10, 125, 133n, 231, 237; pictorial 16, 19, 109, 158, 162, 170, 172–73, 302, 392, 492–505; popular 115, 121, 123, 125, 126; Sweden 246, 247; Värmland 231, 237, 242–46, 247

Repton, Humphrey 433

Resen Mandt, Engelbret Michaelsen 406

resistance: to innovations 209, 239, 291, 483; wartime 174, 175–76, 547, *see also snapphane*

Reykjavík 31, 45, 62, 63, 68, 551, 552, 581

"right of domicile" (Åland) xxv, 451, 456

Riis, Jakob 26

Ritamäki *241*

Ritter, Karl 436

rituals xi, 14, 45, 57, 237, 394, 436, 478n, 518; deritualization 242; ritual symbols 318, *391*
roads 568; Åland 447, 450, 451, *452;* Denmark 187, 548, 549; Finland 436, 459, 505, 561; Greenland 118; Iceland 551; Jutland 25, 31, 36–37, 39, 44n–45n; Norway 329, 335, 401, 554; road deaths 459; roadscapes 505; Scania 159, 173, 186–87; southern Norway 409, 410, 411, 413, 414, 416; Sweden 151, 154, 195, 212, 230, 259, 268, 274, 559, 560; Western Norway 349, *351,* 353, 358, 360, 378, 380, 382, 383, 384; *see also* road book; signs; streets
road book, Norwegian *(NAF veibok)* 315–16, 327, 328, 330, 331, 332, 333, 335
rock carvings 147
rockfalls 377, 379, 380, 389
Rogaland 358
Rögden, Lake 236
Rögnvaldr Jarl the Great of Norway 58
Rom, Per 412
Romani People *(romanifolket; tartere)* 303
Romans 6, 17, 33–34, 162–63, 167
Roma People (Gypsies; *roma; sigøynere*) 303
Rome 45
Rose, Gillian xi
Roskilde 29
Røst 288–90, *290,* 321, 322, 323
Rottneros 228, *229,* 233, 234, 243
Rovaniemi 501
Rumohr, Knut 392–93, *394*
Runeberg, Johan Ludvig 520, 521
runes 12–14, 17, 19, *19,* 22, 27, 33, 42n, 147, 163, *268;* Odin, god of the runes 16
Ruokolahti 458
rural areas (districts) 581, 583, (Åland) 445, (Finland) 471, 474, 458, (Iceland) 63, (Norway) 302, 352, 353, 381, 385, (Sweden) 151, 168, 186, 263, 275; rural buildings 259, 383, *see also* building styles; rural continuity 344; rural environment (Iceland) 63; rural estates 172; rural mansions 233; rural people (Denmark) 5, 22, 27, (Faeroes) 87, (Sweden) 206; rural population 153, 382, 581; rural production 559, 583; rural proletariat 181, 563; rural settlements 163, 544, 548, *see also* villages; rural society 584, (Norway) 416, (Sweden) xix, 184, 559
rural landscape 583; Denmark 10; Finland 423, 496; Norway xxi, 401, 409; Sweden xviii, 153, 211, 251, 258
Russia 5, 7, 144, 300, 301, 307, 309, 324, 329, 330, 333–34, 445, 448, 460, 488, 496, 498, 556, 558, 562; economic sphere in North Norway 327; Finland, Grand Duchy of the Russian Empire 434, 445, 460, 483, 489, 491, 520, 558, 562–63; occupation of Åland xxiv, 444–45, 446; occupation of North Norway 555; Russian Karelia 498; *see also* borderlands; Novgorod; Soviet Union
Ryden, K. C. 517, 527, 529

Saab 558
Saami (Lapps) 577, 579, map *303;* Coast Saami 293–94; East Saami 330, 333; Finland 436, 488, 491, 531, 562, 564; Norway xx, xxi, 20, 283, 284, 287–88, 291–96, 302–303, 306, 307, 308, 310–11, 312, 321, 322, 324–25, 327–35, 555; parliament *(Sámediggi* or *Sametinget)* 307, 332; South Saami 296, 321, 325; Sweden 154, 558; taxation of 308; *see also* language, Saami
Saami landscape 287–88, 325
sagas: *Frithiof's Saga* 396; Icelandic xvi, 57, 68, 318; *see also Gösta Berling's Saga*
Saimaa Canal 497, 562
Saimaa, Lake 469
Säisä, Eino 535
Salangen 323–24
Salmela, M. 434
Salmi 432
Saltdal 321
sämjodelning (Sweden) *see* land division
Sande, Jacob 396
Sanders, Chris 44n
Sandgreen, Otto 125
Sandhåland xxii, 349–58, 359–64, *360, 365, 366,* 368, 369n, maps *351, 353, 356, 357, 362*
Sandoy 81, 82, 94, 97, 100
Sandur 82
Satakunta *495*

satellite imagery 254; Landsat satellite 27
Saudi Arabia 556
Sauer, Carl xxviiin, 161, 511
saunas 460, 488, 491
Saussure, Ferdinand de 313, 314
Savo 227, 235, 238, 432, *495*, 498
Savonlinna 498, 500
Saxby 435
scale xi, xiv, xxvi, 60, 253, 259, 260–62, 275, 404, 476, 505, 512, 514, 517, 518, 519, 520–21, 522, 535; aerial photography 258; maps 151, 161, 213, 262–63, 423
Scandinavia 17, 19, 25, 55, 61, 145, 162, 168, 173, 187, 192, 215, 216–17, 238, 247n, 302, 322, 326, 545, 570, 573, 574, 579, 583; definition ix, 543; name 6, 162; Scandinavian peninsula xv, 3, 7, 558; Scandinavian shield 4
Scandinavism 247n
Scania (Skåne) xv–xvi, xviii, xix, 6, 7–8, 9, 10, 43n, 144, 145, 146, 147, 149, 154, 157–91, *158*, *177*, 216, 231, 243, 246, 263, 558, 559, 573, 574, 577, 581, 583; map *160*; name 162, 187n
scenery xii, xiii, xiv, xv, xxviiin, 8, 10, 16, 108, 162, 173–74, 220, 221, 223, 373, 476, 489; Åland 453; Dalecarlia 211; Finland xxvi, 424, 425, 465, 470, 476–77, 483, 484, 490, 491, 492, 498, 500, 501, 505, 507; Greenland 110, 111; Iceland 64; Jutland xvi, 14, 15, 17, 20–21, 23, 24, 25, 25–27, 35, 39, 40, 41, 42n, 46n; Norway xiv; Scania xix, 7, 167, 170; Sweden 222–23, 232, 243, 246, 251, 259, 266, 276, 277; Wales 40; Western Norway 385, 388
Schama, Simon 432
Schanche, Audhild 287–88
Schleswig–Holstein (Slesvig–Holsten) ix, 9, 17, 18, 29, 42n, 545, 546; *see also* South Jutland
schools: Åland xxv, 453, 455; Denmark 546; Finland 428, 476, 479, 493, 504, 507, 530; Finnmark school of agriculture 331; folk high schools 453; Greenland 111, 119, 128, 134; Scania 158, 162; schoolbooks xxv, 158, 222, 245, 455, 476, 479, 493, 522, 529;

Sweden 222, 231, 240, 243, 245; *see also* children
Scotland 78, 79, 83, 188, 306, 337n, 550, 554
sea bottom 284, 288, 293, 295, 309
seafaring: Åland 443, 446, 450, 455; Denmark xv, 5; Finland 562; Iceland 59, 67; North Norway 288–90, *289*, *290*, 293, 310–11, 321–23, 326, 329; Norway 366, 381, 396, 554; Scania 162; *see also* fishing; sealing; shipping; whaling
sea ice xxv, 7, 174, 309, 444, 447, 450, *488*, 558, 561, 577; drift ice xvi, 55, 69, 310; Ishavet ("Ice Sea"; Arctic Ocean) 310; Østisen ("East Ice") 310; Vestisen ("West Ice") 310
sea-level change *see* shore displacement
sealing 67, 81, 85, 94, 118, 121, 307, 310, 323; *see also* hunting
sea salinity 574
seascape xxv, 446–47, 450; *see also* maritime landscape
seashore 5; Faeroes 84–85, 86, 87, 88, 89, 91–93, 103; Iceland 68–69, 72; Norway 288, 294, 296, 349, 350, 354–55, 358; seashore settlements *(strandsittersteder)* (Norway) 372, 381; *see also* shore displacement
seasonal movement of livestock: Saami reindeer herders 283, 287, 294–95, 311, 324; southern Norway 407–408, *407*; Western Norway xxiii, 376–77, *377*, *379*, 381–82
seasonal resource use: Greenland 128; North Norway xx, 295, 326; southern Norway 407–408; Sweden 148, 198, 204, 217; Western Norway 368, 354–55, 358, 385, 387
seasons 35, 335, 575; breeding season 309; colors of seasons (chromotology) 435, 436; Dalecarlia 198, 204; Denmark 4; Finland 561; harvesting 293, 358, 361; hunting season (Finland) 459, 462; in art 394; Jutland 27, 35; Norway 374, 385, 404; Saami 293, 296n; seasonal fisheries (North Norway) 291, 293, 294–95, 307, 320–21, 323, 554; seasonal food (Greenland) 128; seasonal markets 328; seasonal work 198, 202, 240, 311; Sweden 271; *see also* agriculture, growing season; autumn; spring; summer; winter

seaweed gathering: Faeroes 86, 89, 91–93, 99; Iceland 69; Western Norway 350, 354–55, 364, 368
second homes: Åland (summer houses) 429; Finland (villas) 429, *431;* Greenland (weekend cottages) 128; Scania 187; Sweden (summer houses) 148, 154, 201, 276; Western Norway (leisure cottages) 384
Seglem, Karl 396
self-sufficiency: Åland 446, 447, 449; Denmark 548; Faeroes 81; Finland 564; landscape of 447; Norwegian farms 366, 408
Seljord 402, 404, *405,* 406, 409, *409*
seminatural vegetation 345, 346, 369; Norway 286, 287, 401; southern Norway xxiii, *403*, 414–15, *415*, 416; Sweden 255, 259, 268, 270, 275, 276; Western Norway xxii–xxiii, 361, 364, 368, 372–73, 384–90; *see also* meadows; pastures
semiotics xii, 301, 306, 311, 313–14; Tartu-Moscow school 314
service economy (tertiary occupations) 584; Åland xxv, 440; Denmark 546, 547, 548, 549; Finland 532; Greenland 124; Iceland 551, 552; North Norway xxi, 291, 295; Norway xxi, 295, 336, 555, 556; Sweden 561
seter see summer farms
Setesdal 402, 404
settlements 346, 544, 568, 574, 577; Åland 429, *449;* Dalecarlia 195, 203, 211, 212–13, 214, 215–17; Denmark 548; Faeroes 78–79, 81, 82, 84, 85–94, 96–97, 99, 104; Finland 460, 488; Finnish settlements in Ladoga area 432; Finnish settlements in North Norway *305;* Finnish settlements in Sweden 558; Greenland 109, 119–32, 134n, 550; Iceland 56, 58, 59, 72; North Norway 309, 324, 327, 328, 330, 331; Norway 553, 554; Norwegian settlements in Finnmark, map *304;* Scania 163–64, 166, 179, 182; southern Norway 405, 406; Svalbard 557; Sweden 143, 145, 147, 153, 155, 256, 266, 268, 269, 272; Värmland 225, 226, 227, 232, 234–35, 238, 242; Western Norway 350, 375, 378, 379, 380, 381

Seven Sisters, the (De syv søstre) (mountains) 322–23
Seymour, Susanne 173
Seyðabrævið (Sheep Letter) 77, 84, 87, 89, 90, 95, 96, 97, 99, 100
Shakespeare, William 432
shamanism 14, 236–37
sheep: Denmark 35; Faeroes 77, *78*, 81, 82–83, 86, 88, 89–91, 94, 97, 99, 100–101, 103, 104; Greenland 111, 126; Iceland 61, *61*, 62, 71, 578; Norway 555; southern Norway *405*, 406, 408, 410; Sweden 149, 150–51; Western Norway 350, 351, 352, 356, 357, 364, 365, *365*, 376, 378, *379*, 383, 384, 387, *389*
Sheep Letter *see Seyðabrævið*
Shetland ix, 86, 102, 216, 337n, 554
shielings *see* summer farms
shipping 574; Åland xxv, 440, *442*, 444, 446, 448, 451, 452, 566; Denmark 5, 545, 546, 548; Finland 436, 563, 565; Iceland 551; Norway 333, 381, 389, 554–55; Scania 168, *169*, 170; shipping lanes and land uplift 141
shore displacement xviii, 141, 143, 450, 570, 574, 561, map (Norden) *572;* highest shoreline (HS) 143, 144, 202, 203, 225, 270–71, 272, 404, 570, 572, 573, 579, 581, 583; sea-level change 225, 570, *572; see also* land uplift
signs xxi, 14, 33, 41, 63, 311, 312, 313–15, 318, 516, 522; Åland 453; compass rose 308; Finland 373, 391; graffiti 318; Greenland 122; Iceland 67, 73; North Norway 300–301, 321, 324, 326, 330, 333, 334, 335, 336; road signs xxi, 8, 315, 318; street signs 334; Western Norway 393; *see also* symbols
Sigtuna xxvii
Sildnes, Sissel 334
Siljan, district 200, 203, 211, 214; Lake 192, 200, 201, 202–203, 211, 214
silver mines 146, 554, 560
Sisimiut 116, 131
Sjælland *see* Zealand
Sjöström, Victor 242
Sjugare 201, *203*
Sjúrðardóttir, Gudrun 97
Skagen 13

Skagerrak 143
Skåne *see* Scania
Skånes, Helle 255
Skansen (Stockholm) 198, 223, 246
Skellefteå 560
skerry guard *(skärgård) see* archipelagoes
SKF ball bearings 560
Skibotn 283, 328
Skien 411
Skiftet (Vattuskiftet) 447
Skipper Clement 18–19, 42n
Skjesvatnet, lake 411
Skúvoy 100
Slesvig–Holsten *see* Schleswig–Holstein
Småland xviii–xix, 149, 151
smoke cottages 237, 240, *241*, 248n
Smuts, J. C. 252
Snabeshøy (Snabeshøj) 27, 29, 31, 41, 43n, 45n
Snæfellsjökull (glacier) 69, 70
Snæfellsnes 65, 70, 71
snapphane (-ar) (snaphanse) (Scania) 175–76, 188
Snorri Sturluson 68
snow 543, 577; Canada 113; costs 561; Faeroes 83, 90; Finland 460, 462, *502*, 531, 543, 561; Iceland 69; North Norway 292, 293, 308, 314; southern Norway 406; Sweden 271; Värmland 224–25, 229, 230; Western Norway 374, 377, 378, 394, 397
social class xi, 161, 173; Dalecarlia 201, 206; Finland 493, 505; North Norway 313; Scania 168, 175, 181, 183–84, 185; Sweden 222, 558–559; Värmland 233
society–environment relations x, xii, xxvii, 53, 221, 224, 254, 311, 313, 348, 442, 483, 511, 516, 568; Åland 449; Denmark 546; Greenland 106, 109–10, 113, 120, 125; Iceland 69; Norway 296, 344, 348, 373, 402; Scania 161–62, 166, 167–68; Sweden 251
Sodankylä 561
Söderåsen national park 168
Södermanland 215
Sogn 372–400, *376*, *377*, *378*, *379*, *380*, 386, *389*, *391*, *393*, *395*, map *374*
Sogndal xxvii, 374, 385, 388

Sognefjorden 277, 372, 373, *377*, 382, 388, *391*, 397, map *374*
Sogn og Fjordane 365, 372–400
soil(s) 166, 345, 543, 570–71, 573–74, 581, 583; Åland 448, 452; Dalecarlia 196, 202; Denmark 4, 15, 39–40; Finland 476, 511, 564; Norden map *580;* Norway 329, 353–54, 355, 358, 359, 362, 364, 366, 367, 368, 375, 379, 383, 387, 388, 390, 406; Scania 160, 162, 164, 165, 187; Sweden 141, 143, 145, 147, 155, 235, 258–59, 260, 264, 266, 270–71, 272, 273, 276
Soldan, Björn 501
Solør 227
Solovetsk 330
solskifte see land reorganization
Sønderjylland *see* South Jutland
song(s): Åland 453–55; Denmark 12, 23–24, 35; folk songs 412, 453, 455; Greenland 111, 112, 128, 130, 133n; hymns 71, 128, 175; Iceland 71; Norway 406, 411n, 412; Sweden 151, 167, 175, 201, 228, 268; patriotic 23–24, 111, 133n, 201; *see also* anthems
Sørensen, Henrik 412–13, *413*
Sør-Fugløy 327
Sörmland 231
Sortavala 498
Sorte Mads *see* Black Mads
Sør-Trøndelag 320
Sørvágur xxvii, 78, 91
Sør-Varanger *see* South Varanger
Sound, the *see* Øresund
sources, historical: Denmark 10; Faeroes 97, 100, 103; Finland xxiii, 421, 422–25, 492–93; Norway xxi, 302, 398; Sweden xix, 149–52, 155n, 176, 210, 254, 255, 258–59, 348; *see also* atlases; parish registers; statistics
South Africa 333
southern Norway xx, xxi, xxiii, 283, 284, 285, 287, 291–92, 294, 295–96, 305, 320, 324, 325, 401–18, 573, 577, 581, 583
Southern Ostrobothnia 494, *495*, 497, 529, 530, 531, 532, *533*
South Jutland (Sønderjylland) 17, 174, 546, 581
South Varanger (Sør-Varanger) 330, 333–35

Soviet bloc ix
Soviet Union (USSR) ix–x; and Finland xxiv, 432, 434, 521, 562, 563, 565; and Norway 333, 334
space xiii, xvii, 16, 33, 53, 112, 123, 168, 170, 187, 223, 255, 269, 475, 512–13; Certeau, Michel de 57; Denmark 5, 17, 21, 23; Finland 486, 504, 505, 506, 508, 524, 525, 526, 529, 530, 536; Greenland 114, 120, 123, 130–31; Iceland 53, 56, 57–63, 66, 69, 70, 71, 72–73; North Norway 308; Scania xix, 186; spatial analysis 348; spatial boundaries 3; spatial confinement 133; spatial data 255, 259; spatial distance xiii, 16; spatial distributions 260; spatial expression of power 171; spatial forms 272; spatial framework xxv; spatial imprint 13; spatial integration xxv; spatial margins xvi; spatial organization 256, 258, 544; spatial partitioning 112; spatial patterns 255; spatial phenomena x; spatial practice x, xvii, 54, 72, 178; spatial questions 164; spatial scales xxvi, 253; spatial structure 224; Sweden 246, 254, 269; time and space xix, 59, 63, 168, 223, 242, 254, 255, 269, 348, 472, 536
Spanish Civil War 551
species: animal 67, 118, 267–68; biodiversity 253, 254, 268; fish 85, 550; endangered 267; *hagmark* 251, 259, 266, 267; loss of 275; plant 267, 269, 272, 275, 577; rare 253, 266, 267; sheep 53; species-poor grasslands 270; species-richness 267, 274; threatened 253, 267; whales 84
Spitsbergen *see* Svalbard
Sporrong, Ulf 254–55, 411
spring 38, 67, 292, 309, 386, 387, 393, 407, 444, 462; pastures 365, 377, 388, 389, 390, 407, 408; work 358, 366, 406; *see also* seasons
spring farms: southern Norway 406–408, *407*; Western Norway 376, *377*; *see also* autumn farms; summer farms
Stærke Sejr 27–28, *28*, 29, 38, 41
statistics: Åland 456; censuses 259, 348, 350, 383, 551; Finland 423, 425, 459; Greenland 134n; Iceland 551; Norway 348, 350, 383; Sweden xix, 151 178, 259, 261, 263
Statoil 557; StatoilHydro 557
Stavanger 350, 352, 556, 574, 581
Steffens, Henrik 22, 23, 24
Stenshuvud national park 168
stereotypes 301, 306, 313; Eskimo 110; Finnish regional 524, 30–32; Finnish stereotypical landscapes 483, 496, 507
Stjernsvärd, R. H. 157
stockfish 321, 551, 584
Stockholm 159, 181, 223, 238, 239, 423, 432, 443, 444, 446, 453, 558, 559; University xxviii
Stone Age: Fennoscandian 330; Neolithic (Värmland) 225; southern Norway 404; Sweden 147
stone walls: Faeroes 82, 88; Iceland 64; Norway 401; southern Norway xxiii, 413, 416; Sweden 256, *268*, 269, 274; Western Norway *351*, 356, 365, 380, 383
Storfjord 327, 328
stories 247n; Blicher's 27–28, 34, 35, 45n; children's xvii–xix, 151; Faeroes 78; folktales 63, 71, 78, 268, 411; Greenland 120, 129; Iceland 55–58, 63, 71; North Norway xxi, 312–13; Norway 411; Greenland 120, 129; Iceland 71; origin stories xxi, 55–58, 312–13; Scania 176; Sweden 268; Värmland 235–37, 242–43, 245; *see also Gösta Berling's Saga*; legends; *Nils Holgersson*
Storsjön, Lake 152, 579; district 575, 579, 581
St. Petersburg 429, 432, 489, 562; *see also* Leningrad
Strabo 526
strandflat 366, 553, 574
street(s): Nuuk (Greenland) *108*, 132; street gangs 318; street layout 170; street lighting 185; street names 132, 334; *see also* roads
Streymoy 87
Strindberg, August 157, 187n
Sture family 194
Sturluson, Snorri 68
sublime 222, 488, *488,* 489, 491, 504
suburbs 583; Denmark 548; Greenland 115, 118; Sweden 153

Suðuroy 81, 82, 84, 100
Suðurstreymoy 96–97
Suðursveit 61, 62, 64, 70, 71
Suistamo 432
Sumba 94
summer 31, 85, 151, 292, 330, 422, 432, 458; frost in summer 464; grazing xxiii, 82, 83, 88, 90, 148, 150, 164, 165, 198, 292, 325, 365, 376, 377, 382, 406; landscape xxv, 447, 453, 454; light 308, 309, 436, 575, 579; midsummer xxv, 200, 453–55, midsummer pole 454; recreation xxv, 128, 447; tourism 192, 324, 447; work 198, 201, 202, 295; *see also* seasons
summer farms: 148; Dalecarlia 198, 203–204, 212, 215; Faeroes (*ærgi*) 82–83, 91, 96; Scotland (shielings) 83; southern Norway (*setrer*) 404, 406–408, 407, 410, 414; Sweden (*fäbodar*) 148, 150; Värmland 232; Western Norway xxiii, 376–80, 377, 378, 379, 383, 385, 389, 395; *see also* autumn farms; spring farms
summer houses *see* second homes
Sund 497, 507
Sundblom, Julius 445
Sundborn 201
Sunnmøre 365, 367, 396, 391
Suomenselkä 468, 529
sustainability: ecological sustainability 301, 465; social sustainability 474; sustainable development 252–53, 416; sustainable forestry 472, 474; sustainable future 301; sustainable land use 251, 252; sustainable management xxiii, 254, 277, 292; sustainable production 255, 269, 390
Svabo, J. C. 94
Svalbard (Spitsbergen) 55, 291, 307, 309, 310, 323, 325, 333, 553, 557; Svalbard Treaty (Spitsbergen Treaty) (1920) 310, 557
Svanehom 179, 181, 184, 185, 187, map 183
Svartå 433
Svartdal *see* Hjartdal–Svartdal
Svea 557
Sveen, Arvid 319–20, 321, 323, 324, 325, 327, 331
Svend (Sweyn Forkbeard) (king) 20

Svenstorp 171–72
Svínoy 81
Sweden ix, xiv, xv, xviii–xx, xxiv, xxvii, xxviii, 6, 7, 8, 9, 10, 42n, 65, 139–280, 308, 311, 316, 321, 326, 327, 328, 329, 331, 332, 433, 440, 442, 443, 444–45, 446, 447, 453, 455, 462, 543, 545, 549, 554, 558–61, 562, 563, 564, 565, 566, 570, 573, 575, 577, 579, 581, 583, map 142; *see also* Dalecarlia; Scania; Värmland
Sweden–Finland xxviiin, 144, 216, 329, 331, 423, 443, 444, 545, 563
Sweden–Norway xv, 7, 9, 238–40, 247n, 316, 545, 554, 558
Swedish Matches 560
Swedish Tourist Federation 243
Switzerland xiv, xxviiin, 555
symbols xi, 162, 166–68, 187, 223, 238, 242, 312, 313, 314–15, 325–26, 441, 478n, 483, 484–85, 512–13, 514, 516–19; Åland 453–56, 454; Dalecarlia 195, 214; Denmark 13–14, 16, 22, 42n; Finland 427, 459, 460, 461, 465, 473, 475, 478n, 483, 485, 492, 504, 505, 506, 514, 522, 524, 530, 531, 535; Greenland 116, 122, 125, 126, 132; Iceland 64, 65; North Norway 288, 291, 300, 305, 320–37; Norway 316, 318, 319, 324, 326; Sweden 146, 148, 222; symbolic economy 518; symbolic capital 530; Värmland 228, 243, 244, 245, 247; Western Norway 334, 379, 391–92, 391, 395, 396; *see also* signs

taiga 577
Tällberg 200, 201
Tallinn (Reval) 5, 446
Tammekann, August 526, 527
Tampere 496, 497, 562
Tana *see* Deatnu-Tana
Tana (Deatnu; Teno), River 331, 501
Tannum 147
Tapiola 506
tar production 177, 198, 327, 393, 425, 462, 468, 470, 498
Tarnovius, Thomas Jacobsen 93
Tartu 314
taxation *see* land taxation; Saami, taxation of

tectonic movements 579
Tegnér, Esias 227, 396
Telemark 277, 402–18
Teno (Tana), River 331, 501
terminology: cardinal directions 314; ecology 268; Faeroese 77, 82, 88, 89, 94, 95, 100, 101, 102, 104n; fishing 290; Greenland 112, 116, 120, 132n, 133n; heraldry 337n; ice 450; islands 449; landscape xv, xxiv, 16, 29, 44n, 221, 223–24, 247, 252, 283–84, 286, 288, 292, 323, 421, 440, 483, 506, *see also* landscape concept; legal 56, 96; Norden xv, 543; scientific 293; symbol 315; travel 134
territory ix, xi, xiv, xv, xix, 16, 59, 161–62, 168, 171, 173, 221, 223, 224, 238, 300, 318, 441, 511–12, 515–19, 534; Åland ix, xxiv, 443–44, 446, 447; Faeroes ix; Finland xxv, xxvi, 423, 432, 461, 476, 477, 478n, 484, 486, 489, 490, 492, 504, 513–15, 520–22, 524–27, 530, 535, 537; Greenland ix, xviii, 106; Iceland 62, 63, 72; Norse 102; Norway 301, 323, 326, 330, 334–37, 365, 366; Scania xix, 163, 164–65, 171, 186; Värmland xix, 220, 222, 224, 226, 228, 231, 232, 242–43, 245, 246–47
Thaulow, Fritz 412
theater 15, 20
Theusen, Søren 110
Thingvellir (Þingvellir) 31, 43n, 64–65
Thomas, Bishop (Thomas Simonsson) 201
Thompson, E. P. 44n
Thompson, John B. 242–43
Thorpe, Harry 15, 25
Thule 55, 550
Thyra (Queen) 19, 45n
Tibble 201, *208*
Tidemand, Adolph 392, 412
Tilgmann printing house 425, *428*
Tilley, Christopher 53, 64, 65, 72
timber floating 232, 327, 468, *469*, 470, 554
time geography 248n, 255
time-space compression 472
ting (law assembly; thing) xxiv, 31, 32, 33, 34, 43n–44n, 45n, 163, 214, 226, 444; *see also* parliaments
Tingvalla (Karlstad) 226

toft 213–14; *see also býlingur; heimrust*
Tolkien, J. R. R. 429
tomtåker (arable land closest to toft) (Sweden) 213–14
Topelius, Zachris (Zacharias) 424, 436, 462, 476–77, 479n, 490–92, 503, 520–21, 532
Torneå 326, 329
Torne: Lake 326; Torne River (Torne Älv) 143, 155, 326; Torne Valley 328
Tórshavn 550
tourism: Åland 440, *442*, 447, 453; Dalecarlia 192, 195, 199, 200–201, 202, 203, 209; Denmark 4, 16, 27, 547; Finland xxiv, xxvi, 434, 435–36, 484, 491, 493, 498, 500, 501, 503, 505, 506, 507, 516, 531, 536, 562; Iceland 45n, 552; North Norway 301, 322–22, 324; Norway 324, 402, 412, 553; Svalbard 310, 333, 557; Sweden 154, 186, 232; Western Norway 384, 397; Värmland 223, 227, 228, 229, 241, 242, 243, 247
towns and cities 134n, 166, 185, 188n, 544, 546, 568, 581, 583–84; Åland 440, 445, 451; Denmark xv, 5, 6, *21*, 22, 41, 43n, 544, 546, 548, *see also* Copenhagen; Finland 422, 433, 435, 446, 447, 458, 470, 489, 492, 496, 497, 498, 500, 503, 506, 562, 563, *see also* Helsinki; Greenland 106, 109, 114–23, 124–25, 126, 127–28, 129, 130–32, 550, *see also* Nuuk; Hansa towns 168; Iceland 53, 63, 68, 552, *see also* Reykjavík; Norway 291, 315–16, *316*, 318, 319, 320, 324, 325, 332–33, 334, 382, 554, 555, *see also* Oslo; Scania 159, 163, 164, 166, 167–70, *169*, 174, 186, 187, *see also* Lund; Shetland 337n; Sweden xix, 146, 153, 168–70, 227, 232, 446, 558, 559 *see also* Stockholm; town and country 109, 130, 131, 168, 496; *see also* suburbs; urban areas; urbanization; urban landscape; urban population
townscape 496, 500
trade: Åland 440, 443, 446, 449, 451, 452; Denmark 35, 545, 546, 547, 548, 549; Faeroes 81, 82, 83, 84, 85, 86, 92, 102, 550; Finland 470, 473, 474, 477, 478n, 498, 562, 563, 564–66; Greenland 111, 550; Iceland 551, 552; North Norway 295, 307, 308, 309,

310, 323, 326, 329–30, 331, 333; Norway 410, 554–55, 556; Sweden 162, 165, 168–69, 226, 227, 232–33, 240, 559–60; Western Norway xxiii, 381–82, 395

Trætteberg, Hallvard 319, 320, 321, 325, 330, 332

transitional zones 202, 225, 257, 260, 261, 263, 266, 429, 448, 583; *see also* liminality

Trap, Jens Peter 36

travel 134, 321, 484; travel guides 243, 315, 493, 497; travel routes 31, 37, 55, 62, 67, 151, 201, 327, 333, 381, 432–33, 444, 446, 450, 498; *see also* tourism; travelers

travelers: Denmark 12, 16, 37; Faeroes 94; Finland xxiii, 421, 422, 491, 521; Norway 337n, 372, 392, 397, 406; Sweden xviii, 149, 157, 179, 193, 206, 220, 234

Treaty of Kiel (1814) 545

Treaty of Moscow (1940) 563

Treaty of Paris (1856) xxiv, 445

Treaty of Versailles (1919) 546

Trelleborg 149

trolls xvii, 61, 63, 65, 70, 391, 396

Troms 306, 307, *307*, 312, *316*, *317*, 319, 323–29, 335; North Troms 308, 309, 314, 331, 332; South Troms 314, 326

Tromsø 302, 308, 310, 319, 325, 326, 327, 329, 335; University of 283; University Museum 310

Trondheim xxviii, 329, 337n, 575, 581

Trøndelag 318, 320, 579

tun (clustered farm buildings; farmyard): Faeroes (*tún*) 87; Western Norway 349, 352, 354, 364–65

tunnels 187, 384, 397, 549

Turku (Åbo) 432, 434, 444, 446, *461*, 496

Turunen, Heikki 534

Tuuri, Antti 535

Þingvellir *see* Thingvellir

Þórbergur Þórðarson 64, 69–70

Þorsteinn Jósephsson 61

Ukraine 574

Ullsfjord 326

Ullvi 205

Ultima Thule 55

unemployment: Denmark 547; Finland 470, 474, 531, 563–64; Greenland 550; Iceland 551; Norway 411, 556; Scania 186; Sweden 559

UNESCO 402

United Nations 10

United Nations Conference on Environment and Development (UNCED) 252

United States of America ix, xxviiin, 232, 240, 318, 380, 409, 511, 546, 547, 550, 551, 554, 559, 562; *see also* America (North)

Unjárga-Nesseby 331

Uppåkra 163

Uppland 210, 215, 263

Uppsala 234, 238, 239

urban areas: alienation 534; anti-urbanism (Greenland) 131, 132, (Norway) 412; urban culture 168, 392; urban development 286; urban environment (Iceland) 63, (Sweden) 223; urban growth 401, 548, 562; urban imagination, landscapes of 168; urban influence 185; urban life (Greenland) 109, 126, 131, (Scania) 166; urban parks 433; urban regions 155, 549; urban-rural dichotomy 168, 534, 474; urban settlements (Finland) 488, 496, (Norway) 291, 553, 554; urban society (Finland) 474, (Sweden) xix, 223; urban sprawl (Sweden) 276; urban street gangs 318; urban viewpoint 502–503, 508n; *see also* suburbs; towns and cities; townscape; urbanization; urban landscape; urban population

urbanization 581, 583; Åland 451, 456; Finland 503, 532, 534; Greenland 124; Norway 291, 330; Sweden 164, 167, 171, 186, 222, 227; urbanized areas (Scania) 159, 167, 186

urban landscape: Åland 451; Denmark 134n; Finland 423; Greenland 106, 123; Norway xxi, 283, 295; *see also* townscape

urban population: Åland 566; Denmark 546, 549; Faeroes 550; Finland 562; Greenland 550; Iceland 551, 552; Norway 294, 296, 554; Sweden 153, 154, 222, 227, 558

utility, concept 178–79

Utladalen 393, *395*

Utsjoki 501, 506

Uusimaa *495*, 496, 498, 500, 521, 524, 531, 532, 533

Vadsø 319, 325, 329, 330, 335
Vågå 325
Vågan 320, 321, 323
Vágar 91, 94
Valamo 434, 498
Valdemar the Victorious (Valdemar II) 5, 45n
Valosaari 435, 436
Vänern, Lake 224, 225, 226, 231, 559
Vantaa 458
Varanger 302, 327, 330; Varangerfjorden 331, 333; *see also* South Varanger (Sør-Varanger)
Vardø *317*, 319, 329, 333
Vårdö 429
Vardøhus fortress 329, 334, 335
Varenius, Bernhard 526
Värmeln 225
Värmland xix, 146, 150, 153, 154, 220–50, map *225*
Varsinais-Suomi *495*
Västanfors 195
Västerås 210
Västergötland 145, 146, 581
Västmanland 215
Vatnajökull (glacier) 69, *70*, 71, 581
Vättern, Lake 559
vegetation 255, 344, 345–46, 347, 348, 404, 573, 577–79, map (Norden) *578*; Åland 448, 449; Faeroes 81, 88, 92; Iceland 551, 578; Norway xxii–xxiii, 283, 311, 335, 401; southern Norway 404, 414–15, *415*, 416; Sweden 145, 152, 161, 225; Western Norway xxii, 345–46, 348, 350, 352, 359, 361, 364, 368–69, 369n, 372–73, 374–75, 384–90; *see also* fodder; forest; grass(lands); *hagmark*; heath(s); peat; seminatural vegetation
Vestmanna (Faeroes) 92
Vestmannaeyjar (Westman Islands) (Iceland) 552
Vestvågøy 320, 321, 323
Viborg (Denmark) 22, 25, 29, 30, 31, 41, 43n, 45n, 562
Vidkunsson, Erling 324
Viipuri (Viborg; Vyborg) 429–32, 433, 496, 562, 563, maps *430*, *431*

Vik 384, 395
Vikings 79, 194; Denmark 4–5; Faeroes xvi, xvii, 77–79, 81, 95, 104; Greenland xiv, xvi, xviii; Iceland xvi, 55, 79, 551; Norway 102, 392, 405, 412, 553–54; North Norway 307, 310–11, 320, 321, 323, 325–26; Scania 163, 164–65; Shetland 337n; Sweden 272, 558; *see also* Norse
Vilgerðarson, Flóki 55
villages 583; Åland *449*, 453, 454; Dalecarlia 196, 198, 200, 201, 203–204, 205, 211–15; Denmark 8, 93, 94, 548; Finland *425*, 462, 464, 471, 534; Greenland 117, 122; Iceland 53, 65; Jutland 28, 29, 37, 44n; Norway 328, 382, 383, 394, 397; Scania 8, 149, 159, 163–64, 166, 171, 176–83, 185; Sweden 147, 149, 150, 151, 153, 226, 258, 260, 263, 272, 274, 276; village landscape 166, 256, 272–75; *see also* settlements
Vindelälven, river 143
Vindsjåen, lake 411
Vinland 59
Virestad, 263, maps *257*, *264*
Virgil 166
Virgin Islands 4
Virrat 498
volcanoes 53, 68, 69, 543, 551, 552, 574, 581
Vordingborg 45n

Wadden Sea 574
Wales 40–41
waterfalls and rapids: Iceland 65; Finland 468, 488, 497, 498, 500, 503, 506, 507; Western Norway 384, 391, 393, 394, 397
water power: Finland 433, 468, 564; Norway 557; Sweden 149, 152; *see also* hydroelectric power
welfare state 312, 475, 583; Denmark 548, 549; Finland 460, 472, 478n; 563; Greenland 123–24, 550; Iceland 68, 553; Norway 300, 306, 336, 555; Sweden 558, 559, 563
Wennervik, Lisa-Maja 237
Werenskiold, Erik 412
Wergeland, Henrik 397
West Africa: Danish colonies xv, 4, 545
West, John F. 94–95

West Greenland 110, 132n
West Indians 20, 127
West Indies 127; Danish xv, 4, 545
Western Norway xxii–xxiii, 79, 81, 95, 216, 283, 295, 296, 328, 344–400, 543, 551, 575; cultural influence on Faeroes 77–79, 81, 88, 95
Westman Islands (Vestmannaeyjar) 552
wetlands 163, 198, 259, 270–71, 272, 274, 276, 375, 573, 577
whaling: Faeroes 77, 81, 84–86, 91–93, 99, 103; Greenland 118; Norway 310, 333; *see also* hunting
White Sea 308, 309, 310, 323, 329–30
Widgren, Mats 226
wilderness x, 579; Finland 421, 426, 428–29, 432, 459–60, 462, 464, 465, 470, 472, 487, 498, 500, 503; Iceland 53, 58, 62, 553; Greenland 113; Scania 166–68; Sweden 154; Värmland 231, 241, 243
Wille, Hans Jacob xxiii, 406
Williams, Raymond 40–41, 512
wind power 549, 560
Wingård, Johan af 239–40
Winnipeg 59
winter 321, 436, 462, 543–44, 561, 575, 577; darkness 291, 308, 335, 436, 599; fisheries 83, 85, 90, 289, *290*, 291; fodder xxiii, 82, 165, 309, 333, 350, 365, 385–87, *386*, 406, 408, 410, 450; grazing 83, 88; landscape *230*, 422, 435, 447, 450, 451, 462, *502*; recreation 502; rye and barley 149, 406; tourism 198, 201, 503; transport 331, 408, 450, 561; *Winter by the Sognefjell* (J. C. Dahl) *391*; work 198; *see also* ice; sea ice; seasons; snow
Winter War 563

wolves 324, 377, 395, 462, 470
women 35, 44n, 64, 66, 173, 233, 247n, 328–29; equal rights (Sweden) 558; Greenlandic women 113, 116, 117, 122, 123, 126, 130; North Norwegian women 291, 292, 294, 295, 301, 328–29, 333; witches 237, 333; women and girls on summer farms 150, 198, 377, 407; women and landownership (Dalecarlia) 204, 210–11; women farmers (Norway) xx, 291, 292, 294, 295, 555; women in workforce 548, 552, 556, 558; women migrant workers 198; women represented in heraldry 328–29; women's costumes (Dalecarlia) 199, 202; women's suffrage (Finland) 460, 563
Wood, Denis 178, 526–27
Wordsworth, William 435
World Commission on Environment and Development (Brundtland Commission) 252
World Heritage 143–44, *145*, 147, 149
World War I 445, 546
World War II x; Denmark 547; Faeroes 547; Finland 432, 474, 521, 563; Greenland 547; Iceland 4, 9, 551–52; Norway 302, 308–309, 325, 330, 332, 333, 334, 410, 555
Wright, Magnus von 424

Yates, Frances 54
Ylinen, V. *428*

Zealand (Sjælland) xv, xix, 4, 6, 7, 8, 9, 12, 16, 22, 29, 40, 43n, 45n, 162, 167, 168, 174, 186–87, 549, 581
Zilliacus, Benedikt 432
Zorn, Anders 200, 201
Zukin, S. 513

Index compiled by Michael Jones and Catriona Turner